Surgical Atlas of Cardiac Anatomy

Xiaodong Zhu
Editor

Surgical Atlas of Cardiac Anatomy

人民卫生出版社
PMPH PEOPLE'S MEDICAL PUBLISHING HOUSE

Springer

Editor
Xiaodong Zhu
Department of General surgery
Fuwai Hospital
Beijing
China

Jointly published with People's Medical Publishing House

ISBN 978-94-017-9408-4 ISBN 978-94-017-9409-1 (eBook)
DOI 10.1007/978-94-017-9409-1
Springer Dordrecht Heidelberg New York London

Library of Congress Control Number: 2014955151

Printed on acid-free paper

Springer is part of Springer Science+Business Media (www.springer.com)

Foreword

Patients with congenital heart defects begin a life of suffering the moment they are born into the world. For years, cardiac surgeons have been trying hard to find ways to make up for and correct such defects using various means. Through such efforts, these surgeons have alleviated the patients' sufferings and brought happiness back to the patients and their families, much like Angel *Nüwa* in the ancient Chinese folktale who benefits mankind by repairing the wall of heaven. Dr. Xiaodong Zhu, an expert in cardiac surgery who has been engaged in clinical works for more than half a century and was named Famous Doctor of the Chinese Academy of Medical Sciences and PUMC in 1993, is such an angel. Dr. Zhu is particularly skillful at hemodynamics and cardiac anatomy. In this book, *Surgical Atlas of Cardiac Anatomy*, pathological features of heart malformations are explained with Dr. Zhu's clinical experiences covering decades.

The book is based primarily on real images in cardiac surgical anatomy with a focus on congenital heart defects. Anatomy is one of the most fundamental disciplines on which clinical practices in surgical departments are based, and it is of great significance particularly for the understanding and treatment of congenital heart defects. In such disciplines, practical experiences are much more important than are theoretic knowledge and instruments. My hope is that this unique book, which analyzes different kinds of cardiac malformation specimens in depth, will inspire colleagues to further develop novel methods in the diagnosis and treatment of congenital heart defects.

Another feature of this book is that it contains massive amounts of precious medical records. Practice makes perfect. This saying is particularly true for surgeons. However, the current autopsy rate in hospitals in China is still far from ideal. This book contains invaluable images of more than 200 well-preserved specimens of heart malformations including some postsurgery specimens collected by the Fuwai Cardiovascular Hospital over the years.

There is no royal road to learning, and cardiac anatomy and surgery are no exceptions. I would like to thank Dr. Zhu for opening a new horizon for the diagnosis and treatment of congenital heart defects with his new book, and I encourage all cardiac experts, scholars, and ambitious young surgeons who are interested in making strides in these areas to read this book and make a difference in this field.

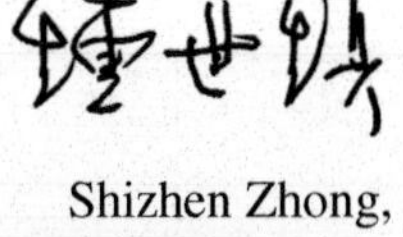

Shizhen Zhong,
Senior Academician of Chinese Academy of Engineering,
Professor of Institute of Clinical Anatomy,
Southern Medical University

Preface

Cardiac anatomy is the foundation for cardiac surgery and the prerequisite for modern cardiac imaging. Gaining a thorough understanding of actual pathological specimens of the heart cannot be achieved simply by studying anatomical models or by observing operations. The author wrote *The Basics of Cardiac Surgery*: *An Illustrated Guide* in the 1980s. It covered both cardiac anatomy and surgeries, aiming to integrate anatomical concepts with surgical techniques. Although the book was recommended and highly valued by colleagues, it was able to describe typical lesions and their surgical significance only by schematic diagrams rather than images of actual specimens due to the technical limitations that then prevailed.

In recent years, the rapid development of technologies for cardiac surgeries in China has enabled us to gain a considerably more insightful understanding of pathological specimens through which we have accumulated a wealth of clinical experiences. Familiarity with actual heart specimens is absolutely necessary for the cardiac surgeon to fully recognize the different types of heart malformations and to identify pathological changes quickly and accurately during surgery. To facilitate the cardiac surgeon's achievement of these goals, this book provides hundreds of three-dimensional images of actual heart specimens taken from multiple perspectives, many also accompanied by schematic illustrations. All specimens come from a collection established over the course of more than 50 years by the Department of Pathology, Fuwai Hospital, and the National Center for Cardiovascular Disease, Beijing.

Specimens of the same disease acquired from different people at different times and places can be quite diverse. The emphasis of this book is on the understanding of this pathology rather than the evaluation of the constantly evolving surgical techniques. All anatomical images of hearts in this book are of actual specimens, and the legends have been written to be as clinically practical as possible.

The book includes 28 chapters, with the first half concentrated on the normal heart structures and the second on congenital heart defects. Because the contents of the book were written according to the various numbers of available specimens, the lengths of the chapters vary accordingly. For example, we have numerous specimens of certain diseases such as Tetralogy of Fallot but only a few of other diseases such as tricuspid valve disease. Nonetheless, we had a sufficient store and variety of samples to provide substantial information on most of the common congenital heart defects in the book.

The majority of the book is composed of images. Each chapter begins with a text overview followed by images numbered by the pathological specimens. Most of the specimens are displayed from their anatomy perspective and some from the surgical perspective. In all images, *S* (*superior*) denotes the head side of the patient, *I* (*inferior*) the foot side, *L* (*left*) the left side, *R* (*right*) the right side, *A* (*anterior*) the front side, and *P* (*posterior*) the back side. As noted in the first chapter, common terms for the anatomy are shortened after their first use to acronyms, which are used in their place in subsequent chapters. The reader is advised to refer to the chart of acronyms for reference.

We would like to thank all the pathology experts, particularly Prof. Hong Zhao, who have worked diligently for many years to collect, store, and organize all the heart specimens used in the book. Also, we would like to express our highest respect and gratitude to the donors of these specimens, without whom our knowledge of cardiac anatomy and our surgical techniques would not have been advanced. We are grateful for the encouragement and support from colleagues in the National Center for Cardiovascular Disease, the Department of Pathology, and the Department of Cardiac Surgery, Fuwai Hospital. Moreover, we are indebted to Prof. Shizhen Zhong from the Chinese Academy of Engineering, who is a master in anatomy, has been supportive and helpful, and has provided photos of six specimens used in the book.

The images in this book have been inspected and proofread many times. However, mistakes may still occur. The readers are invited to inform us in such case. Until now, only a few books on cardiac anatomy have been available in China, and even fewer provide clinical specimens for illustration. It would be the greatest delight for us if the book could benefit doctors, colleagues, and medical students in cardiac surgery and related fields.

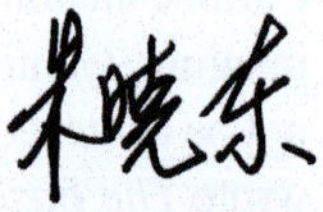

Beijing, China Xiaodong Zhu

Translator's Word

As one of the founders of cardiac surgery in China, Dr Zhu Xiaodong has made great contributions to the establishment and development of cardiac surgery. With more than 50 years of experience in this area, on the occasion of his approaching 80th birthday, Dr Zhu began to compile this book with the intention of sharing information that he thought would improve patients' outcomes after cardiac surgery.

Specimens shown in this book cover normal heart, various cardiac malformations, and malformations that received surgical treatment. In addition to filling a gap in the knowledge of cardiac surgery in China, this book also has significant value for the field of cardiac surgery around the world. Therefore, Dr Zhu authorized its conversion into English in order to share his expertise with peers around the world.

Zheng Zhe

About the Author

Dr. Xiaodong Zhu was born in Kaifeng, Henan. He graduated from the Medical Department of Harbin Medical University in 1956 and then worked in the Chest Hospital of the Chinese People's Liberation Army. His life as a cardiac surgeon started as a resident in the Fuwai Cardiovascular Hospital, Chinese Academy of Medical Sciences, in 1958. He then pursued his M.D. in the Department of Cardiac Surgery, PUMC, in 1962. He visited the United Kingdom and Australia during the mid-1970s and early 1980s, respectively, to advance his education in cardiac surgery. Dr. Zhu was the Director of Fuwai Cardiovascular Hospital and the Director of the Cardiovascular Disease Research Institute from 1992 to 1996. He was the Chairman for the fourth and fifth sessions of the Chinese Society for Thoracic and Cardiovascular Surgery from 1996 to 2006. Currently, he is the Chief Expert in the Department of Cardiac Surgery, Fuwai Cardiovascular Hospital.

Dr. Zhu has long devoted his life to clinical cardiac surgery as well as to carrying out diverse research projects in cardiac catheterization technology, hemodynamics, cardiac anatomy, and biomedical engineering including artificial heart valves and heart assistance devices. He developed the biological valves (pericardial valve) and successfully performed its clinical use first in 1976 for which he received honors and awards from the state, the Ministry of Health, the Academy of Medical Sciences, and the Government of Beijing. He won the Young Medical Science and Technology Expert award by the Ministry of Personnel in 1988 due to his prominent contributions, was named Famous Doctor of the Chinese Academy of Medical Sciences and PUMC in 1993, and was elected as Academician of the Chinese Academy of Engineering in 1996.

In his more than 50 years as a physician, Dr. Zhu has completed as chief surgeon more than 4,000 cardiac surgeries including operations for congenital heart defects, coronary artery diseases, valvular heart diseases, and aortic artery surgeries. In 1990, he led a medical team to aid Tibet, successfully performing open-heart surgery using cardiopulmonary bypass at the altitude of 3,600 m. Dr. Zhu values the training of both the team members and leaders. Under his leadership, the Department of Cardiac Surgery, Fuwai Cardiovascular Hospital, was designated by the State Education Commission in 1990 as the only nationwide key cardiac surgery department. Dr. Zhu also actively participates in the establishment of national technical collaborations promoting new advancements in cardiac surgery techniques across the nation. In the 1980s, he promoted heart valve surgeries in China. In the 1990s, he helped many domestic hospitals establish and develop the Departments of Surgery for Coronary Artery Disease. In more recent years, he served as the founding member for several hospitals specializing in cardiovascular disease in China.

Dr. Zhu was accepted as a member in the International Society of Surgery in 1988. He has been invited to academic conferences abroad many times, visiting Manila in 1980, Switzerland in 1985, and Indonesia, Iran, Azerbaijan, Japan, the USA, and Australia in the 1990s. During these visits, he demonstrated the advancements and achievements of cardiac surgery that were being accomplished in China, which were highly appreciated by colleagues from all over the world.

Dr. Zhu has published more than 130 papers in academic journals. He was the main author of ***The Basics of Cardiac Surgery*: *An Illustrated Guide*** (1st and 2nd editions), ***A Guide for Cardiac Surgery***, and ***Cardiac Surgery***. The book ***Cardiac Surgery*** won an award from the General Administration of Press and Publication in 2008 for its originality. Dr. Zhu was also the chief translator of ***Surgery for Congenital Heart Defects***. In addition, he participated in the writing of several academic monographs.

Contents

Translators

Chief Translator
Shengshou Hu

Associate Chief Translator
Zhe Zheng, Liang Tao

Proof Reader
Shiju Zhang, Betty Lee Ligon, Zhou Zhou

Translator

Chapter 1 Cardiac Anatomy: Nomenclature and Abbreviations
Mingyao Luo

Chapter 2 Position and Observation of the Heart and Surgical Approaches
Mingyao Luo

Chapter 3 The Right Atrium
Mingyao Luo

Chapter 4 The Right Ventricle
Yan Zhang

Chapter 5 The Coronary Artery
Yan Zhang

Chapter 6 The Conduction System of the Heart
Yan Zhang

Chapter 7 The Left Atrium
Peng Zhang

Chapter 8 The Left Ventricle
Peng Zhang

Chapter 9 The Cardiac Skeleton and the Aortic Root
Bai Lei

Chapter 10 General Considerations of Cardiac Embryology
Dingxu Gong

Chapter 11 Atrial Septal Defect and Cor Triatriatum
Meice Tian

Chapter 12 Atrioventricular Septal Defect
Liang Zhang

Chapter 13 Ventricular Septal Defect
Yi Chang

Chapter 14 Anomalous Pulmonary Venous Drainage
Hongguang Fan

Chapter 15 Aortic Arch Coarctation and Interrupted Aortic Arch
Hongguang Fan

Chapter 16 Congenital Anomalies of the Tricuspid Valve
Hongguang Fan

Chapter 17 Right Ventricular Outflow Tract Obstruction
Bai Lei

Chapter 18 Mitral Atresia and Hypoplastic Left Ventricle
Changwei Zhang

Chapter 19 Single Ventricle
Changwei Zhang

Chapter 20 Persistent Truncus Arteriosus
Fang Fang

Chapter 21 Aortopulmonary Window
Fang Fang

Chapter 22 Aneurysm of the Sinus of Valsalva
Liang Zhang

Chapter 23 Tetralogy of Fallot
Liang Zhang, Jianguo Yang

Chapter 24 Double-Outlet Right Ventricle
Changwei Zhang, Bo Wang

Chapter 25 Complete Transposition of the Great Arteries
Fang Fang, Xiao Wang

Chapter 26 Corrected Transposition of the Great Arteries
Meice Tian, Zhengdong Hua

Chapter 27 Pulmonary Atresia
Bai Lei, Jing Jin

Chapter 28 Double-Outlet of Left Ventricle
Bai Lei, Daqing Hu

Cardiac Anatomy: Nomenclature and Abbreviations

1

1.1 Nomenclature of Cardiac Anatomy

1.1.1 Right Atrium (RA)

The right atrium (RA) is the right upper chamber of the heart. It normally receives systemic venous drainage from the superior vena cava (SVC) and the inferior vena cava (IVC) (the two largest veins in the body collectively known as the *venae cavae*) and coronary venous drainage from the coronary sinus (Fig. 1.1). The RA can be divided into the body part, called the *auricula*, and the sinus part, called the *sinus venarum*. The auricula forms the front part of the RA and includes the right atrial appendage (RAA) and lateral wall, with numerous muscle bundles inside that form a network of hills and furrows, giving it a trabeculated surface. Located between the superior and inferior venae cavae, with an internodal tract passing through it, is a special muscle bundle called the *crista terminalis*. The sinus venarum is located just behind the crista terminalis and contains the orifices of the vena cava and coronary sinus.

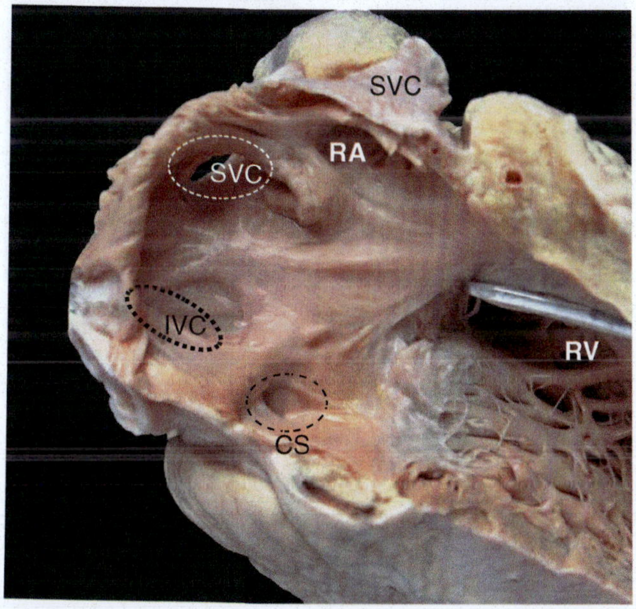

Fig. 1.1 The characteristics of the right ventricle (*RV*). The RV is a cardiac cavity directly connected with the SVC, the IVC, and the coronary sinus. *CS* coronary sinus, *IVC* inferior vena cava, *RA* right atrium, *RV* right ventricle, *SVC* superior vena cava

X. Zhu (ed.), *Surgical Atlas of Cardiac Anatomy*,
DOI 10.1007/978-94-017-9409-1_1, © Springer Science+Business Media Dordrecht and People's Medical Publishing House 2015

1.1.2 Right Atrial Appendage (RAA)

The RAA is a small conical, triangle-shaped muscular pouch. It is attached to the right atrium of the heart and often is short and blunt (Fig. 1.2).

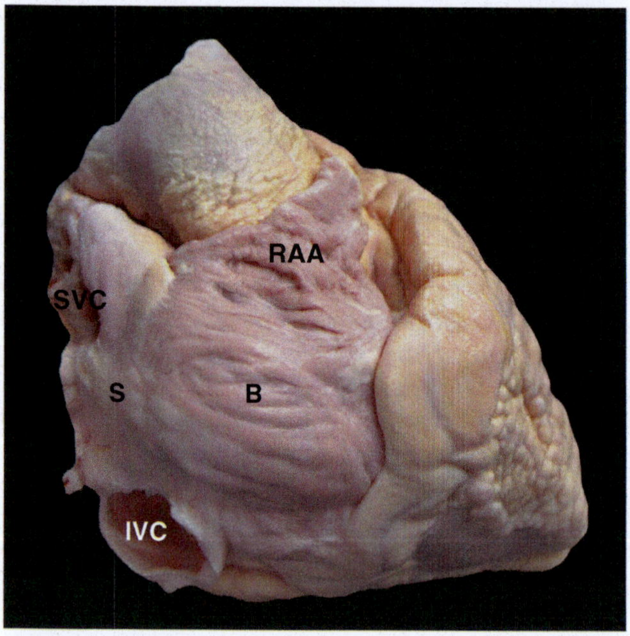

Fig. 1.2 The characteristics of the right atrial appendage (*RAA*). The RAA is short, blunt, wide-based, and triangle-shaped, with numerous muscle bundles inside that give it, like the RA, a trabeculated surface. *B* the body part of RA, *RAA* right atrial appendage, *S* the sinus venarum of RA, *SVC* superior vena cava, *IVC* inferior vena cava

1.1.3 Left Atrium (LA)

The term "left" when used for the left atrium (LA) is a morphological concept, rather than a simple direction. The LA is the cardiac chamber that receives drainage from the pulmonary veins through four ostia in its posterior wall. The LA has a smooth inner wall, whereas in the left atrial appendage (LAA), muscle bundles form a network of hills and furrows (Fig. 1.3). A crescent reductus can be seen in the interatrial septum, the wall of tissue separating the right and left atria, when looking through the LA, in contrast to the fossa ovalis, which can be seen when looking through the RA.

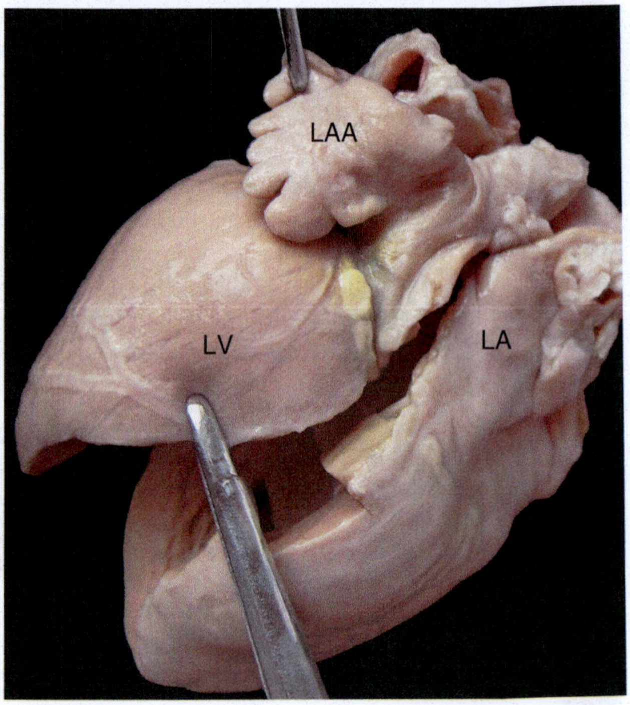

Fig. 1.3 Some muscle bundles form a network of hills and furrows in the LAA. This network is different from the inner wall of other parts of the LA

1.1.4 Left Atrial Appendage (LAA)

The left atrial appendage (LAA) is a small pouch located high in the LA. It pumps oxygenated blood from the lungs into the left ventricle. In contrast to the short and blunt RAA, the LAA has a long, narrow, and tubular shape and an obvious indentation; it usually is multilobed (Fig. 1.4).

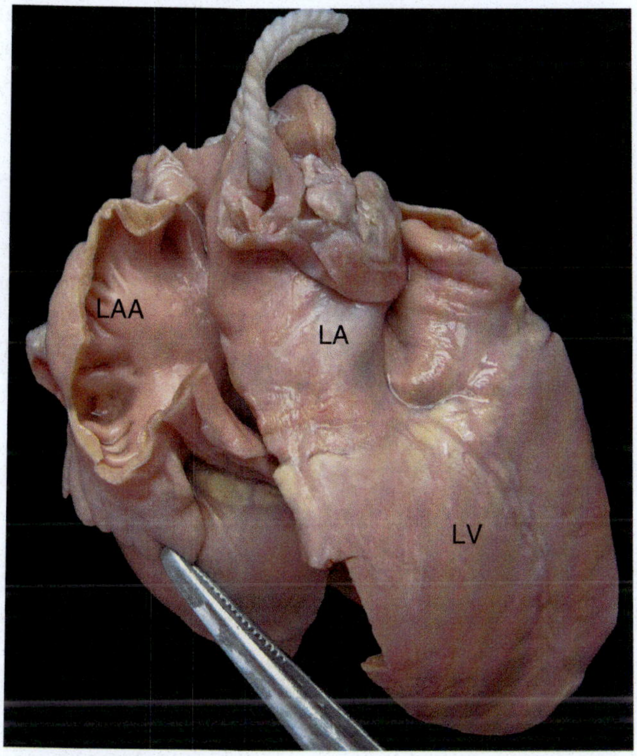

Fig. 1.4 In contrast to the RAA, the LAA has an obvious indentation that is just like the neck of the LAA. *LAA* left atrial appendage, *LA* left atrium, *LV* left ventricle

1.1.5 Right Ventricle (RV)

The right ventricle (RV) is a cardiac chamber located in the front side of the heart. It contains numerous muscle bundles and looks coarsely trabeculated. The term "right" is also a morphological concept. The morphological characteristic is that crista supraventricularis muscles widely separate the atrioventricular (tricuspid) valve from the semilunar (pulmonary) valve in the normal heart (Fig. 1.5). The inner surface of the RV consists of three parts, the orifice portion, the trabecular portion, and the outlet portion.

The outlet portion of the RV is composed of muscles of the infundibulum. The muscle of the inferior border of the infundibular septum is called *crista supraventricularis*, or *trabecula septomarginalis*. It contains several different muscle bundles, the parietal band, the moderator band, and the septal band (Fig. 1.6). The septal band often extends apically to be continuous with the moderator band, which is an important trabeculation located among the septum, anterior papillary muscle of the tricuspid valve (TV), and the free wall. The crista supraventricularis is found between the TV and the pulmonary valve (PV).

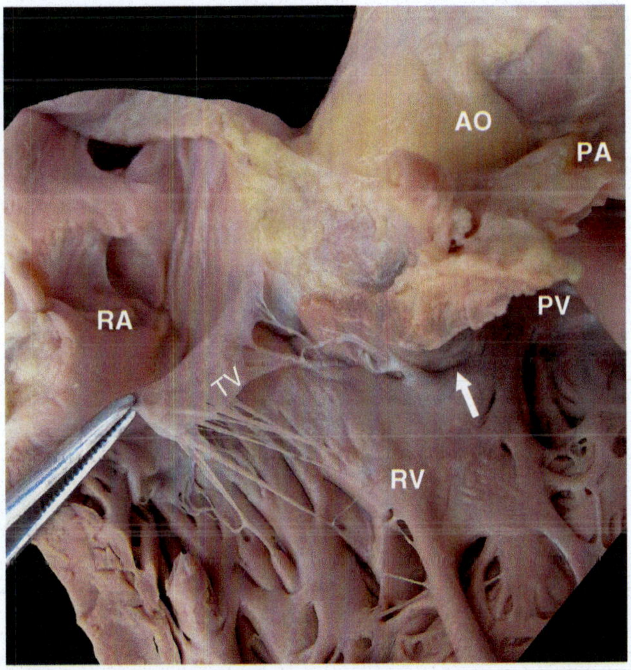

Fig. 1.5 The characteristics of the right ventricle (*RV*). The RV has numerous muscle bundles inside with coarse trabeculae. The atrioventricular or TV, with some chordae tendineae attached to the interventricular septum. The tricuspid valve is separated from the PF by the *crista supraventricularis muscles*. *AO* aorta, *PV* pulmonary valve, *RA* right atrium, *RV* right ventricle, *TV* tricuspid valve, *SVC* the *white arrow* show the muscle that separate pulmonary valve and tricuspid valve

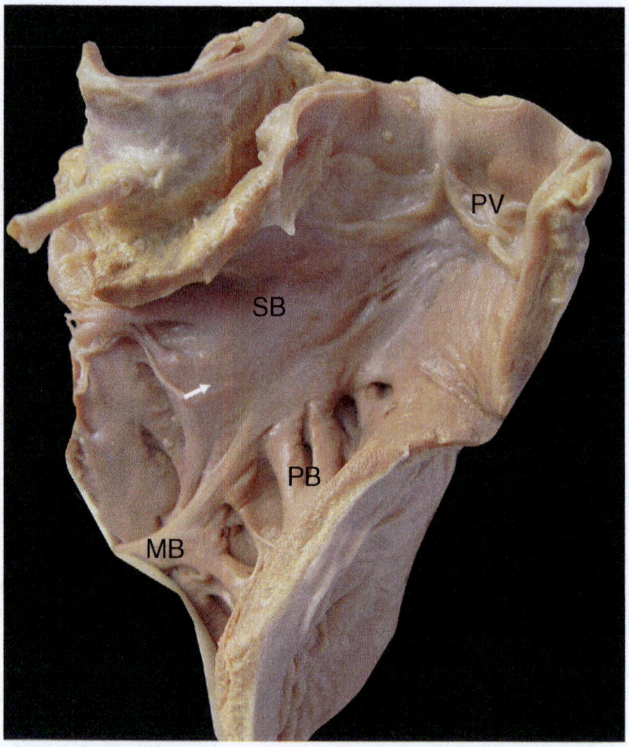

Fig. 1.6 The muscle bundles between the TV and the PV. *MB* moderator band, *PB* parietal band, *PV* pulmonary valve, *SB* septal band

1.1.6 Tricuspid Valve (TV)

The TV is the atrioventricular valve of the RV and has three leaflets: the anterior, posterior, and septal leaflets. The anterior leaflet is usually the largest one, with a strong anterior papillary muscle; the septal leaflet is usually smaller than the anterior leaflet but often larger than the posterior leaflet; it has a small chorda that mostly connects with the interventricular septum.

1.1.7 Left Ventricle (LV)

The left ventricle (LV) is a cardiac chamber that is divided into the orifice portion, the trabecular portion, and the outlet portion. The term "left" is also a morphological concept with regard to the LV. The mitral valve (MV) is the atrioventricular valve of the LV. In a normal heart, the MV is separated from the aortic valve (AV) by fibrous tissue instead of muscles.

1.1.8 Mitral Valve (MV)

The mitral valve (MV) is the atrioventricular valve of the LV and has two leaflets: anterior and posterior. These two leaflets connect with the anterior and posterior papillary muscles, respectively, by chordae. Normally, no chordae are attached to the interventricular septum.

1.1.9 The Posterior Cross of the Heart

The *posterior cross of the heart* refers to the cross in the posterior aspect of the heart, located at the connection of the interatrial groove and the interventricular groove. The coronary sinus runs from the left to the right of the atrioventricular groove, forming the transverse line of the cross (Figs. 1.7 and 1.8). The posterior margins of the interatrial septum and the interventricular septum compose the upright line. The posterior margin of the interventricular septum is the same line as the posterior interventricular groove, with the posterior descending branch of the coronary artery passing through it.

1.1.10 Interatrial Septum (Atrial Septum)

The interatrial septum is the wall of tissue separating the left and right atria of the heart (the upper chambers of the heart). As noted previously, the fossa ovalis (FO) in the interatrial septum can be seen when looking through the RA cavity, whereas a crescent can be seen when looking through the LA. When the interatrial septum does not develop normally, a hole in the septum, called an *atrial septal defect* (ASD), results.

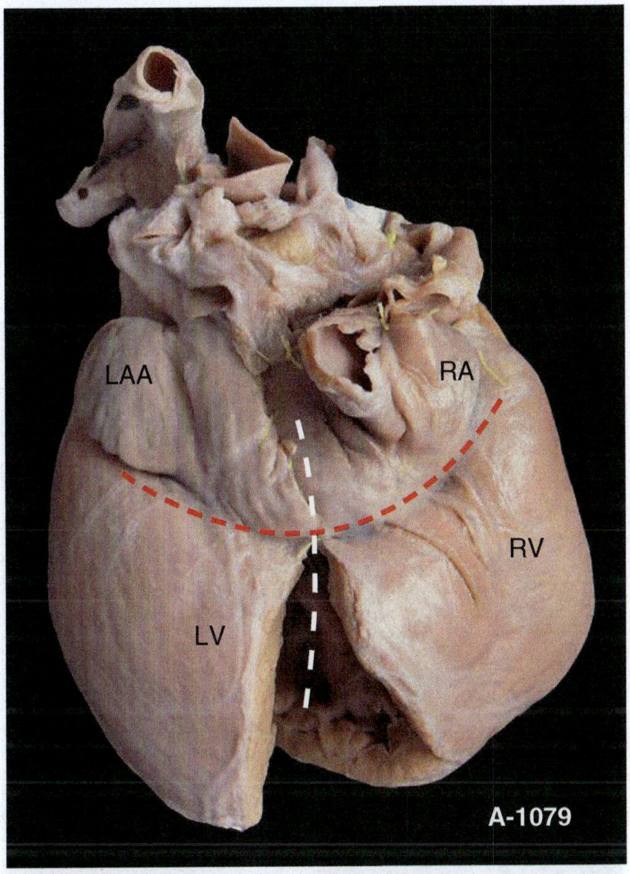

Fig. 1.7 The transverse axis refers to the left and right atrioventricular groove; the longitudinal axis refers to the interatrial groove and the interventricular groove. *LAA* left atrial appendage, *LV* left ventricle, *RA* right atrium, *RV* right ventricle

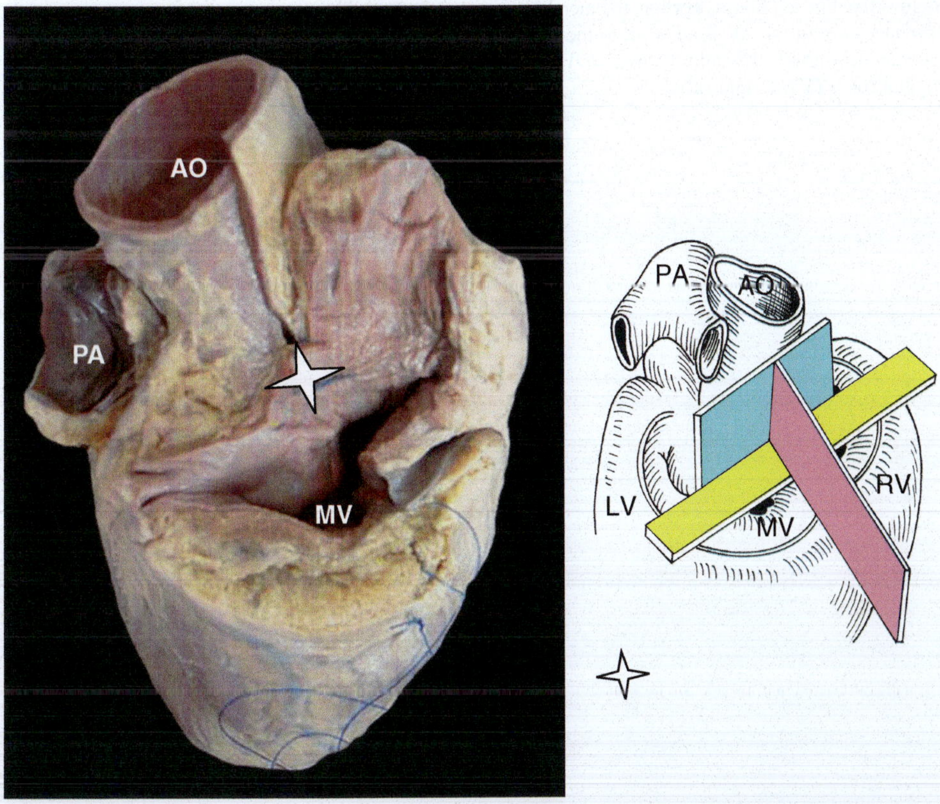

Fig. 1.8 3-D cross section of the heart. *AO* aorta, *PA* pulmonary artery, *MV* mitral valve

1.1.11 Interventricular Septum (Ventricular Septum)

The interventricular septum is another wall just like the interatrial septum but thicker. It separates the left and right ventricles of the hearts (the lower chambers of the heart). Two atrioventricular valves, the MV and the tricuspid valve (TV), are located at its two sides, respectively (Fig. 1.9). The interventricular septum can be divided into the muscular ventricular septum and the membranous ventricular septum (Fig. 1.10). The muscular ventricular septum is thick and is located at the anterior and lower part of the septum. The membranous ventricular septum is thin and fibrous and is located at the posterior and upper part of the septum.

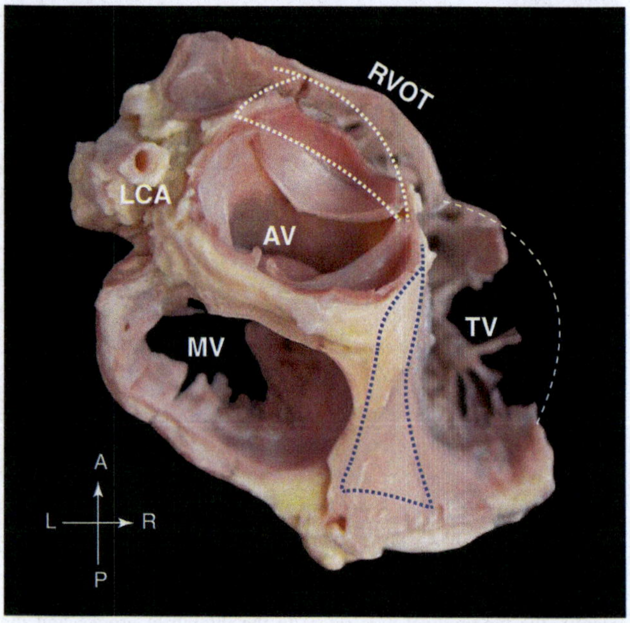

Fig. 1.9 Cross section of the interventricular septum (IVS) (from the upper view angle). *Blue dash line* shows the posterior portion of the IVS; *white dash line* shows the anterior portion of the IVS; between them is the thin and fibrous portion, the membranous septum. The right coronary cusp of the AV is adjacent to the IVS. *AV* aortic valve, *LCA* left coronary artery, *MV* mitral valve, *ROVT* right outflow of the ventricular tract, *TV* tricuspid valve

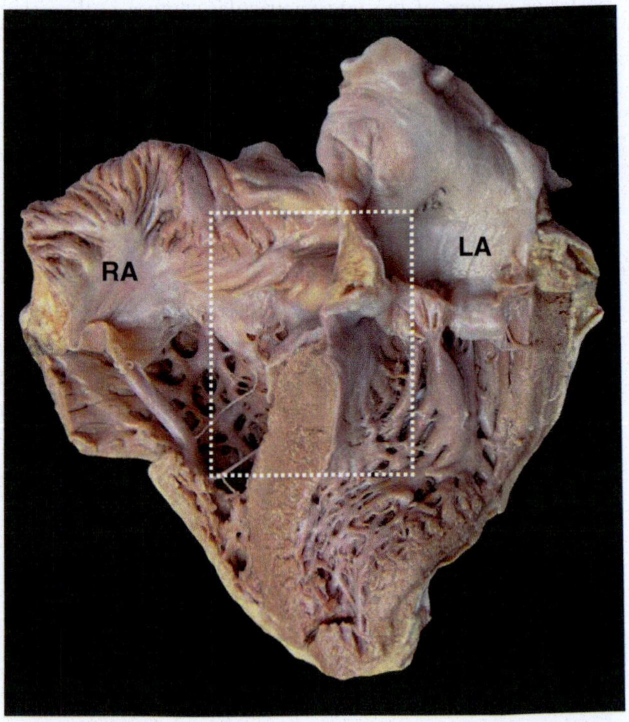

Fig. 1.10 Cross section of the posterior portion of the interventricular septum (IVS). From the anterior view angle, the posterior wall of the four chambers of the heart can be seen clearly. It also shows that the annulus of the MV lies a little higher than that of the TV

1.1.12 Atrioventricular Septum

The septum between the RA and the LV is called the *atrioventricular septum* (Fig. 1.11). Generally, no communication exists between the LA and the RV. The atrioventricular septum is located behind the membranous ventricular septum and below the atrial septum. Its left and right portions are located under the annulus of the MV and the annulus of the TV, respectively.

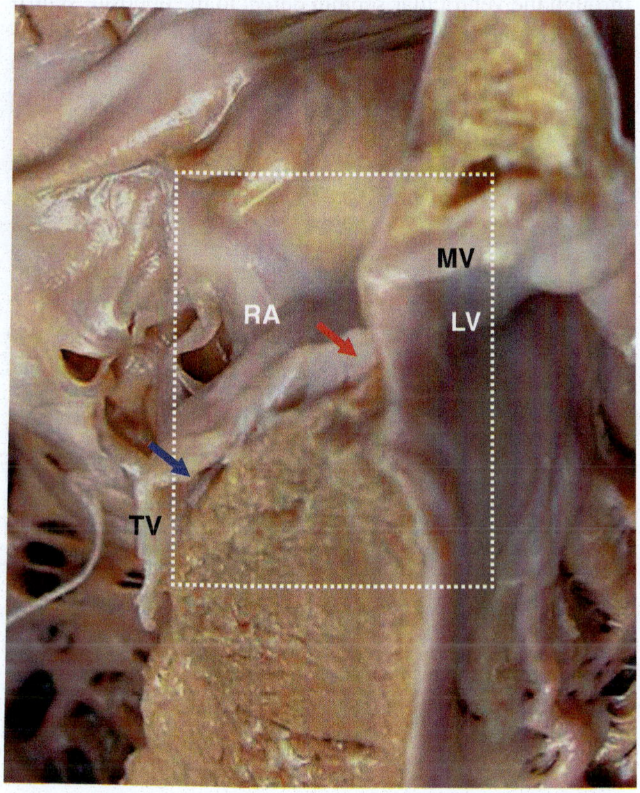

Fig. 1.11 The atrioventricular septum. The *red arrow* shows the membranous septum between the right atrium (*RA*) and the left ventricle (*LV*), *SVC* the *blue arrow* show the root of septal leaflet of tricuspid valvc, *MV* mitral valve, *TV* tricuspid valve

1.1.13 Aorta

The aorta is a large artery arising from the base of the heart and extending down to the abdomen. It can be divided into five sections: ascending aorta, arch of aorta, descending aorta, thoracic aorta, and abdominal aorta. It gives rise to the main coronary arteries and the brachiocephalic trunk.

1.1.14 Main Pulmonary Artery (Pulmonary Trunk)

The main pulmonary artery is the portion of the pulmonary artery that originates from the base of the RV. It is short and wide, and it branches into the left and right pulmonary arteries. The function of the pulmonary artery is to transport deoxygenated blood from the RV to the lung.

1.2 Malformations of the Heart

1.2.1 Atrial Septal Defect (ASD)

As one of the most common congenital heart diseases, atrial septal defect (ASD) is a defect in the interatrial septum that enables oxygenated blood to pass from the LA to the RA directly, or vice versa. It can lead to right-sided volume overload of the heart, pulmonary hypertension, Eisenmenger syndrome, and other conditions.

1.2.1.1 Ostium Secundum Atrial Septal Defect

As the most common type of ASD, ostium secundum ASD is formed by dysplasia of the septum secundum in the embryonic period. It usually involves the foramen ovale and the posterior part of the interatrial septum, often close to the caval vein but away from the TV annulus (Fig. 1.12).

1.2.1.2 Ostium Primum Atrial Septal Defect

This type of ASD is formed by dysplasia of the ostium primum in the embryonic period. Because it often involves the endocardial cushion, it is sometimes called an *endocardial cushion defect*. It usually is found quite close to the TV annulus; the foramen ovale would not be involved chiefly. In this type of ASD, the atrioventricular septum is narrower and shorter, the MV annulus is lower, and the triangle of Koch is smaller (Fig. 1.12). Both types of ASD can be present at the same time in a patient.

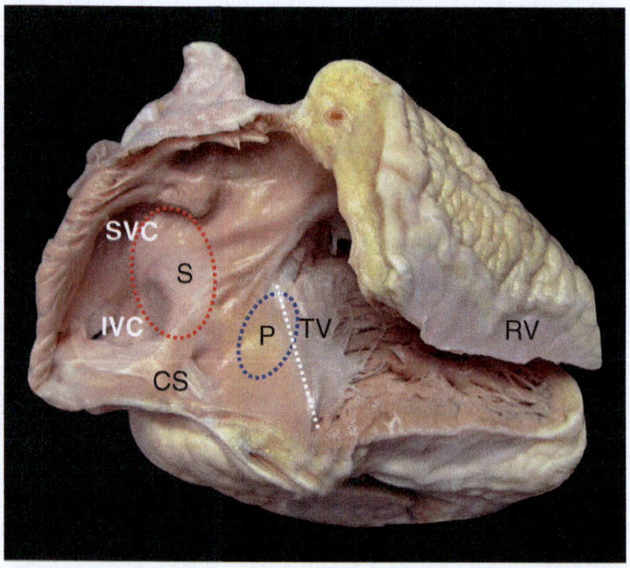

Fig. 1.12 The classification of the two types of ASDs. The *red dash line* shows the area in which an ostium secundum atrial septal defect (*S*) usually occurs. The *blue dash line* shows the area in which an ostium primum atrial septal defect (*P*) usually occurs. *CS* coronary sinus, *IVC* inferior vena cava, *RV* right ventricle, *SVC* superior vena cava, *TV* tricuspid valve

1.2.2 Atrioventricular Septal Defect

Even though a defect in the ostium primum is classified as an ASD, it is more commonly considered as an atrioventricular septal defect. When an ostium primum ASD involves the interventricular septum, the heart's four chambers then can communicate with each other. The two atrioventricular valves are usually connected together by some fibrous tissues, and some chordae tendineae of one valve may pass through the ventricular septum defect to the opposite ventricle. Accordingly, a common atrioventricular valve is necessary for the diagnosis of atrioventricular septum defect, but an isolated ASD and an isolated ventricular septal defect (VSD) without a common atrioventricular valve cannot be considered as an atrioventricular defect.

1.2.3 Single Ventricle

Single ventricle refers to a special cardiac deformity in which there is only one functional ventricle. It may manifest in one of two ways: (1) only a single ventricle is present anatomically or (2) a main functional ventricle and a poorly developed ventricle are present simultaneously. However, if the maldevelopment of one ventricle results from mitral/tricuspid atresia, it cannot be classified as single ventricle and should be diagnosed as mitral/tricuspid atresia.

1.2.4 Left Ventricle Dysplasia

In LV dysplasia, the LV develops poorly, with a very small cavity, but the main structure is present. It can be complicated by some mitral deformities.

1.2.5 Tetralogy of Fallot (TOF)

Tetralogy of Fallot (TOF) is a common congenital heart disease that involves four anatomical abnormalities of the heart. Stenosis exists in the right outflow of the ventricular tract [(ROVT)] and in the main pulmonary artery to various degrees. A large VSD is usually seen around the infundibulum or below the supraventricular crista. The AV is situated above the large VSD and is connected to both the LV and the RV. A diagnosis of TOF requires that some fibrous connection must exist between the AV and the MV.

1.2.6 Double-Outlet Right Ventricle (DORV)

In the condition known as *double-outlet right ventricle* (DORV), the PV arises from the RV as usual. However, the

AV is above the interventricular septum; as a result, more than 75 % of the aortic blood may come from the RV. A VSD often coexists. The fibrous connection between the AV and the MV may be replaced by some muscular tissues. Moreover, pulmonary hypertension or stenosis may appear concomitantly.

1.2.6.1 Taussig-Bing Deformity

A special type of DORV is the Taussig-Bing syndrome. Its anatomical feature is the PV overriding a VSD and the AV arising totally from the RV.

1.2.7 Double-Outlet Left Ventricle (DOLV)

When the aorta arises from the LV normally while a large portion of the pulmonary artery comes from the LV, the condition is called *double-outlet left ventricle* (DOLV). Like the DORV, DOLV often is complicated by a VSD, pulmonary stenosis, or hypertension.

1.2.8 Complete Transposition of the Great Arteries

In complete transposition of the great arteries, the two great vessels are connected to the wrong ventricles. In other words, the AV is connected to the RV, and the PV is connected to the LV. The semilunar valves, atria, ventricles, and atrioventricular connection are normal.

1.2.9 Corrective Transposition of the Great Arteries

Corrective transposition of the great arteries is a cardiac deformity in which the great vessels are transposed. As the two great vessels are connected to the opposite ventricles, respectively, so the atrioventricular connections also are inverted (i.e., the LA connects with the RV and the RA connects with the LV). Accordingly, deoxygenated blood of the RA can flow into the pulmonary artery (PA) through the LV, and oxygenated blood of the LA can also flow into the aorta through the RV. Hence, this deformity is called *corrective transposition of the great arteries*.

1.2.10 Pulmonary Artery Atresia (PAA)

In pulmonary artery atresia (PAA), deoxygenated blood from the RV cannot flow to the pulmonary artery because the main pulmonary artery does not exist or has no lumens. When the PA has a very small lumen and some blood can flow in from the RV, the condition is *pulmonary stenosis*, which must be differentiated from PAA.

1.2.11 Obstruction of the Right Outflow of the Ventricular Tract (ROVT)

Obstruction of the ROVT refers to the muscular thickening of the ROVT, without overriding of the aorta and pulmonary atresia.

1.2.12 Persistent Truncus Arteriosus

Persistent truncus arteriosus is a rare cardiac deformity of infants. The embryological structure called the *truncus arteriosus* fails to divide properly into the pulmonary trunk and aorta. Only one semilunar valve is present in the heart, and a subvalvular VSD can be found. The main pulmonary artery or the two branched pulmonary arteries can originate from the ascending aorta or the aortic arch.

1.2.13 Aortopulmonary Window

The term *aortopulmonary window* refers to a congenital heart defect in the arterial wall of the connection portion of the aorta and main pulmonary artery. Through this "window," blood flow of these two great vessels can communicate with each other directly. In this case, no canals or ducts are present between the aorta and main pulmonary artery.

1.3 Nomenclature and Acronyms

The abbreviations shown in the table above will be adopted in the following chapters of this book. For example, "VSD" will be taken to stand for "ventricular septal defect," and no more explanation or information about its full name will be given when the abbreviation is used.

AO	Aorta
A-P	Aortopulmonary
ASD	Atrial septal defect
AV	Aortic valve
AVSD	Atrioventricular septal defect
CS	Coronary sinus
CTGA	Corrected transposition of great arteries
DORV	Double-outlet right ventricle
DOLV	Double-outlet left ventricle
FO	Foramen ovale
IVC	Inferior vena cava
LA	Left atrium
LAA	Left atrial appendage

LBB	Left bundle branch
LCA	Left coronary artery
LPA	Left pulmonary artery
LSVC	Left superior vena cava
LV	Left ventricle
LVOT	Left ventricular outflow tract
LVOTO	Left ventricular outflow tract obstruction
MPA	Main pulmonary artery
MV	Mitral valve
PA	Pulmonary artery
PAPVD	Partial anomalous pulmonary drainage
PDA	Persistent ductus arteriosus
PV	Pulmonary valve
RA	Right atrium
RAA	Right atrial appendage
RBB	Right bundle branch
RCA	Right coronary artery
RPA	Right pulmonary artery
RV	Right ventricle
RVOT	Right ventricular outflow tract
SVC	Superior vena cava
TAPVD	Total anomalous pulmonary venous drainage
TGA	Transposition of the great arteries
TOF	Tetralogy of Fallot
TV	Tricuspid valve
VSD	Ventricular septal defect

Position and Observation of the Heart and Surgical Approaches

2

2.1 Position of the Heart

The heart is a muscular organ in the mediastinum, a cavity located anterior to the vertebral column, posterior to the sternum, above the diaphragmatic muscle, and between the two lungs (Figs. 2.1 and 2.2). Its function is to pump blood through the blood vessels by repeated, rhythmic contractions. The portion of interpleural space between the heart and the sternum is called the *anterior mediastinum*. It contains no important structures and has only a lamellar sternum in front of it; hence, it is the most frequently selected surgical approach to the heart. In contrast is the *posterior mediastinum*, the interpleural space between the posterior wall of the heart and the vertebral column; it contains many important organs, such as the trachea, the esophagus, the aortic arch, and branches of the descending artery. The trachea branches into two main bronchi on the level of the sternal angle, and enlargement of the LA may raise the left main bronchus. The lower segment of the esophagus is situated close to the posterior wall of the heart. The aortic arch is directed backward on the left side of the trachea and passes downward on the left side of the body. It is also continuous with the descending aorta at the lower border of the fourth thoracic vertebra. A large aortic arch aneurysm may result in dyspnea due to compression of the trachea. The *superior mediastinum* refers to the cavity located above the heart and between the manubrium sterni and the upper thoracic vertebra. It holds several vital organs, including the thoracic gland, the aortic arch and its vital branches, the anonymous veins, the azygous vein and the upper portion of the SVC, the diaphragmatic nerve, and the vagus nerve. Located behind the heart is the vertebral column.

X. Zhu (ed.), *Surgical Atlas of Cardiac Anatomy*,
DOI 10.1007/978-94-017-9409-1_2, © Springer Science+Business Media Dordrecht and People's Medical Publishing House 2015

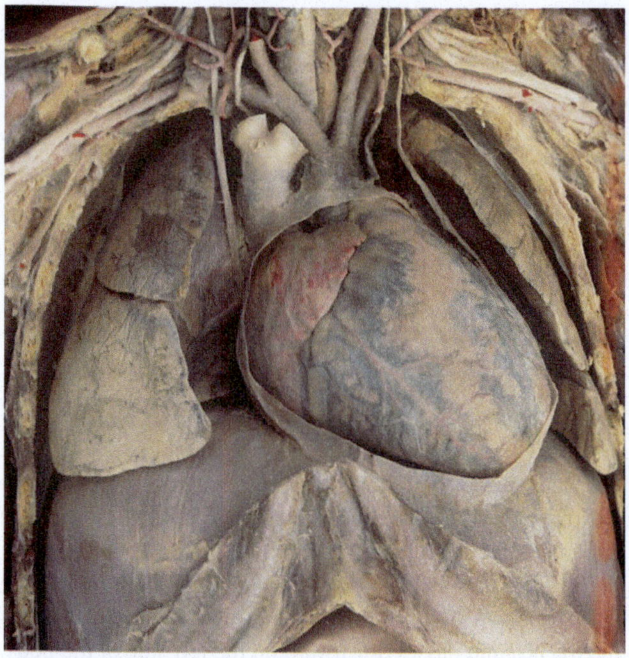

Fig. 2.1 The position of the heart. The heart is located in the mediastinum, anterior to the vertebral column, posterior to the sternum, above the diaphragmatic muscle, and between the two lungs. It is protected by a perfect bony framework

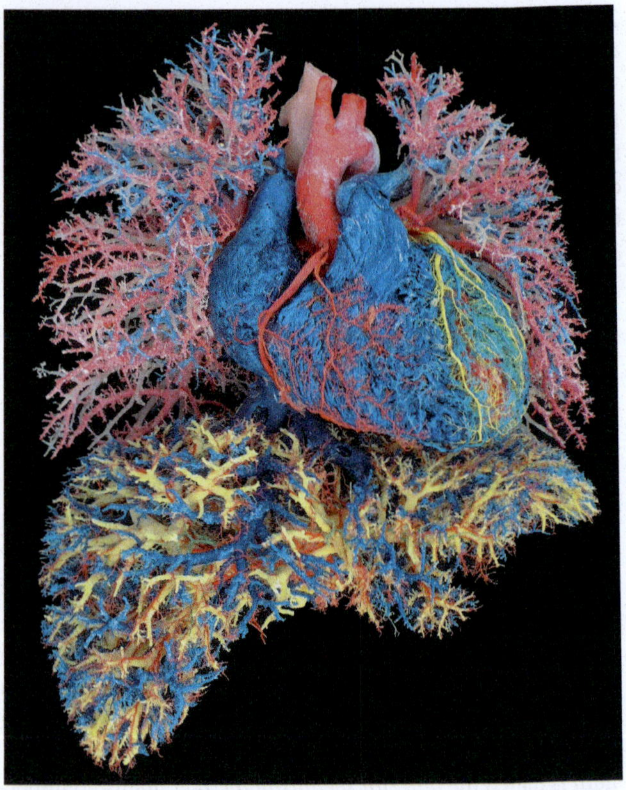

Fig. 2.2 Stereoimaging of the vessels of the heart, lung, and liver

2.2 Observation of the Heart

The anatomy of the heart can be described by observing it from different visual angles that provide views of the axes of the heart (Figs. 2.3 and 2.4). The visual angle used most frequently provides a view from the anterior aspect and is called the *anatomical angle* (Figs. 2.5 and 2.6). The *left anterior oblique angle* allows the four chambers of the heart to be seen simultaneously (Figs. 2.7, 2.8, 2.9, and 2.10), whereas the *right anterior oblique angle* provides the clearest visualization of the RA and the RV (Fig. 2.11). Moreover, the operator usually observes the heart from the right shoulder side (Fig. 2.12), so this visual angle may be named the *operator angle* (Fig. 2.13). In some special conditions, it will be described by some special angles, such as from the side of the head or back (Figs. 2.14 and 2.15).

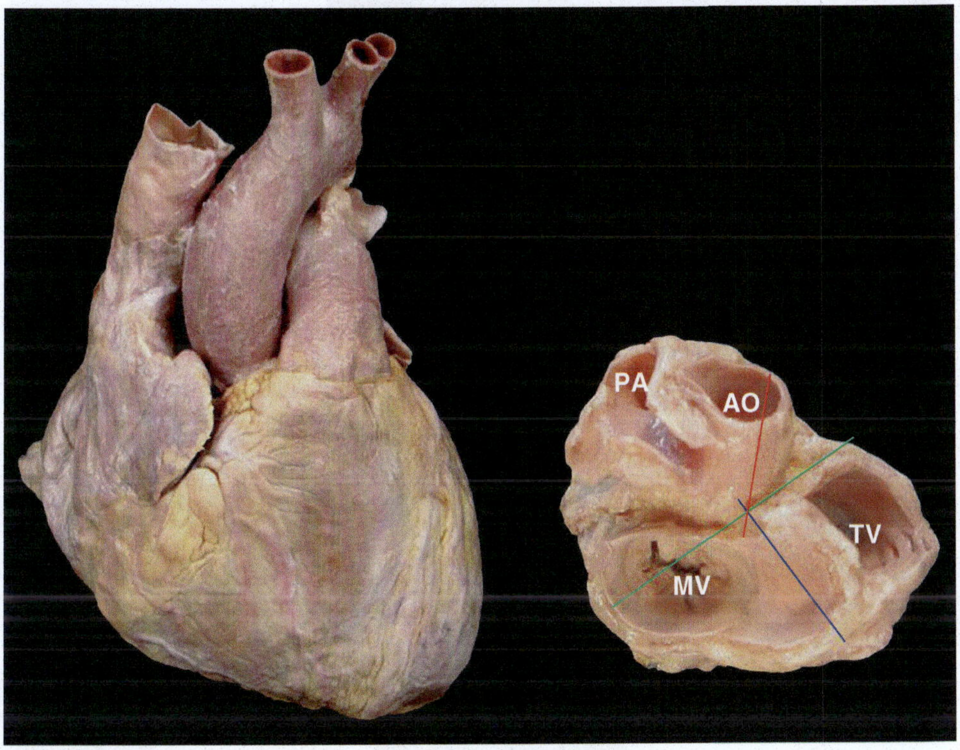

Fig. 2.3 The axes of the heart. The *blue line* parallel to the interatrial septum is the anteroposterior axis; the *red line* is the cephalocaudal axis; the *green line* is the Y-axis. These axes cross at the central fibrous trigone. *PA* pulmonary artery, *AO* ascending aorta, *TV* tricuspid valve, *MV* mitral valve

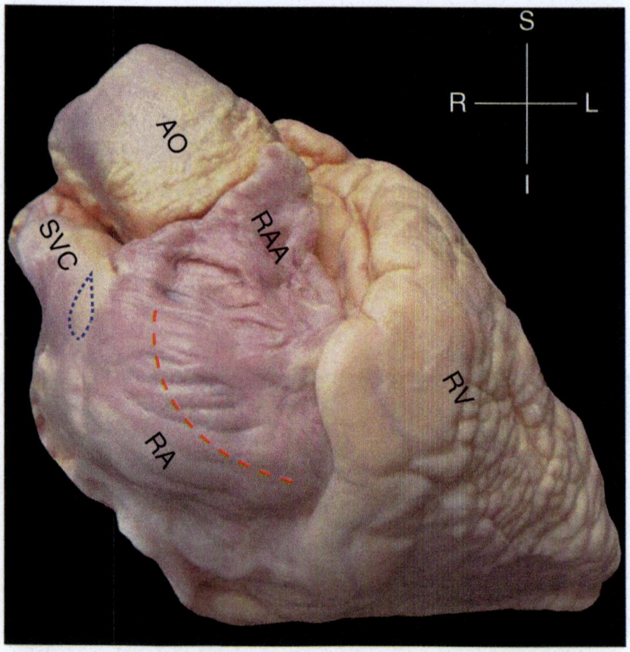

Fig. 2.4 Observation from the *anatomical angle*. The right atrioventricular groove, with the right coronary artery passing through it, and the anterior interventricular groove, with the left anterior descending artery. The right margin of the heart is composed of the RA, and the inferior margin is formed by the RV. The aortic arch, main pulmonary artery, LAA, and apex constitute the left margin. The *red dash line* indicates the incision made in the RA during surgery, and the *blue dash line* indicates the sinoatrial node. *AO* ascending aorta, *RA* right atrium, *RAA* right atrium appendage, *RV* right ventricle, *SVC* superior vena cava

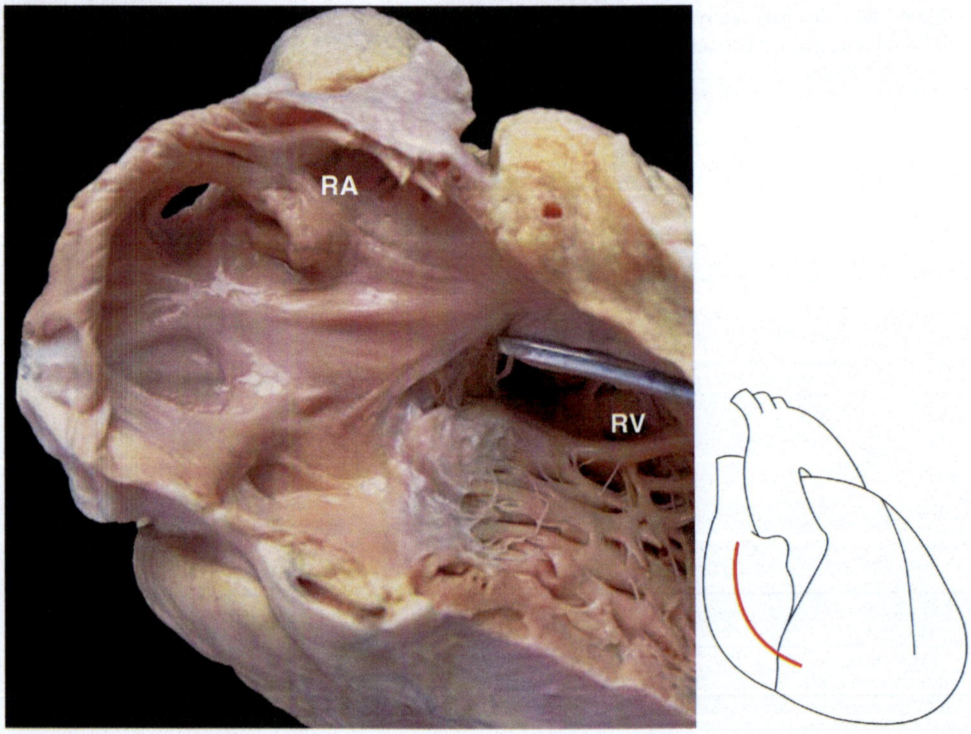

Fig. 2.5 Observation from the right side. Through the RA incision, inner wall of the RA and septal leaflet of the tricuspid valve can been seen. *RA* right atrium, *RV* right ventricle

Fig. 2.6 Observation by the right anterior oblique position. The RVOT was exposed. The tricuspid valve and the RA can be seen clearly. *FO* fossa ovalis, *RA* right atrium, *RV* right ventricle

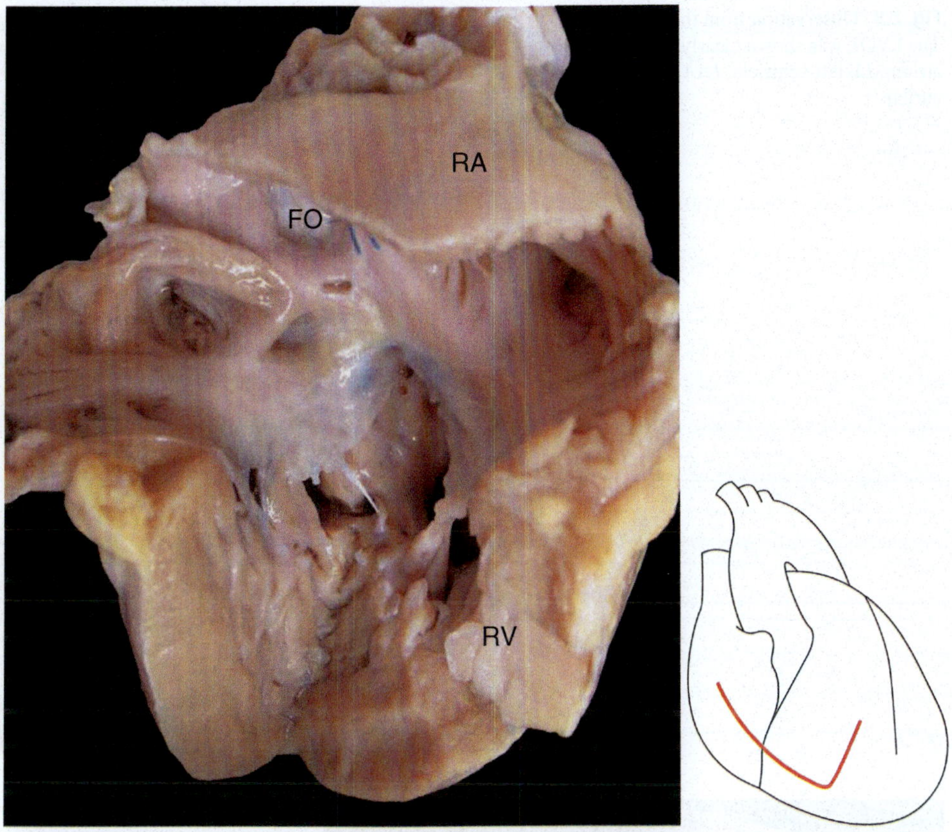

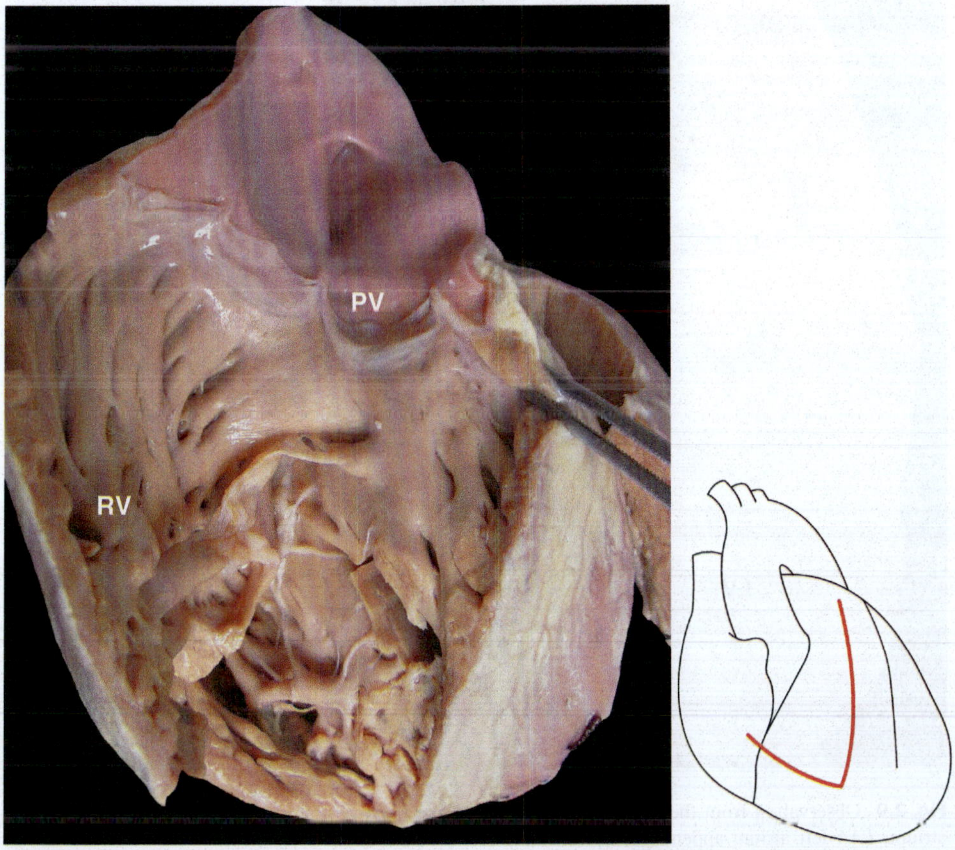

Fig. 2.7 Observation from the *anatomical angle*. The anterior wall of the RV was incised, and the RVOT was shown. *PV* pulmonary valve, *RV* right ventricle

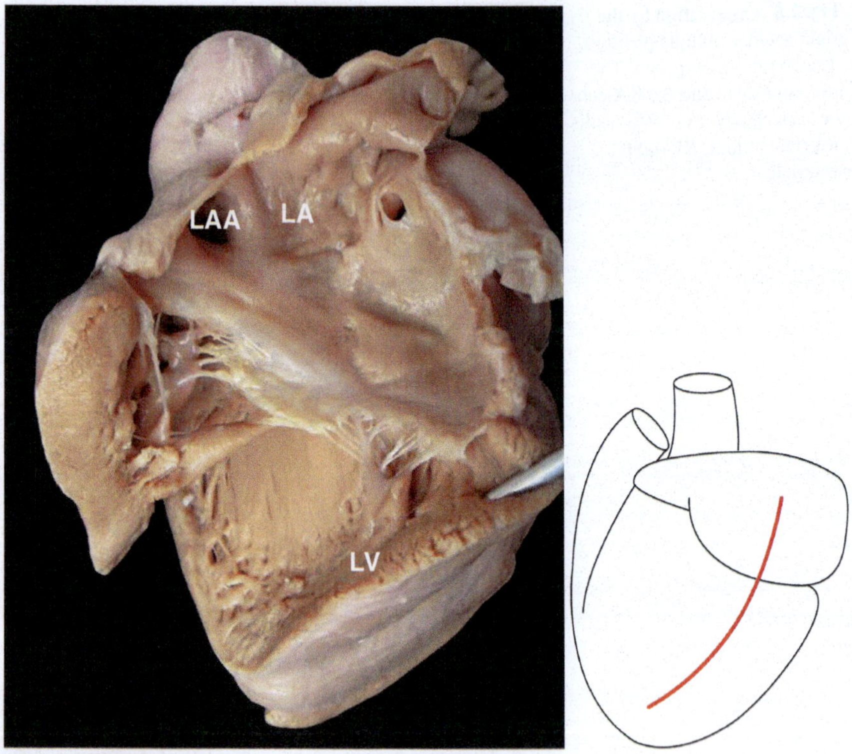

Fig. 2.8 Observation from the left side. The LVOT was shown clearly. *LA* left atrium, *LV* left ventricle, *LAA* left atrium auricle

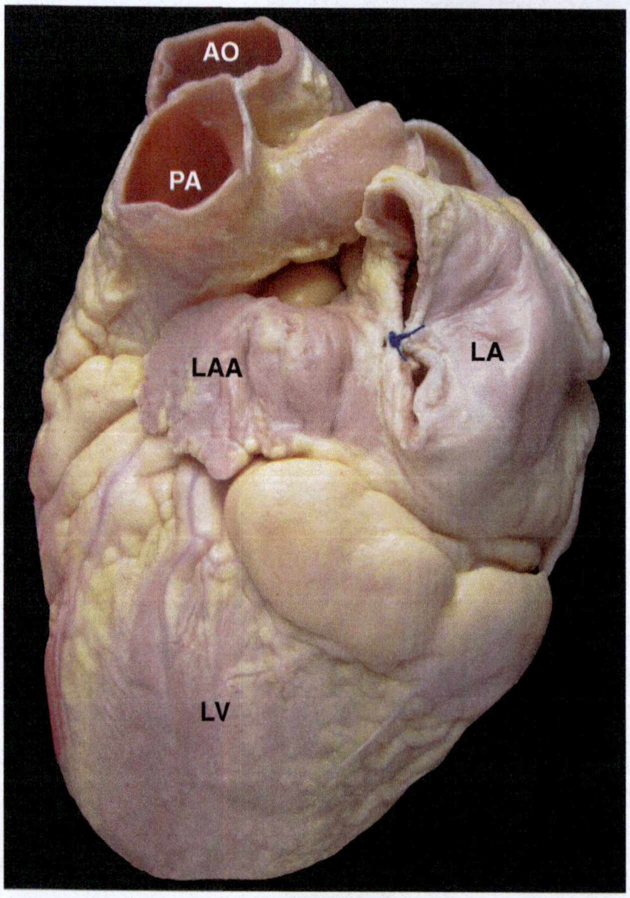

Fig. 2.9 Observation from the *left side*. *AO* ascending aorta, *LA* left atrium, *LAA* left atrium appendage, *LV* left ventricle, *PA* pulmonary artery

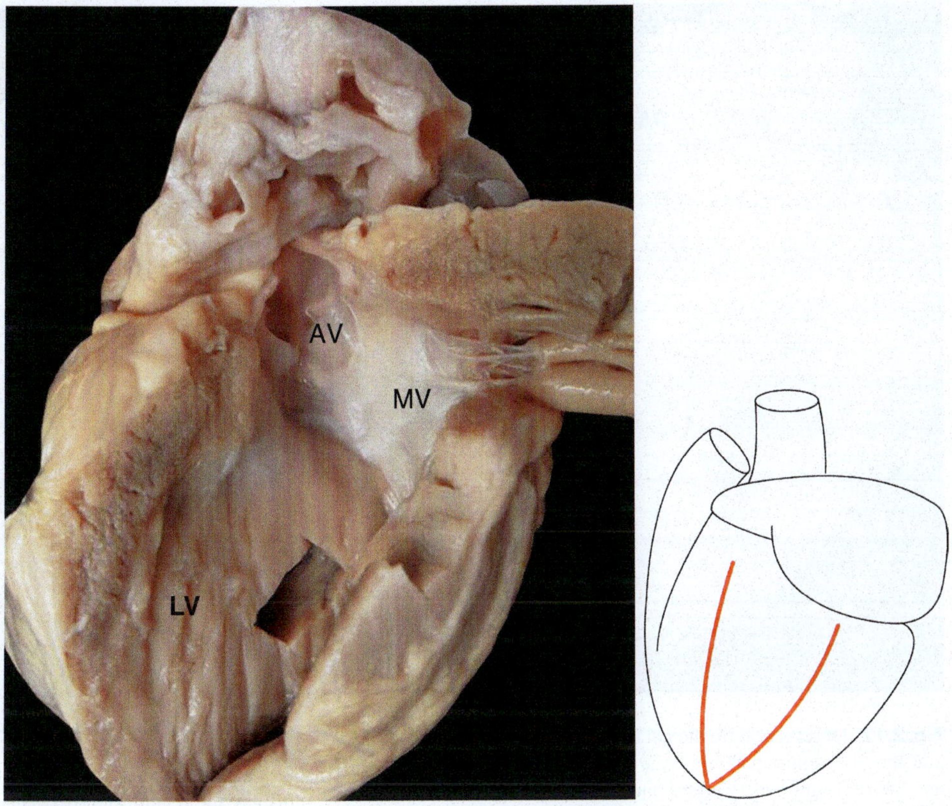

Fig. 2.10 Observation by the *left anterior oblique position*. The LVOT is exposed, and the MV, AV, and interventricular septum can be seen. *AV* aortic valve, *MV* mitral valve, *LV* left ventricle

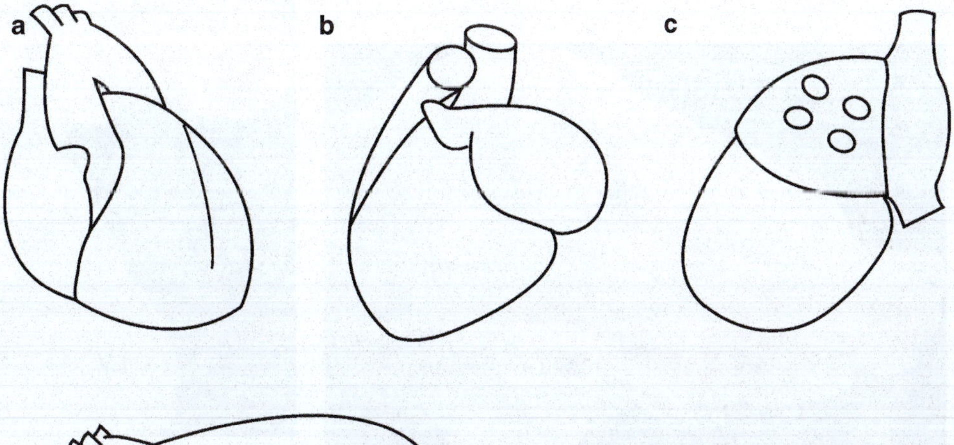

Fig. 2.11 Different visual angles. Most expressions will be described by the anatomical angle in this book. (**a**) anterior angle, (**b**) the left anterior oblique position, (**c**) posterior angle

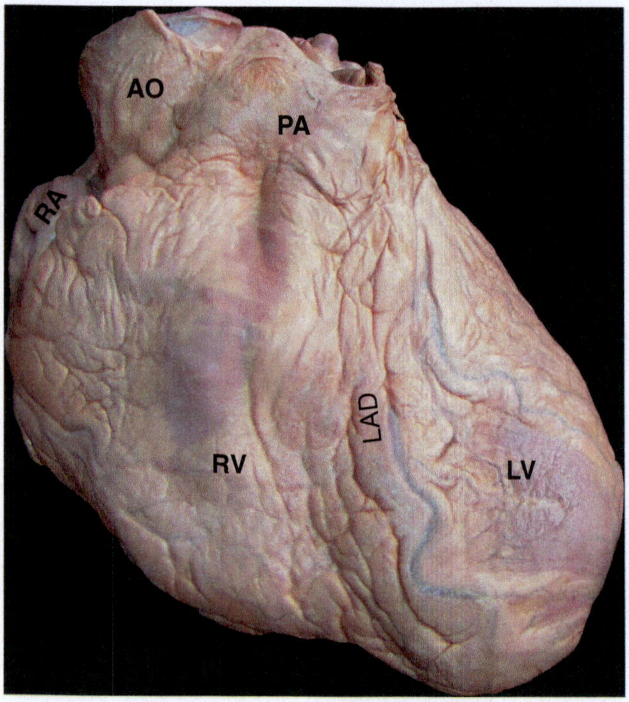

Fig. 2.12 Observation by the *left anterior oblique position*. The LAD and the interventricular septum are seen perfectly. Moreover, the LV and RV, main pulmonary artery, and the LAA, except the RA and aortic root, can be seen well. This is the most frequently used visual angle. The LAD going along the interventricular groove is often considered a feature of the anterior margin of the LV. With fewer vessels, the apex behind the LAD is usually used for an incision in the LV. *AO* ascending aorta, *PA* (main) pulmonary artery, *LAA* left atrial appendage, *LAD* left anterior ascending artery, *LV* left ventricle, *RA* right atrium, *RV* right ventricle

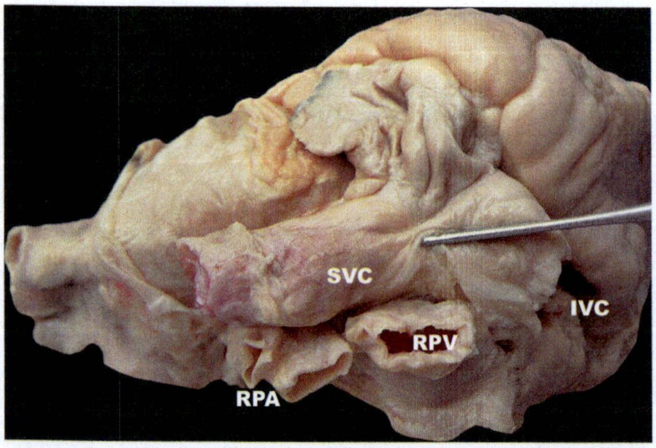

Fig. 2.13 Observation from the *operator angle*. Observed from the right shoulder side, the SVC and the RA can be seen well, as can the ascending aorta and the right pulmonary artery. *IVC* inferior vena cava, *RA* right atrium, *RPA* right pulmonary artery, *RPV* right pulmonary vein, *SVC* superior vena cava

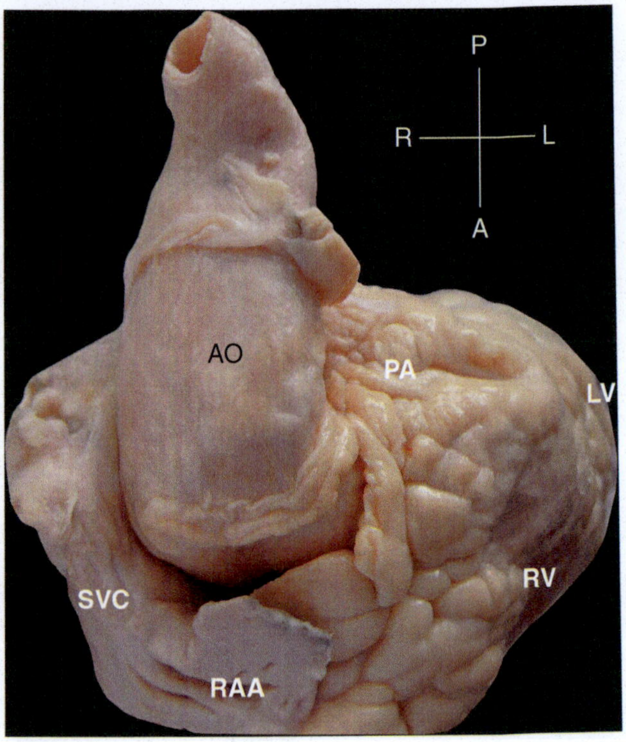

Fig. 2.14 Top view from the anterior side of the heart. This view can display the relationship between the aorta, the pulmonary artery, and the SVC. The AV will be exposed if the aorta wall is sectioned. The ascending aorta locates in the pericardial cavity. *PA* (main) pulmonary artery, *RAA* right atrium appendage, *RV* right ventricle, *SVC* superior vena cava

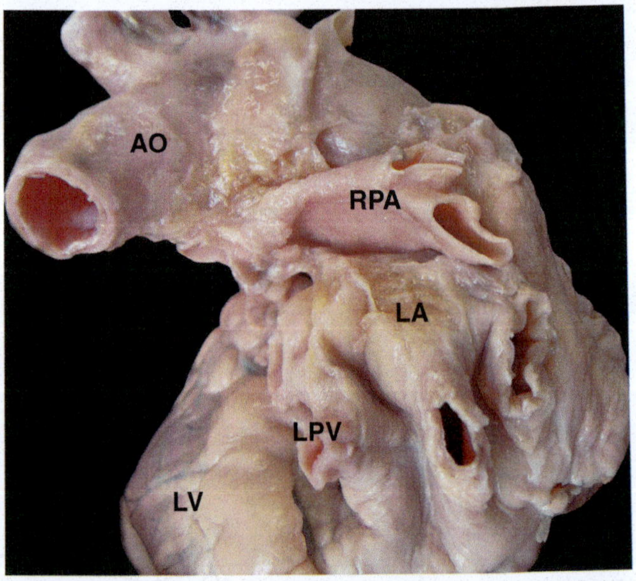

Fig. 2.15 Dorsal view of the heart. This view is used only to describe the back of the heart, but it cannot be displayed during an operation. The location of the right pulmonary artery origin is very important for the interventional treatment of atrial fibrillation. *AO* aortic arch, *LA* left atrium, *LPV* left pulmonary vein, *LV* left ventricle, *RPA* right pulmonary artery

2.3 Surgical Approaches to the Heart

2.3.1 Thoracotomy

To gain access to the heart, the cardiac surgeon must first perform a thoracotomy, which is an incision into the pleural space of the chest. Thoracotomy is a major surgical maneuver and is considered one of the most difficult surgical incisions to deal with postoperatively. A thoracotomy can be made by one of several different approaches, the most common of which are median sternotomy, right thoracotomy, left thoracotomy, and minimally invasive thoracotomy (Figs. 2.16 and 2.17).

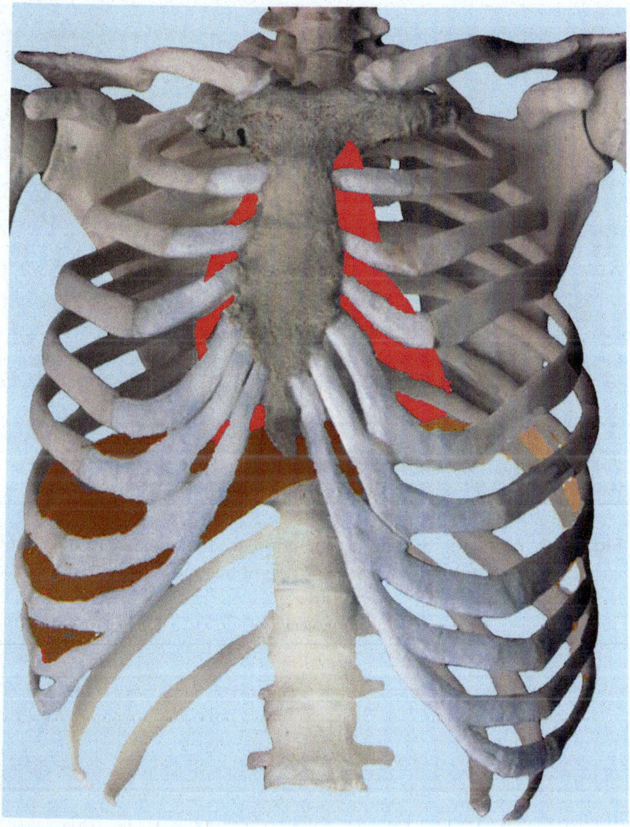

Fig. 2.16 Projection image of the heart in the thorax (provided by Shizhen Zhong)

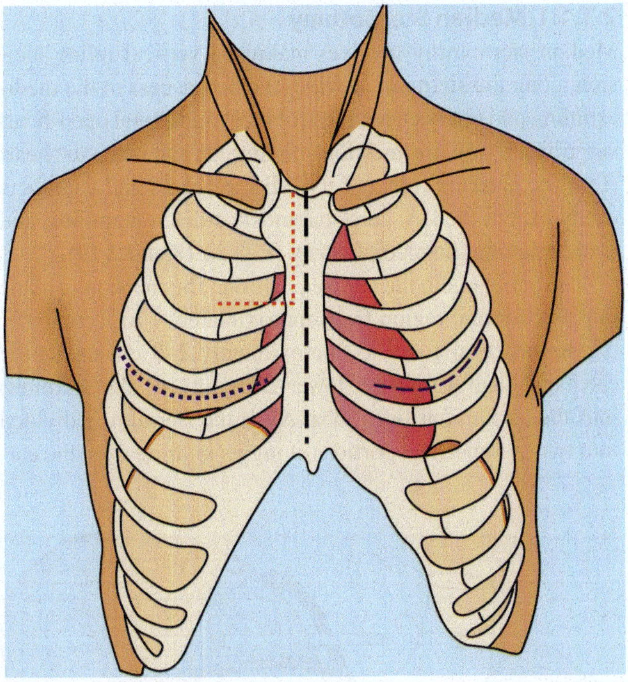

Fig. 2.17 Several common methods for thoracotomy. The patient is in the dorsal position. The *dash line* shows a median sternotomy, a left-and-right thoracotomy in the fourth intercostal space, and the reverse L-shaped mini-invasive thoracotomy

2.3.1.1 Median Sternotomy

Median sternotomy involves making a vertical inline incision along the sternum, provides the best access to the mediastinum, and has been adopted for most traditional open-heart surgeries. Using this incision, the surgeon can reach the heart from the anterior side, observe all the valves and chambers easily except the LA, carry out most cardiac operations, and avoid opening the pleural cavity (Figs. 2.18 and 2.19).

The patient is in the supine position. The incision is made in the presternal region from approximately 1 to 2 cm under the suprasternal fossa to approximately 2–3 cm under the cartilago ensiformis (the lower extremity of the sternum) into the abdominal linea alba. After the sternum is divided into two symmetrical portions along its midline from the cartilago ensiformis toward the suprasternal fossa, hemostasis must be established by using cauterization to control hemorrhaging vessels in the sternal periosteum. Additionally, bone wax can be used to control bleeding from the incision of the medullary canal. When accretio cordis or enlargement of the ascending aorta exists, it is of vital importance to cut the sternum carefully, or the RV or aorta may be injured severely. For infants and young children, the thymus gland is developed well and often needs to be partially resected. Then the pericardium is reached and should be incised vertically. With this incision, all of the heart chambers can be manipulated during an operation. Moreover, during a coronary artery bypass grafting procedure, a median sternotomy allows easy harvesting of the internal mammary artery.

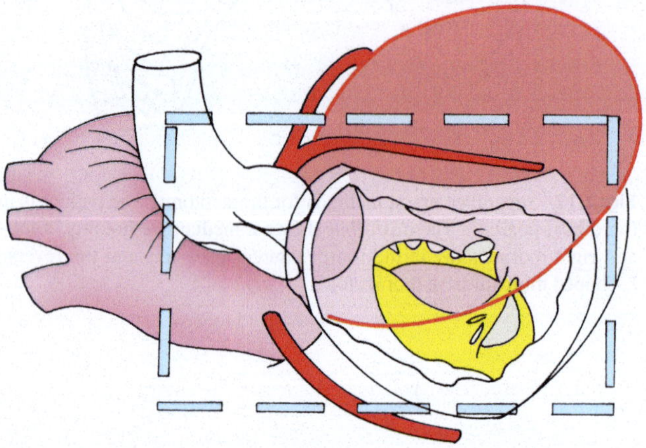

Fig. 2.18 The exposed extent in median sternotomy. The *gray dash line* shows the extent of exposure, and the *yellow site* shows the TV. *TV* tricuspid valve

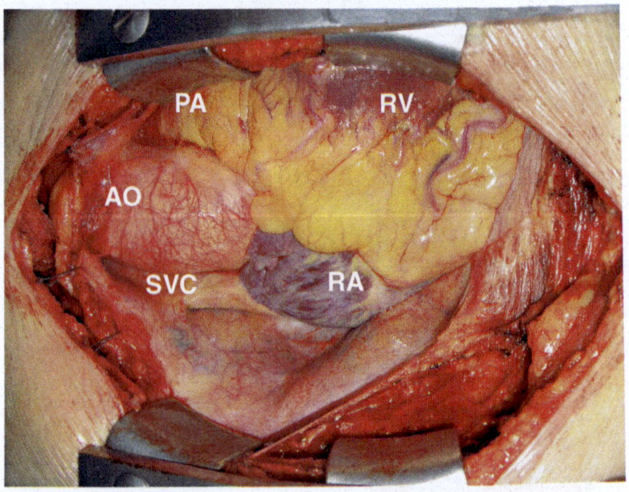

Fig. 2.19 A photograph in an operation by median sternotomy. *AO* aorta, *PA* main pulmonary artery, *RA* right atrium, *RV* right ventricle, *SVC* superior vena cava

2.3.1.2 Right Thoracotomy

A right thoracotomy is usually used for operations on the right chambers. The patient is in the left recumbent position with the right upper extremity suspended properly (Fig. 2.20). The incision is made in the fourth right intercostal space. After the thoracotomy is completed, the right lung should be pushed backward; the diaphragmatic nerve appears downward on the outer wall of the pericardium. The pericardium should be incised along this nerve to expose the RA, the RV, the interatrial groove, and the aortic root (Fig. 2.21). For infants and young children, the ascending aorta and RV outflow tract can be seen well by retracting the pericardium through this incision, but doing so may be dangerous for adults. In this position, the interatrial septum is in a horizontal level, and, therefore, it is convenient to discharge gas from the LA during the repair procedure of the ASD.

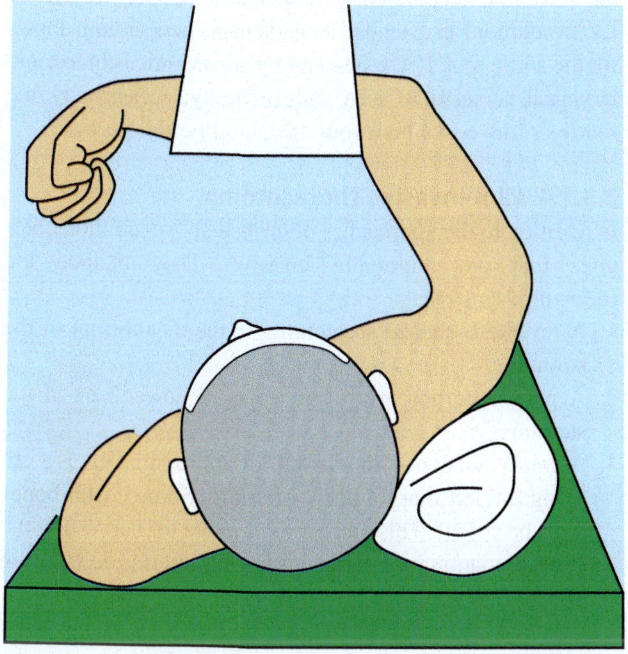

Fig. 2.20 The position for the right thoracotomy. The patient is in the left recumbent position, and his right upper extremity has been suspended properly

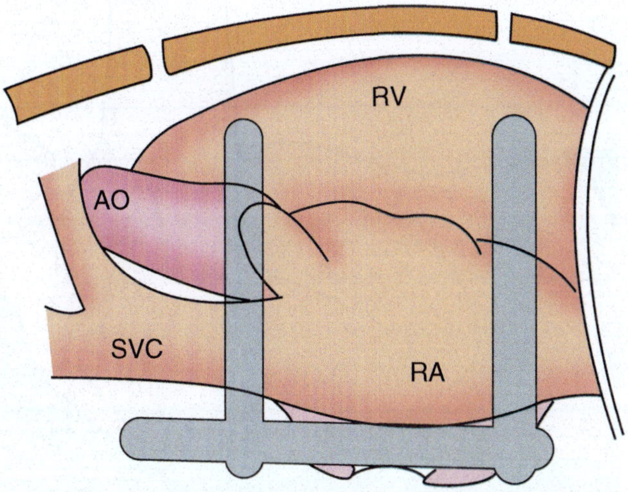

Fig. 2.21 The exposed extent by the right thoracotomy. *AO* aorta, *RA* right atrium, *RV* right ventricle, *SVC* superior vena cava

2.3.1.3 Left Thoracotomy

For the left thoracotomy, the patient is in the right recumbent position (Fig. 2.22), and the incision is made in the fourth left intercostal space, usually from the midaxillary line to the sternum. Depending on the need of the operation, the length of the incision can be modified accordingly. Regardless, care should be taken to avoid injury to the breast mammary gland.

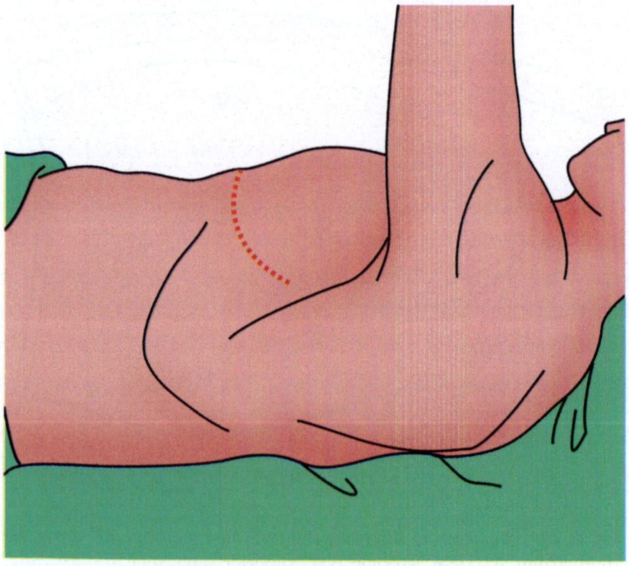

Fig. 2.22 The position for the left thoracotomy

With a left or right thoracotomy through the intercostal space, the extent of exposure is limited because of the fixed costal bones. Once the intercostal muscle and pleural membrane are incised, the pleural cavity has been reached. Subsequently, the left pericardium, the diaphragmatic nerve, and the cranial nerve can be exposed, after the left lung has been pushed up. Then an incision should be made in the pericardium about 1 cm in front of and parallel to the diaphragmatic nerve. This procedure allows for very good exposure of the LAA, LA, left pulmonary vein, main pulmonary artery, and LV. In addition, an extended incision in the pericardium allows for the aortic root, RAA, descending aorta, aortic isthmus, and laryngeal nerve to be seen with difficulty; occasionally, the fourth or fifth costal bone must be cut off for this purpose.

2.3.1.4 Mini-Invasive Thoracotomy

In addition to the approaches described above are thoracotomies that are minimally invasive. They include the following:

1. A reverse L-shaped sternotomy in the manubrium of the sternum.
2. A partial sternotomy in the upper or lower part of the sternum.
3. A small incision with a length of approximately 5–8 cm along the left margin of the sternum. Some costal bones must be cut off. This incision is suitable for harvesting the internal mammary artery and the left anterior descending artery.

2.3.2 Anatomical Surgical Features

2.3.2.1 Pericardium

The pericardium is a double-layered membranous sac that contains the heart and the roots of the great vessels. Its physiological function is to anchor and protect the heart and to guarantee that it beats normally. The double layers are the outermost fibrous pericardium (parietal pericardium) and the inner serous pericardium (epicardium) (Fig. 2.23). The air-tight space between the two layers is called the *pericardial cavity*; it contains pericardium fluid to help lubricate the heart. Occasionally, too much fluid accumulates in the cavity (pericardial effusion) and can result in pericardial tamponade. In the case of an adherent pericardium (the pericardium is connected to the heart), the beats will be restricted and the heart function will be impaired.

In the folding area of the parietal pericardium and the visceral layer of the pericardium, the roots of the great vessels, including the aorta, vena cava, pulmonary artery, and pulmonary vein, are surrounded by the pericardium (Fig. 2.24). This area is called the *base of the heart*. Separating the LA and the root of the pulmonary vein is very difficult.

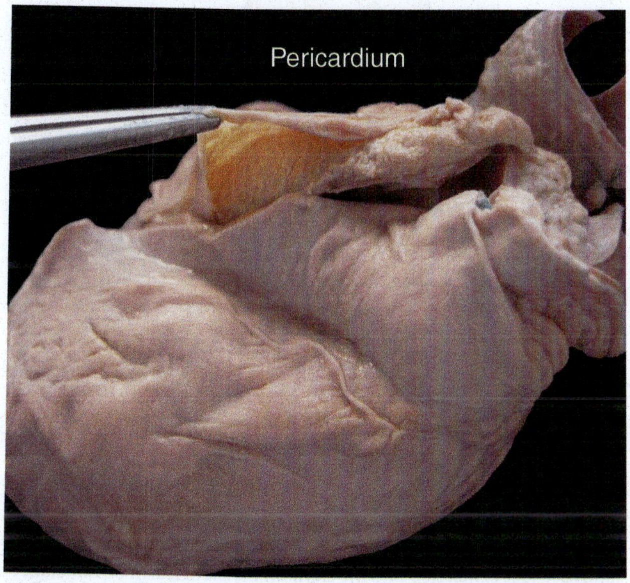

Fig. 2.23 The inner serous pericardium. The inner serous pericardium is so close to the myocardium that it is also called the epicardium. In a pericardectomy procedure, the myocardium is likely to be injured, resulting in hemorrhage

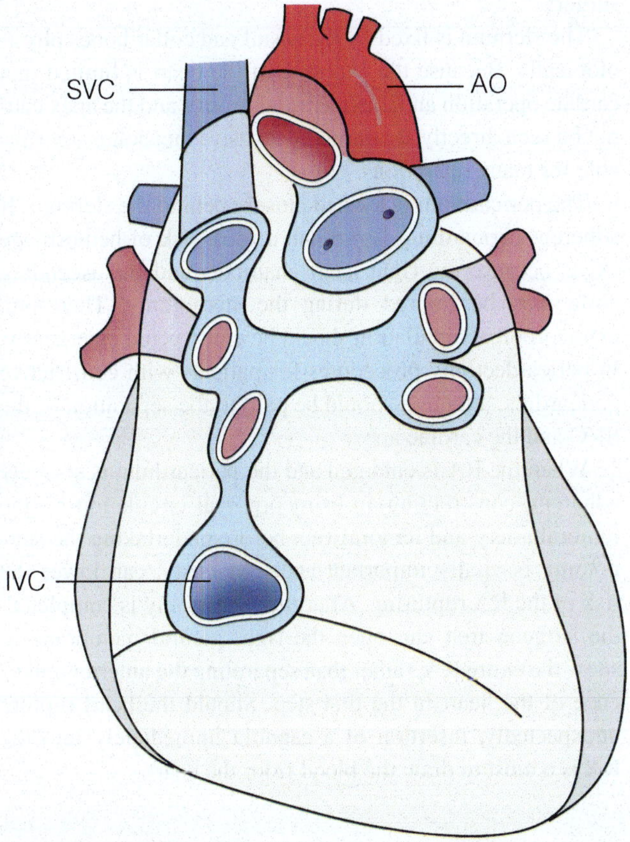

Fig. 2.24 The folding area of the two layers of the pericardium. The folding area is also the transition zone of the two layers of the pericardium. The space between these two layers is the pericardial cavity. The folding area contains and fixes the base of the heart, including the roots of great vessels

2.3.2.2 Sternum

The sternum is a narrow, shallow, and long bone located vertically in front of the pericardium. The incision of the sternum should be made along its midline as precisely as possible, or it will influence the surgical exposure of the heart and the healing of the bone (Fig. 2.25).

The incision of the sternum can easily become infected. The reasons include the slight amount of muscle surrounding it, impairment of the periosteum vessels by coagulation, excessive bone wax inserted into the medullary canal, and failed fixation of the sternum. Moreover, the internal mammary artery may be injured when the steel wires are inserted into the intercostal space to fix the sternum after cardiac surgery.

The sternum is fixed by the costal and collar bones physiologically. Because the exposure of the heart is limited in a cardiac operation and the interatrial groove and the apex cannot be seen directly, a minimally invasive incision is not suitable for many operations.

The pericardium is located closely behind the sternum. If adherent pericardium is present, a great risk of hemorrhage exists because the right heart chambers and the ascending aorta may be injured during the sternotomy. Therefore, extracorporeal circulation should be constructed emergently. In pericardectomy procedures for patients with constrictive pericarditis, attention should be paid to the separation of the SVC and the cardiac apex.

When the RA is enlarged and the pericardium is severely adherent, constructing an extracorporeal circulation by the femoral artery and femoral vein before undertaking the sternotomy is vitally important and, obviously, can lower the risk of the RA rupturing. After the sternotomy is completed, the surgeon first can open the right pleural membrane to show the entire RA, rather than separating the anterior adherence of the heart in the first step. Should the heart rupture unexpectedly, insertion of a cannula immediately into the RA is a must to draw the blood from the heart.

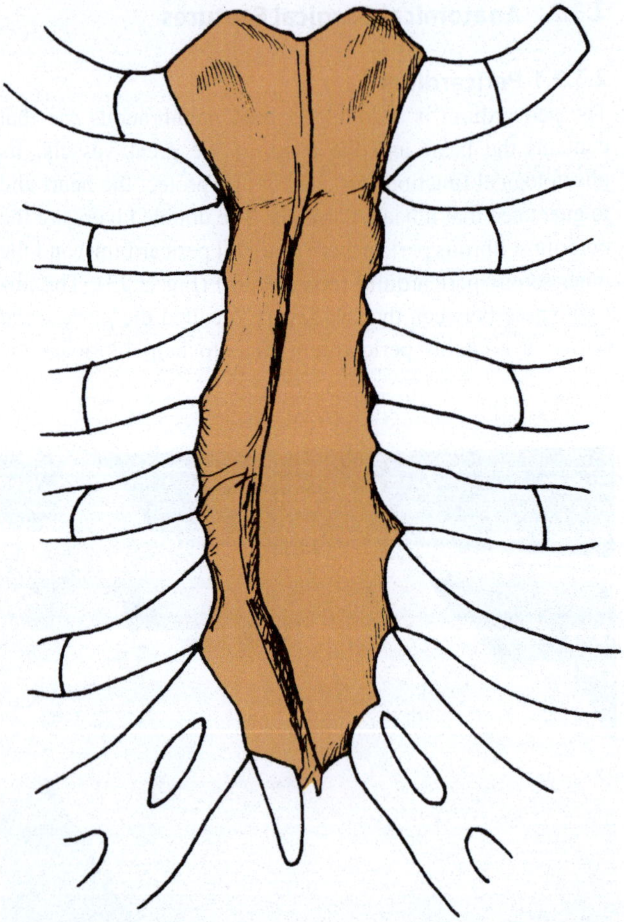

Fig. 2.25 A deflected median sternotomy. The sternum is narrow, shallow, and long, so the sternotomy may be deflected. If so, when a sternum spreader is inserted into the sternum incision, it will be unbalanced between the two sides and not provide good exposure of the heart

The Right Atrium

3

3.1 General Considerations

3.1.1 Location

The right atrium (RA) is the right upper chamber of the heart and is located above the RV (Fig. 3.1). After the pericardium is opened, the lateral free wall can be seen directly (Fig. 3.2). The free wall is the common area for cannulation and surgical approaches into the heart. The vena cava orifices, aortic sinus, and LA are located around the RA (Fig. 3.3). The anterior wall of the RA is located very close to the right coronary cusp (Fig. 3.4). The atrioventricular node is in front of the septal leaflet of the TV, in the posteroinferior portion of the membranous ventricular septum (Fig. 3.5). The interatrial groove is located between the right and left atria (Fig. 3.6).

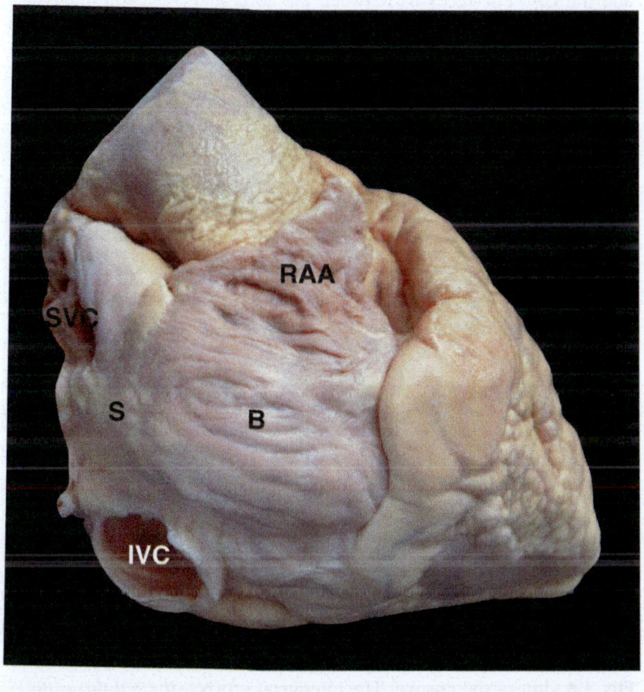

Fig. 3.1 Appearance of the right atrium (RA). The RA is located above the RV and behind and to the right of the root of the aorta. The main structure of the RA, including the sinus venosus and body of the RA, can be seen perfectly from the right lateral side. Usually, the lateral free wall of the body portion is chosen to make the incision. *B* body of RA, *IVC* inferior vena cava, *RAA* right atrium appendage, *S* sinus venosus portion of RA, *SVC* superior vena cava

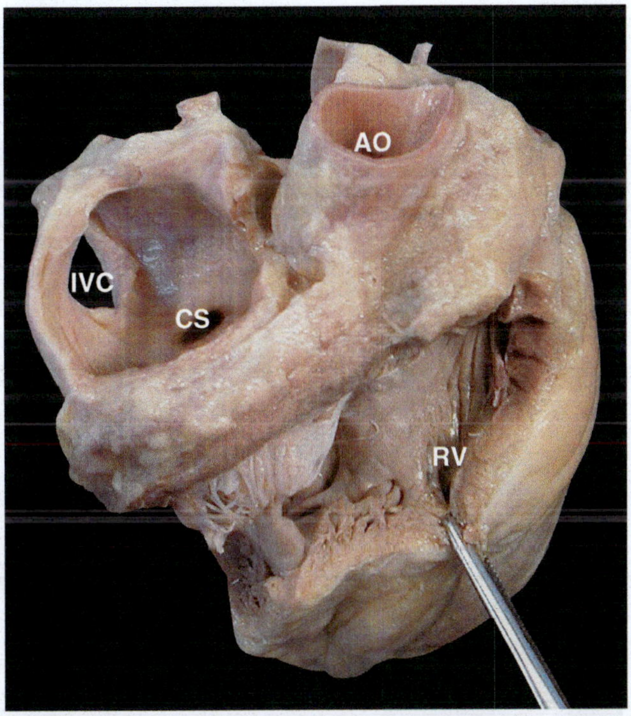

Fig. 3.2 Location of the right atrium in the heart (anatomy position). After the outer free wall of the RA is incised, the sinus venosus portion, including the coronary sinus, the IVC, and the posterior wall of the RA, is shown well. *AO* aorta, *CS* coronary sinus, *IVC* inferior vena cava, *RV* right ventricle

X. Zhu (ed.), *Surgical Atlas of Cardiac Anatomy*,
DOI 10.1007/978-94-017-9409-1_3, © Springer Science+Business Media Dordrecht and People's Medical Publishing House 2015

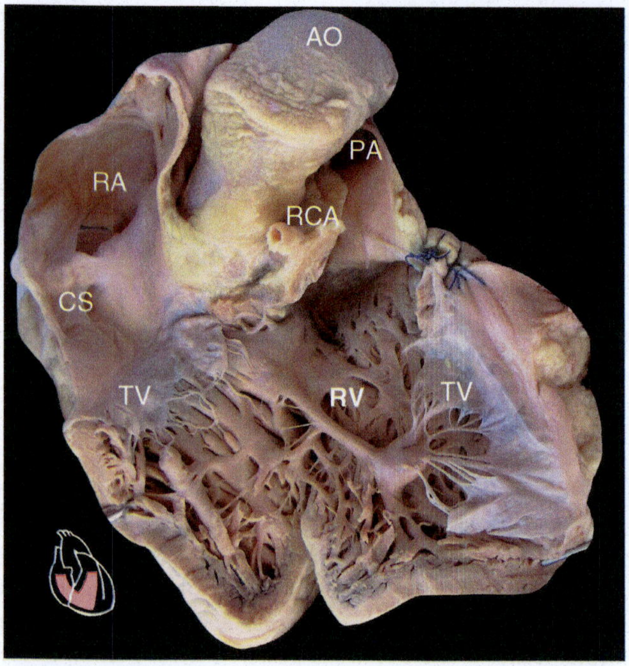

Fig. 3.3 Frontal plane of the heart (anatomy position). The elementary structures and connection, transition of RA, and RV are revealed. *PA* pulmonary artery, *RCA* right coronary artery, *TV* septal leaflet of tricuspid valve

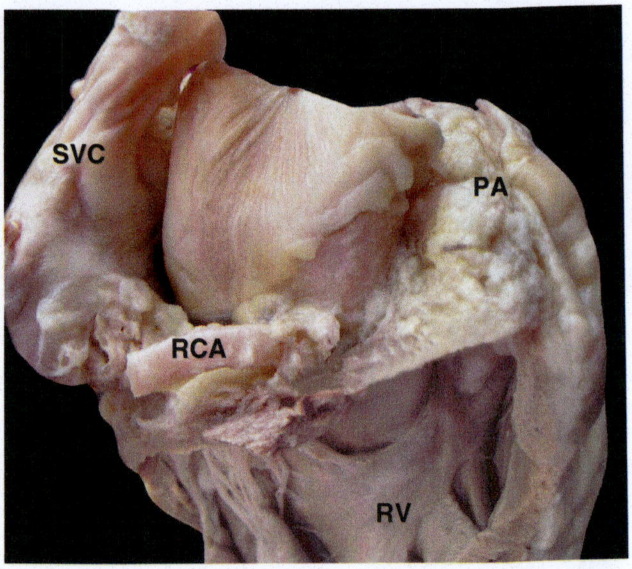

Fig. 3.4 Relationship between the ascending aorta and the anterior wall of the right atrium. Retracting the aortic root through the recessus between the aorta and the SVC, the operator can see that the anterior wall of the RA is very close to the right coronary cusp and that the posterior half portion of the aortic root has been surrounded by the RA and the LA. The recessus between the aorta and SVC communicates with the transverse sinus of the pericardium cavity normally. This is also the space where the aortic blocking forceps are placed during cardiac surgery. *RCA* right coronary artery, *RV* right ventricle, *SVC* superior vena cava

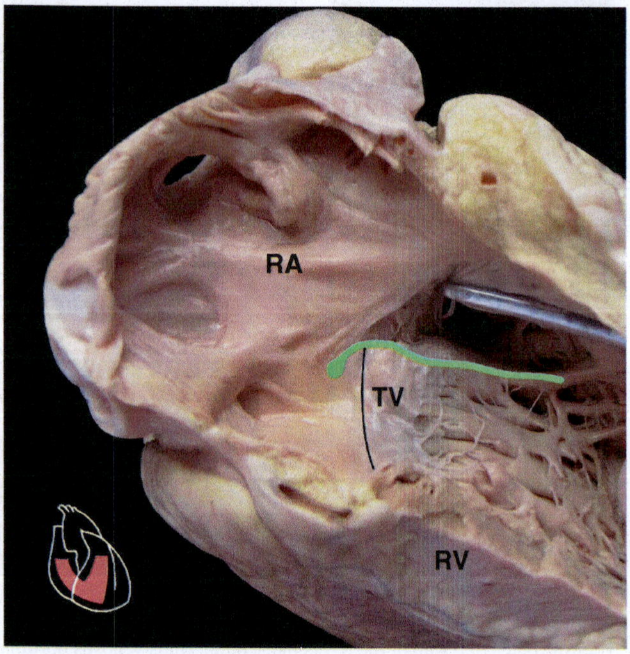

Fig. 3.5 The relationship between the atrial septum and the septal leaflet of the tricuspid valve. The atrioventricular node (*green mark*) is located in front of the septal leaflet of the TV and in the posteroinferior portion of the membranous ventricular septum

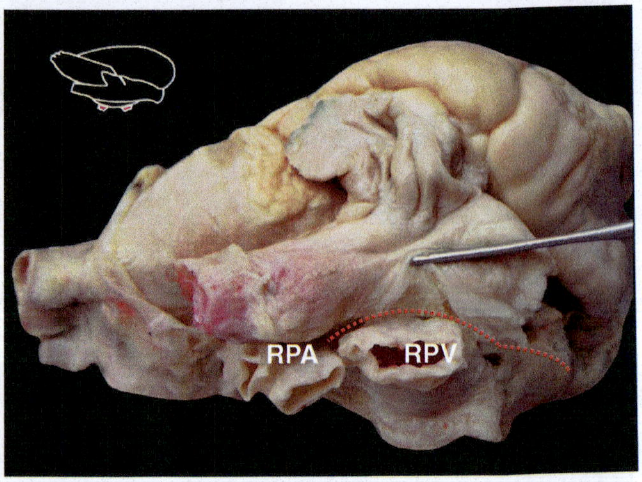

Fig. 3.6 Interatrial groove. The interatrial groove (*the red dash line*) is seen from the position of the operator. It is located between the RA and the right pulmonary vein. After loose connective tissues in the groove have been stripped, the LA turns out. The dash line also indicates an important surgical approach to the LA, such as for operations on the MV. *MV* mitral valve, *RPA* right pulmonary artery, *RPV* right pulmonary vein

3.1.2 Morphological Characteristics

The morphology of RA differs from that of the LA by having the following characteristics: (1) a short, blunt, and thin-walled RAA, a piece of triangle-shaped tissue that has more muscle bundles than does the LAA and is a common area for intracardiac detection and cannulation; (2) the orifice of the superior and inferior venae cavae; and (3) the orifice of the coronary sinus vein (Figs. 3.7, 3.8, 3.9, and 3.10). According to the embryonic development, the RA can be divided into two portions, the sinus venosus portion, which is the main portion and is located posteriorly, and the body of the RA. The sinus venosus portion receives systemic venous drainage from the superior and inferior venae cavae and coronary venous drainage from the coronary sinus. The body of the RA is located anteriorly and contains the RAA and the lateral free wall. A muscle bundle called *crista terminalis*, which has an internodal bundle passing through it, is located between these two portions (Figs. 3.11 and 3.12).

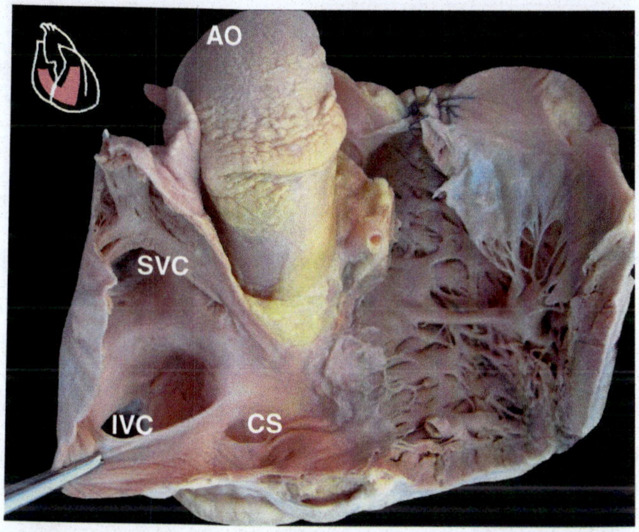

Fig. 3.7 Morphological characteristics of the RA. In contrast to the LAA, the RAA is blunt and large-based. The confirmatory morphological features of the RA are the orifices of the vena cava and the coronary sinus. The smooth wall in front of the vena cava is the sinus venosus portion of the RA, in which the orifices of the vena cava and the coronary sinus are located. This photo shows the structures of the RA and the AV after the anterior wall was resected. *CS* coronary sinus, *IVC* inferior vena cava, *SVC* superior vena cava

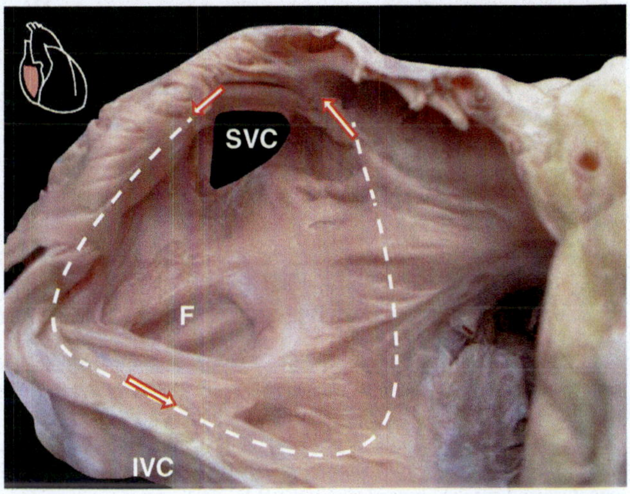

Fig. 3.8 Conducting pathway of an atrial flutter. The circular conducting pathway is shown after the anterior wall of the RA has been removed. The wave passes downward along the crista terminalis; follows a course that runs inferior to the coronary sinus and IVC orifice, in front of the FO; and returns to the crista terminalis (Wilcox BR, Cook AC. Surgical Anatomy of the Heart, 3 edition. Cambridge University Press; Cambridge, UK; 2004)

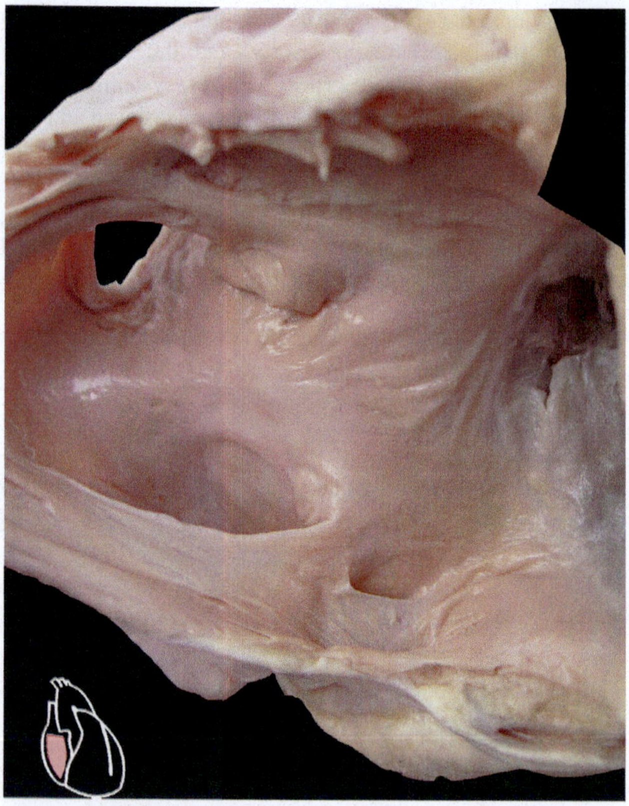

Fig. 3.9 Orifice of the coronary sinus. Located anteroinferior to the orifice of the IVC, the orifice of the coronary sinus is considered to be the posterior margin of the triangle of Koch. This photo shows the relationship of the orifices of the vena cava and the coronary sinus, the FO, and the septal leaflet of TV. This area is very important for atrial septum puncture and surgical procedures in the RA

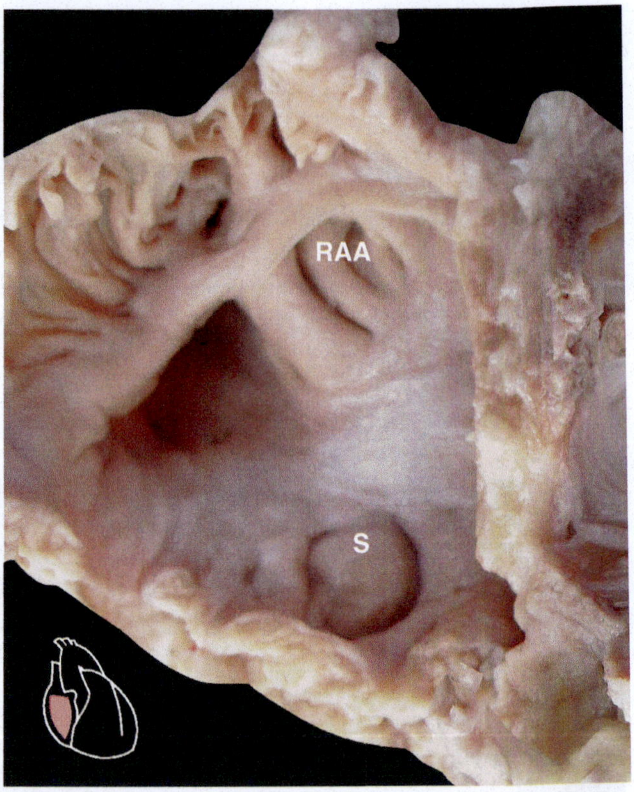

Fig. 3.10 Right atrial appendage (RAA). The anterior wall of the RA was resected, and the RAA was opened. Compared to the LAA, the RAA is triangular shaped, without obvious incisure in the base. *RAA* right atrium appendage, *S* atrium septum

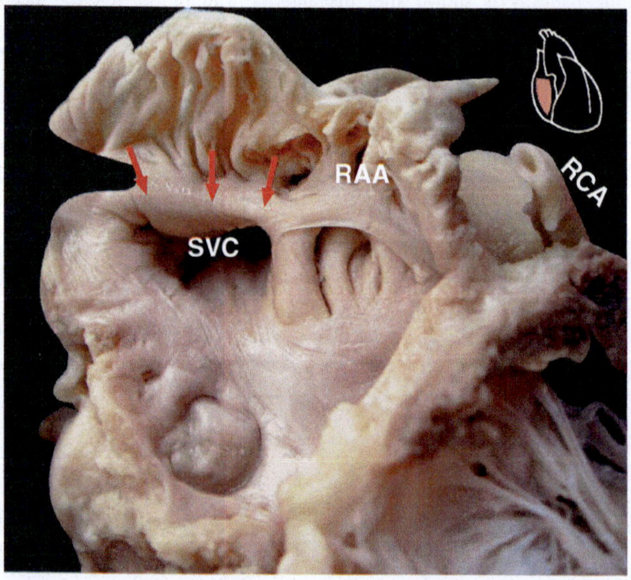

Fig. 3.11 Muscle bundle in front of the superior vena cava. A muscle bundle located in front of the orifice of the SVC passes downward to the caval valve. It is the demarcation line of the sinus venosus and the body of the RA. One of the conducting pathways between the sinoatrial node and the atrioventricular node passes along this muscle, also called *crista terminalis*. In front of the crista terminalis is the anterior free wall of the RA, which has a trabeculated inner surface

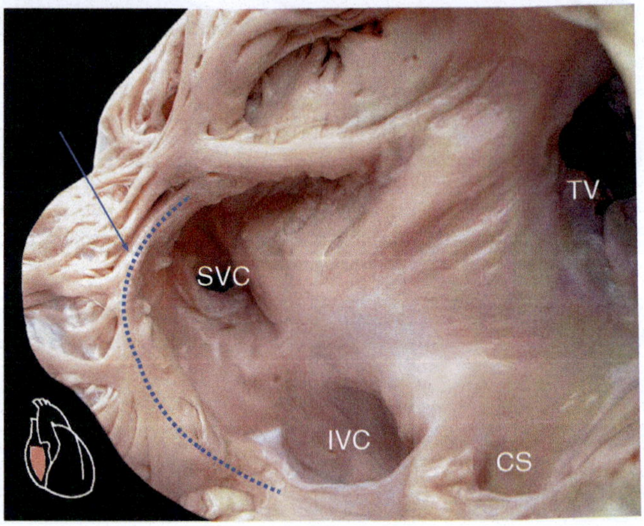

Fig. 3.12 Crista terminalis. The term *crista terminalis* refers to the borderline between the posterior, smooth sinus venosus portion and the anterior, rough portion of RA. It is a vertical ridge of the smooth myocardium on the lateral inner wall of the RA. It is located from the front (*right*) side of the orifice of the SVC inferiorly to the right side of the valve of the IVC (eustachian). The crista terminalis also passes forward and forms the Thebesian valve, which connects with the atrioventricular node by the Todaro tendon. The sinoatrial node is found within the upper part of the crista terminalis; it usually is described as being located in the endocardium in front of the orifice of the SVC. The sinoatrial node sends out three conducting pathways to the atrioventricular node, one of which passes along the crista terminalis. The blue dash line shows the crista terminalis. *CS* coronary sinus, *IVC* inferior vena cava, *SVC* superior vena cava

3.1.3 Atrial Septum, Atrioventricular Septum, and Triangle of Koch

3.1.3.1 Atrial Septum

The actual extent of the atrial septum is the fossa ovalis and the tissues surrounding it (Figs. 3.13 and 3.14a, b). The thinnest part in the middle of the atrial septum is called the *fossa ovalis* (FO) (Figs. 3.15 and 3.16). The anterior margin of the FO is a horseshoe-shaped muscular ring called the *Vieussens ring*. This ring can be divided into the superior and the inferior borders. The inferior border of the ring is contiguous with the valve of the IVC (also called the *eustachian valve*) (Fig. 3.17), so it is sometimes called the *papillary muscle* of this valve. The valve of the IVC extends forward and forms the Todaro tendon (Fig. 3.18). The anterior border of the atrial septum is very close to the aortic sinus; hence, the aorta may be injured during an atrial septum puncture procedure (Fig. 3.19)

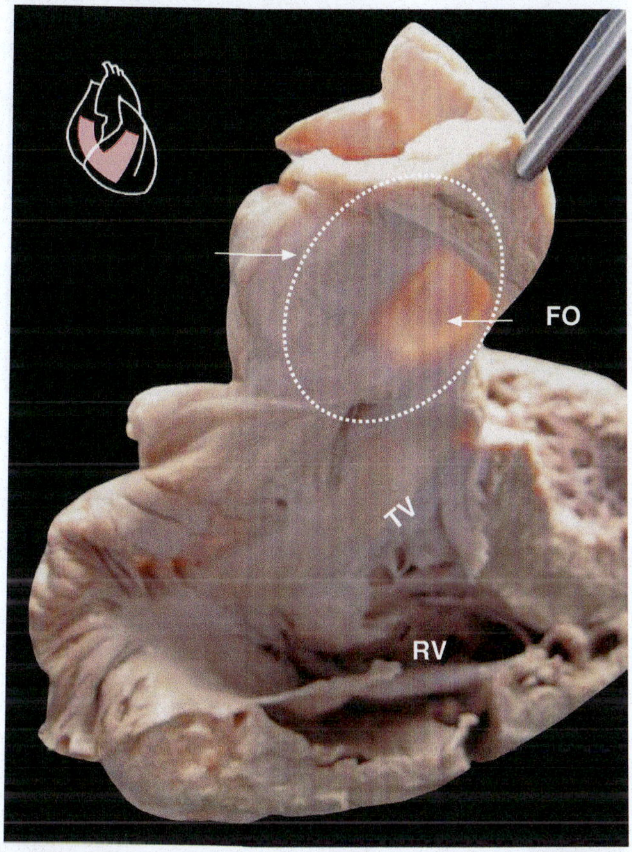

Fig. 3.13 Atrial septum. The atrial septum is indicated by the *white dash line*. The thinnest part (photic zone) is the fossa ovalis (FO). The muscular ventricular septum is located between the atrial septum and the septal leaflet of TV

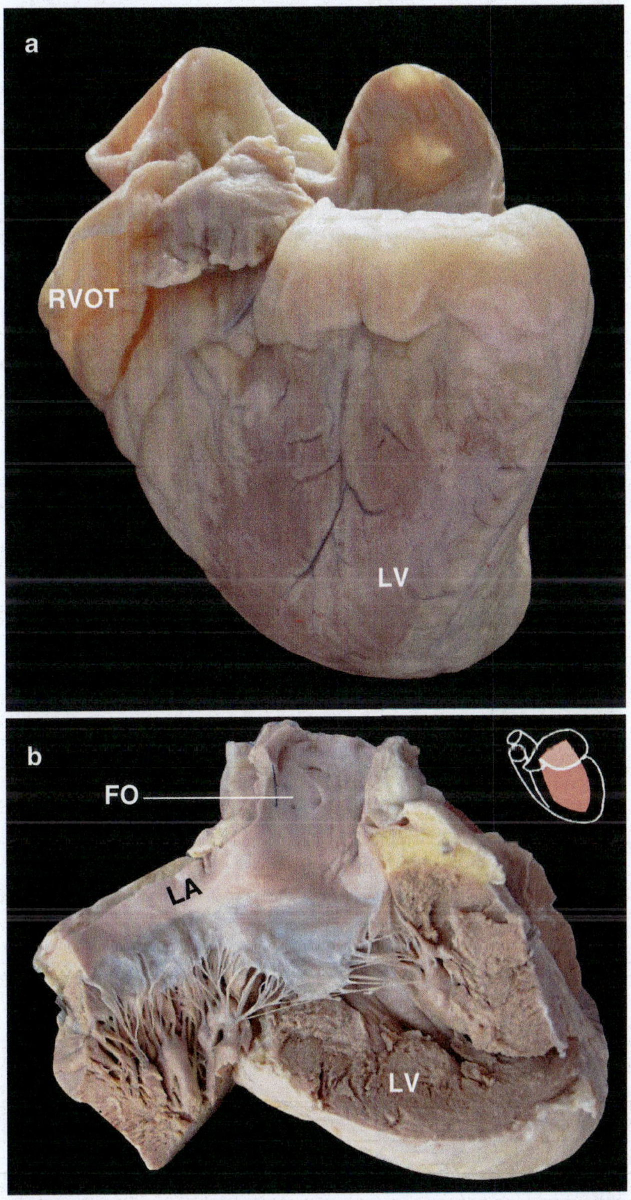

Fig. 3.14 (**a**) Left side of the atrial septum. With the lateral wall of the LA removed, the atrial septum is observed from the left side. Shown is the spatial relationship between the atrial septum and the LA. The photic zone is the FO. *LV* left ventricle, *RVOT* right ventricular outflow tract. (**b**) Folding zone in the left atrium corresponding to the fossa ovalis. With the LA and the LV open, the folding zone, which corresponds to the FO, can be seen in the atrial septum. This folding is one of the best characteristics of morphological LA in complicated malformations of the heart

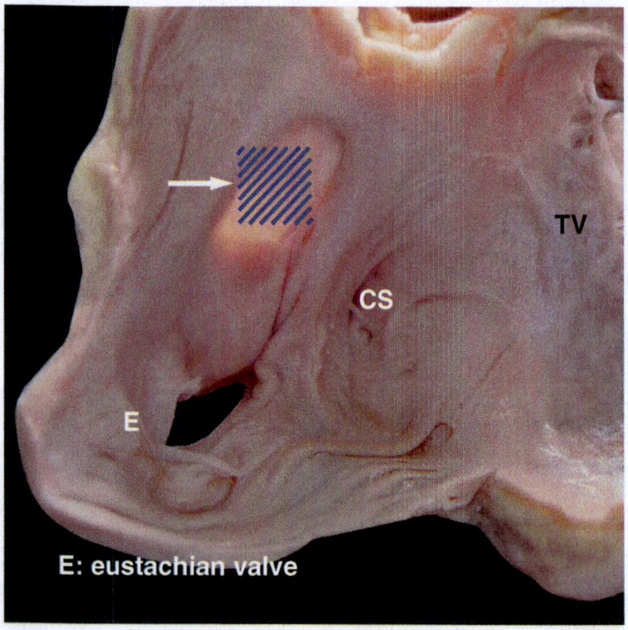

Fig. 3.15 Puncture site of the atrial septum. The blue oblique lines show the FO, the thinnest part of the atrial septum. It is located far from the AV and the atrioventricular node. Therefore, it is the best site for puncture of the atrial septum, such as interventional balloon dilation therapy for the MV

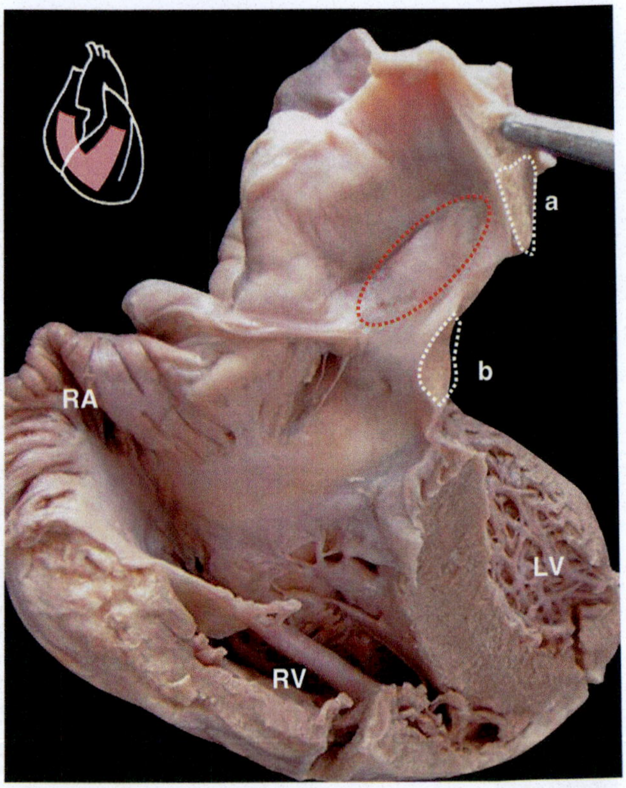

Fig. 3.16 Fossa ovalis. The anterior margin of the FO is muscular and can be divided into the superior border and the inferior border. Internodular conducting pathways pass through it. The *red dash line* indicates the FO. *a* superior border, *b* inferior border, *LV* left ventricle, *RV* right ventricle

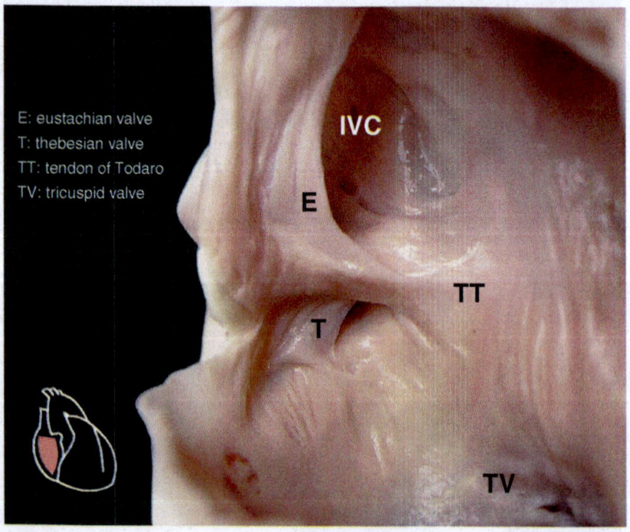

Fig. 3.17 Valve of the inferior vena cava (IVC). The IVC is also called the *eustachian valve*. It is a crescent-shaped fold of endocardium arising from the anterior margin of the IVC orifice. The medial horn of the valve joins the Thebesian valve, a semicircular valvular fold forming the upper rim of the orifice of the coronary sinus. This photo shows the margins of the triangle of Koch. *E* eustachian valve, *T* Thebesian valve, *TT* Todaro tendon, *TV* tricuspid valve

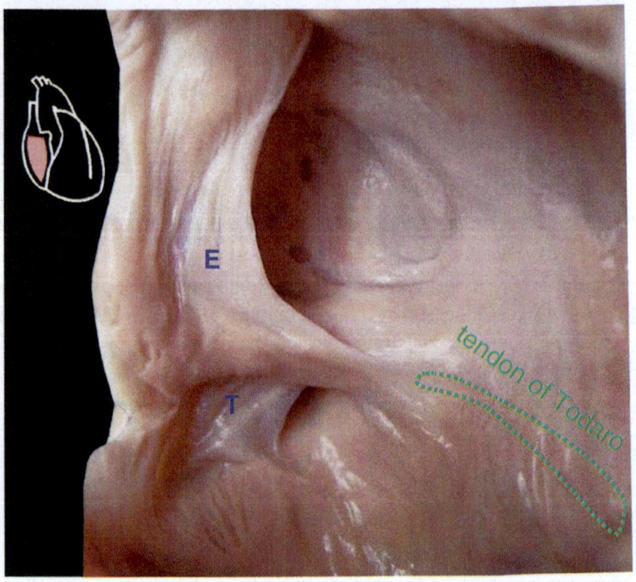

Fig. 3.18 Todaro tendon. The Todaro tendon is a fibrous tendon found between the Thebesian valve and the atrioventricular node and indicated by the *green dash line* in the photo. *E* eustachian valve, *T* Thebesian valve

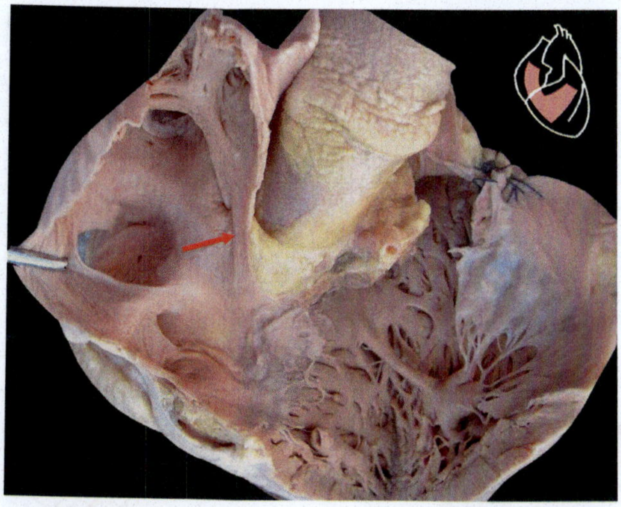

Fig. 3.19 Risk for atrial septum puncture. The FO is the correct and safe puncture site. If the puncture site is too low, the aortic sinus and the atrioventricular node may be injured. The *red arrow* shows the common *wrong* site for puncture, which usually results in severe complications

3.1.3.2 Atrioventricular Muscular Septum

The muscular septum located between the RA and the LV is called the *atrioventricular muscular septum* (Figs. 3.20a, b and 3.21); it may be considered the inferior wall of the RA. Observed from the RA, the inferior wall is seen to be attached to the ventricular septum, just as an oblique triangle is located between the atrial septum and the septal leaflet of the TV. Located in front of the inferior wall of the RA and behind the membranous septum, it is the meeting point of the mitral annulus, the tricuspid annulus, and the central fibrous body. The left portion of the atrioventricular muscular septum is located below the mitral annulus, whereas the right portion is above the TV (Figs. 3.21 and 3.22).

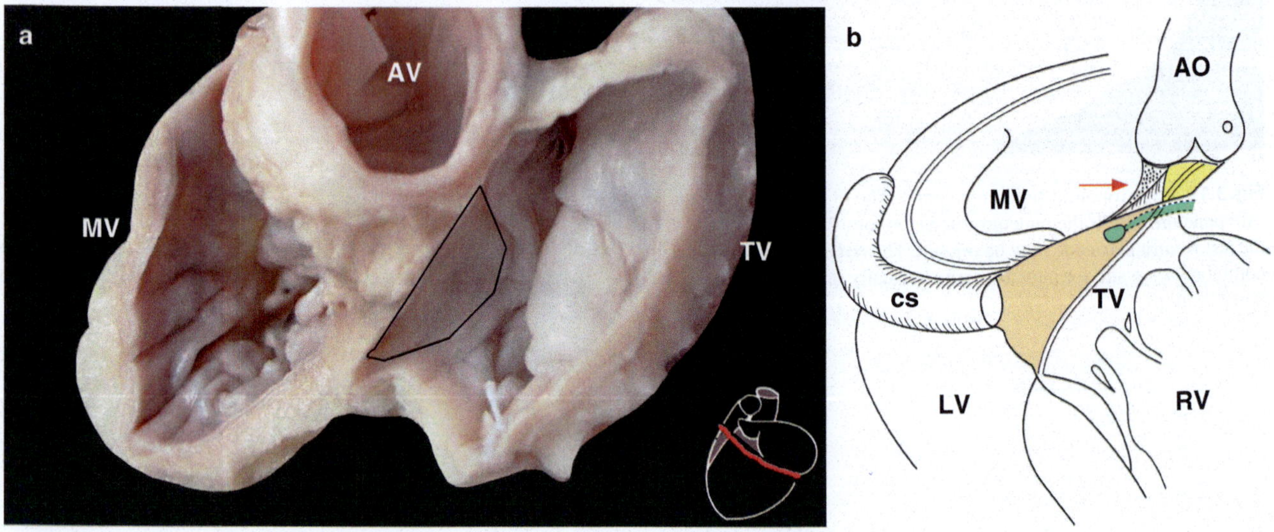

Fig. 3.20 (**a**) Inferior wall of right atrium (RA). The ventricular septum extends backward. Its right part is located in the inferior wall of RA (triangle of Koch), whereas the left part remains under the mitral annulus, which belongs to the LV. The atrioventricular muscular septum is found between the RA and LV. This picture also shows that the mitral annulus is at a higher level than is the tricuspid annulus. The *black line* indicates the atrioventricular muscular septum. *AV* aortic valve, *MV* mitral valve, *TV* tricuspid valve. (**b**) Diagrammatic drawing of the atrioventricular muscular septum. This drawing shows the relationship of the inferior wall of the RA, the orifice of the coronary sinus, the atrioventricular node, and the septal leaflet of the TV. The mitral and tricuspid annuli are not at the same level. The atrioventricular node (*green mark*) is located anteroinferiorly. The yellow zone indicates the atrioventricular muscular septum, the right atrium side of which is the triangle of Koch. The red arrow indicates the central fibrous body. *LV* left ventricle, *MV* mitral valve, *RV* right ventricle

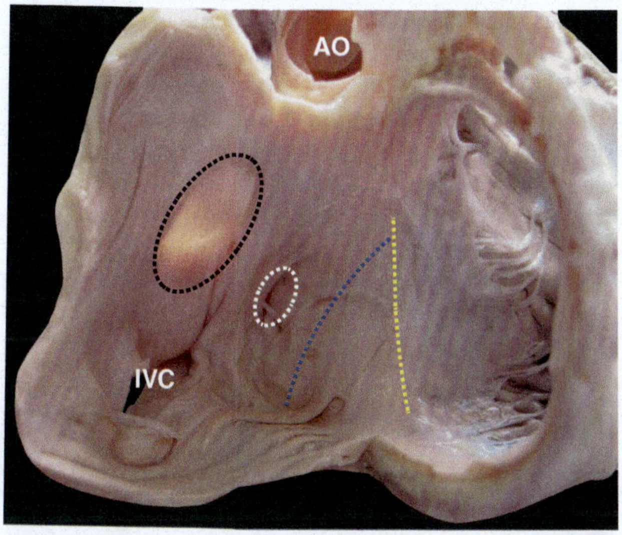

Fig. 3.21 Atrioventricular muscular septum. The atrioventricular muscular septum is located in the area between the *blue* and *yellow dash lines*. The white dash line indicates the coronary sinus

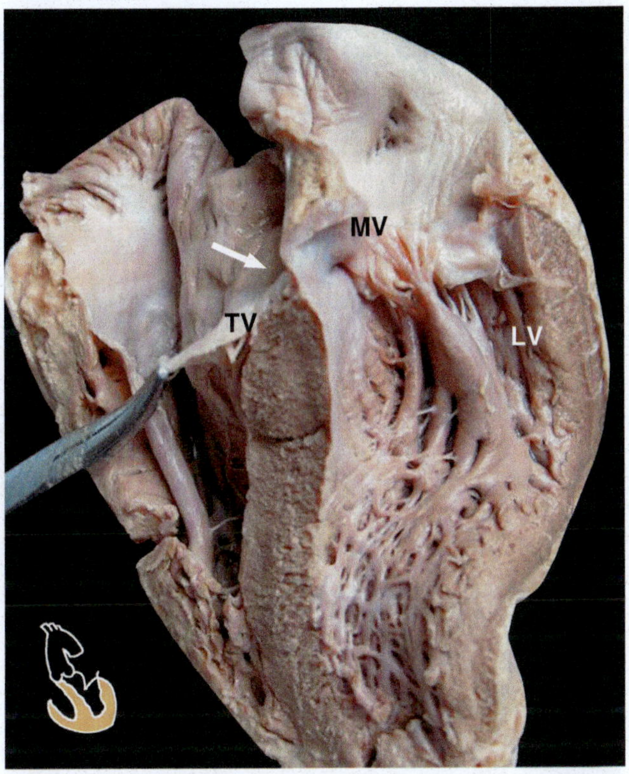

Fig. 3.22 Frontal plane of the posterior portion of the ventricular septum. The four chambers are seen after the two atriums and the anterior half of the two ventricles have been resected. The mitral and tricuspid annuli are not at the same level. The *white arrow* indicates the atrioventricular muscular septum. *LV* left valve, *MV* mitral valve, *TV* tricuspid valve

3.1.3.3 Triangle of Koch

The triangle of Koch (also called the *Koch triangle* and *triangle of the atrioventricular node*) is composed of the Todaro tendon, the septal leaflet of the TV, and the orifice of the coronary sinus (Figs. 3.23 and 3.24). It can be considered the RA side portion of the atrioventricular muscular septum. It is an important area and has special functions. Most importantly, the atrioventricular node is located in this triangle (see Fig. 3.5). The atrioventricular conduction pathway passes through the central fibrous body and along the posteroinferior border of the membranous septum, reaching the lower chambers of the heart.

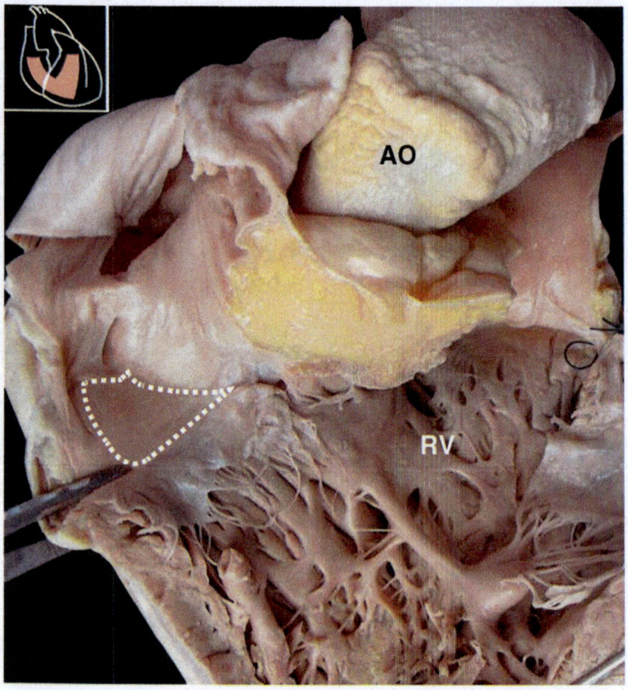

Fig. 3.23 Triangle of Koch. The *white dash line* indicates the border of the triangle of Koch

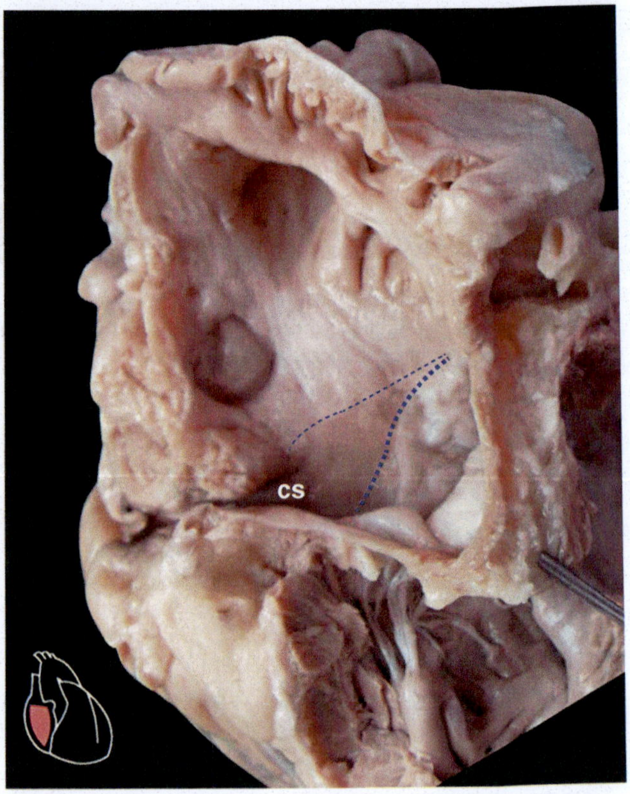

Fig. 3.24 The relationship between the triangle of Koch and the anterior border of the FO is shown. The *blue dash line* indicates the triangle of Koch. The anterior border of the FO is located posterosuperiorly

3.2 Surgical Incision

An incision in the RA usually is made for surgical procedures of the RV, such as operations on the TV and repair of the VSD or TOF. It is very important to pay special attention to protect the triangle of Koch, or the conduction pathway may be injured. In order to protect the sinoatrial node and the crista terminalis, the RA incision should not be made too close to the SVC or too posteriorly. Any surgical procedures in the RA should be performed very carefully to avoid injuries to the atrioventricular node. If the RA turns posteriorly because of right-handed rotation of the heart, exposing the RA may be very difficult and the TV may be injured by the retractor.

The surgical approach through the RA and interatrial septum often is used for surgical procedures of the LA because it easily exposes the MV. However, the atrial septum is not large, so the incision cannot be too prolonged (Figs. 3.25 and 3.26).

Sometimes, an incision of the atrial septum can be extended toward the top of the atrium properly. Opening the top of the upper chambers may provide better exposure, but caution is warranted because this long incision may injure the coronary collaterals of the sinoatrial node, resulting in arrhythmia. Additionally, suture of this extended incision at the end of the operation is much more complicated than usual.

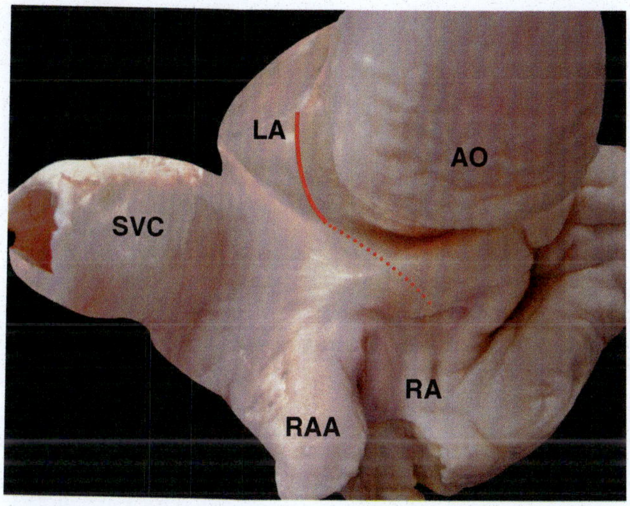

Fig. 3.25 Atrial septum incision extending to the left atrium (LA). The *red dash line* indicates the combined incision of the RA and LA. The atrial septum was opened from the top of the LA. *AO* aorta, *LA* left atrium, *RA* right atrium, *RAA* right atrium appendage, *SVC* superior vena cava

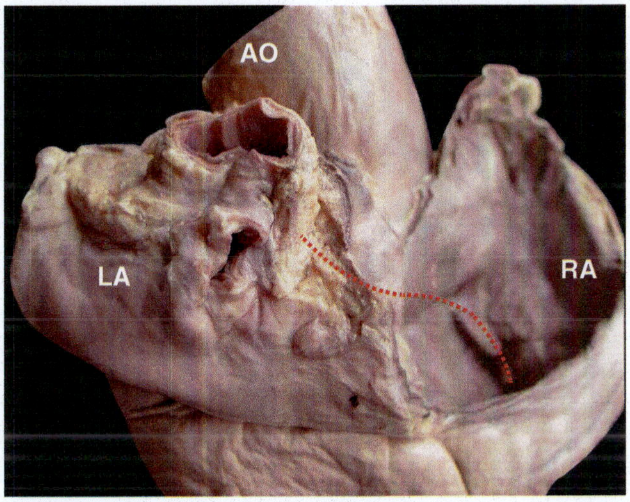

Fig. 3.26 Exposure of the LA by an upper atrial septum incision. This picture shows the exposure of the LA by an upper atrial septum incision after the sinus venosus of the RA has been resected. Special attention should be paid to decide the length and extent of this incision. The incision of the atrial septum can be made only approximately 1.2 cm longer toward the LA. If the LA incision is too long, the retractor may injure the circumflex branch of the coronary artery

The Right Ventricle

4

4.1 General Considerations

4.1.1 Position

The two ventricles of the heart are separated by an interventricular groove. The right ventricle (RV) is located anterosuperior to the LV. The left anterior descending artery (ADA) runs in the anterior interventricular sulcus and supplies blood for the anterior portion of the interventricular septum. The posterior descending artery (PDA) runs in the posterior interventricular sulcus to the apex of the heart and supplies blood for the posterior portion of the interventricular septum. The right atria are separated from the RV by the right atrioventricular groove. The RCA travels down the right atrioventricular groove, toward the crux of the heart.

4.1.2 Structural Characteristics

In contrast to the LV, the RV is thin walled and compliant, and it can sustain only the low pressure of pulmonary circulation. The LV is circular shaped with thick walls and bears the high pressure of the systemic circulation. The ventricular septum is considered as part of the LV. In the systolic phase, the LV makes a concentric movement. Because the contractile force of the RV is weaker than that of the LV, the RV bears a lower pressure of pulmonary circulation. Hence, when the pulmonary artery pressure is elevated or the blood flow in the right heart increases, the RV is prone to failure. Therefore, surgery in this area should be avoided to reduce damage to the right ventricular structure. A right atrium and main pulmonary artery incision can be chosen in certain right ventricular cavity operations.

Other characteristics of the RV include the TV as the inflow entrance, the pulmonary valve as the outlet, a trabecular portion with many muscle bands, and the absence of a continuous fibrous connection between the tricuspid and pulmonary valves. The tricuspid and pulmonary valves are separated by infundibular muscle bundles that include a septum band on the right-side surface of the ventricular septum and a parietal band on the inner surface of the insolated wall of the RV. Compared to the LV, the RV has more muscle tissue between the inflow and outflow tracts (Fig. 4.1a, b).

X. Zhu (ed.), *Surgical Atlas of Cardiac Anatomy*,
DOI 10.1007/978-94-017-9409-1_4, © Springer Science+Business Media Dordrecht and People's Medical Publishing House 2015

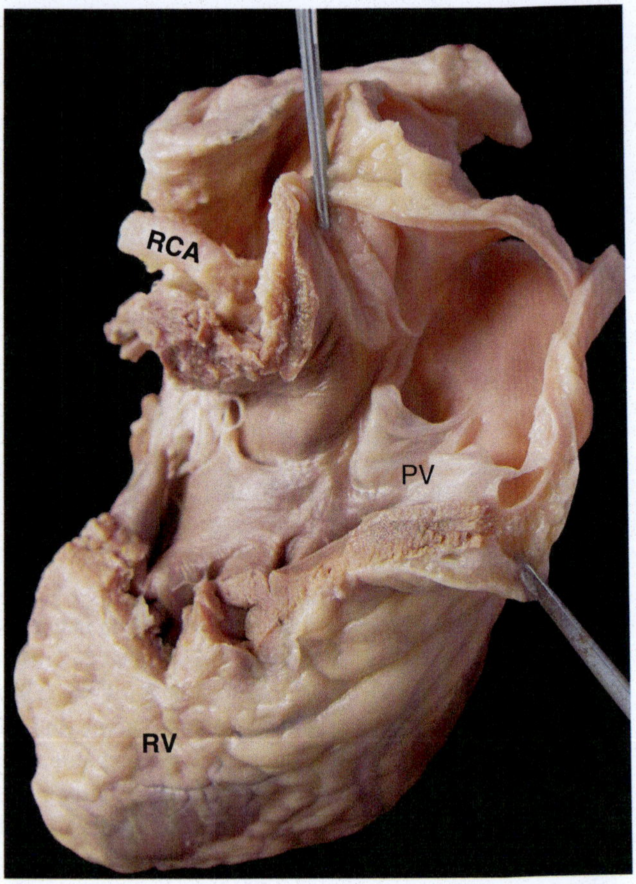

Fig. 4.1 (**a**) Pulmonary valve adjacent structures. The pulmonary and tricuspid valves are separated by the infundibular muscle. It is the important characteristics for determining if the arterial location relationship is normal. The left coronary artery lies in the back of the main pulmonary artery. (**b**) Muscle band under the main pulmonary artery (*white arrow*). The inner wall in this part of the heart usually is filled with small trabeculae muscles in the normal heart, but infundibular muscular hypertrophy can lead to pulmonary subvalvular stenosis

4.2 Right Ventricular Wall and Surgical Approaches

As the most anterior part of the heart, the anterior wall of the RV is easily exposed. Therefore, it is the first choice for making the right ventricular incision; however, to avoid injury to the anterior papillary muscle and the moderator band, the cardiac surgeon must be certain of the structure of RV before cutting it open. The anterior right ventricular wall can be divided into three regions: the conal part, the sinus, and the trabecular, as shown in Fig. 4.2. An incision made in the conus arteriosus (indicated by "C" in Fig. 4.2) is appropriate (Fig. 4.2). An incision can be made at the upper half of the right ventricular anterior wall; however, because the tricuspid anterior papillary muscle attachment point and the complex trabecular structure are found in the lower right corner, an incision should *not* be made in the region(Fig. 4.3).

Right ventricular incisions are limited by the branches of the conus arteriosus from the right coronary artery, injury to which should be avoided. An infundibular incision is useful to display the pulmonary valve and the subvalvular structure, but not the membranous portion.

A larger right ventricular incision or violent pull will damage the right ventricular muscle, affecting the right ventricular systolic function. In some cases, a massive right ventricular branch from the right coronary artery crosses the right ventricular anterior wall, and a right ventricular incision may injure this giant branch, leading to localized myocardial infarction (Fig. 4.4).

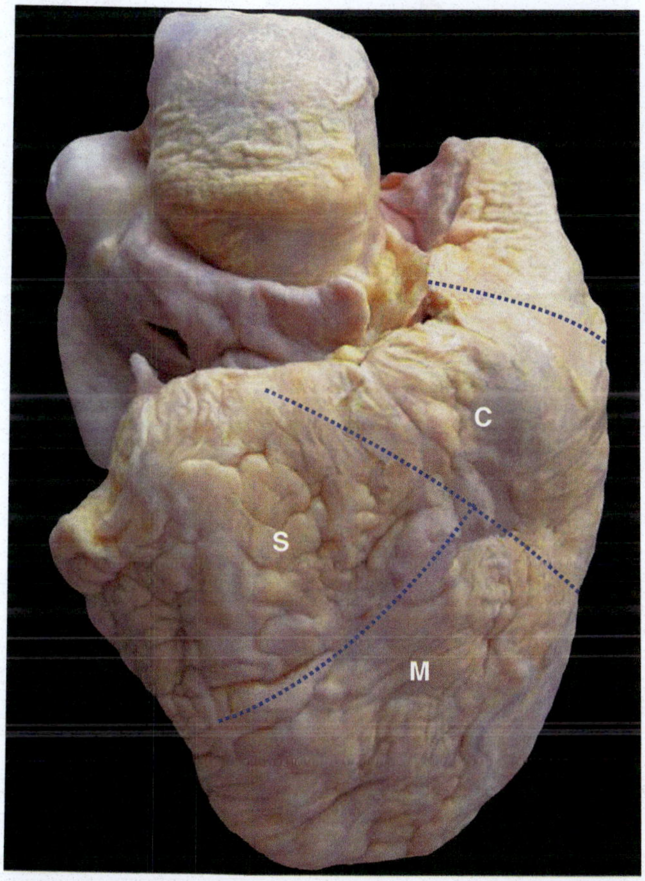

Fig. 4.2 Right ventricle. This specimen displayed by the frontal anatomical position shows the right ventricular anterior wall. Between the interventricular sulcus and the LV, the left anterior descending coronary artery runs within the interventricular groove. Between the atrioventricular groove and the right atrium, the right coronary artery runs within the atrioventricular groove. The aortic root is covered by the upper edge of the right ventricle. The anterior right ventricular wall can be divided into three parts: the conus arteriosus (*C*), the sinus (*S*), and the trabecular muscle (*M*)

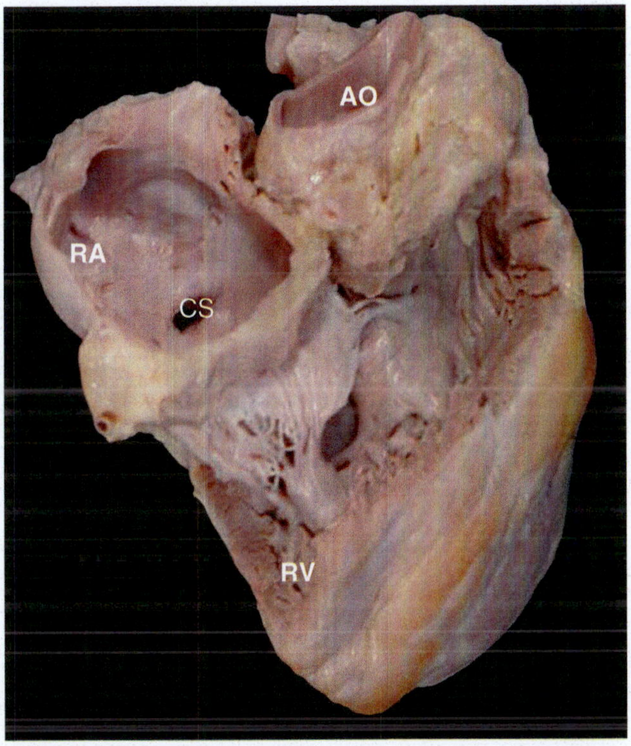

Fig. 4.3 Right ventricle without anterior wall. Viewed from its face side, without the anterior wall, retention of the tricuspid annulus, anterior leaflet of the TV, and the magnitude of the anterior papillary muscle can be seen. In front of the TV, the conical papillary muscle is connected with the supraventricular crest; between the right ventricular outflow tract and the aorta is the reflection of the infundibulum. *AO* aortic root, *CS* coronary sinus, *RA* right atrium, *RV* right ventricle

Fig. 4.4 Right ventricle used as the operation incision area. The principle for right ventricular incision is to avoid the important branches of the coronary arteries and the tricuspid anterior papillary muscle. The right bundle branch runs along the moderator band, cutting of which can cause a blockage of the right bundle branch and a decrease in the right ventricular contractility. In figure (***b***) for the anterior papillary muscle, the *dotted box* shows the area where a right ventricular incision can be made

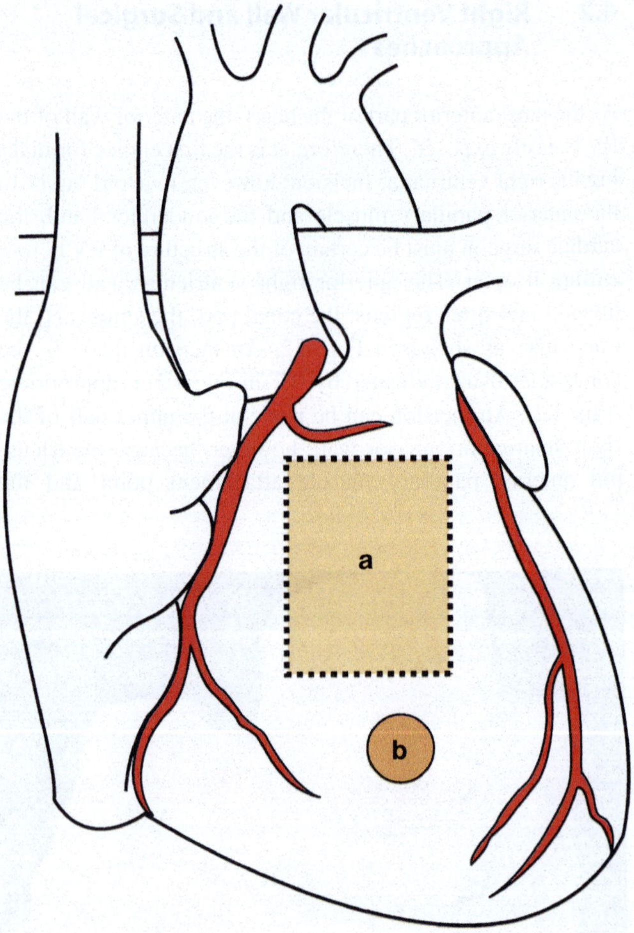

4.3 Right Ventricular Cavity Divisions

The right ventricular cavity is divided into three portions (Figs. 4.5, 4.6, and 4.7):

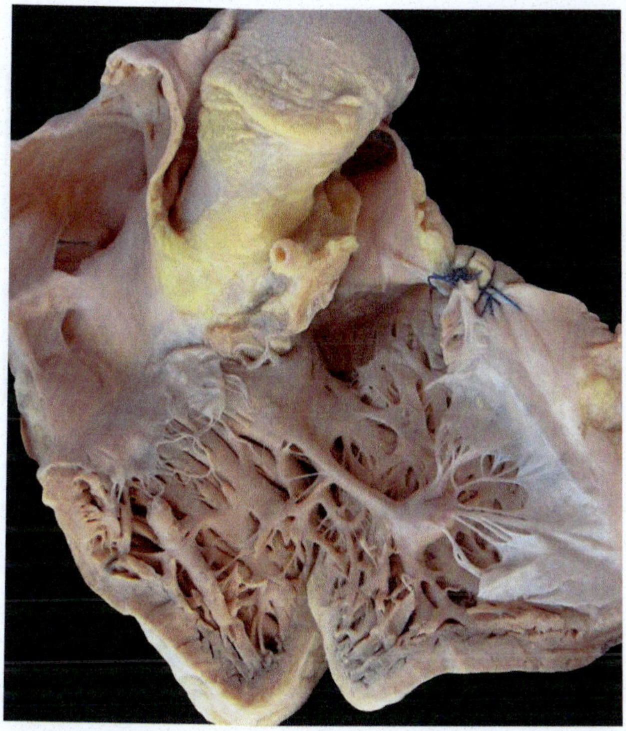

Fig. 4.5 Right ventricular cavity. The right ventricular cavity can be divided into three parts: (1) the right ventricular inflow part (namely, the sinus), which has a smooth inner wall and an upper edge formed by the TV annulus, also including membranous ventricular septal; (2) the apical area of trabecular portion; and (3) the right ventricular infundibulum, also referred to as the outflow portion, the distal end of which is connected to the main pulmonary artery

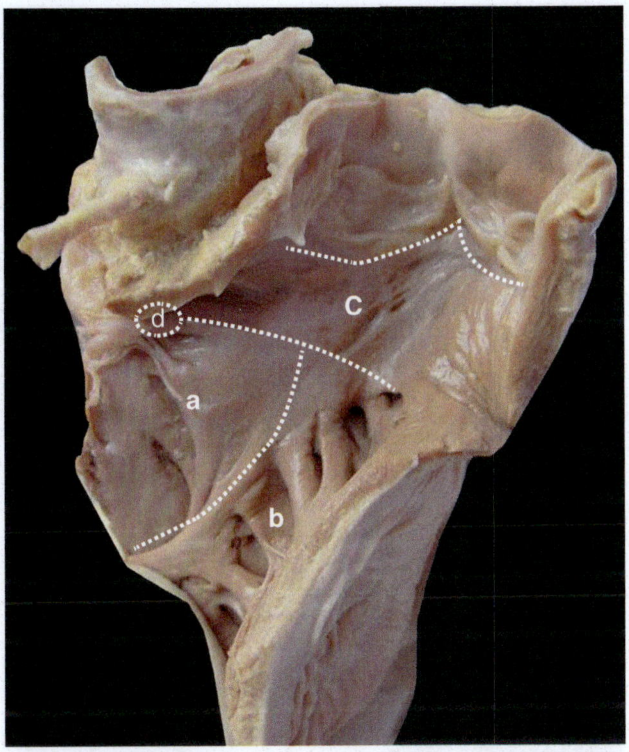

Fig. 4.6 Interventricular septum viewed from the right ventricle. From this view, the interventricular septum can be divided into four portions: (*a*) the ventricular inflow tract or sinus, which has the smooth inner wall; (*b*) the trabecular portion, where many bundles are present; (*c*) the right ventricular outflow tract as the infundibulum, a supraventricular crest connects with the sinus; (*d*) the membranous septum, which is very small, is in front of the tricuspid septal leaflet, and is positioned below the conical papillary muscle

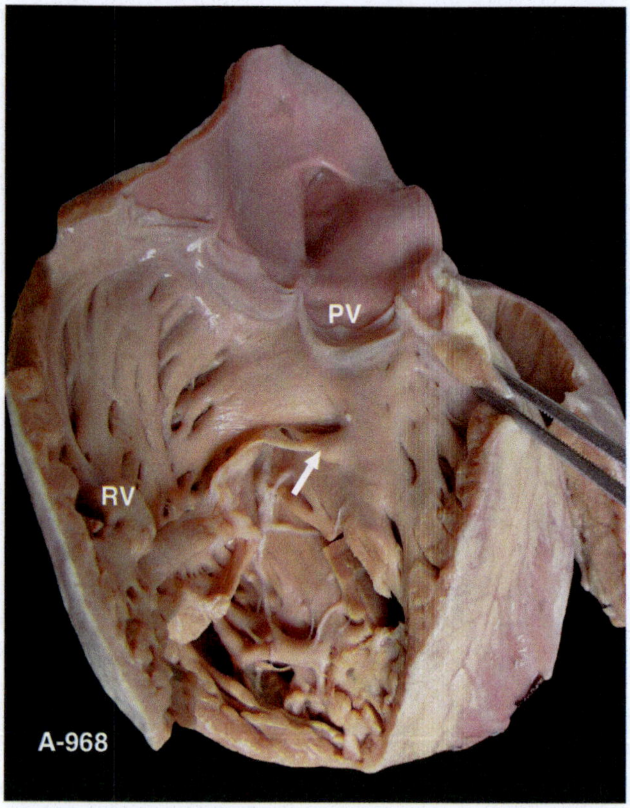

Fig. 4.7 Right ventricular flow tract. The right ventricular flow tract between the pulmonary valve and the TV is seen, along with infundibular muscular tissue under the pulmonary valve; there is no continuous fiber between the pulmonary valve and TV. *Arrow* cone papillary muscle, *PV* pulmonary valve

4.3.1 Sinus Portion

The right ventricular sinus portion is the right ventricular inflow tract (Fig. 4.8). It is formed by embryonic, primitive, muscular ventricular septal growth. The inner wall is smooth. The TV apparatus can be seen in this portion.

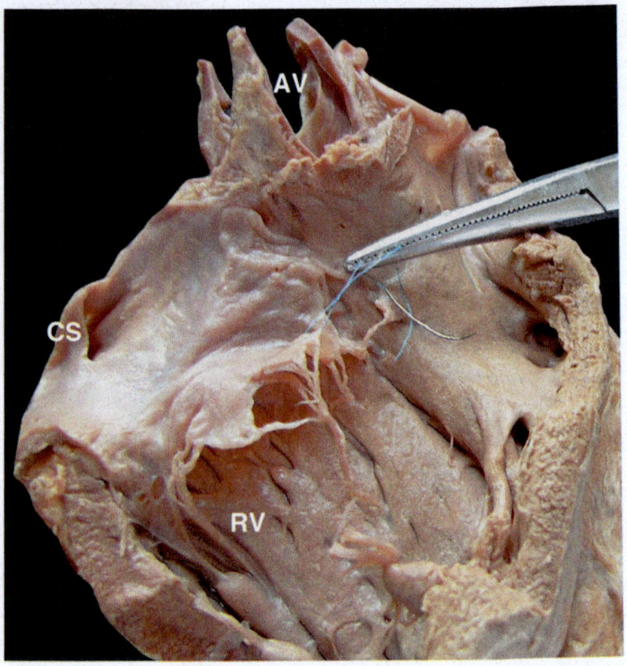

Fig. 4.8 Right ventricular sinus. From the right side of the ventricular septum, the right ventricular sinus portion held on by the forceps is the tricuspid septal leaflet. Beneath the septal leaflet is relatively smooth muscle. In front of the septal leaflet is the conical papillary muscle. The coronary sinus (*CS*) can be seen from above the septal leaflet

4.3.2 Trabecular Portion

The right ventricular trabecular portion, as the right ventricular cavity bottom, is a result of embryonic ventricular cavity spongy absorbing expansion (Fig. 4.9). The inner wall in this portion of the heart chamber is filled with trabeculae containing tricuspid valve papillary muscles. A ventricular septal defect is a rare occurrence in this portion of the heart, although they often are seen in the intricate muscle band and can be missed easily during surgery.

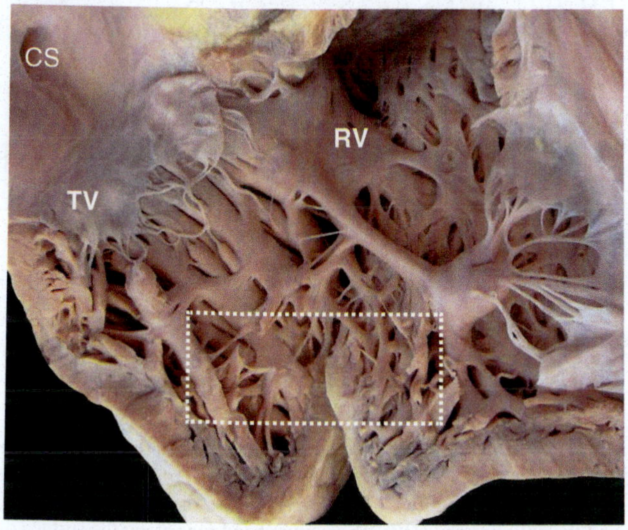

Fig. 4.9 Trabecular portion. Along the right ventricular longitudinal incision, an opened TV revealed the entire right ventricular cavity. The *dotted area* is the trabecular portion of apical regions

4.3.3 Infundibulum

The right ventricular infundibulum develops from the embryonic cone. It is located between the pulmonary valve and the supraventricular crest, as an integral part of the infundibular muscle structure. The surrounding muscle bundles form the right ventricular outflow tract and a smooth inner wall, with the supraventricular crest dividing the inflow part and the trabecular portion (Fig. 4.10).

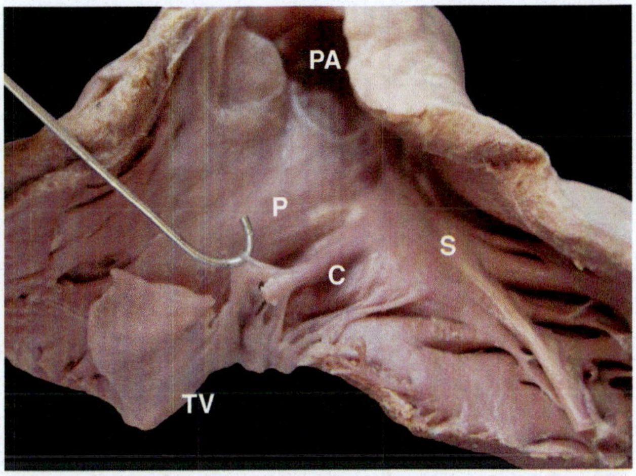

Fig. 4.10 Infundibulum. The infundibulum is also the right ventricular outflow tract. The right ventricular anterior wall has been removed from the specimen. In order to highlight the conical papillary muscle (*C*), the chorda of the tricuspid septal leaflet was lifted with a hook. The infundibulum muscle band, including the parietal band (*P*) and the septal band (*S*), also was revealed. The pulmonary valve ring issued from the infundibular muscular. The conical papillary muscle is an important marker between the right ventricular infundibulum and the conical portion. It can be easily overlooked and injured in TV operations. In membranous ventricular septal defect repair, it is not easy to damage the conduction bundle when suturing at the basal area of the conical papillary muscle. *PA* pulmonary artery, *TV* tricuspid valve

4.3.3.1 Infundibular Septum

The infundibular ventricular septum is a part of the infundibulum and includes a variety of muscle bands, such as the muscle on the septum, which is called the *septal band*, and the muscle around the right ventricular anterior wall, called the *parietal band* (Figs. 4.11 and 4.12). In addition, the muscle running along the septal and parietal bands becomes the moderator band. The infundibular septum is located between the left and the right ventricular outflow tracts and close to the AV.

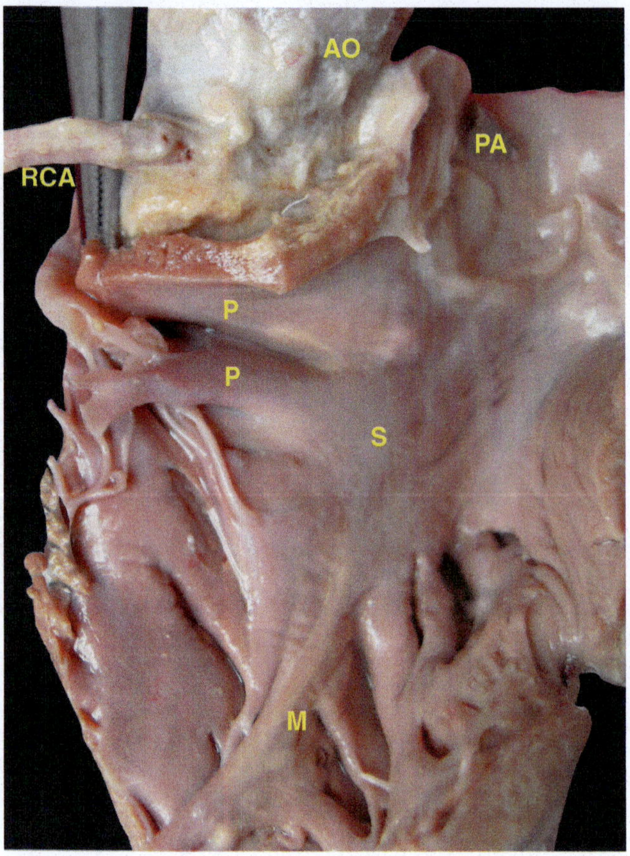

Fig. 4.11 Infundibular muscle bundle. The right ventricular anterior wall has been removed from this specimen. The infundibulum is seen wrapped around the aortic root. The infundibular muscle displays the parietal band (*P*), the septal band (*S*), and the moderator band (*M*). *AO* aorta, *PA* the main pulmonary artery, *RCA* right coronary artery

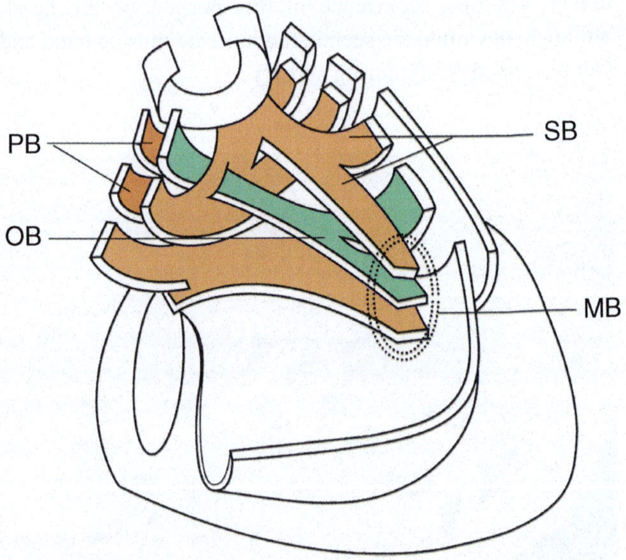

Fig. 4.12 Bundle structure diagram. This is a schematic diagram of the infundibular muscle layers. The inner band according to the running direction is divided into the septum band (*SB*), the parietal band (*PB*), and the oblique band (*OB*). The septum band runs along the surface of the infundibular septum. The parietal band surrounds the anterior right ventricular wall. *MB* moderator band

4.3.3.2 Membranous Area

The membranous area is very small but has large clinical significance (Figs. 4.13 and 4.14). Above the membranous portion and to the back are the tricuspid ring and the atrial part of the membranous septum (Fig. 4.15); the lower part is the muscular ventricular septal crest, the front part is the infundibular muscle, and the upper portion is composed of the septal valve and the adjacent portion of the aortic valve ring (Fig. 4.16).

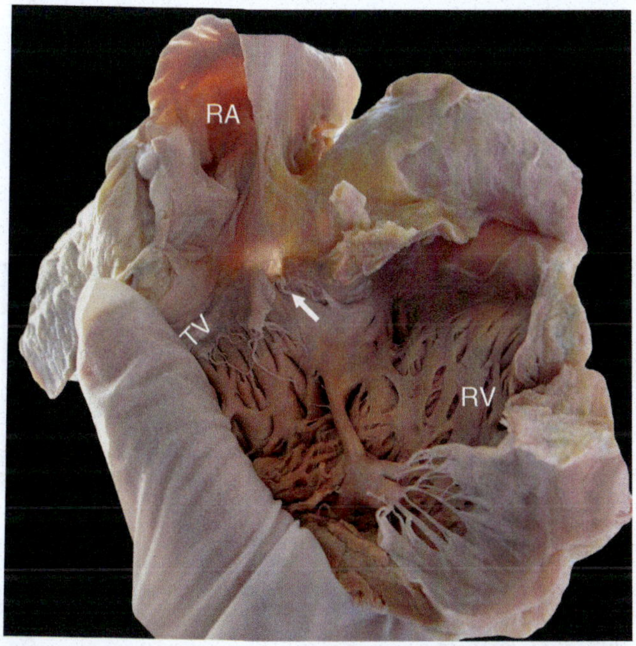

Fig. 4.13 Membranous septum. The membranous septum is located under the junction of the right coronary AV and the noncoronary valve, in front of the septal leaflet of the TV. The translucent area indicates the membranous septum

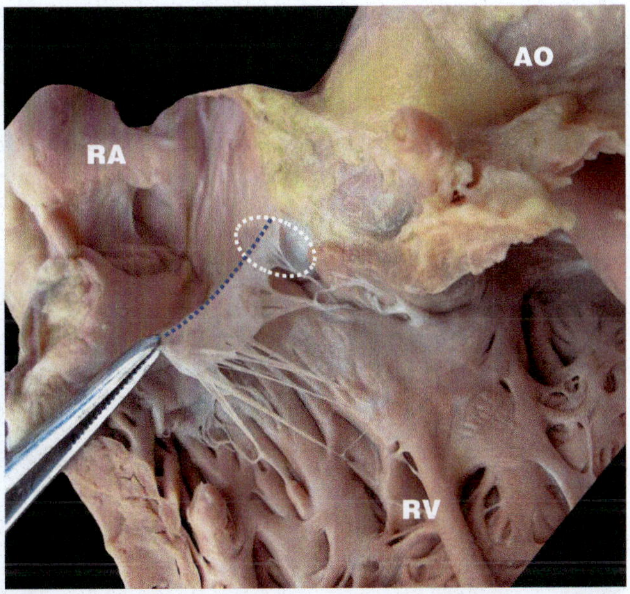

Fig. 4.14 Membranous septum division. The tricuspid septal leaflet attachment edge (*blue dotted*) across the membrane can be divided into two parts. The lower part is the membranous part, and the upper is the left ventricular-right atrial septum. *AO* ascending aorta, *RA* right atrium, *RV* right ventricle

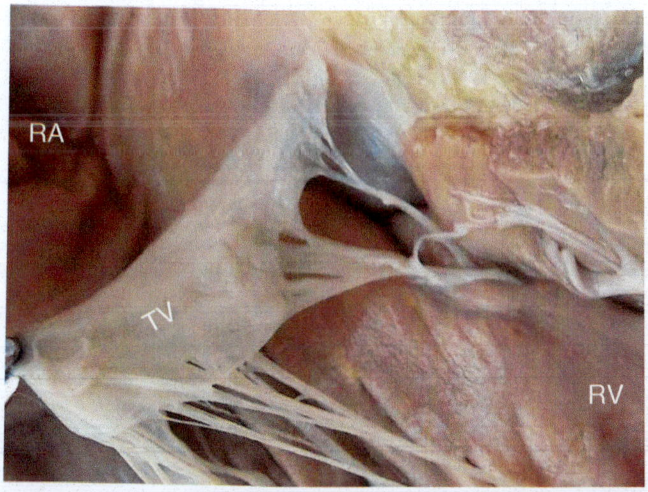

Fig. 4.15 Membranous septum and tricuspid. This is a specimen of the membranous septum with slight observation; the *oval light gray area* shows the membranous septum in front of the tricuspid septal leaflet

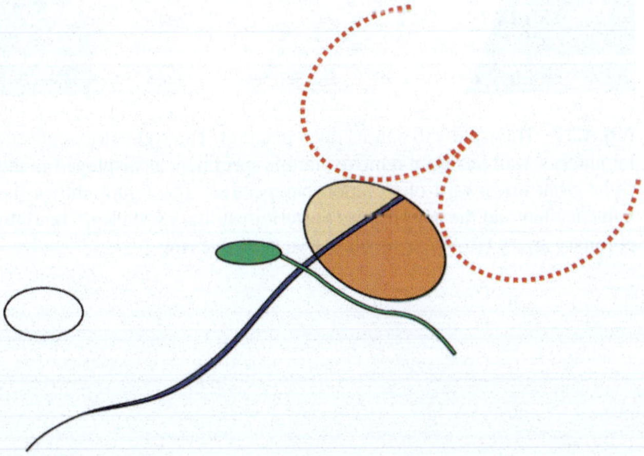

Fig. 4.16 Partition of the membranous septum. From the right ventricular view, the membranous septum is divided into two parts by the tricuspid annulus: A, membranous septum of left ventricular-right atrial; B, membranous ventricular septum. The *red dotted line* is the AV ring, the *blue line* is the tricuspid septal leaflet attachment edge, and the *green* represents the atrioventricular node. *CS* coronary sinus, *P* noncoronary artery valve ring, *R* right coronary artery valve ring

4.4 Tricuspid Valve Device

4.4.1 Characteristics

The front half part of the tricuspid annulus tissue is fragile and soft. The tricuspid annulus is the attachment point for the TV and is part of the fibrous skeleton of the heart; it is not located on the same horizontal plan as the mitral annulus. The tricuspid annulus can be divided into the fixed part (the tricuspid ring) and the free part (the atrioventricular node) (Figs. 4.17 and 4.18). The portion of the tricuspid annulus on which the septal leaflet is attached to the interventricular septum is known as the *fixed part* because it has no flexibility (Fig. 4.19). The portion of the tricuspid annulus found in the right atrioventricular groove is called the *free part*. As the attachment edge of the anterior leaflet of the TV, it can be elongated along with the enlargement of the ventricular cavity. It is also the part for tricuspid annular shrinkage. Due to the fragility of tricuspid annulus tissue and the great mobility of the annulus plastic part, the suture may easily be torn, which can result in failure of the valve repair or in valve leakage after cardiac valve replacement. The valve ring is slightly triangular, and one angle is equivalent to the septal leaflet connected with the central fibrous body.

The normal TV generally has three leaflets (the septal leaflet, the anterior leaflet, and the posterior leaflet) and papillary muscles (Fig. 4.20).

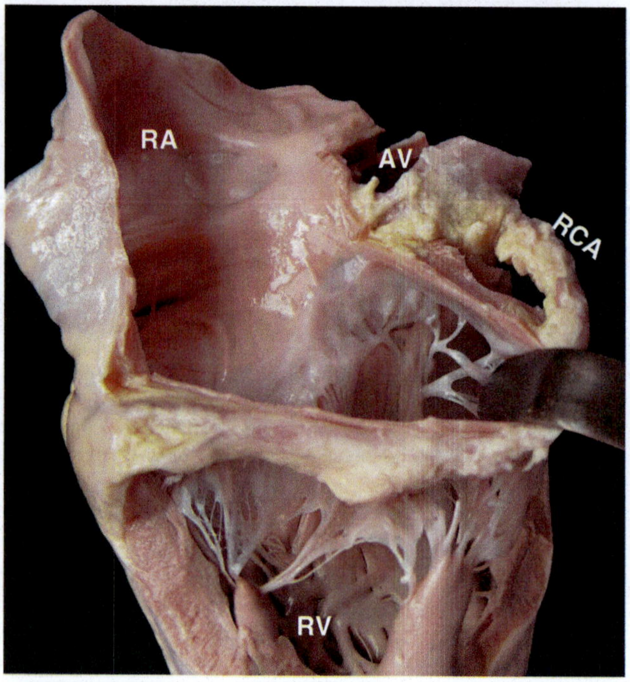

Fig. 4.17 Tricuspid valve in its open position. The right atrioventricular anterior wall has been removed in this specimen, as displayed in the right atrial downward observation perspective. The figure shows the valve leaflets and the anterior and posterior papillary muscles. The right coronary artery (RCA) is visible above the valve ring

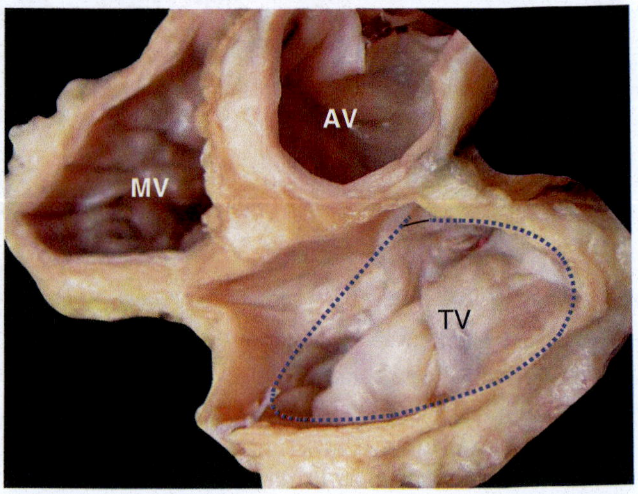

Fig. 4.18 Tricuspid annulus. The *dotted line* shows the tricuspid annulus. In the central fibrous body, the tricuspid and mitral annuli are connected, but the tricuspid annular plane is below that of the MV ring. From an external oblique angle, the tricuspid annulus can be divided into two parts: the septal part of the tricuspid ring (its position is fixed, and it has no flexibility) and the atrioventricular node (the free part of the tricuspid annulus), which corresponds to the right atrioventricular ring part. The more active and flexible free part is the major part of the TV repair operation

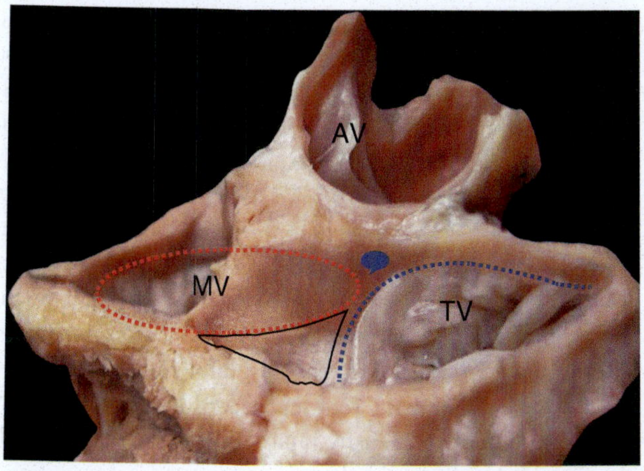

Fig. 4.19 The relationship between the central fibrous body and the atrioventricular valve. The *red dotted line* shows the mitral annulus; *blue dotted lines* show the tricuspid annulus. The target star shows the central fibrous body

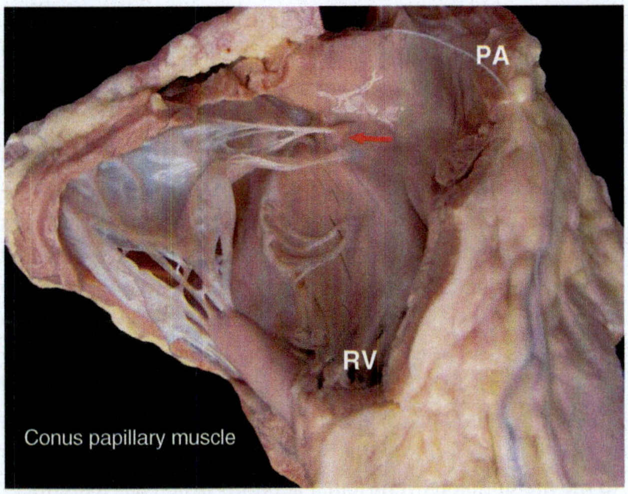

Fig. 4.20 Conical papillary muscle. This specimen has been removed from the right ventricular anterior wall. Looking down from the apical to the tricuspid annulus, the three leaflets and all papillary muscle can be seen. The *red arrows* indicate the conical papillary muscle. Pay attention to the adjacent relationship around the papillary muscle

4.4.1.1 Leaflets

The *septal leaflet* of the tricuspid annulus is attached forward to the middle of the membranous septum, which is divided into two parts. The upper half is found between the left ventricular septum and the right atrium, and the lower half is attached to the ventricular septum (Fig. 4.14). An atrioventricular bundle passing through the atrioventricular ring runs below the membranous septum. The septal leaflet is attached to the ventricular septum by many small tendons from the septum and a small part of the tendons from the papillary muscle. Its function is inferior to that of the anterior leaflet.

The *anterior leaflet* is the largest of the three leaflets and maintains the main part of the TV's function. The junction between the anterior leaflet and the septal leaflet is adjacent to the membranous ventricular septum, which has a predilection for congenital cardiac malformation. Due to pulmonary hypertension or other causes for enlargement of the tricuspid annulus, backflow often appears at the junction between the septal and the anterior leaflets.

The *posterior leaflet* is the smallest of the three leaflets.

4.4.1.2 Chordal and Papillary Muscles

The main chordal and papillary muscle of the TV is the *anterior papillary muscle* and its tendons. The tendons are attached mainly to the anterior leaflet, with a few attached to the posterior leaflet. The anterior papillary muscle (Fig. 4.21) is the largest of the papillary muscles of the TV, which has many large bundles connected to the ventricular septum, the thickest of which is called the *moderator band* (Fig. 4.22). The moderator band, in which runs the right bundle branch conduction system, is connected to the anterior papillary muscle and the supraventricular crest through the free chamber. In the case of right ventricular outflow tract muscular stenosis, dredging can easily injure the moderator band and the anterior papillary muscle.

The *posterior papillary muscle* is small, with its tendons connected to the posterior leaflet (Figs. 4.20 and 4.23). Beneath the junction of the anterior and posterior leaflets, the conal papillary muscle grows from the lower edge of the supraventricular crest, which is the important marker for surgical intervention (Figs. 4.6 and 4.10).

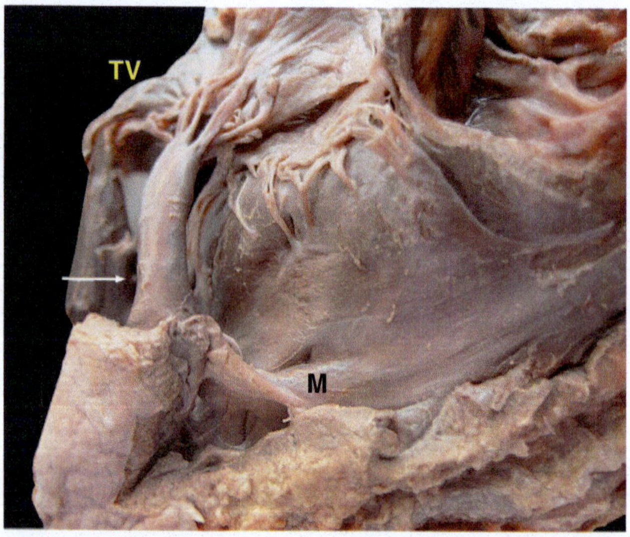

Fig. 4.21 Tricuspid anterior papillary muscle. A tricuspid anterior papillary muscle (*white arrow*) with a strong tendon can assure the normal function of the anterior leaflet of the TV. In tricuspid valvuloplasty, it is critical to maintain and to restore the function of anterior leaflet of the TV. *M* moderator band

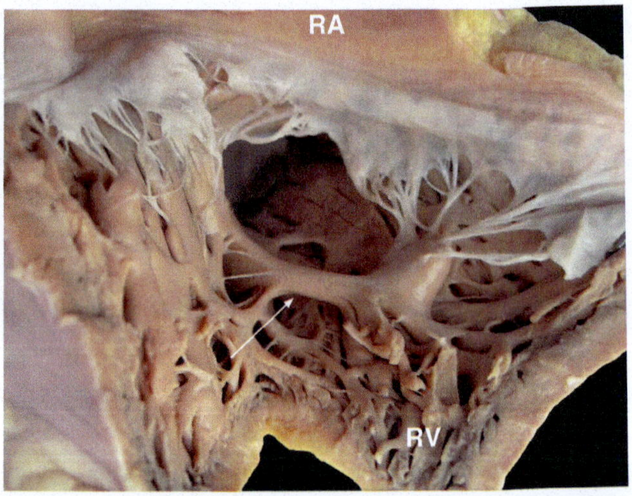

Fig. 4.22 Moderator band. The moderator band is located between the supraventricular crest and the tricuspid anterior papillary muscle, free from the right ventricular cavity. Right ventricular dilation may injure the moderator band by pining it down. The *white arrows* refer to the moderator band

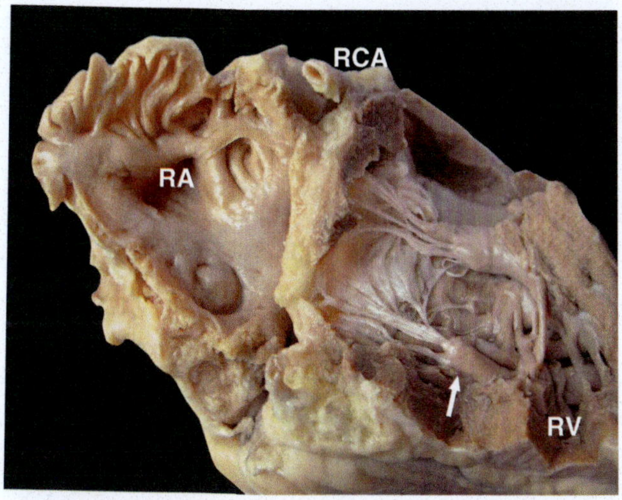

Fig. 4.23 Posterior papillary muscle. The *white arrow* points to the posterior papillary muscle that anchors the posterior leaflet chorda. In tricuspid valvuloplasty, the posterior leaflet is discarded and the function of the anterior leaflet is ensured. *RA* right atrium, *RCA* right coronary artery, *RV* right ventricle

4.4.2 Tricuspid Valve Surgery

Valvuloplasty is a surgical procedure performed primarily to guarantee the physiological function of the anterior leaflet (Fig. 4.21). When the TV annulus is narrowed, the anterior leaflet can cover the TV orifice. Tricuspid ring dilation is performed to retract the tricuspid annulus of the right ventricular free wall (i.e., anterior leaflet of the tricuspid valve attachment edge portion) (Fig. 4.24). The junction of the septal and the anterior leaflets is adjacent to the membranous septum, through which the atrioventricular bundle runs at the posterior inferior edge. The atrioventricular bundle may be damaged during tricuspid valvuloplasty.

TV replacement also should prevent atrioventricular bundle injury (Fig. 4.25). If a bioprosthetic valve is to be implanted, the biological valve foot should be away from the right ventricular outflow tract (Fig. 4.26).

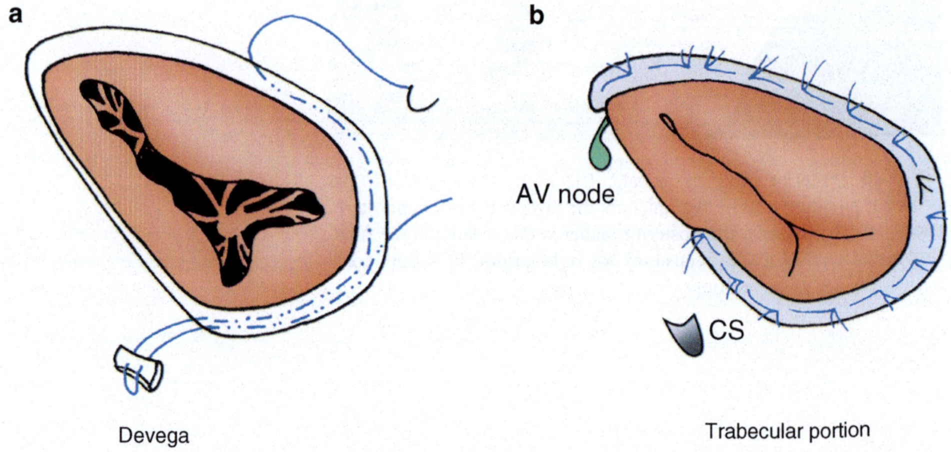

Fig. 4.24 Tricuspid annuloplasty diagram. The figure shows enlargement of the tricuspid annulus. The leaflets cannot close normally, leading to regurgitation. The subvalvular structure is normal. First, make the ring of the anterior leaflet and the posterior leaflet constriction with sutures. Close the valve leaflets (**a**) and then use the appropriate size of the annuloplasty ring to make a reinforcement (**b**)

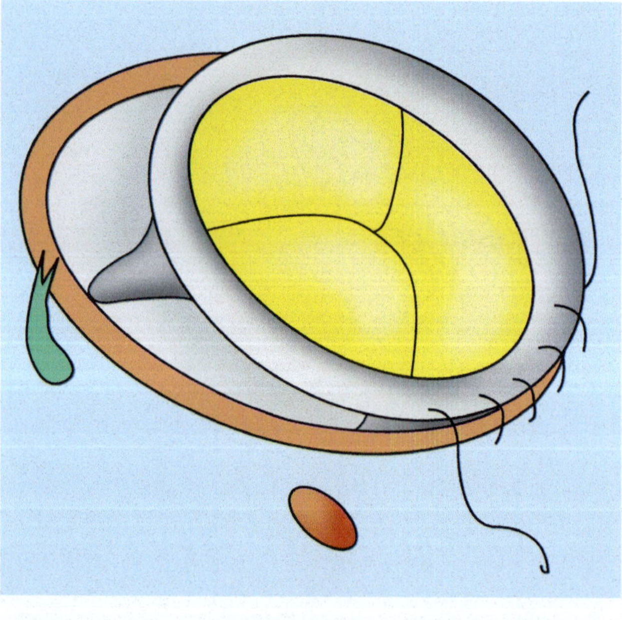

Fig. 4.25 Tricuspid valve replacement chart. The TV is located in the low pressure area. Bioprosthetic valves are often used for replacement of a TV. This picture shows a pericardial valve replacement. Note: avoid injuring the atrioventricular node and bundle with the suture; the biological valve foot should not obstruct the right ventricular outflow tract

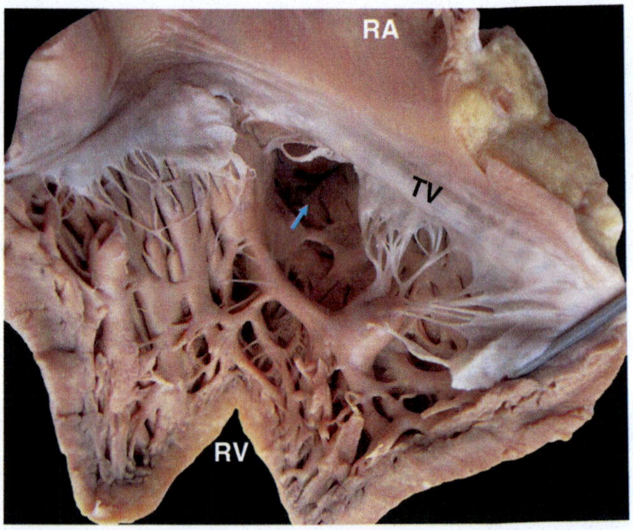

Fig. 4.26 The relationship between the sinus portion of the right ventricular and the infundibulum. When the TV is to be replaced with a stented bioprosthetic valve, the valve foot should not block the entrance of the infundibulum. *TV* tricuspid valve

The Coronary Artery

5

5.1 Coronary Artery Anatomy

The coronary artery anatomy includes: (1) the left main coronary artery (LMCA), also known as the left coronary artery (LCA); (2) the left anterior descending artery (LAD) and its main branches; (3) the left circumflex artery (LCx) and its main branches; (4) the right coronary artery (RCA); and (5) the posterior descending artery (PDA) (Figs. 5.1a, b). The coronary vein is parallel to the artery (Figs. 5.2a–c). It merges into the coronary sinus, and all drain into the RA. The LCA originates from the left aortic sinus, whereas the RCA originates from the right aortic sinus above the AV (Figs. 5.3 and 5.4).

X. Zhu (ed.), *Surgical Atlas of Cardiac Anatomy*,
DOI 10.1007/978-94-017-9409-1_5, © Springer Science+Business Media Dordrecht and People's Medical Publishing House 2015

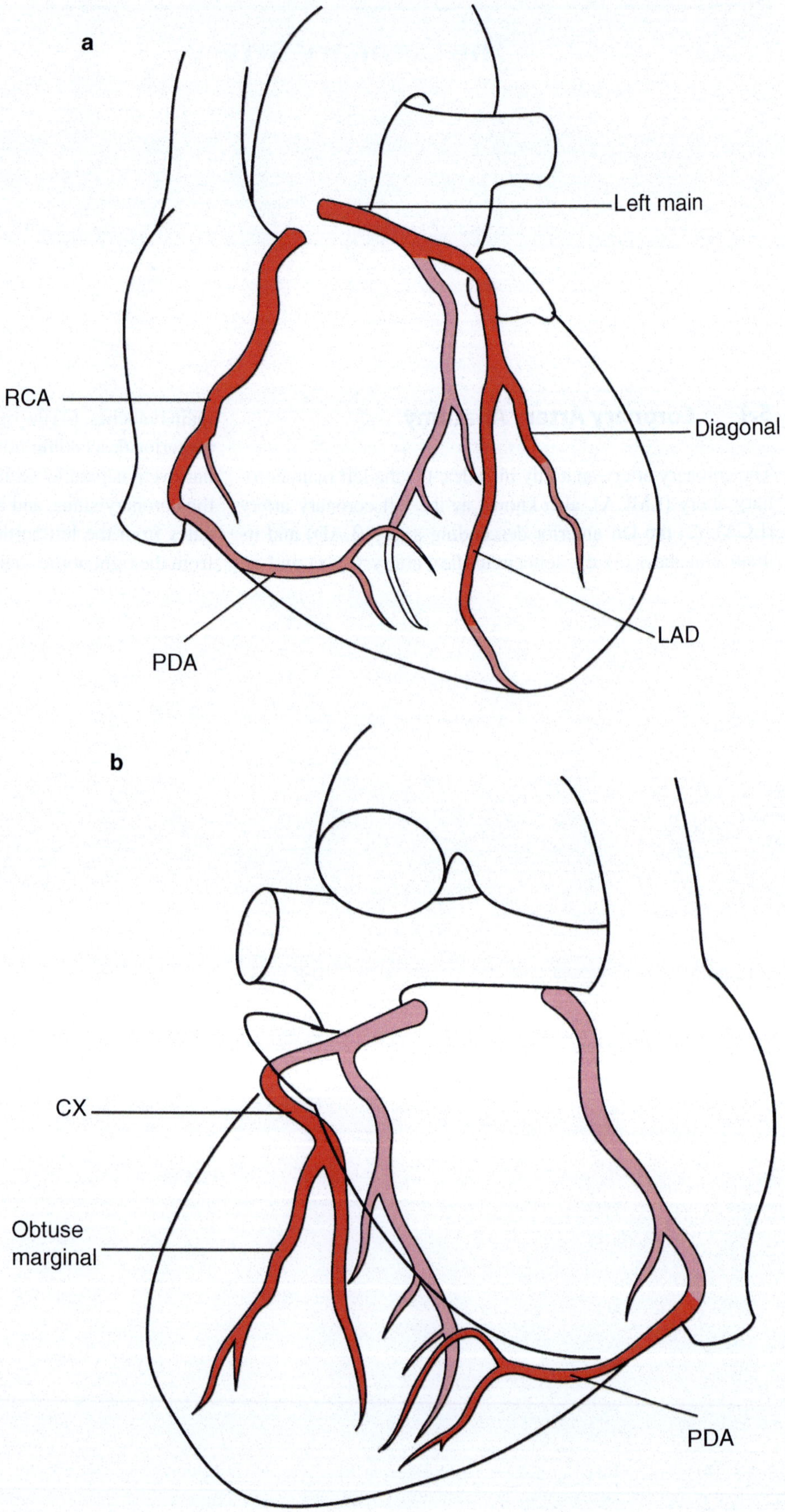

Fig. 5.1 (**a**) Coronary artery anatomy position, projecting from the front view. Deep *red shade* shows the anterior of the heart; *light shade* shows area behind the heart. The LAD is located in the left front portion. The heart is easily exposed by either a median sternotomy or a small incision on the left side of the chest. Minimally invasive techniques are used commonly for coronary artery bypass at this branch. Because it contains no giant branch of the LAD, the LV provides an ideal site for making a left ventricular incision. (**b**) Back projection of the coronary artery. Deep *red shade* indicates the LCx. The obtuse marginal branch (OM), the posterior descending branch, runs from the back of the heart. Exposing the branches at the back of the heart is more difficult in bypass surgery. LCx bypass is covered by the left atrial appendage. Caution must be exercised to avoid tearing the LAA, which may damage the LCx and cause fatal hemorrhage during the operation

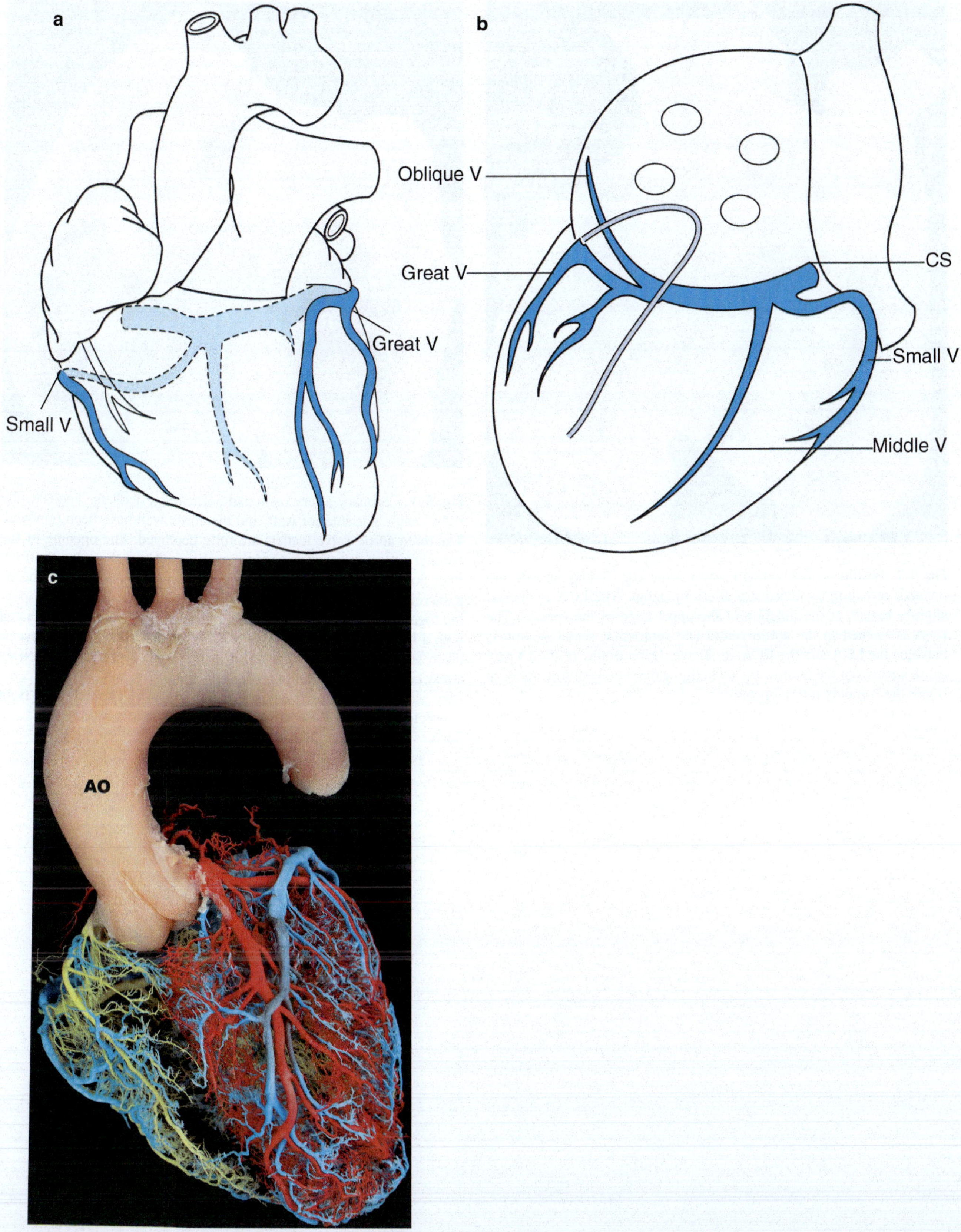

Fig. 5.2 (**a**) Front positive, coronary venous projection. The *deep blue* branches indicate the great cardiac vein. On the right is the small cardiac vein. The vessels colored in *light blue* are the cardiac dorsal heart vein and coronary sinus. (**b**) Back projection of the heart coronary vein. The middle vein (middle V), the oblique vein (oblique V) of Marshall on the atrial posterior wall, and the coronary sinus (CS) in the atrioventricular groove are seen entering together into the RA. The oblique vein without full degradation becomes the thick left SVC, which may influence the cardiac operation under direct vision. The coronary sinus is the site for retrograde perfusion and for insertion of the catheter for interventional treatment. (**c**) Coronary artery. *AO* ascending aorta (from Zhong Shizhen academician)

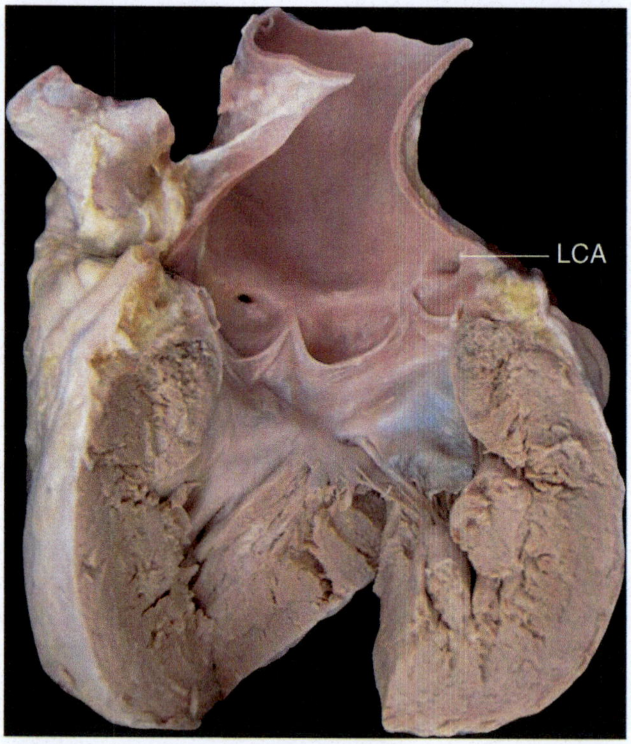

Fig. 5.3 Position of the coronary artery ostia. The LV and the aorta are opened, revealing two coronary artery openings. The LCA is opened slightly higher, at the junction of the upper edge of the sinuses. The RCA is opened in the aortic sinus. The ventricular septal section is between the LCA and the RCA, as shown. Hypertrophy of the LV can cause stenosis of the outflow tract. Dredging the septal outflow tract can injure the coronary artery opening

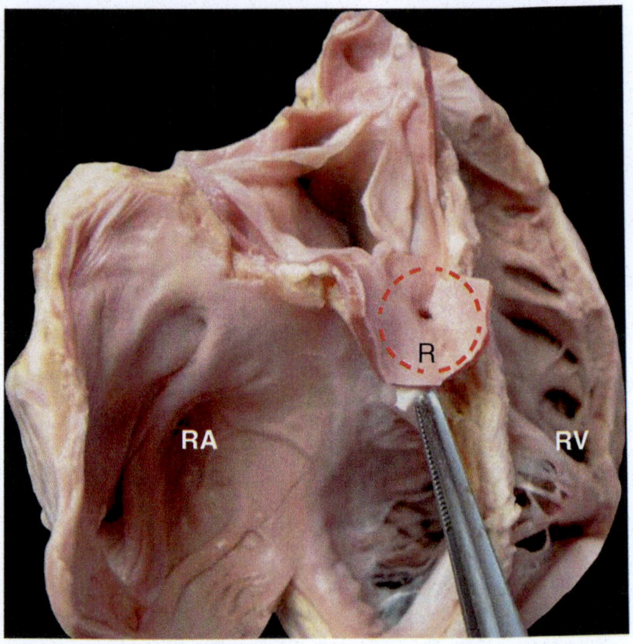

Fig. 5.4 Coronary artery ostia and aortic valve leaflets. The RA, the right ventricular anterior wall, and the aortic wall have been removed. The three aortic valve leaflets are fully deployed. The opening of the coronary artery is in the sinus pipe junction and near the valve junction. In an aortic root replacement operation, the coronary artery often needs to be reimplanted in an artificial vascular graft. A coronary artery opening together with the surrounding of the aortic wall is often dissociated with a button-type cutting. If the AV is to be preserved, care must be taken to avoid damaging the valve leaflet when freeing the coronary artery ostium. The *dotted line* represents the button tangent. *L* left coronary ostium, *R* right coronary artery ostium, *RA* right atrium, *RV* right ventricle

5.1.1 The Left Main Coronary Artery

The LMCA originates from the left coronary sinus in the ascending aorta (Fig. 5.5). The LMCA is short, running along the main pulmonary artery after the LAA and extending under the left atrioventricular groove (Figs. 5.6 and 5.7). The LMCA is embedded in fat and is covered by the LAA (Figs. 5.6 and 5.7). The LMCA bifurcates into two or three branches: the LAD, which runs along the anterior interventricular sulcus; the LCx, which runs along the left atrioventricular groove; and a branch between these two branches that becomes the intermediate branch (Fig. 5.8). The LAD, which is adjacent to the pulmonary artery (Figs. 5.6 and 5.9), crosses the apex and meets with the posterior interventricular groove, also called the PDA, in the posterior interventricular groove. The anterior main descending artery enters the apex as a single vessel and later grows into a diagonal branch with numerous septal branches (Fig. 5.10). The proximal LAD is covered by fat tissue on the heart's surface. In most people, the proximal LAD is located in the inner layer of the pericardium. In a minority of patients with the anterior descending branch running within the myocardium, the ventricular contraction may compress the anterior descending artery and block the blood flow. The LCx runs along the left atrioventricular groove and has numerous left ventricular branches that bifurcate from it before it arrives at the apex (Fig. 5.11a, b). The ventricular branches supply the blood of the posterior wall of the LV (Fig. 5.12a).

The LMCA angioplasty is as shown in Fig. 5.12b.

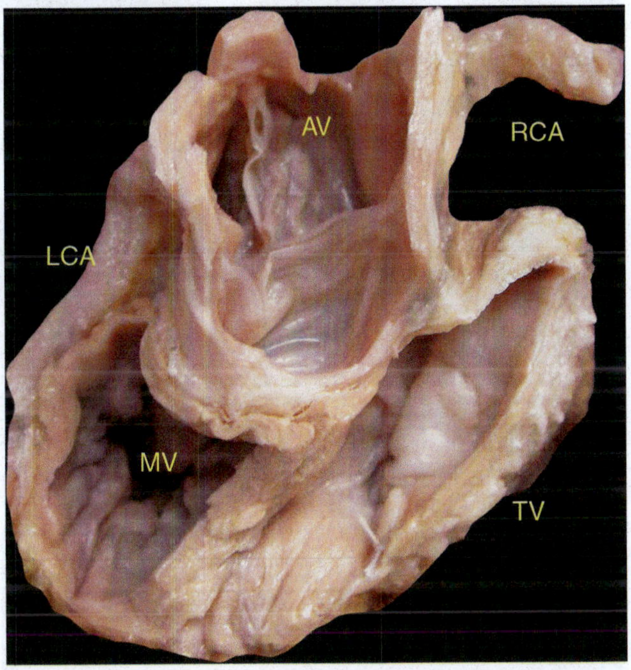

Fig. 5.5 Right and left coronary artery initial segments. A horizontal cross section of the atrioventricular ring is shown. Viewed from the back side, the initial segments of the LCA and the RCA are seen. The photo also shows the relationship between the three valves and coronary artery root. In MV surgery, peeling calcified nodules or cutting too deeply in the posterior lobe attachment edge may damage the LCx. *AV* aortic valve, *MV* mitral valve, *TV* tricuspid valve

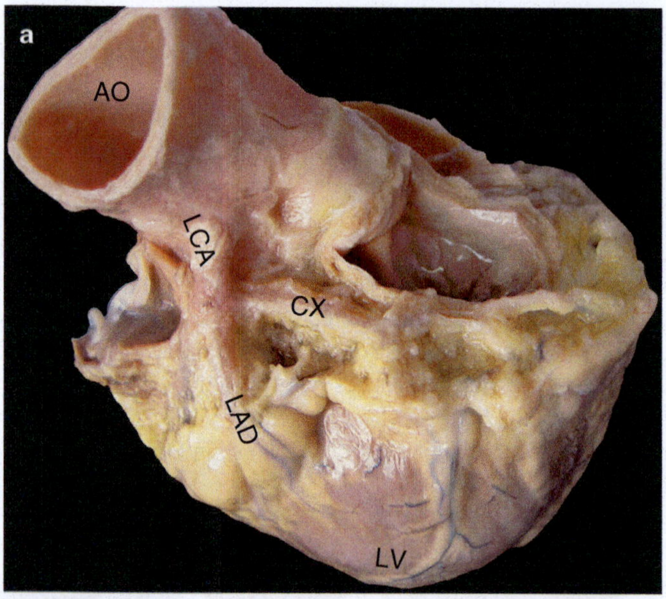

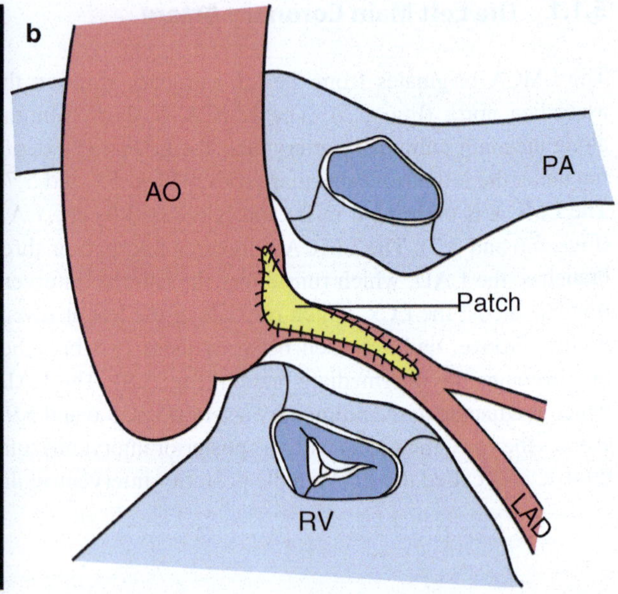

Fig. 5.6 (**a**) Left circumflex coronary artery and lateral wall of the left ventricle. From the left side the left ventricular branch from the circumflex lies along the left ventricular wall. Note there is no intermediate branch between the circumflex and anterior descending branch. There are fewer vessels in the apical area between the circumflex and anterior descending arteries, and no thick branches. It is safe to make a left ventricular incision. (**b**) Left main coronary artery angioplasty

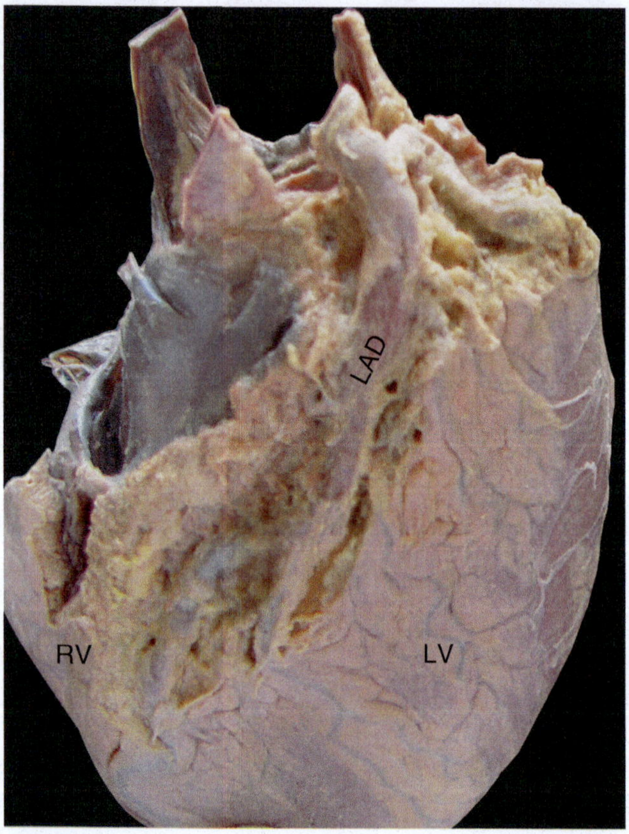

Fig. 5.7 Left anterior descending branch. Left-side view shows the LAD and its diagonal branch. The RV is in front of the anterior descending artery. The LV is at the back of the anterior descending artery. The adjacent structure around the LAD root is seen here clearly. There is no large branch in the left apex. Note that the descending distal vessels have a segment of dark lesions, revealing arteriosclerosis

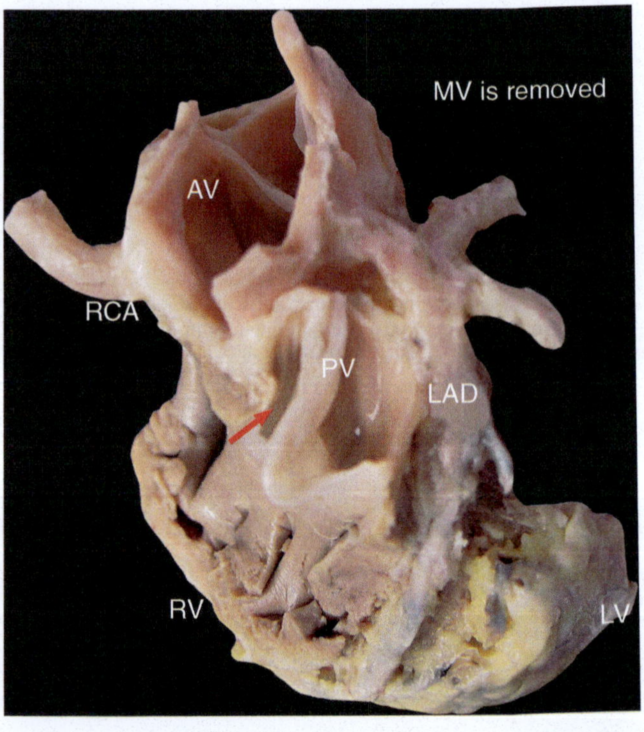

Fig. 5.8 Left coronary artery and the main pulmonary artery. Left-side view showing that the left and right atria and most of the ventricle have been removed and only the root of the aorta and the main pulmonary artery remain. The LCA lies posterior to the main pulmonary artery and is adjacent to the pulmonary artery. The anterior commissure of the AV is adjacent to the posterior commissure of the pulmonary valve. The LAD runs along the anterior interventricular sulcus. The *red arrow* indicates the base of the pulmonary valve. *Surgical considerations*: When dissociating the left coronary artery ostium while making an aortic root replacement, the pulmonary valve can be damaged. Doing left ventricular outflow tract surgery may involve cutting deeply into the septum. *Arrow* points into the right ventricle. *AV* aortic valve, *LV* left ventricle valve, *PV* pulmonary valve, *RV* right ventricle

Fig. 5.9 Branches of the left coronary artery. The left and right atria were stripped in the specimen. The left side of the heart is displayed here, showing the pulmonary valve, MV, aortic root, and LCA. The LMCA (*1*) emits about 2 cm to form a plurality of branches. The LAD, the diagonal branch, and the septal branch issue from the anterior descending branch. The LMCA extends backward along the left atrioventricular groove, or LCx (*4*). The LCx and the LAD bifurcation often include a strong middle branch (*3*). The LMCA and the beginning portion of the LAD initially are seen posterior to the main pulmonary artery. When doing the Ross operation of stripping the pulmonary autograft root, one can easily injure the LAD or the first septal branch. *AV* aortic valve, *LV* left ventricle, *MV* mitral valve, *PV* pulmonary valve

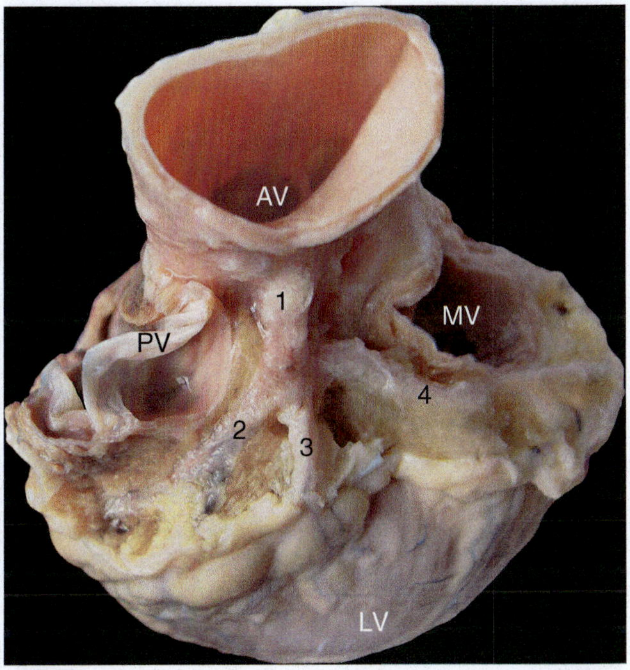

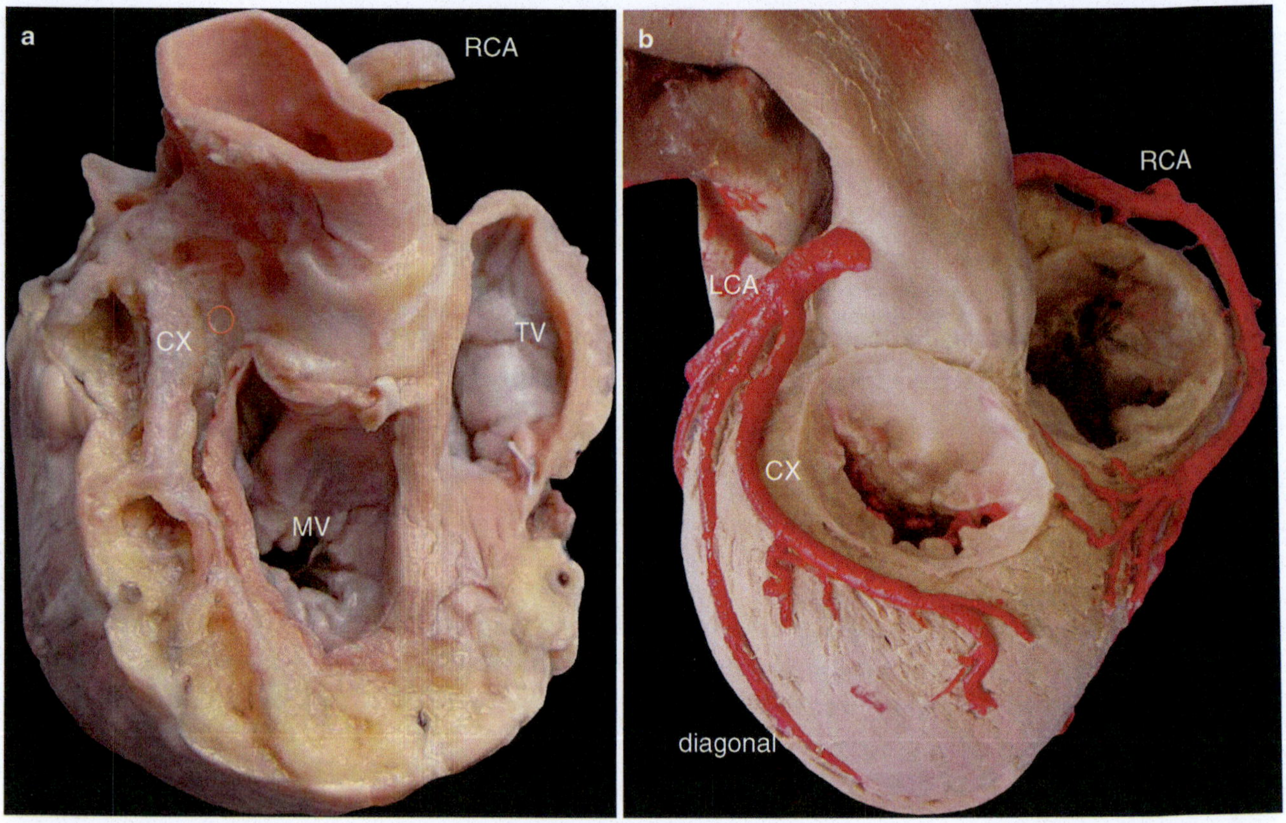

Fig. 5.10 (**a**) Horizontal cross section at the atrioventricular valve level. The LCx lies along the left atrioventricular groove (outside the lateral mitral annulus), near the left fibrous trigone (*red*) of the heart. Multiple left ventricular branches from the LCx lie along the left ventricular posterior wall. Excessive resection of leaflet tissue in a MV operation may damage the LCx. Deep suture in the left fibrous trigone may injure the coronary artery. (**b**) Coronary artery from the dorsal side (provided by academician Zhong Shizhen)

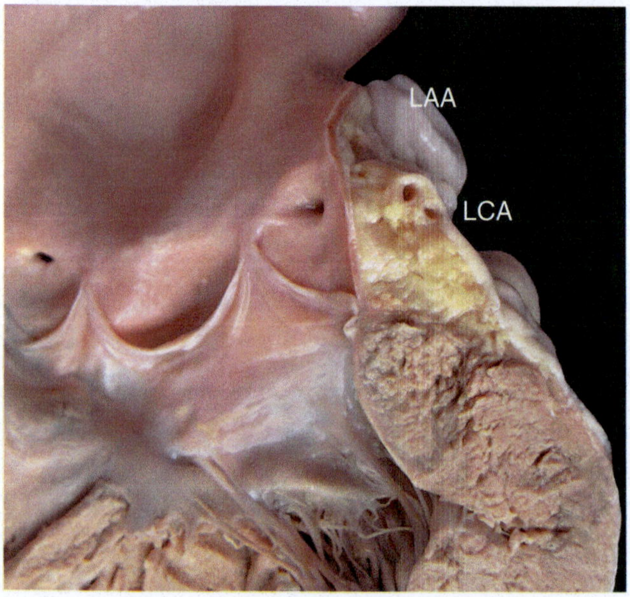

Fig. 5.11 Left circumflex artery beneath the left atrial appendage. Shown in ventricular longitudinal display, the end of the LCx is located beneath the adipose tissue of the LAA. Making an incision in the LAA or LA may damage the LCx. *LAA* left atrial appendage, *LCA* left circumflex coronary artery

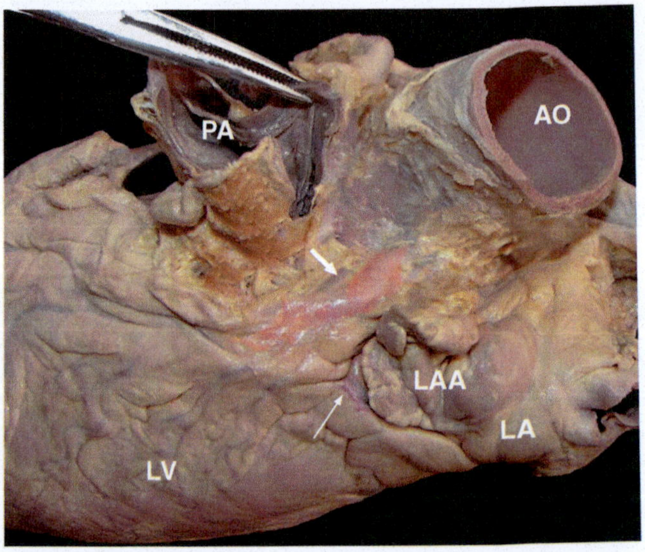

Fig. 5.12 Left atrial appendage and coronary artery, left-side view. The *thick arrow* indicates the LAD; the *small arrow* indicates part of the LCx, most of which is hidden in the LAA. *AO* ascending aorta, *LA* left atrium, *LV* left ventricle, *PA* pulmonary artery

5.1.2 The Right Coronary Artery

The RCA originates from the right coronary aortic sinus and continues downward along the right atrioventricular groove and then into the PDA (Fig. 5.13). It supplies blood to the left ventricular posterior wall and the interventricular septum. The infundibular (or conus) branch, which arises from the proximal portion of the RCA, courses anteriorly over the base of the ventricular infundibulum. The acute marginal artery branch from the RCA courses along the acute margin of the heart. In the posterior interventricular sulcus, a branch of the RCA forms a U-shaped segment, from which the atrioventricular nodal artery originates. This coronary artery layout is called the *right dominant system*. If the PDA originates from the left LCx, then it is called the *left dominant system*.

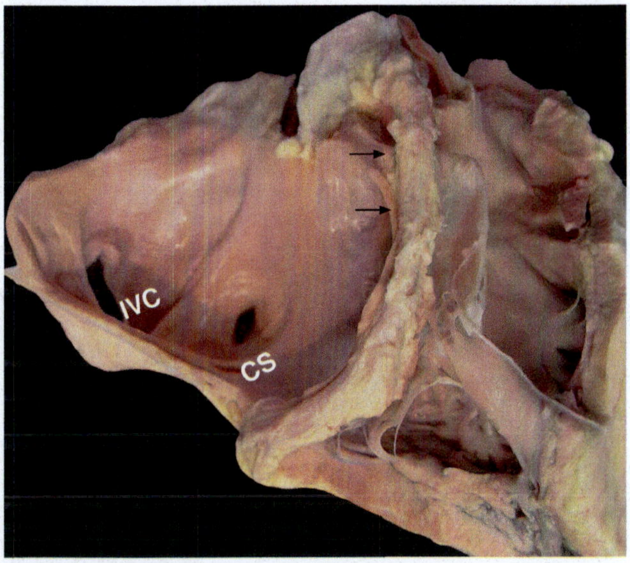

Fig. 5.13 Right coronary artery. Right-side view shows that the anterior wall of the RA and the anterior right ventricular wall have been removed, but the right atrioventricular ring remains. The free segment of the atrioventricular ring is parallel to the RCA trunk (*arrow*). The right ventricular branch and the acute marginal branch arise from the RCA. The RCA crosses the crux, giving off the posterior descending coronary. *CS* coronary sinus, *IVC* inferior vena cava. *Surgical considerations*: Occasionally, the right ventricular branch is quite large, and incision of the RV may injure the right ventricular branch, affecting the right ventricular function. The RCA is located anterior to the heart, and the median sternotomy incision in severe adhesions of the pericardium may injure the RCA

5.2 Coronary Artery Exposure in CABG

In the event of severe angiosclerosis or embolization (Fig. 5.14), surgical intervention on the coronary artery must be considered. In order to increase the blood supply of the distal artery, a bypass operation, referred to as *coronary artery bypass grafting* (CABG), is established (Fig. 5.15). Establishing a coronary artery anastomosis is the key to the success or failure of the operation, and achieving a full exposure and stability of the anastomosis site is extremely important.

The LAD and its diagonal branch are easily exposed. A traction line is used to lift the left side of the pericardium; full drainage of the LV is necessary. A wet gauze pad is used under the LV. An assistant pulls the heart slightly to the right and downward.

Because the RCA and the initial part of the PDA are located in the rear of the heart, exposing them is very difficult. Usually, the right edge of the pericardium is lifted by the traction line; the heart diaphragm surface is lifted with a wet gauze pad, and the RV is pulled to the left (Fig. 5.16). Also at the distal and medial sites (a distance of about 1 cm from the site of the anastomosis for a traction-line, bypassed coronary artery), the traction line is lifted and the coronary artery is straightened such that it reaches the superficial parts.

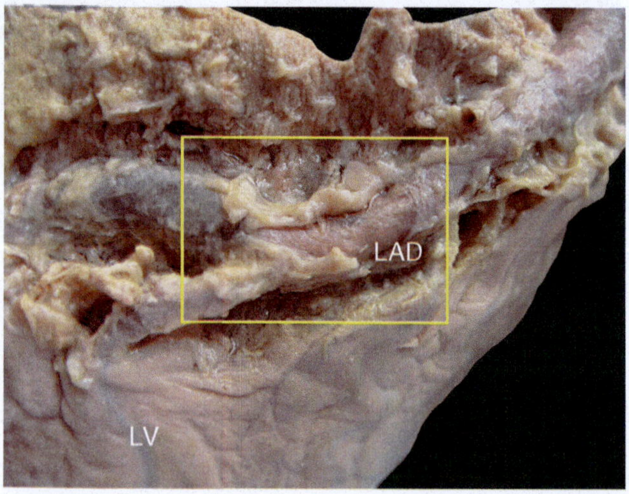

Fig. 5.14 Coronary artery disease. The specimen shows atherosclerosis in the LAD. The vascular wall is significantly thicker (*yellow box*) than normal. Its distal portion is visibly black, which indicates thrombus blockage in the vessel. *LV* left ventricle, *LAD* left anterior descending artery

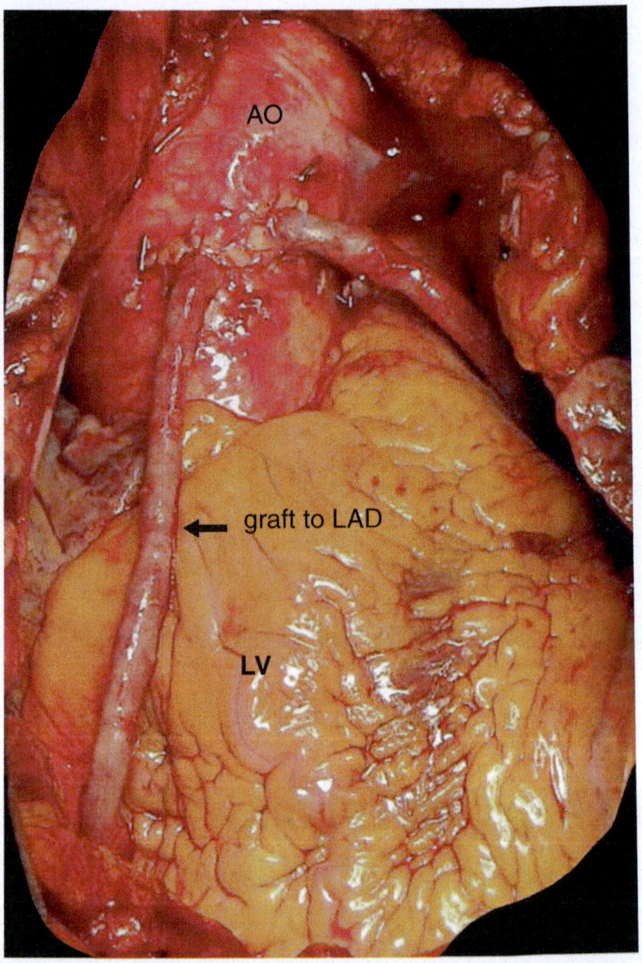

Fig. 5.15 Coronary artery bypass graft (CABG) completed. This photo shows the operation with the saphenous vein as the bridge material for two bypasses: one is from the aorta to the left ventricular branch; the other one is from the aorta to the left descending artery branch. The *arrow* points to the bridge (graft to LAD). *AO* ascending aorta, *LV* left ventricle

Exposing the LCx of the left ventricular posterior wall also is difficult. The traction lines at the right margin of the pericardium should be released, or the right pleural cavity may be opened. The left ventricular apex should be lifted to the right. In the oblique sinus, the pericardium is lifted by traction line in order to better expose the left ventricular branch.

Usually, the coronary artery running within the myocardial surface pericardium is easy to expose (Fig. 5.16). However, in some patients, the left anterior descending branch runs in the ventricular muscle layer, known as the *muscle bridge*. The myocardial vasomotor directly affects the LAD blood flow and can lead to serious myocardial ischemia. A LAD branch in the muscular bridge is unlikely to be recognized at the heart surface. Opening the myocardium to search for the LAD may damage the coronary artery branches and even lacerate the wall and penetrate the heart chamber. Usually, the first diagonal branch of the anterior descending artery can be found on the heart surface first. After the incision is made, a thin metal probe is used to retrograde into the diagonal branch of the LAD, and, finally, a probe guide is used to find the LAD.

After the target vessel is exposed, the surgical area must be fixed by using a special stabilizer. The LAD artery lesion often is severe in the patient with a ventricular aneurysm resection, and only the diagonal branch can be bypassed.

Single coronary artery malformation sometimes can be seen in clinical cases (Figs. 5.17 and 5.18).

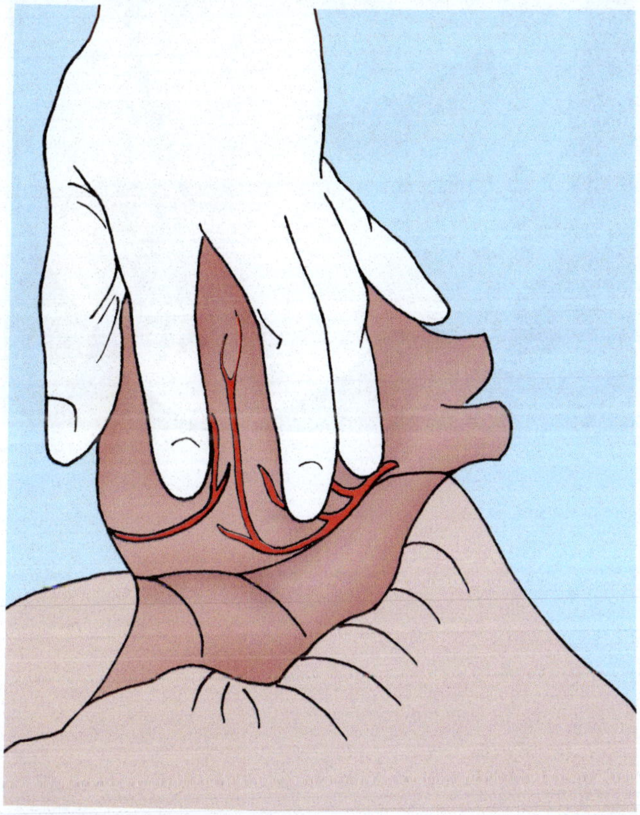

Fig. 5.16 Exposure of the coronary artery in the posterior wall of the left ventricle in operation. The patient is in the supine position. With the surgeon's hands on the apex, the diaphragmatic surface beneath the heart is fully exposed, and the selected target vessels are fixed by the fingers

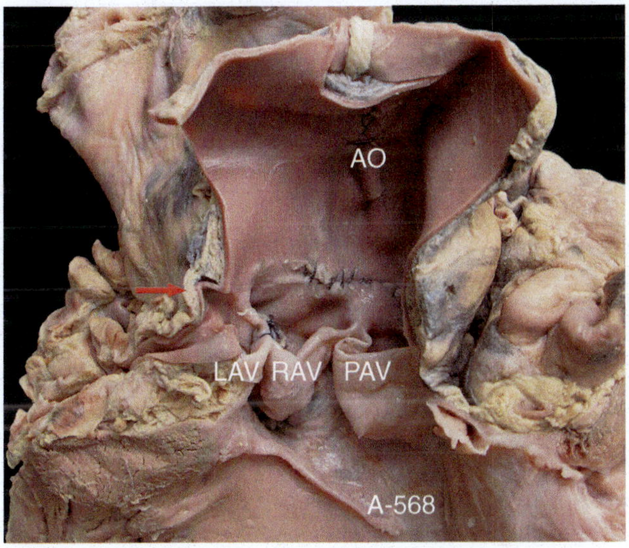

Fig. 5.17 Single coronary artery malformation associated with aortic valve prolapse (568 specimens). The ascending aorta and coronary artery have been dissected. The RCA opening is absent. Only the left coronary ostium is shown here. The patients with AV prolapse had undergone aortic valvuloplasty. Illustration shows sutures at the junction of the LCA and the right coronary valve. *LAV* left coronary valve, *NAV* no coronary valve, *RAV* right coronary valve

Fig. 5.18 Single coronary artery malformation (568 specimens). Because the left ventricular lateral wall and the RCA originate from the *left main* trunk, the LAD and the LCx in this plane cannot be seen. *MV* mitral valve

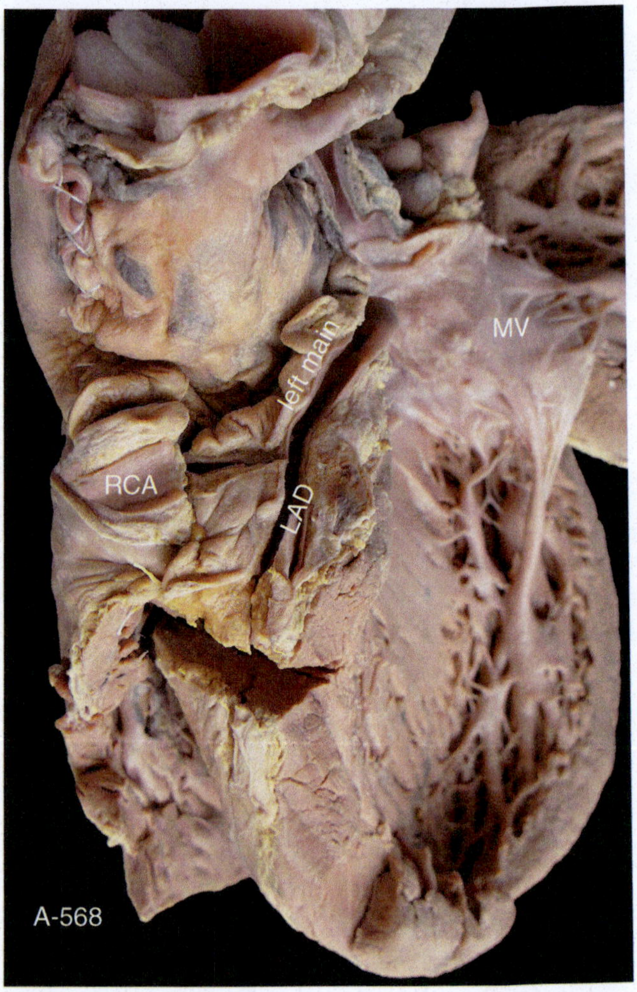

The Conduction System of the Heart

6

6.1 General Considerations

The conduction system of the heart, which provides the heart its automatic rhythmic beat, consists of cardiac muscle cells and conducting fibers that initiate impulses and conduct them rapidly through the heart, initiate the normal cardiac cycle, and coordinate contractions of the cardiac chambers. The events that occur in the cardiac cycle must be coordinated if the heart is to pump efficiently and the systemic and pulmonary circulations are to operate in synchrony.

6.1.1 Sinoatrial Node

Located in the front and upper side of RA, where the SVC enters, the sinoatrial (SA) node has a spindle shape (Fig. 6.1). Its upper end is coarse, stretches downward between the superior and IVC, and is covered only by visceral pericardium. The lower end is deep and contacts the endocardium. Nutrition of the SA node is supplied by a branch from the RCA in most (55–60 %) patients or the LCx in others (40–45 %).

6.1.2 Atrioventricular Node

The atrioventricular (AV) node is located in the lower right atrial septal plane, downward of the coronary sinus opening (Fig. 6.2). It is cylindrical and approximately 3 mm long and approximately 8–10 mm wide. Blood to the atrioventricular node is supplied by the atrioventricular node artery from the RCA. The AV node is surrounded by an abundance of conductive fiber called the *atrioventricular node area*.

6.1.3 The Internodal Conduction Bundle

An electric conduction pathway exists between the SA node and the AV node (Fig. 6.3). Impulses from the SA node to the AV node run through three channels or pathways called *internodal tracts*, which are divided into the anterior beam, the middle beam, and the posterior beam. The *anterior beam* extends from the SVC, descending anteriorly into the atrial septum, near the FO to the AV node. Branches from the anterior internodal beam to the LA are called the *Bachmann bundle*. The middle beam extends from the rear of the SVC and descends into the atrial septum, connecting with the anterior beam near the front of the FO. The posterior beam extends from the SVC and runs along the outer edge of the right atrial wall, continuing downward to the right atrial wall within the crista terminalis, along the valve attachment of the IVC, reaching the AV node.

When an abnormality or damage is associated with the three beams, arrhythmias, such as complete block or nodal rhythm, can occur. In the case of an atrial septal defect, the closure of an open surgery may damage the internodal conduction bundle.

6.1.4 The Bundle of His and Its Branches

The AV node sends forth a coarse beam called the *bundle of His* (Figs. 6.4 and 6.5). It is approximately 1–2 mm in diameter and 10–20 mm in length. Connecting with the distal part of the AV node, the bundle of His runs through the right fibrous trigone to the tricuspid septal leaflet until it reaches the lower edge of the membranous septum, where it divides into two branches (left and right) (Fig. 6.4). Because the bundle of His is affixed tightly to the posterior and inferior margin of the membranous septum, it is pushed to the bottom and backward in the cases of membranous VSD and endocardial cushion defect. Its left and right branches will shift and cause different degrees of conduction block (Fig. 6.6). Surgery can cause injuries to the lower edge of the membranous septum, which can result in a complete conduction block.

The right and left bundle branches cross the muscular ventricular septum on either side. The right bundle branch is

X. Zhu (ed.), *Surgical Atlas of Cardiac Anatomy*,
DOI 10.1007/978-94-017-9409-1_6, © Springer Science+Business Media Dordrecht and People's Medical Publishing House 2015

a single beam that extends from the stem portion of the bundle of His (Fig. 6.7). It traverses along the right subendocardium beneath the crista supraventricularis, reaching the base of the anterior papillary muscle by coursing through moderator bands. In the process, former Purkinje fibers reach the visceral pericardium. In the case of right ventricular dilatation, the right bundle branch can be damaged, leading to atrioventricular block. The left bundle branch originates from the right side of the interventricular septum, arriving at the LV through the ventricular septum (Fig. 6.8). It divides into two beams, which are called the *left anterior fascicle* and *left posterior fascicle*, to the left side of the septum. The left anterior fascicle fans out toward the end of the Purkinje fiber mesh on the ventricular surface, and the left posterior fascicular branch fans out toward the posterior papillary muscle.

6.2 Specimen Demonstrations

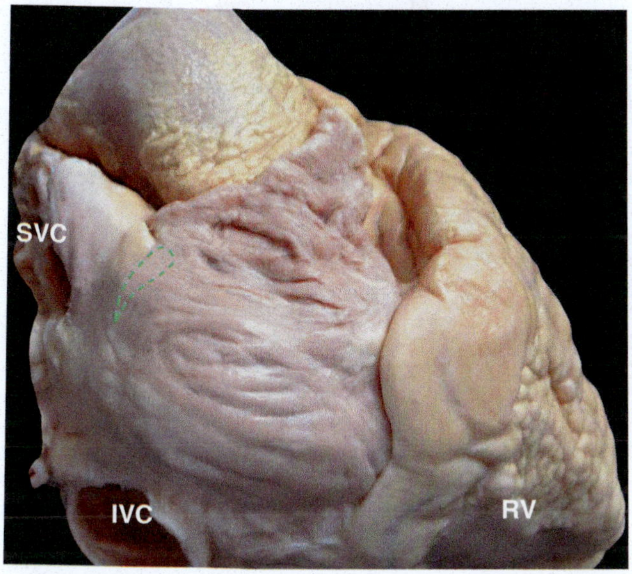

Fig. 6.1 Sinoatrial node. This is a specimen of a healthy heart. The sinoatrial node (*green label*) is located at the entrance of the SVC. It is spindle shaped and covered by the epicardium. Impulses sent by the sinus node traverse through internodal tracts to the atrioventricular node, which controls the rate and rhythm of the heart. *IVC* inferior vena cava, *SVC* superior vena cava, *RV* right ventricle. *Surgical tips*: High right atrial incision may damage the sinoatrial node. A right atrial to left ventricular roof incision may damage the coronary sinus branch and may cause arrhythmia

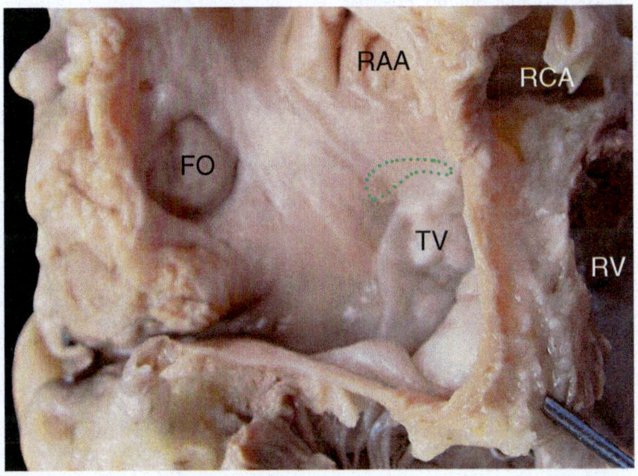

Fig. 6.2 Atrioventricular node. A right-side view is shown. Removal of the right atrioventricular anterior wall reveals the right atrial cavity. Opening the tricuspid annulus reveals the tricuspid septal leaflet (*TV*). The *green dotted line* indicates the atrioventricular node. The atrioventricular bundle from the node courses through the tricuspid annulus to the ventricular septum, where it separates into the right side (right bundle branch) and the left side (left bundle branch). *FO* fossa ovalis, *RAA* right atrial appendage, *RCA* right coronary artery, *RV* right ventricle. *Surgical tip*: in TV replacement surgery or valvuloplasty, a deep cut or stitches in the junction of the septal leaflet and the anterior leaflet can result in conduction block

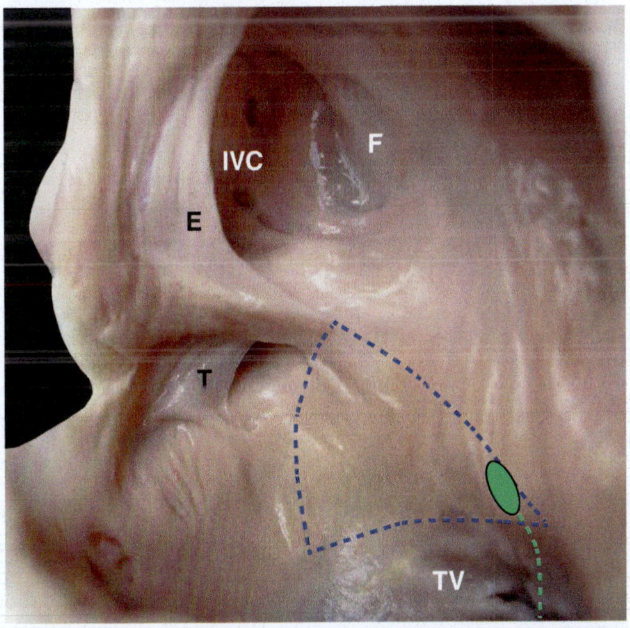

Fig. 6.3 Triangle of Koch and atrioventricular node position. The *blue dotted* area indicates the triangle of Koch, and the *green area* indicates the atrioventricular node. *E* eustachian valve, *F* fossa ovalis, *IVC* inferior vena cava, *T* Thebesian valve

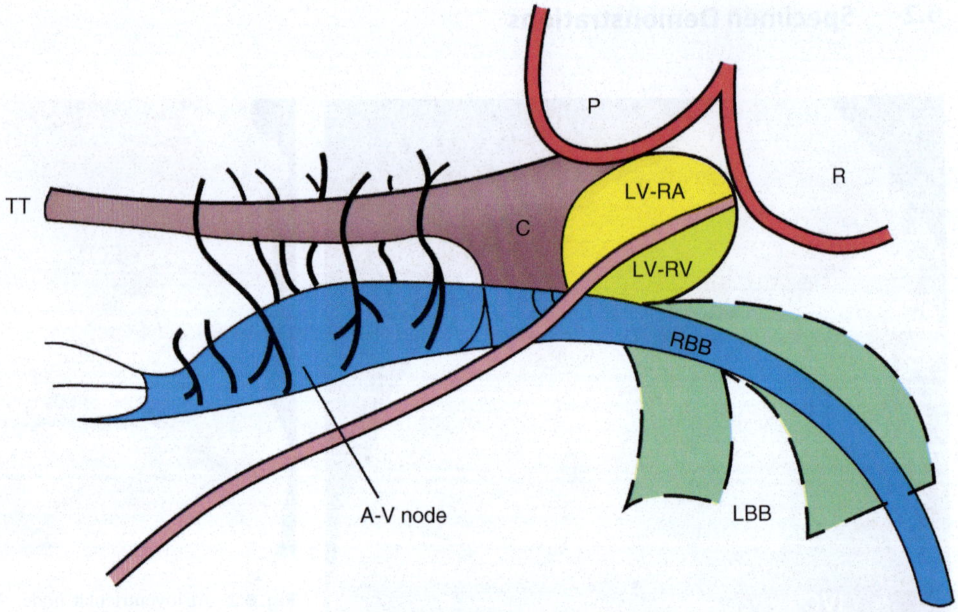

Fig. 6.4 The central fibrous body and the schematic diagram of the relationship with the atrioventricular bundle. Positive perspective demonstrates the central fibrous body and the position of the atrioventricular bundle projection. The *yellow area* represents the membranous septum, which is divided into two parts by the tricuspid annulus. The top portion is the septum of the left ventricle-right atrial septum (LV-RA); the lower portion is the member septum of the left ventricle-right ventricle (LV-RV). *A-V node* atrioventricular node, *C* the central fibrous body, *LBB* left bundle branch, *P* posterior of the aortic valve ring, *R* right aortic valve ring, *RBB* right bundle branch, *TT* Todaro tendon

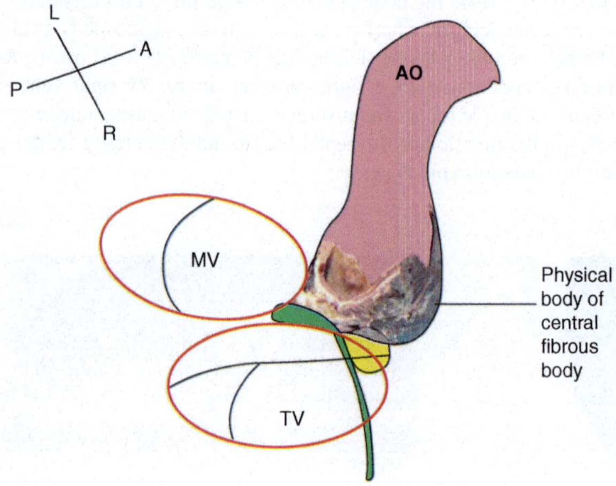

Fig. 6.5 Map of the central fibrous body. *AO* ascending aorta, *MV* mitral valve, *TV* tricuspid valve

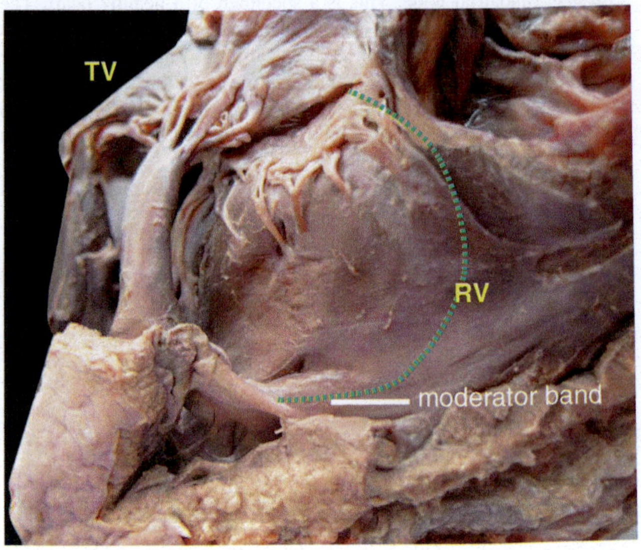

Fig. 6.6 Right bundle branch. These are specimens from the right ventricular cavity. The *green dashed lines* represent the right bundle branch. Running along the lower edge of the membranous septum from the front of the septal leaflet of the TV, the right bundle branch is separated by the atrioventricular bundle at the right side of the RV. It runs along the lower edge of the supraventricular crest into the moderator band and extends to the anterior papillary muscle. Injury to the moderator band during surgery can cause right bundle branch block. *RV* right ventricle, *TV* tricuspid valve

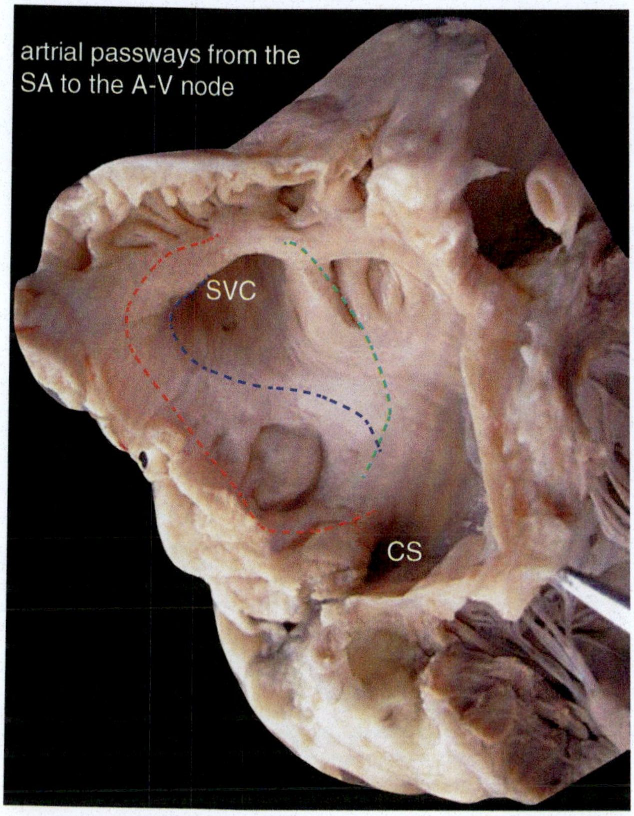

Fig. 6.7 Internode beam tract. Anatomy from the right-side view reveals three impulse pathways to the atrioventricular node between the sinoatrial node and the atrioventricular node by electrophysiological confirmation. The three pathways are called the internodal tracts. The *green dotted line* indicates the anterior internode tract, along the FO front walk down; the *blue dotted line* represents the intermediate tract, from the SVC down also to the marginal branch of the FO. The *red dotted line* indicates the lateral internode tract, coursing along the crista terminalis of the right atrial wall downward through the IVC into the Todaro tendon to the atrioventricular node. *CS* coronary sinus, *FO* fossa ovalis, *IVC* inferior vena cava, *SVC* superior vena cava. *Surgical tip*: a right atrial incision that cuts off the crista terminalis or the marginal branch of the oval fossa may lead to atrial arrhythmias

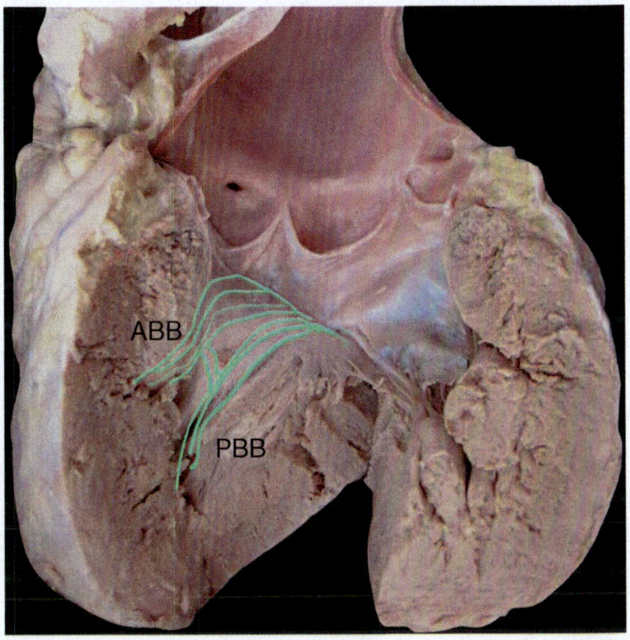

Fig. 6.8 Left bundle branch. This figure shows the interventricular septum from the left ventricle view. The *green line* represents the left bundle branch. These conductions spread in the left-sided endocardium of the septum and are divided into two groups: the left anterior bundle branch (*ABB*) and the left posterior bundle branch (*PBB*)

The Left Atrium

7

7.1 Structure and Characteristics of the Left Atrium

The left atrium (LA) lies behind the ascending aorta and is superior to the LV (Figs. 7.1, 7.2, and 7.3). The LA is demarcated from the LV by the atrioventricular groove, within which lie the LCx and the coronary sinus.

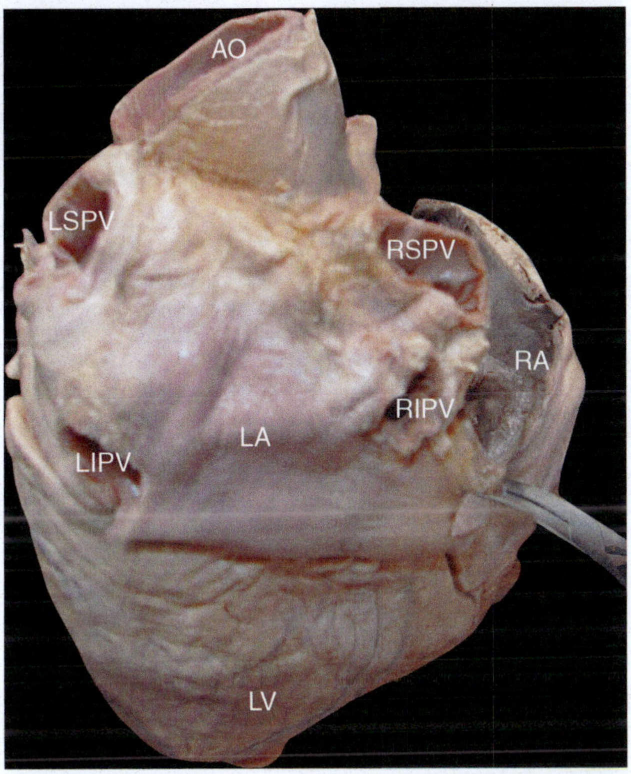

Fig. 7.1 Position of the LA. Figure shows anatomical view of the dorsal part of the heart. The LA lies behind the aorta and above the ventricles. The RA is on the right side of the LA. The left atrioventricular groove is between the LA and the LV. Four orifices of the pulmonary veins lie in the posterior wall of the LA. *AO* aorta, *LA* left atrium, *LIPV* orifice of left inferior pulmonary vein, *LSPV* orifice of left superior pulmonary vein, *LV* left ventricle, *RA* right atrium, *RIPV* orifice of right inferior pulmonary vein, *RSPV* orifice of right superior pulmonary vein

X. Zhu (ed.), *Surgical Atlas of Cardiac Anatomy*,
DOI 10.1007/978-94-017-9409-1_7, © Springer Science+Business Media Dordrecht and People's Medical Publishing House 2015

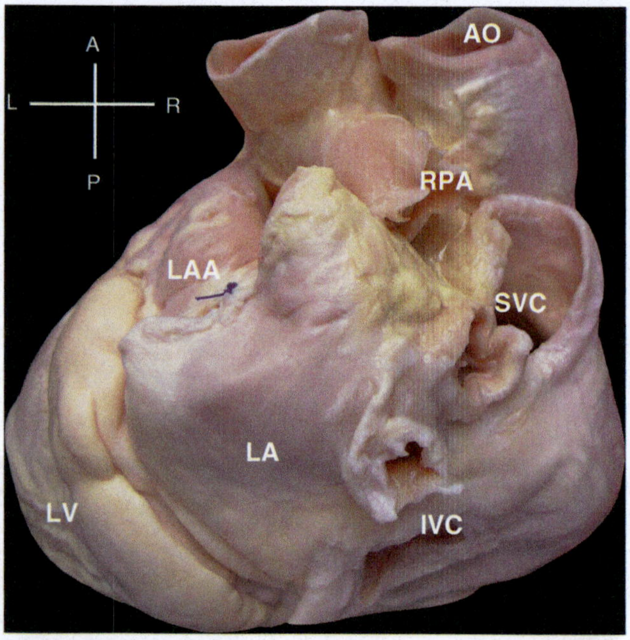

Fig. 7.2 Neighboring of the LA. Overview of the heart from the left posterior side shows neighboring relations of the LA and superior vena cava (*SVC*), inferior vena cava (*IVC*), right pulmonary vein (*RPV*), left atrial appendage (*LAA*), left ventricle (*LV*), right pulmonary artery (*RPA*)

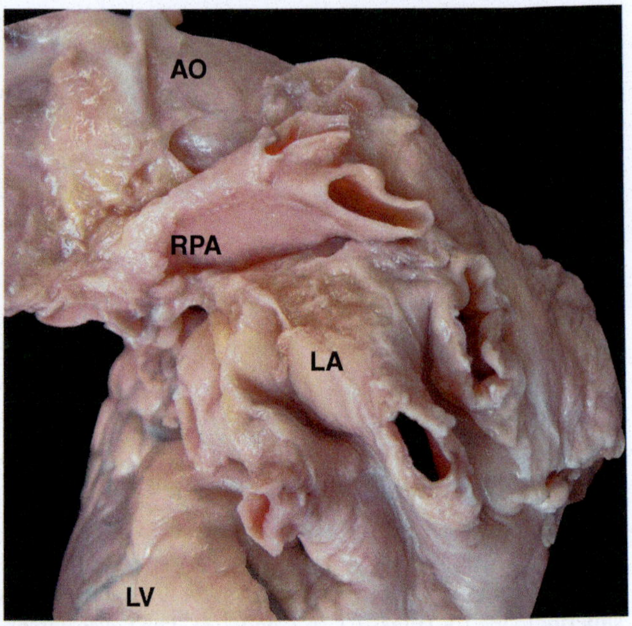

Fig. 7.3 Right pulmonary vein above the LA. The right pulmonary vein (*RPA*) extends across the roof of the LA. *AO* ascending aorta, *LA* left atrium, *LV* left ventricle

7.1.1 Structure

Similar to the RA, the LA consists of trabecular and sinus portions (Fig. 7.4). The trabeculated part, confined within the LA appendage, is developed from preliminary LA of the embryonic heart (Fig. 7.5). The LAA is a triangular protrusion with several exterior incisurae and interior cavernous trabeculae (Fig. 7.6). The sinus portion is developed from the dorsal bulge of the embryonic LA, which becomes the common pulmonary venous trunk, joining the pulmonary veins and eventually becoming enlarged (Fig. 7.4). The LA has a smooth interior, with four pulmonary veins entering dorsally. Posteroinferiorly, the LA adjoins the posterior annulus of the MV and coronary sinus, where it is called the *vestibule of the left artery* (Fig. 7.7). In an abnormally enlarged LA, the posteroinferiorly enlarged vestibule results in the displacement of the MV annulus. The LA volume reduction surgery is performed in the left atrial vestibule.

The posterior wall of the LA joins the entries of the four pulmonary veins (Fig. 7.1); being deep and immobile, it is hard to access during surgery. The inner wall of the LA is smooth. The LA closely adjoins the esophagus dorsally. When enlarged, the LA pushes the esophagus rightward and backward. An oblique ligament located lateral to the posterior wall of the LA and extending inferiorly and rightward is the relic of regressed embryonic oblique veins. If incompletely regressed, it remains the persistent left superior vein, which connects the cephalobrachial vein superiorly and the coronary sinus inferiorly. One should be alert to its possible presence when inspecting the enlarged coronary sinus orifice. The LAA is well exposed through the left lateral thoracotomy, which is the ideal route for LA exploration.

The roof of the LA is located beneath the aortic arch and the right pulmonary artery (Fig. 7.6). Dorsally, the LA neighbors the tracheal bifurcation, which can be lifted by an enlarged LA.

The interatrial septum lies between the left and right atria. The semilunar fold on the left side of the interatrial septum, which is the relic of the primary interatrial septum, is an important marker for identifying the LA. The anterior wall of the LA closely neighbors the ascending aorta, where it is separated from the left and part of the noncoronary sinus. The anterior wall of the LA continues downward to the base of the anterior leaflet of the MV, where, together with the aortic annulus, it forms the converging part and is the site of incising the LA and the anterior leaflet of the MV during an aortic annular enlargement procedure (Figs. 7.8 and 7.9).

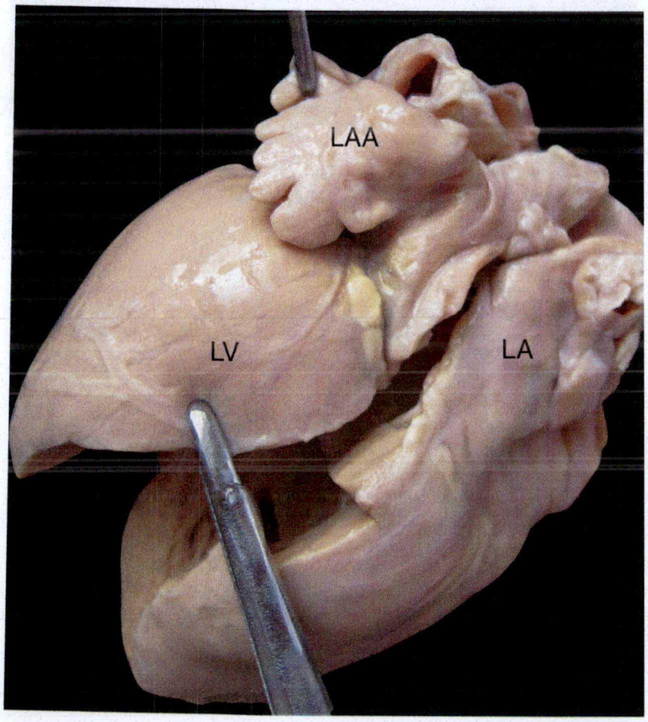

Fig. 7.4 Incisurae on the neck of the left atrial appendage. The anatomical view from the left side displays the LA and LAA. In contrast to the RAA, the neck of the LAA is slender and narrow with exterior incisurae and interior trabeculae. The wall of the LAA is thin and fragile. Cautions should be taken to avoid incision avulsion during cannulation and finger exploration, which, in severe cases, can result in massive bleeding and injury to the left circumflex artery. *LV* left ventricle

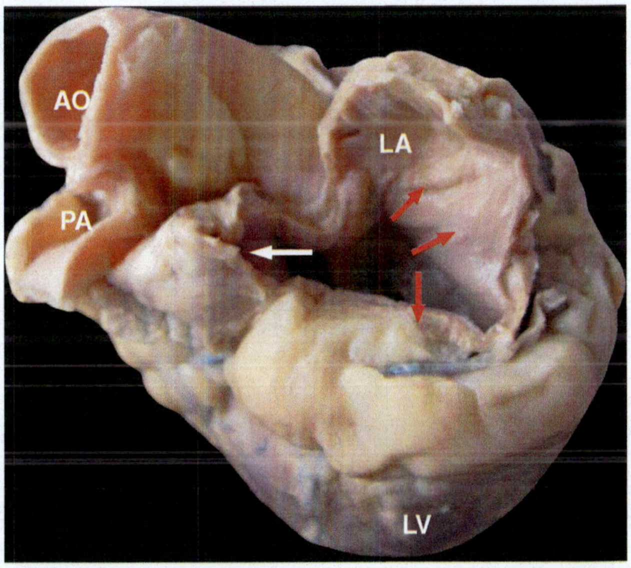

Fig. 7.5 Compartments of the left atrial cavity. The specimen is viewed from the left side. The lateral wall of the LA is removed. The *white arrow* indicates the cavity of the LAA, which is trabecular. The *red arrow* indicates the main part of the LA and its smooth inner wall, where the pulmonary veins emerge in the embryonic period. *AO* aorta, *LA* left atrium, *LAA* left atrial appendage, *LV* left ventricle, *PA* pulmonary artery

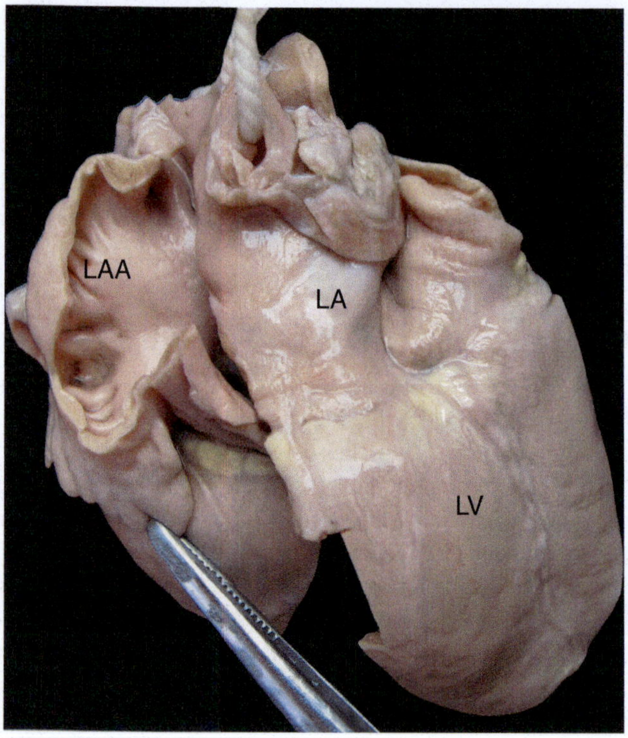

Fig. 7.6 Trabeculae of the left atrial appendage. The LAA is incised, revealing the trabeculated portion. *LV* left ventricle, *LA* left atrium

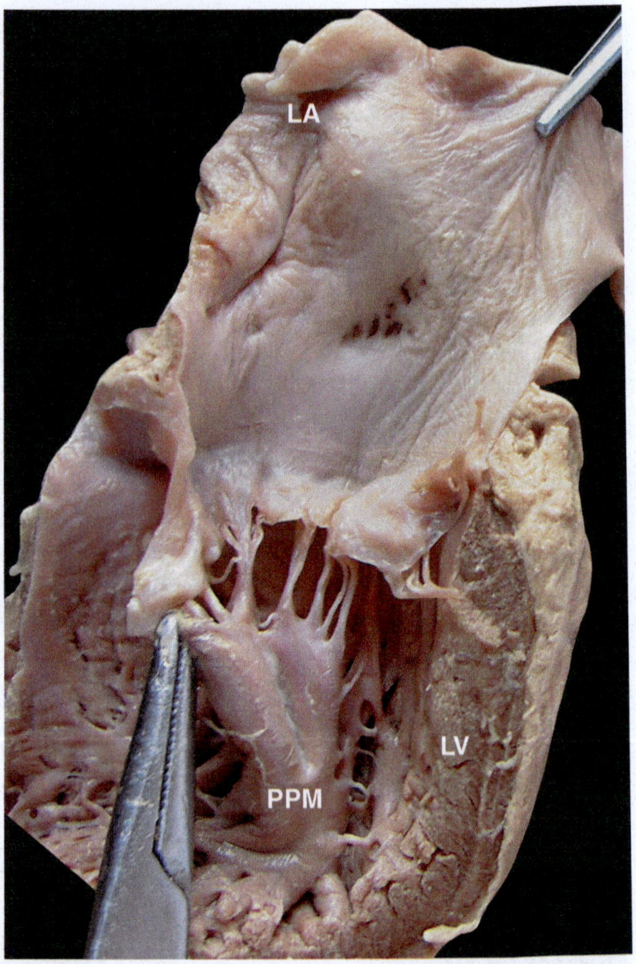

Fig. 7.7 Inner wall of the left atrium. The sagittal section viewed from the left side shows the reticular fold on the center of the left side of the interatrial septum, corresponding to the FO on the right atrial side of the septum. The posterior leaflet and the PPM of the MV are also displayed. *FO* fossa ovalis, *PPM* posterior papillary muscle, *LV* left ventricle, *LA* left atrium

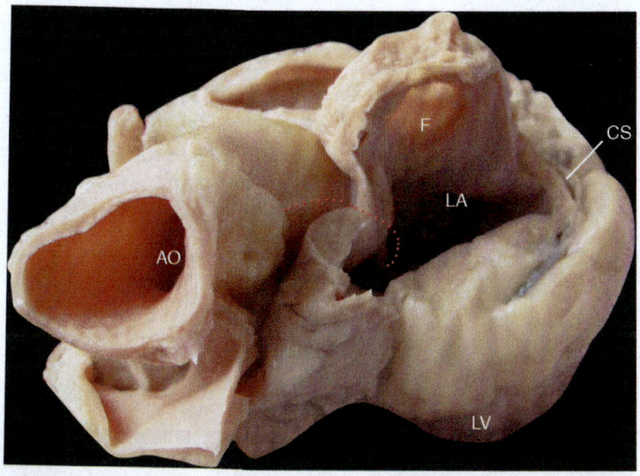

Fig. 7.8 Left atrium overview from the left side with the lateral wall removed. The semitransparent area (*F*) of the interatrial septum corresponds to the FO. The interatrial septum is close to the MV. The incision on the septum directly exposes the MV, which is the most frequently used route in MV procedures. In balloon mitral valvuloplasty, the FO is punctured from the RA side and the MV is accessed. The *red dotted line* indicates the position of the joint incision in the aorta and the LA. *AO* aorta, *CS* coronary sinus, *FO* fossa ovalis, *LA* left atrium, *LV* left ventricle, *MV* mitral valve, *RA* right atrium

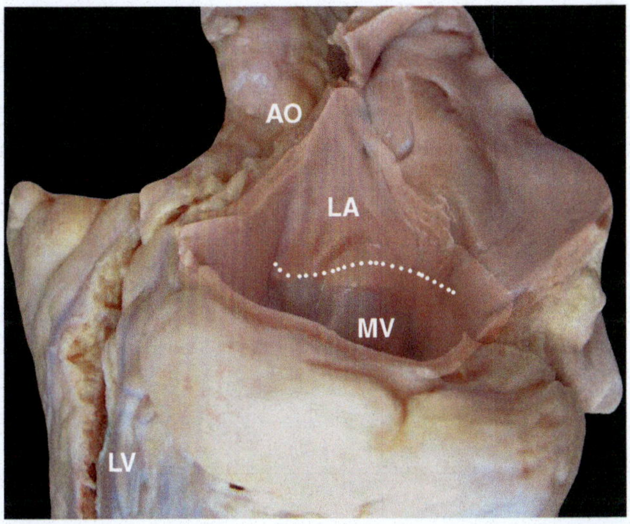

Fig. 7.9 Demarcation line between the mitral and aortic annuli. The specimen is placed in its anatomical position and viewed from the left side. The lateral wall of the LA and the main pulmonary artery are excised. The *white dotted line* corresponds to the demarcation line of the aortic (noncoronary and left coronary leaflets) and mitral annuli, also indicating the common borders of the posterior wall of the aorta, anterior wall of the LA, and the anterior leaflet of the MV. When an aortic root enlargement procedure is needed in the case of a small aortic root, the posterior wall of the aorta can be incised to transect the demarcation line until the base of the anterior leaflet of the MV is reached. *AO* aorta, *LA* left atrium, *LV* left ventricle, *MV* mitral valve

7.1.2 Surgical Characteristics

In the valvular procedure, especially the valvuloplasty, maximum surgical exposure is required for optimal inspection of the structures of the MV and for assessing the MV in its physiological position. Therefore, the following characteristics of the LA should be taken into consideration: *first*, the LA is deep and difficult to expose; *second*, the LA is immobile; *third*, the space for the incision in the LA is limited and hardly expandable; and, *finally*, the left atrial wall is thin and fragile and is liable to tension injury.

7.2 Surgical Routes to Left Atrium

There are different choices of surgical routes to the LA, including through (a) the right interatrial groove, (b) the interatrial septum via the RA, (c) the roof of the LA, (d) the LA and the LAA, and (e) the left ventriculotomy reversely and (f) combined incisions (Figs. 7.10 and 7.11). The most frequently used route is through the right interatrial groove and the interatrial septum.

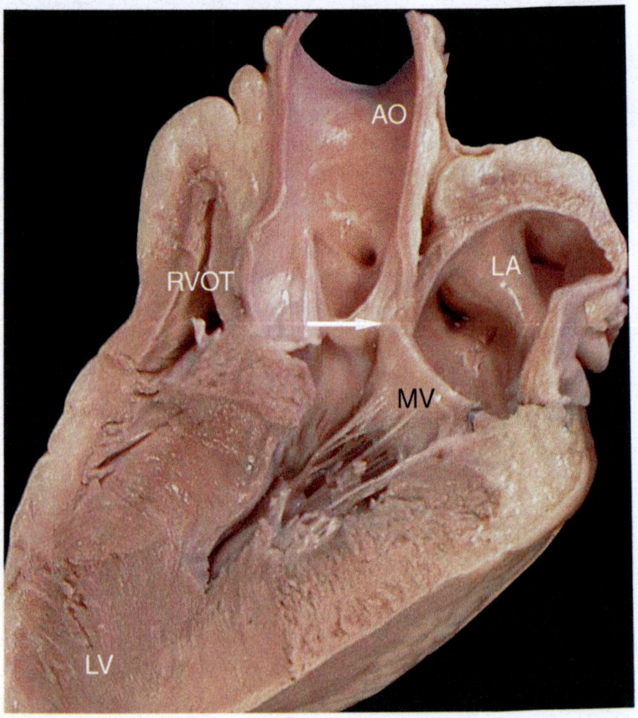

Fig. 7.10 Confluent point of the aorta, the left atrium, and the mitral valve. The sagittal section of a specimen is observed from the left side. *Arrow* indicates the confluent point of the three structures. *AO* aorta, *LA* left atrium, *MV* mitral valve, *RVOT* right ventricular outflow tract

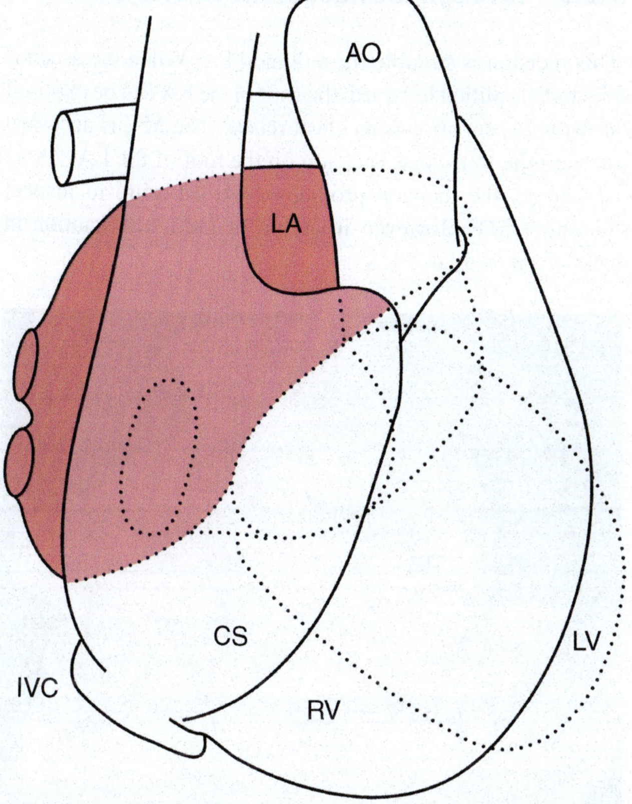

Fig. 7.11 Frontal projection showing the position of the left atrium. The front view of the heart in its anatomical position demonstrates the position of the LA relative to other cavities of the heart. The pink part indicates the LA, on the right of which are the orifices of the pulmonary veins. The roof of the LA can be seen between the aorta and the SVC. The *dotted line* shows the left ventricle (*LV*) and the interatrial septum (A). *IVC* inferior vena cava, *AO* ascending aorta, *CS* coronary sinus, *RV* right ventricle

7.2.1 Through the Right Interatrial Groove

The interatrial groove is exposed when the RA is pushed rightward. A longitudinal incision anterior to the right pulmonary veins is made to enter the LA. The incision can be extended caudally to the rear of the IVC and cranially to the rear of the SVC. In the case of a small LA, the interatrial groove is hidden behind the RA, resulting in poor exposure through this incision (Fig. 7.12). The right side of the interatrial groove is the most convenient site for LA drainage.

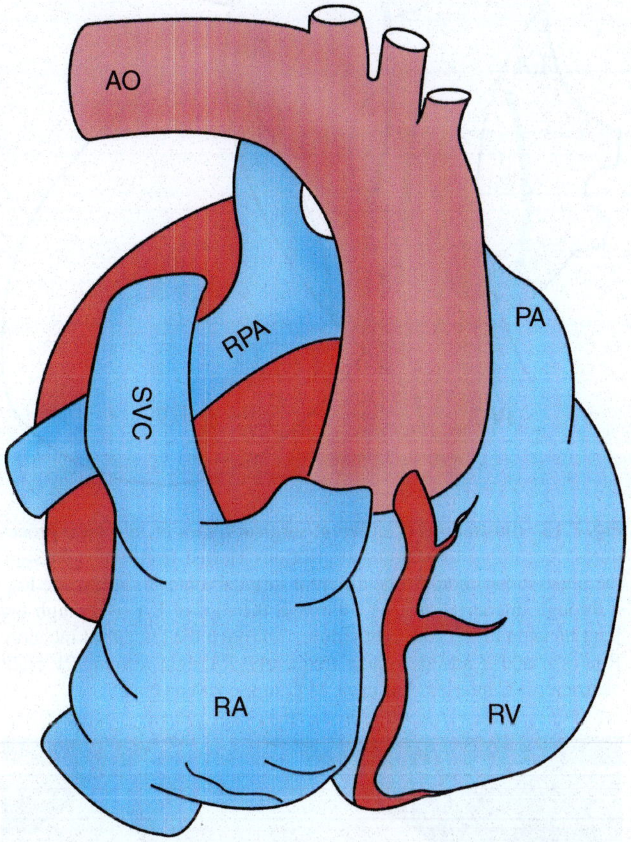

Fig. 7.12 Illustrations of the roof of the left atrium. The illustrations demonstrate the frontally overlooked heart. The *red part* indicates the LA. Note that the site of incision in the roof of the LA lies among the aorta, the SVC, and the RPA. *RPA* right pulmonary artery, *SVC* superior vena cava, *RA* right atrium, *RV* right ventricle, *PA* pulmonary artery, *AO* ascending aorta

7.2.2 Through the Interatrial Septum

The left and right atria are separated by merely the interatrial septum; thus, incising the interatrial septum is the most direct way to access the MV. After the RA is incised, a longitudinal incision is made in the interatrial septum (Figs. 7.13, 7.14, and 7.15). The lesions of the TV can be managed simultaneously. However, it is difficult to extend the incision.

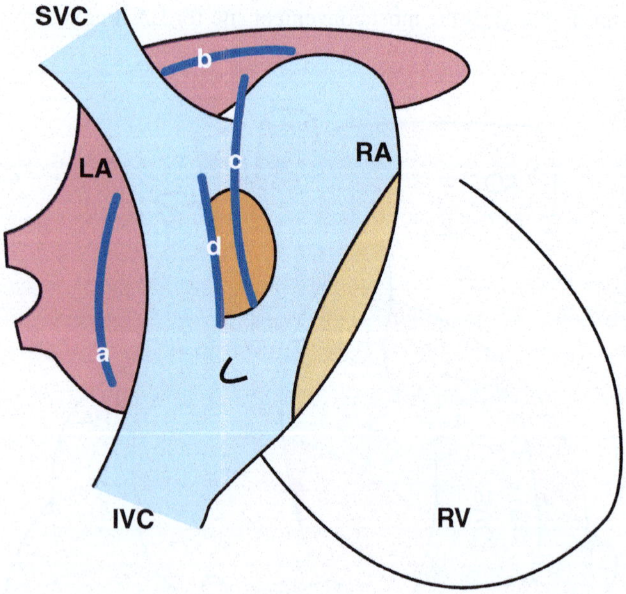

Fig. 7.13 Illustrations of different surgical routes of left atrial procedures. According to anatomical positions viewed from the right side, the illustrations demonstrate different surgical routes to access the LA. *a* through interatrial groove; *b* through the roof of LA; *c* through the joint incision in left and right atrium; *d* through the interatrial incision. Routes *a* and *d* are used most frequently. *RA* right atrium, *RV* right ventricle, *IVC* inferior vena cava, *SVC* superior vena cava

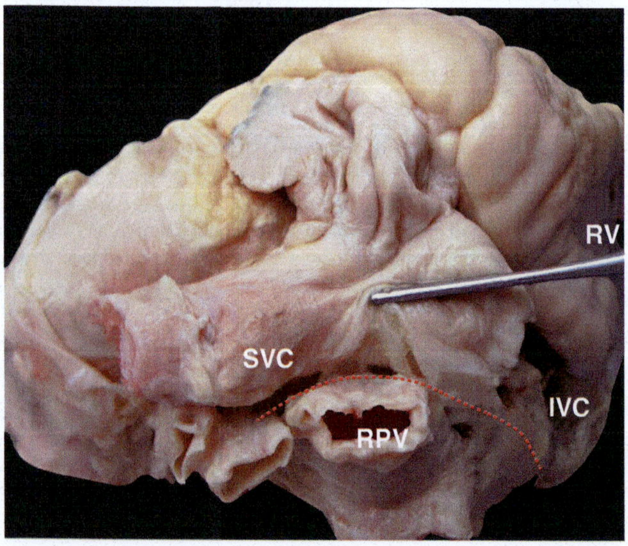

Fig. 7.14 Site of incision in the interatrial groove. Observed from the right side, the *dotted line* indicates the surgical incision to access the LA through the interatrial groove. The lower end of the incision approximates the orifice to the IVC; the upper end could circle to the back of the SVC. *IVC* inferior vena cava, *RV* right ventricle, *SVC* superior vena cava, *RPV* right pulmonary vein

7.2.3 Through the Roof of the Left Atrium

This incision is suitable for a dilated LA. When the ascending aorta is pulled leftward, the roof of the LA can be exposed at the aorto-superior vena caval recess. The MV is accessed through the transverse incision on the roof of the LA. As an advantage, the incision provides good exposure to inspect the entire MV. However, it should be used with caution in case of a non-dilated LA.

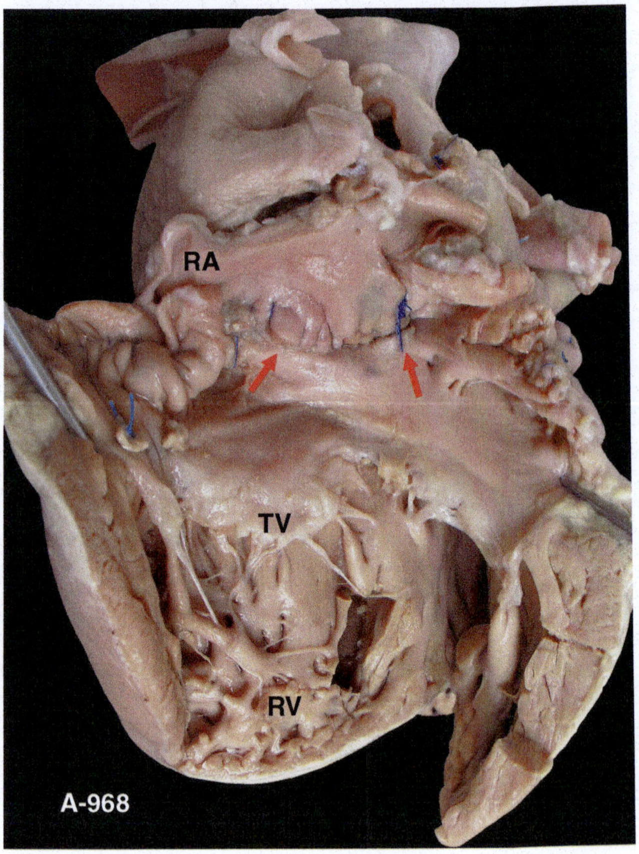

Fig. 7.15 Specimen NO. 968 showing the heart after a left atrium procedure using the interatrial septal route. Viewed from the RA, the *arrow* indicates the suture of the interatrial septum incision. *RA* right atrium, *TV* tricuspid valve, *RV* right ventricle

7.3 Comparison of Different Surgical Routes and Related Risks

Although there are different routes to the LA, all are not optimal. The incision in the LA or LAA is usually used for cannulation and manometry. The reverse route through the left ventriculotomy is used merely in procedures combined with left ventriculoplasty; however, although the various combined incisions provide good exposure, they should be avoided as much as possible due to the potential for massive injury (see Table 7.1).

In the LA volume reduction procedure for a giant LA, the vestibule of the LA should be resected or folded and sutured to reduce the volume of the LA. The vestibule, lying between the mitral annulus and the orifices of the pulmonary veins, is thin walled; therefore, caution should be exercised to avoid injury to the MV and bleeding from the left atrial wall (Fig. 7.7). Regardless of whether it is performed operatively or percutaneously, the radiofrequency ablation of the atrial fibrillation is conducted in the posterior wall of the LA surrounding the orifices of the pulmonary veins (Fig. 7.16). Usually, for satisfactory effect, the lesion resulting from ablation in the wall of atrium should have full thickness. The procedure has high risks for injury to the left atrial wall and bleeding from the suture due to the difficulty of establishing hemostasis after cardiac arrest.

Table 7.1 Comparisons of routes to the LA

Route	Advantages	Disadvantages	Indications
LAA and free wall	Simple	Limited length of incision	Left thoracotomy
Right interatrial groove	Incision is extendable	Unsuitable for small LA	MV procedure
Right atrium-interatrial septum	Close to MV, TV can be managed simultaneously	Limited extendibility	MV and TV procedure
Roof of LA	Good exposure of overall view of MV	Exclusively for enlarged LA	MV procedure
LA and aorta	Relatively good exposure	Massively traumatic	Aortic annular enlargement procedure
Interatrial septum and roof of LA	Good exposure	Liable to injury to sinus node artery	Small LA
Interatrial groove and roof of LA	Good exposure	Massively traumatic	Small LA

LA left atrium, *LAA* left atrial appendage, *MV* mitral valve, *TV* tricuspid valve

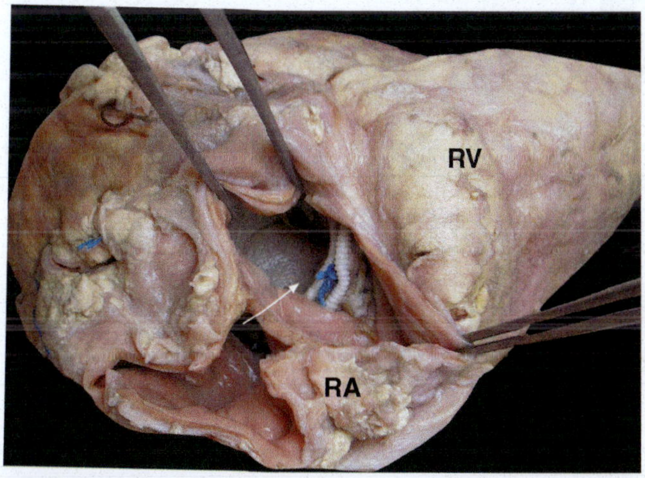

Fig. 7.16 Mitral valve replacement procedure observed through the interatrial septal incision. The nearest route to access the MV is through the interatrial septum. The suture border of the implanted artificial cardiac valve prosthesis (*arrow*) can be seen in the picture. *RA* right atrium, *RV* right ventricle

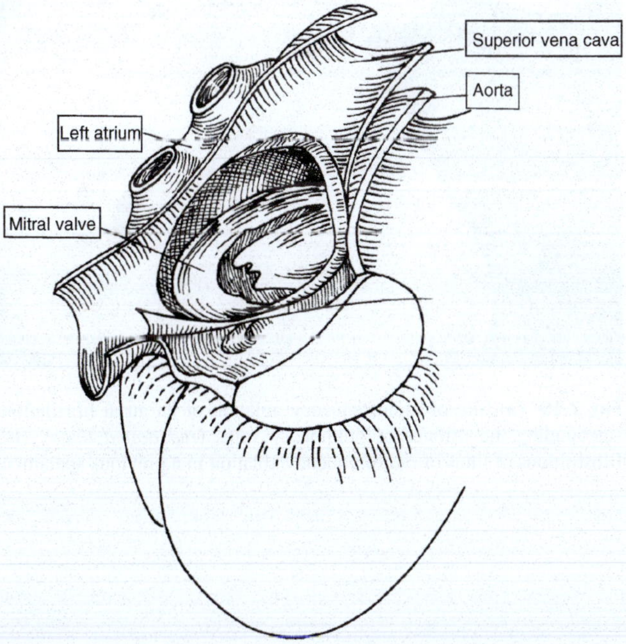

Fig. 7.17 Illustrations of the left atrium with the interatrial septum totally excised

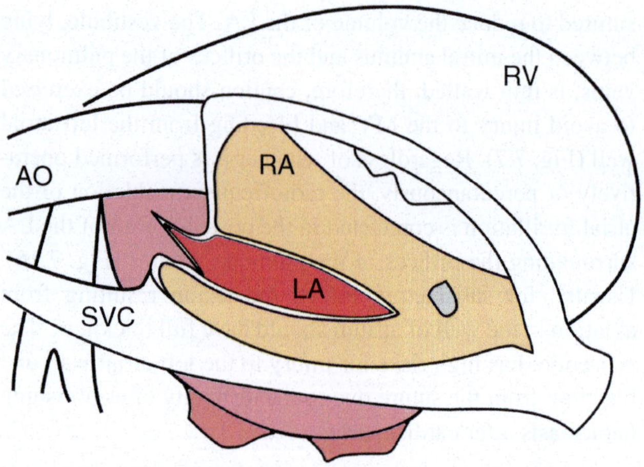

Fig. 7.18 Joint incision of left and right atria. *LA* left atrium, *RA* right atrium, *RV* right ventricle. Surgical implications: The MV can be exposed sufficiently through this incision, but cautions should be taken to avoid avulsion of the incision in the left atrium. Suturing the incision is demanding. Moreover, the incision requires cutting off of a coronary artery supplying the sinus node, which can result in arrhythmia in some cases. *AO* ascending aorta, *SVC* superior vena cava

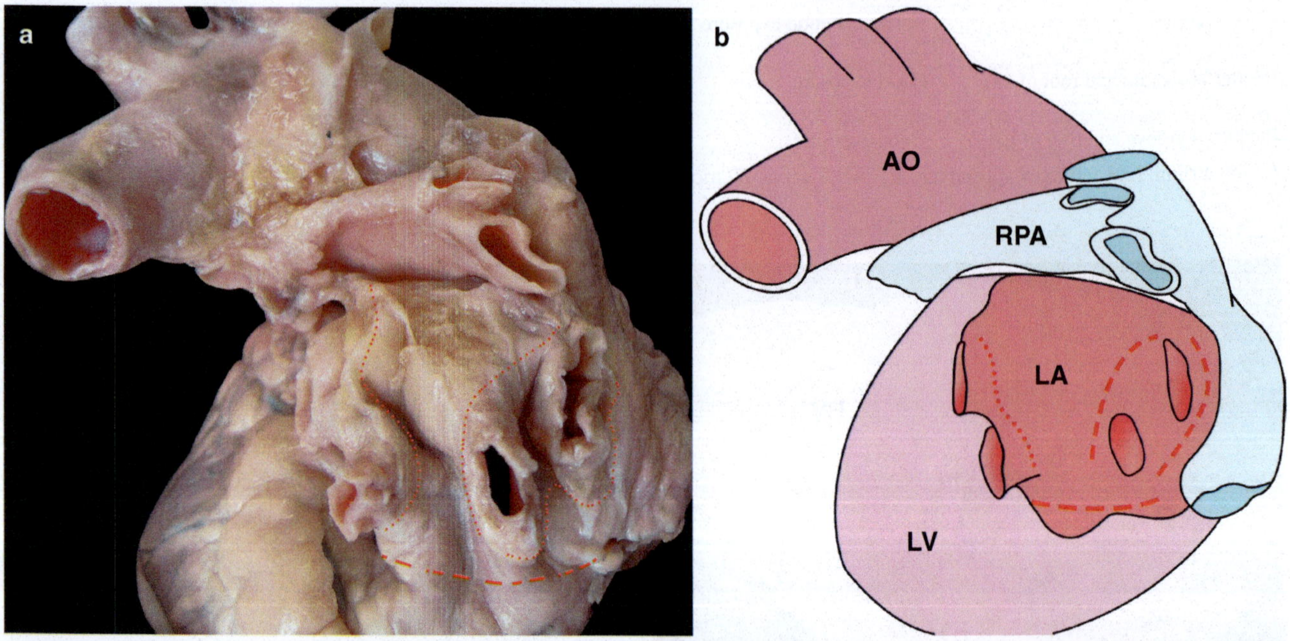

Fig. 7.19 (**a**) Site of radiofrequency ablation of an atrial fibrillation surrounding the orifices of pulmonary veins (*red dotted lines*). (**b**) Illustrations of lines of radiofrequency ablation in a previous specimen (surrounding the orifices of pulmonary veins). *LA* left atrium, *RPA* right pulmonary, *LV* left ventricle, *AO* ascending aorta

8 The Left Ventricle

8.1 General Considerations

8.1.1 Wall of the Left Ventricle

The ventricular wall is separated from the LA by the atrioventricular groove, from which the coronary sinus and the LCx extend. The left ventricular anterior wall is demarcated from the RV by the interventricular sulcus, an important marker of the LV and in which is found the LAD (Fig. 8.1).

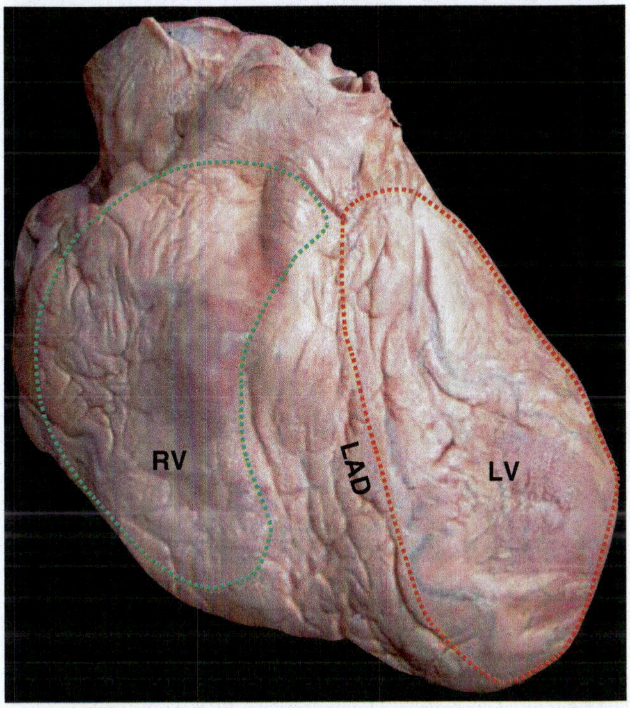

Fig. 8.1 Left anterior oblique view of heart showing the left ventricular lateral wall. The portion behind the LAD is the lateral wall of the LV (area encircled by *red dotted line*). The *green dotted line* indicates the RV. *LAD* left anterior descending artery, *LV* left ventricle, *RV* right ventricle

X. Zhu (ed.), *Surgical Atlas of Cardiac Anatomy*,
DOI 10.1007/978-94-017-9409-1_8, © Springer Science+Business Media Dordrecht and People's Medical Publishing House 2015

8.1.1.1 Characteristics

The left ventricular wall is formed by the twisted myocardial bands that anchor to the cardiac fibrous skeleton at the upper end, forming the left ventricular orifices, namely, the left atrioventricular valve and the AV, that constitute the apex. The myocardial bands consist of the superficial, middle, and deep layers of spiral muscles, which twist and form the entire left ventricular wall (Fig. 8.2). Some of the myocardial bands take part in forming the RV and the interventricular septum, which is evidently thicker in the left ventricular wall than in the right counterpart (Fig. 8.3).

The left ventricular wall is mortar shaped and formed by thick and strong muscles, in contrast to the thin right ventricular wall (Figs. 8.4, 8.5, and 8.6). This difference of structures between two ventricles reflects their physiological functions: the LV bears the resistance of systemic circulation.

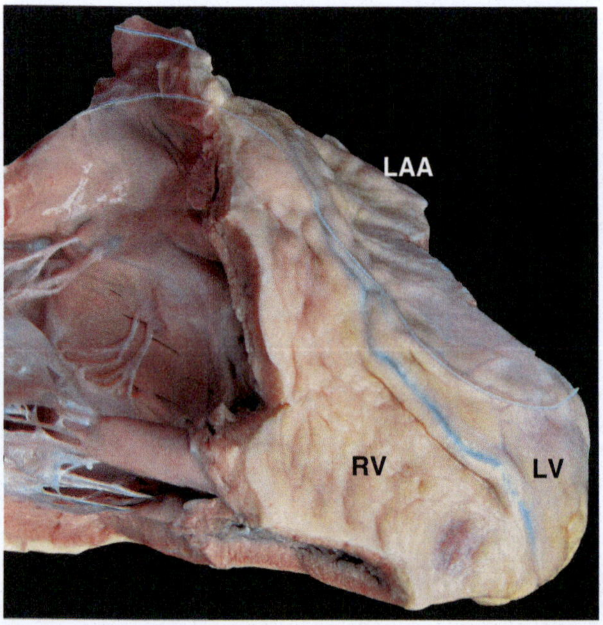

Fig. 8.2 Comparison of the morphologies of the left and right ventricles. The front wall of the RV is removed. Viewing the front of the heart, the interventricular septum convexes to the right, rendering the LV a cone-shaped cavity, while the RV attaches to the right of the LV. *LAA* left atrial appendage, *LV* left ventricle, *RV* right ventricle

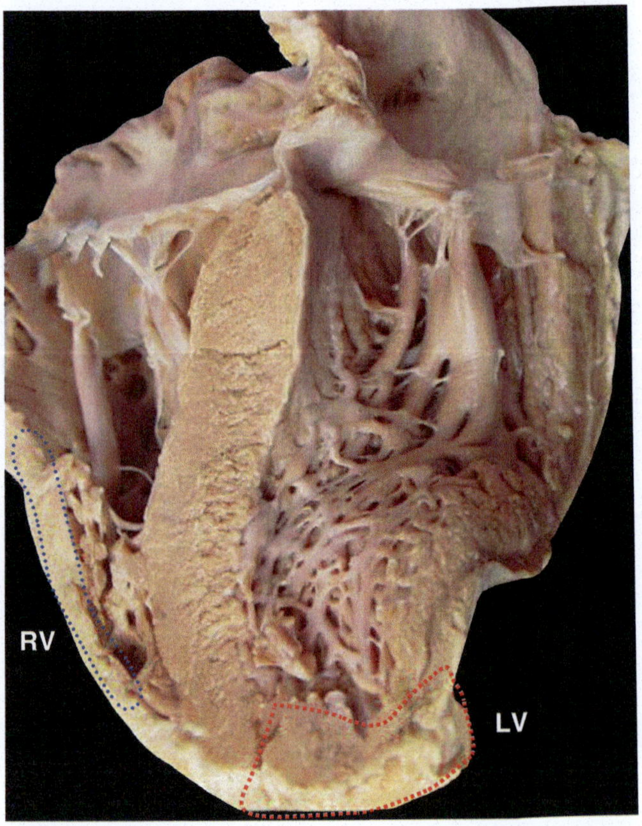

Fig. 8.3 Comparison of left and right ventricular walls. The coronal section demonstrates the walls of the left and right ventricles. The area enclosed in the *red dotted line* indicates the left ventricular wall. The *blue dotted line* encloses the right ventricular wall. The LV works in systemic circulation, so its wall is apparently thicker than the right ventricular wall

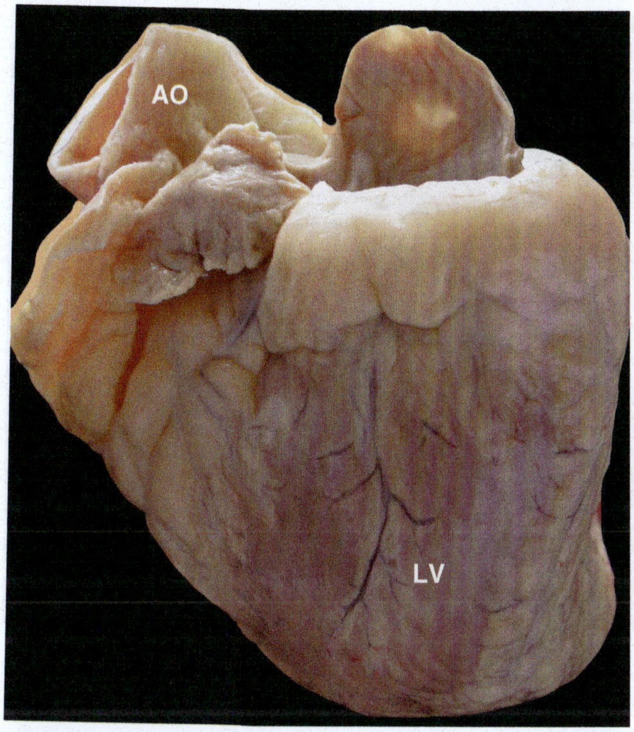

Fig. 8.4 Mortar-shaped left ventricle with coronary vessels on the surface. The wall of the LA is removed, leaving the LAA and the interatrial septum. *LA* left atrium, *LAA* left atrial appendage

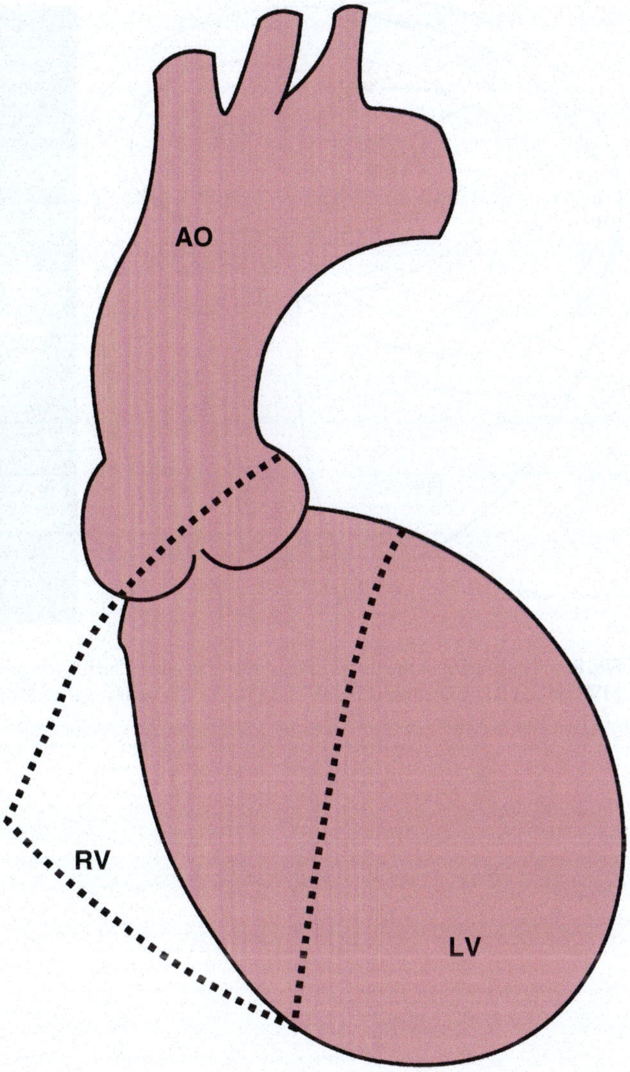

Fig. 8.5 Projection of the front view of the left ventricle. The shapes and relationship of the left and right ventricles are demonstrated. The *blue dotted line* depicts the RV closely adjoining the LV

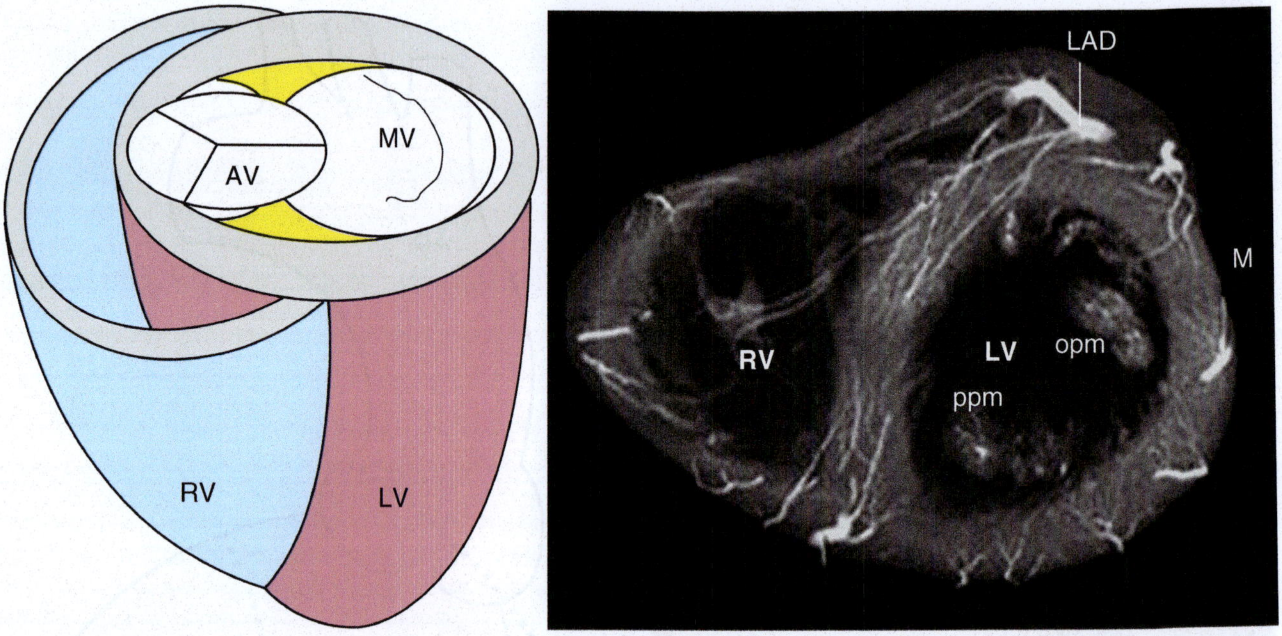

Fig. 8.6 Illustration of the transaction of the ventricles. The LV is a mortar-shaped muscular pump. The upper orifices include the inlet orifice (the MV) and outlet orifice (the AV). *AV* aortic valve, *MV* mitral valve, *RV* right ventricle

8.1.1.2 Wall Portions

The left ventricular wall is divided into three portions:

(a) Interventricular septum

This portion is composed of thick and strong myocardial tissue. The interventricular septum separates the two ventricles; it concaves in the left side and convexes in the right side. The upper end of the septum connects to the right coronary sinus of the aorta. The muscular part of the septum is connected with the central fibrous body at the membranous part, which is small. The conduction bundle passes underneath it (Figs. 8.7 and 8.8).

(b) Diaphragmatic side of the left ventricular wall

This portion connects with the diaphragm and is hard to expose.

(c) Free ventricular wall in the left anterior of the free ventricular wall in the left anterior of the LV, separated from the RV by the interventricular sulcus; within it passes the LAD.

The LCx extends into the atrioventricular groove. The avascular area behind the interventricular sulcus is the preferable site for performing a ventriculoplasty.

8.1.2 Cavity of the Left Ventricle

The left ventricular cavity can be divided into the sinus portion, the trabeculated portion, and the outlet portion or tract.

8.1.2.1 Sinus Portion

The sinus portion is the inlet tract of the LV and has a smooth inner wall.

8.1.2.2 Trabeculated Portion

The trabeculated portion is at the apex and has trabeculae of different sizes. Two groups of trabeculae of larger sizes form the papillary muscles of the MV.

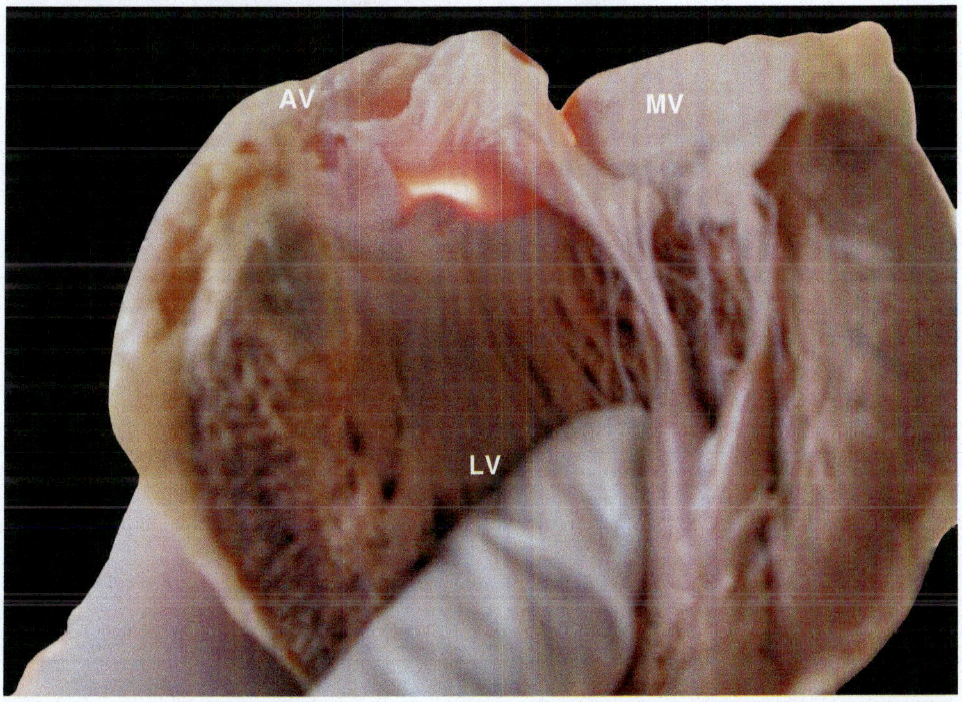

Fig. 8.7 Membranous interventricular septum. The left ventricular lateral wall is removed. The membranous septum (the semitransparent area) is observed from the left ventricular cavity

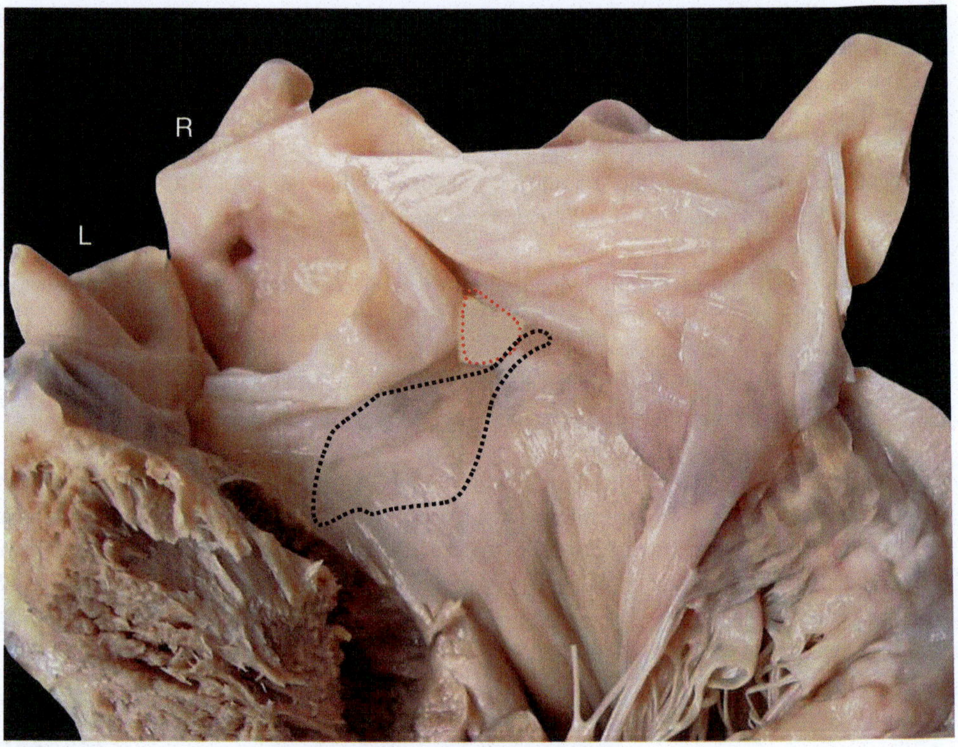

Fig. 8.8 Membranous interventricular septum and left bundle branch. The interventricular septum is displayed from the left ventricular cavity. The membranous septum (*red dotted line*) lies beneath the commissure between the right and noncoronary cuspids. The septum beneath the left and right aortic cuspids is incised. The left bundle branch (*blue dotted line*) passes subendocardially on the left side of the interventricular septum beneath the membranous septum. *L* left coronary cuspid, *R* right coronary cuspid

8.1.2.3 Outlet Tract

The left ventricular outlet tract (LVOT) is encircled by a muscular interventricular septum and fibrous continuity of the mitral-aortic valve. The right coronary sinus of the aorta is tightly adhered to the interventricular septum (Figs. 8.9 and 8.10). The septum is smooth with a fan-shaped left branch of the conduction bundle extending in the superficial layer (Fig. 8.8). There is neither complete infundibular muscle nor muscle bands as in the supraventricular crest of the RV. The continuity of the anterior leaflet of the MV and of the leaflets of the AV appears to be like a screen, separating the inlet portion from the LVOT and forming the membranous portion of the left ventricular outlet tract (Fig. 8.9). The interventricular septum forms the muscular portion of the LVOT. The existence of the fibrous continuity of the anterior leaflet of the MV and the leaflets of the AV is one of the important identifications of congenital cardiac deformities.

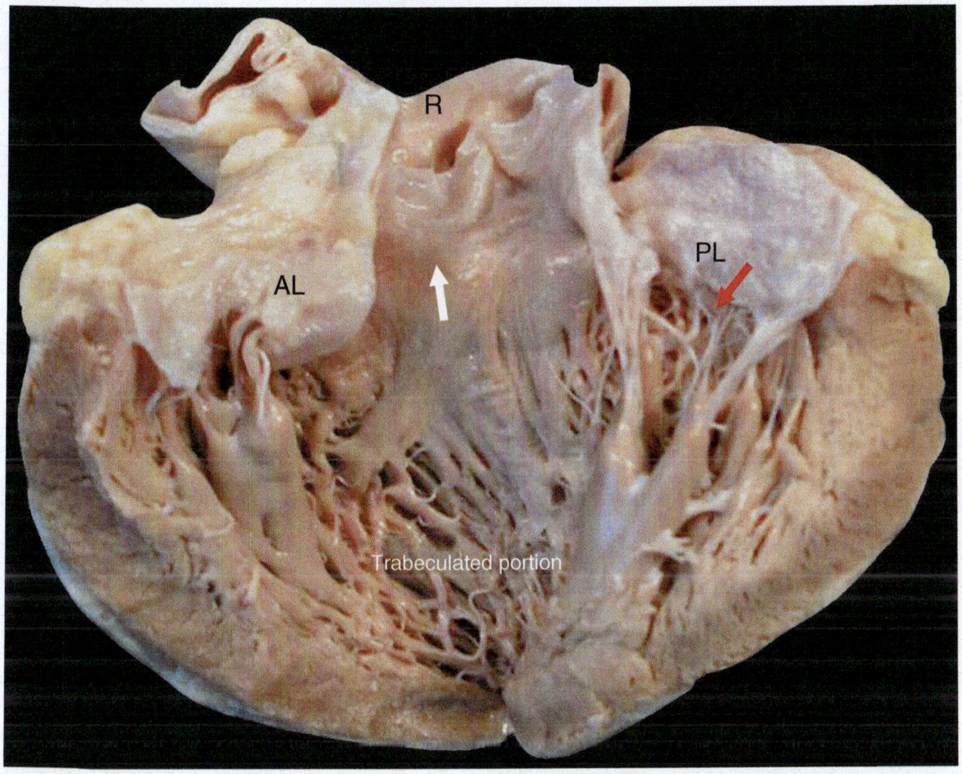

Fig. 8.9 Inlet and outlet tracts of the left ventricle. The specimen is dissected along the LVOT to display the overall view of the LV. The *red arrow* indicates the inlet tract (i.e., the MV). The *white arrow* indicates the outlet tract. The anterior wall of the outlet tract is the interventricular septum, whereas the posterior wall is the anterior mitral leaflet. *AL* anterior mitral leaflet, *LV* left ventricle, *LVOT* left ventricular outlet tract, *PL* posterior mitral leaflet, *R* right coronary cuspid. Surgical implications: In hypertrophic obstruction of the LVOT, one of the surgical procedures is to excise some of the interventricular septal muscles at the site indicated by the white arrow, where, by accident, the interventricular septum could be perforated to enter the RV and the AV and the papillary muscles of the MV would be injured. Another surgical procedure is MV replacement, thus eliminating the anterior movement of the anterior mitral leaflet in systole, which exacerbates the obstruction of the LVOT

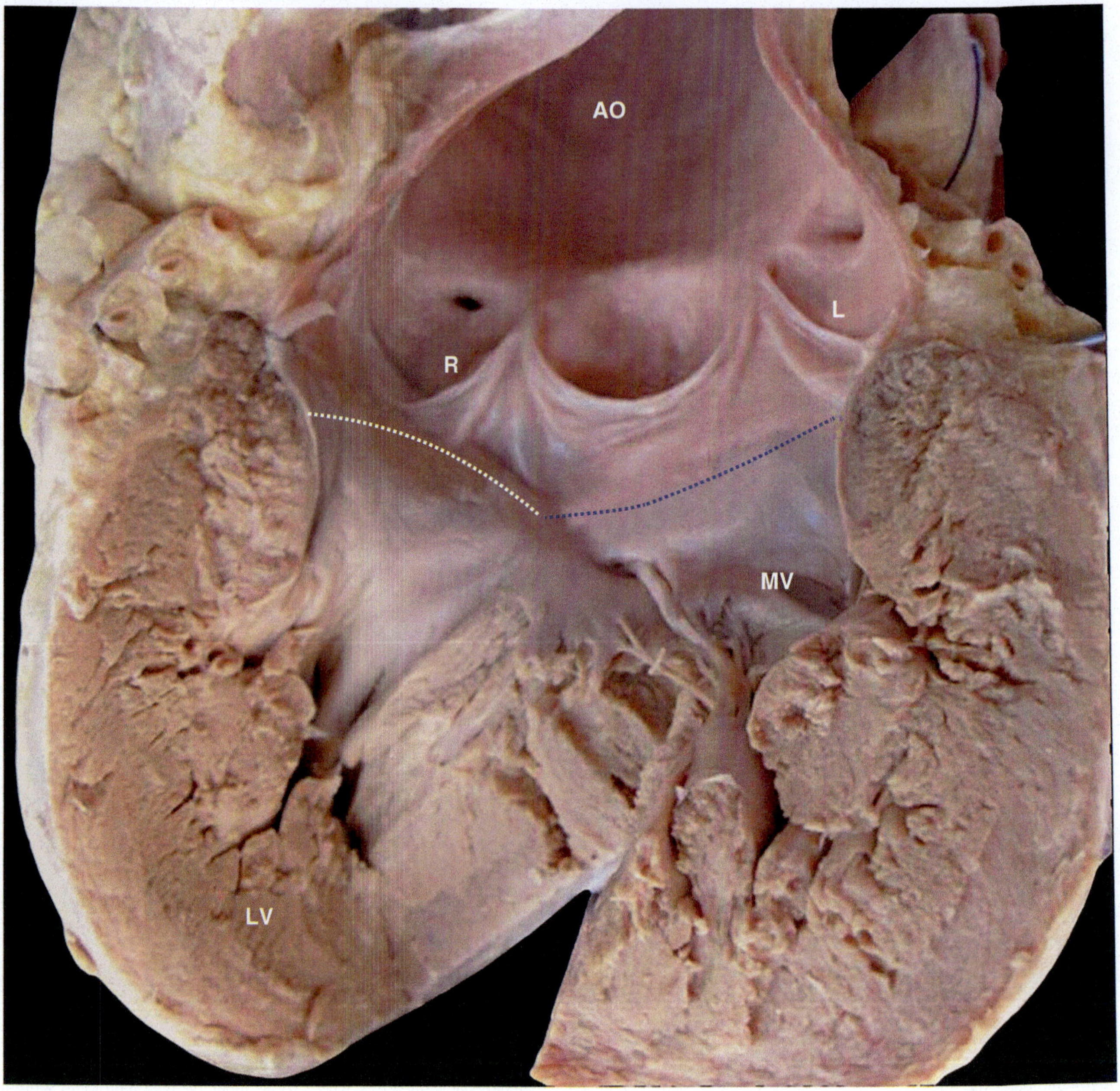

Fig. 8.10 Upper border of the left ventricular outlet tract. The LVOT is dissected from the apex to the aorta. The upper border of the outlet tract is composed of the attaching margin of the aortic cuspids on the interventricular septum (*white dotted line*) and the joint line of the aortic-mitral fibrous continuity (*blue dotted line*). *L* left coronary cuspid, *MV* anterior mitral leaflet, *R* right coronary cuspid

8.2 Mitral Valve Device

Generally, the MV device includes an annulus, leaflets, chordae tendineae, papillary muscles, and the left ventricular wall (Figs. 8.11 and 8.12).

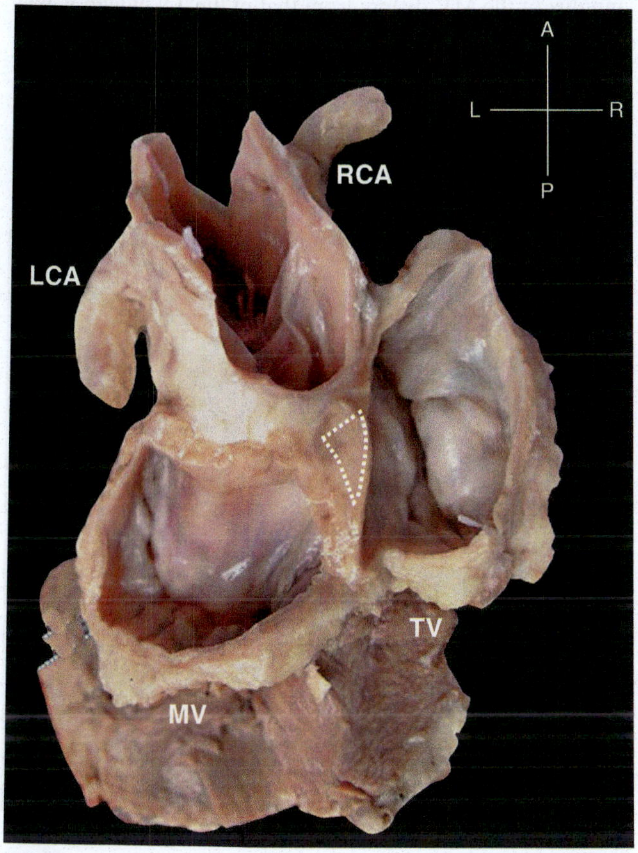

Fig. 8.11 Mitral valve and the central fibrous body. This is a posterior view of the specimen at the plane of the atrioventricular valves. The confluent point of the aortic, mitral, and tricuspid valves is the central fibrous body (area enclosed by *white dotted line*). It is the core support point of the heart. The anterior mitral leaflet is close to the central fibrous body. The specimen displays mainly the surroundings of the MV. *LCA* left coronary artery, *MV* mitral valve, *RCA* right coronary artery

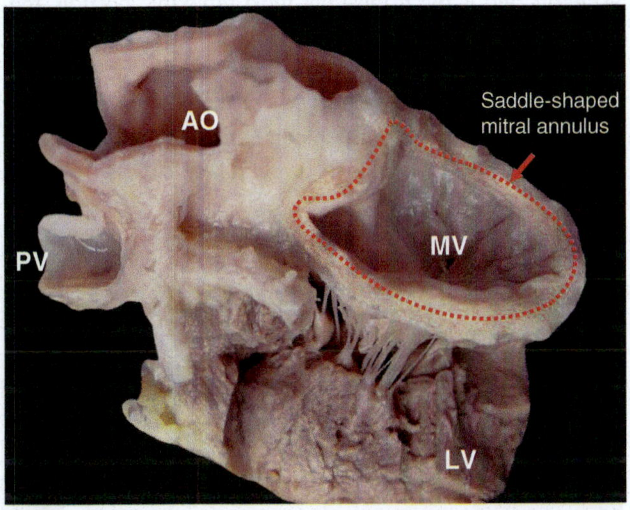

Fig. 8.12 Mitral valve device. The specimen is not a normal heart. The LV is obviously hypertrophic, but the entire MV device (i.e., the mitral annulus, mitral leaflets, chordae tendineae, and papillary muscles) is still well displayed. The *dotted line* indicates the saddle-shaped (the front and back are upheaved and lateral sides are slightly depressed) MV. *AO* ascending aorta, *LV* left ventricle, *MV* mitral valve, *PV* pulmonary valve

8.2.1 Mitral Annulus

The annulus is a band of fibrous tissue attached to the border of the left atrioventricular orifice. The mitral annulus is not a complete and uniform fibrous band (Fig. 8.13).

The anterior third of the mitral annulus is the continuity of the anterior leaflet and the AV, which is part of the cardiac skeleton, and is the firmest but most inflexible portion, rendering it unsuitable for annuloplasty (Figs. 8.14, 8.15, 8.16, and 8.17).

The posterior two-thirds is where the posterior leaflet of the MV is attached, corresponding to the upper border of the LV free wall (Fig. 8.18). The fibrous tissue is thin with considerable extensibility; thus, it is an important site for annuloplasty. The MV annulus is no longer considered to be fixed and flat structured. Actually, it is saddle shaped, and its size and shape change with the heart cycles. In systole, the posterior annulus moves anteriorly, reducing the area of the MV orifice and enlarging the LVOT. The saddle shape of the annulus of the MV differs from that of the TV, which is bow shaped and is in a plane. The annuli of the mitral and tricuspid valves join each other in the central fibrous body, but they are not in the same plane: the annulus of the MV is higher than that of the TV (Fig. 8.16).

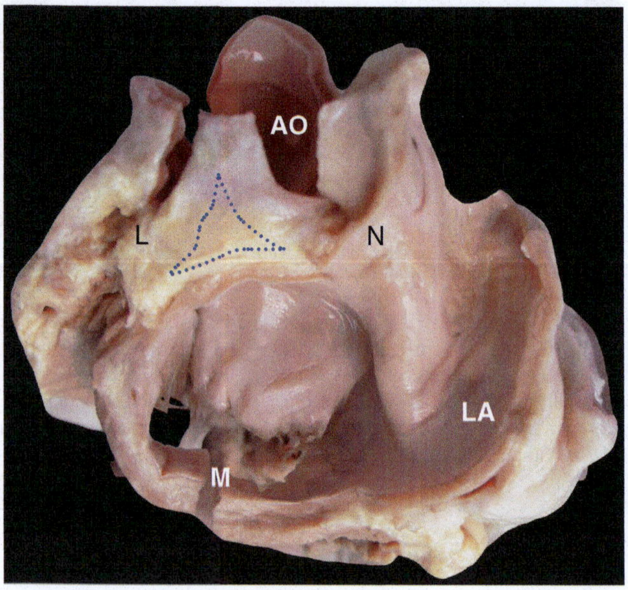

Fig. 8.13 Joint zone of aortic valve and mitral valve. The aortic (AO) and mitral (M) valves are overviewed. The joint zone of the noncoronary (N) and left coronary cuspid (L) is a dense fibrous structure, the so-called intervalvular triangle, the base of which is continued with the anterior mitral valve

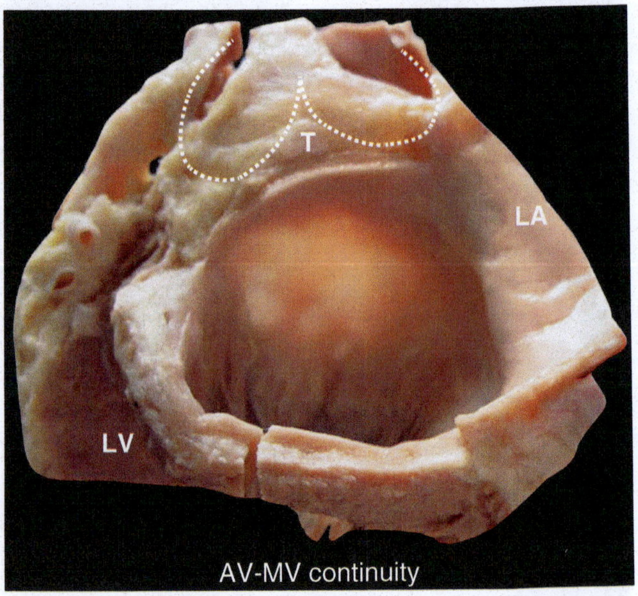

Fig. 8.14 Aortic intervalvular triangle. The semitransparent membrane is the anterior MV. The dotted line indicates the aortic annulus. T indicates the fibrous triangle between the noncoronary cuspid and left coronary cuspid, the intervalvular triangle

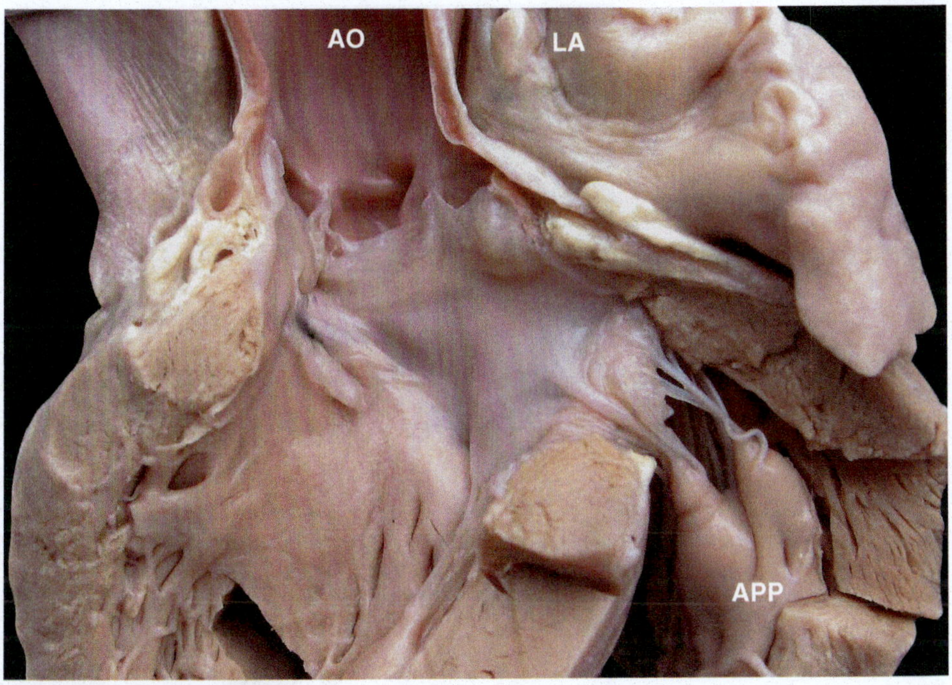

Fig. 8.15 Surroundings of the anterior mitral leaflet. The specimen was obtained from a healthy child; all valves are normal. The aortic-mitral fibrous continuity and the APP are clearly displayed. The aortic-mitral fibrous continuity is an important identification marker of whether or not there is congenital anomaly of the great vessels, which does not exist in the DORV, where the MV and the AV are separated by muscular tissue. *APP* anterior papillary muscle, *AV* aortic valve, *DORV* double-outlet right ventricle, *MV* mitral valve

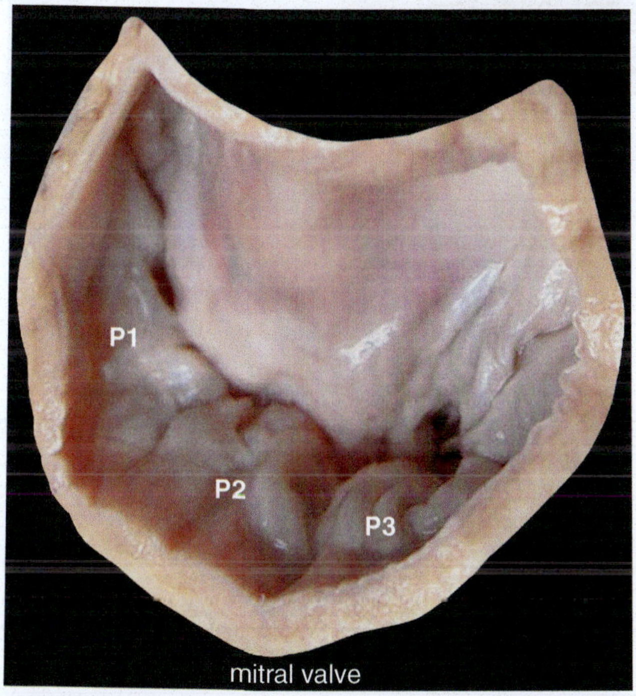

Fig. 8.16 Mitral annulus and the posterior mitral leaflet. The posterior mitral leaflet does not have a fixed width. Rather, it is divided into three scallops as P1, P2, and P3. The posterior mitral leaflet inserts on a flexible portion of the mitral annulus, the length and shape of which can be changed in the mitral valvuloplasty

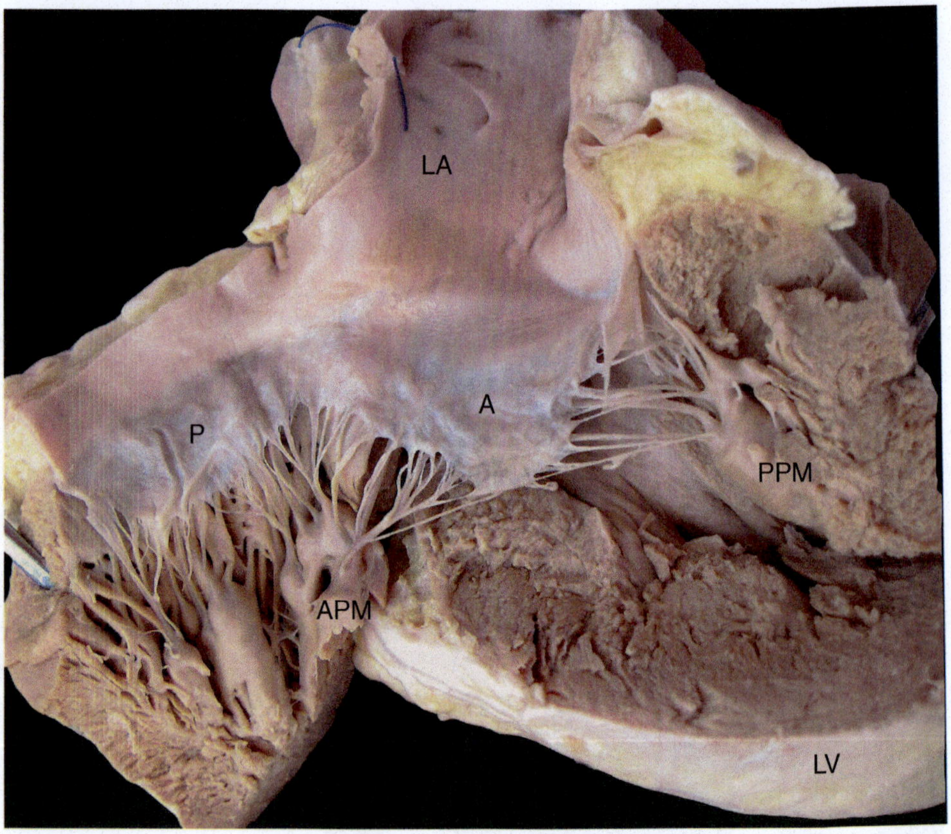

Fig. 8.17 Unfolded mitral leaflets. The heart is dissected through the left atrium, showing the anterior leaflet, posterior leaflet, chordae tendineae, and papillary muscles of the MV. *A* anterior mitral leaflet, *APM* anterior papillary muscle, *P* posterior mitral leaflet, *PPM* posterior papillary muscle

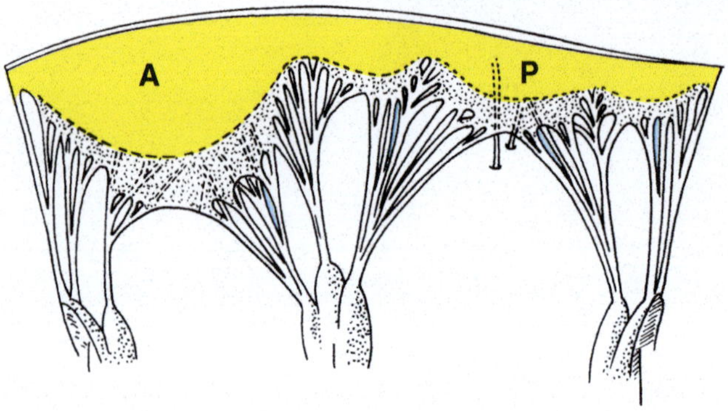

Fig. 8.18 Illustration of the mitral leaflets. In the unfolded mitral leaflets, the *dotted area* in the picture indicates the coaptation zone of the anterior and posterior leaflets

8.2.2 Mitral Leaflets

The mitral leaflets are the flake-like structures derived from the endocardial protrusion and are composed of connective tissue covered with endocardium and inserted on the fibrous ring of the cardiac skeleton. The leaflets consist of basal, marginal, and middle parts, although they are not completely divided but are as membranous tissue in a continuum with variable widths. The base of the leaflet is attached to the mitral annulus. The inferior part of the leaflets droops into the LV to variable extents. The leaflets of the MV are divided into the anterior and posterior leaflets by two prominent incisures, called the *anterior commissure* and the *posterior commissure*, respectively (Figs. 8.11 and 8.12).

8.2.2.1 Anterior Leaflet

The anterior leaflet is wide and short, almost rectangular. It is composed of the smooth part and the rough part. The former forms the main part of the anterior leaflet, and the latter forms the marginal part (Fig. 8.19). Along with the subaortic muscles, the anterior leaflet of the MV forms the LVOT (Fig. 8.15). The marginal part is the coaptation zone (Fig. 8.12).

8.2.2.2 Posterior Leaflet

The posterior leaflet is narrow and long. It is divided by indentations into three fan-like parts; thus, it is called a *tri-scalloped structure* (Figs. 8.18 and 8.20).

8.2.3 Subvalvular Structure

The subvalvular structure includes chordae tendineae and papillary muscles. Chordae tendineae, originating from the tips of papillary muscles in the majority and from the ventricular wall in the minority of people, insert on the free edge and the ventricular side of the leaflets. Chordae tendineae are composed of coarse chords and dispersed fine chords. Chordae tendineae can be classified into three classes:

8.2.3.1 Primary

Primary chordae tendineae originate from papillary muscles and insert on free edge of leaflets. They are coarse, and their main function is to prevent the eversion of the leaflets (Fig. 8.21).

8.2.3.2 Secondary

Secondary chordae tendineae also originate from papillary muscles, but they insert on the ventricular side of the leaflets. Their function is to reinforce the tractive effect.

8.2.3.3 Tertiary

Tertiary chordae tendineae originate from the posterior wall of the LV and insert on the ventricular side of the leaflets.

8.2.4 Papillary Muscles

Papillary muscles are not isolated muscular columns but rather connect with chordae tendineae. The posterior papillary muscles collect chordae tendineae and insert them into the posterior portions of both the anterior and posterior leaflets. The anterior papillary muscle is located beneath the anterior commissure, originating from the left ventricular lateral wall, and thus is also called the *left papillary muscle* (Fig. 8.11). Blood supplying the anterior papillary muscle comes mainly from the LCx. Blood supplying the posterior papillary muscle comes from the LCx or the posterior descending branch of the RCA.

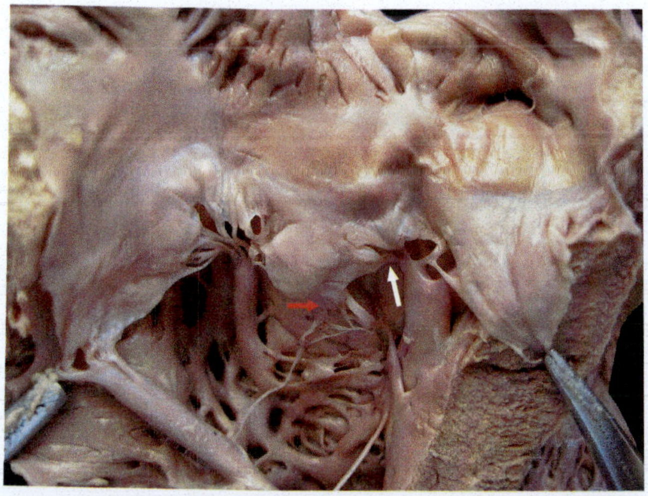

Fig. 8.19 Posterior mitral leaflet and chordae tendineae. The posterior mitral leaflet is scallop shaped. Some of the chordae tendineae (*white arrow*) originate from papillary muscles and some from the ventricular wall (*red arrow*). When ruptured, the chordae tendineae inserting on the anterior leaflet can be substituted by chordae tendineae of the posterior leaflet

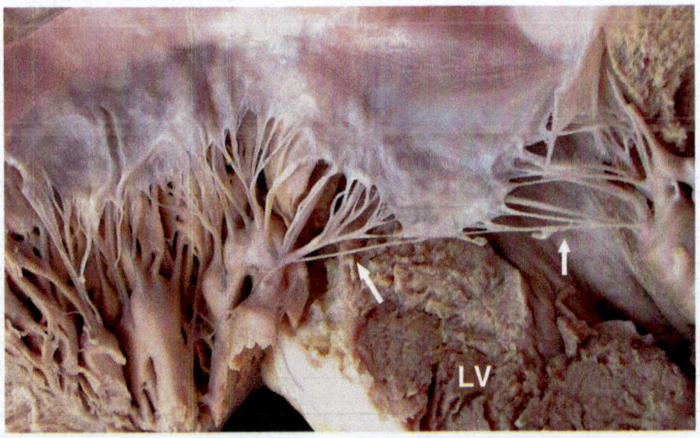

Fig. 8.20 Chordae tendineae of the anterior mitral leaflet. All of the chordae tendineae of the anterior leaflet are coarse, inserting on the free edge and originating from the tips of papillary muscles

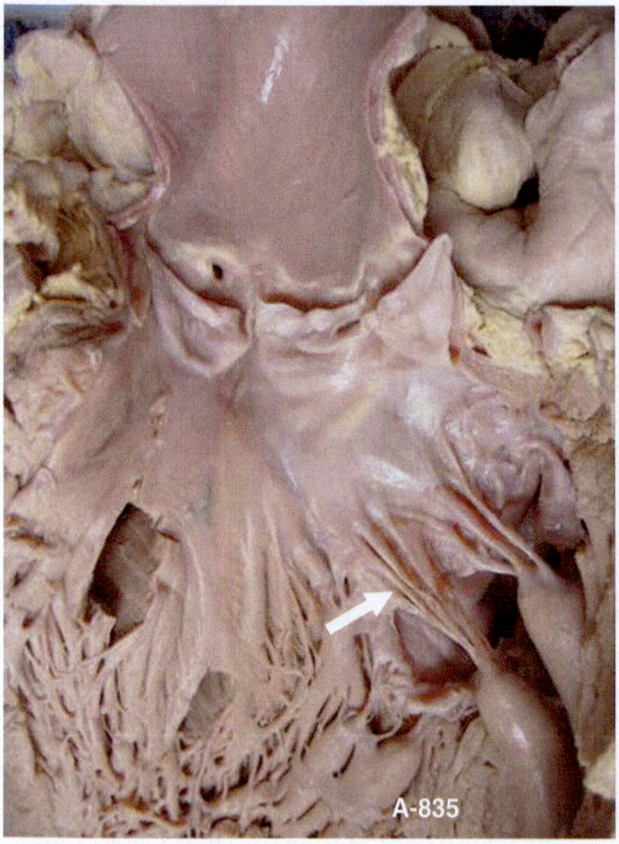

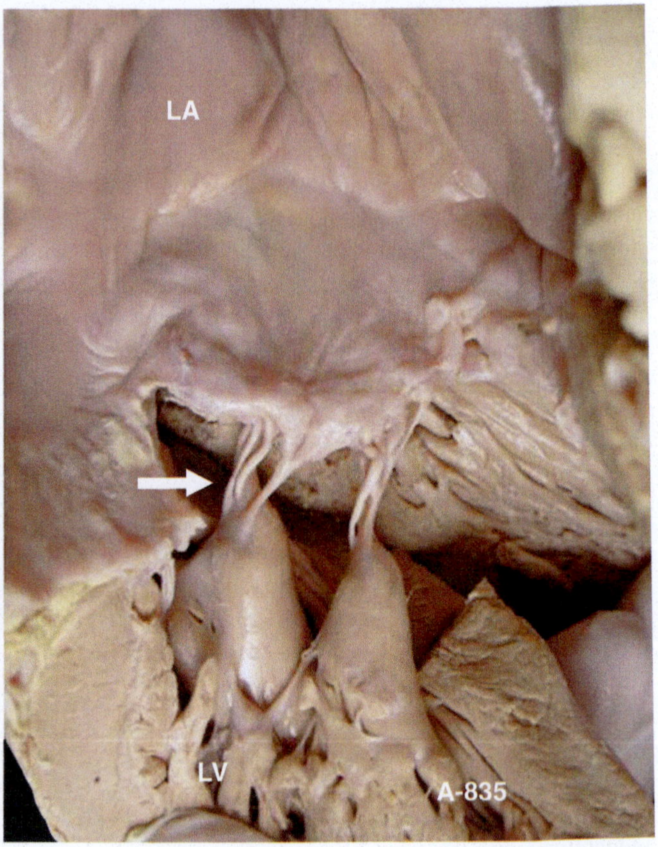

Fig. 8.21 Mitral chordae tendineae displayed from the atrial and ventricular side. Some chordae tendineae insert on the ventricular side of the leaflets. Major chordae tendineae insert on the free edge of the leaflets. From the left atrial side, the roughness of the coaptation zone of leaflets is displayed

8.3 Surgical Implications

8.3.1 Mitral Valvuloplasty

The basic purposes of mitral valvuloplasty are to preserve and to restore the area and function of the anterior mitral leaflet, as much as possible. As much as one half to two thirds of the posterior leaflet and annulus can be sacrificed to preserve the area of the anterior leaflet. In some cases, the chordae tendineae inserting on the posterior leaflet can be transferred to the anterior leaflet to support the normal movement of the anterior leaflet. As long as there is enough area and movement of the anterior leaflet, the function of the MV can be well preserved. The size of the annuloplasty ring is chosen according to the area of the anterior leaflet. The length of anterior leaflet annulus takes only one third of that of the whole annulus, where it is relatively fixed. Therefore, it is unsuitable for surgical repair; hence, for the most part, the posterior leaflet is the choice for valvuloplasty, to leave the anterior leaflet intact. This strategy requires minor sacrifices to safeguard major interests. An improper annuloplasty ring might impair the LVOT (Figs. 8.22, 8.23, and 8.24).

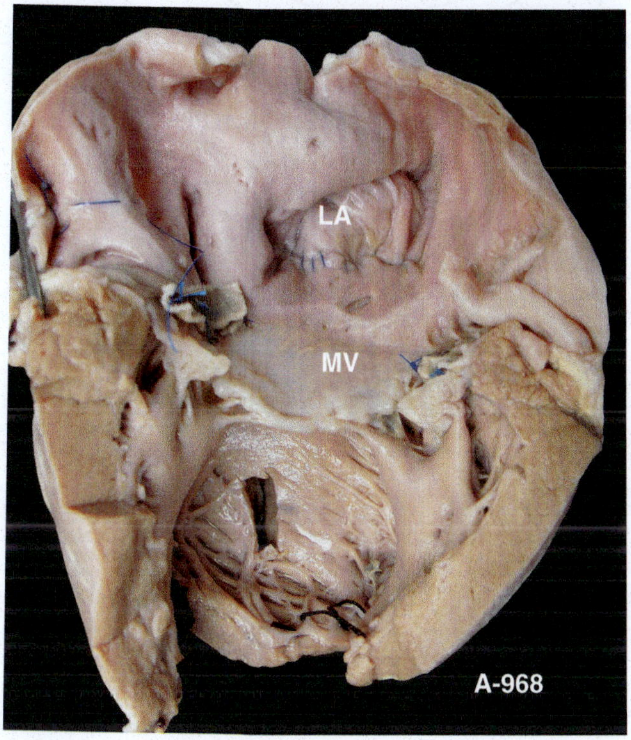

Fig. 8.22 Operative specimen showing mitral valvuloplasty. The anterior and posterior commissures are sutured and compressed

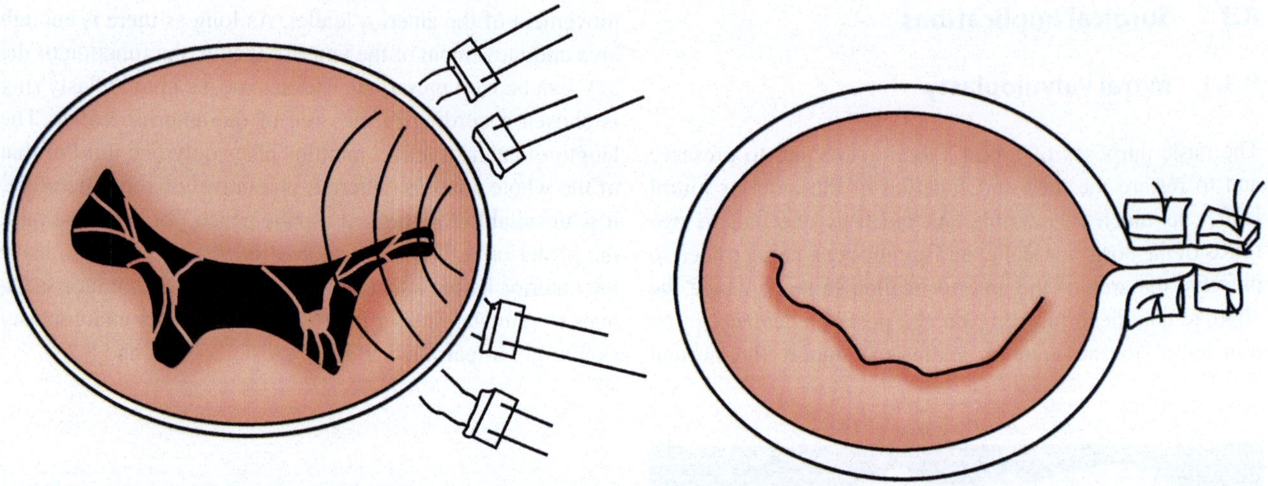

Fig. 8.23 Annular compression in mitral incompetence. The annulus at the posterior commissure is compressed using pledgeted horizontal mattress sutures. In some cases, the simple compression sutures may come loose. A plastic ring is used to reinforce in this situation

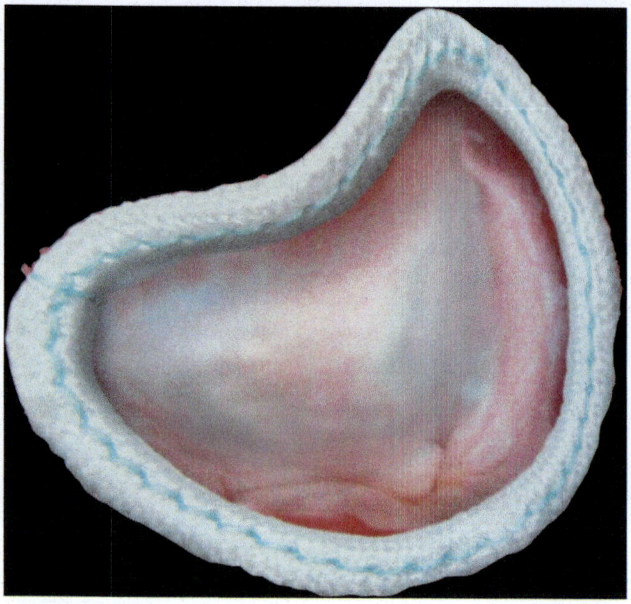

Fig. 8.24 Artificial plastic ring used after compression of the mitral annulus

8.3.2 Mitral Valve Replacement

When the MV is not suitable for repair in a severely diseased heart, as shown in the specimen below, MV prosthesis replacement is indicated (Figs. 8.25, 8.26, 8.27, 8.28, and 8.29). Either a bioprosthesis or a mechanical prosthesis can be chosen, depending on the clinical situation. The suture should be sound to avoid periannular leak. In MV replacement with a bioprosthesis, if the pedicle of the prosthesis is winded by the suture, severe regurgitation may occur. Over-excising of the papillary muscles in MV replacement could result in the rupture of the left ventricular posterior wall, which is a fatal complication. The opening of the rupture is small despite a rather large area of damaged myocardium in the interior of the ventricular cavity. Therefore, a simple pledgeted mattress suture is unreliable. Suturing should be done when the heart is arrested; the prosthesis is taken down if necessary, the opening of rupture is closed by patch, and the prosthesis is reimplanted (Figs. 8.30, 8.31, and 8.32).

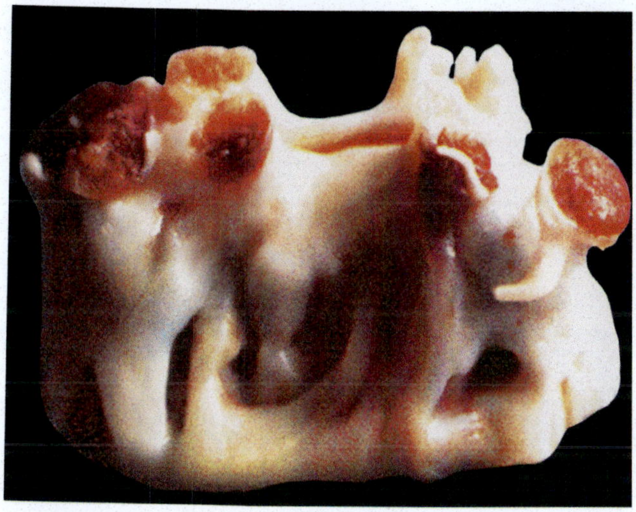

Fig. 8.25 Specimen of rheumatic mitral disease. The specimen was excised in the surgical procedure. The mitral leaflets are apparently thickened and fused. The leaflets are immobile, excluding valvuloplasty. The valve replacement is indicated

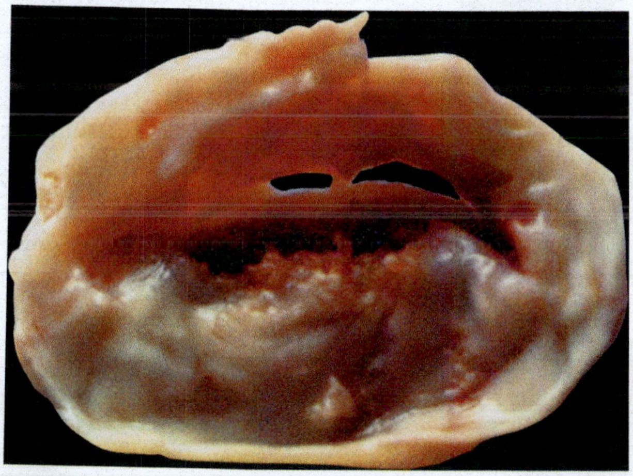

Fig. 8.26 Calcified mitral valve

Fig. 8.27 Massive thrombi retrieved from the left atrium in patient with mitral stenosis

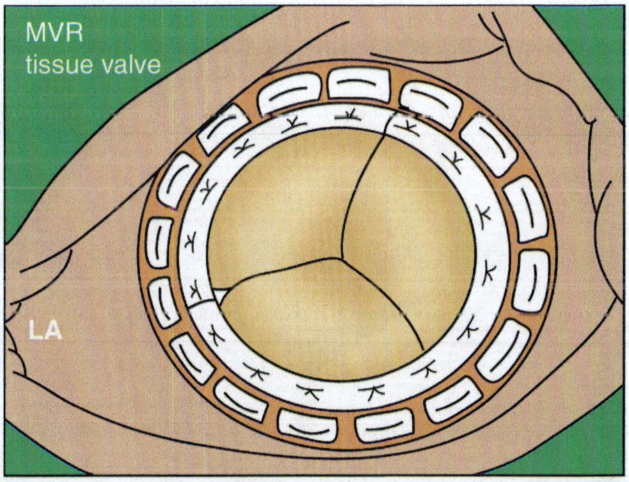

Fig. 8.28 Mitral bioprosthesis. This is a pedicled pericardial bioprosthesis. In addition to protecting the artificial valve from mechanical damage, catching of the pedicle by sutures is avoided to prevent prosthetic incompetence

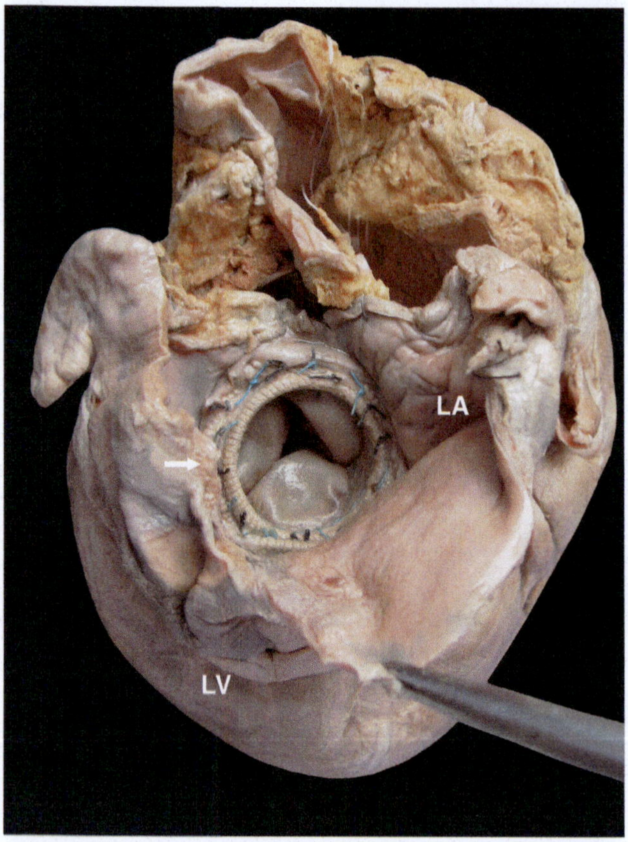

Fig. 8.29 Post-operative specimen following mitral bioprosthesis. The autopsy specimen of a patient dying of early post-operative heart failure following MV replacement displays bovine pericardial bioprosthesis. *Arrow* indicates bioprosthesis (viewing from the ventricular side), showing the prosthetic leaflets are still intact

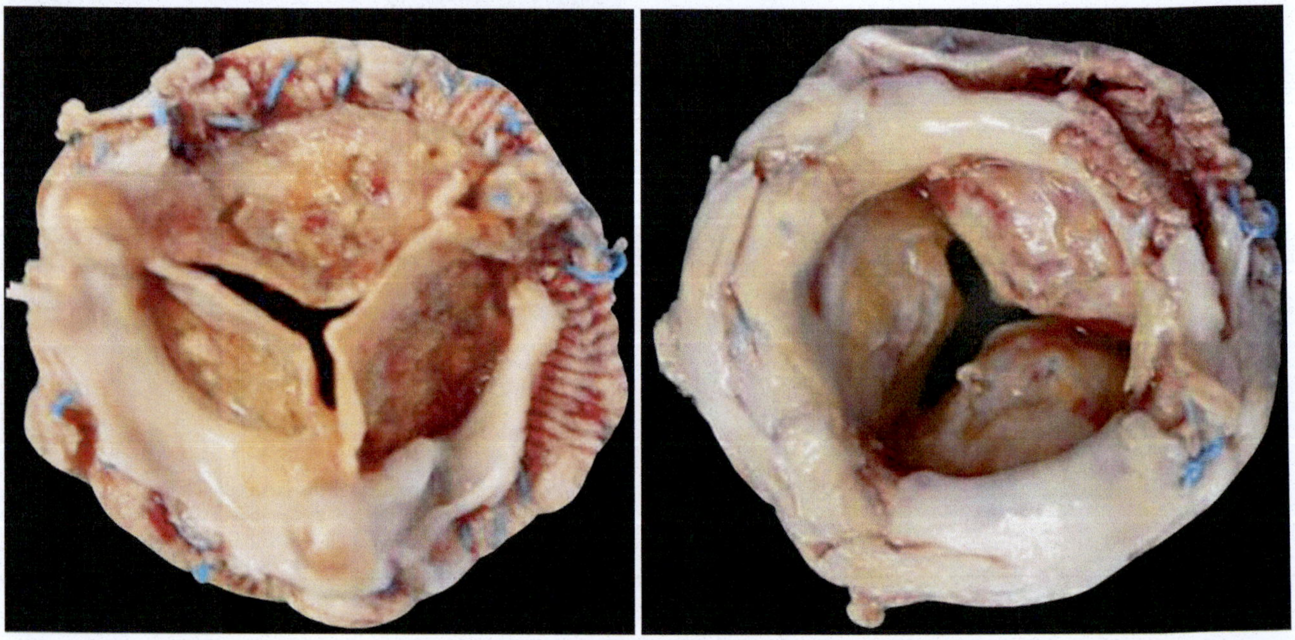

Fig. 8.30 The BN bovine pericardial bioprosthesis used in mitral valve replacement. There is severe calcification, but no avulsion is identified 23 years postoperatively

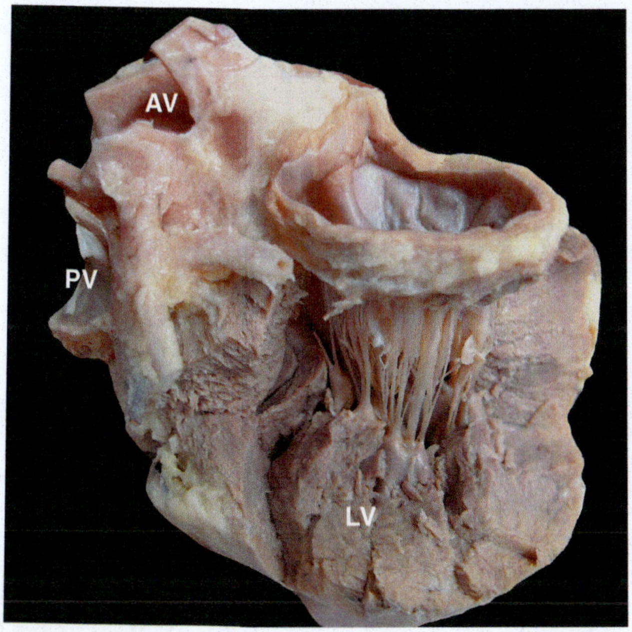

Fig. 8.31 Highly hypertrophic left ventricular wall and shortened chordae tendineae. *AV* aortic valve, *LV* left ventricle, *PV* pulmonary valve

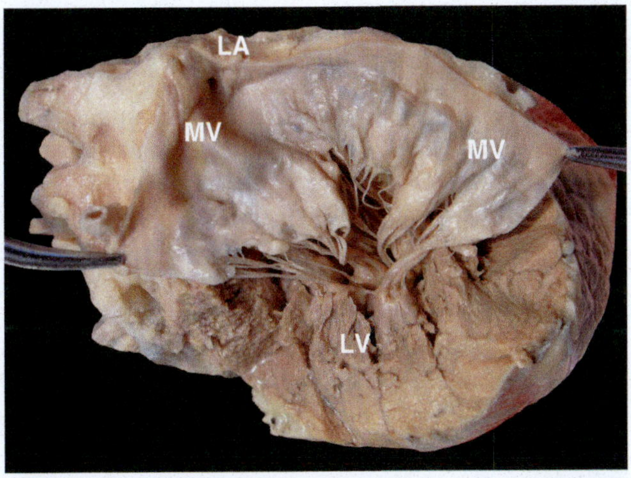

Fig. 8.32 The same specimen as the previous figure, showing the posterior mitral leaflet and shortened chordae tendineae

8.3.3 Procedure of Hypertrophic Subaortic Stenosis

The exaggerated, hypertrophic myocardium in the hypertrophic cardiomyopathy can result in muscular obstruction of the LVOT. High-speed blood stream in systole moves the anterior mitral leaflet anteriorly (called *systolic anterior movement*, SAM). A MV replacement can alleviate the obstruction. Usually, the obstructive hypertrophic myocardium is locally excised to relieve the obstructed outlet tract. Care should be taken to avoid making an excessive excision, which can result in perforation of the interventricular septum into the RV (Figs. 8.33, 8.34, and 8.35).

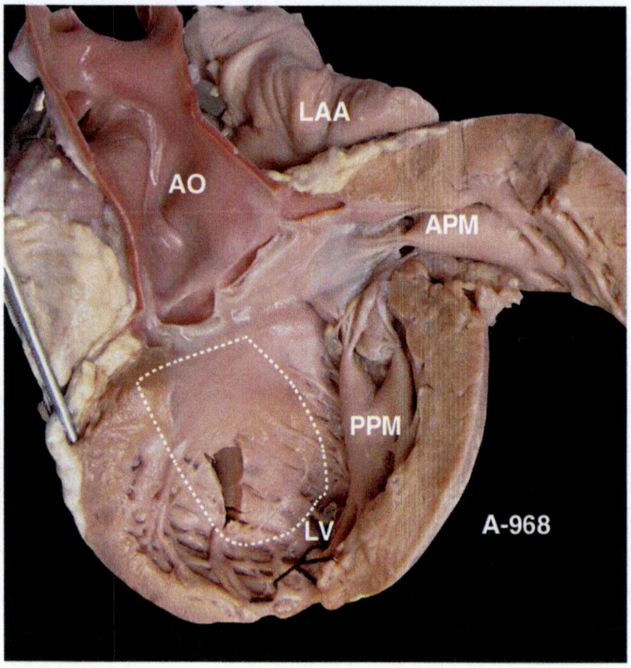

Fig. 8.33 Normal heart demonstrating site of excision in repair of the left ventricular outlet tract obstruction (area encircled in *dotted line*)

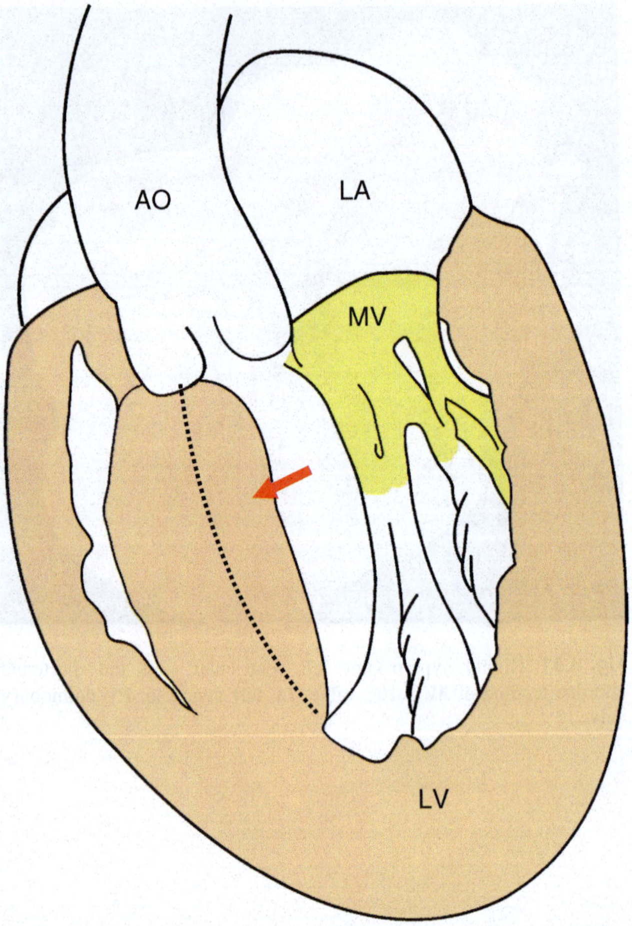

Fig. 8.34 Excision of hypertrophic subaortic septum. The *dotted line* and *arrow* indicate the excessive myocardium. The surgical procedure aims to relieve subaortic obstruction, simultaneously avoiding injuries to the aortic cuspids and mitral papillary muscles and perforation of the interventricular septum

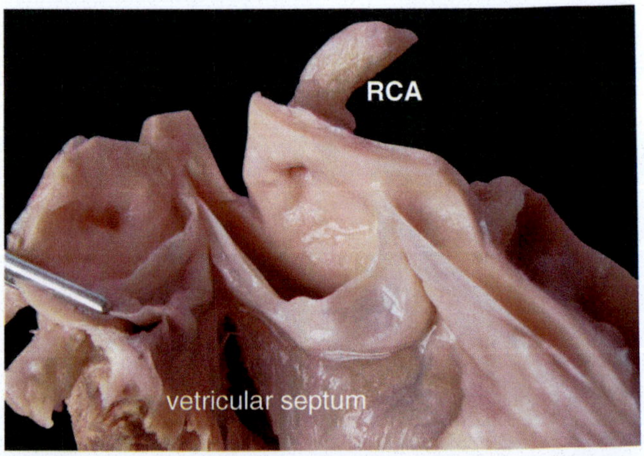

Fig. 8.35 Site of incision in hypertrophic interventricular septum. The incision on the left half of the interventricular septum beneath the AV is displayed. The interventricular septum is notched to relieve obstruction, leaving a groove of 1.0 cm wide, 1.0 cm deep, and 5 cm long. Excessive excision is avoided to prevent interventricular septal perforation and injury to the AV and submitral structures. To avoid atrioventricular block, the excision should not cross over the middle point of the right coronary cuspid

8.3.4 Left Ventriculoplasty

In dilated cardiomyopathy, the LV is over-dilated and weak in contractility. The mitral annulus can dilate as well, so as to result in mitral incompetence. In severe cases, heart transplantation is indicated. In less severe cases, the volume of the left ventricular cavity can be surgically reduced to improve left ventricular function. The ventriculotomy usually is performed at the apex behind the LAD. Efforts should be made to avoid injuring the papillary muscles of the LV (Fig. 8.36).

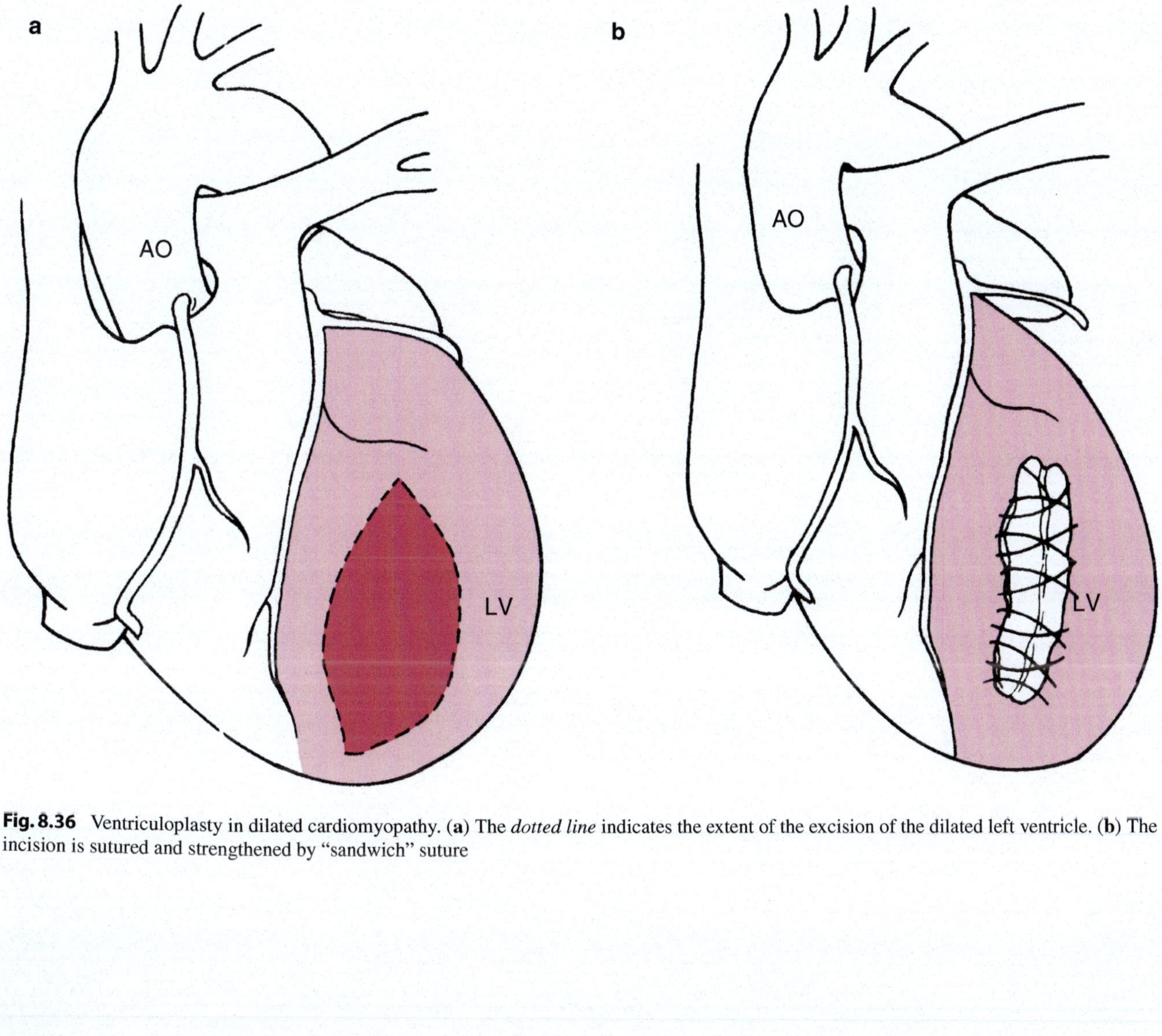

Fig. 8.36 Ventriculoplasty in dilated cardiomyopathy. (**a**) The *dotted line* indicates the extent of the excision of the dilated left ventricle. (**b**) The incision is sutured and strengthened by "sandwich" suture

9 The Cardiac Skeleton and the Aortic Root

9.1 General Considerations

9.1.1 The Cardiac Skeleton

The fibrous skeleton of the heart, namely, the *cardiac skeleton*, is a framework supporting all of the chambers of the heart. Cardiac surgery involving the chambers of the heart and the aortic root also involves the cardiac skeleton (Fig. 9.1). The cardiac skeleton is composed of the left and right AV valve rings and the aortic annulus and consists of four bands of dense connective tissue. The cardiac valve leaflet and the myocardium are accreted to the cardiac skeleton, so as to maintain the shapes of the chambers of the heart and the three-dimensional structure (Fig. 9.2). The aortic annulus is in the central portion of the cardiac skeleton and adjoins the PV ring with the loose connective tissue.

The core of the cardiac skeleton is composed of the central fibrous triangle (body) and the left fibrous triangle (Fig. 9.3). The central fibrous triangle is the site at which the aortic annulus, the mitral ring, and the tricuspid annulus meet; it is fibrous and cartilaginous. The membranous interventricular septum lies in its front and at its end is the Todaro tendon, running from the interatrial septum inferiorly and to the right to the inferomedial part of the IVC valve. The atrioventricular band passes through the central fibrous body, following the lower margin of the membranous interventricular septum into the ventricular (Fig. 9.4).

The left fibrous trigone is the site at which the left aortic and the mitral annuli meet. Its lateral end adjoins the left coronary artery (Fig. 9.5).

9.1.2 The Aortic Root

The aortic root may be defined generally as a supra-coronary aorta inside the pericardium – the AV with its crown-shaped annulus, interleaflet triangles, sinotubular ridge, and sinus of Valsalva. For practical purposes, the aortic root should be between the aortic annulus and the sinotubular ridge, concluding at the sinus of Valsalva and left and right coronary artery ostia (Fig. 9.6).

Because the aortic root is the central part of the heart, the aortic annulus is the important part of the supporting frames of the heart. All cardiac surgery should refer to the aortic root; therefore, the topography of the aortic root is very important for the surgery.

The *sinus of the Valsalva* refers to the expanded portions of the aortic root confined proximally by the concave attachments of each cusp and named accordingly (right, left, and noncoronary). Part of the wall of the right sinus of Valsalva is the myocardium of the ventricular septum. The base of the sinus of Valsalva is imbedded in other tissues, increasing the difficulty for exposing it during surgery on the aortic root. It is surrounded by the left and right atria, and the atrial septum faces the middle of the noncoronary sinus of Valsalva. The right sinus of Valsalva adjoins to the RA, and its front adjoins to the major pulmonary artery. These imbedding contact surfaces are connected by the loose connective tissue, which could be divided by blunt dissection. The right sinus of Valsalva straddles the infundibular septum and adjoins the RVOT (Fig. 9.7).

The *ostia of the left and right coronary arteries* originate from the posterior left sinus and the upper portion of the anterior sinus of Valsalva, respectively (Fig. 9.8). It is difficult to stop bleeding of the anastomosis between the left coronary artery and a vascular prosthesis because this anastomosis is covered by the vascular prosthesis after the heart rebeats, rending it difficult to expose. Traction on the vascular prosthesis aiming to expose the anastomosis may result in more bleeding. The ostium of the RCA lies superficially, but the incision of the aortic root is too low to impair it. After the sinus of Valsalva is opened, one can find the ostia of the coronary arteries, the AV, the aortic annular, and the position of the sinus of Valsalva in its adjoining relationship (Figs. 9.8 and 9.9). An AV device includes a cusp of the aorta, an aortic annular, and a subvalve structure.

The *AV ring* is not a completely uniform fiber cable; the bottom of the right coronary cusp is attached at the septum

X. Zhu (ed.), *Surgical Atlas of Cardiac Anatomy*,
DOI 10.1007/978-94-017-9409-1_9, © Springer Science+Business Media Dordrecht and People's Medical Publishing House 2015

muscle. The bottom of the left coronary cusp and the noncoronary cusp is connected with the anterior leaflet of the MV. The connecting wires form the last half of the aortic annulus. It includes the right fibrous triangle, with the triangular and left fibrous triangle, and is a very dense and solid part. The aortic annulus shows a wave type to form the upper edge of the LVOT.

The AV presents the three half-moon leaves and is attached to a wave-type aortic ring. The structure is different from the atrioventricular valve, having no accompanying chordae tendineae and relying mainly on the aortic sinus tension among the three leaflets to the free edge of the ventricular diastolic to prevent insufficiency. The free margin of the valve lies at the same level, and the basal adhesion is in a wave-like valve ring (Fig. 9.10).

The valve and its corresponding aortic wall constitute an upwardly opening bag-shaped depression, called the *aortic sinus*. The three leaflets are identical with regard to size, location, and other characteristics; the free edges contact each other, and the free edge of the leaflet midpoint *locally thickened free limbal nodule formation often* [*this is not clear*]. During the left ventricular ejection, the AV opens. The valve is attached to the sinus. During left ventricular diastole, the AV closes, with the valve leaflet contacting the free edge in order to prevent the backward flow of blood.

The tissue beneath the AV is the LVOT, around which are (Figs. 9.11, 9.12, 9.13, 9.14, 9.15, 9.16, 9.18, and 9.19):

1. The anterior wall, composed of the muscle structure of the ventricular septum, which lies under the left and right aortic valves.
2. Fibrous triangle between the noncoronary ring and the left coronary ring; it stretches with the anterior leaflet of the MV, continuing a screen shape, which forms the posterior wall.
3. Below the leaflet of the left coronary; a valve there may be a ventricular free wall myocardium, which is of little clinical significance.
4. Below the junction of the right coronary and noncoronary leaflets is the membranous septum. It is a thin membrane-like tissue. The scope is very small, but it is in the central part of the fibrous skeleton of the heart, and its surrounding anatomy is complex and is of considerable surgical significance.

9.2 Surgical Significance

9.2.1 Incision of the Aortic Root

The anterior wall and the right diagonal back are dissociated, so the incision can be made there. For thick tissue, the surgeon should choose the anterior wall; for thin tissue, the surgeon should choose the right diagonal back. The right diagonal back incision starts from the upper left and extends to the right bottom. Sometimes, the aortic artery can be traversed for a clearer view. If the aortic artery is not fully grown, it is widened with a patch. An incorrect incision leads to a more difficult operation; the bleeding around the aortic ring should be prevented (Fig. 9.27).

9.2.2 Aortic Valve Plasty

The juncture adhesion should be treated as a commissural incision, but the leaflet cannot be repaired. The plasty of a loose and prolapse leaflet is not ideal. A single leaflet could be repaired. The constriction is suitable for an enlarged aortic ring, but the method used for a normal leaflet with a wave-like ring is different. The aortic regurgitation is the main complication after operation (Fig. 9.20).

9.2.3 Aortic Valve Replacement

Mechanical valves have a level suture edge, whereas bioprosthetic ones have a wave-like suture edge; neither of them can match with the aortic ring, and close attention should be paid to them. Sometimes, the root is too thin to plant a valve, so the aortic artery must be widened. We choose different patterns according to the extent of the expansion (Figs. 9.21 and 9.22).

9.2.4 Stenosis (Figs. 9.28, 9.29, 9.30, and 9.31)

9.2.4.1 Moderate Stenosis

The root incision should be extended to the back sinus, and the root should be widened by a tricuspid patch, the bottom of which should be parallel with the ring.

9.2.4.2 Severe Stenosis

The root incision should be extended to the juncture between the left sinus and the back sinus, the ring to the base of the anterior leaflet of the MV should be cut off, and a combined incision of the aortic artery and the LA should be made. Simultaneously, the aortic ring should be widened sufficiently. The key point is divergence of the aortic ring, the base of the anterior leaflet of the MV, and the lower limb of the anterior wall of the LA. After the aortic and mitral valves are replaced, the anterior leaflet of the MV should be repaired, the incision of the LA should be sutured, and the aortic root should be widened by a patch. The Ross-Konno procedure, which involves cutting off the right sinus, extends to the ventricular septum. It also involves part of the anterior wall of the RV, avoiding the RCA. After AV repair is completed, a three-vane patch is used to repair the ventricular septum, the wall of the RV, and the incision of the aortic artery (Figs. 9.39 and 9.40).

9.2.5 Aortic Root Dilatation and Aneurysm

We define dilatation as a diameter of the base of the aortic ring >25 mm, diameter of sinotubular junction >30 mm, and diameter of aortic artery > 50 mm, which can be considered as an aneurysm and should be treated surgically. We classify the operations as replacement and plasty.

The root *replacement* includes replacement of the AV and the ascending aortic artery and transplantation of the coronary.

The *plasty* involves plasty of the AV and the aortic sinus and includes replacement of the ascending aortic artery. The representative operation is the "David procedure."

According to the property of the aortic regurgitation, we classify it into two types:

Type one: Normal leaflet, central regurgitation caused by a dilated ring (Fig. 9.33)

Type two: Prolapsed, crimped, or punched leaflet (Figs. 9.34, 9.35, and 9.36a, b)

According to the dilatation of the root, plasty or artificial vascular replacement is used. In the root replacement, the AV should be retained as much as possible. The Bentall operation is a combination of AV replacement, ascending aortic artery replacement, and transplantation of the coronary artery and is used in the cases for which the valve cannot be retained (Fig. 9.23).

In recent years, transcatheter AV implantation (TAVI) has been developed and is rapidly being implemented. Its basic principle is the inlaying of the leaflet into a variable metal stent after chemical treatment. First, the stent is mounted onto a catheter and then it is delivered to the target, such as the pulmonary artery or aorta. Then the stent is released, allowing the new aortic valve to work. Finally, the catheter is removed to finish the operation. The purpose is to replace the valve without surgery, which is especially useful for the elderly and/or seriously ill patient (Figs. 9.24, 9.25, and 9.26).

9.3 Specimen Demonstrations

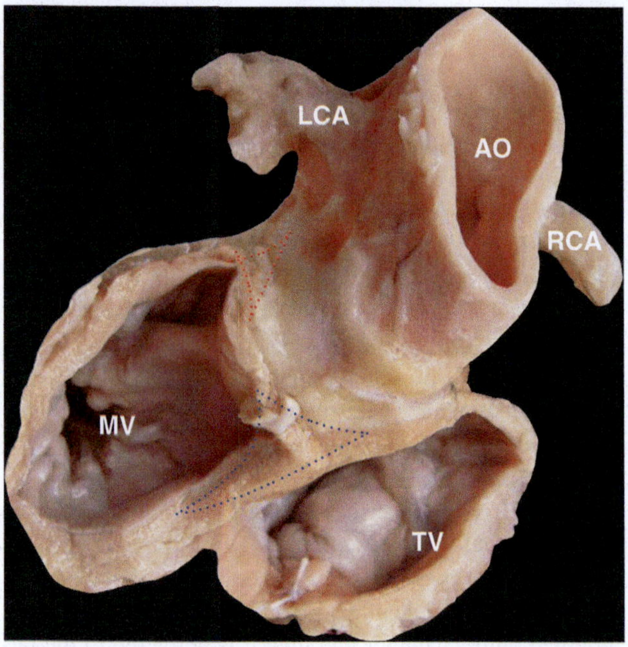

Fig. 9.1 Fibrous skeleton of the heart. The angle of the view is from the back, overlooking the ascending aorta and mitral and tricuspid valves. The juncture of the aortic root, the MV, and the TV is called the *central fibrous body* or *right fibrous triangle central fibrous body*; the MV and the left sinus are called the *left fibrous triangle*. Generally speaking, the fibrous skeleton of the heart contains the aortic ring, the MV, and the TV; it is the central point of support. *AO* aortic artery, *LCA* left coronary artery, *MV* mitral valve, *RCA* right coronary artery, *TV* tricuspid valve

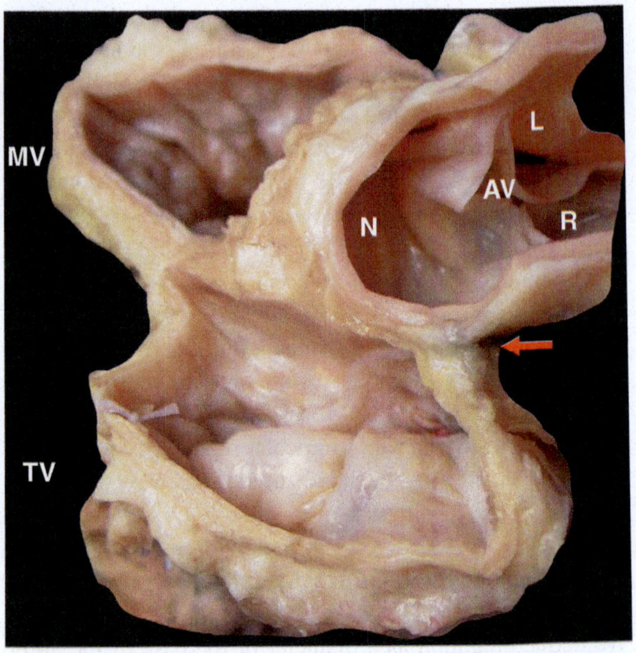

Fig. 9.2 Aortic valve and central fibrous body. This specimen shows the aortic ring, leaflet, and contiguity of the MV and the TV; the middle point of the left leaflet and right leaflet head on; the middle part of the anterior leaflet of the MV; the bottom of the back leaflet and right leaflet head on; and the septal leaflet of the TV. These identifications are significant in distinguishing a congenital deformity

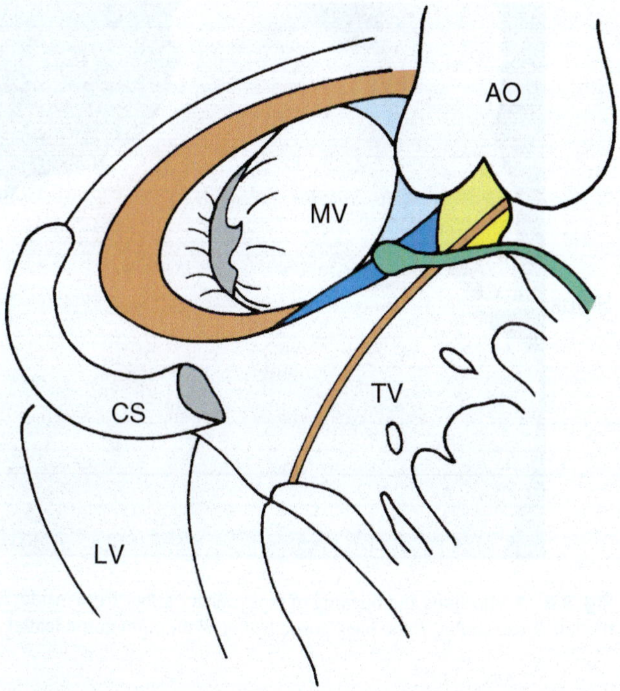

Fig. 9.3 Sketch of fibrous skeleton of the heart. *Blue* fibrous triangle, *yellow* perimembranous ventricular septum, *green* conduction bundle, *CS* coronary sinus

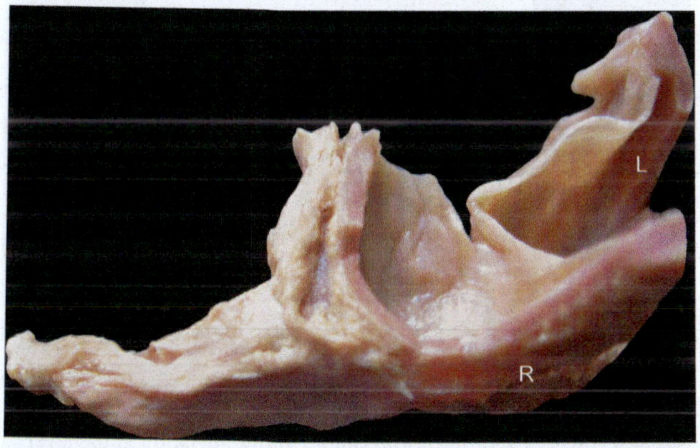

Fig. 9.4 Central fibrous body. The angle of the view is the right side of the central fibrous body. The aortic ring is linked to the front edge of the central fibrous body. *L* left aortic leaflet, *R* right aortic leaflet

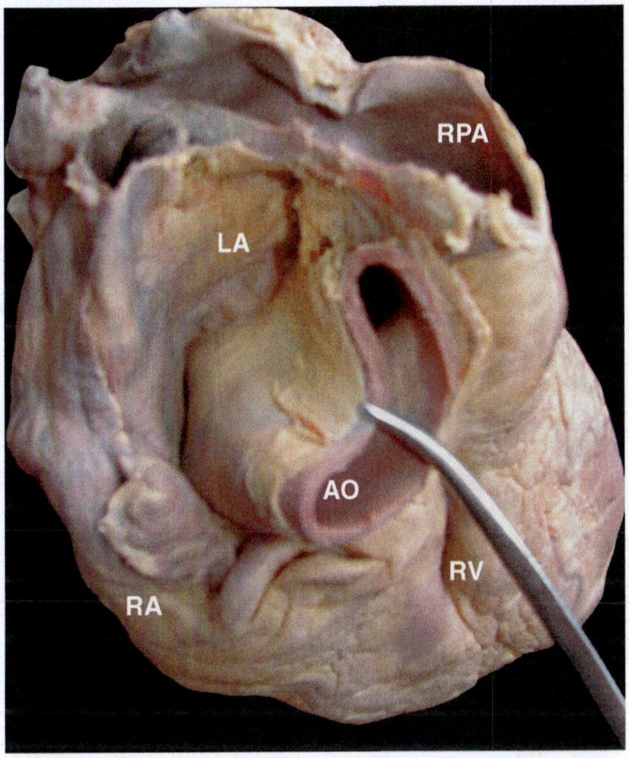

Fig. 9.5 Contiguity between the aortic root and the atrium. The rear part of the ascending aorta is surrounded by the left and right atria. The anterior wall of the LA and the lateral wall of the aorta join together at the ring of the MV

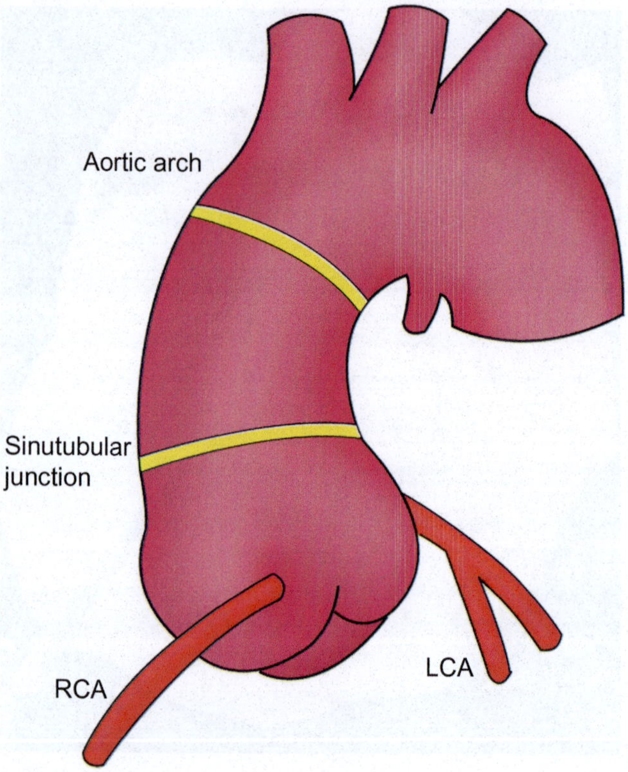

Fig. 9.6 Range of the aortic root. The root contains the sinus of Valsalva, the AV, and the initial part of the coronary artery. The juncture of the aortic sinus and the ascending aorta is called the *sinotubular junction*; it is significant to surgery. The sinotubular junction is the upper edge of the root, whereas the aortic ring is the bottom. The bulging wall of the aorta is called the *aortic sinus*, emitting the left and right coronary arteries

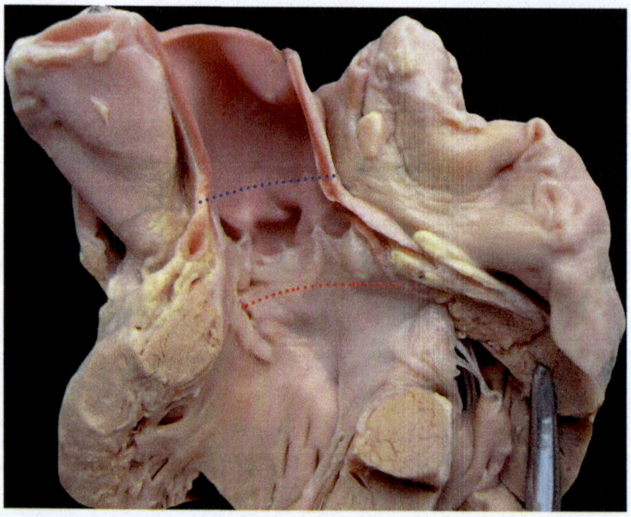

Fig. 9.7 Upper limb and inferior limb of aortic sinus. *Blue dashed line* sinotubular junction, *red dashed line* the junction of the AV and the ventricular septum

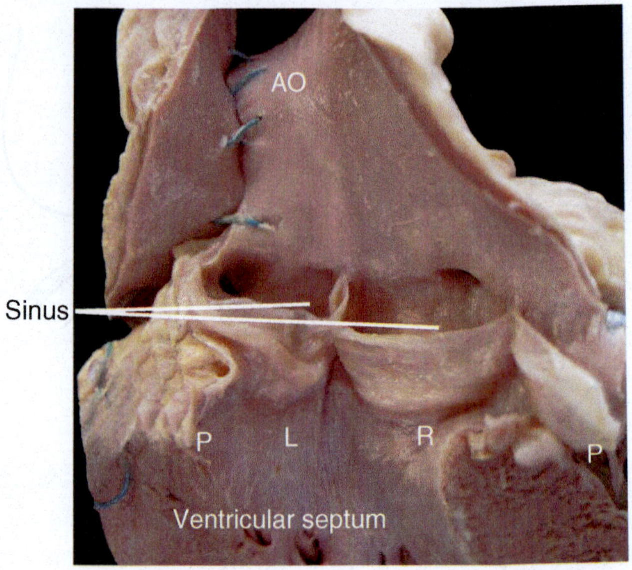

Fig. 9.8 Aortic sinus and opening of the coronary artery from inside. *L* the left aortic leaflet, *P* the back aortic leaflet, *R* the right aortic leaflet

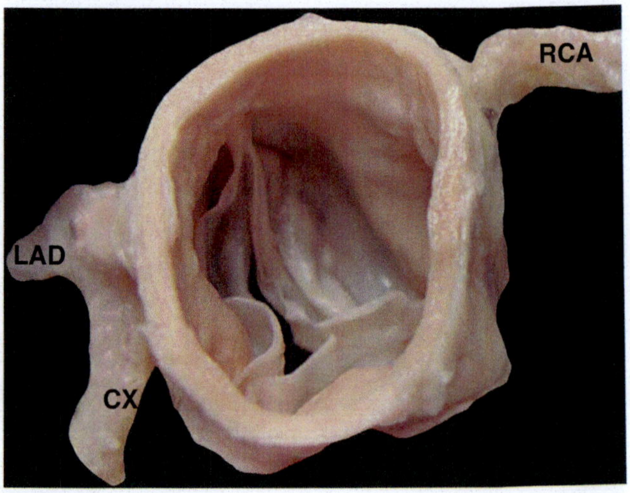

Fig. 9.9 Overview of the aortic sinus. Cutting the aorta at the level of the sinotubular junction reveals three leaflets and the origination of the coronary artery. Attention should be paid to the structure underneath the right sinus, which is muscular tissue of the ventricular septum. *CX* circumflex, *LAD* left anterior descending branch, *RCA* right coronary artery

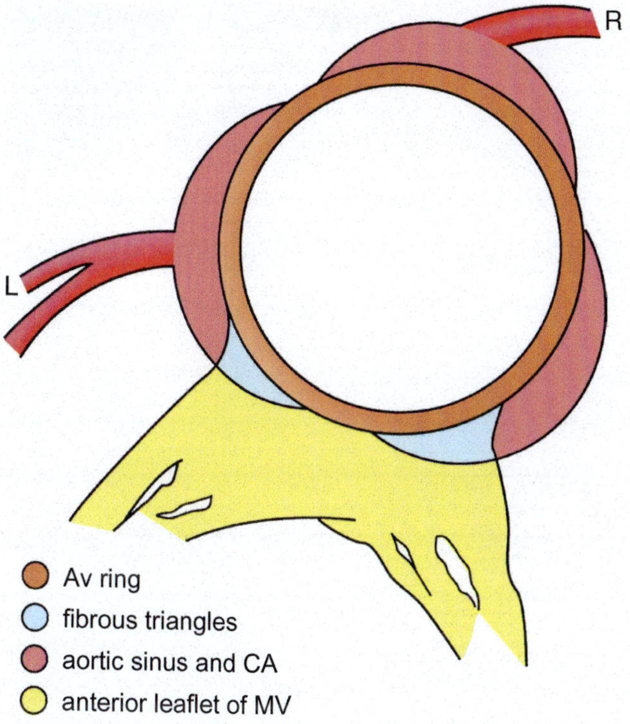

Fig. 9.10 Overview of the mitral valve and fibrous trigone. The contiguity of the aortic sinus and the anterior leaflet of the MV are shown. *Blue part* fibrous triangle, *yellow part* anterior leaflet of the MV

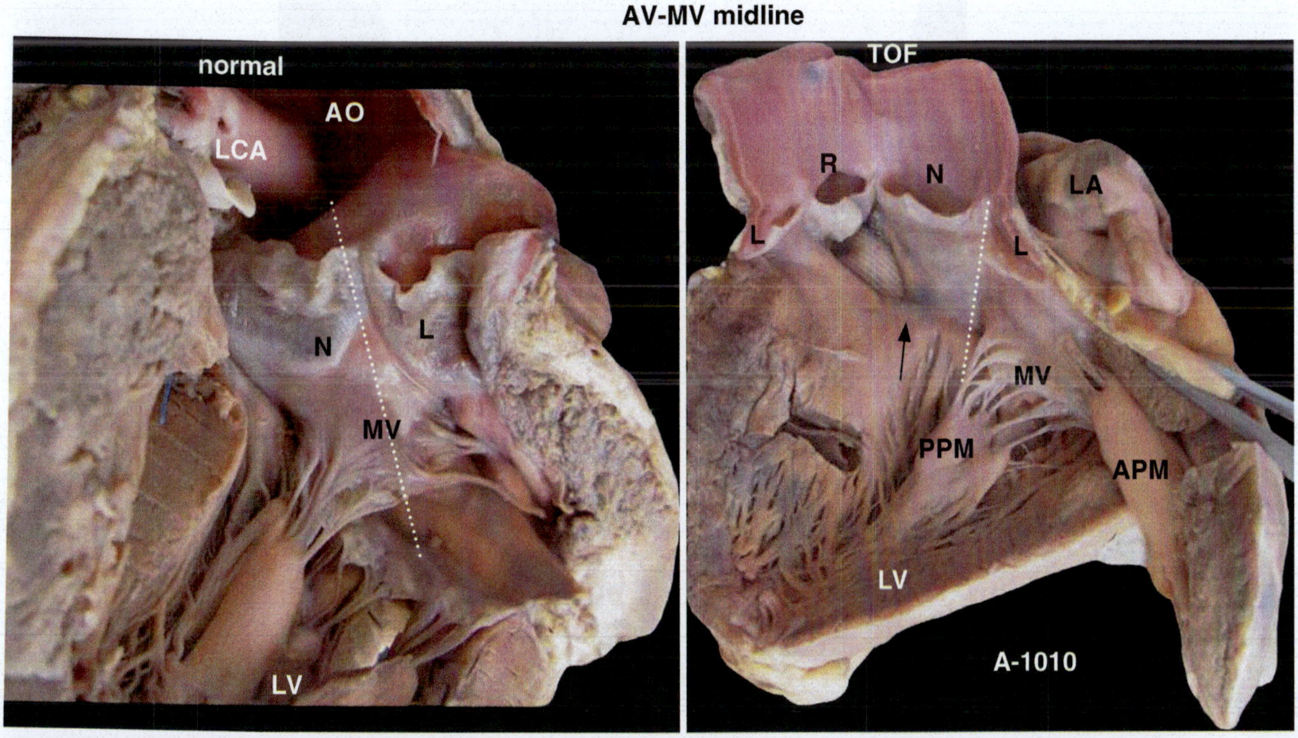

Fig. 9.11 Excursion of *AV-MV midline* in the tetralogy of Fallot. In the normal heart, the line goes from the middle point of three aortic valves to the middle point of the anterior leaflet of the MV. When the main artery is excursed, the line is far away from the middle point. *TOF* tetralogy of Fallot

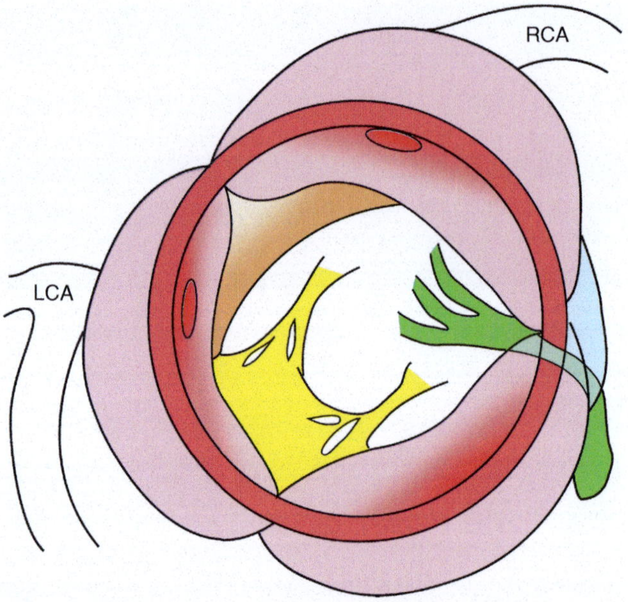

Fig. 9.12 Overview of the ventricular septum and the left bundle branch through the aortic sinus. *Green* left bundle branch, *yellow* leaflet of MV

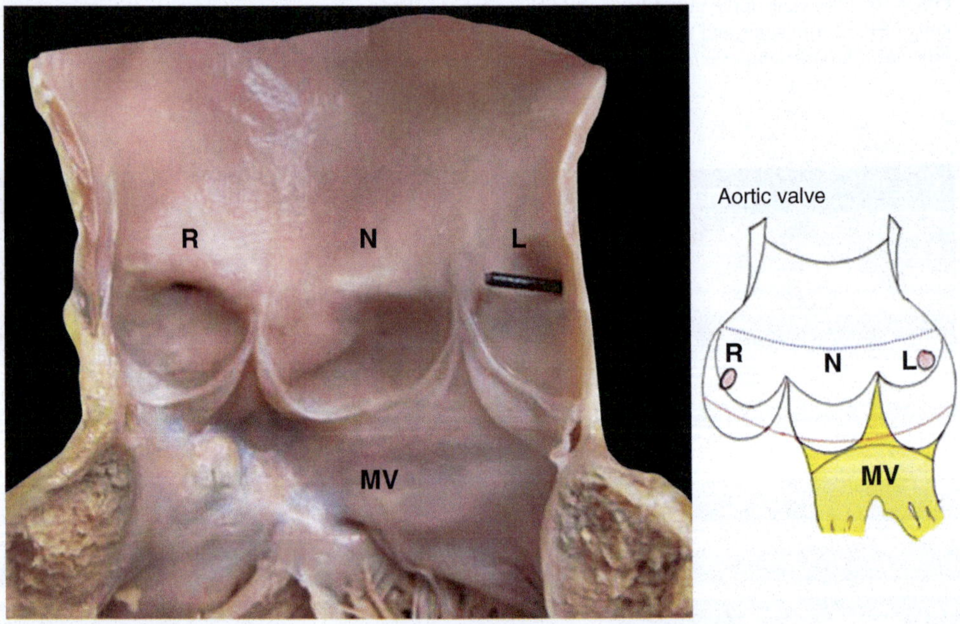

Fig. 9.13 Device of the aortic valve. Observe the aortic sinus, the opening of coronary artery, three aortic valves, and the ring and the structure underneath from the LV

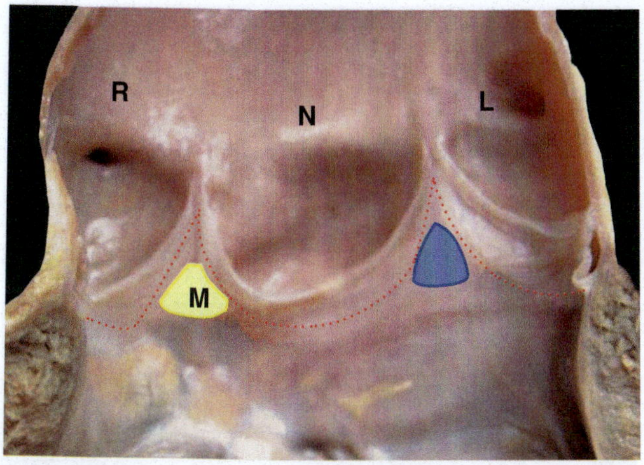

Fig. 9.14 Wave-like aortic ring. Observe the anterior leaflet of the MV extending to the AV from the front when cutting the aorta. The view of the anterior leaflet of the MV is from the ventricle. *Red dashed line* wave-like aortic ring, *yellow dashed line* perimembranous ventricular septum, *blue dashed line* the triangle in the center of the valves

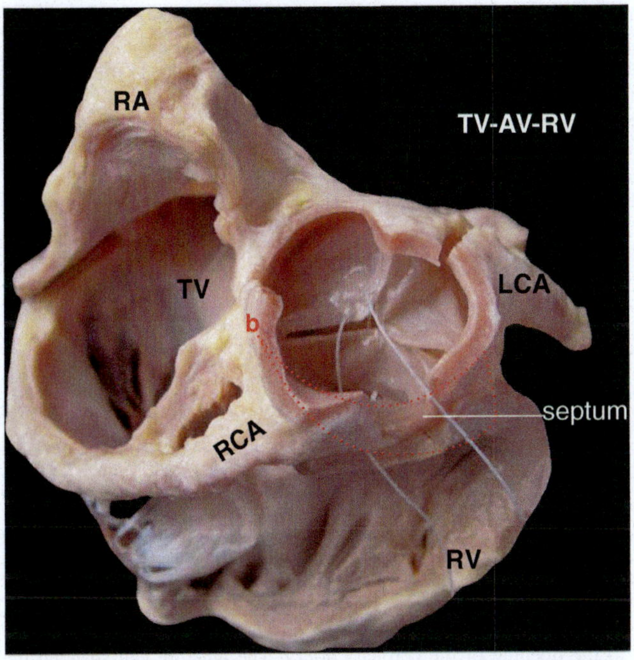

Fig. 9.16 The right aortic leaflet and ventricular septum. The angle of view is an anterior overview. The RA and RV have been cut off. The *red dashed line* is a projection of the forepart of the left ventricular septum. The band of the RVOT can be seen. In the normal heart, the right leaflet bestrides on the ventricular septum and is easily damaged in the operation. In a congener stripping operation, there must be muscular tissue under the right leaflet. *B* perimembranous ventricular septum

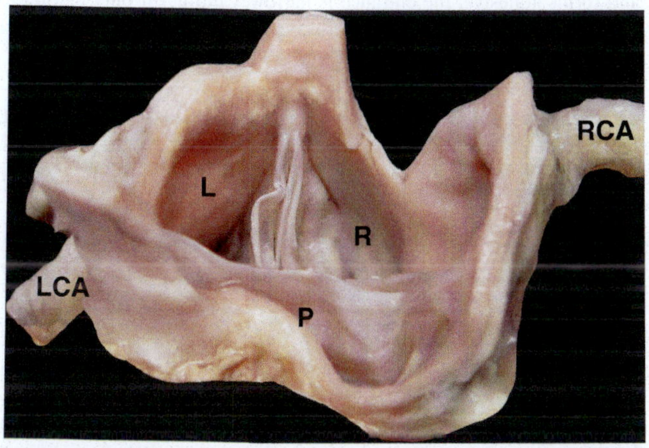

Fig. 9.15 Leaflet of the aortic valve. The figure shows the normal occlusion of the valves. The normal leaflets are thin and well distributed, just slightly thick at the middle point of the free edge. The junction of the leaflets is high and close to the sinotubular junction, suspending the leaflets and keeping the contact surface from regurgitation. If the ring is dilated or loose, it will lead to wrong involution and reverse flow of blood. *L* the left sinus, *LCA* left coronary artery, *P* the back sinus, *R* the right sinus, *RCA* right coronary artery

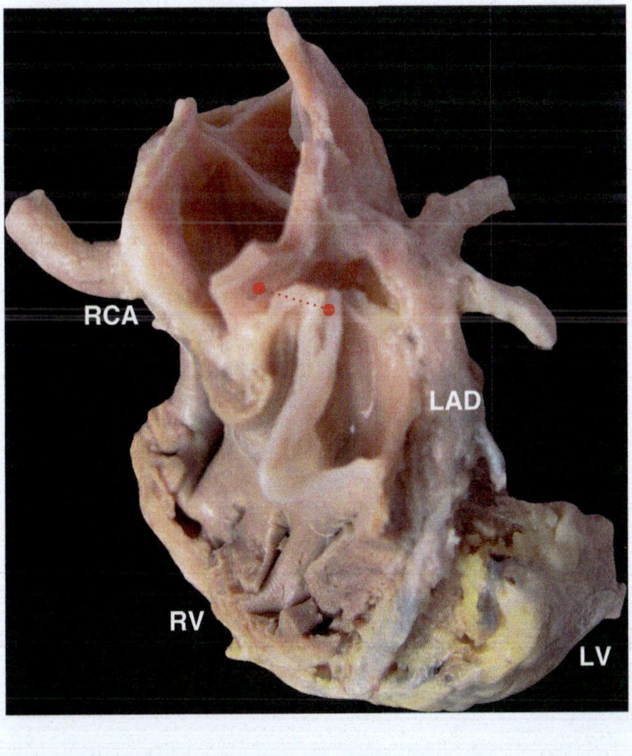

Fig. 9.17 The right aortic sinus and pulmonary valve. The angle of view is the front. The right and left atria, the MV, and the TV have been cut off. Contiguity can be seen between the root and the pulmonary leaflets. The *red line* is the juncture of the aortic left and right valve heads on the juncture of the pulmonary left and right valves. The left main coronary artery lies behind the PV. The replacement of the PV may damage the LCA

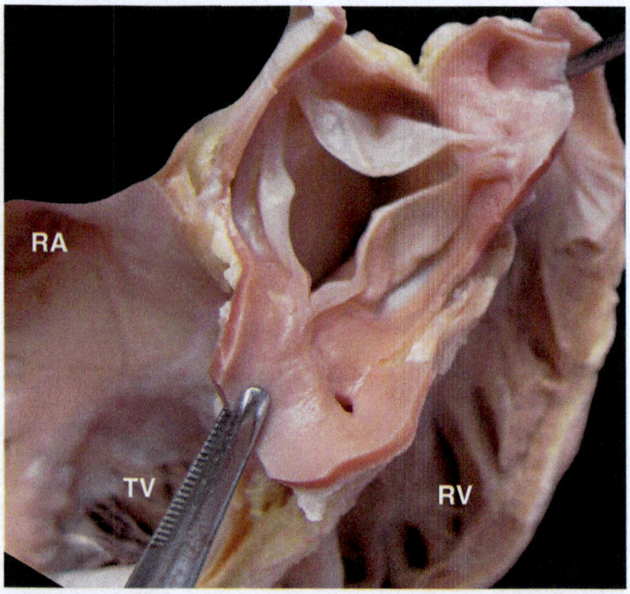

Fig. 9.18 Juncture of the aortic valve. The top of the juncture is close to the opening of the coronary artery. The juncture can suspend the leaflets

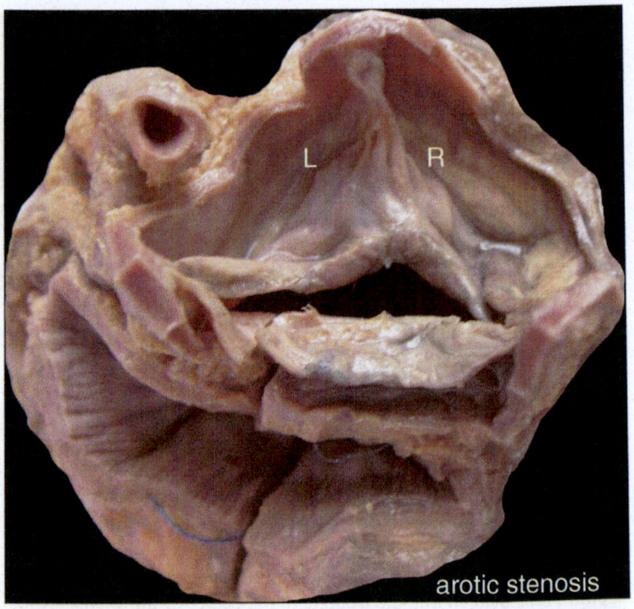

Fig. 9.20 The leaflet is thickening; the juncture of the right leaflet and back leaflet has mixed together totally; the AV incrassates, is calcified, and forms adhesive; the entrance is narrow. Cutting off only the juncture will lead to regurgitation; therefore, a replacement is indicated. If the ring is small, it will need to be widened

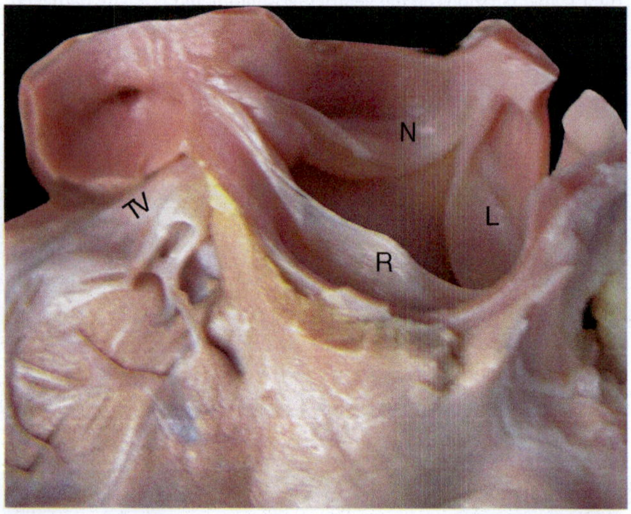

Fig. 9.19 Contiguity between the right aortic leaflet and the septal leaflet of the TV. In the normal heart, the juncture of the right leaflet and the back leaflet head on the forepart of the septal leaflet of the TV, but in some cases of congenital heart disease, such as excursion of AV-MV midline in TOF, the contiguity changes, which is an indication for surgery. *L* left leaflet, *N* back leaflet

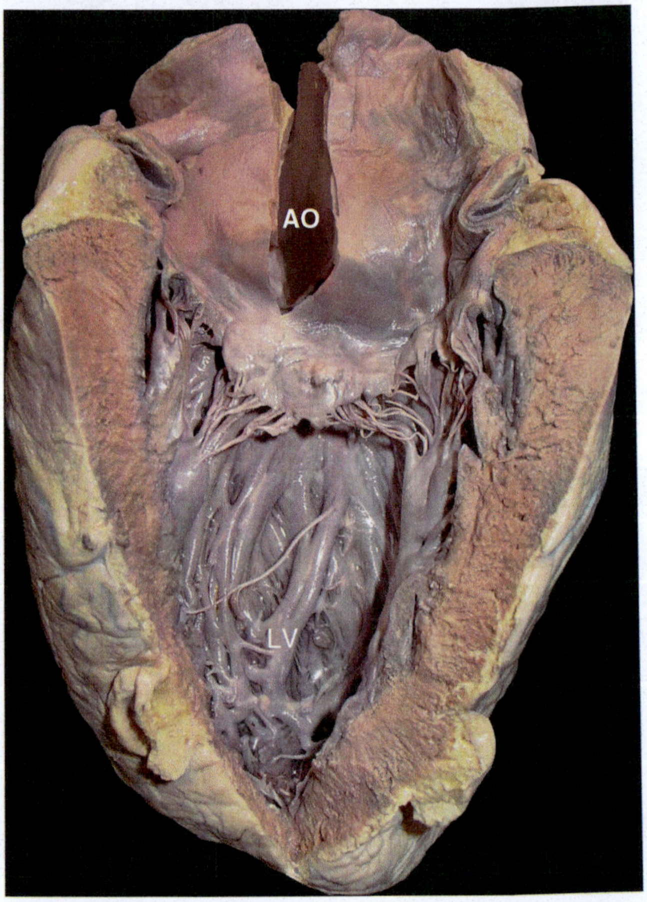

Fig. 9.21 How the aortic ring expands. By extending the incision of the aorta to the base of the anterior leaflet of the MV, a large artificial valve can be planted

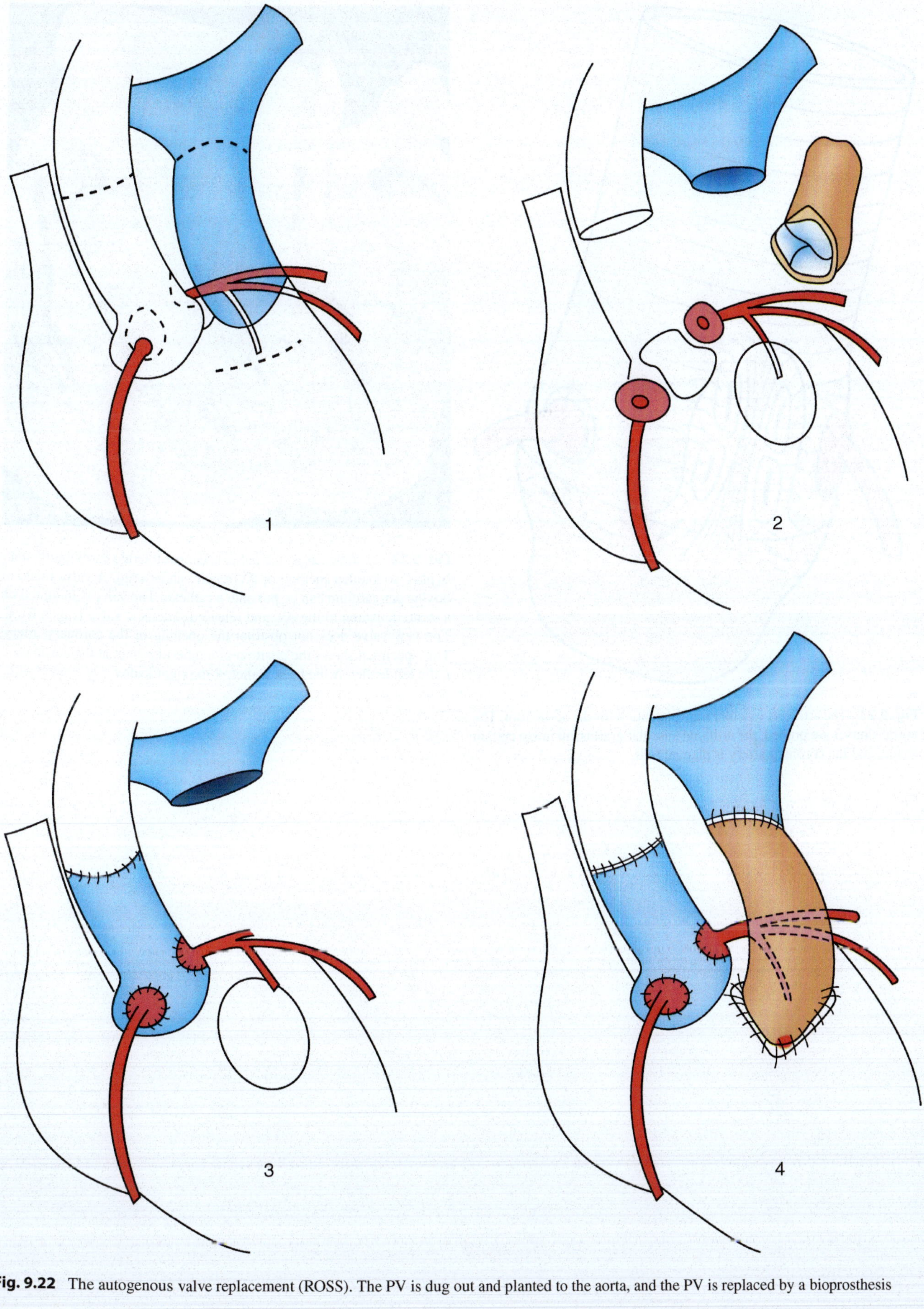

Fig. 9.22 The autogenous valve replacement (ROSS). The PV is dug out and planted to the aorta, and the PV is replaced by a bioprosthesis

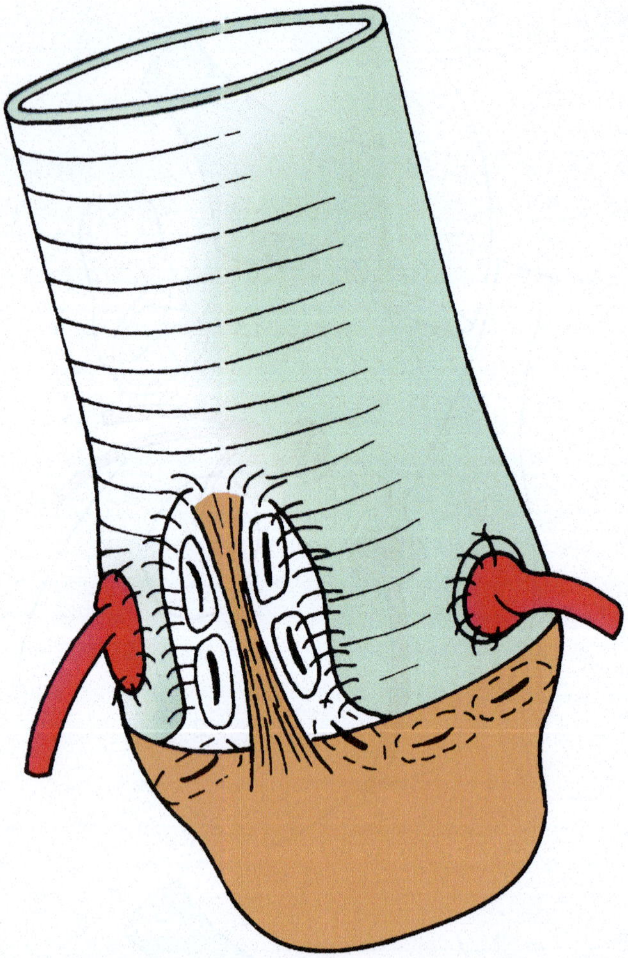

Fig. 9.23 The artificial vascular transplantation for the aortic root. The aortic sinuses are pruned, the artificial vascular graft is cut into a certain shape, and the coronary artery is planted to it

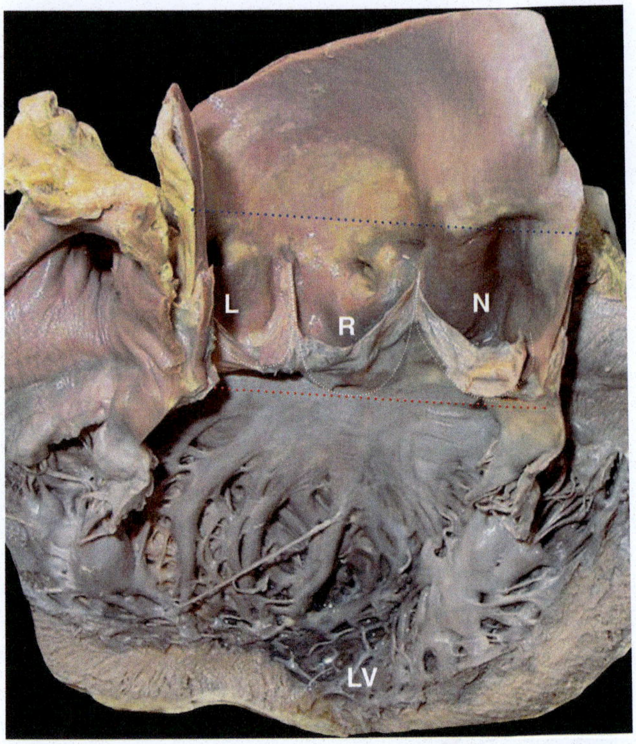

Fig. 9.24 In some cases, the patients cannot tolerate cardiopulmonary bypass, so another method of AV repair can be used. A valve made of bovine pericardium can be put onto a catheter. The valve is mounted on a stent, delivered to the AV, and released. The new valve is then fixed. The new valve does not obstruct the opening of the coronary artery. This specimen shows the position of a tube-like stent at the aortic root. *L* the left leaflet, *N* the back leaflet, *R* the right leaflet

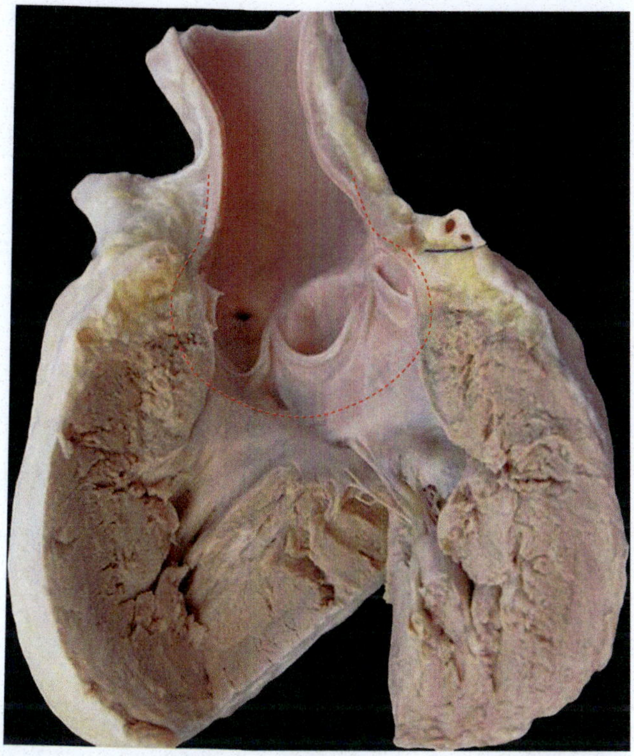

Fig. 9.25 The inset position for an umbrella stent

Fig. 9.26 The aortic valve in the umbrella stent. Fixed in three sinuses, the stent is like a net, so it does not obstruct the opening of coronary artery

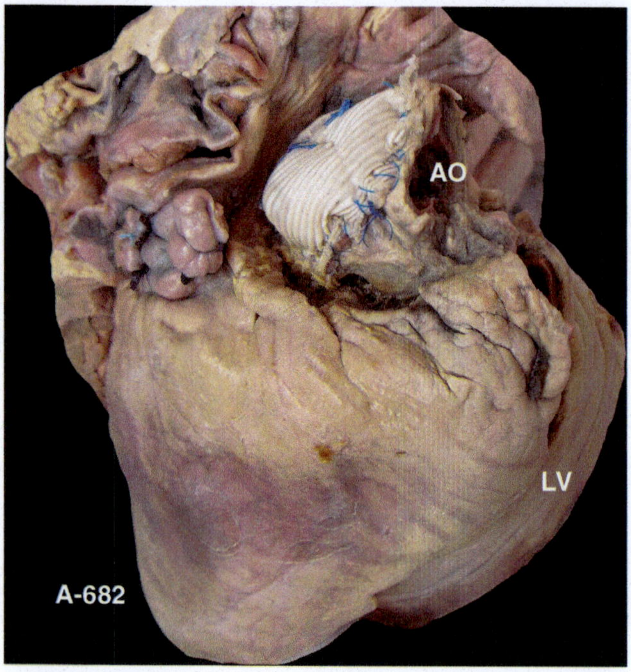

Fig. 9.27 Specimen 682. The incision is widened by artificial vascular tablet after AV repair is completed. *AO* the ascending aorta, *LV* the left ventricle

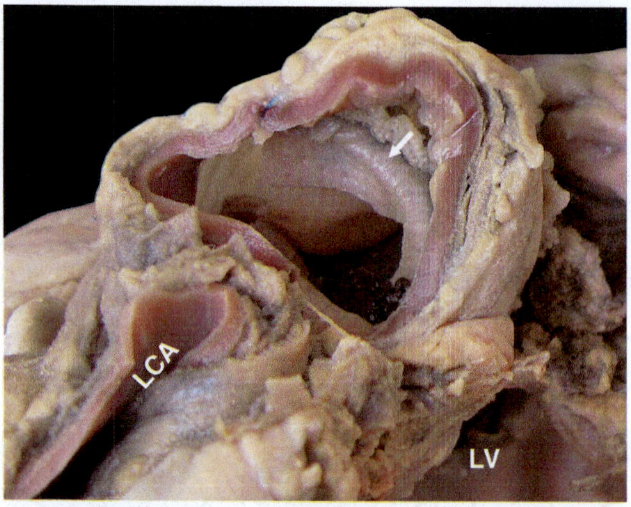

Fig. 9.28 Specimen 682. The lesson from AV replacement. The artificial valve tilts. The leaflet has been broken. The *arrow* shows a planted pericardium valve. The case is a pericardium valve replacement. The artificial valve does not fix well, and the ring tilts obviously with an angle of 40°, so half the ring is not tight, and regurgitation continues to occur. When the ring is small, it should be widened, rather than tilting the artificial valve. *LCA* left coronary artery, *LV* the left ventricle

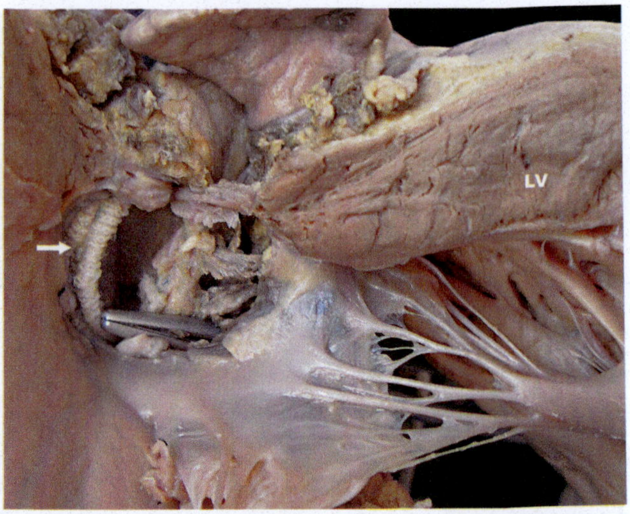

Fig. 9.29 Specimen 682. View from left ventricular outflow tract (*Arrow*). The artificial valve tilts, and there is remaining regurgitation under the artificial valve. *LV* the left ventricle

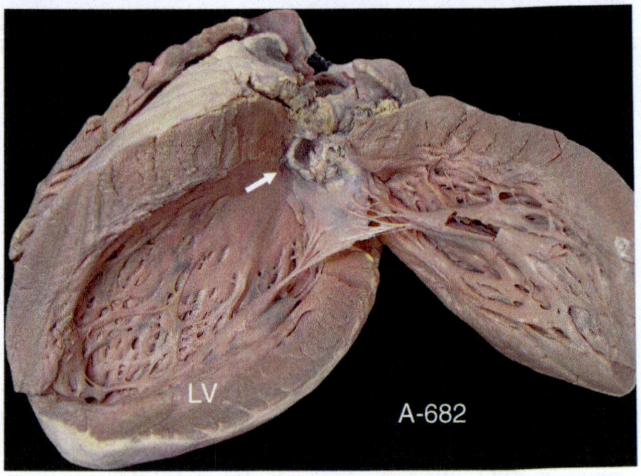

Fig. 9.30 Specimen 682. The view is from the ventricle, the artificial valve (*Arrow*)

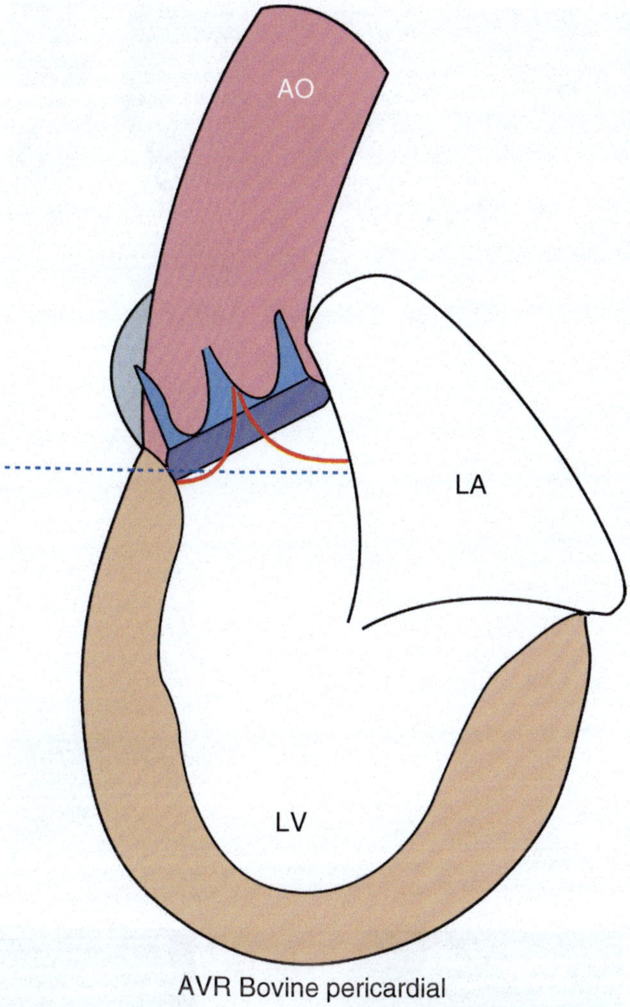

Fig. 9.31 Specimen 682. The incorrect position of implantation. The *dashed line* is the correct position. When the diameter of the aortic ring is small or the diameter of the artificial valve is large, the incision can be widened rather than filled by force, because it will lead to stenosis, and it is not easy to suture. Hence, the perivalvular leakage is common

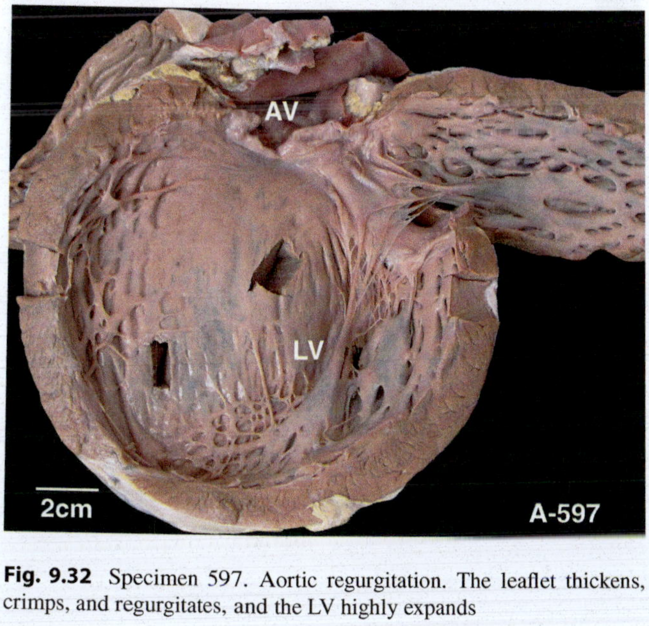

Fig. 9.32 Specimen 597. Aortic regurgitation. The leaflet thickens, crimps, and regurgitates, and the LV highly expands

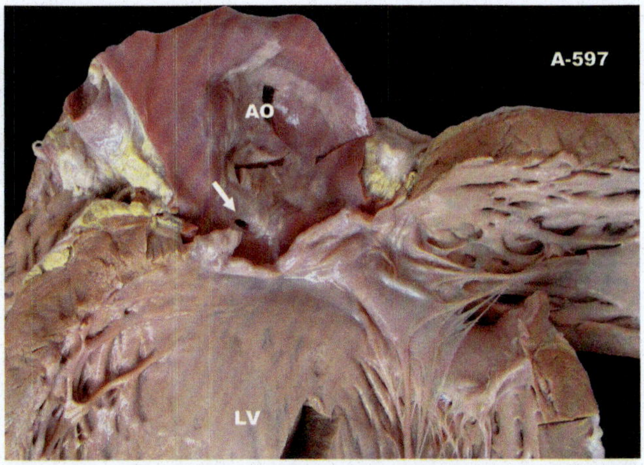

Fig. 9.33 Specimen 597. The leaflet crimps and cannot involute; the *arrow* shows the opening of the coronary artery

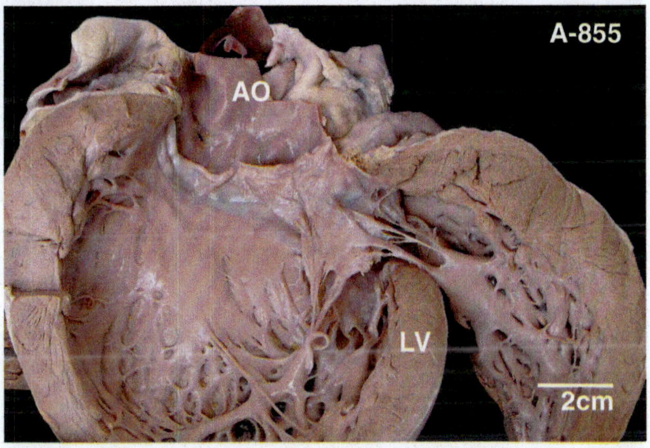

Fig. 9.34 Specimen 855. The leaflet is infected, crimps, and regurgitates

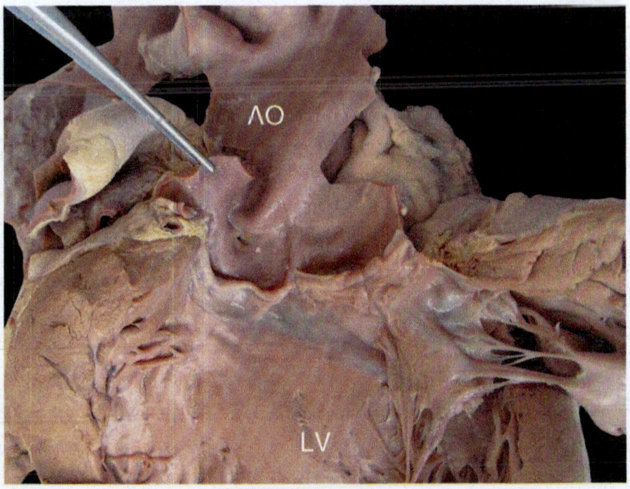

Fig. 9.35 Specimen 855. Close-up view

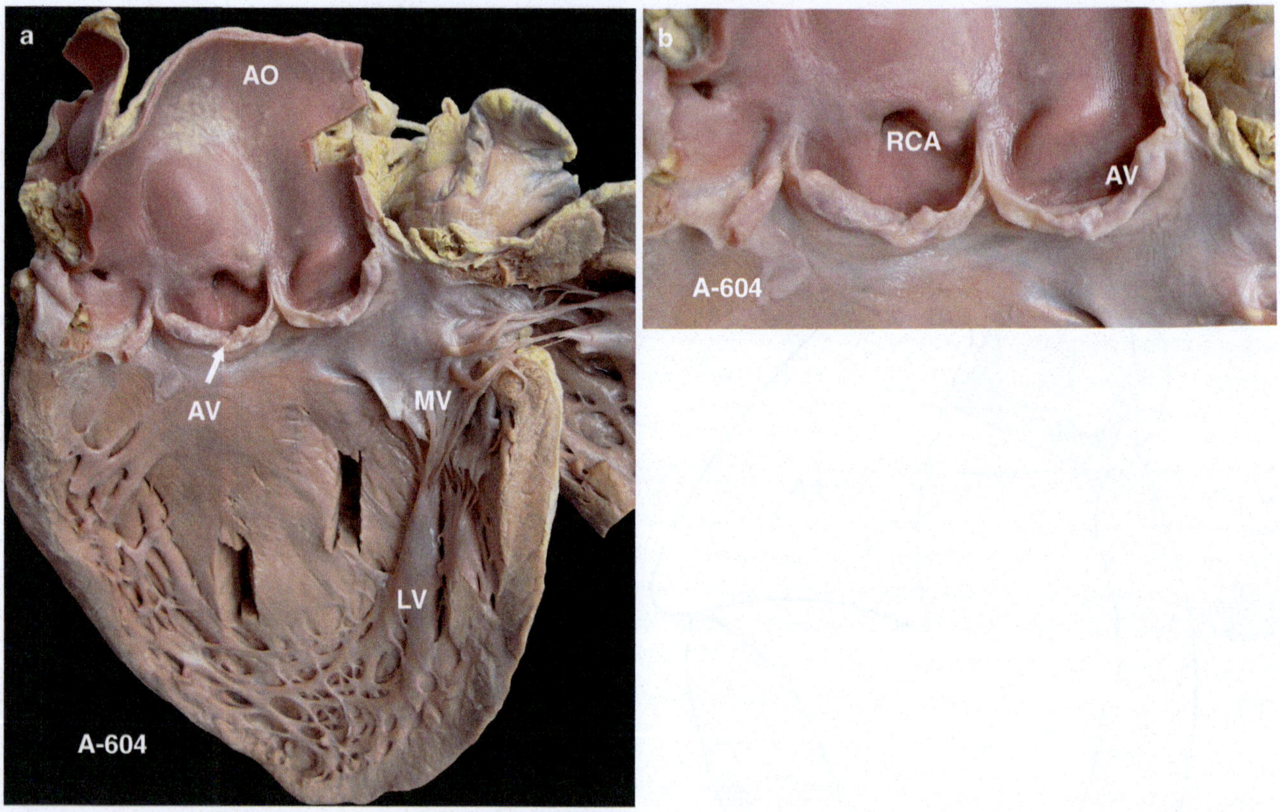

Fig. 9.36 (**a**) Specimen 604a. The leaflet crimps. (**b**) Specimen 604a, close-up view

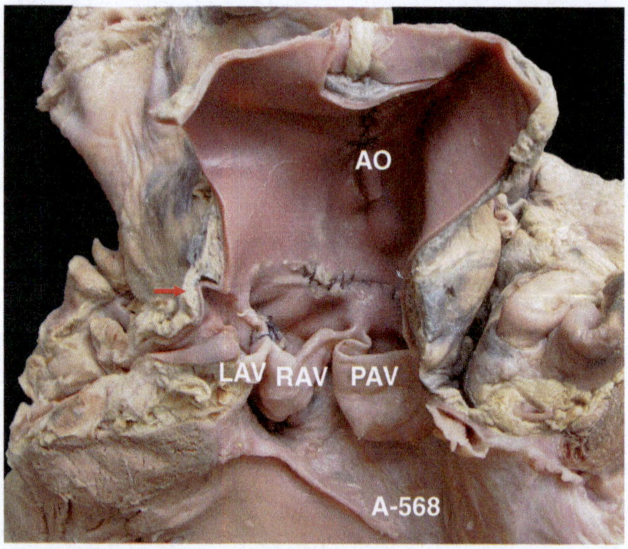

Fig. 9.37 Specimen 568. Prolapsed aortic valve with single coronary artery after plasty. *LAV* left leaflet, *RAV* right leaflet, *PAV* posterior leaflet

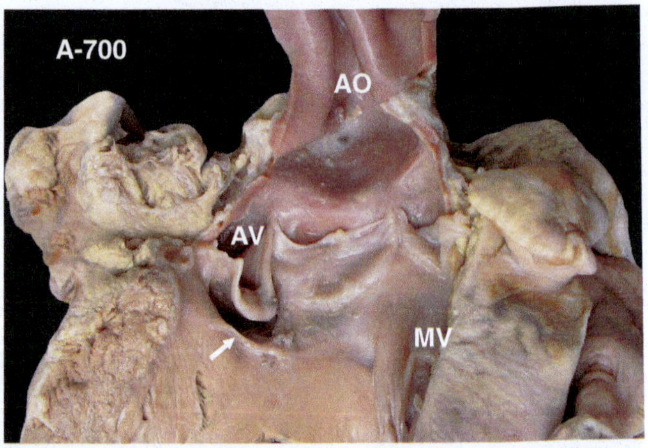

Fig. 9.38 Prolapsed aortic valve and VSD (*arrow*). *AV* aortic valve, *MV* mitral valve

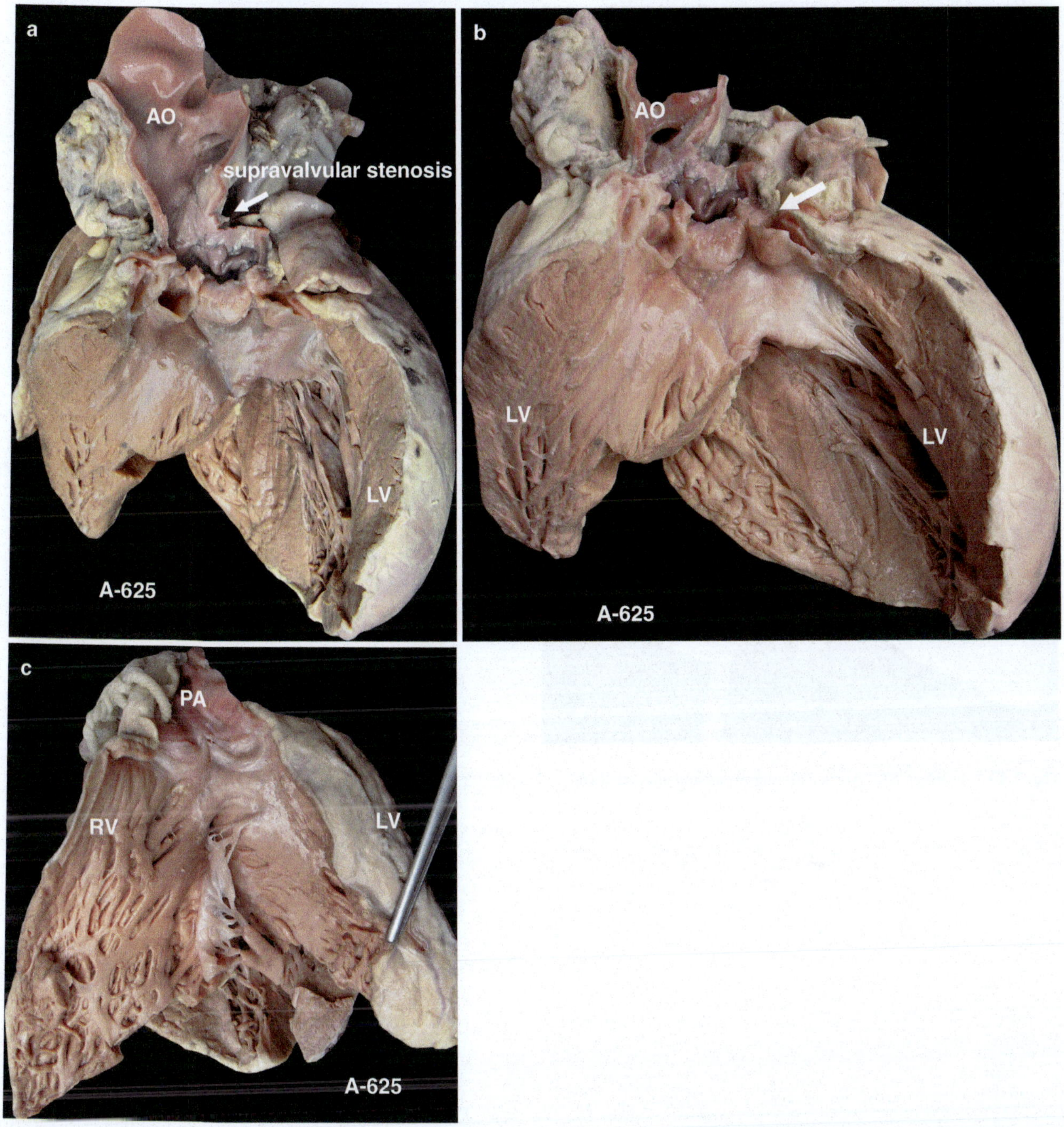

Fig. 9.39 (**a**) Specimen 625. Aortic stenosis and supravalvular stenosis (*arrow*). The ascending aorta has expansion after stenosis. The LV has thickening, and the RV has thinning, and a small ring. *AO* ascending aorta, *LV* the left ventricle. (**b**) Specimen 625. The leaflet thickening (*arrow*), part of the juncture mixing together, a small ring, and the LV highly thickened. (**c**) Specimen 625. View from the right ventricular outflow tract. The LV is thickening, and the RV is thinning. *RV* right ventricle

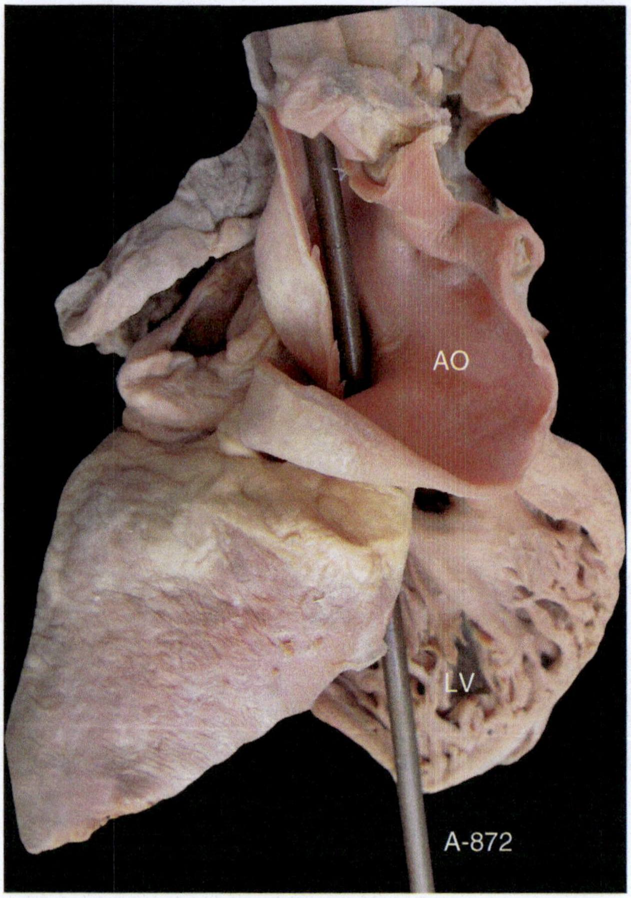

Fig. 9.40 Specimen 872. Aortic valvular expansion after stenosis

10 General Considerations of Cardiac Embryology

10.1 General Considerations

The primitive cardiovascular system begins forming during the fourth week of human embryonic development. The heart begins to beat, and blood begins to circulate through the primitive heart tube. The blood originates from the cephalic dorsal aorta and flows back to the heart tube through the common cardinal vein. The tube-like primitive heart tube will develop into the complete cardiovascular system step by step. Abnormity during the development of the embryo circulatory system leads to various cardiovascular abnormalities.

The primitive heart tube gradually develops into three layers: endocardium, myocardium, and epicardium. The pericardial sac originates from the cranial end of the primitive heart tube. After the cranial end folds toward the ventral side, the pericardial sac approaches that side of the heart tube, after which the pericardium surrounds the heart tube from the ventral to the dorsal side. Hence, the pericardial reflection is located on the dorsal side around the proximal portion of the main vessels. A more detailed description of the development of the primitive heart is provided in the discussion below.

The development from the primitive heart tube to the mature heart includes four aspects. The *first aspect* is the determination of heart chambers and the roots of the main arteries. This process includes the elongation of the primitive heart tube, segmentation, and looping. After all of these developmental processes are completed, the three-dimensional structure of the heart is formed. Abnormal determination of the heart does not always result in abnormal blood flow. For example, when malposition of the great arteries occurs, the AV lies in front of the pulmonary artery but the aorta still can receive blood flow from the LV as normal. The *second aspect* involves the connections of the heart chambers and the roots of the main arteries. If abnormalities in these connections result in an abnormal circulating system, an active correction is required. The *third aspect* is the septation of the heart chambers and the roots of the main arteries. This process includes the septations of the conus arteriosus, the ventricles, the atrioventricular canal, and the atria. The *fourth aspect* involves the maturation of the heart chambers, which includes the enlargement of the heart chambers, the formation of the heart valves, and the development of the cardiac conducting system.

X. Zhu (ed.), *Surgical Atlas of Cardiac Anatomy*,
DOI 10.1007/978-94-017-9409-1_10, © Springer Science+Business Media Dordrecht and People's Medical Publishing House 2015

10.2 Heart Chambers and Roots of Main Arteries

10.2.1 Formation of the Primitive Heart Tube

The primitive cardiac cells (cardiac progenitors) develop to form a primitive single chamber, the linear heart tube. The cranial end of the primitive heart tube is called the *truncus arteriosus* (TA). The primitive ventricle is caudal to the bulbus cordis and the vitelline veins (from the yolk sac). As described above, the pericardial sac moves from the caudal side to the dorsal side of the primitive heart tube and further develops to enclose the primitive heart tube. The pericardial reflex line lies outside the primitive heart tube (Fig. 10.1).

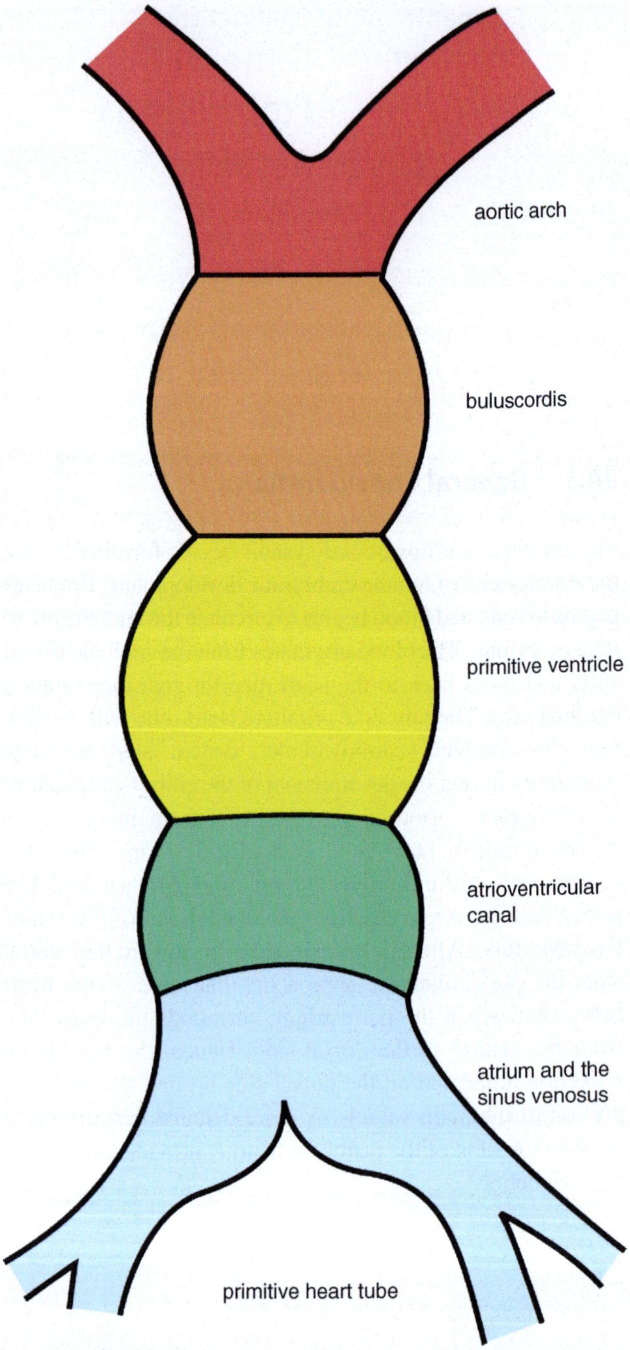

Fig. 10.1 Primitive heart tube. The primitive heart tube initially is bulged and segmented. The bulbus cordis develops into the truncus and conus, and the AV canal is shortened into the atrioventricular foramen. The sinus venosus is merged into the atrium

10.2.2 Segmentation of the Primitive Heart Tube

The primitive heart tube elongates and loops in the pericardial sac and develops many constrictions outlining future structures. The cranial most area is the *bulbus cordis*, and the ventricles are caudal to it. The AV canal is the caudal-most structure of the tubular heart. The atrium and sinus venosus are not formed yet; they only connect to the AV canal and the vitelline veins. The primitive atrium and sinus venosus lie outside the caudal end of the pericardial sac.

In the development of the primitive heart tube, the bulbus cordis divides into two segments: the *truncus*, or front, side and the *conus*, or posterior, side. These two segments together form the *conotruncus*. The primitive atrium and sinus venosus merge into one chamber, which forms a part of the primitive heart tube. The primitive heart tube is divided into six segments at this point: conus, conotruncus, ventricles, AV canal, atrium, and sinus venosus. The atrium and sinus venosus move upside into the pericardial sac. The conotruncus connects to the first aorta arch at the cepharal end, and the sinus venosus receives the umbilical veins, the vitelline veins, and the common cardinal veins at the caudal end.

10.2.3 Looping of the Primitive Heart Tube

The heart tube elongates and bends in the pericardial sac. This looping process occurs in three directions: right, caudal (posterior), and ventral.

10.2.3.1 Right Direction

The ventricular segment of the bulbus cordis bends to the right to form the bulboventricular loop. The most extensive bending of the ventricle and the bulbus cordis occurs here as the tube forms a convex curve to the right front side. The internal chamber of this connection is called the *ostium bulbi*, which functions as the outlet of the primitive ventricle. The cepharal end of the ostium bulbi is the conus, which develops into the outlet of the ventricles. The caudal end of the ostium bulbi develops into the RV. The D-looping process of the bulboventricular loop places the ventricles of the mature heart in the right positions.

10.2.3.2 Caudal Direction

As the bulbus cordis bends to the right, the atrium and the sinus venosus at the caudal end of the AV canal bend to the dorsal side of the ventricles.

10.2.3.3 Ventral Direction

The atrium and the sinus venosus move to the cepharal end and then to the dorsal end of the conotruncus (backward of the ventricles), where the atrium of the mature heart localizes. During this upside bending process, the atrium and the sinus venosus elongate, enlarge, and, for the most part, merge into the pericardium.

After the bending and movement are completed, the heart has a smooth lining except for the area just proximal and just distal to the bulboventricular foramen, where trabeculae form. The primitive ventricle will eventually develop into the LV, and the proximal portion of the bulbus cordis will form the RV. The distal part of the bulbus cordis, an elongated structure, will form the outflow tracts of both ventricles, and the truncus arteriosus will form the roots of both great vessels. The bulbus cordis gradually acquires a more medial position due to the growth of the RA, forcing the bulbus to be in the sulcus between the two atria (Fig. 10.2).

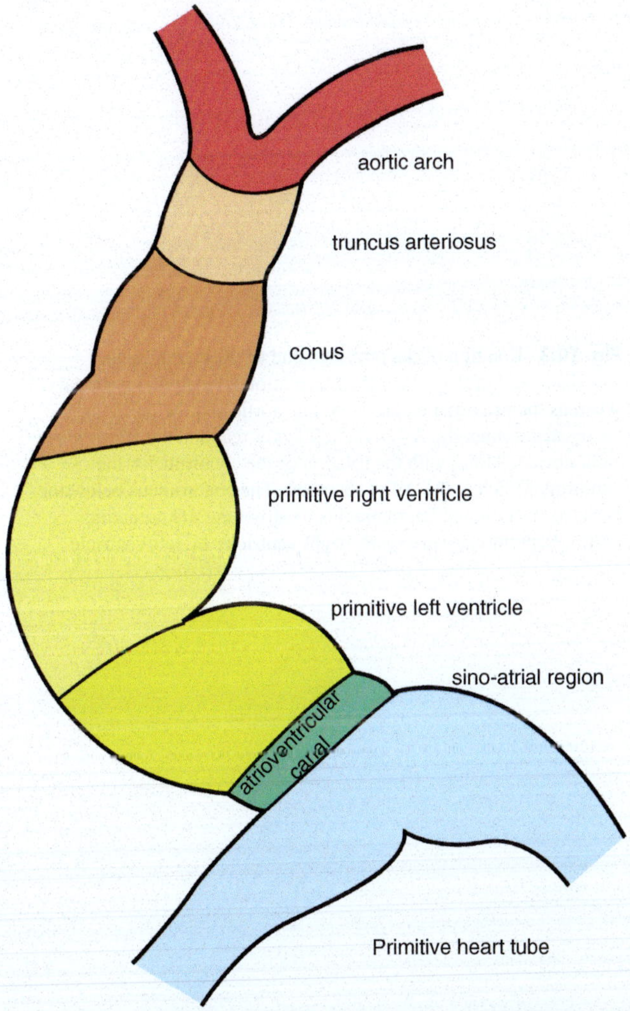

Fig. 10.2 Bending and D-looping of the heart tube. The conus and ventricles elongate and bend to the right, which is called *D-looping*. The ostium bulbi moves to the right side, and the primitive ventricle lies below. The middle part of the primitive ventricle sprouts into the interventricular ostium, which divides the left and right ventricles. The primitive atrium moves backward and to the upside

10.2.4 Rotation of the Proximal End of the Conotruncus

The conotruncus septation divides the conotruncus into two parallel chambers. The distal end of the conotruncus connects to the great arteries; the main artery on the right side becomes the aorta, and the main artery on the left side develops into the pulmonary artery. The proximal end of the conotruncus is the ostium bulbi, and this structure rotates clockwise approximately 90°–110° along the longitudinal axis (Fig. 10.3). At this point, the distal end of the conotruncus and the aortic arch are relatively stable. Due to this rotation, the previous parallel blood flow becomes spiral blood flow. The conotruncus below the aortic valve rotates to the left rear, and the conotruncus below the pulmonary valve rotates to the right front. Concomitantly, the conotruncus itself becomes shortened and is absorbed, with most of the portion below the aortic valve disappearing (Fig. 10.4).

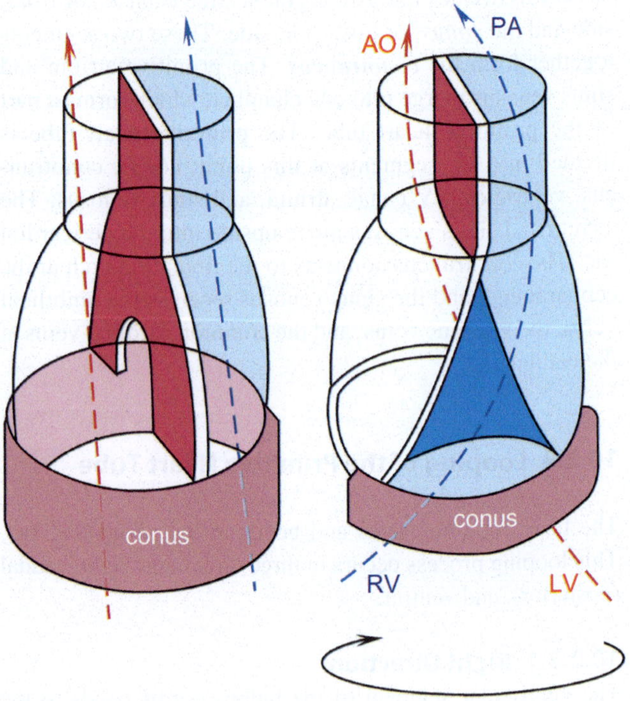

Fig. 10.3 Rotation of the proximal end of the conotruncus. The distal end of the conotruncus septation is relatively stationary, whereas the proximal end rotates. The aortic and the pulmonary valves also rotate: the AV to the left lower rear and the PV to the right upside. Along with the rotation of the conotruncus, the conotruncus below the AV is absorbed. The conotruncus below the PV is also shortened but maintains its structure. *AO* ascending aorta, *PA* pulmonary artery, *RV* right ventricle, *LV* left ventricle

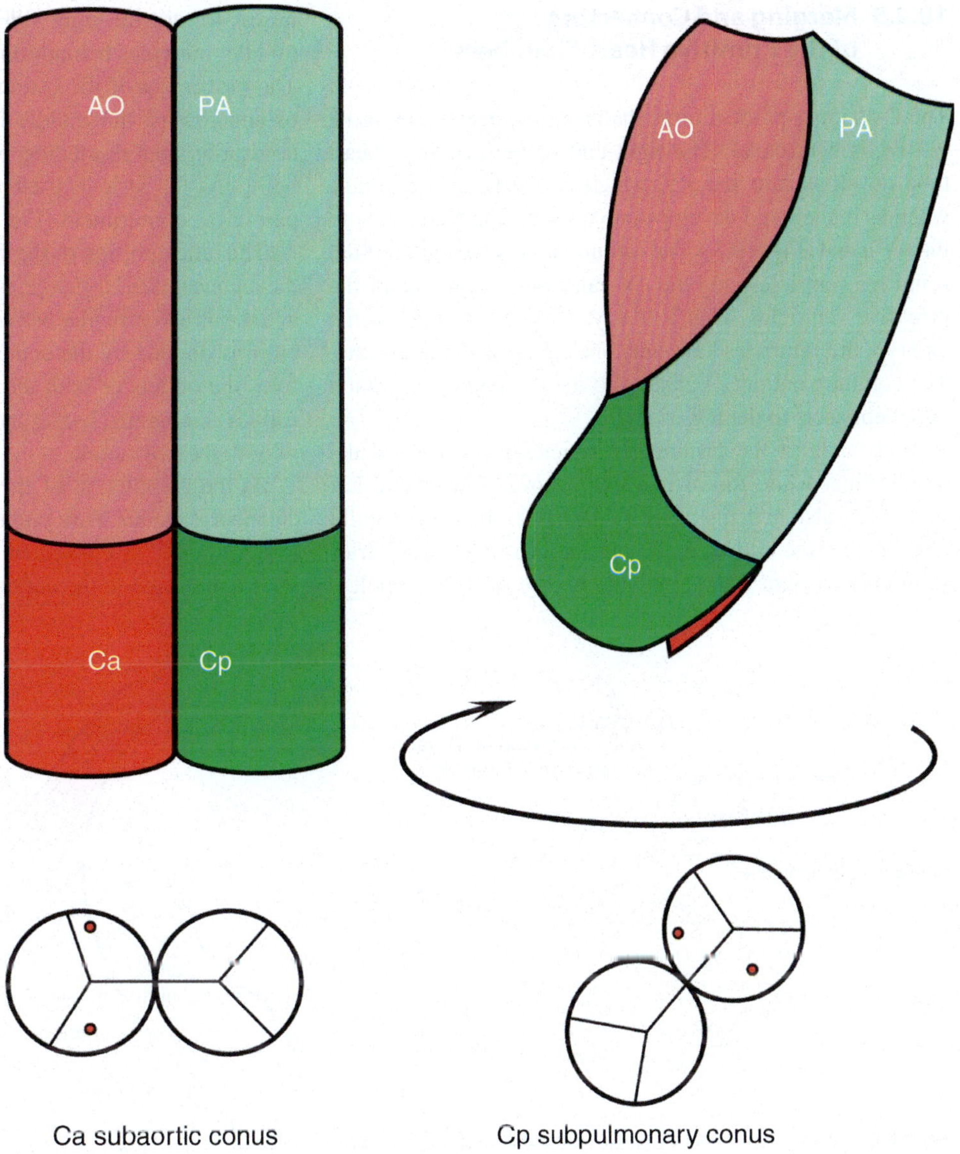

Fig. 10.4 Rotation and absorption of the proximal end of the conotruncus. At first, the conotruncus is divided into two parallel chambers. The aorta and the subaortic conus (*Ca*) are located on the right side, whereas the pulmonary artery and the subpulmonary conus (*Cp*) are on the left side. Then, the aorta and the pulmonary artery remain stationary as the conus rotates. The conotruncus below the AV is absorbed at the same time. *AO* ascending aorta, *PA* pulmonary artery

10.2.5 Merging and Connecting of the Primitive Heart Chambers

The D-looping process turns the primitive heart tube into a left recumbent canal. The right end of the primitive heart tube develops into the RV, which connects to the conus, whereas the left end develops into the LV, which connects to the AV canal (Fig. 10.5). The ostium bulbi is the outlet of the primitive ventricles, and the AV canal is the entrance of the primitive ventricles. That is to say, at the early stage of the embryo, the ventricles have only one entrance and one outlet. The common entrance connects to the LV, and the common outlet connects to the RV.

Both ends of the primitive ventricles move toward the middle line, while the ostium bulbi moves toward the left side. After also arriving at the middle line, the ostium bulbi rides above (the cranial end) both the ventricles. The atrioventricular ostium (the common entrance of the ventricles) moves toward the right side and then, after arriving at the middle line, rides behind (dorsal side) both of the ventricles. The movement of the ostium bulbi and the atrioventricular ostium toward the middle line is the consequence of the absorption and shortening of the conotruncus. The ostium bulbi and the atrioventricular ostium merge at the bulging part of the conotruncus (Fig. 10.6).

The reflex portion of the ventricles and the conotruncus is the *conoventricular flange*, which separates the ostium bulbi and the atrioventricular ostium. The cranial end of the ostium bulbi is divided by the conus septation into two halves that form the outlets of both the chambers. The atrioventricular ostium is separated by the endocardial cushion into entrances of both the ventricles.

At the completion of the septation process of the heart chambers, the atria are connected to the inlets of both ventricles, the AV is connected to the outlet of the LV, and the PV is connected to the outlet of the RV.

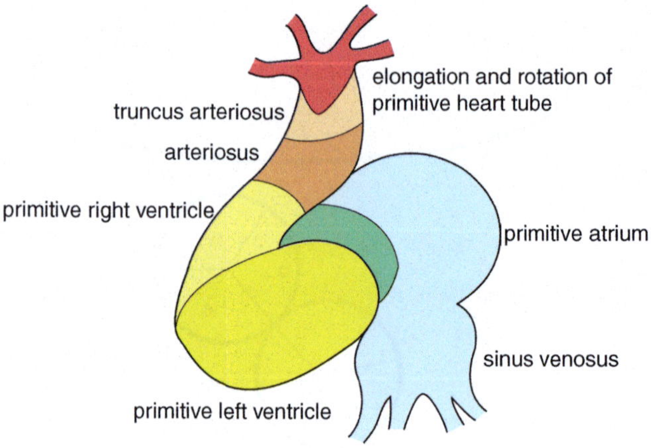

Fig. 10.5 Upward movement of the atria and the venous sinus. This figure shows that the primitive atria have moved to the dorsal side of the ventricles

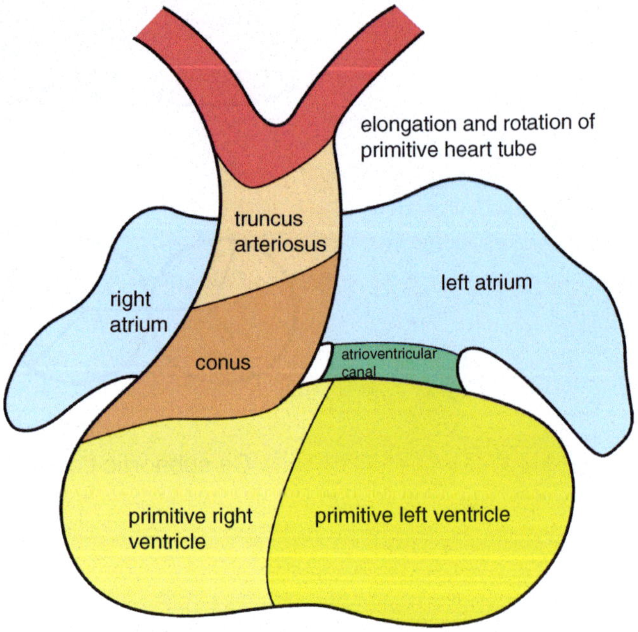

Fig. 10.6 Merging of the AV ostium, ventricular ostium, and conus ostium. The foramen between both ventricles, an intact ring-shaped structure, is called the *interventricular ostium*. As the elongation and bending of the heart tube occur, the heart chambers meet, and then the conus ostium moves to the upside of the interventricular ostium. The integrity of the cranial end of the interventricular ostium is then destroyed. The atrioventricular ostium meets the interventricular ostium from the dorsal side at the same time. In this way, the former intact ventricular ostium becomes a curved edge

10.2.6 Septation of Heart Chambers and Main Artery Roots

The primitive heart tube is a single-chambered canal. The cranial end is connected to the aortic arch, and the caudal end is connected to the sinus venosus. As the heart tube bends, its inner wall forms septations to result in a double-chambered tube, and two parallel blood flow systems are formed. The aortic arch at the cranial end is divided into the ascending aorta and the main pulmonary artery. The sinus venosus at the caudal end moves toward the right side to connect to the right heart system (Fig. 10.5). Because different heart segments have different patterns of development and septations, we describe this process in four sections.

10.2.6.1 Septation of the Conotruncus

The conotruncus is a single-chambered straight tube initially. Two parallel bulges, termed *truncus crests*, are formed in its inner wall as the truncus develops. These two crests grow and meet the *truncus* to form the conotruncus septum. Hence, the single-chambered conotruncus is separated into two parallel tubes: the ascending aorta and the main pulmonary artery (from the right side to the left side). The conotruncus septum is composed of the distal truncus septum and the proximal conus septum.

The proximal end of the truncus septum is very short. It and the endometrial tissue of the truncus together form two semilunar valves, the AV and the PV. Before the rotation of the conotruncus occurs, the aorta is located on the right side, and the pulmonary artery and PV are on the left side. After the proximal end of the conotruncus rotates, the AV moves to the left below the back side, and the PV moves to the right front upside. The ascending aorta and the main pulmonary artery change their parallel positions into a spiral-twining relationship.

The conus septum is composed of two longitudinal conus crests and is connected to the truncus septum. It separates the conus segment into the subpulmonary conus (the outlet of the RV) and the subaortic conus (most part absorbed). After the rotation is completed, the subpulmonary conus is located in the right front side, and the subaortic conus is located in the left back side.

Then, the subpulmonary conus and its septum are partially absorbed and shortened. The subpulmonary conus is not entirely absorbed, so in the mature heart, a complete subpulmonary muscular conus structure is still present. Most of the subaortic conus is absorbed, so no complete subaortic muscular conus structure is observed in the mature heart. A result of the absorption of the subaortic conus is that the AV moves downward to meet the fibrous tissue of the MV annulus.

The conus septum makes up the border of the interventricular ostium, which grows downward to join the muscular part of the interventricular septum. The conus septum grows backward to join the atrioventricular endocardial cushion to form the membrane part of the interventricular septum. The subpulmonary conus maintains its tubular muscular structure and is located between the PV and the TV.

10.2.6.2 Septation of the Ventricles

The ventricular septum is composed of the intrinsic part, the outlet part, and the membrane part (Fig. 10.7). The intrinsic part is also called the *ventricular sinus part*, and it is developed from the muscular septum of the primitive heart. The outlet part is also called the *infundibulum part*, and it develops from the conus part. The membrane part has a small area and is the last closed part of the ventricular septum; it is composed of the muscular septum, the conus septum, and the endocardial cushion.

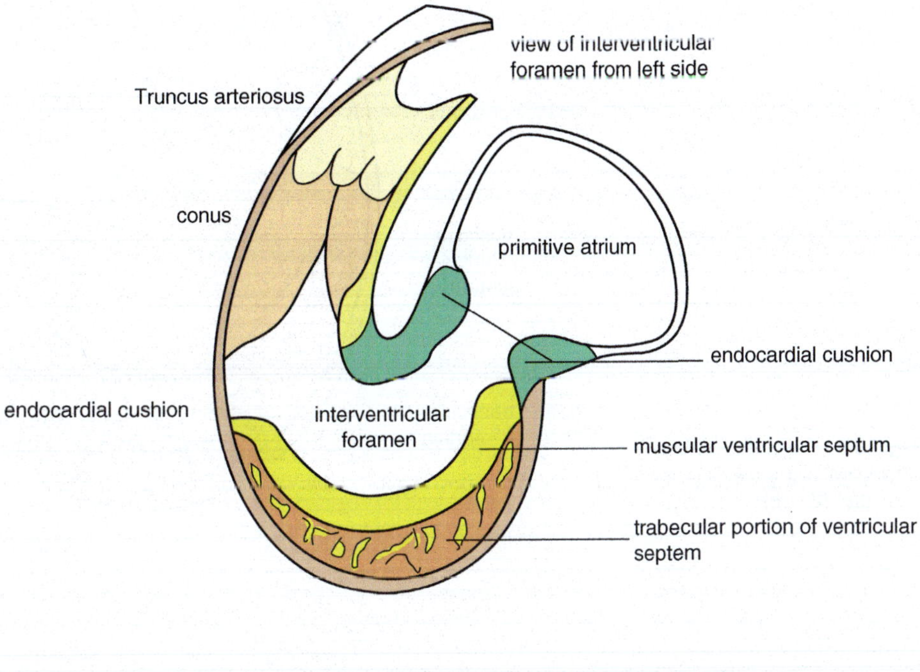

Fig. 10.7 The nearby structures of the interventricular foramen (view from the left side)

Muscular septum. The longitudinal muscular crest present in the middle segment of the inner wall of the primitive heart is called the *muscular septum*. It divides the heart chamber into the left and right ventricles. The ostium above the muscular septum is named the *interventricular foramen*.

Two mechanisms guarantee the formation of the muscular septum. *First*, the rising of the muscular crest narrows the interventricular foramen; this part of the muscular septum is very smooth and accounts for only a small area of the muscular septum. *Second*, the enlargement of the ventricular chambers enables the base of the muscular septum to grow. Because the enlargement of the ventricles comes from the irregular sinking and absorption of the inner muscles of the ventricles, this part of the muscular septum is uneven and forms many muscular trabeculae. This uneven septum is also called trabeculated portion of the muscular septum, which occupies a great area of the muscular septum to reduce the interventricular foramen. The muscular septum grows toward the cranial end to join the conus septum and backward to join the endocardial cushion.

Infundibulum septum. The infundibulum septum is composed of the subpulmonary conus, and most of the subaortic conus is absorbed.

Membrane septum. The development and formation of the membrane septum is the last step in the closure of the interventricular foramen. The muscular septum and the infundibulum septum account for most of the ventricular septum, and the membrane septum occupies only a small part of the ventricular septum.

10.2.6.3 Septation of the Atrioventricular Foramen

After the primitive heart tube bends to the right side, a canal named the *atrioventricular canal* is connected to the atria and the ventricles. The atrioventricular canal is shortened to form the atrioventricular foramen. The atrioventricular foramen initially is located behind the LV. As the primitive heart tube develops, the atrioventricular foramen moves from the left side to the middle line to ride on the back of the ventricles. Then, the dividing ventricles are just below (the ventral side) and the dividing atria are just above (the dorsal side) the atrioventricular foramen. The atrial septum and the ventricular septum are at the same level, which separates the atrioventricular foramen. Subsequently, the endocardium of the atrioventricular foramen bulges to form the endocardial cushion, which grows along the plane of the atrioventricular septum. The common atrioventricular foramen is separated into left and right atrioventricular foramen; the left atrioventricular foramen becomes the MV foramen, and the right atrioventricular foramen becomes the tricuspid foramen (Fig. 10.8). The left side and the right side of the inner wall of the atrioventricular foramen sprout a lateral atrioventricular endocushion, which, together with the central endocushion, forms the leaflets of the atrioventricular valve.

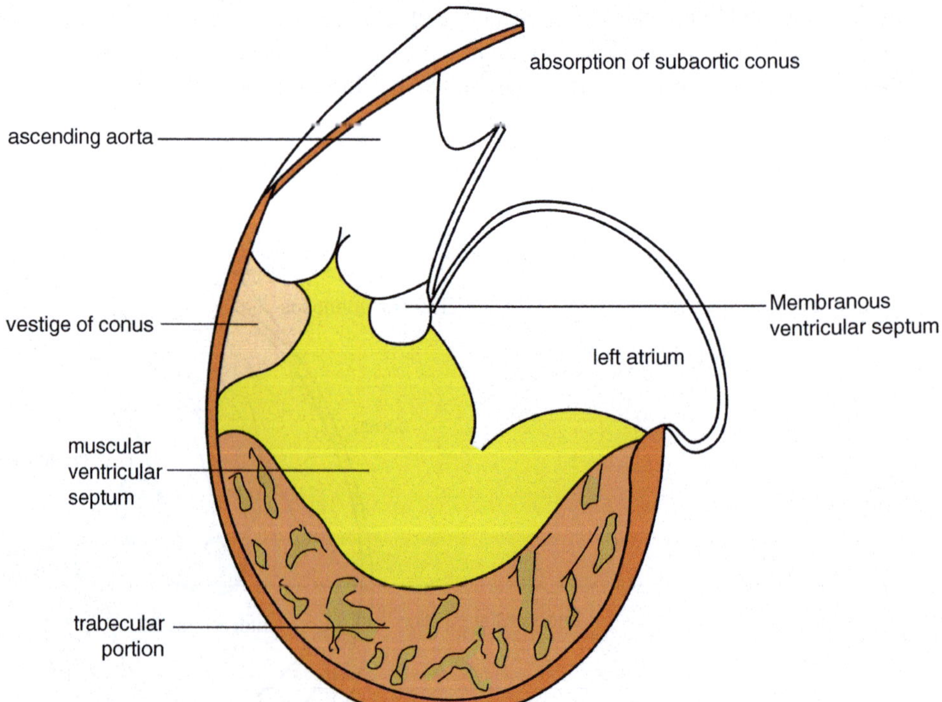

Fig. 10.8 The closure of the interventricular foramen (view from the left side)

10.2.6.4 Septation of the Atrium

As the primitive atrium moves to the back and upward of the ventricles, the cranial end of the inner wall of the primitive atrium sprouts a longitudinal crest that grows downward to form a septum named the *first septum*. The lower border of this crest extends to the atrioventricular foramen and joins the anterior and posterior atrioventricular endocushions to divide the primitive atrium into two halves. The sunken part of the first septum joins the central atrioventricular endocushion to form a communicating foramen named the *foramen primum*. The foramen primum closes completely, and simultaneously, the top is absorbed to form another foramen called the *foramen secundum*.

After the appearance of the foramen secundum, another septum (called the *second interatrial septum*) appears at its right side, which closes against the right side of the first septum. The second interatrial septum grows from the cranial end to the sinus venosus at the back of the atria, which also has a crest-shaped hollow. The second interatrial septum covers the foramen secundum, and its lower border forms the edge of the FO. The uncovered left first interatrial septum forms the base of the FO. The first interatrial septum at this location is membrane-like and covers the left side of the FO like a valve. When the pressure at the RA becomes high, the blood flow opens this valve toward the LA, and the blood flows from the second interatrial foramen to the LA. In this way, the systemic circulation of the fetus operates. After the birth of the fetus, the pressure of the LA becomes higher than that of the RA, and the FO is closed by the membrane-like interatrial septum to avoid blood shunt.

In the process of the formation of the interatrial septum, the sinus venosus moves to the right side to enter the RA, and it enlarges to become a portion of the RA. The sinus venosus includes the superior and inferior venae cavae and the inlet of the coronary sinus.

10.3 Development of the Heart Chambers

10.3.1 Development of the Atria

10.3.1.1 Formation of the Right Atrium and the Auricles

When the atrial septum forms, the inlet of the sinus venosus is separated into the RA. Valve-like bulges (named *sinus venosus valves* (SVVs)) are observed in the border of the inlet of the sinus venosus. The left and right SVVs join together to form a pseudo-atrial septum. In the third month of the embryonic development, most of the left SVV and the pseudo-interatrial septum are absorbed and fuse with the atrial septum. The right SVV is not absorbed completely, and a longitudinal muscular crest (termed *crista terminalis*) is still available. The crista terminalis is the borderline of the sinus venosus and the primitive RA. As the enlargement of the sinus venosus and fusion with the right atrial chamber occur, the sinus venosus itself becomes the major component of the mature RA. It is located behind the atrium and has a smooth inner wall and three venous openings (the inlet of the SVC, the inlet of the IVC, and the inlet of the coronary sinus). The primitive RA also forms the right auricle and the trabeculated anterior RA wall.

10.3.1.2 Formation of the Left Atrium and Auricle

The posterior wall of the primitive LA sprouts a diverticulum called *the pulmonary venous trunk*. The venula in the pulmonary bud also converge into four venous trunks, which join the pulmonary venous trunk to form the main body and posterior wall of the LA. The primitive LA in turn forms the left auricle and the trabeculated left lateral atrium wall.

The small venous network in the pulmonary bud is called the *pulmonary venous plexus*. Before draining into the LA, the venous plexus has communications with the systemic venous system. If the pulmonary venous trunk cannot drain into the LA and some communication veins are present, then abnormal pulmonary drainage occurs.

10.3.2 Development of the Ventricles and Valve Devices

10.3.2.1 Ventricles

The process the ventricular septum forms is also the process the heart chambers develop. The ventricles include the infundibulum portion, the sinus portion, and the trabeculated portion. The outlet of the RV has an intact tubular conus structure, whereas the outlet of the LV does not. The outlet of the LV and the sinus portion of the LV are separated only by the large valve leaflet of the MV. The enlargement of the ventricle chamber forms the sinus portion of the ventricle, which has a smooth inner wall. The trabeculated portion of the ventricle comes from the absorption of the muscles of the inner wall. The inner wall of the trabeculated portion has trabecular muscles and papillary muscles.

10.3.2.2 Atrioventricular Valve Devices

The valve annulus and the valve leaflet of both the left and the right heart come from the atrioventricular endocardial cushion. The central endocardial cushion forms the anterior leaflet of the MV and the septal leaflet of the TV. In the process of the absorption and enlargement of the inner wall muscles of the ventricles, some muscles become papillary muscles, and tendon chords are formed between the valve leaflets and the papillary muscles.

10.4 Development of the Conducting System

The conducting system of the heart is composed of muscular bundles with special conducting ability. The formation of this system is related to the development of the cardiac annular tissues in the primitive heart tube. The so-called cardiac annuli are the constricted rings among different segments of the primitive heart tube. They change their locations during the looping and bending of the heart tube, and the locations of the conducting system are in accordance with the final positions of the cardiac annuli (Fig. 10.9).

The sinoatrial tissue in the primitive atrium and the sinus venosus thicken at the inlet of the SVC to form the sinoatrial node. The SVV on the right side grows into the crista terminalis, the eustachian valve, and the Thebesian valve.

The atrioventricular annular tissue of the primitive atrium and ventricle grows from the backside of the atrioventricular groove to the anterior side. This tissue named as compact node lies on the top of the posterior interventricular septum and stretches to the posterior border of the interventricular foramen. It develops into the atrioventricular node, which lies in the posterior-inferior border of the central fibrous body. When the atrioventricular node does not develop well, the atrioventricular annular tissue can form a pre-atrioventricular node. The atrioventricular bundle from the atrioventricular node courses across the posterior-inferior side of the central fibrous body to bifurcate into the left and right bundles above the ventricular septum. The atrial septal tissues nearby the atrioventricular node and some atrial wall cells also have some ability to conduct the pulses of the heart. This area is called the *transitional cell zone*, and the Todaro tendon from the central fibrous body passes inside this area.

The right bundle is a long, slim structure that courses along the lower border of the supraventricular crest on the right ventricular side. The right bundle is thought to be a special tissue coming from the bulboventricular annulus. The left bundle lies spread on the left ventricular side, and it divides into the anterior and the posterior bundles (Fig. 10.10). The possible embryonic origins of this structure are the special tissues from the bulboventricular annulus and the special tissues from the atrioventricular annulus.

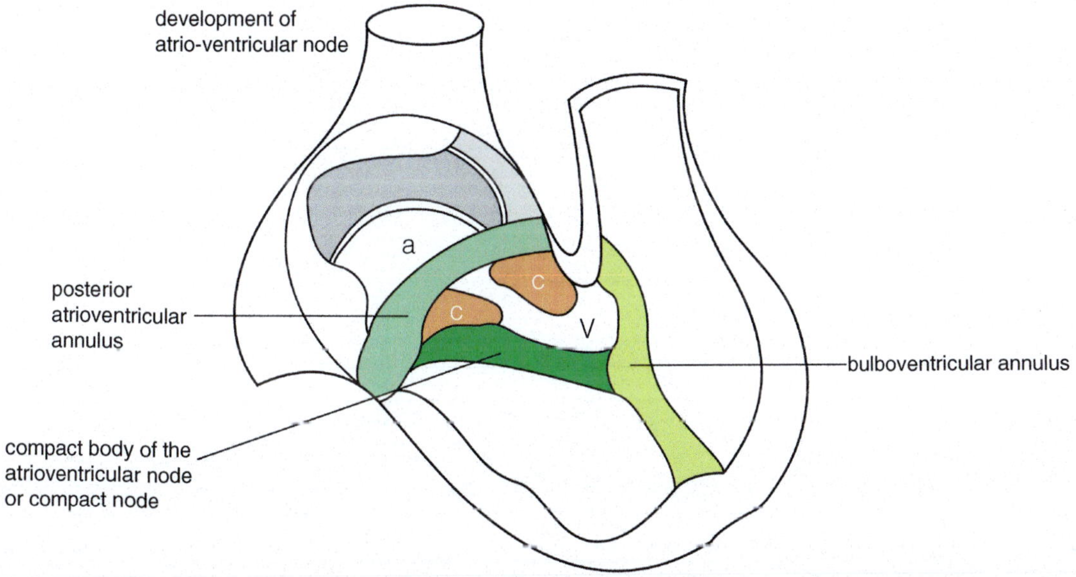

Fig. 10.9 The developmental process of the conducting system. As viewed from the left side, the first atrial septum (*a*) and the ventricular foramen (*v*) are not closed yet, and the central endocardial cushion is not joined (*c*). The special tissue of the posterior atrioventricular annulus develops into the ventricles, and it stretches along the upper border of the ventricular septum. This special tissue is called the *compact body of the atrioventricular node*, and it finally develops into the atrioventricular node and the atrioventricular bundles. The atrioventricular node that comes from the posterior atrioventricular annulus is also called the *posterior atrioventricular node*

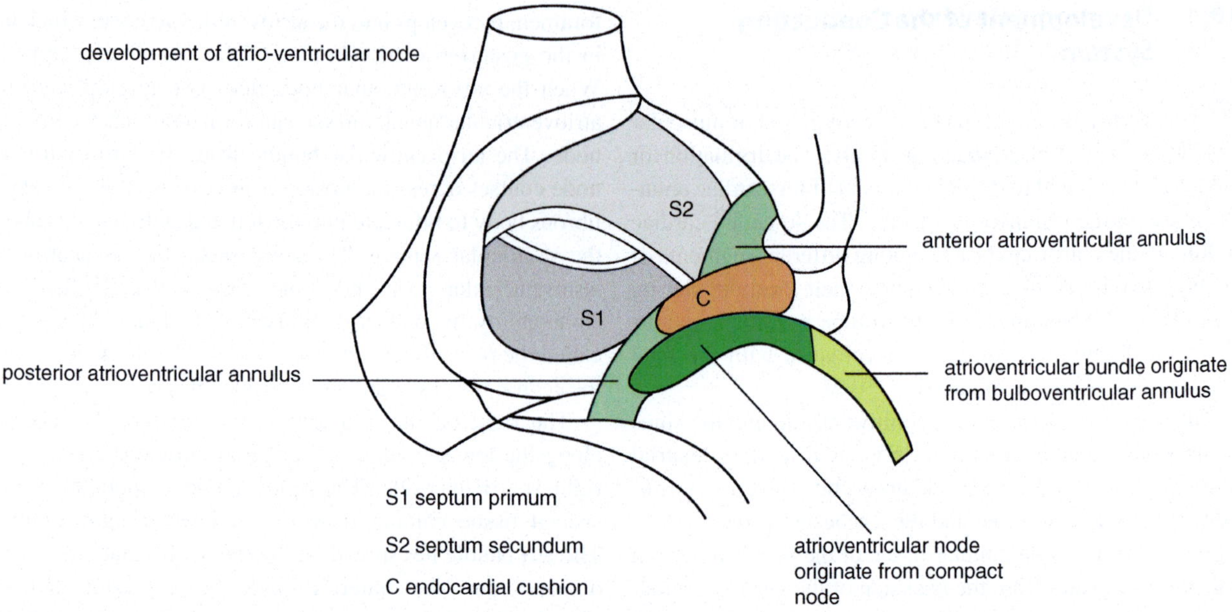

Fig. 10.10 The completion of the atrioventricular node and atrioventricular bundles. At the end stage of the embryo, the endocardial cushion is merged to form the left and right atrioventricular foramen. The first atrial foramen and the ventricular foramen are already closed. The compact body of the atrioventricular node develops into the atrioventricular node and the atrioventricular bundles. If the atrioventricular node is not well developed, the atrioventricular annulus tissue can form the pre-atrioventricular node. *S1* septum primum, *S2* septum secundum, *C* Cendocardial cushion

10.5 Embryonic Development and the Classification of Cardiac Abnormalities

Congenital heart abnormalities are due to defects that occur in the developmental process from the primitive heart tube to the mature heart. Hence, the classification and nomenclature of the congenital heart defects along the developmental process of the heart can clearly reflect the features of all of the abnormalities.

When classifying the abnormalities, the relative locations of the heart chambers and the valves of the great arteries and their connection mode should be considered. The abnormal 3-D structure can affect the surgical approach, and the abnormal connections of the great arteries and the ventricles should be corrected.

11 Atrial Septal Defect and Cor Triatriatum

11.1 Atrial Septal Defect

11.1.1 Formation of Atrial Septal Defect

An atrial septal defect (ASD) is an abnormality that occurs during the separation of the primitive atrium, leaving an unclosed interatrial foramen between the left and right atria.

A crescent-shaped thin wall called the *septum primum* is formed and divides the posterior and superior portions of the inner wall of the primitive atrium. It then grows downward toward the atrioventricular orifice and finally meets and merges with the central endocardial cushion. Until then, the single-lumen primitive atrium is initially separated into left and right atria. Before the septum primum completely merges with the endocardial cushion, the foramen separating them is called *foramen primum* or *ostium primum*. After the closure of the septum primum occurs, its roots are spontaneously absorbed into a perforation to maintain communication between the left and right atria; the perforation is called the *foramen secundum* or *ostium secundum* (Fig. 11.1). At the same time, the right side of the foramen secundum is covered by another septum, called the *septum secundum*, which grows from anterior to posterior. The septum secundum is crescent shaped with a depressed inferior border, which forms the edge of the FO. A certain interspace is kept between the septum primum and septum secundum and is called the *foramen ovale* (Fig. 11.2). During development, blood from the umbilical vein flows from the RA into the LA through the foramen ovale to provide blood supply for the fetus.

The septum primum is thin and comprises the floor of the FO, acting like a flap. The septum secundum is composed of thick muscular tissue and forms the anterior and superior border, also called the *Vieussens ring*, of the FO. If the septum primum fails to close, an ostium primum defect is formed (Fig. 11.3). In another case, if the foramen secundum is too large or left uncovered by the septum secundum, it is called *ostium secundum defect* (Fig. 11.3).

During the development of the atrial septum, the venous sinus and pulmonary veins are also in constant development and shifting. The venous sinus moves to the RA, dilates, and becomes part of it, forming the opening orifices of the SVC, the IVC, and the coronary sinus (Fig. 11.1) in the RA, respectively. The right border of the venous sinus forms the crista terminalis of the RA, the lower portion of which forms the Eustachian and the Thebesian valves. If shifting of the venous sinus is insufficient, then development of the atrial septum is affected and a sinus venosus ASD, also called a *superior vena cava defect*, is present and located at the posterosuperior portion of the atrial septum (Fig. 11.3).

11.1.2 Ostium Secundum Defect

11.1.2.1 Central-Type ASD

Central-type ASD, called the *fossa ovalis defect* (Fig. 11.4), is the most common type of the ostium secundum defects. It is located at the center of the atrial septum, which is also the position of FO. This type of ASD is surrounded by a complete atrial septal structure; it usually can be sutured directly or repaired by a patch under direct vision, without damaging adjacent important tissue.

11.1.2.2 Inferior Vena Cava-Type ASD

IVC-type ASD is less common. It is located at the posterior and inferior portion of the atrial septum (Fig. 11.2). It has no complete atrial septum border and continues into the orifice of the IVC, with the posterior wall of the atrium forming the posterior border of the defect. The inferior extremity of the IVC valve connects with the border of the defect. If the valve of the IVC is too large, it may be mistaken as the border of the defect; blood from IVC drains into the LA, resulting in right-to-left shunting.

11.1.2.3 Superior Vena Cava-Type ASD

SVC-type ASD is also less common. It is located at the posterior and superior portion of the atrial septum, without having a clear margin with the orifice of the SVC. The FO may still be in its normal position. This type of ASD is usually

X. Zhu (ed.), *Surgical Atlas of Cardiac Anatomy*,
DOI 10.1007/978-94-017-9409-1_11, © Springer Science+Business Media Dordrecht and People's Medical Publishing House 2015

associated with anomalous right superior pulmonary venous drainage. Direct suturing may cause stenosis of the entrance of the SVC.

11.1.2.4 Mixed-Type ASD

This type of ASD is a combination of more than two types of the giant ostium secundum defects mentioned above. The defect is usually too large for direct suturing; the stitches may experience too much tension and be torn, resulting in a residual shunt. Therefore, it may require patch repair.

11.1.3 Simple Ostium Primum Defect

The term *simple ostium primum defect* refers to a residual ASD anterior to the orifice of the vena coronaria without involvement of the atrioventricular valve or the fissure on the MV. The inferior border of the defect is the junction of the left and right atrioventricular rings, with its anterior close to the aortic wall and its posterior margin close to the atrioventricular node. A simple ostium primum defect is not commonly seen.

11.1.4 Simple Left Ventricle-Right Atrium Shunt

Simple left ventricle-right atrium shunt refers to a type of ASD located at the atrial part of the membranous ventricular septum, superior to the parietal septum of the triple valve, and inferior to the anterior of the MV. The defect is usually small in size, connected with the AV, and without abnormality of the atrioventricular valve. Because the atrioventricular valve is not involved but the LV and RA are, no consensus has been reached on the classification of this type of abnormality. In most cases of left ventricle-right atrium shunt ASD, direct suturing can be performed without a patch; however, care must be taken not to damage the AV.

11.2 Anatomical Characteristics of the Atrial Septum

The structure of the atrial septum and its relationship with peripheral tissue have great significance for the treatment of ASD. The FO (Fig. 11.1) lies in the center of the atrial septum, with its floor composed of membranous tissue. Thicker muscular tissue constitutes the edge of the FO, except for its posterior and inferior border, called the *limbus fossae ovalis*. The superior and inferior borders are termed the *superior limbus* and the *inferior limbus*, respectively. The superior limbus of the FO continues with the atrial wall muscle bundle of the SVC, and the inferior limbus with the Eustachian valve. The anterior and middle internodal bundle lies in the anterior border of the FO.

The anterior border of the atrial septum faces the AV, which is about the midpoint of the posterior aortic sinus. In cases such as transposition of the great vessels, the aorta moves anteriorly. The junction of the anterior and inferior portions of the atrial septum with the AV ring is the central fibrous body (also termed the *right fibrous triangle*), which courses along the inferior border of the atrial septum and extends posteriorly to form a long fibrous band called the *Todaro tendon*. At its inferior portion are the atrioventricular node and the orifice of the coronary sinus. The inferior border of the atrial septum is the MV ring, which forms the upper boundary of the midfeather of the atrial septum. The attachment edge of the parietal septum of the TV is much lower than the MV, which forms the lower boundary of midfeather. In cases of atrioventricular channel abnormalities, the MV ring descends to the plane of the parietal septum of the TV.

An ostium primum defect (Fig. 11.11a) lies anterior and inferior to the orifice of the coronary sinus, which is close to the atrioventricular node and the AV, whereas an ostium secundum defect lies more posteriorly and is not associated with lowering of the MV ring. This latter type of defect could be identified during intracardiac exploration as being posterior and superior to the orifice of the coronary sinus, which is a distance from the conduction tissue and, therefore, relatively safe for surgery.

11.3 Diagnosis of ASD

ASD is one of the most common congenital heart diseases and is often associated with other malformations. Diagnosis should be confirmed by exploring the RA, if necessary. The procedure for performing an intracardiac exploration and the specific areas to which attention should be paid are described below.

11.3.1 Confirmation of Presence

First, the presence of an ASD should be clearly confirmed, which can be accomplished only when the MV can be touched by a finger reaching through the defect; otherwise, a pulmonary vein trunk that drains into the RA or an enlarged orifice of the coronary sinus can be mistaken for an ASD.

11.3.2 Size and Type

Both the size and type of an ASD should be confirmed. A defect that lies posteriorly and superiorly to the coronary sinus is an ostium secundum defect; an ostium primum defect lies anteriorly and inferiorly to the orifice of coronary sinus.

11.3.3 Presence of Anomalous Pulmonary Venous Drainage

Whether there is anomalous pulmonary venous drainage should be confirmed, which is easily done when there is partial anomalous pulmonary venous drainage (PAPVD) but may be difficult when there is total anomalous pulmonary venous drainage (TAPVD). If the LA chamber is found to be too small by intracardiac exploration and no pulmonary vein orifice is identified, TAPVD should be considered.

11.3.4 Other Anomalies

11.3.4.1 Left SVC

The existence of a left SVC should be made clear. The coronary sinus is enlarged when there is a left SVC, which could also be seen from outside the heart going longitudinally and posteriorly to the left auricle.

11.3.4.2 Splits

Another anomaly is seen if there are splits on the left or right atrioventricular valve, or if a common atrioventricular valve has been formed.

11.3.4.3 Ventricular Septal Defect

The determination of whether a defect coexists with ventricular septal defect should be made.

11.4 Cor Triatriatum

Cor triatriatum (Fig. 11.16a) usually refers to a condition in which the LA is divided into two chambers by a fibromuscular septum. The chamber connecting with the pulmonary vein is called the *accessory left atrium*, and that with the MV is called the *real left atrium*. Blood from the accessory LA may flow through the pores on the septum into the real LA and may also flow into the RA through the ASD. During the embryonic stage, pulmonary veins first converge into a common trunk. Simultaneously, the primitive atrium divides into left and right atria. The common pulmonary vein trunk then fuses with the LA into an LA chamber. If the common pulmonary vein trunk fails to fuse with the LA, cor triatriatum occurs. Clinically, attention should be paid to determining whether there is communication between the accessory and real left atria through the septum. Attention also should be paid to the existence of an ASD and whether all pulmonary veins are connected with the accessory LA.

11.5 Specimen Demonstrations

11.5.1 Normal Atrial Septum and Patent Foramen Ovale

The aim is to introduce the anatomical characteristics of the normal atrial septum and patent foramen ovale. FO is an important mark on the right side of the atrial septum, which includes the FO, orifices of the superior and inferior venae cavae, and the orifice of the coronary sinus. Attachment edge of the septal leaflet of the TV forms the boundary of the RA. Common positions of an ASD are shown in Fig. 11.2. Specimen 1039 shows a tiny split in the FO, which is a patent foramen ovale but with limited shunting and surgical significance.

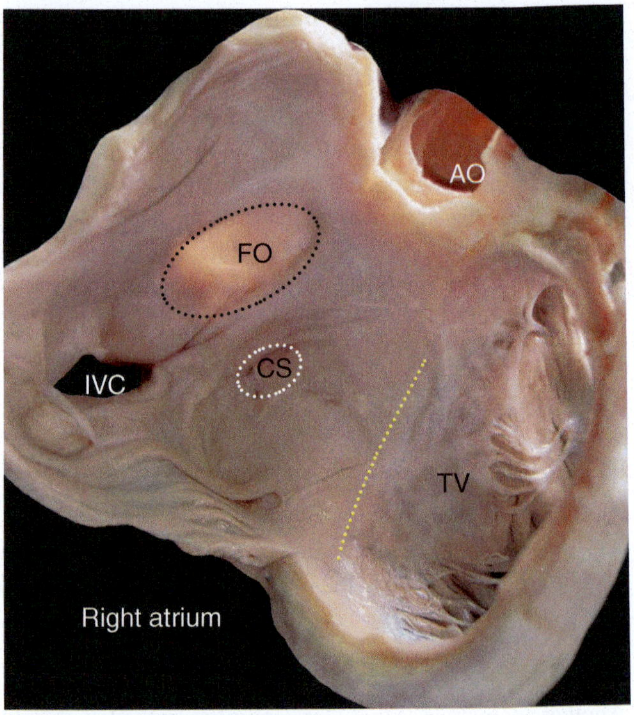

Fig. 11.1 Anatomical landmarks of the right atrium. Identification of the RA is usually made according to its morphological features, rather than the side on which it is located. If the atrium is connected with the superior and inferior venae cavae, with an orifice of the coronary sinus, it could be confirmed as the RA, with no consideration of which side on which it is located. This figure shows a dissected RA. The orifices of the IVC and coronary sinus can be seen. The AO lies anterior to the RA. The *yellow dashed line* marks the TV ring. *AO* aorta, *CS* coronary sinus, *IVC* inferior vena cava, *RA* right atrium, *TV* tricuspid valve

Fig. 11.2 Positions of different types of atrial septal defects. An ASD can form at any position of the atrial septum. In this figure, the RA is unfolded. Using the coronary sinus as the boundary, all defects superior to it are part of the ostium secundum defect. Defects anterior and inferior to the coronary sinus are all ostium primum defects, which lie next to the atrioventricular node. *C* central-type ASD, *CS* coronary sinus, *I* IVC-type ASD, *RA* right atrium, *S* venous sinus or SVC-type ASD

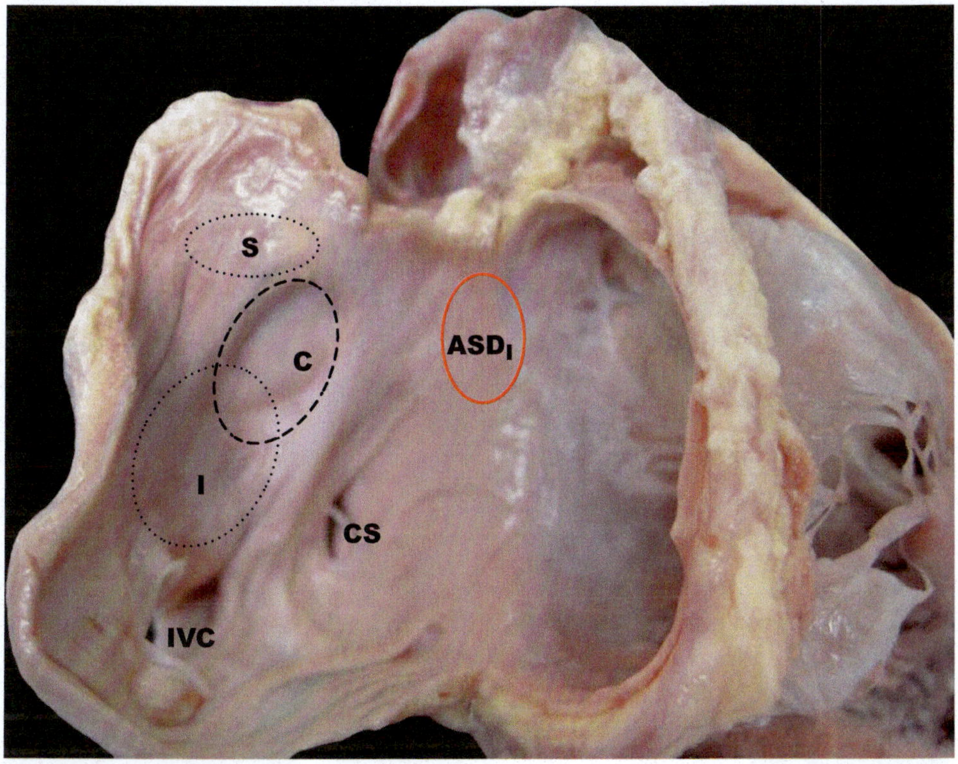

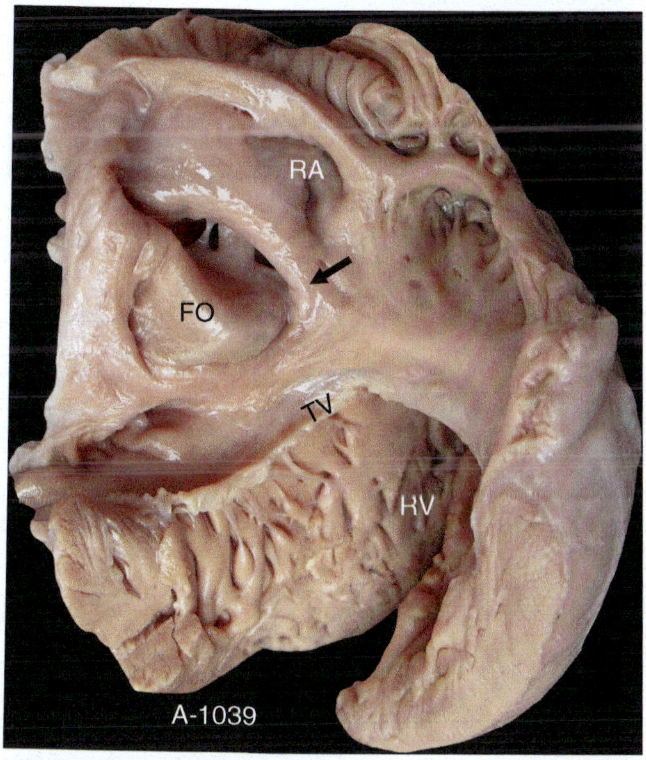

Fig. 11.3 Patent foramen ovale. FO shows the position of a fossa ovalis with a hole, through which it communicates with the LA. A muscle bundle lies at the anterior border of the FO. It is also called the *anterior limbus* (as the *arrow* indicates), which extends to the TV ring. The atrioventricular node is located at the junction of the anterior limbus of the FO and the TV ring. The anterior bundle of the internodal tract courses along the anterior limbus of the FO to reach the atrioventricular Node

11.5.2 Surgical Specimen of Central-Type Ostium Secundum Defect

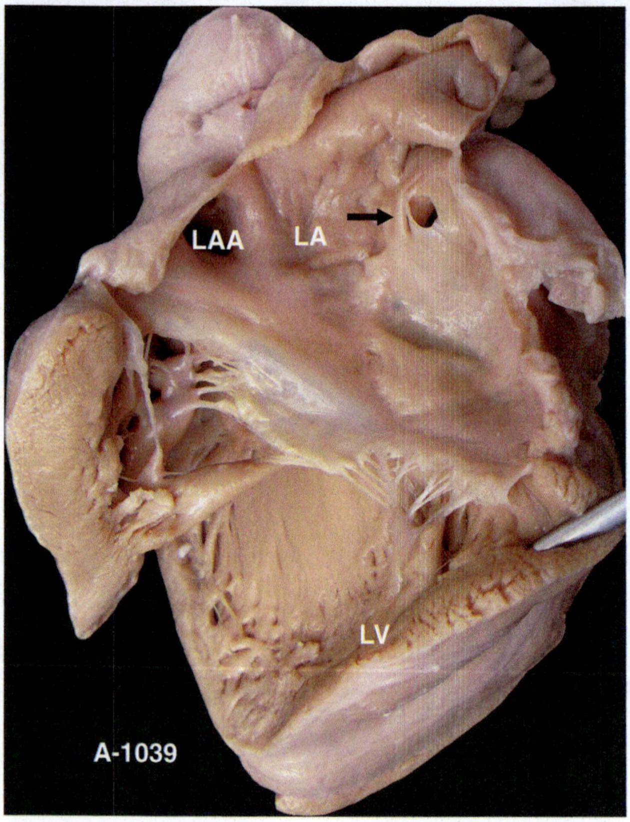

Fig. 11.4 Observing the fossa ovalis from the left atrium side. The *arrow* shows the fold anterior to the fossa ovalis, which is the trace of the absorption of the septum primum. *LA* left atrium, *LAA* left auricle, *LV* left ventricle

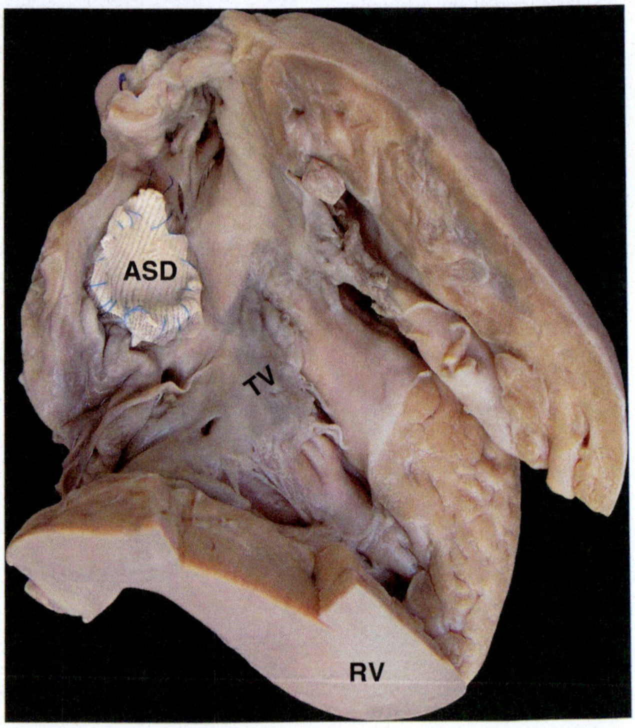

Fig. 11.5 Central-type atrial septal defect associated with double orifice of coronary sinus malformation. Repair with Dacron patch has been performed. Viewed from the RA side, two coronary sinus orifices could be seen inferior to the ASD. The defect is surrounded by complete edge. Viewed from the right side, the tissue anterior and inferior to the patch is close to the atrioventricular node, which requires attention during surgery to avoid damage. This type of ASD is also an indication for transcatheter interventional occlusion. *RA* right atrium, *TV* tricuspid valve

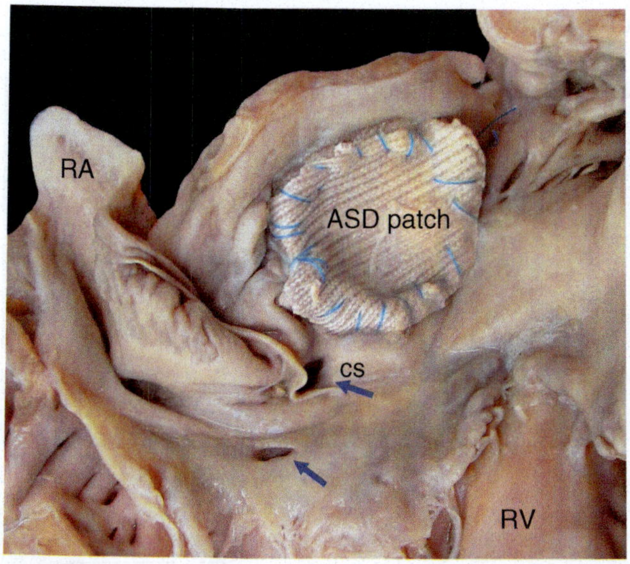

Fig. 11.6 A close observation of a central-type ASD. Same specimen as 11.5 above. Suturing of the patch is neat and tight. *Arrows* indicate the two CS orifices. *CS* coronary sinus

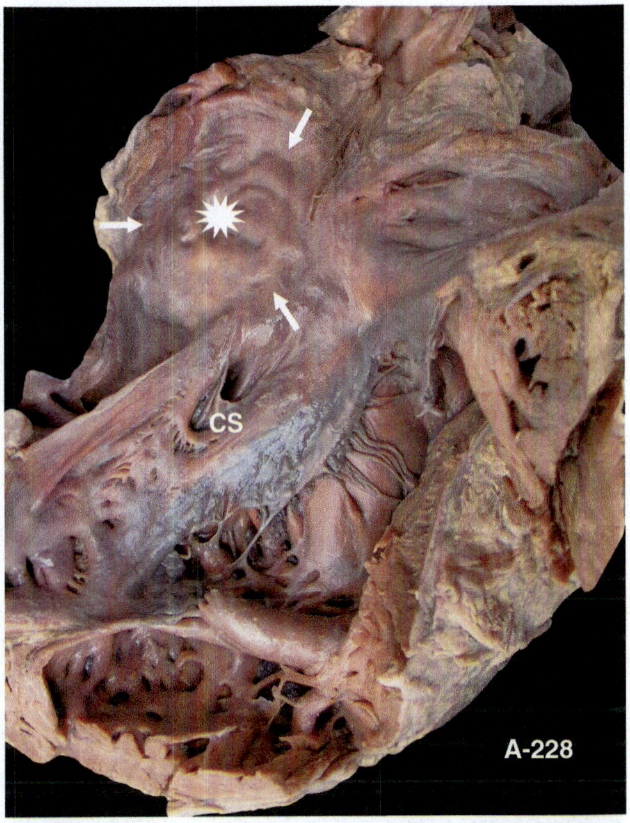

Fig. 11.7 Long-term result of atrial septal defect repaired with Dacron patch: endocardium completely covers the patch. The three *arrows* in the figure indicate the edge of the Dacron patch, with the *asterisk* symbol indicating the surface of the patch. Judging by its appearance, this part is not different from the rest of the inner atrial wall. Because of the covering of endocardium, the Dacron patch is not exposed and, therefore, poses no risk of thrombosis. *CS* coronary sinus

11.5.3 Atrial Septal Defect Associated with Ventricular Septal Defect

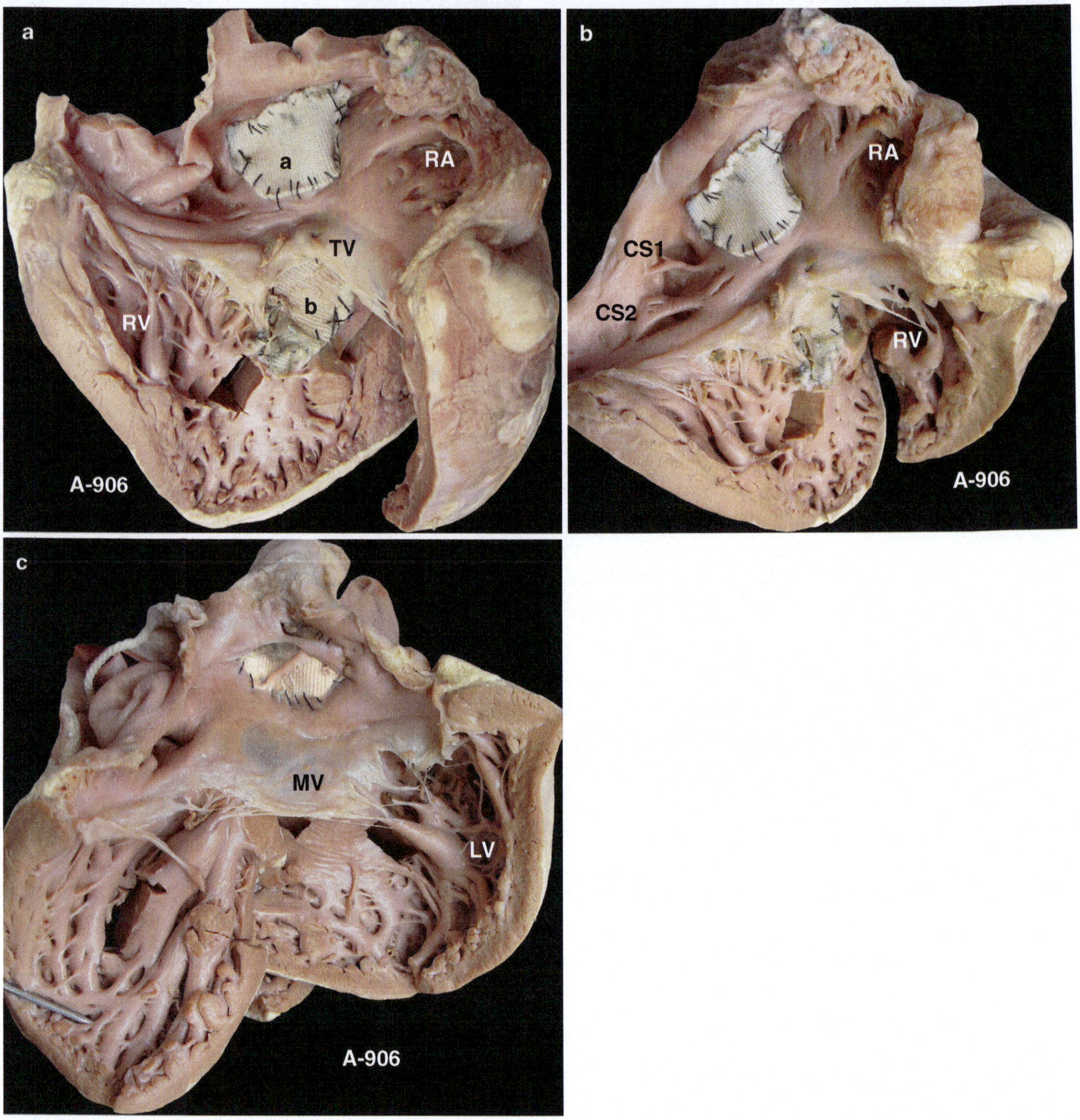

Fig. 11.8 (**a**) This specimen is a case of an ostium secundum defect (*a*) associated with a ventricular septal defect (*b*). This is a specimen after surgical repair, with the ASD superior to the CS, which is an ostium secundum defect. The VSD belongs to the perimembranous type, which is beneath the anterior of the septal leaflet of the TV. *CS* coronary sinus, *RA* right atrium, *RV* right ventricle *TV* tricuspid valve, *VSD* ventricular septal defect. (**b**) Same specimen, showing the entrance of double coronary sinuses (*CS1*; *CS2*). The patch of ASD and VSD could be seen with no abnormality of the TV. *VSD* ventricular septal defect, *TV* tricuspid valve. (**c**) Viewing from the dissection plane of the left atrium and ventricle. Suturing of the ASD patch is neat and tight. The VSD is blocked by the MV. *MV* mitral valve, *VSD* ventricular septal defect

11.5.4 Lutembacher Syndrome

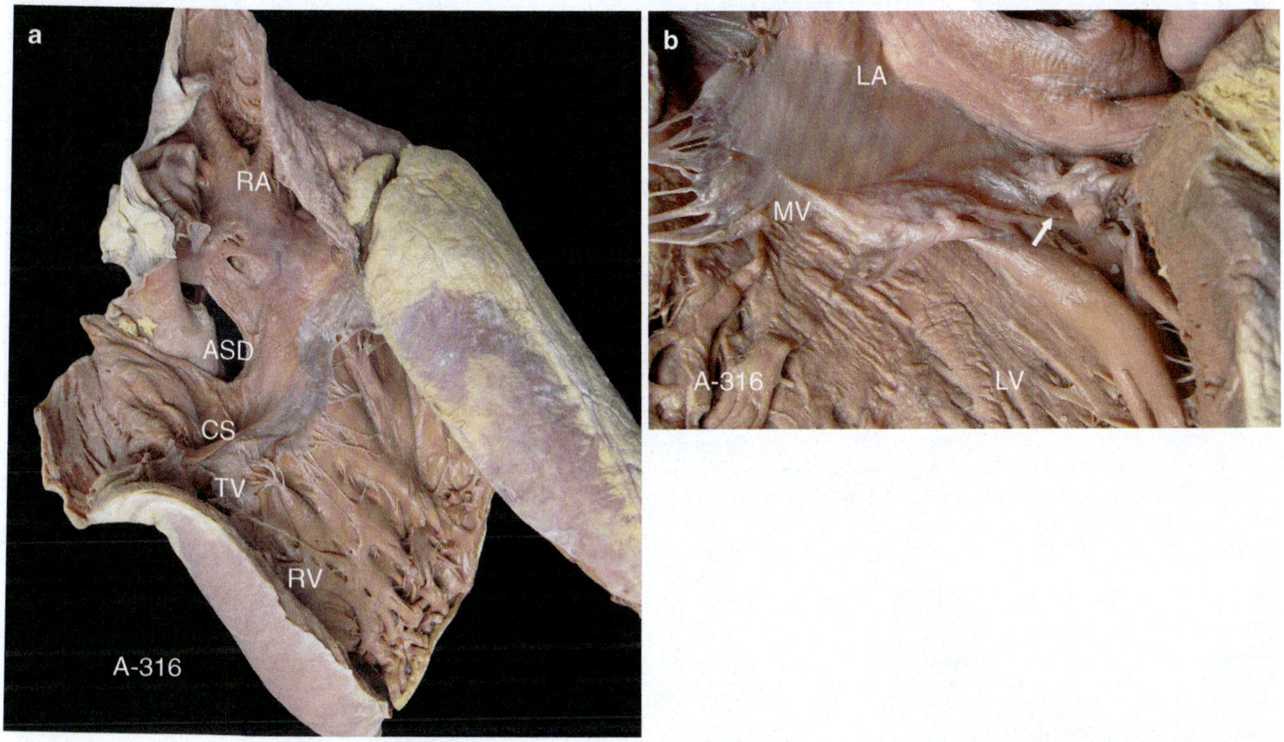

Fig. 11.9 (**a**) Atrial septal defect associated with mitral stenosis (*Lutembacher syndrome*). Mitral stenosis usually is an acquired disease, with the leaflet thickened and adhered. Mild enlargements of the RA and RV are seen. This figure shows a view from the right side, in which only a central-type ASD could be seen with an intact ventricular septum. *RA* right atrium, *RV* right ventricle, *TV* tricuspid valve. (**b**) Same specimen shown from the left side. Leaflets of the MV are thickened and adhered at the posterior commissure (*arrow*), which is similar to rheumatic disease, without calcification. *MV* mitral valve

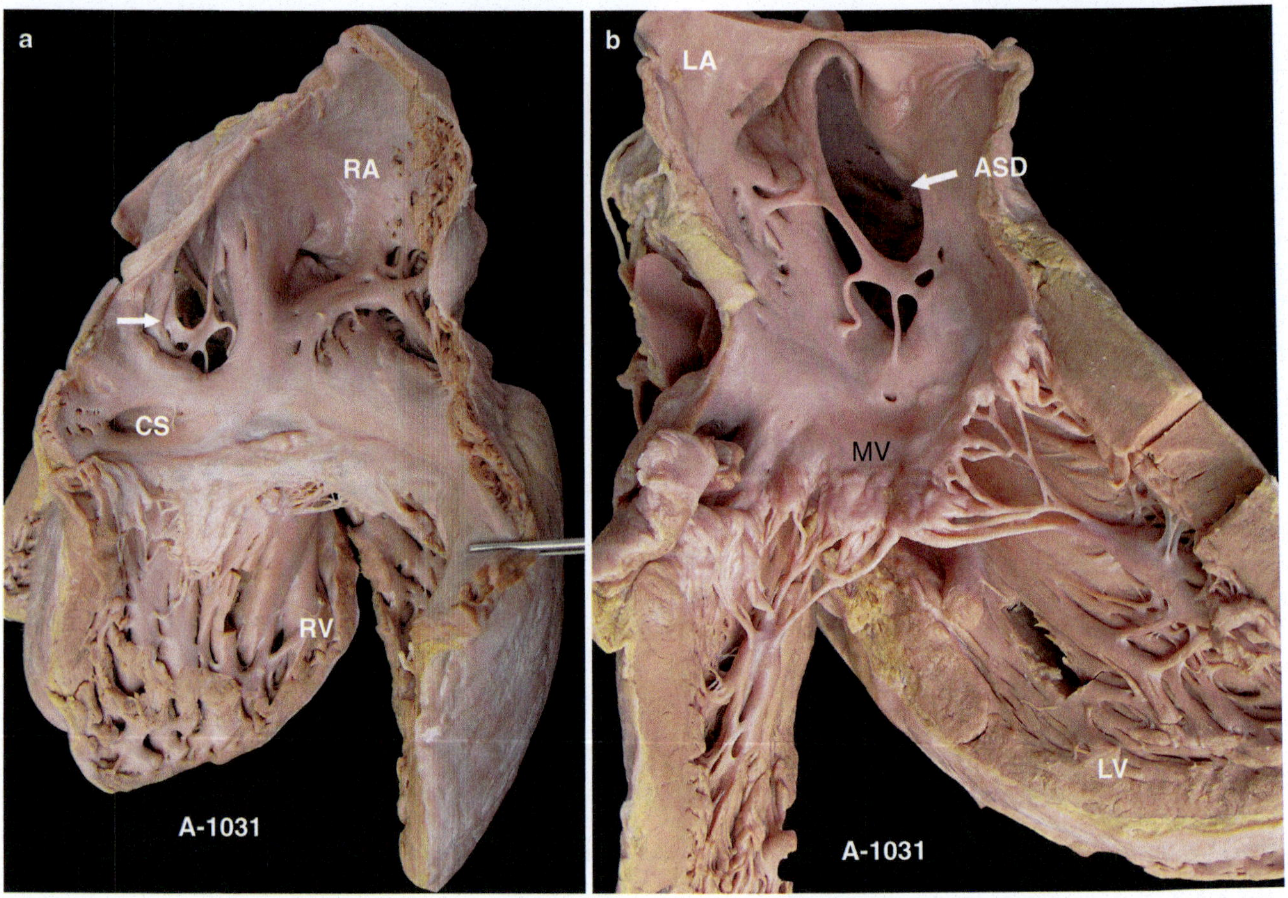

Fig. 11.10 (a) A central-type atrial septal defect shown from the side of the right atrium. Reticular cord could be seen in the middle of the defect (*arrow*), which is the trace of absorption of the atrial septum tissue. This reticular cord structure poses no interference with surgery but presents difficulty for transcatheter interventional occlusion. *RA* right atrium. (b) Specimen 1031, viewed from the left atrium side. *Arrow* shows the reticular structure in the defect. Attention: this case of ASD still has a certain distance from MV, which should not be mistaken for an ostium primum defect. *MV* mitral valve

Fig. 11.11 (a) Specimen 966. Ostium primum defect associated with perforation of aortic valve. This is an ASD specimen after surgical repair. The ASD is low in position, and the TV ring is involved (*arrow*); therefore, it could be confirmed as an ostium primum defect. To make sure the suturing is tight and to avoid damage to the conduction system, suturing of the inferior border of the defect is intentionally more to the right side, and the coronary sinus is separated to the LA side. The patch is composed of three pieces for purposes of tailoring, with the peripheral part tightly sutured. *LA* left atrium, *TV* tricuspid valve. (b) Same specimen viewed from the left atrium side. The giant ASD has been closed by patch. The inferior border of the defect is the MV ring, which is evidence of an ostium primum defect. The MV is normal without any split, which helps in excluding an AVSD. *AVSD* atrioventricular septal defect, *LA* left atrium, *LV* left ventricle, *MV* mitral valve. (c) View from the left side. The *red probe* is put in the coronary sinus, demonstrating that the coronary sinus has been separated to the LA side. The aim of this type of repair is to make sure the suturing edge is far from the atrioventricular node, in order to avoid damage to the conduction system. *CS* coronary sinus, *LA* left atrium. (d) Specimen 966. View of the perforation of the aortic valve from the outflow of left ventricle, which may be a congenital perforation (*arrow*). The inferior border of the ASD is the MV ring, indicating the MV ring had lowered to the same plane as that of the TV. The outflow of the LV is normal without stenosis. *LV* left ventricle, *MV* mitral valve, *TV* tricuspid valve. (e) Same specimen (966). A close observation of the aortic valve with perforation (*arrow*). This leaflet has no secondary changes such as thickening or transformation

11.5.5 Postsurgical Ostium Primum Defect

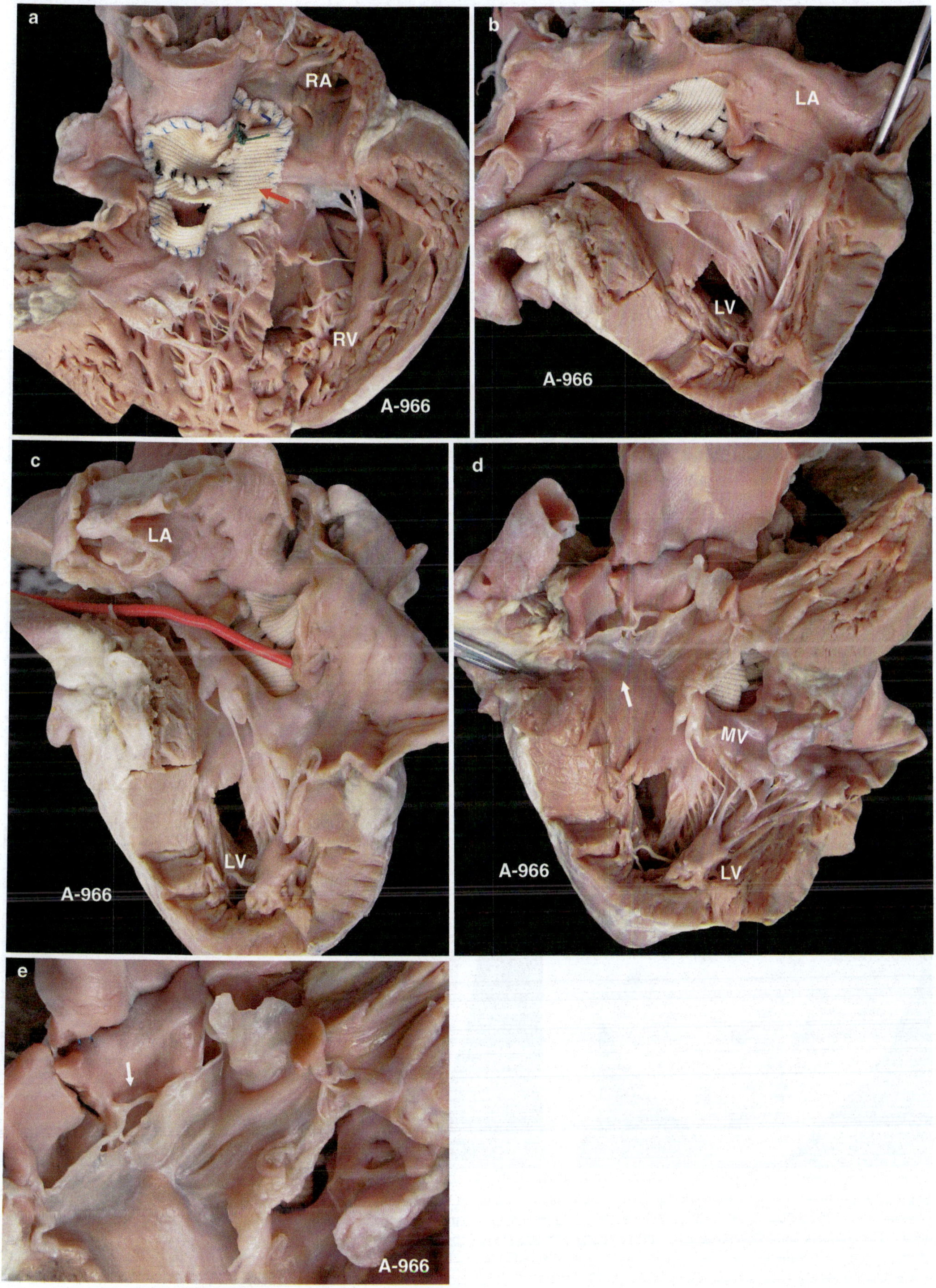

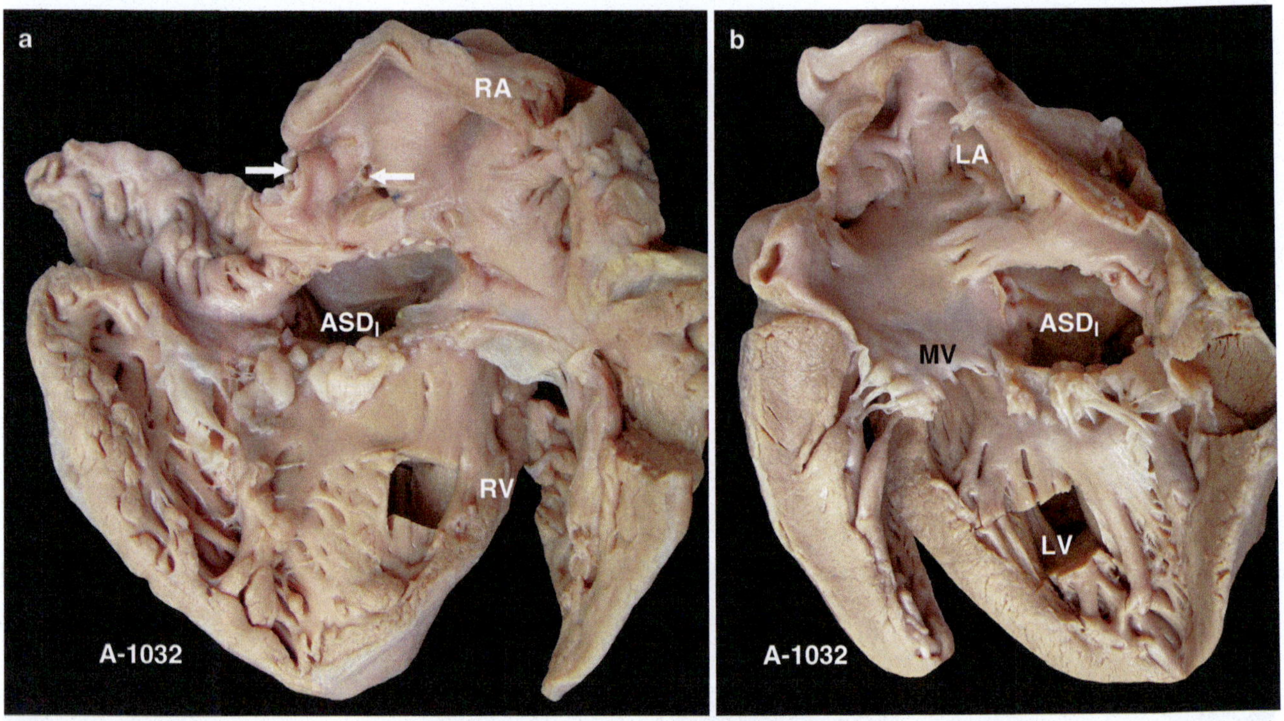

Fig. 11.12 (**a**) Typical ostium primum defect. Surgical patch has been removed. The inferior border of the defect is next to the TV ring, which forms the inferior edge of ASD. The FO could still be seen in the RA (between *two arrows*). The atrioventricular valve has no split. The LA could be seen through the defect. *LA* left atrium, *RA* right atrium, *TV* tricuspid valve. (**b**) Specimen 1032. Ostium primum defect shown from left atrium side. MV ring constitutes the inferior border of the defect, with the leaflet free of split. *MV*, mitral valve

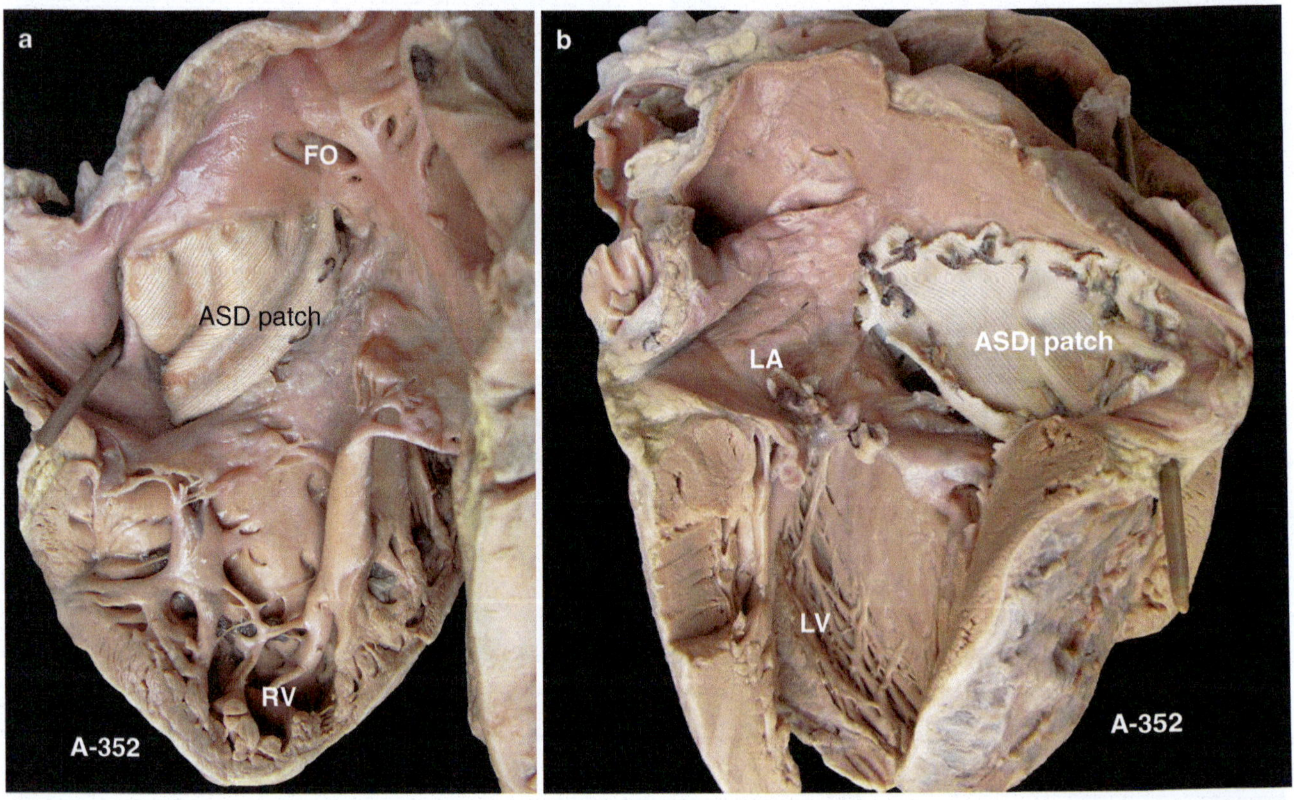

Fig. 11.13 (**a**) Specimen 352. Remnant leak after repair of giant ostium primum defect. The existence of an FO could rule out ostium secundum defect. The coronary sinus (has probe in it) still lies in the RA. Knotting of the suturing line of the Dacron patch is partly in the LA side, with the rest in the RA side. *LA* left atrium, *RA* right atrium. (**b**) Specimen 352. Ostium primum defect. Left-side view shows remnant leak anterior and inferior to the Dacron patch of the defect. There is no fibrous connection between the mitral and tricuspid valves. A single mattress suture with clout is added to the mitral commissure due to regurgitation. A metal probe is inserted into the coronary sinus from RA. *RA* right atrium

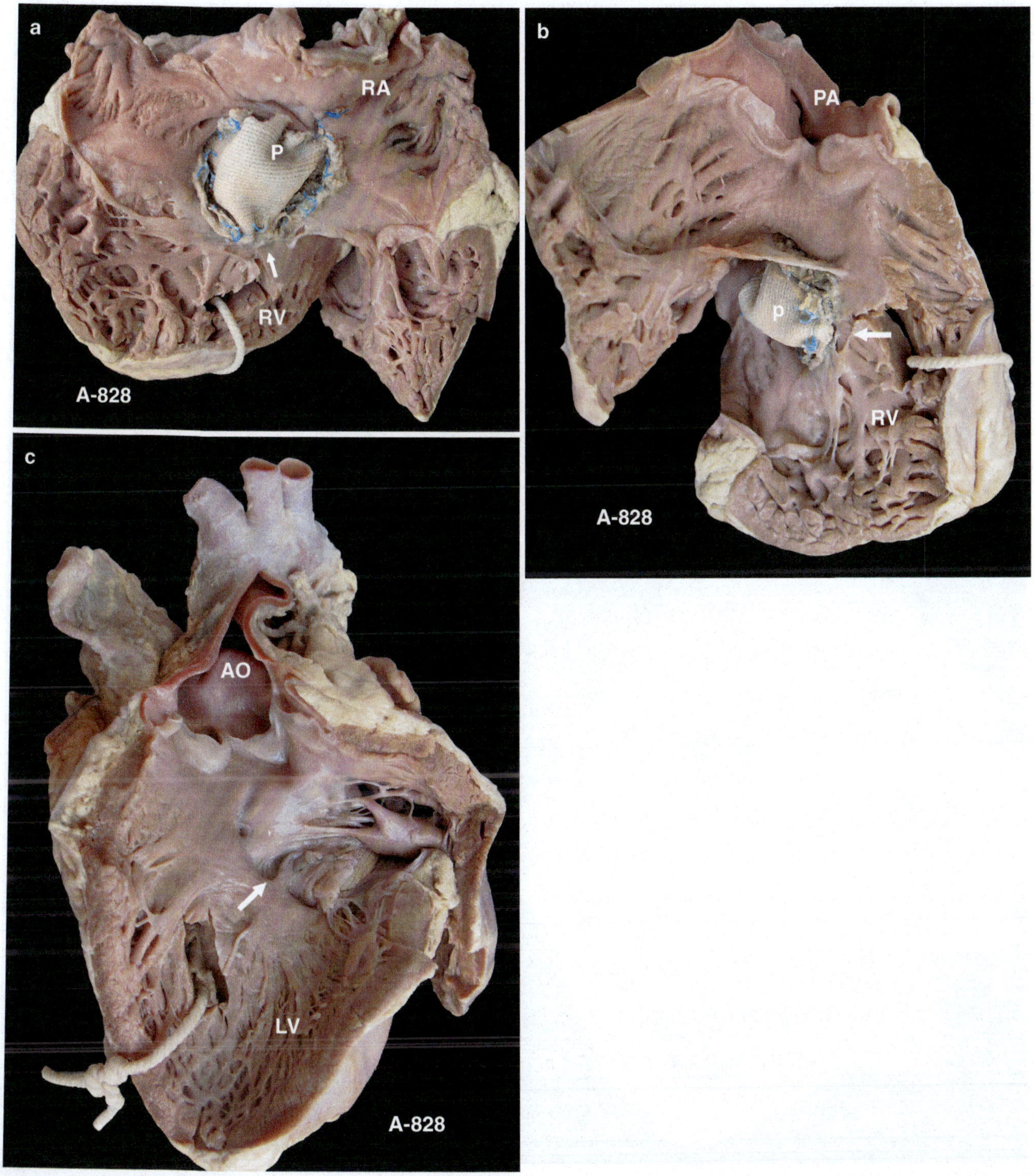

Fig. 11.14 (**a**) Specimen 828. A giant ostium primum defect. The inferior border of the defect has reached the TV ring. Deep split (*arrow*) could be seen in the TV leaflet. Coronary sinus has been separated into the LA by Dacron patch. *LA* left atrium, *P* patch for ASD, *TV* tricuspid valve. (**b**) Specimen 828. The inferior border of atrial septal defect shown from outflow of right ventricle. The *arrow* indicates the split in the TV leaflet. The repair of the ASD is tight and the ventricular septum intact. The pulmonary artery is mildly enlarged. *P* patch for ASD, *TV* tricuspid valve. (**c**) Specimen 828. View from outflow of the left ventricle. The ventricular septum is intact with a tiny crack in the leaflet of the MV (*arrow*), but without split. The MV is not lowered, with a normal outflow of the LV. *LV* left ventricle, *MV* mitral valve

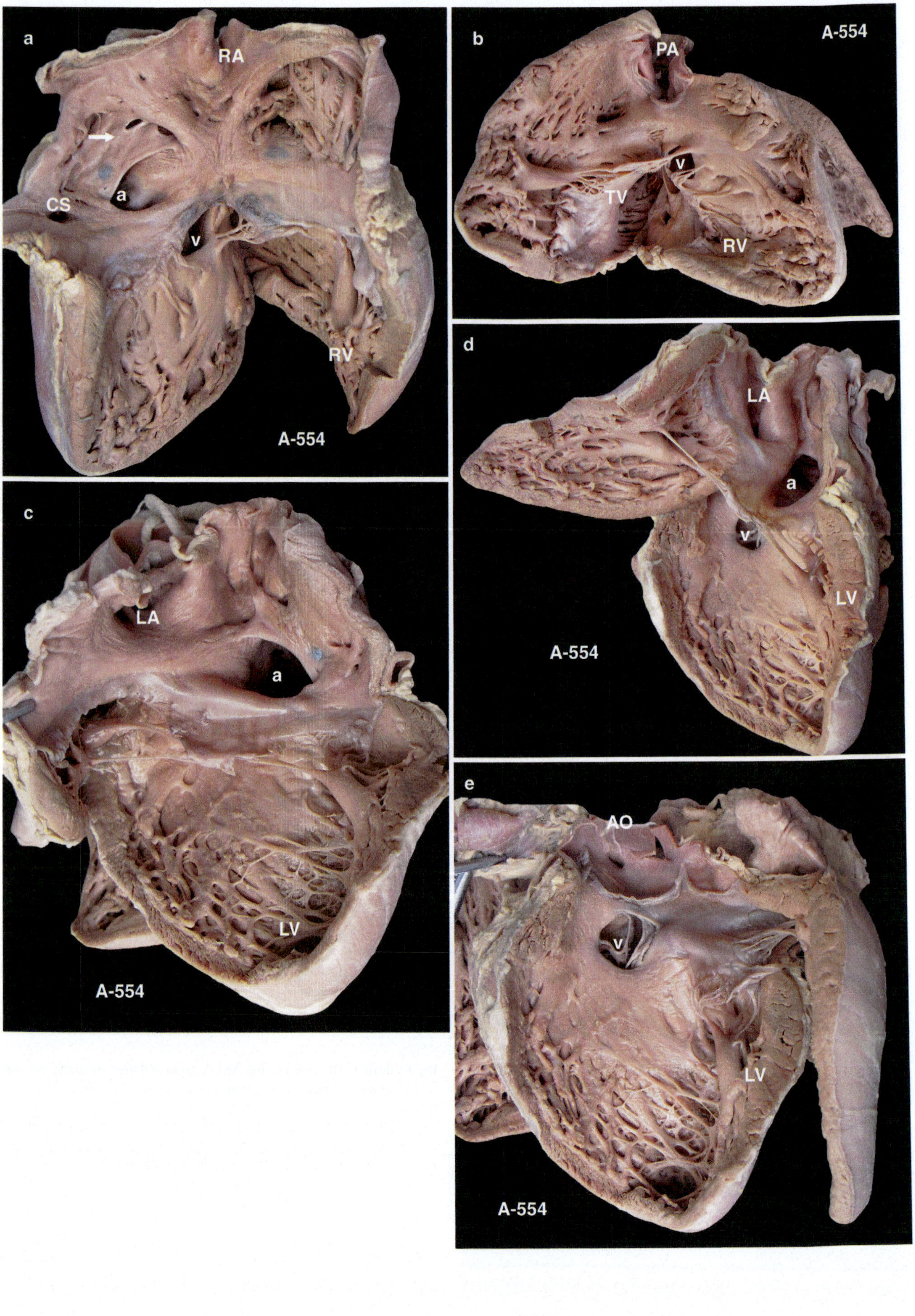
a
RA
CS
a
v
RV
A-554
b
A-554
PA
v
TV
RV
c
LA
a
LV
A-554
d
LA
a
v
LV
A-554
e
AO
v
LV
A-554

Fig. 11.15 (**a**) Ostium secundum defect with independent ventricular septal defect which is hard to identify. ASD (*a*) has a relatively low position, so it is suspected to be an ostium primum defect, with the FO superior to it. However, the inferior border of the ASD does not completely connect with the TV ring. The two are separated by muscular tissue, which means the foramen primum has closed and, therefore, an ostium secundum defect can be confirmed. *ASD* atrial septal defect, *CS* coronary sinus, *TV* tricuspid valve, *V* ventricular septal defect (VSD). (**b**) Showing ventricular septal defect from outflow of the right ventricle. The defect lies posterior to the papillary muscle of the conus, which belongs to perimembranous subpulmonary VSD. The VSD has no communication with the ASD. *ASD* atrial septal defect, *PA* pulmonary valve, *TV* tricuspid valve, *VSD* ventricular septal defect. (**c**) Ostium secundum defect viewed from the left side (*a*). Some certain muscular edge exists between the inferior border and the MV ring, with the leaflet generally normal. *LA* left atrium, *LV* left ventricle, *MV* mitral valve. (**d**) Atrial septal defect and ventricular septal defect from the inflow of the left ventricle. The ASD and the VSD are viewed from the inflow of the LV. The MV has no split, and no fibrous connection exists between the left and right atrioventricular valves, which is distinct from an AVSD. *ASD* atrial septal defect, *AVSD* atrioventricular septal defect, *LV* left ventricle, *MV* mitral valve, *VSD* ventricular septal defect. (**e**) View from outflow of left ventricle. Distance is kept between the VSD and both the MV and the AV. The VSD has an intact border peripherally, through which chordae tendineae of the right atrioventricular valve can be seen. *AV* aortic valve, *MV* mitral valve, *VSD* ventricular septal defect

11.5.6 Cor Triatriatum

This specimen is a typical case of cor triatriatum. The LA is divided into two parts by the intact fibromuscular septum within it. The upper chamber connecting with the pulmonary vein is called the *accessory left atrium*, and the lower chamber connecting with the MV is termed the *real left atrium*. The two chambers can communicate through the pores in the middle of the septum. Pulmonary venous blood in the accessory left atrium can flow through the ASD into the RA

Fig. 11.16 (**a**) Specimen 846. Cor triatriatum viewed from the left side of the heart. The left atrial chamber is revealed by cutting open the LA along the left atrioventricular groove. This cavity connects with the MV and is the real LA (*LA1*), which is smaller, with a muscular septum membrane above it. Superior to the muscular septum is another left atrial chamber, which receives blood from the pulmonary vein. This chamber, called the *false left atrial chamber*, is formed by the converging and enlargement of the pulmonary veins. There is a pore (*arrow*) in the septum between the two chambers through which they can communicate. Blood in the pulmonary vein can flow from the false left atrial chamber into the real one through the communicating pore. *LA* left atrium, *MV* mitral valve. (**b**) The same viewing angle with the last figure. The real LA is fully exposed, through which (*LA1*) a red probe is sent into the MV opening. The *arrow* indicates the communicating pore between the true and false left atria. *LA* left atrium, *MV* mitral valve. (**c**) Specimen 846. The roof of the LA has been cut open to reveal the false left atrial chamber, which receives blood from the pulmonary veins and communicates with the real chamber through the communicating pore (*red probe*). The white dashed line indicates the range of the muscular septum between the two chambers. LA2 is the false left atrial chamber. (**d**) Similar viewing angle as the last figure. Besides the shown false left atrial chamber (*LA2*), a LAA can also be seen communicating with the real left atrial chamber. (**e**) View from the left side of the heart. The metal probe is sent into the false left atrial chamber (*LA2*) from an opening of the pulmonary vein, with the real left atrial chamber (*LA1*) beneath it. A LAA communicates with the real left atrial chamber. (**f**) Specimen 846. Cor triatriatum associated with atrioventricular septal defect. As demonstrated, the defect has been repaired. In this case, the direction of blood flow from the pulmonary vein is pulmonary vein → false left atrial chamber → ASD → real left atrial chamber → LV. *ASD* atrial septal defect, *CS* coronary sinus, *LV* left ventricle, *RA* right atrium, *RV* right ventricle. (**g**)Specimen 846. Schematic diagram of cor triatriatum associated with total atrioventricular septal defect

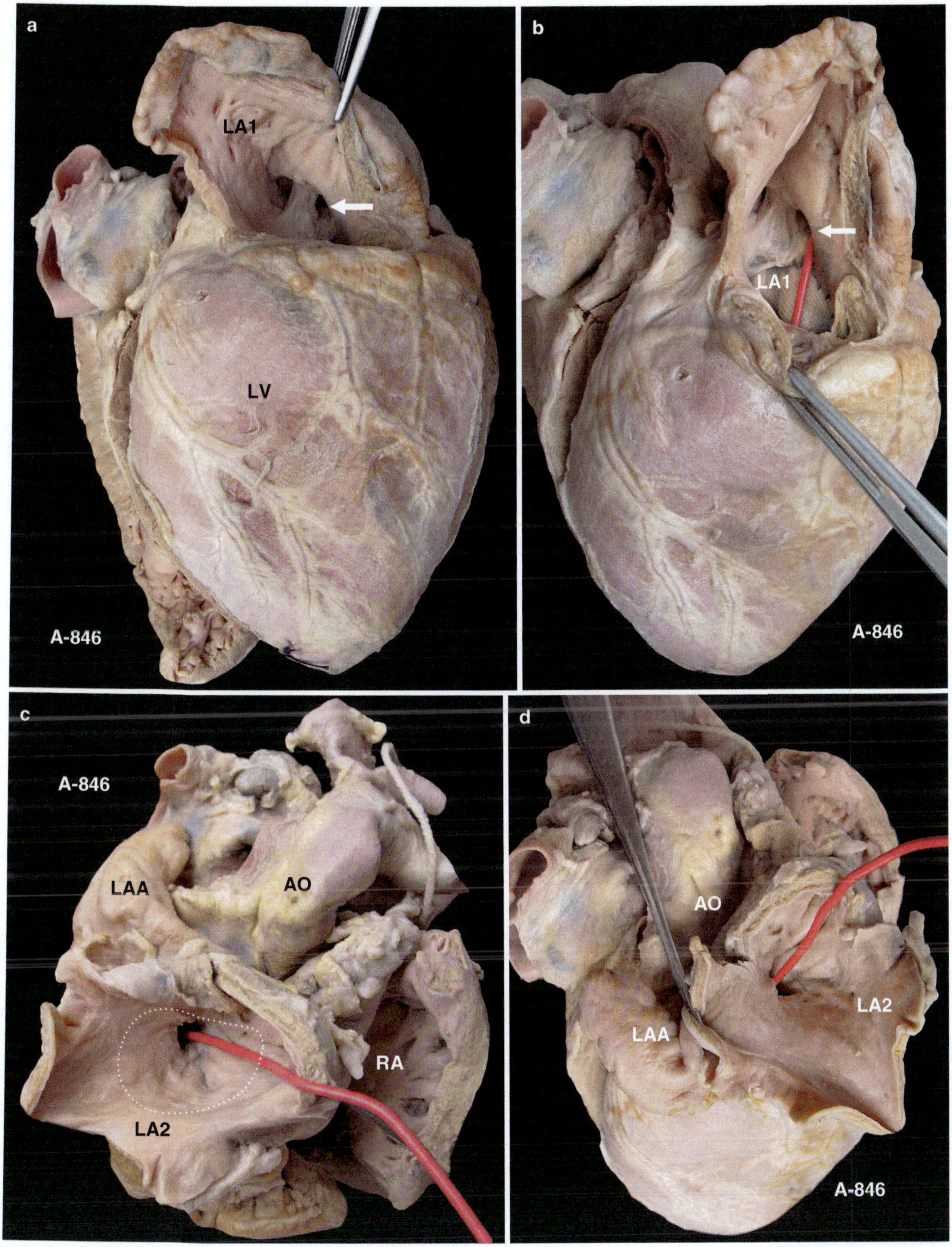
a
LA1
LV
A-846
b
LA1
A-846
c
A-846
LAA
AO
RA
LA2
d
AO
LAA
LA2
A-846

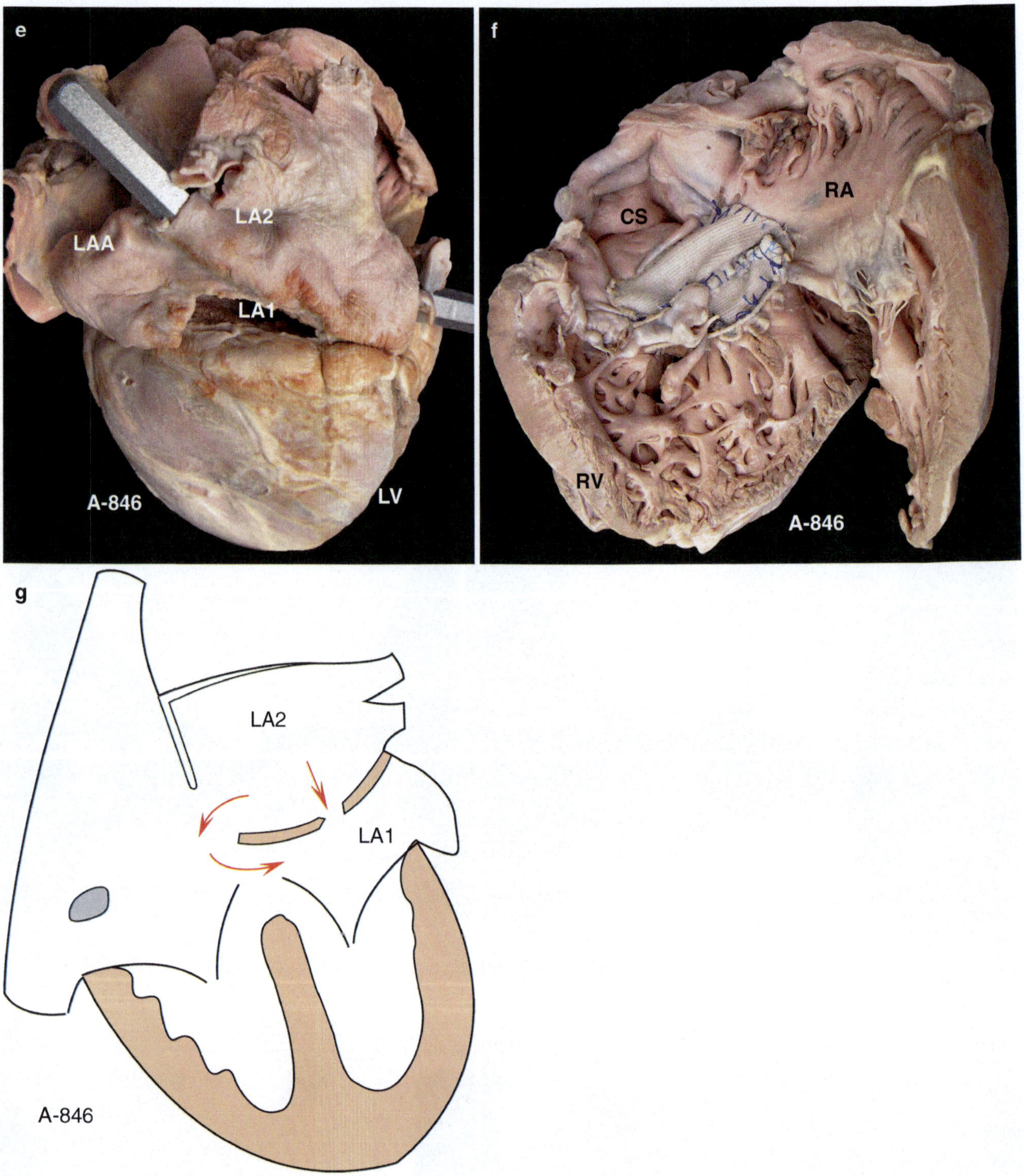

Fig. 11.16 (continued)

Atrioventricular Septal Defect

12

12.1 General Consideration

Atrioventricular septal defects (AVSDs) have also been called *endocardial cushion defects* and *atrioventricular canal defects*, excluding isolated ASDs and isolated VSDs.

12.2 Embryology

In common circumstances, the superior and inferior endocardial cushions grow continuously toward each other while protruding wing shaped to both sides to form the left and right nodules. They gradually join to produce the medial endocardial cushion and to divide the atrioventricular canal into two channels. The left and right nodules of the endocardium form the anterior leaflet of the MV and the septal leaflet of the TV, respectively. The main body of the endocardial cushion may be part of the septum. The malformation of the atrioventricular cushion will make the atrioventricular annulus discontinued, and the annulus of the MV will descend to the level of the septal leaflet of the TV. In those cases, the four chambers can communicate with each other.

12.2.1 Ostium Primum Atrial Septum Defect (ASD)

The inferior rim of the septum primum fuses with the endocardial cushions to close the ostium primum. The maldevelopment of the endocardial cushions or the septum primum results in an ostium primum ASD.

12.2.2 Ventricular Septum Defect (VSD)

The endocardial cushions, muscular ventricular septum, and conal septum merge to close the interventricular foramen. Abnormal development of the endocardial cushions may result in variable sizes of VSDs, which influence the membranous septum or conal septum and lead to obstruction of the LVOT.

12.2.3 Translocation of the Atrioventricular Node

The atrioventricular node and the atrioventricular bundle derive from the tissue of the primary annulus, coursing posteriorly to the endocardial cushion. As the interventricular foramen gradually diminishes, the atrioventricular bundle continues along the posterior and inferior edge of the membranous septum. When the endocardial cushions develop abnormally, the atrioventricular node and the His bundle are displaced posterior to the defect and right posterior and inferior to the entry of the coronary sinus. The left bundle branches occasionally are influenced, and particular electrocardiographic characteristics can be observed.

12.3 Classification of Atrioventricular Defects

12.3.1 Partial Atrioventricular Septal Defect

A partial AVSD is similar to the ostium primum ASD, which has no VSD and the atrioventricular valves are variably cleft. A partial AVSD differs by having a fiber connection between the MV and the septal leaflet of the TV, both of which directly attach to the ventricular septum (Fig. 12.1a–k).

12.3.2 Complete Atrioventricular Septal Defect

A complete AVSD includes an ostium primum ASD and a complete cleft that divides the atrioventricular valve into the common superior valve and the common inferior valve. The basic parts of those two common valves are not attached to

X. Zhu (ed.), *Surgical Atlas of Cardiac Anatomy*,
DOI 10.1007/978-94-017-9409-1_12, © Springer Science+Business Media Dordrecht and People's Medical Publishing House 2015

the ventricular septum directly but by chordae tendineae, with the result that both leaflets float above a VSD (Figs. 12.2a–e, 12.3a–g, and 12.4a–e).

12.3.3 Transitional Atrioventricular Septal Defect

This type of AVSD is on a spectrum between the partial and the complete types, as only a small VSD exists inferior to the leaflets.

12.4 Surgical Management

A partial AVSD is similar to the ostium primum ASD. It requires closure of the interatrial communication and repair of the cleft of the MV (Fig. 12.5a–f). Surgical treatment of a complete AVSD is directed basically toward remolding the ventricular septum, raising the atrioventricular annulus back to the normal level, and closing the ostium septal ASD (Figs. 12.6a–f, 12.7a–f, 12.8a–d, 12.9a–d, 12.10a–e, 12.11a–e, 12.12a–c, 12.13a–c, 12.15a–e, 12.16a–d, 12.17a–e, and 12.18a–d).

12.4.1 Special Considerations

Operations should be performed so as to avoid damage to the conduction system in the posterior and inferior edge of the VSD and the AV anterior and superior to the VSD. The chordae tendineae are always attached to the edge of the defect, like barriers, so using a patch for repair is difficult. Because visualization is poor, measuring the size of the defect also is difficult, especially when it is not clear if a defect exists below the inferior common valve.

It is important to avoid having an obstruction of the LVOT. In the case of a huge complete AVSD, a LVOT obstruction is due partly to the conal septum being affected; the medial ventricular septum also is affected. As a result, the atrioventricular annulus descends and is concaved to the apex. Repairing only the defects without raising the atrioventricular annulus to an ordinary level may leave an important LVOT obstruction.

The MV regurgitation should be corrected. If the cleft of the MV is not accompanied by marked regurgitation, dealing with the MV may not be necessary. The coronary sinus may be used as an anatomical landmark to explore the cardiac anatomy and to locate the conduction system.

12.4.2 Technical Difficulties and Risks

12.4.2.1 Residual Mitral Valve Regurgitation

The cleft of the MV may lead to various degrees of regurgitation. The malformation of the MV and the descent of the annulus may result in a more difficult repair.

12.4.2.2 Complete Atrioventricular Block

The incidence of complete atrioventricular block is high in treating ostium primum ASDs, and it is necessary to keep a certain distance from the defect while suturing the posterior inferior rim (Fig. 12.12a).

12.4.2.3 Left Ventricle Outflow Tract Obstruction

An LVOT obstruction is due to the downward movement of the mitral annulus and may cause postoperative left-heart dysfunction.

Fig. 12.1 (**a**) Specimen 759. Partial atrioventricular septal defect. This defect resembles an isolated ostium primum ASD, without interventricular communication. Fibrous connections exist between the mitral and tricuspid valves, and common *atrioventricular* valves are formed. Those characteristics may distinguish an AVSD from an ostium primum ASD. This figure is viewed from the right atrial and ventricular sides. The MV can be seen across the defect. The mitral annulus has descended to the same level as that of the TV. The *white arrow* marks the FO, which verifies that the defect belongs to a type of ostium primum. *MV* mitral valve, *RA* right atrium, *RV* right ventricle, *TV* tricuspid valve. (**b**) Specimen 759. Emphasized fossa ovalis and cleft in the tricuspid valve. This figure is typical of a partial AVSD, viewed from the RA. It is important to investigate the malformation during surgery. Common atrioventricular valves are beneath the defect. The *white arrow* show fossa ovalis. (**c**) Specimen 759. Viewed from the right atrium. Cleft in MV is seen across the defect. *MV* mitral valve, *RA* right atrium, *RV* right ventricle. (**d**) Specimen 759. Orifice of the coronary sinus viewed from the right atrium. The Thebesian valve is lifted by using forceps. The inferior edge of the defect is the tricuspid annulus. Seeing through the defect, a marked cleft in the MV is present, but coaptation is good, and there is no significant regurgitation. *MV* mitral valve, *RA* right atrium, *RV* right ventricle. (**e**) Specimen 759. Relation between partial AVSD and venae cavae. The metal bougie is put in the superior and inferior vena cavae. *D* partial atrioventricular defect, *RA* right atrium, *RV* right ventricle. *TV* tricuspid valve. (**f**) Specimen 759. Viewed from exterior of left heart. LA and LV are normal. *LA* left atrium, *LV* left ventricle. (**g**) Specimen 759. Partial AVSD viewed from the left atrial aspect. Cleft in MV can be seen. *MV* mitral valve. (**h**) Specimen 759. Overhead view of the left atrium. *Dash line* marks the mitral annulus. Note that the MV leaflets have connected with the contralateral TV leaflets to form a common atrioventricular valve, so this specimen is confirmed to be a case of a partial AVSD. *MV* mitral valve, *TV* tricuspid valve. (**i**) Specimen 759. Viewed from the left ventricular inflow tract. The MV is exposed. Mitral and tricuspid annuli are at the same level. Seen through the defect, common atrioventricular valves are formed by fusion of the left and right atrioventricular valves, and there is a cleft in the MV. *MV* mitral valve. (**j**) Specimen 759. Viewed from the left ventricular outflow tract. The MV leaflets and chordae are exposed. The *yellow arrow* marks a slightly inferior displacement of the mitral annulus, but there is no important LVOT obstruction. The *red arrow* indicates that an area of the ventricular septum is deficient, with chordae attached, and no interventricular communication is present. *MV* mitral valve. (**k**) Specimen 759. Diagram of partial atrioventricular septal defect. Diagnosis is AVSD, common atrium. A common atrioventricular valve and cleft MV are present

12.5 Specimen Demonstration

12.5.1 Partial Atrioventricular Septal Defect

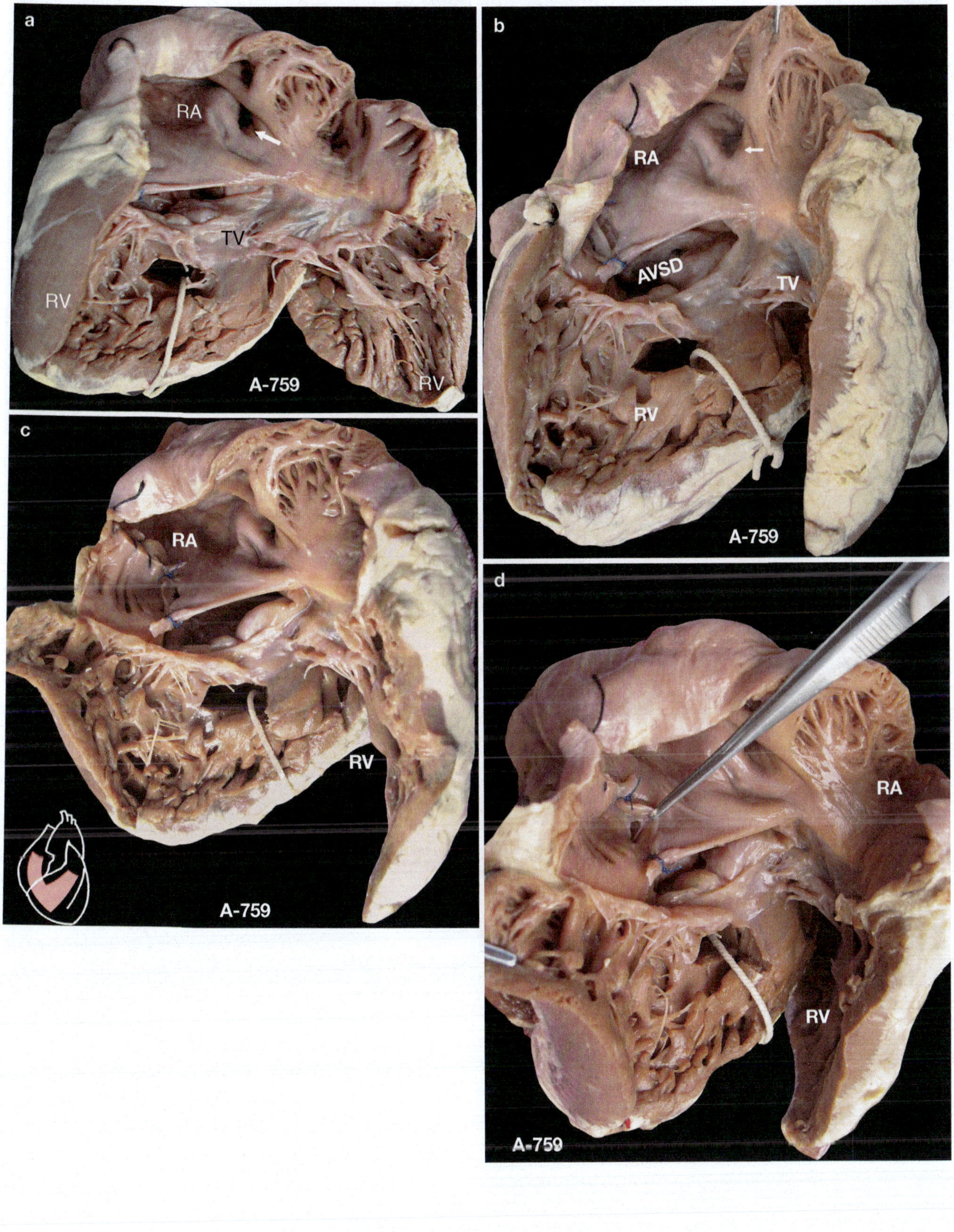

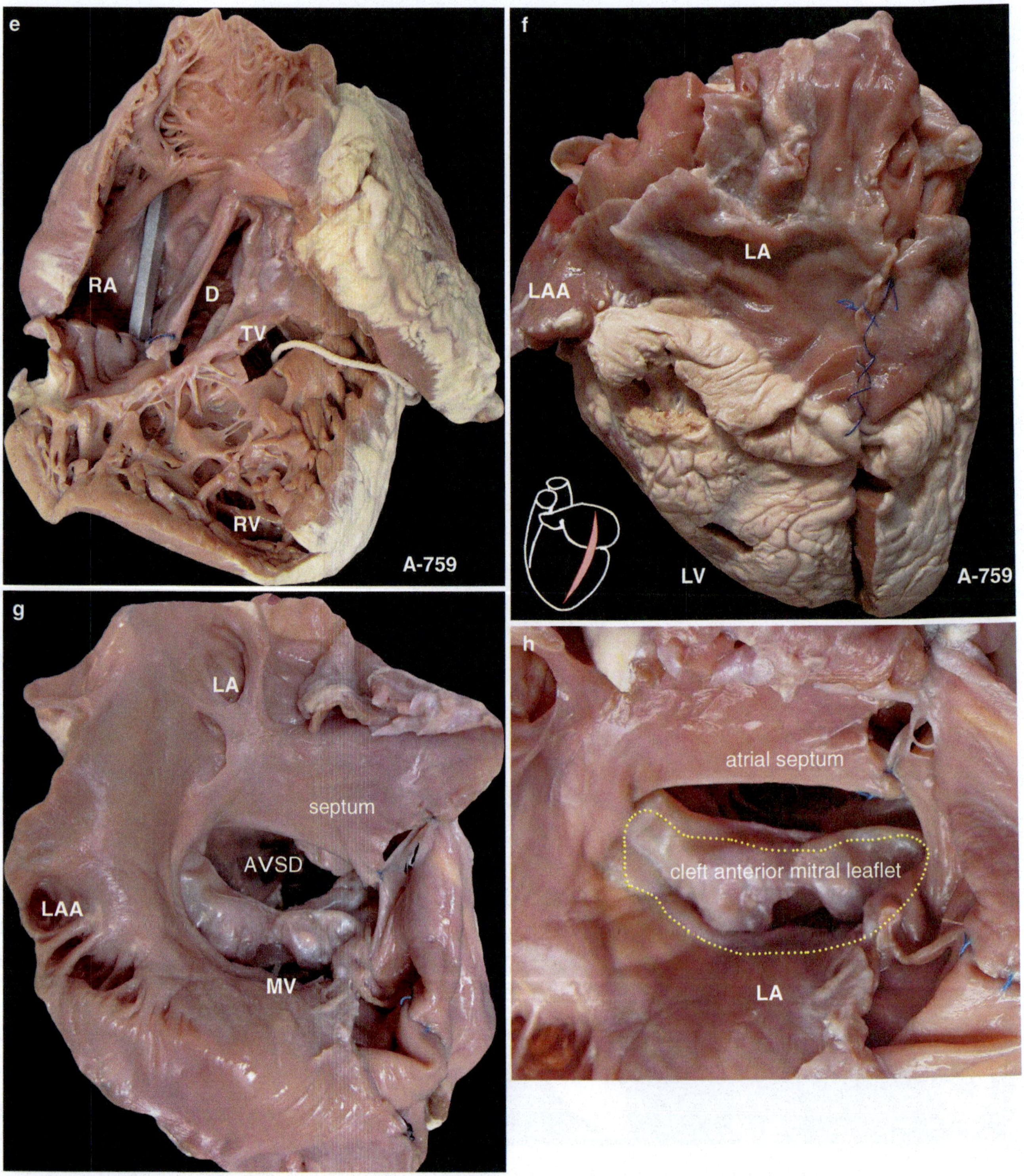

Fig. 12.1 (continued)

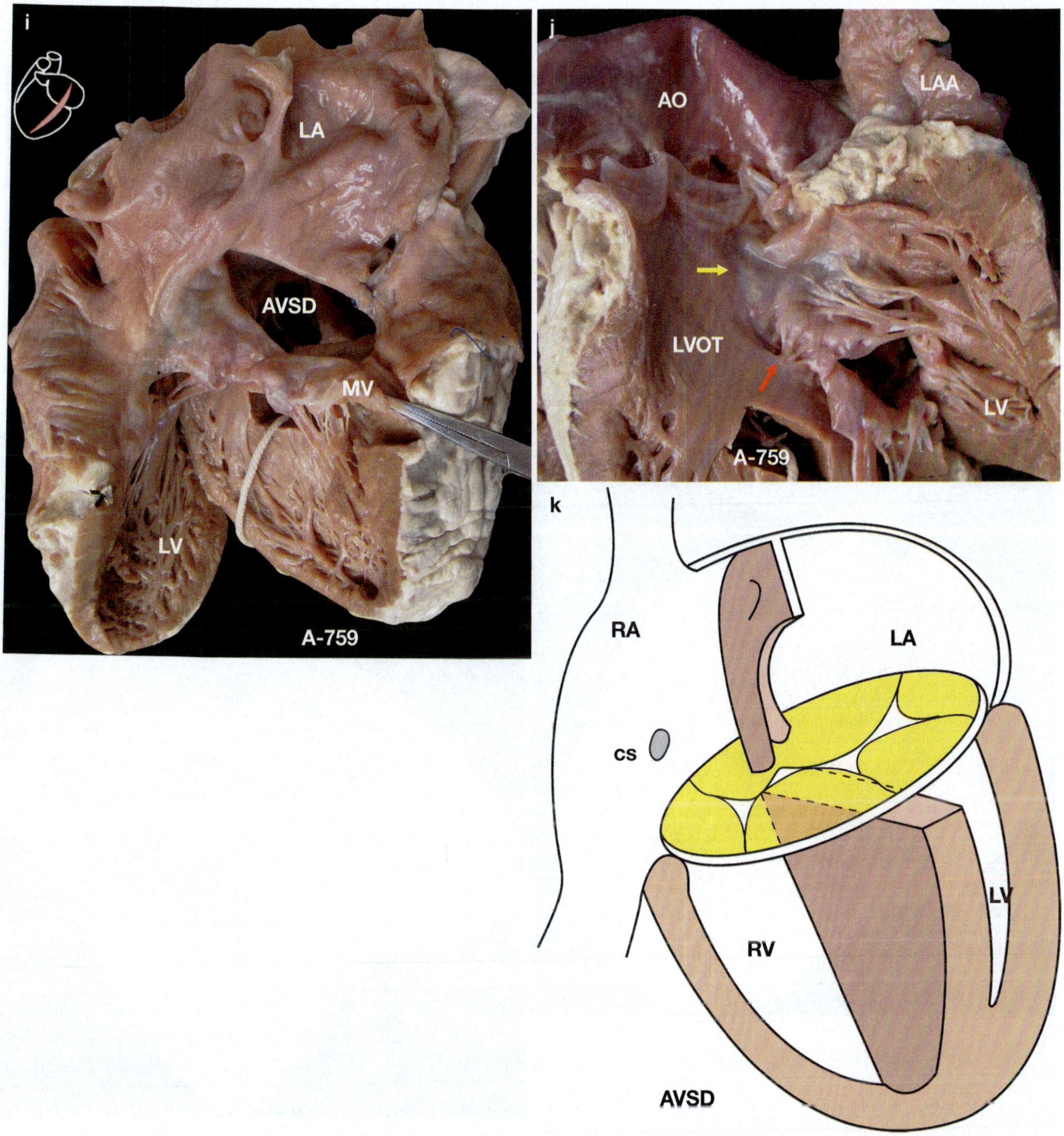

Fig. 12.1 (continued)

12.5.2 Complete Atrioventricular Septal Defect

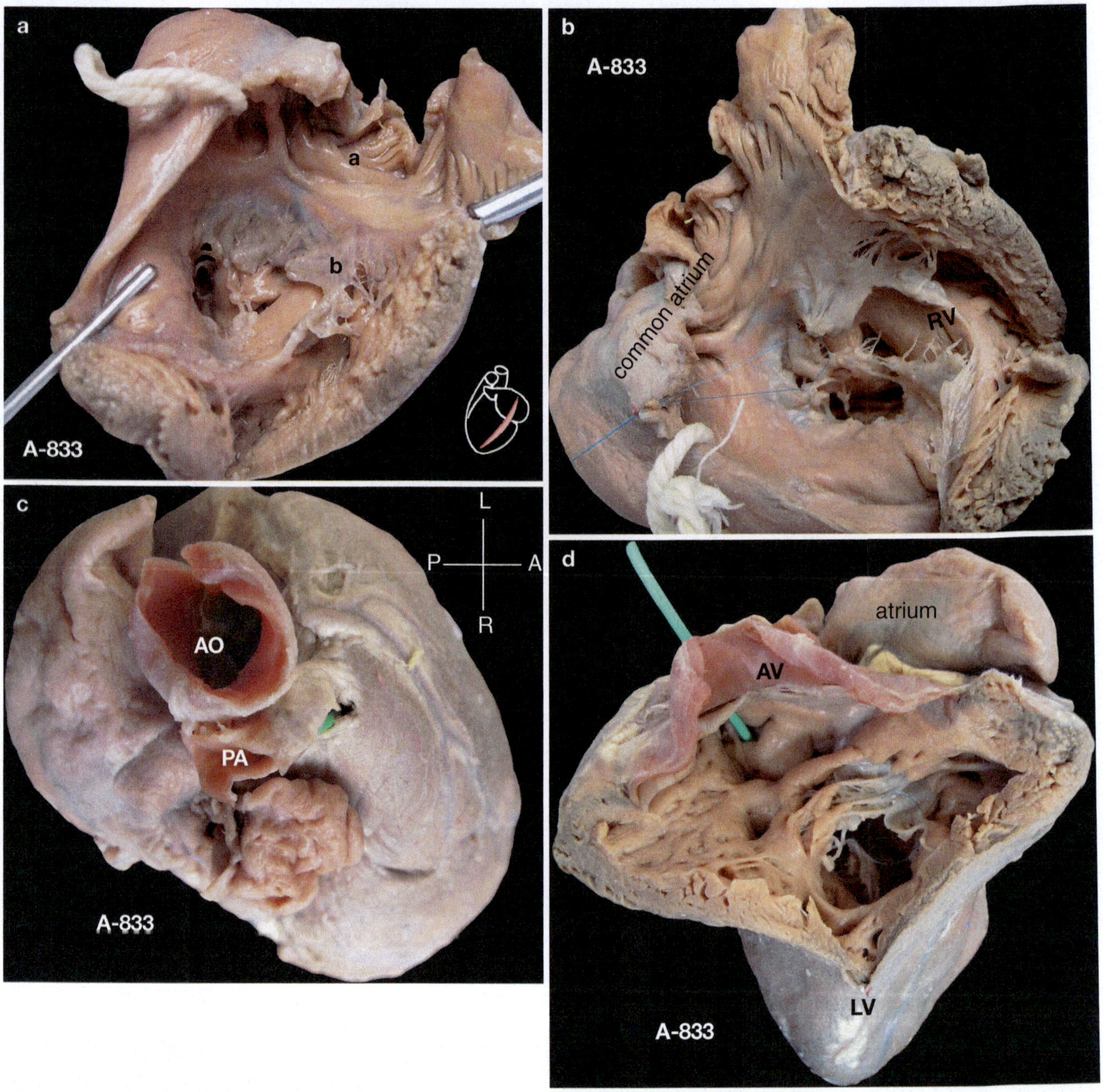

Fig. 12.2 (**a**) Specimen 833. Complete atrioventricular valve septal defect, common atrium, accompanied with pulmonary stenosis. Overhead view of RA and RV. No definite atrial septum is present. *a*: atrium, *b*: common atrioventricular valves. (**b**) Specimen 833. Ventricular structure seen beneath the common atrioventricular valves by overlooking. The RV is small, and the LV beneath the AV is large; together they function as a single ventricle. A common atrium is present due to deficiency of the atrial septum. *AV* aortic valve, *LV* left ventricle, *RV* right ventricle. (**c**) View of the heart from cephalic side. The large aorta, enormous common atrium behind the aorta, and the hypoplastic pulmonary artery, which has tiny lumen on the right of the aorta, are shown. The aorta and pulmonary artery are side by side. *AO* ascending aorta, *PA* pulmonary artery. (**d**) Viewed from the left ventricular outflow tract. The AV can be seen. There is a narrow fissure in the muscle below the AV, through which a bougie of 3 mm is placed. The bougie is in the pulmonary artery, resembling a functional pulmonary atresia. *LV* left ventricle. (**e**) Specimen 833. Diagram. Complete AVSD, common atrium, hypoplastic pulmonary artery, and hypoplastic RV are seen. *AO* ascending aorta, *LV* left ventricle, *PA* pulmonary artery, *RV* right ventricle

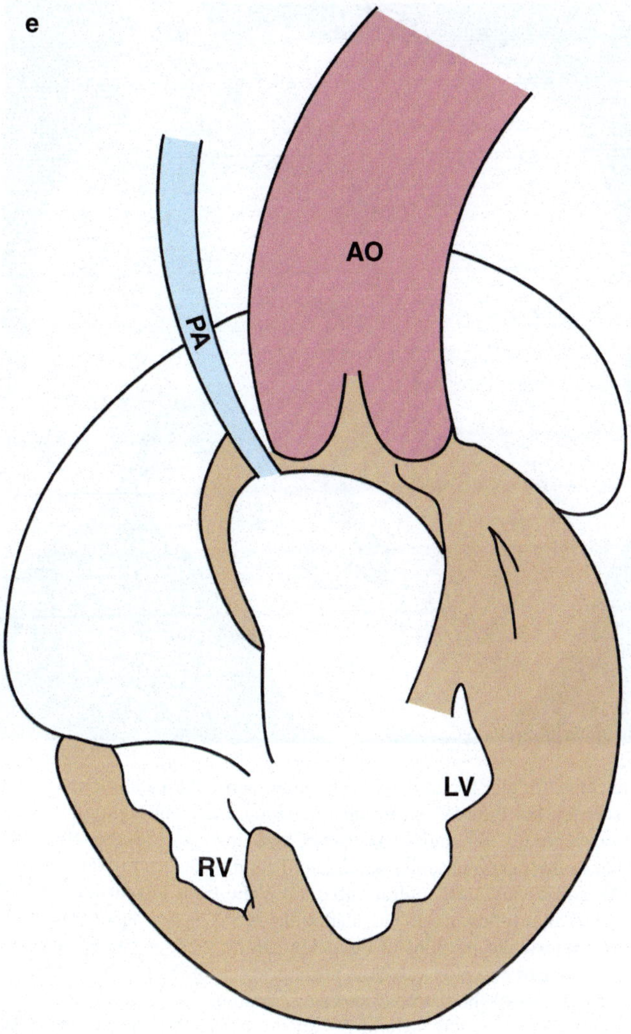

Fig. 12.2 (continued)

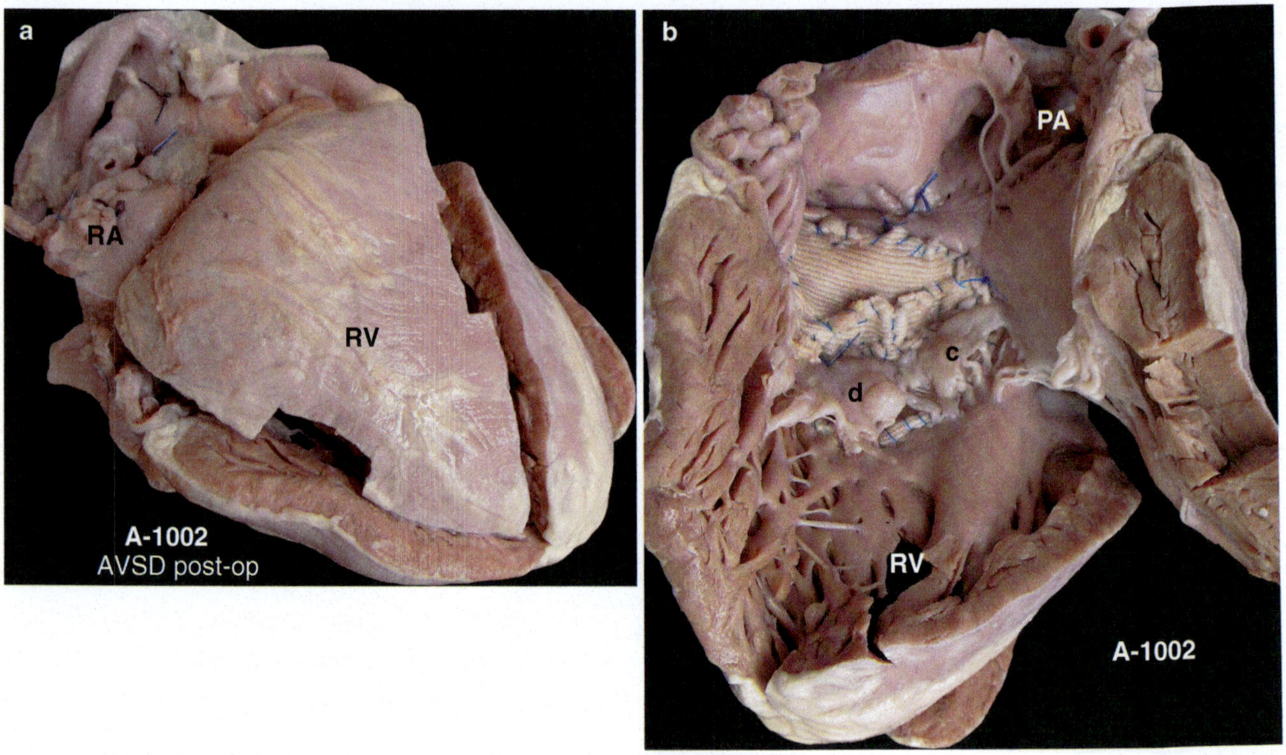

Fig. 12.3 (**a**) Complete atrioventricular valve septal defect (specimen 1002). External view, postoperation, is shown. *RA* right atrium, *RV* right ventricle. (**b**) Specimen 1002. Viewed from right ventricle outflow tract. The atrial septum and TV are exposed. The defect is repaired with a Dacron patch by using a two-patch technique. *C*: superior common valve, *d*: inferior common valve. Those two common valves are sutured together, and coaptation is good. The inferior edge of the ventricular patch can be seen under the TV. It typically is a two-patch technique that uses two Dacron patches separately to repair the ASD and the VSD. *ASD* atrial septal defect, *TV* tricuspid valve, *VSD* ventricular septal defect. *PA* pulmonary artery, *RV* right ventricle. (**c**) Same view as the previous figure. The *dash line* shows atrioventricular valve node and the bundle of His, which has been displaced to the posterior edge of the AVSD. *PV* pulmonary valve, *RA* right atrium, *RV* right ventricle. (**d**) Left ventricular view to observe the repair of the atrioventricular septal defect. The interventricular patch is steady and smooth, but the MV is torn apart (*arrow*), inducing regurgitation. *LV* left ventricle, *MV* mitral valve. (**e**) Close-up view from the left ventricle of the interventricular patch and the slit of the mitral valve leaflet. *AV* aortic valve, *MV* mitral valve. (**f**) View from left atrium and left ventricular inflow tract. The leaflet of the MV is torn apart. The atrial and ventricular portions of the patch can be seen, as can the superior and inferior common valves. *A* atrial portions of the patch, *B* ventricular portions of the patch, *C* superior common valves, *D* inferior common valves. *LA* left atrium, *LV* left ventricle, *MV* mitral valve. (**g**) Specimen 1002. Left-side view of complete atrioventricular septal defect after two-patch technique procedure. The interventricular patch and the slit of the MV leaflet of a postoperative complete AVSD treated with two-patch technique can be seen. *Red arrow* show the rupture of mitral leaflet. *MV* mitral valve

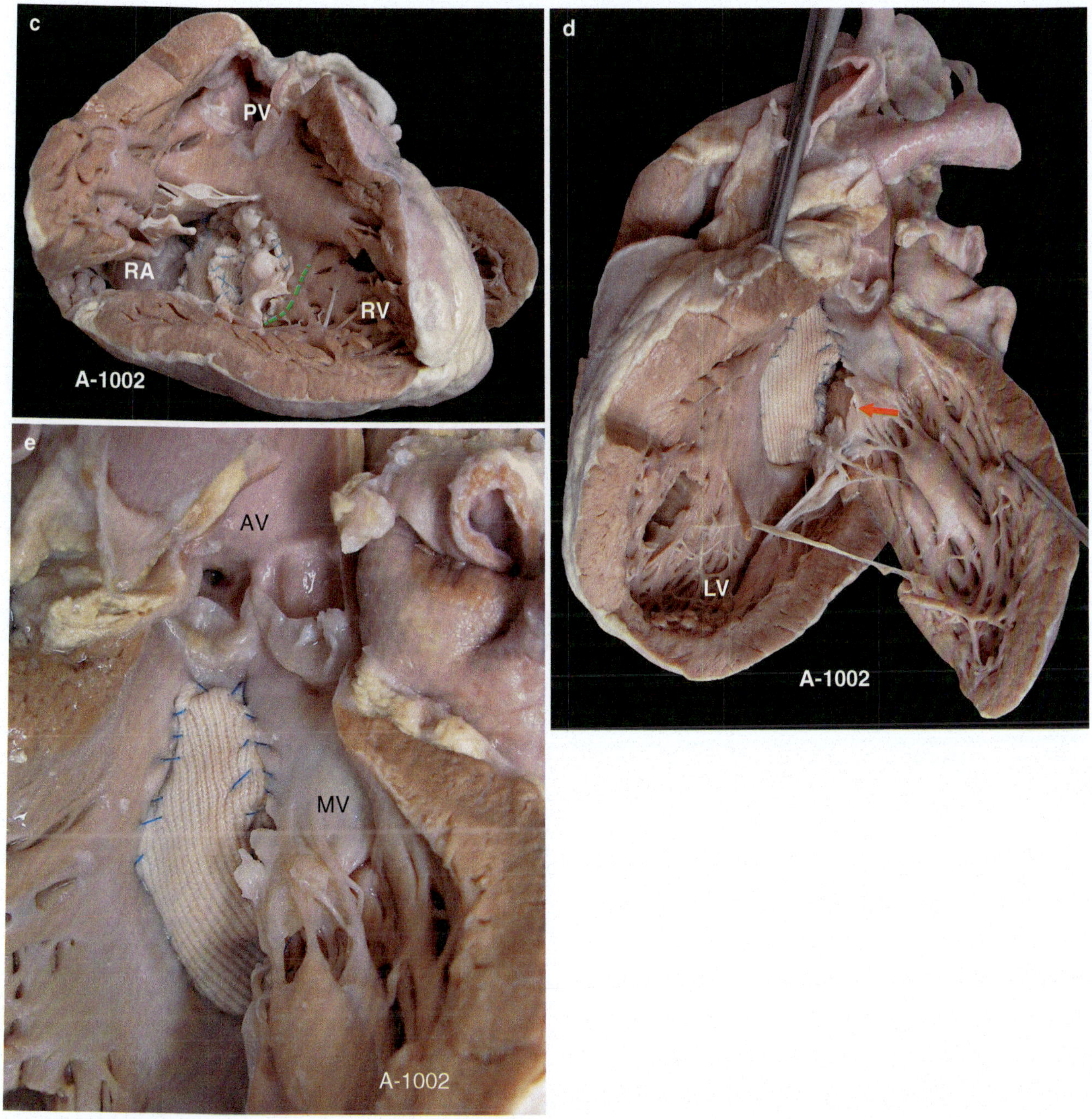

Fig. 12.3 (continued)

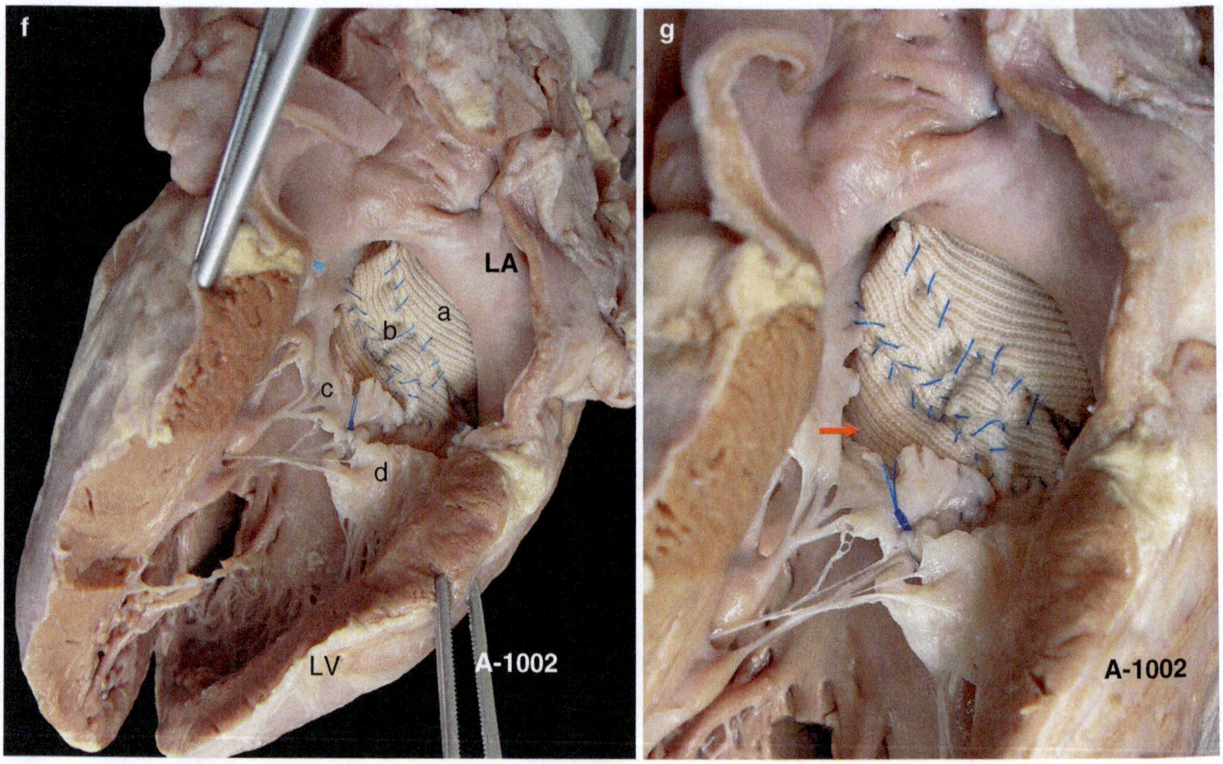

Fig. 12.3 (continued)

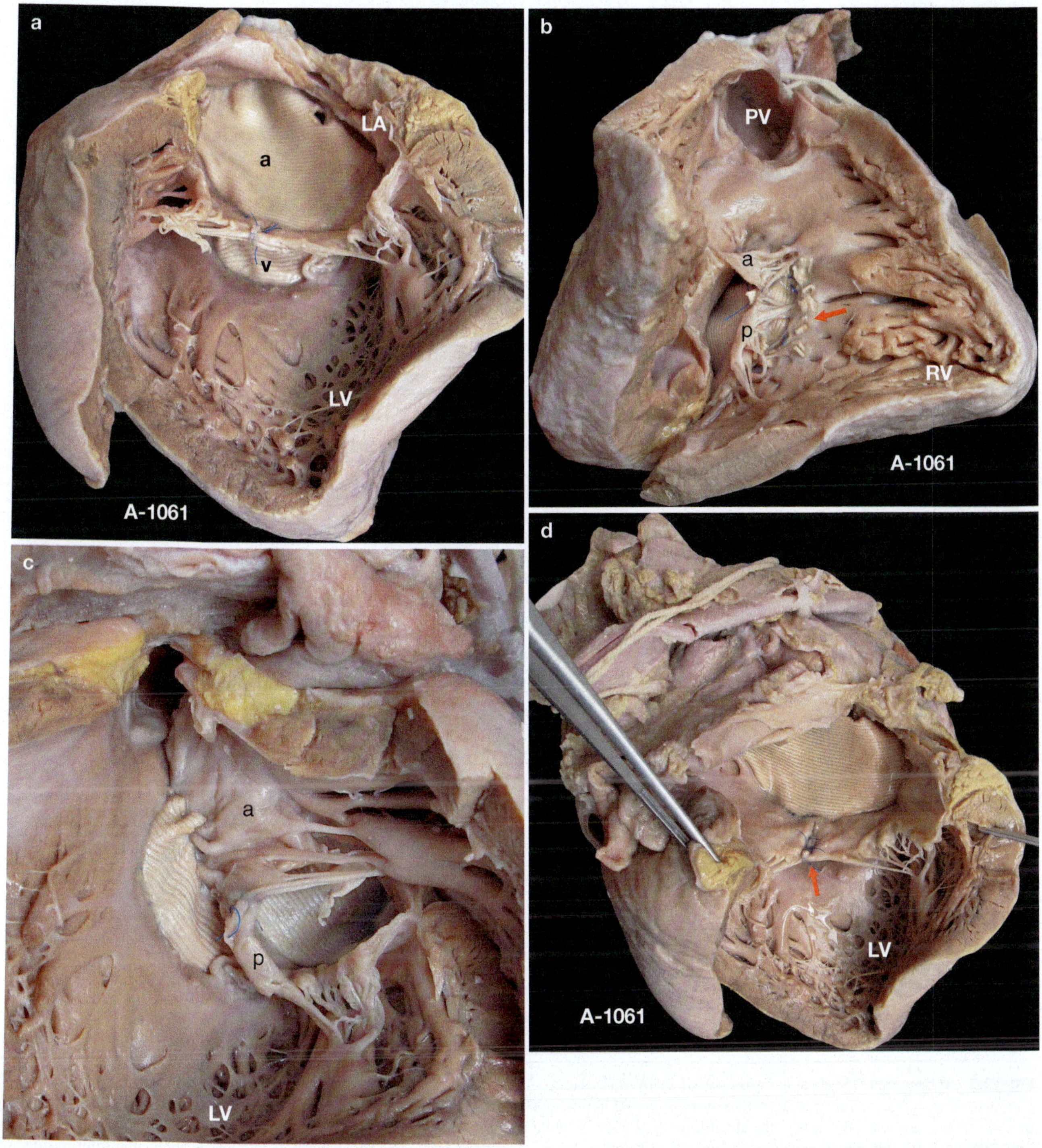

Fig. 12.4 (**a**) Specimen 1061. Left-side view of complete atrioventricular valve septal defect after two-patch technique procedure. The interatrial patch is superior, and the interventricular patch is inferior. Common atrioventricular valves are on the commissure of those two patches. *A* interatrial patch, *V* interventricular patch. *LA* left atrium, *LV* left ventricle. (**b**) Specimen 1061. View from right ventricular outflow tract. The superior common valve and inferior common valve, sutured to the atrial defect patch and the ventricular patch, respectively, are present. The *arrow* shows the suture line on the inferior edge of the ventricular patch. *PV* pulmonary valve, *RV* right ventricle, *A* Superior common valve, *p* inferior common valve. (**c**) Specimen 1061. Exposure of atrioventricular septal defect patch and common valves from left ventricular view. The patch is steady and smooth. *a* Superior common valve, *p* inferior common valve. *LV* left ventricle. (**d**) Specimen 1061. View from left atrium and left ventricular inflow tract. Only the ASD patch and the suture line of the cleft in the MV can be seen, with the VSD being covered (*arrow*). *ASD* atrial septal defect, *LV* left ventricle, *MV* mitral valve, *VSD* ventricular septal defect. (**e**) Diagram of repair of complete atrioventricular septal defect with two-patch technique. The ASD patch (*green*) and the VSD patch (*red*) are sutured together. Common valves are sewn to the attachment line of the two patches. *ASD* atrial septal defect, *LA* left atrium, *LV* left ventricle, *RA* right atrium, *RV* right ventricle, *VSD* ventricular septal defect

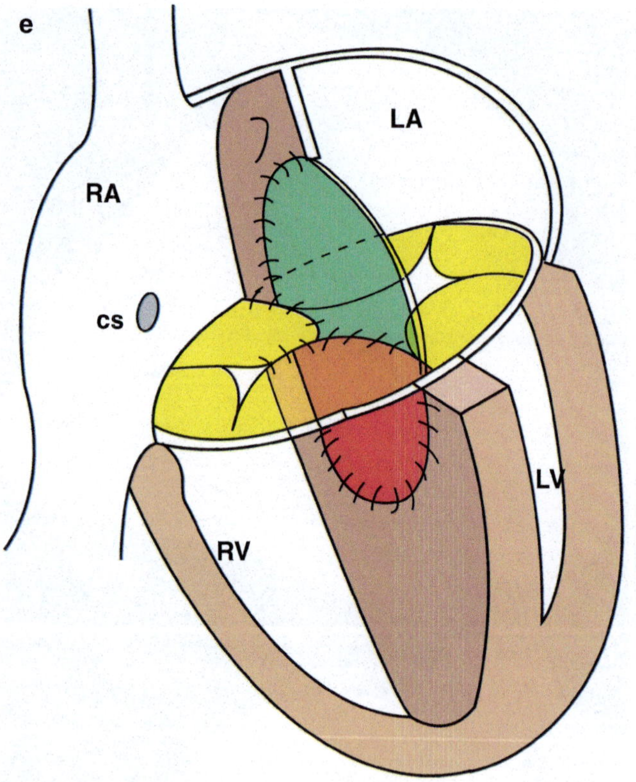

Fig. 12.4 (continued)

Fig. 12.5 (**a**) Specimen 757. Postoperative partial atrioventricular septal defect. The defect is similar to an ostium primum septal defect, but the cleft in the common valve is near the annulus, with deficiency of the septal leaflet of the TV, which indicates that the left and right atrioventricular valves have been connecting with each other to form common valves. This specimen is an AVSD with intact ventricular septum. Meanwhile, note that there is endocardial hematoma above the mitral annulus caused by the operation. The *arrow* shows that the orifice of the coronary sinus is left of the LA, so the suture line is safely far away from conduction system. *CS* coronary sinus. *LA* left atrium, *RA* right atrium, *RV* right ventricle, *TV* tricuspid valve. (**b**) Specimen 757. Partial atrioventricular septal defect viewed from the left heart. The orifice of the coronary sinus is left of the LA (bougie). The interatrial patch and the suture line of the cleft in the MV are present. *CS* coronary sinus, *LA* left atrium, *LV* left ventricle, *MV* mitral valve, *PA* pulmonary artery, *RV* right ventricle. (**c**) Specimen 757. Viewed from the right ventricular outflow tract. The PV is normal. The *arrow* shows the inferior edge of the AVSD patch. (**d**) Specimen 757. Viewed from the left ventricular outflow tract. The ventricular septum is intact, and the cleft in the MV has been sewn (*arrow*). Due to the downward displacement of the MV, the LVOT is slightly obstructed. *LVOT* left ventricular outflow tract, *MV* mitral valve, *AO* ascending aorta, *LV* left ventricle. (**e**) Specimen 757. Full view of mitral valve. *LA* left atrium, *LV* left ventricle. (**f**) Specimen 757. Repair of partial atrioventricular septal defect. The endocardial hematoma (*arrow*) results in damage to the left ventricular endocardium by operation. *LA* left atrium, *LV* left ventricle, *MV* mitral valve

12.5.3 Postoperative Partial Atrioventricular Septal Defect

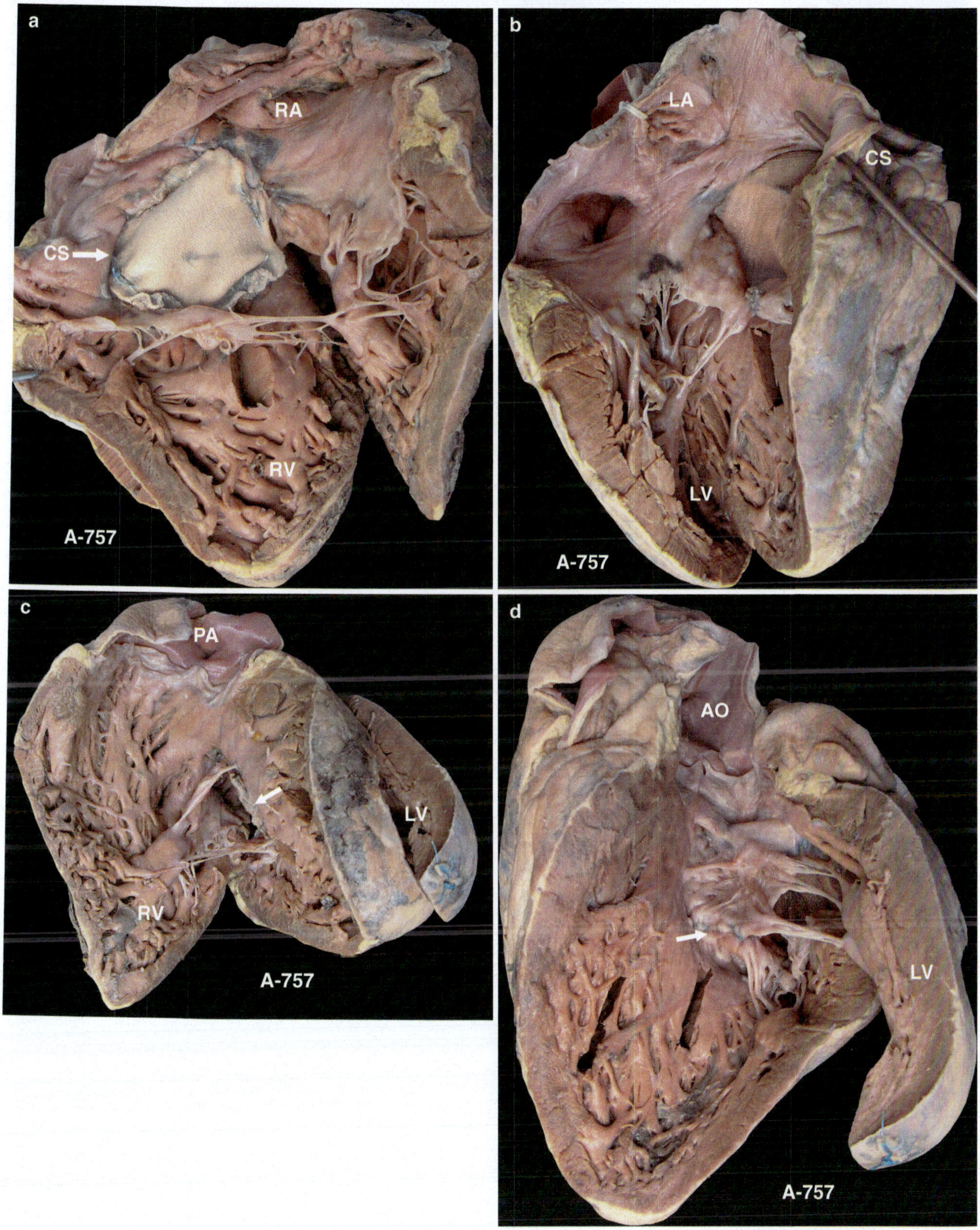

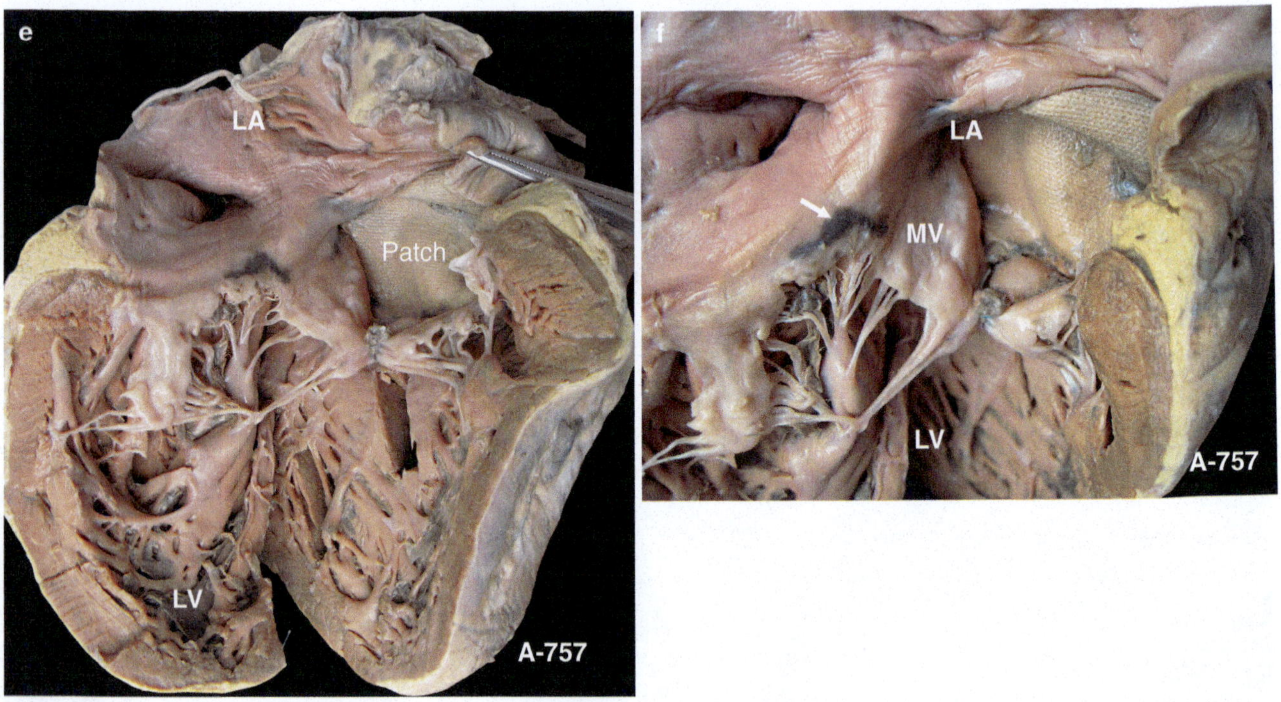

Fig. 12.5 (continued)

12.5.4 Complete Atrioventricular Septal Defect, Small Ventricular Septal Defect

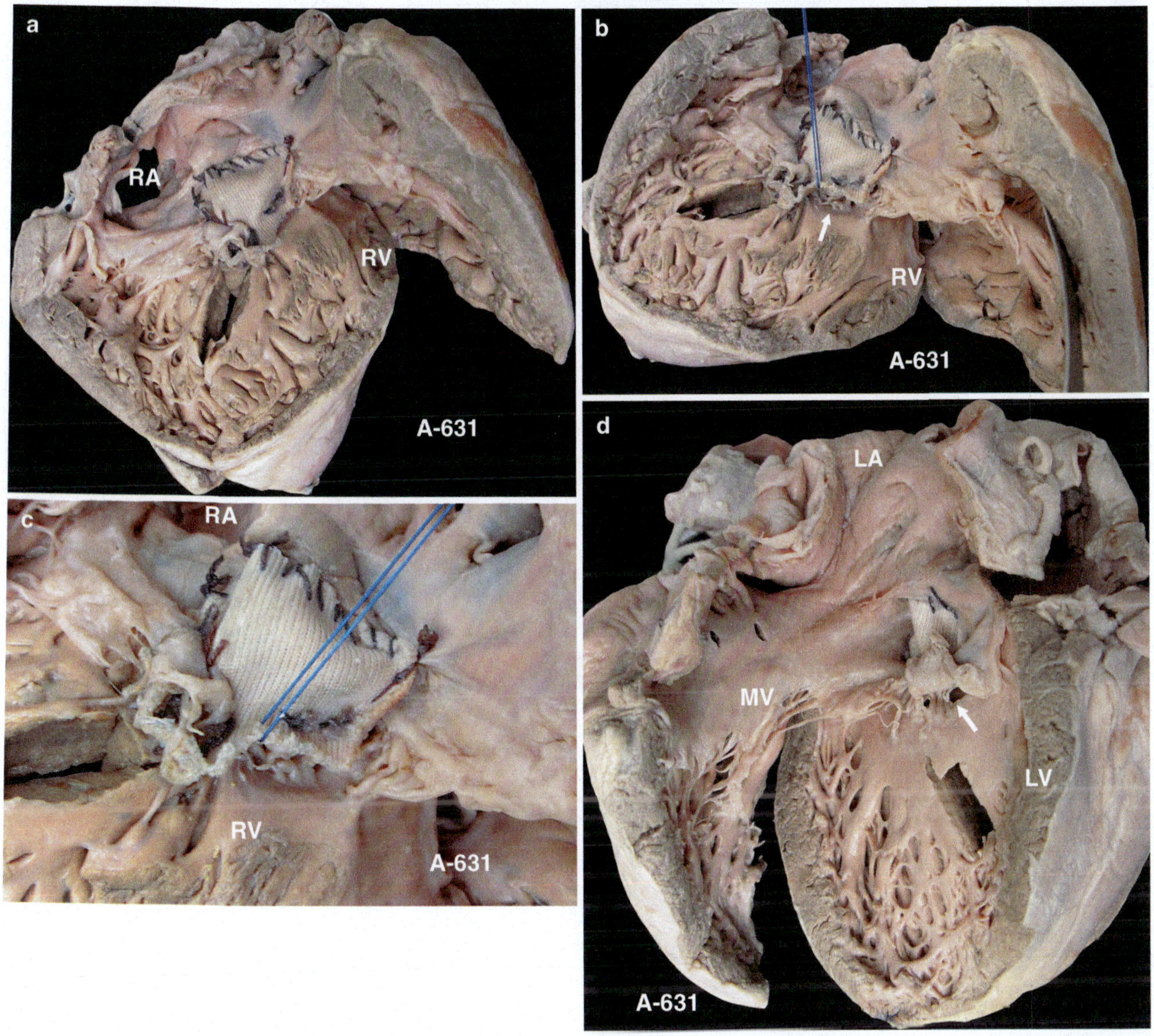

Fig. 12.6 (**a**) Specimen 631. Complete atrioventricular septal defect. Exposure of the atrial defect patch from the RVIT. The VSD is as small as a narrow crack. Numerous chordae are attached to the crest of the ventricular septum and almost obstruct the defect. Coaptation of the TV is poor, which may cause regurgitation postoperatively. *RVIT* right ventricular inflow tract, *TV* tricuspid valve, *VSD* ventricular septal defect. (**b**) Specimen 631. Viewed from the right atrium. A stay suture is placed at the inferior edge of the atrial defect patch to show the narrow deficiency (*arrow*) of the ventricular septum below the valves, the rim of which is attached by many chordae that nearly close the VSD. Hence, the VSD is left without repair. *VSD* ventricular septal defect. *RV* right ventricle. (**c**) Specimen 631. Viewed closely from the right atrium. The cleft in the TV has not been sewn, and the superior common valve is still far away from the inferior, certainly leaving regurgitation behind. *TV* tricuspid valve. *RA* right atrium, *RV* right ventricle. (**d**) Specimen 631. Exposure of the atrial defect patch from the left ventricular inflow tract. The interventricular portion of the AVSD is small and attached by chordae (*arrow*), so it is not closed. *LA* left atrium, *LV* left ventricle, *MV* mitral valve. (**e**) Specimen 631. Adequate exposure of the interventricular portion of atrioventricular septal defect from left ventricular outflow tract. Many chordae are attached to the crest of the ventricular septum (inferior edge of the VSD) (*arrow*). *AO* ascending aorta. *LA* left atrium. *LV* left ventricle. *MV* mitral valve. (**f**) Specimen 631. Diagram of complete atrioventricular septal defect. The ASD is dominant, and the VSD is small with chordae attached to its inferior edge. *AVSD* atrioventricular septal defect, *VSD* ventricular septal defect. *LA* left atrium, *LV* left ventricle, *RA* right atrium, *RV* right ventricle

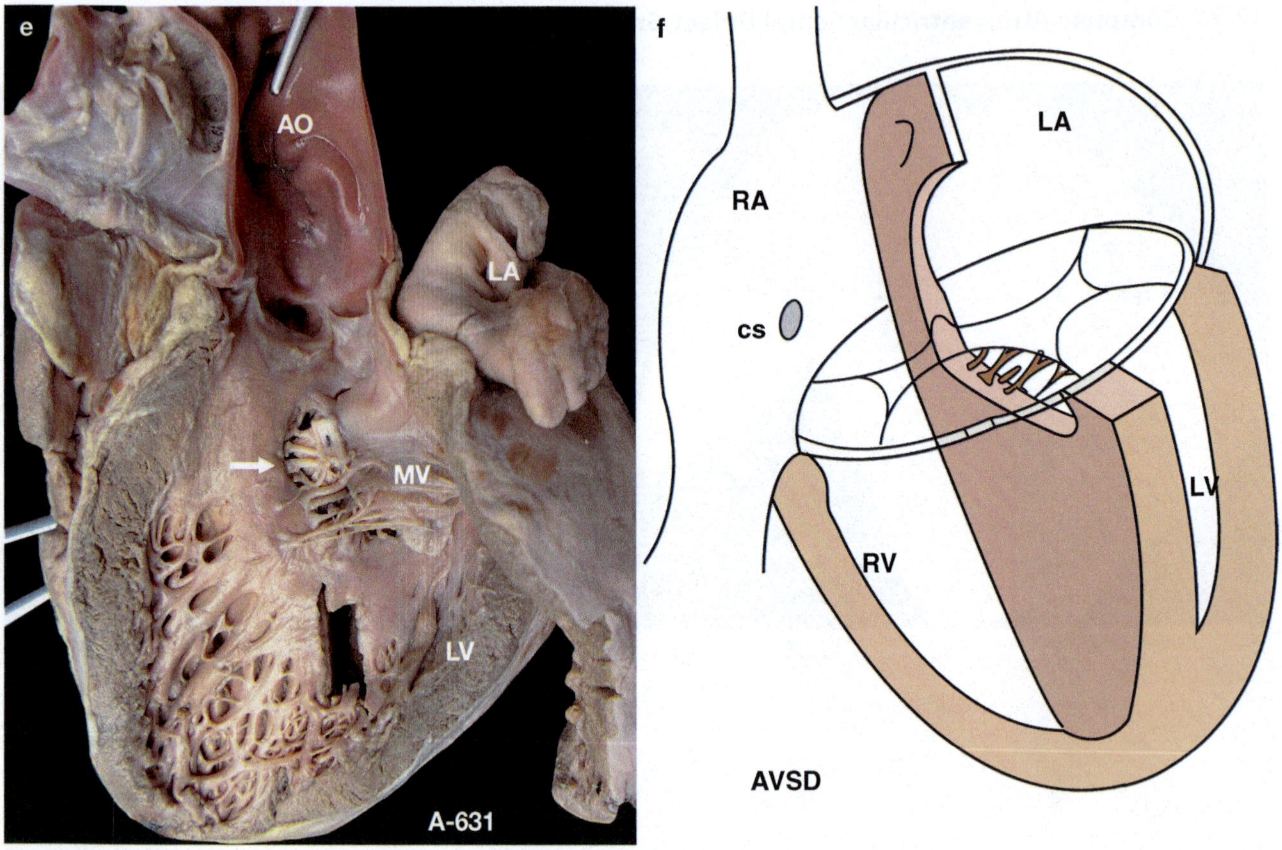

Fig. 12.6 (continued)

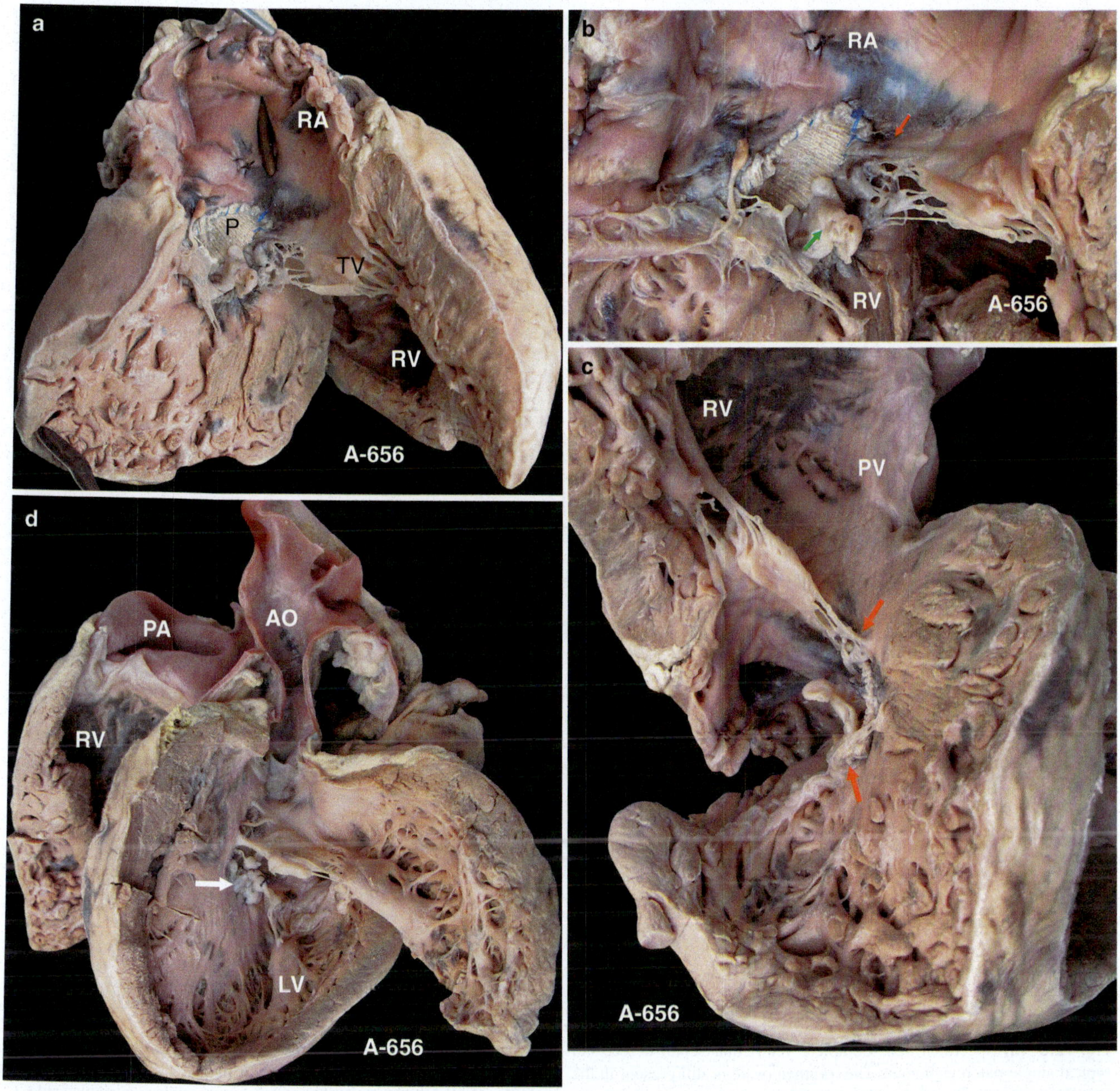

Fig. 12.7 (**a**) Specimen 656. Postoperative complete atrioventricular septal defect. The atrial defect patch (*P*) and a large endocardial hematoma in the portion of the atrioventricular node are revealed from the RA and RV views. *RA* right atrium, *RV* right ventricle, *TV* tricuspid valve. (**b**) Specimen 656. Close view from right atrium. *Red arrow* shows superior common valve, and *green arrow* shows inferior common valve, both of which do not coaptate well and may lead to TV regurgitation. The atrioventricular node is anterior and inferior to the patch, where there is a large endocardial hematoma (*purple area*). *RA* right atrium, *RV* right ventricle, *TV* tricuspid valve. (**c**) Specimen 656. View from right ventricular outflow tract. The cleft in the TV is shown with an *arrow*. The superior and inferior common valves are not sutured together. Attention is paid to the endocardial hematoma. *PV* pulmonary valve, *TV* tricuspid valve. *RV* right ventricle. (**d**) Specimen 656. View from left ventricular outflow tract. The interventricular portion of the AVSD (*arrow*) and the cleft in the MV are observed. The VSD is small and narrow, closed by the chordae on the edge of the defect. *AO* ascending aorta, *LV* left ventricle, *MV* mitral valve, *PA* pulmonary artery, *RV* right ventricle, *VSD* ventricular septal defect. (**e**) Specimen 656. Exposure of the atrioventricular septal defect patch from the left atrium and the left ventricular outflow tract. The patch is confined to the interatrial portion. There is no shunt through the VSD, so it is not treated. *LA* left atrium, *LV* left ventricle, *VSD* ventricular septal defect. (**f**) Specimen 656. View from left ventricle. This view shows the atrial defect patch and the superior, inferior common atrioventricular valves (*arrow*). *LA* left atrium, *LV* left ventricle, *VSD* ventricular septal defect.

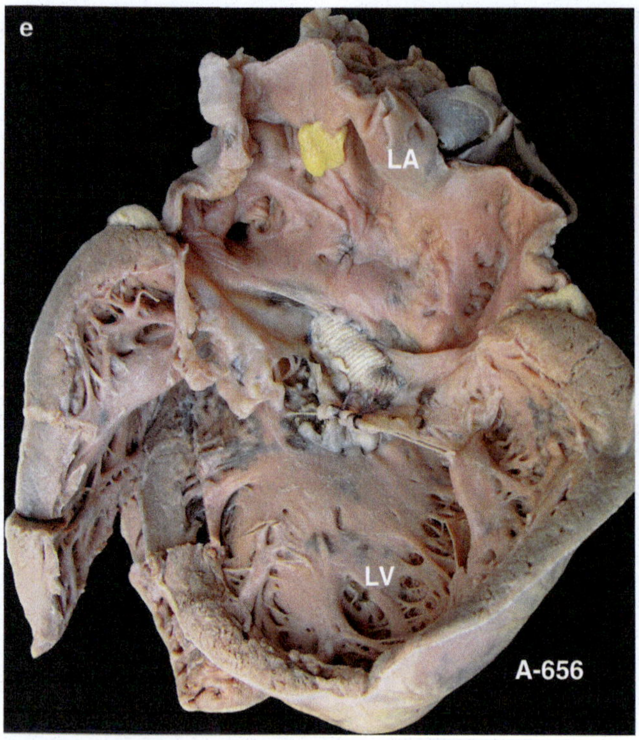

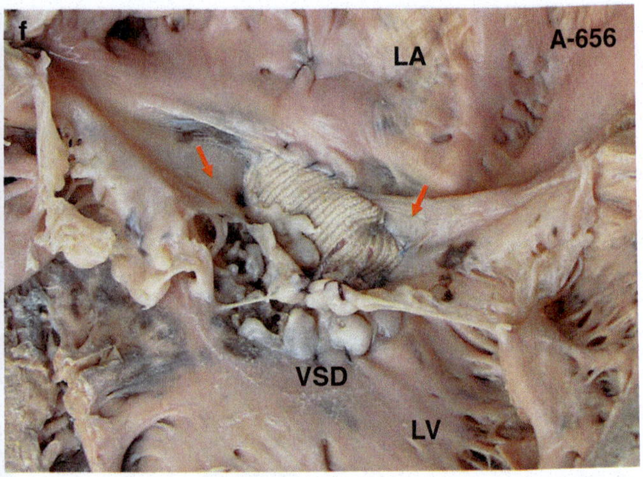

Fig. 12.7 (continued)

Fig. 12.8 (**a**) Specimen 823. Complete ventricular septal defect and typical single-patch technique. The foramen ovale is still present in the RA. The septal leaflet of the TV is completely torn apart, with a large VSD beneath the atrioventricular valve. The "single-patch technique" uses a single patch to repair the ASD and VSD at the same time, suturing the left and right atrioventricular valves to the same patch at the appropriate height to ensure good coaptation of the valves without regurgitation. Presence of the foramen ovale can help to identify the AVSD. *AVSD* atrioventricular septal defect, *RA* right atrium, *TV* tricuspid valve, *VSD* ventricular septal defect. *FO* fossa ovalis, *RV* right ventricle. (**b**) Specimen 823. View from right ventricular outflow tract to expose the interventricular portion of AVSD. The patch is sutured using interrupted pledgeted mattress sutures *red arrow*. (**c**) Specimen 823. Viewed from left ventricular outflow tract to expose the interventricular portion of AVSD. Note that (1) the VSD patch is on the right ventricular side, (2) the chordae of the MV are attached to the edge of the VSD, (3) the cleft in the MV, and (4) slight LVOT obstruction. *AO* ascending aorta, *LVOT* ventricular outflow tract, *LV* left ventricle, *MV* mitral valve, *VSD* ventricular septal defect. (**d**) Specimen 823. View from left atrium and left ventricular inflow tract. The AVSD is closed with a single Dacron patch, while the superior common valve (*red arrow*) and the inferior common valve (*green arrow*) of the MV are sutured together and remain on the patch. Repair of the malformation is satisfactory. *LA* left atrium, *LV* left ventricle. *MV* mitral valve

12.5.5 Complete Atrioventricular Septal Defect With Typical Single-Patch Technique

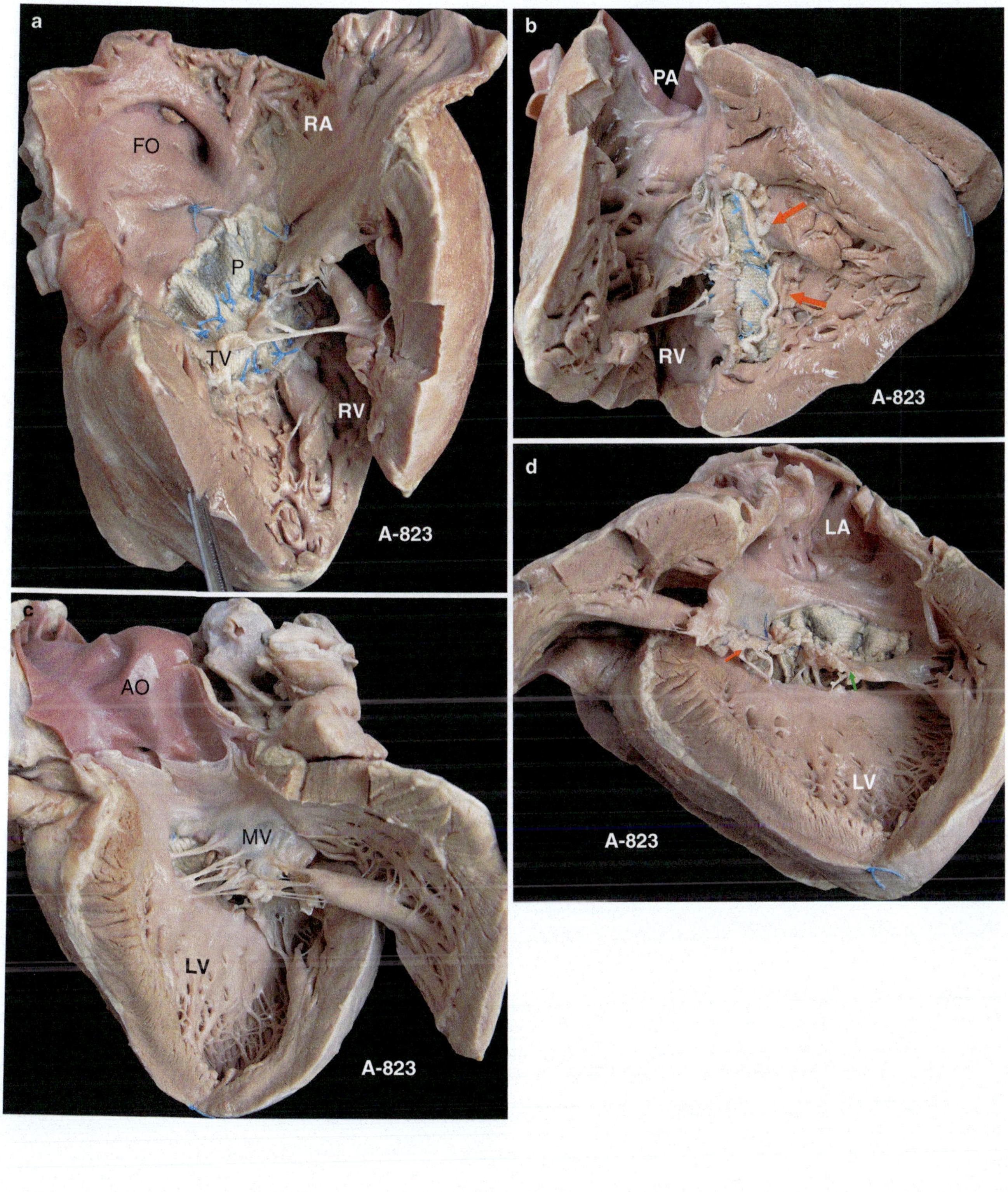

12.5.6 Complete Atrioventricular Septal Defect, Small Ventricular Septal Defect

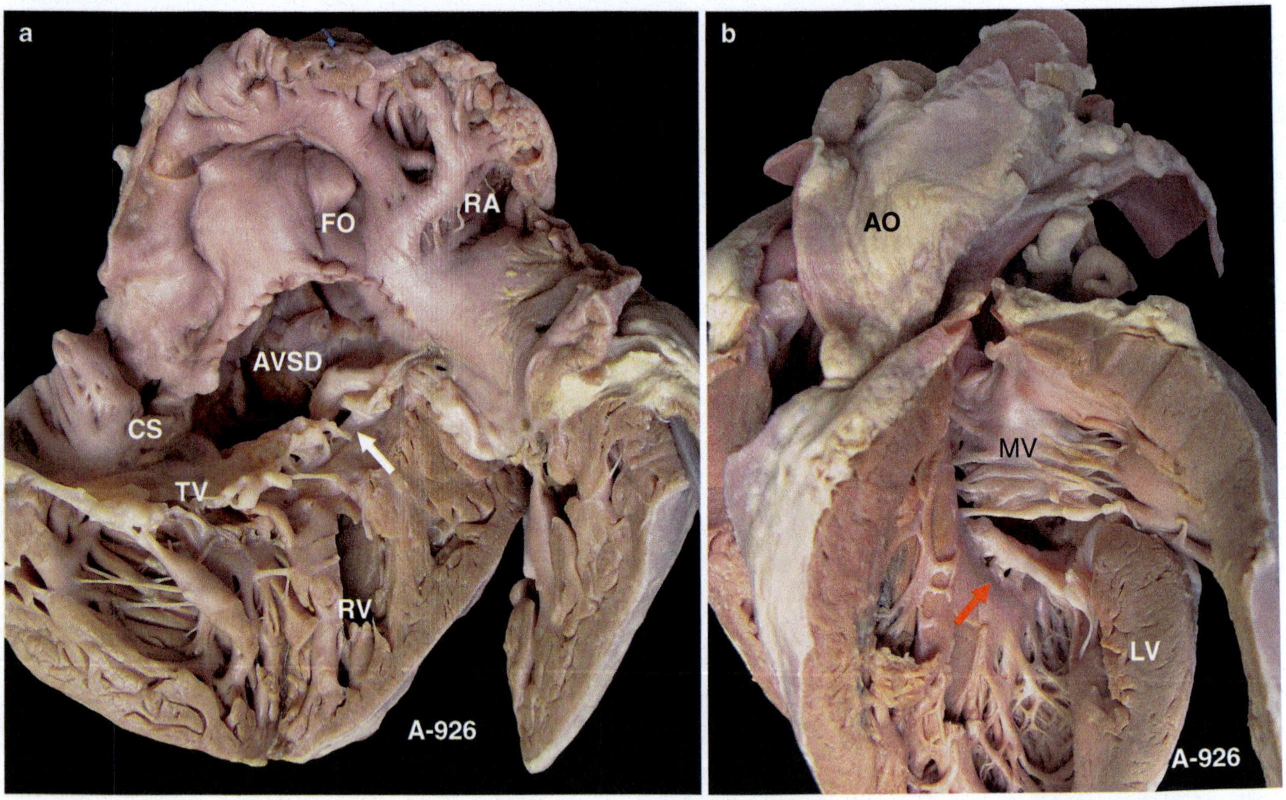

Fig. 12.9 (**a**) Specimen 926. Complete atrioventricular septal defect viewed from the right atrium and the right ventricular inflow tract. The patch has been removed. Only the remainder of the suture line is left in the atrial septum. The foramen ovale is still present above the defect (*white arrow*). The VSD is small and narrow, attached by fence-shaped stripes below the TV. *TV* tricuspid valve, *VSD* ventricular septal defect. *AO* ascending aorta, *AVSD* atrioventricular septal defect, *CS* coronary sinus, *RA* right atrium, *RV* right ventricle. (**b**) Specimen 926. View from left ventricular outflow tract. The cleft in the MV that is connected with the TV to form common atrioventricular valves is apparent. The ASD is prominent in this specimen of AVSD, with only a small VSD (*arrow*) that is attached by chordae. There is no functional shunt. *AO* ascending aorta, *LV* left ventricle. (**c**) Specimen 926. View from the left atrium and the left ventricular inflow tract. The ASD is verified as being predominant in this specimen of complete AVSD, and the VSD is only a narrow slit. The *arrow* indicates the superior common valve formed by mitral and tricuspid valves. *LV* left ventricle. (**d**) Specimen 926. Primary exposures of the common valves. The left superior leaflet has bridged on the right side of the ventricular septum, and chordae are attached to the crest of the ventricular septum under the inferior common valve. *A* Superior common valve, *p* inferior common valve, *AVSD* atrioventricular septal defect, *LA* left atrium, *LV* left ventricle.

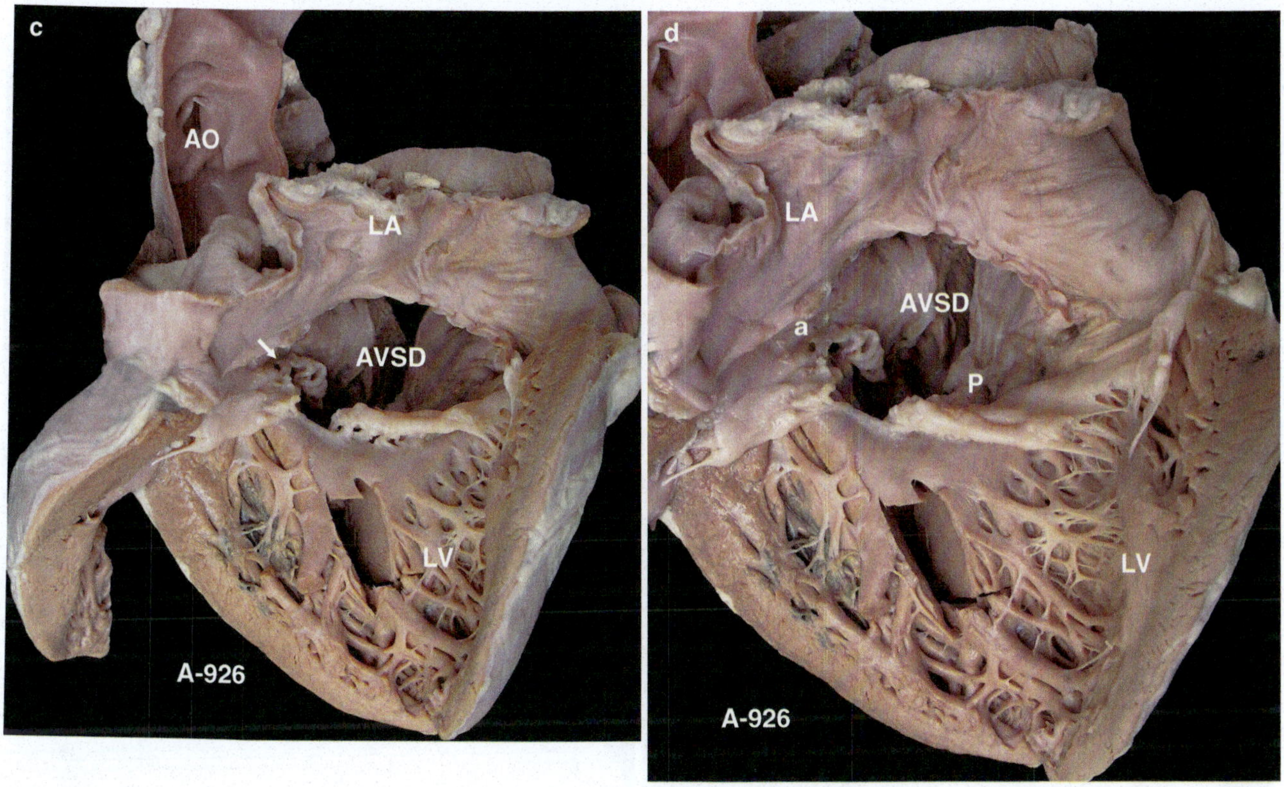

Fig. 12.9 (continued)

12.5.7 Repair of Complete Atrioventricular Septal Defect With Single-Patch Technique

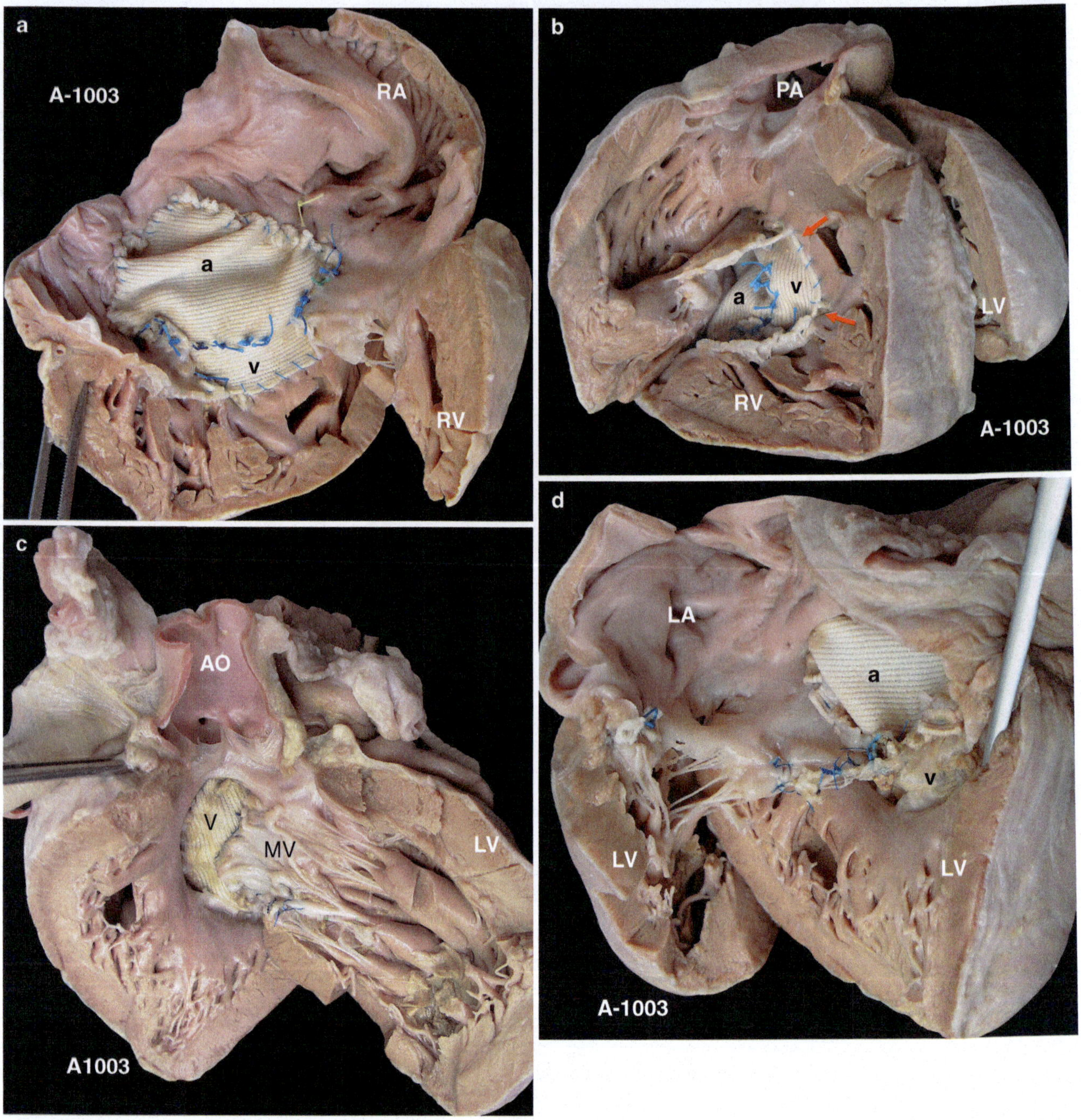

Fig. 12.10 (**a**) Specimen 1003. Right atrium view of repair of complete AVSD with single-patch technique. Clearly seen is that a Dacron patch is placed on the atrioventricular septum. The remaining knots on the patch indicate the position where the MV is sutured on the other side of the patch. The coaptation of the TV is poor, but the defect patch is steady. *a* the ASD portion, *v* the VSD portion. *RA* right atrium, *RV* right ventricle. (**b**) Specimen 1003. View from the right ventricular outflow tract to observe the AVSD patch. Two *arrows* show the position of the common atrioventricular valves, which are too far away from each other and fail to coaptate, causing postoperative TV regurgitation. *a* the ASD portion, *v* the VSD portion, *LV* left ventricle, *PA* pulmonary artery, *RV* right ventricle. (**c**) Specimen 1003. View from left ventricular outflow tract to expose the VSD portion of the AVSD. The "*v*" indicates the patch on the right ventricular side, which is smooth and steady. The MV has been sutured to the patch, and the correction is satisfactory. *AO* ascending aorta, *LV* left ventricle. (**d**) Specimen 1003. View from the left atrium and left ventricular inflow tract. The left atrioventricular valve has been sutured to the lower third level of the Dacron patch, and the cleft in the MV has also been closed. This figure shows a typical repair of a complete AVSD with a single-patch technique. *a* interatrial portion, *v* interventricular portion, *LA* left atrium, *LV* left ventricle. (**e**) Specimen 1003. Diagram of repair of complete AVSD with single-patch technique. The *green part* represents the single patch. *CS* coronary sinus, *LA* left atrium, *LV* left ventricle, *RA* right atrium, *RV* right ventricle

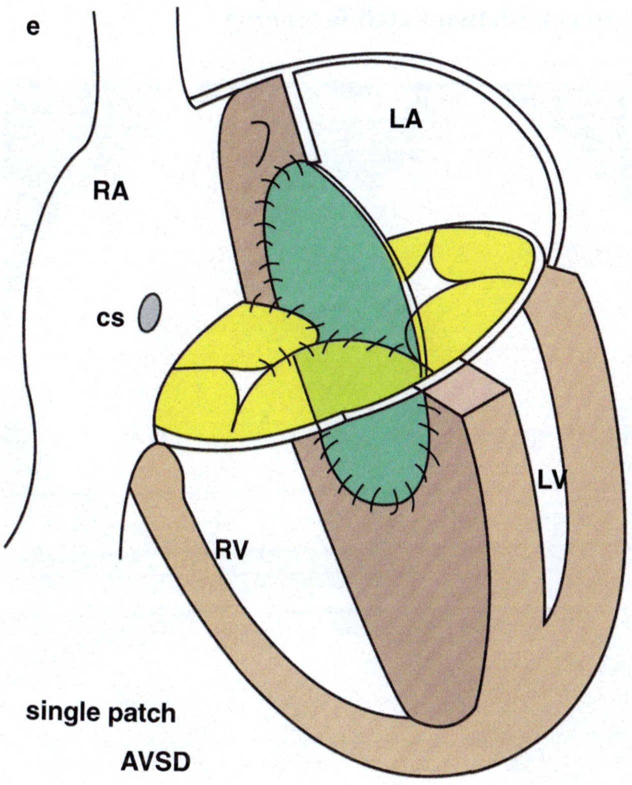

Fig. 12.10 (continued)

12.5.8 Repair of Complete Atrioventricular Septal Defect With two-Patch Technique

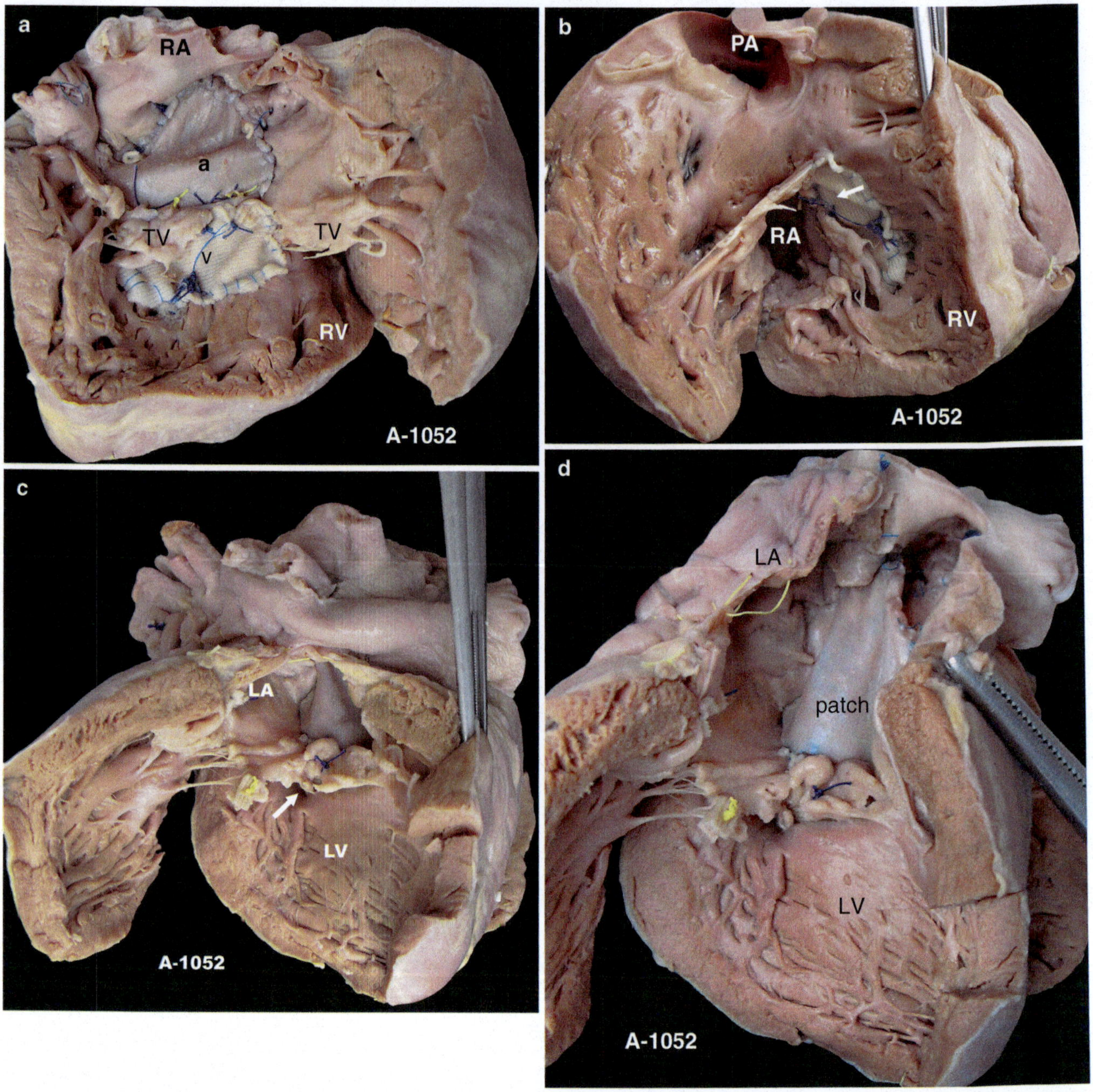

Fig. 12.11 (**a**) Specimen 1052. Specimen of postoperative complete atrioventricular septal defect repaired with two-patch technique. The interatrial defect is repaired with a pericardial patch (*a*), and the interventricular defect is closed with a Dacron patch (*v*). Both join together at the level of the atrioventricular annulus, and the common atrioventricular valves are sewn to the Dacron patch. Generally, the two-patch technique is more accurate and secure than is the single-patch technique. *RA* right atrium, *RV* right ventricle, *TV* tricuspid valve. (**b**) Specimen 1052. View from right ventricular outflow tract. Only the VSD patch can be seen (*arrow*). *PA* pulmonary artery, *RA* right atrium, *RV* right ventricle. (**c**) Specimen 1052. Exposing the atrioventricular septal defect patch from the left ventricular outflow tract. Only the ASD portion is exposed, and the VSD patch is sheltered by the leaflets of the MV. *Arrow* shows the inferior edge of the VSD. *LA* left atrium, *LV* left ventricle. (**d**) Specimen 1052. Left atrium view of a specimen of a complete atrioventricular septal defect to expose the interatrial pericardial patch (*patch*). *LA* left atrium, *LV* left ventricle. (**e**) Specimen 1052. View from left ventricular outflow tract of specimen of complete atrioventricular septal defect. The interventricular polyester patch is observed. The LVOT is narrowed. *AV* aortic valve, *LV* left ventricle, *MV* mitral valve

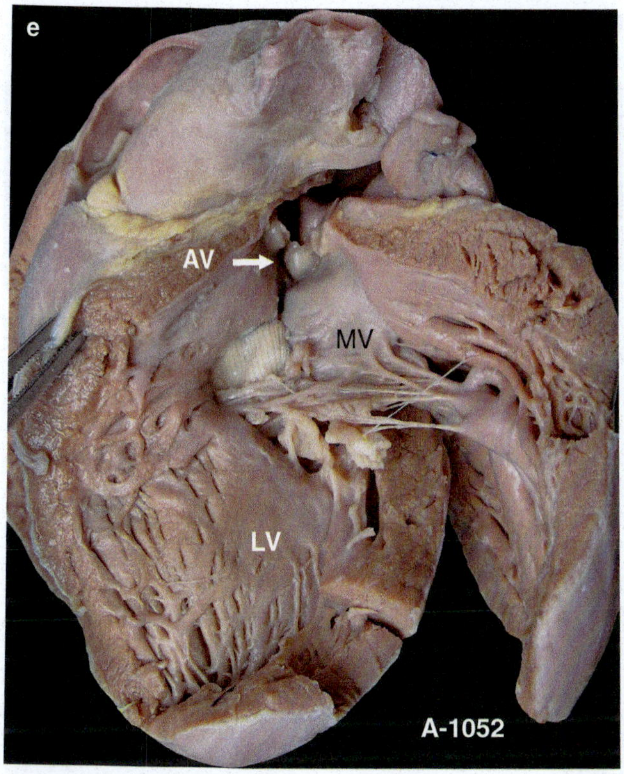

Fig. 12.11 (continued)

12.5.9 Complete Atrioventricular Septal Defect, With Large Ventricular Septal Defect

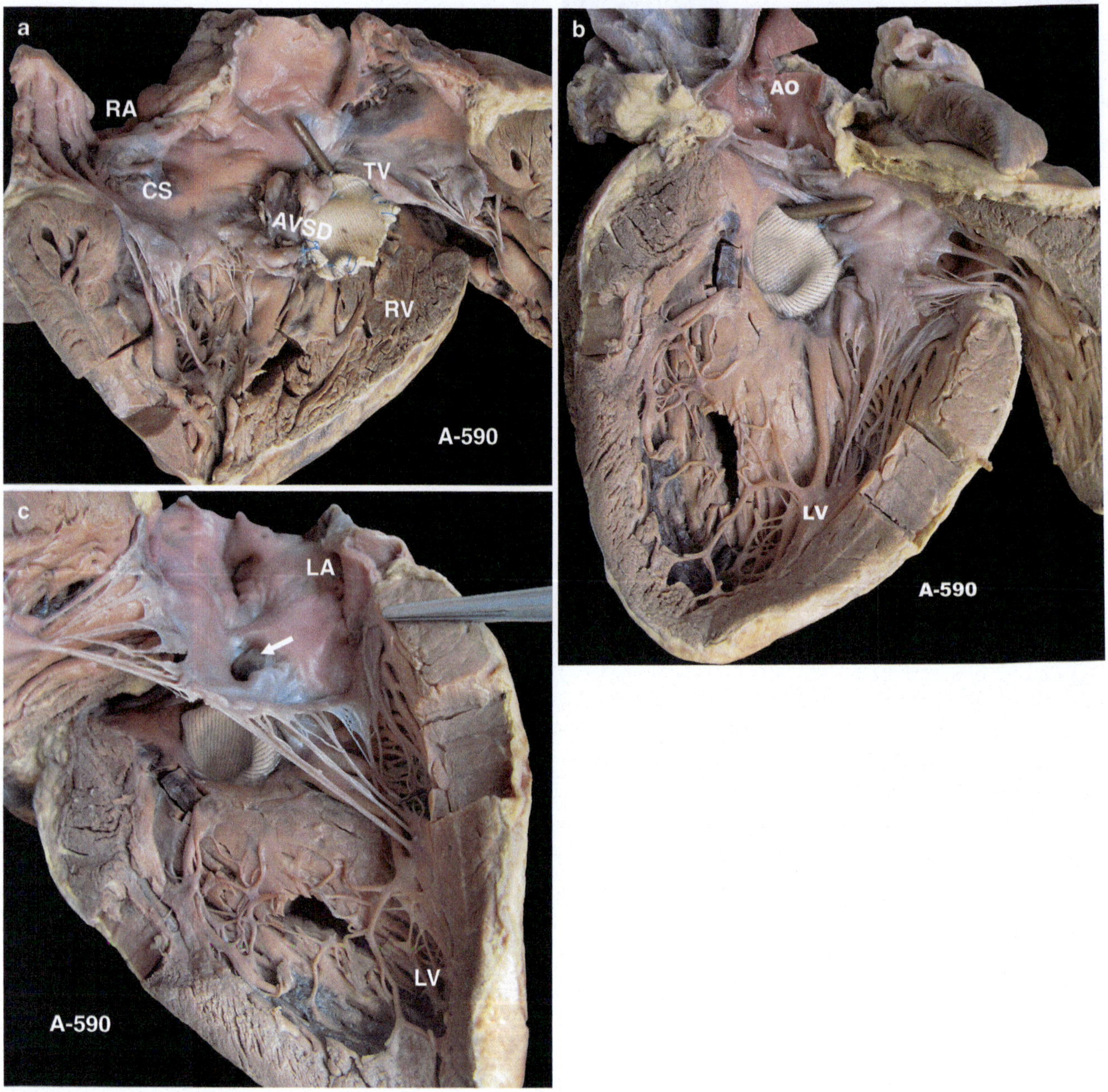

Fig. 12.12 (**a**) Specimen 590. Complete atrioventricular septal defect. The ASD is small, and the VSD is predominant. A residual shunt is present inferior to the patch (bougie). A large endocardial hematoma can be seen in the location of the atrioventricular node near the posterior edge of the patch, which may impart the conductive function of the atrioventricular node and the bundle of His. Note that the coronary sinus is superior and posterior to the defect. *AVSD* atrioventricular septal defect, *CS* coronary sinus, *RA* right atrium, *RV* right ventricle, *TV* tricuspid valve. (**b**) Specimen 590. Complete atrioventricular septal defect. Exposing the large VSD patch from the left ventricular outflow tract view. The MV has been fixed on the Dacron patch. *AO* ascending aorta, *CS* coronary sinus, *LV* left ventricle, *RA* right atrium. (**c**) Specimen 590. View from left ventricular inflow tract. The large VSD patch is superior to the MV. *Arrow* indicates a small ASD near the mitral annulus. *LA* left atrium, *LV* left ventricle, *RV* right ventricle

12.5.10 Repair of Complete Atrioventricular Septal Defect With Two-Patch Technique

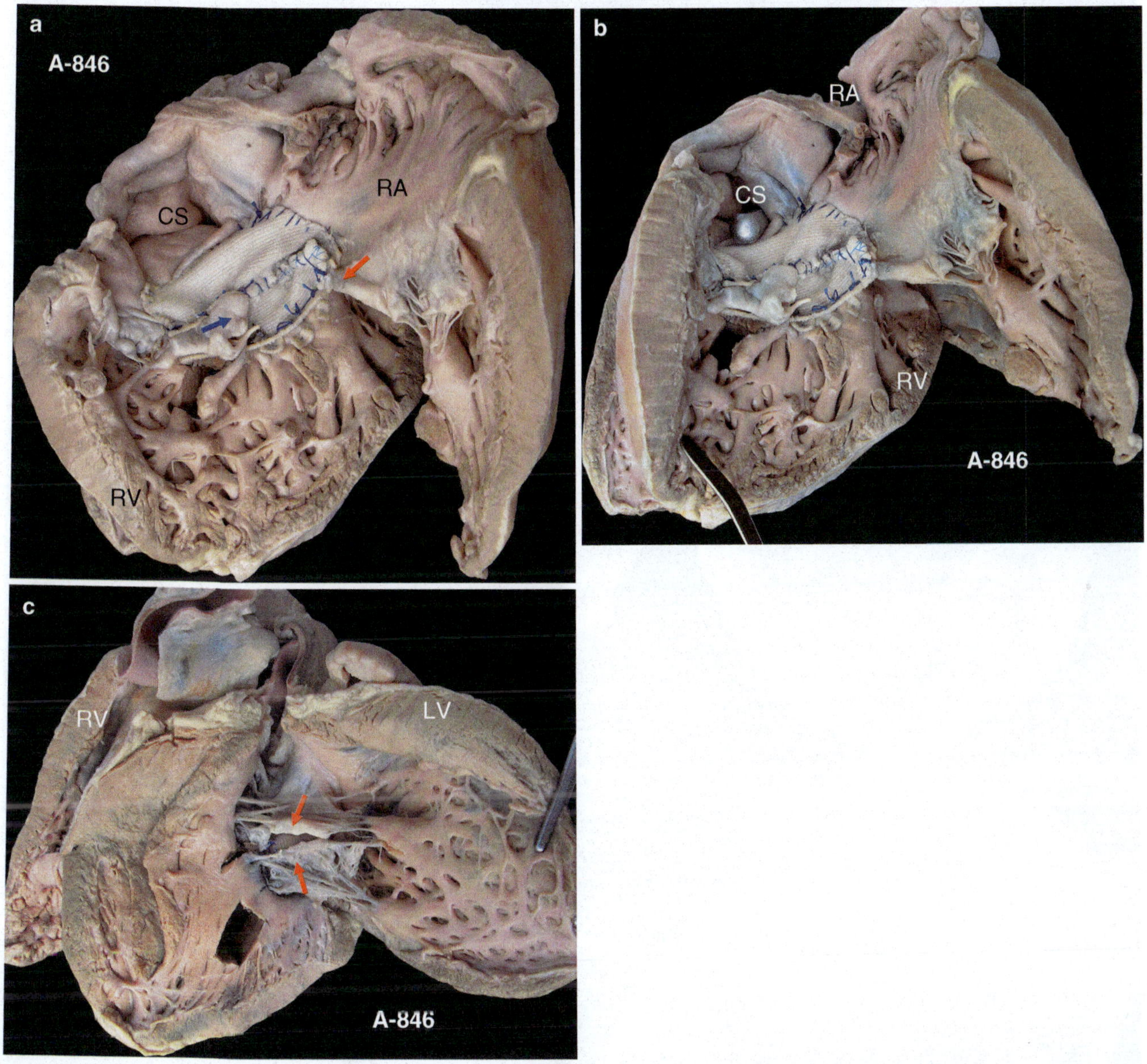

Fig. 12.13 (**a**) Specimen 846. Repair of complete atrioventricular septal defect with two-patch technique. The TV leaflets do not coaptate well, and the left SVC is led to a large coronary sinus. The defect has been repaired with a two-patch technique. The *red arrow* shows the superior common valve, and the *blue* one shows the inferior common valve. The two common valves are not sutured together, which may leave regurgitation of the TV. *CS* coronary sinus, *RA* right atrium, *RV* right ventricle. (**b**) Specimen 846. Large orifice of the coronary sinus in the right atrium. Left SVC is present. *RV* right ventricle, *SVC* superior vena cava. (**c**) Specimen 846. Left ventricular outflow tract view of specimen of complete atrioventricular septal defect to expose the ventricular portion. Cleft in the MV that has not been closed with suture is shown with an *arrow*. Meanwhile, MV chordae are attached to the muscular crest of the VSD

12.5.11 Partial Atrioventricular Septal Defect

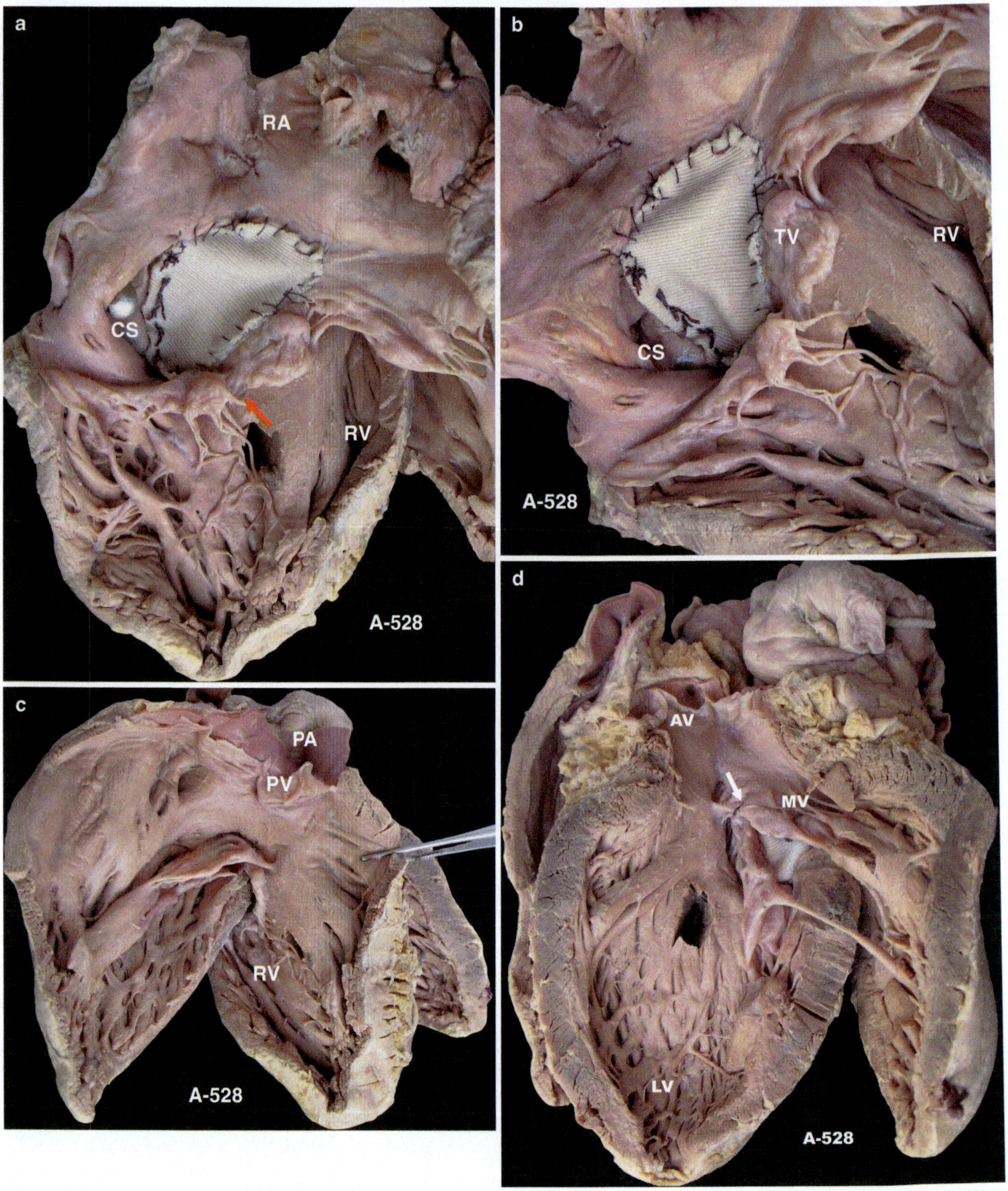

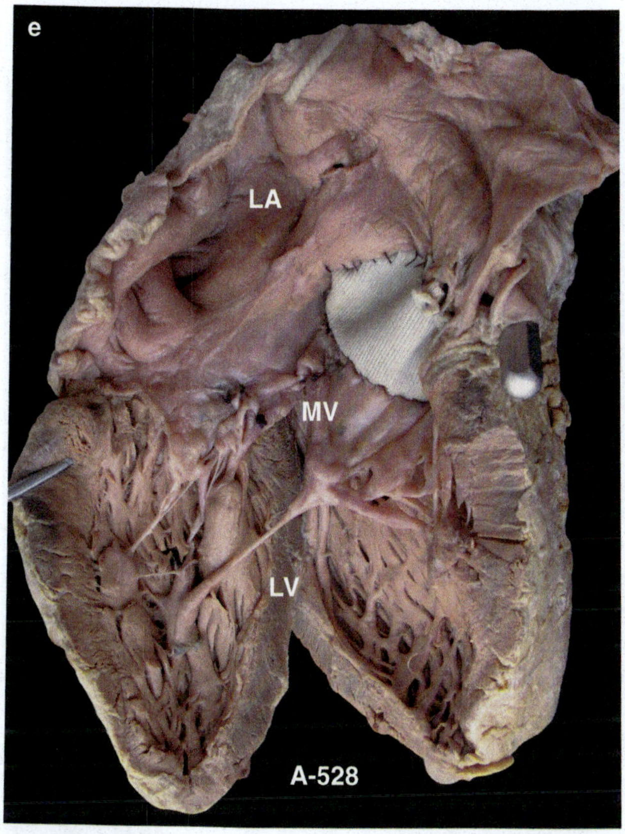

Fig. 12.14 (continued)

Fig. 12.14 (**a**) Specimen 528. Partial atrioventricular septal defect. The ventricular septum is nearly intact, but the left and right atrioventricular valves connect with each other to form a common atrioventricular valve. The *arrow* indicates the cleft in the TV. Viewed from the RA, the defect has been repaired with a Dacron patch, and the coronary sinus is superior and posterior to the patch. There is a metal bougie placed in the coronary sinus. (**b**) Specimen 528. Right atrium view of specimen of partial atrioventricular septal defect. The inferior edge of the ASD is the TV annulus. The cleft in the TV has reached the ventricular septum and continued to the MV on the other side. *CS* coronary sinus, *RV* right ventricle, *TV* tricuspid valve. (**c**) Specimen 528. Viewed from the right ventricular outflow tract. No interventricular communication is present. There is a fissure in the MV. The pulmonary artery is normal. *PA* pulmonary artery, *PV* pulmonary valve, *RV* right ventricle. (**d**) Specimen 528. Viewed from the left ventricular outflow tract. (1) The cleft in the MV has reached the root of the leaflet; (2) the MV continues with the TV on the other side (*arrow*) to produce a common atrioventricular valve, which differs from ostium primum ASD; (3) a narrow fissure is present on the ventricular septum below the MV, which is the remnant of the closure of the interventricular foramen at the embryo stage. *AV* aortic valve, *LV* left ventricle. (**e**) Specimen 528. View from left atrium to expose the partial atrioventricular septal defect patch. The cleft in the MV has been sewn, and the metal bougie is established in the coronary sinus

12.5.12 Complete Atrioventricular Septal Defect With Primary Atrial Septal Defect

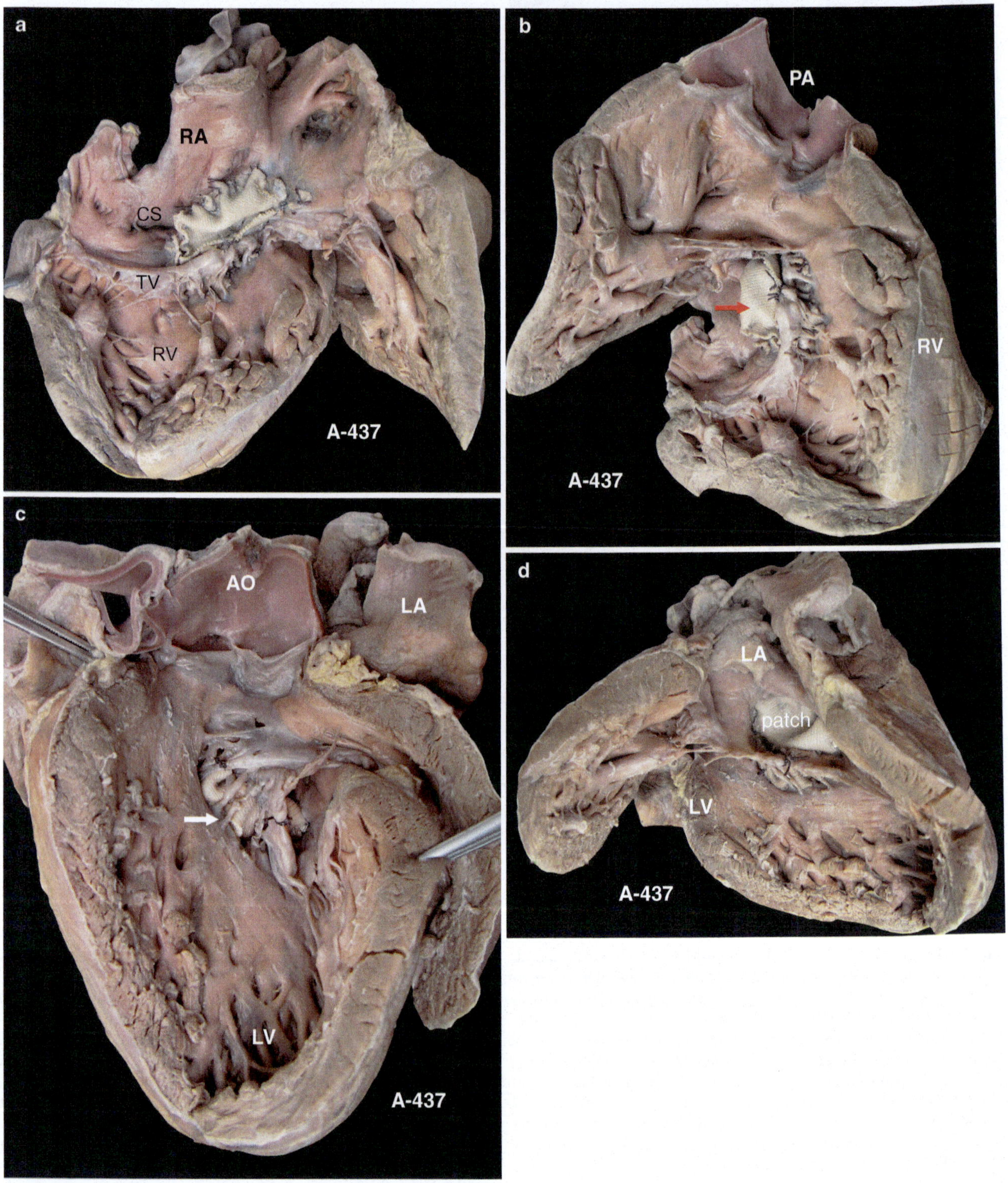

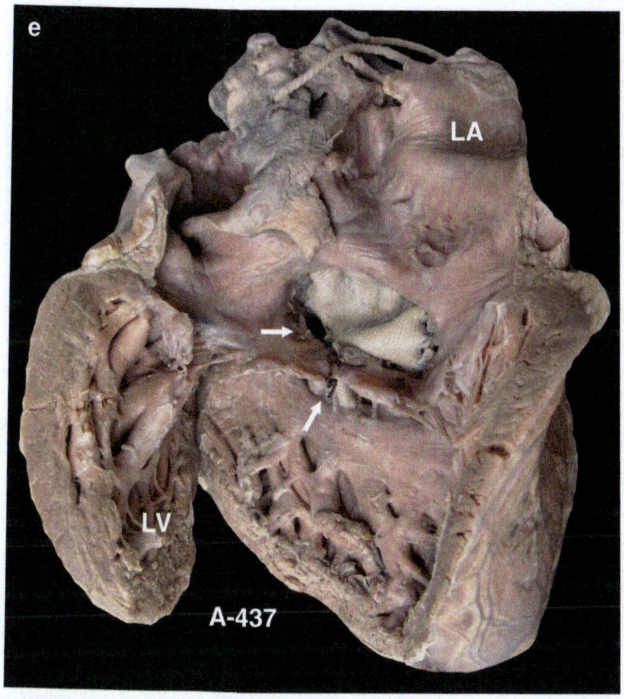

Fig. 12.15 (continued)

Fig. 12.15 (**a**) Specimen 437. Complete atrioventricular septal defect. In the specimen, the ASD is primary and the VSD is small. This figure shows the ASD patch from the RVIT, and the inferior edge of the VSD patch is visible beneath the TV. *CS* coronary sinus, *TV* tricuspid valve, *RA* right atrium, *RV* right ventricle. (**b**) Specimen 437. View from the right ventricular outflow tract to show complete atrioventricular septal defect. The *arrow* points at the ASD patch. Below the TV is a narrow VSD, the muscular crest of which is attached by fence-like chordae. *PA* pulmonary artery, *RV* right ventricle. (**c**) Specimen 437. Complete atrioventricular septal defect viewed from left ventricular outflow tract. *Arrow* indicates the edge of the VSD where chordae and patch are present. *AO* ascending aorta, *LA* left atrium, *LV* left ventricle. (**d**) Specimen 437. View from left atrium and left ventricular inflow tract of complete atrioventricular septal defect. The MV is completely expanded as the cleft is sown. Above the MV, the ASD is repaired with a Dacron patch (*patch*); and below it is another patch to close the VSD. The common atrioventricular valve is sutured to the attachment line of those two patches. *LA* left atrium, *LV* left ventricle. (**e**) Specimen 437. View from left atrium. There is residual shunt anterior to the ASD patch (*arrow*). *LA* left atrium, *LV* left ventricle

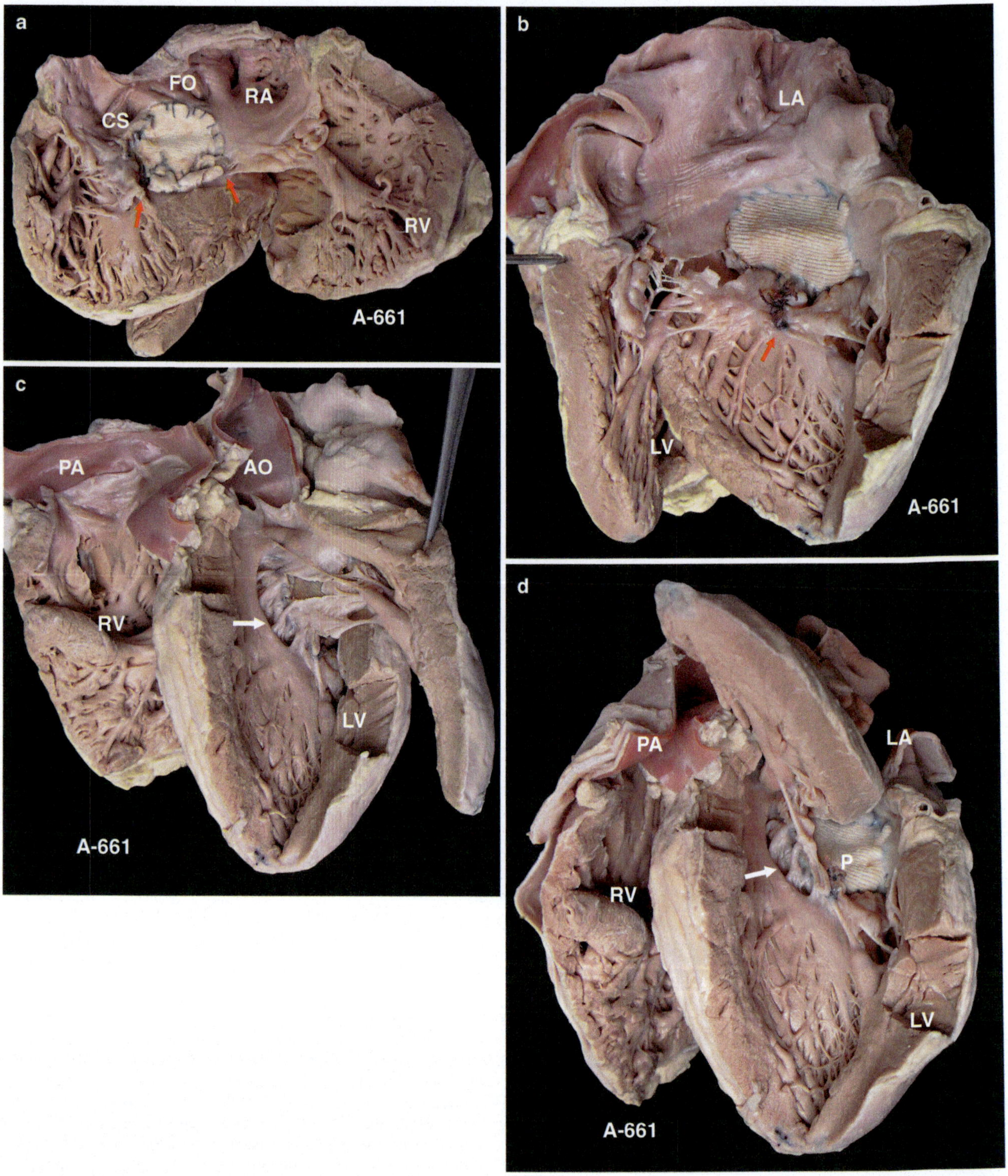

Fig. 12.16 (**a**) Specimen 661. Complete atrioventricular septal defect. The ASD is primary, and the VSD is small. Viewed from the RA and the RVIT, the inferior edge of the ASD is the atrioventricular valve. The defect is repaired. The leaflets of the TV are totally separated (*arrow*). FO and coronary sinus are present in RA. *CS* coronary sinus, *RV* right ventricle. (**b**) Specimen 661. Viewed from left atrium and left ventricular inflow tract. The ASD patch is steady, and the cleft in the MV is sewn (*arrow*). *LA* left atrium, *LV* left ventricle. (**c**) Specimen 661. Viewed from the left ventricular outflow tract to expose the VSD (*arrow*) under the MV and dilatation of the pulmonary artery (*PA*) as well. *AO* ascending aorta, *LV* left ventricle, *RV* right ventricle. (**d**) Specimen 661. The lateral wall of the LV has been lifted up, and both the ASD patch (*P*) and the VSD (*arrow*) may be seen at the same time. Due to the shelter of the chordae, the VSD is not treated. *LA* left atrium, *PA* pulmonary artery, *RV* right ventricle

12.5.13 Complete Atrioventricular Septal Defect With Right Ventricular Outflow Tract Obstruction

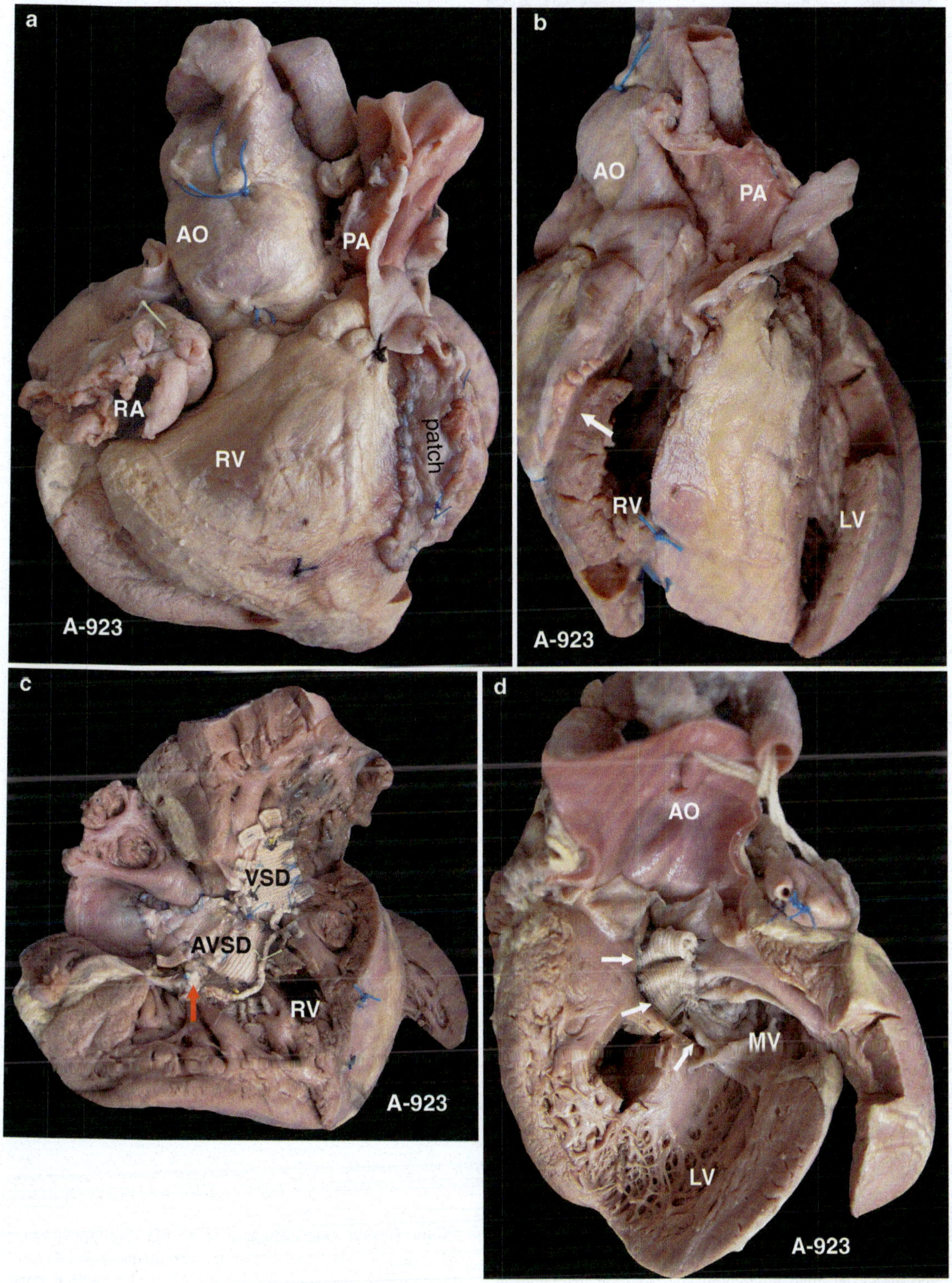

Fig. 12.17 (**a**) Specimen 923 Complete atrioventricular septal defect with right ventricular outflow tract obstruction. The RVOT is enlarged by pericardial *patch*, and the defect is repaired with two-patch technique. *AO* ascending aorta, *PA* pulmonary artery, *RA* right atrium, *RV* right ventricle. (**b**) Specimen 923. Left anterior oblique view. *Arrow* shows that the RVOT is enlarged by a transannular patch. *AO* ascending aorta, *LV* left ventricle, *PA* pulmonary artery, *RV* right ventricle. (**c**) Specimen 923. Complete atrial septal defect. The anterior wall of the RV is lifted up to show that two patches have been applied to repair the AVSD and VSD. The *arrow* points to the common atrioventricular valve, which is also the position through which the bundle of His passes. (**d**) Specimen 923. View from the left ventricular outflow tract. The interventricular part of the complete AVSD is exposed. The *arrow* shows the VSD patch. MV indicates the cleft of the MV. *AO* ascending aorta, *LV* left ventricle. (**e**) Specimen 923. View from left ventricular inflow tract to show the cleft of the mitral valve. The interatrial patch and the interventricular patch can be seen. The complete AVSD is repaired with a two-patch technique. *LA* left atrium, *LV* left ventricle, *MV* mitral valve

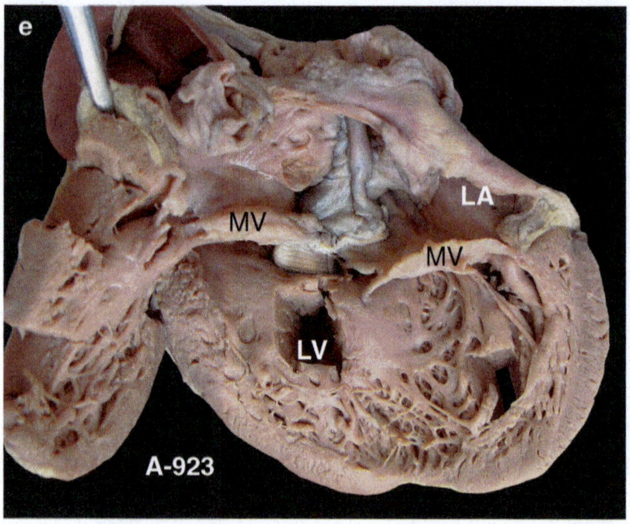

Fig. 12.17 (continued)

Fig. 12.18 (**a**) Specimen 1079. Complete atrioventricular septal defect with double-outlet right ventricle. This figure is viewed from the RA and the RV and shows the small and hypoplastic common atrioventricular valves, which stay in the AVSD (*D*). The *arrow* shows the muscular crest of the ventricular septum that is the inferior rim of the VSD. (**b**) Specimen 1079. View from the right atrium to expose complete atrioventricular septal defect. The *red arrow* indicates the patent foramen ovale. The *dash line* shows the muscular bundle anterior to the SVC in the RA. The tip of the forceps is placed in the coronary sinus, with the AVSD located anteriorly and inferiorly. (**c**) Specimen 1079. Mitral valve and AVSD viewed from the left atrium and left ventricle. Below the common atrioventricular valves is the muscular crest of the ventricular septum (inferior edge of VSD). *a* superior common valve, *p* inferior common valve. (**d**) Specimen 1079. Complete atrioventricular septal defect with double-outlet right ventricle. The pulmonary artery (*PV*) valve and aortic valve (*AV*) both derive from the morphological RV. Inverted transposition of the ventricle coexists in this specimen, and morphological RV is on the left side

12.5.14 Complete Atrioventricular Septal Defect With Double-Outlet Right Ventricle

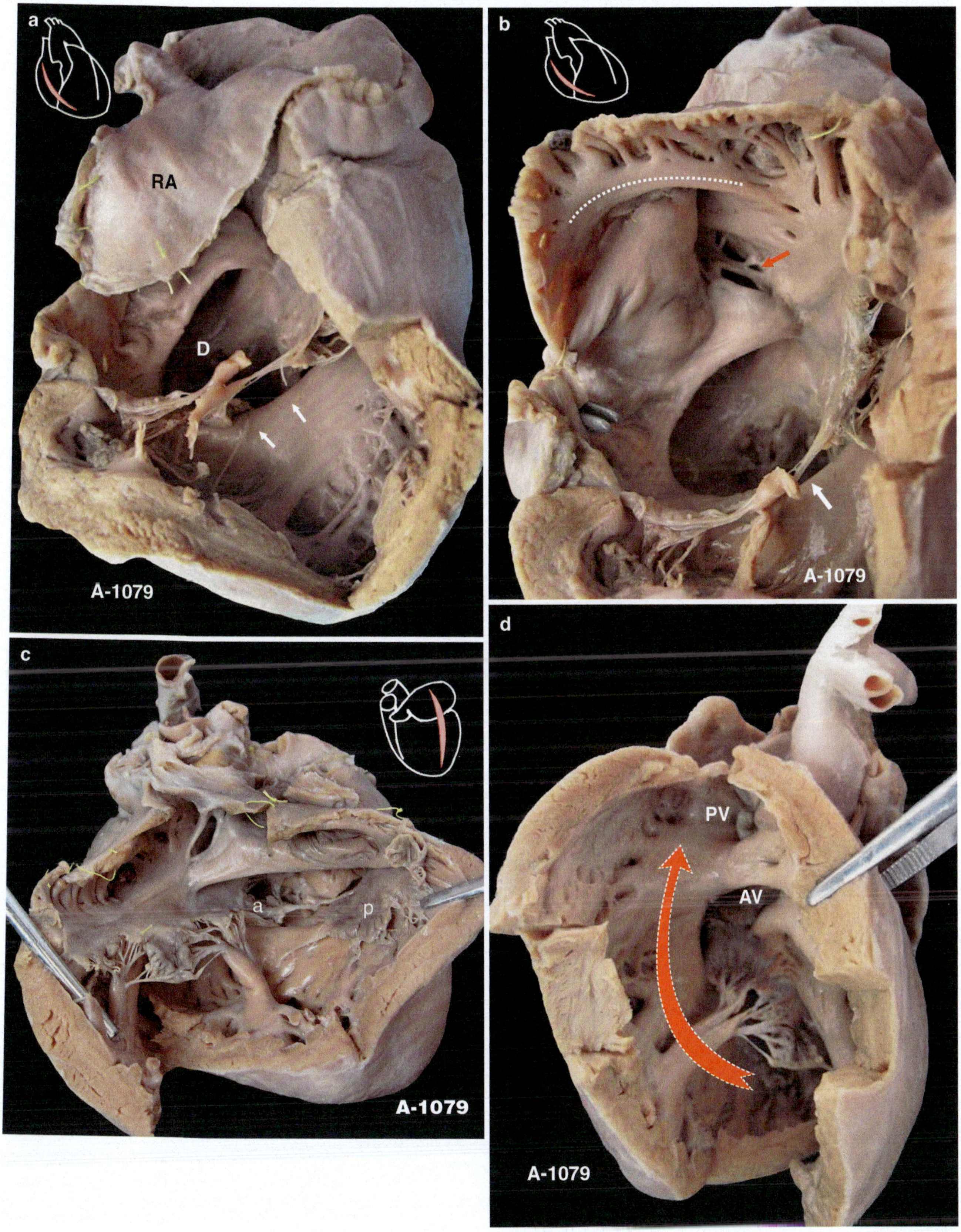

13 Ventricular Septal Defect

This chapter focuses primarily on the primary ventricular septal defect (VSD). The VSD that occurs as a component of other major cardiac malformations is discussed in other chapters.

13.1 Classification of VSDs

Several systems are used to classify VSDs, and the differences among them are not significant. In 2000, Jacobs introduced the consensus of the Congenital Heart Surgery Nomenclature and Database Project (Box 3.1), which divides VSDs into four subtypes as follows (Fig. 13.1):

Type I subarterial VSD
Type II perimembranous VSD
Type III AV-type or inlet-type VSD
Type IV muscular VSD (Fig. 13.5)

In China, no consensus has been reached for classifying VSDs. Box 13.1 lists the classification system used most frequently in Fuwai Hospital, Beijing.

A simple membranous VSD (Figs. 13.4 and 13.3) can be barely distinguishable from an infracristal VSD. Generally, the former is small and bordered by intact white fibrous tissue, whereas the infracristal VSD and the juxtatricuspid VSD are relatively larger. Both simple membranous and infracristal VSDs are quite common occurrences, whereas a muscular VSD is rarely seen.

Because the hypoplastic membranous portion of the atrioventricular septum, also known as the *left ventricular-right atrial defect*, is relevant to the development of the endocardial cushion and has no agreed-upon classification, it can also be classified as a VSD.

Box 13.1 Classification of VSDs

Subtype		Definition
Infundibular VSD	Subarterial	The upper margin of the defect consists of part of the pulmonary and aortic annuli
	Intracristal	The defect is surrounded by muscular tissue; the infundibular septum is separated from the tricuspid annulus by a band of muscle
Membranous VSD	Simple membranous	The margins of the defect are fibrous tissue
Perimembranous VSD	Infracristal	The defect involves membranous septum and partial supraventricular crest, located anterior to the papillary muscle of the conus artery
	Juxtatricuspid	The defect involves membranous septum and partial sinus septum, located posterior to the papillary muscle of the conus artery
	AV canal type	A large defect exists in the sinus septum
Muscular VSD		The defect is in the sinus septum and the trabecular septum, with entire muscular borders

X. Zhu (ed.), *Surgical Atlas of Cardiac Anatomy*,
DOI 10.1007/978-94-017-9409-1_13, © Springer Science+Business Media Dordrecht and People's Medical Publishing House 2015

13.2 Anatomical Key Points of VSDs

1. *Small VSD* (Figs. 13.14, 13.6, and 13.11) is a small defect bordered by white fibrous tissue with a diameter less than 5 mm. The predilection site in the embryonic period is the commissural line of the ventricular septum, especially the inferior edge of the supraventricular crest (equivalent to the commissural line of the anterior and posterior ventricular septum in the embryonic period) and the bulbar ridge (equivalent to the commissural line of the conus artery ridge in the embryonic period). These small VSDs can be sutured directly without patch due to their strong edge. In many cases of small single membranous VSDs, the defects are bordered by the TV and fibrous tissue. The external aperture of the defect seems to be surrounded by a fibrous canal when it is viewed from the RV. The suture line should be placed on the base of the VSD, rather than on the aperture of the canal, to avoid having a loose stitch. Sometimes a fibrous stripe is found across the defect. If it is mistaken for the edge of the defect, a residual defect may be left.
2. *Large subarterial VSD* (Figs. 13.12, 13.15, and 13.16) is tear shaped rather than circular and has no muscular upper margin. Parts of the aortic and pulmonary annuli form the upper margin. The front of the right coronary cusp can be seen through the defect. The AV is easily injured, and the conduction system remains at a distance from the defect. The AV should be protected during the operation.
3. Above the *large infracristal VSD* (Figs. 13.8, 13.9, and 13.2) is the right coronary cusp of the AV, and the posterior edge of the defect, which is either fibrous or muscular, extends to the tricuspid annulus. A conduction bundle runs along the posterior-inferior edge of the defect. The papillary muscle of the conus artery begins posterior to the defect, and the supraventricular crest lies anterior to the defect. The defect usually is large and should be repaired with a patch rather than direct suture. The conduction system should be protected when repairing the posterior inferior edge of defect.
4. The defect beneath the septal leaflet of the TV and the *AV canal type of VSD* (Figs. 13.13, 13.19, 13.20, 13.7, and 13.24) are usually long and slender, located below the septal leaflet of the TV. The tricuspid annulus forms the upper margin of the defect and is located posterior to the papillary muscle of the conus artery. The anterior margin remains at a distance from the AV. Most of the defect can be obscured by the septal leaflet of the TV.

13.3 Surgical Technique: Difficulties and Risks

The purpose of VSD repair is to close the defect and to eliminate left-to-right shunt. Transcatheter closure of ventricular defects is limited, so open-heart surgery is the main method. Any fault or risk can appear during operation, including:

1. A residual, fistula, small membranous defect may be divided into two holes by a fibrous stripe or be surrounded by the fibrous canal below the TV. The hole may be mistaken for the entire defect during operation, resulting in residual shunting and murmur.

   In a large VSD repair or an infant VSD repair with patch, the suture line may be too tight and shallow, cutting the muscular edge and leading to residual shunting. Residual shunting also may appear if continuous stitches are not tightened (Figs. 13.21, 13.22, 13.23, 13.17, and 13.18).
2. AV injury leads to aortic insufficiency. In the case of a large membranous VSD, especially a subarterial VSD, the AV may be injured during surgery, resulting in avulsion and perforation of the cusp. Traction from the suture line may cause deformation of the valve. When these situations arise, reoperation should be performed decisively to repair the injured AV (Fig. 13.10).
3. Complete atrioventricular block may occur. An atrioventricular bundle usually runs along the posteroinferior edge of a membranous VSD (including an infracristal VSD), close to the right ventricular side. Sutures that are too deep may injure the perforating branch or the left and right bundle branch of the bundle of His, resulting in heart block. Violent traction of local tissue may cause temporary or permanent atrioventricular block (Fig. 13.25).
4. Traction of the TV leads to tricuspid insufficiency. A large membranous and perimembranous VSD may be obscured by the chordae of the TV and the papillary muscle of the conus artery. Part of the leaflet or the chordae may be sutured onto the patch so that the TV closure is disturbed, leading to tricuspid insufficiency.

13.4 Specimen Demonstrations

Location of VSDs
Type 1 subarterial/supracristal/conal septal defect/infundibular
Type 2 perimembranous/paramembranous/conoventricular
Type 3 inlet/AV canal type
Type 4 muscular

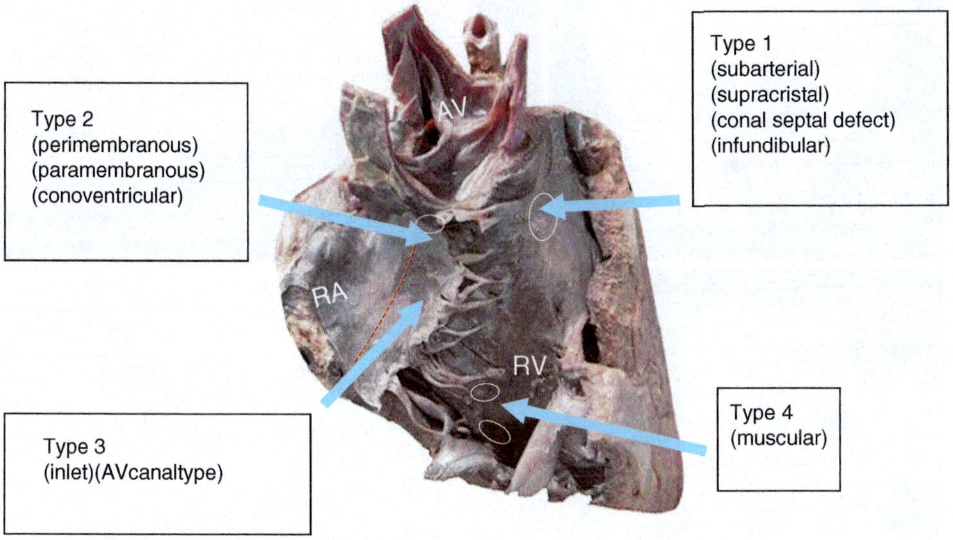

Fig. 13.1 International classification of VSD in contrast to others. The figure demonstrates four common types of VSDs viewed from the RV of the normal heart

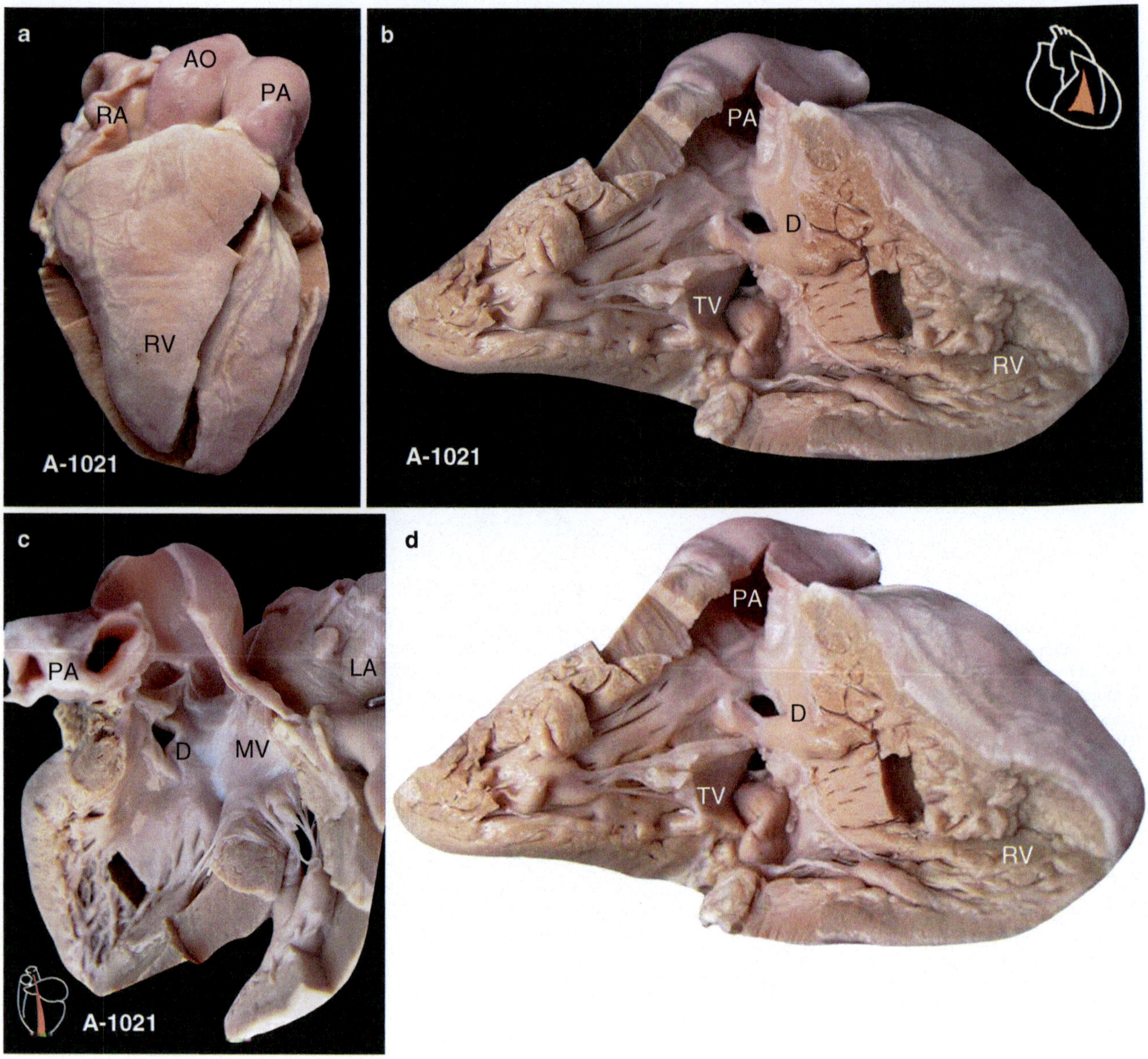

Fig. 13.2 (**a**) Specimen 1021. Small intracristal VSD with normal heart contour. (**b**) Specimen 1021. Right ventricle-pulmonary artery cut open. The anterior wall of the RV is elevated to show the TV, pulmonary artery, and VSD. The defect has intact muscular margins and infundibular muscular tissue that can be seen between the posterior inferior border of the defect and the tricuspid leaflet, so the defect is inside the infundibular septum, representing precisely a small intracristal VSD. (**c**) Specimen 1021. VSD viewed from left ventricle. Defect is away from the membranous septal, so it is not a membranous VSD. The figure also shows the fibrous continuity between the AV and the anterior leaflet of the MV. *LA* left atrium, *MV* mitral valve, *PA* pulmonary artery. (**d**) Specimen 1021. Surgical interpretation for a small intracristal VSD. (*1*) Intracristal VSD can be repaired with a patch or direct suture, avoiding injury to conducting tissue because it is bordered by muscle. (*2*) The VSD occlude is easily released through the TV

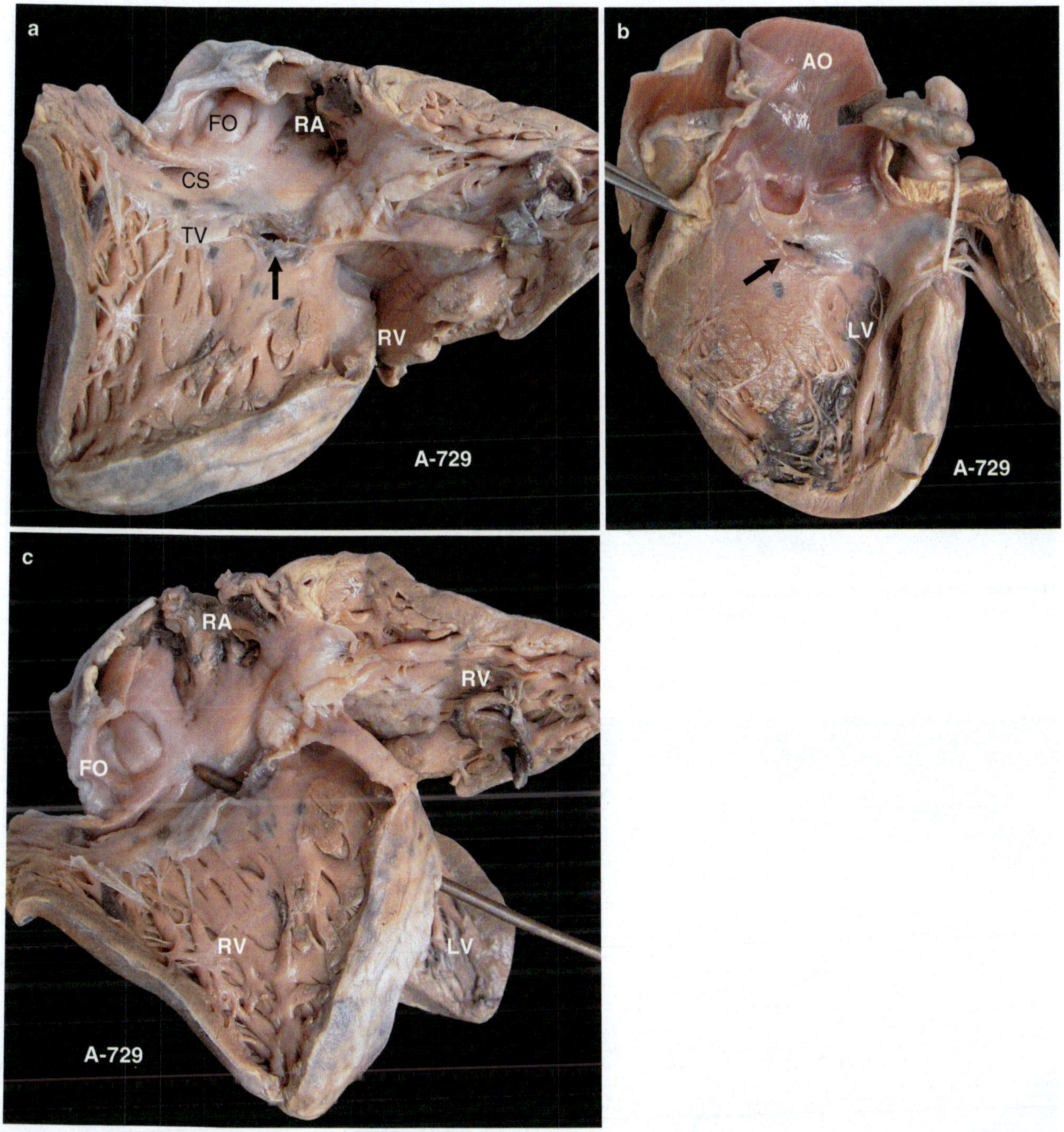

Fig. 13.3 (**a**) Specimen 729. Membranous VSD. The defect is located at the anteroseptal commissure of the TV (*arrow*). *CS* coronary sinus, *FO* fossa ovale, *TV* tricuspid valve. (**b**) Specimen 729. Membranous VSD viewed from the left ventricle. As the *arrow* indicates, the defect was beneath the right aortic cusp and the noncoronary cusp. (**c**) Specimen 729. The probe penetrated the membranous VSD from LV to RV. *LV* left ventricle; *RV* right ventricle

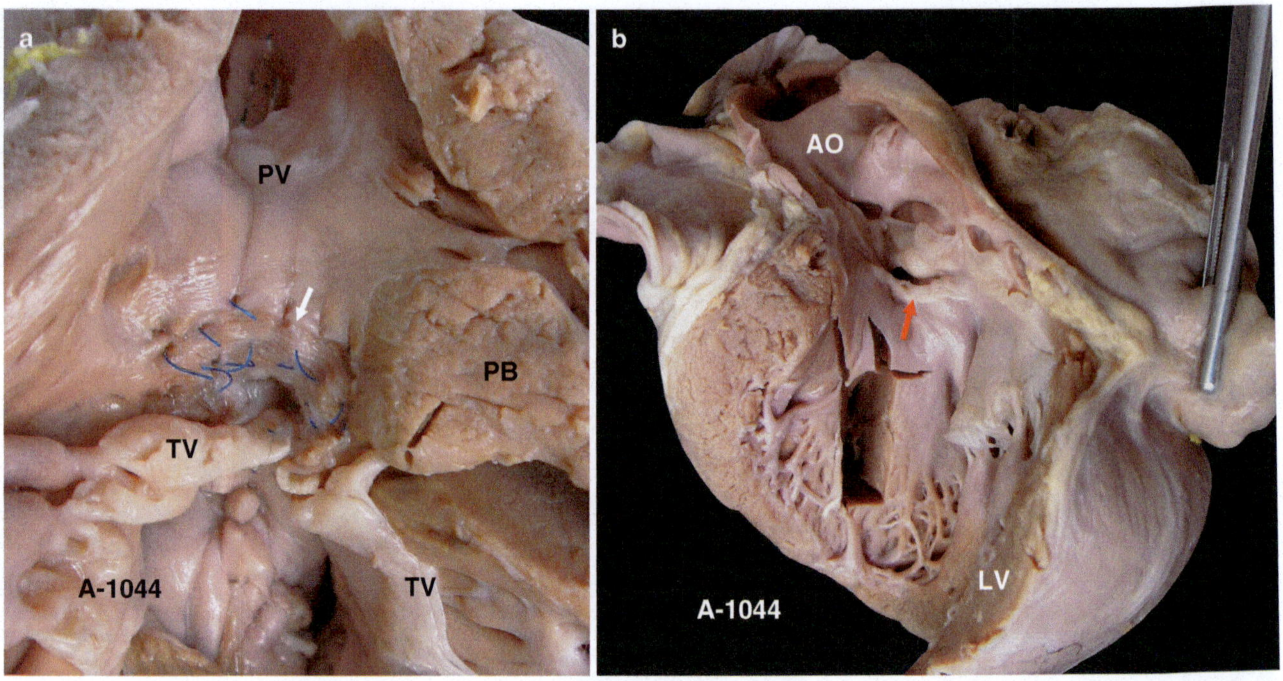

Fig. 13.4 (**a**) Specimen 1044. Small membranous VSD viewed from the right ventricle. *White arrow* indicates the patch. Surgical suture of the VSD has been partly removed. *PB* parietal band, *TV* tricuspid valve. (**b**) Specimen 1044. VSD (*arrow*) viewed from left ventricle

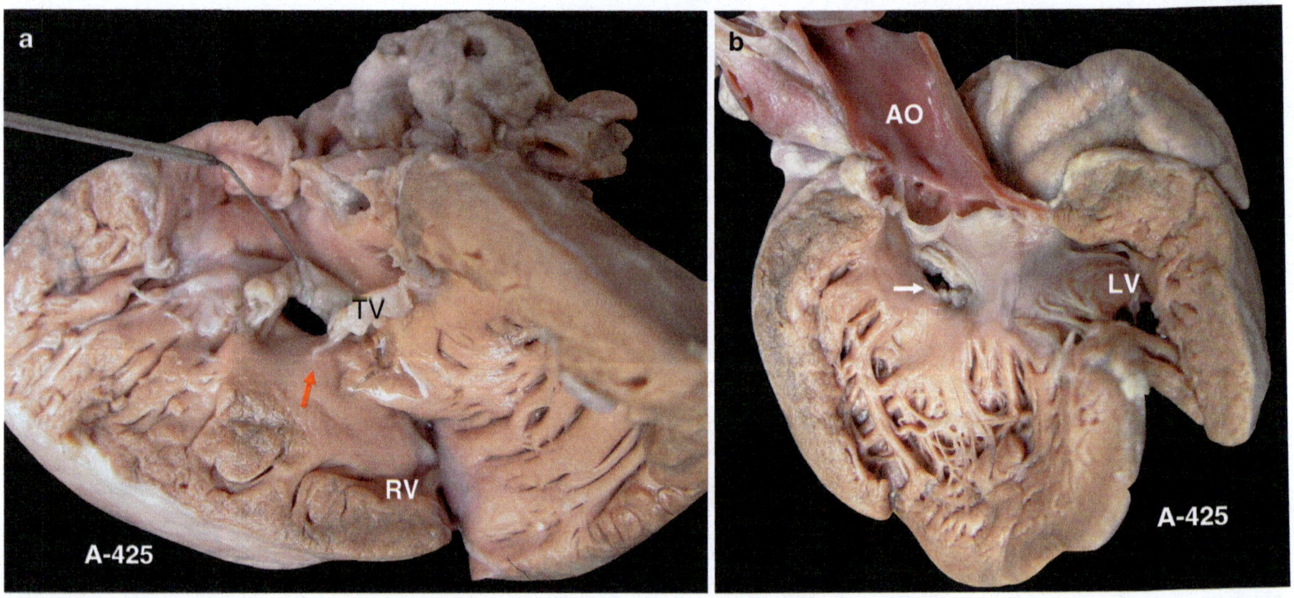

Fig. 13.5 (**a**) Specimen 425. VSD beneath septal leaflet of tricuspid valve. The septal leaflet of the TV is elevated to demonstrate the VSD. The upper margin is attached to the edge of the septal leaflet. The TV entrance is obscured. The *arrow* indicates papillary muscle of the conus artery. *TV* tricuspid valve. (**b**) Specimen 425. Viewed from left ventricle (*arrow*). VSD is located beneath the septal leaflet of the TV and is away from the AV. *AO* ascending aorta, *LV* left ventricle

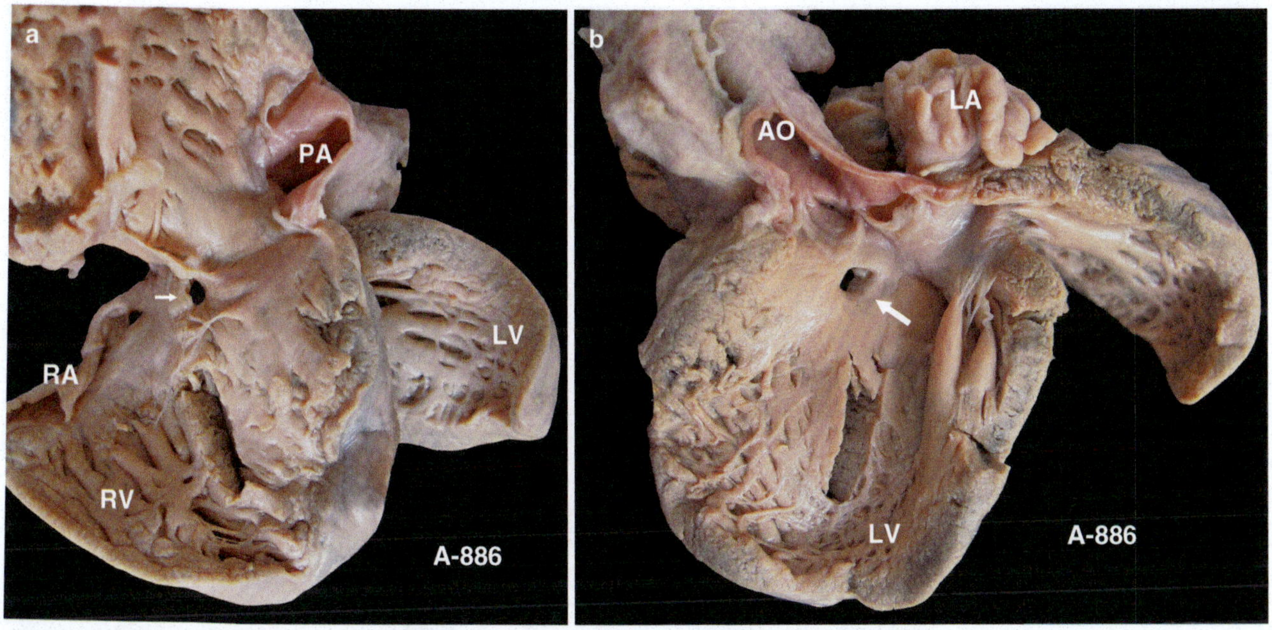

Fig. 13.6 (**a**) Specimen 886. Typical small membranous VSD. The defect is near the TV orifice and can be seen clearly via a right atrial incision in the operation. (**b**) Specimen 886. Defect viewed from left ventricle. Note the spatial relationship between the VSD and the AV

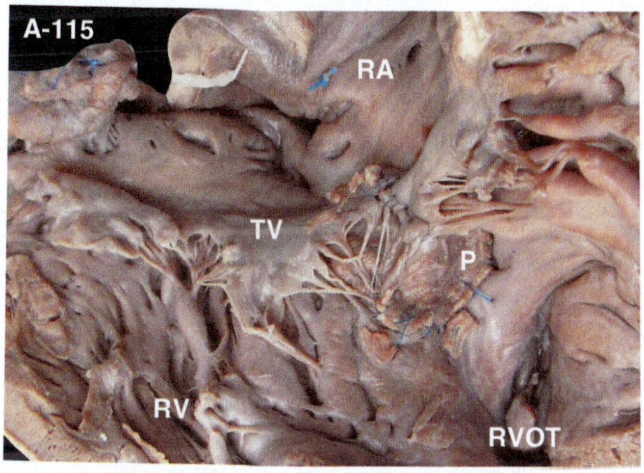

Fig. 13.7 Specimen 115. Perimembranous VSD. The defect lies at the commissure of the inferior edge of the supraventricular crest and the tricuspid annulus. The septal leaflet lies over the VSD. The defect is usually large and needs to be repaired with a patch. The bundle of His lies along the posteroinferior margin of the VSD, and the anteroinferior margin is a safe zone for repair without causing injury to the bundle of His or AV. *P* patch, *TV* tricuspid valve

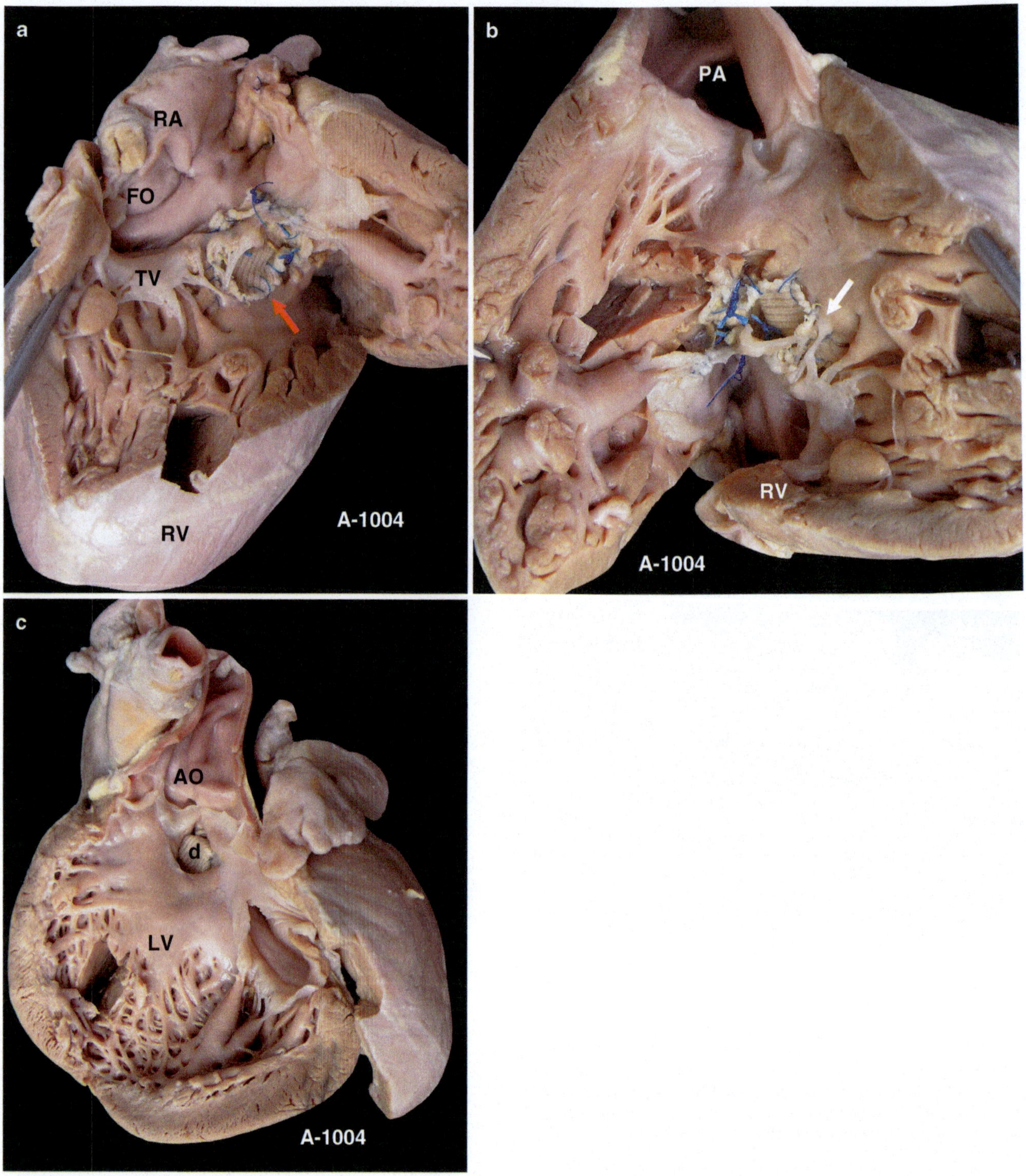

Fig. 13.8 Specimen 1004. Perimembranous VSD. The defect extends beyond the membranous septum, so the RA and RV are dissected to illustrate the patch. The *red arrow* indicates the safe zone in front of the papillary muscle of the conus artery, which remains a distance from the conduction bundle. *FO* fossa ovalis, *TV* tricuspid valve. (**b**) Specimen 1004. VSD viewed from the right ventricle outlet tract. The *arrow* indicates papillary muscle of conus artery. (**c**) Specimen 1004. VSD viewed from the left ventricle. The defect lies beneath the co-commissure of the right coronary cusp and the noncoronary cusp. (**d**) is the tightly sutured patch

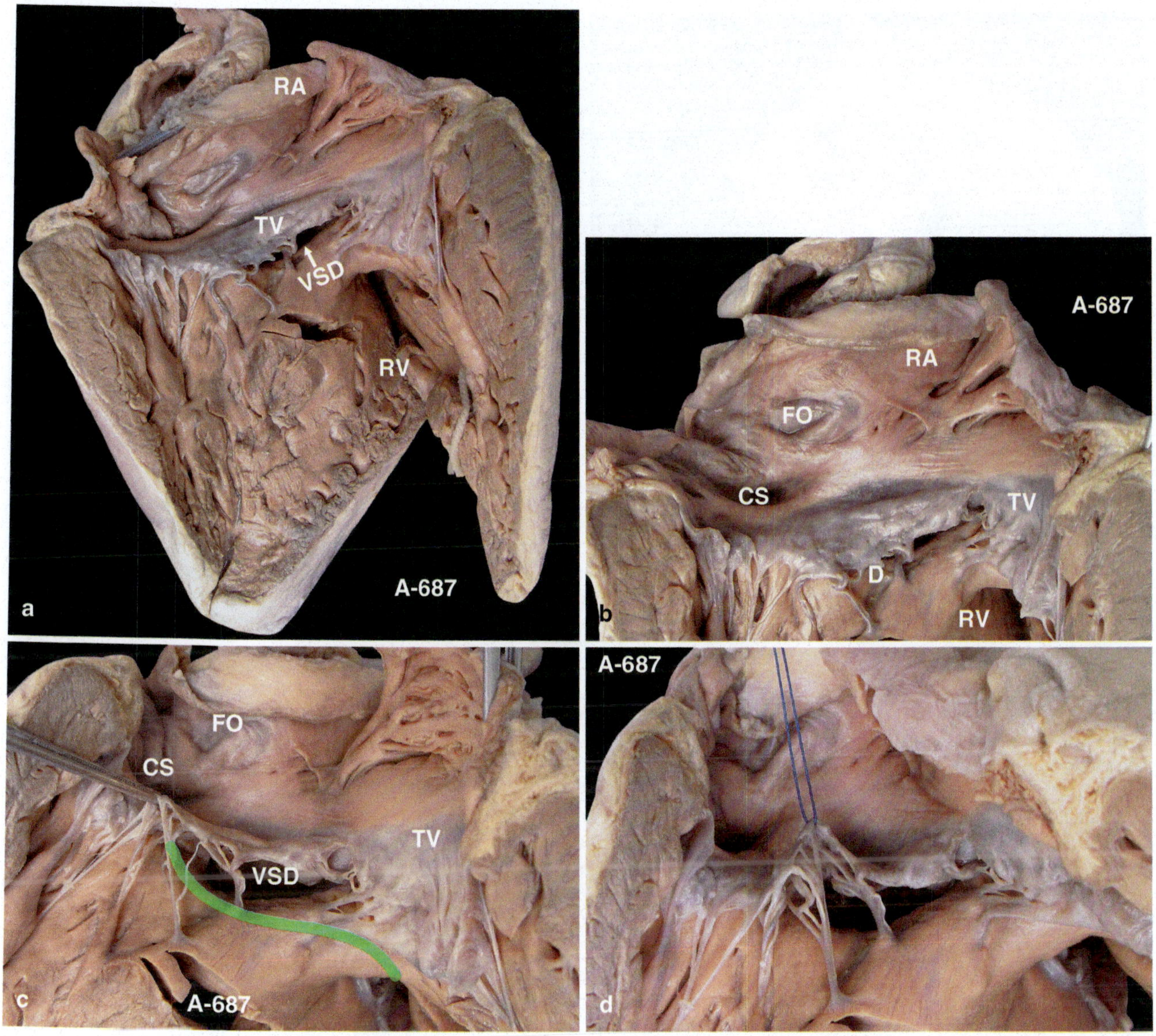

Fig. 13.9 (**a**) Specimen 687. AV canal type of VSD. The defect involves the entire area beneath the septal leaflet. The location is similar to that of the primum atrial septal defect. It lies beneath the coronary sinus (*arrow*), and the atrioventricular node displaces posterior to the VSD. The conduction bundle is particularly vulnerable to injury during operation. (**b**) Specimen 687. The rear half of VSD obscured by septal leaflet of the TV. *CS* coronary sinus, *FO* fossa ovalis. (**c**) Specimen 687. Septal leaflet of tricuspid valve elevated to demonstrate the entire defect. The *green line* indicates the course of the right bundle branch, which usually is injured when repairing the posterior rim of the defect. (**d**) Specimen 687. Anterior margin of AV canal type of VSD is demonstrated. Residual fistula usually appears here. (**e**) Specimen 687. The defect viewed from left ventricle (*arrow*). The TV can be seen through the defect. (**f**) Specimen 687. Interpretation of Specimen 687. (*1*) The defect is located infra-anterior to the coronary sinus and below the septal leaflet of the TV, similar to the endocardial cushion defect. (*2*) The defect is semicircular, and the patch needs to be properly trimmed. (*3*) *P* indicates posterior margin of the defect. Surgeons should avoid injuring the conduction bundle. (*4*) *A* indicates anterior margin of the defect, which needs to be sutured tightly. *CS* coronary sinus, *TV* tricuspid valve

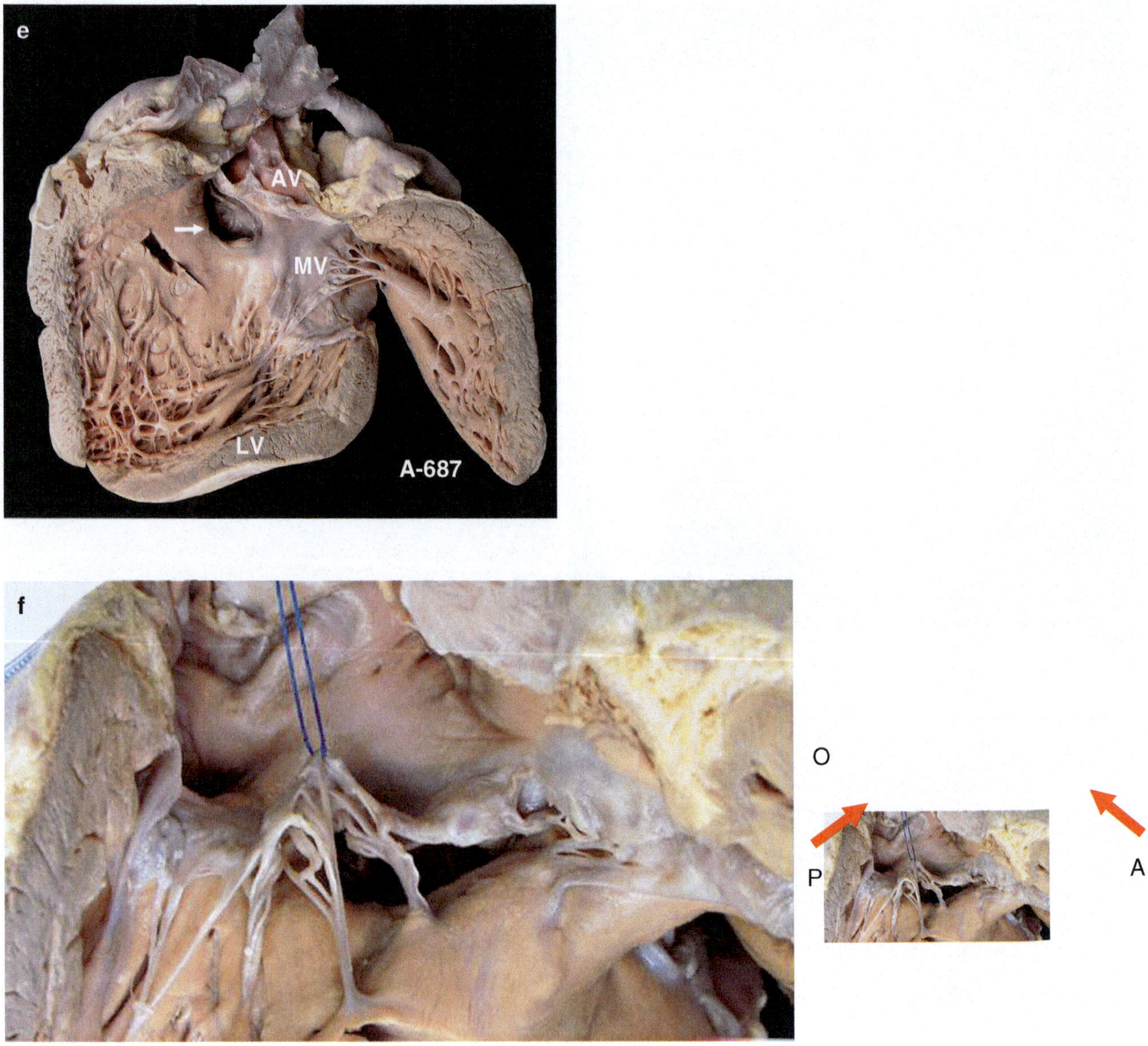

Fig. 13.9 (continued)

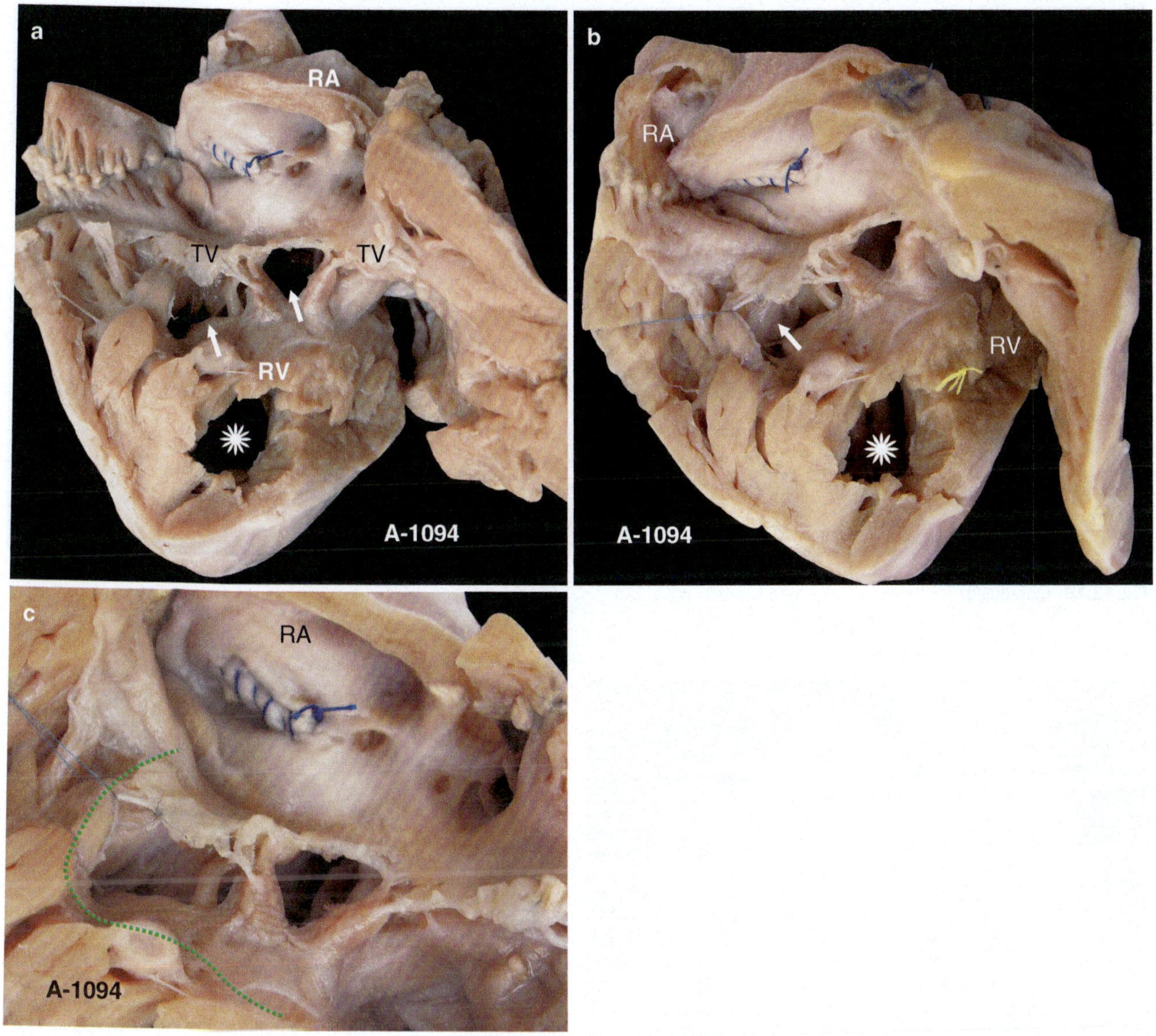

Fig. 13.10 (**a**) Specimen 1094. Aortic valve canal type of ventricular septal defect viewed from right ventricle inlet tract. The chordate below the septal leaflet attaches to the margin (*arrow*) of the defect. *Asterisk* indicates a factitious hole in the pathological examination but not a lesion. *RA* right atrium, *TV* tricuspid valve, *RV* right ventricle. (**b**) Specimen 1094. This figure focuses on the posterior margin of the defect. Chordae attached to the posterior commissure of the TV have been retracted by a string. *Arrow* indicates that the conduction bundle proceeds down along the posterior margin of the defect. *RA* right atrium, *RV* right ventricle. (**c**) Specimen 1094. Close observation of posterior margin of the defect. The *green dotted line* indicates the course of the right bundle branch. The suture line used in repairing the patent foramen ovale can be seen in the RA. *RA* right atrium. (**d**) Specimen 1094. Septal band of supraventricular crest viewed from the right ventricle outlet tract. The defect beneath the septal leaflet is unclear. The *red arrow* indicates the section of the septal band. *PA* pulmonary artery, *RV* right ventricle. (**e**) Specimen 1094. Defect viewed from the left ventricle inlet tract. This AV canal type of VSD lies beneath the MV, as the *arrow* indicates. *MV* mitral valve, *LA* left atrium, *LV* left ventricle

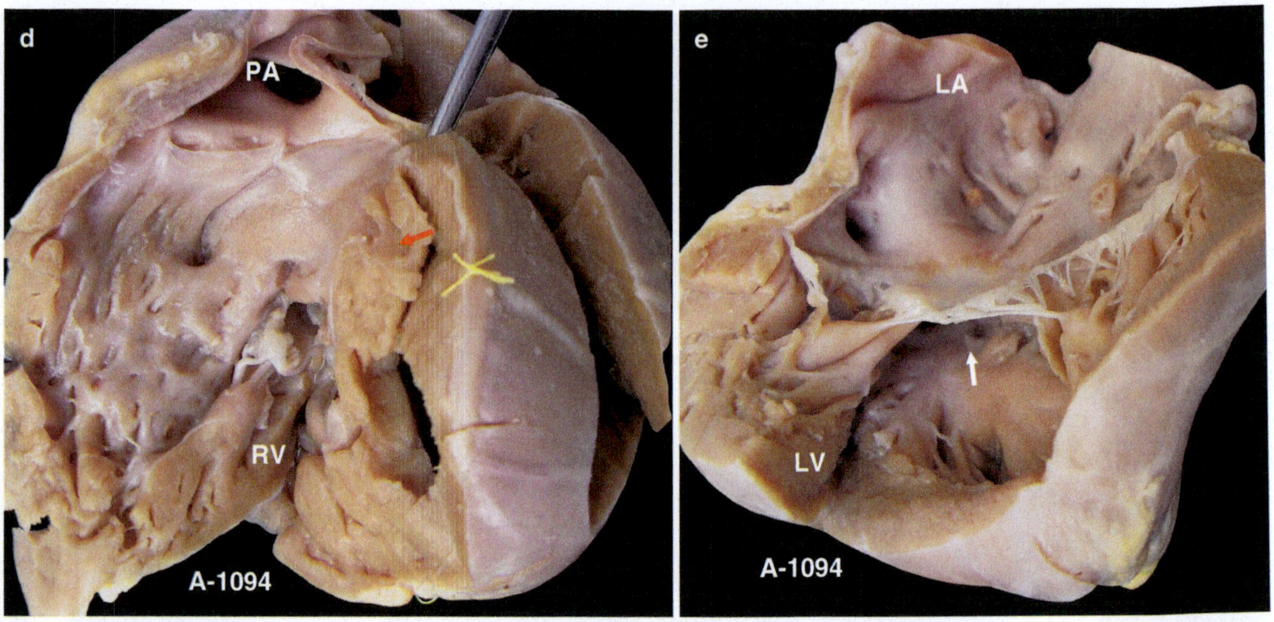

Fig. 13.10 (continued)

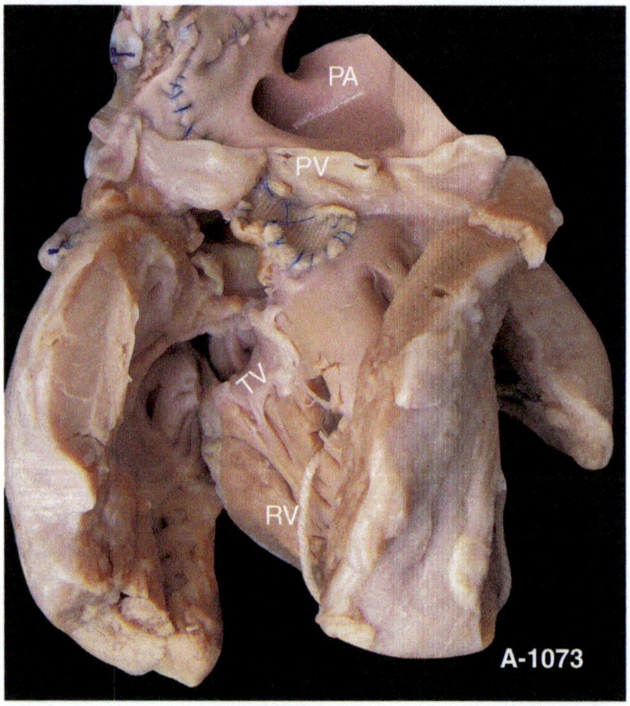

Fig. 13.11 Specimen 1073. Typical subarterial VSD. The inferior margin of the defect is muscle of the infundibular portion, and the superior margin is fibrous tissue of the junction of the aortic annulus with the pulmonary annulus. The AV can be seen through the large subarterial VSD. PV indicates the pulmonary valve, beneath which is the patch. The subarterial VSD is far from the conduction bundle, so the operation seldom causes injury to it, but the AV is vulnerable to injury that leads to aortic insufficiency. The upper margin of the defect is under the annulus, which may be closed, resulting in residual fistula. *AV* aortic valve, *PV* pulmonary valve, *RV* infundibular portion of right ventricle

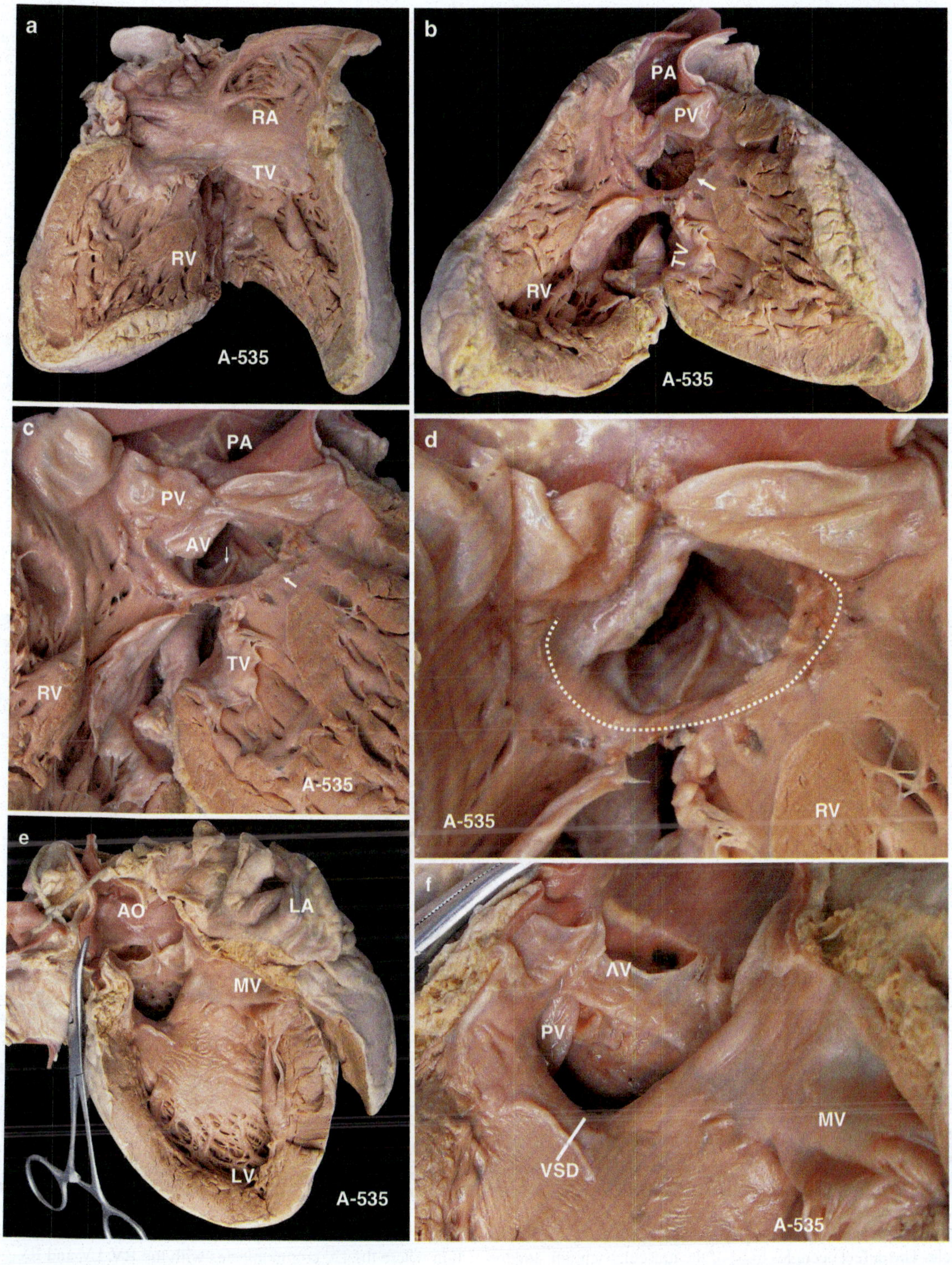

Fig. 13.12 (**a**) Specimen 535. Subarterial VSD. The defect can be exposed through the RA and the TV. The right ventricular or pulmonary incision is better than is the right atrial approach. Some surgeons repair a subarterial VSD through the aorta. *RA* right atrium, *TV* tricuspid valve, *RV* right ventricle. (**b**) Specimen 535. Subarterial VSD far away from the TV. The arrow point to upper margin of the defect where is the pulmonary annulus. *PA* pulmonary artery, *PV* pulmonary valve, *TV* tricuspid valve, *RV* right ventricle. (**c**) Specimen 535. The aortic valve (*fine arrow*) in the LV seen through the VSD. *Coarse arrow* indicates inferior margin of the defect. *LV* left ventricle, *PA* pulmonary artery, *PV* pulmonary valve, *TV* tricuspid valve, *RV* right ventricle. (**d**) Specimen 535. Close observation of the VSD and its relations to surrounding tissues. The *dotted line* indicates the inferior margin of the defect. *RV* right ventricle. (**e**) Specimen 535. Subarterial VSD viewed from the left ventricle. *AV* aortic valve, *PV* pulmonary valve, *MV* mitral valve, *LV* left ventricle. (**f**) Specimen 535. Close observation of the VSD from the leftside. Aortic and pulmonary annuli form the upper margin of the defect. *AV* aortic valve, *PV* pulmonary valve, *MV* mitral valve. (**g**) Specimen 535. Interpretation of Specimen 535. *PV* pulmonary valve, *RV* right ventricle. (*1*) There is no intact muscular margin around the subarterial VSD; the inferior margin of the defect is a muscle, which is a safe zone for repair. (*2*) The upper margin of the defect consists of a semilunar annulus. The stitch here is liable to be loose or injure the valve. The *yellow dotted line* indicates the pulmonary annulus, and the *red dotted line* indicates the aortic annulus. (*3*) The VSD is tear shaped rather than circular, and its tip is the intersection of the aortic and pulmonary annuli, so the patch should be trimmed accordingly

Fig. 13.12 (continued)

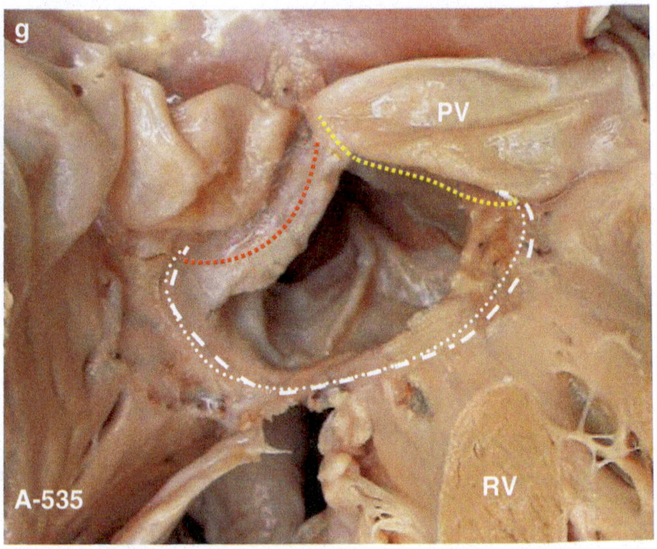

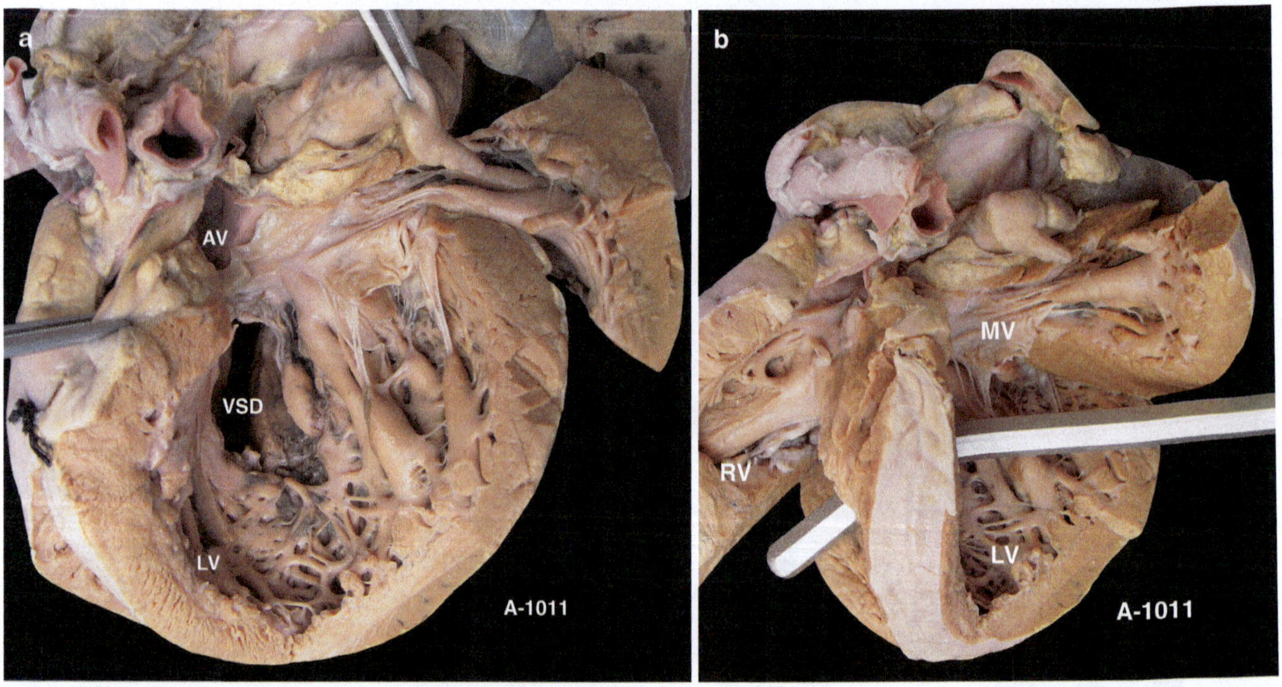

Fig. 13.13 (**a**) Specimen 1011. Large muscular VSD viewed from left the ventricle. The defect occupies most of the muscular septum, and the papillary muscle in the RV can be seen through the defect. *RV* right ventricle, *AV* aortic valve, *LV* left ventricle. (**b**) Specimen 1011. Large muscular VSD. The probe penetrating the defect demonstrates the location where the LV communicates with the RV. LV and RV outlet tracts can be seen from the front of the specimen at the anatomical site. *LV* left ventricle, *RV* right ventricle, *MV* mitral valve

13.4.1 Suture Surpassing Conduction Bundle with a Patch

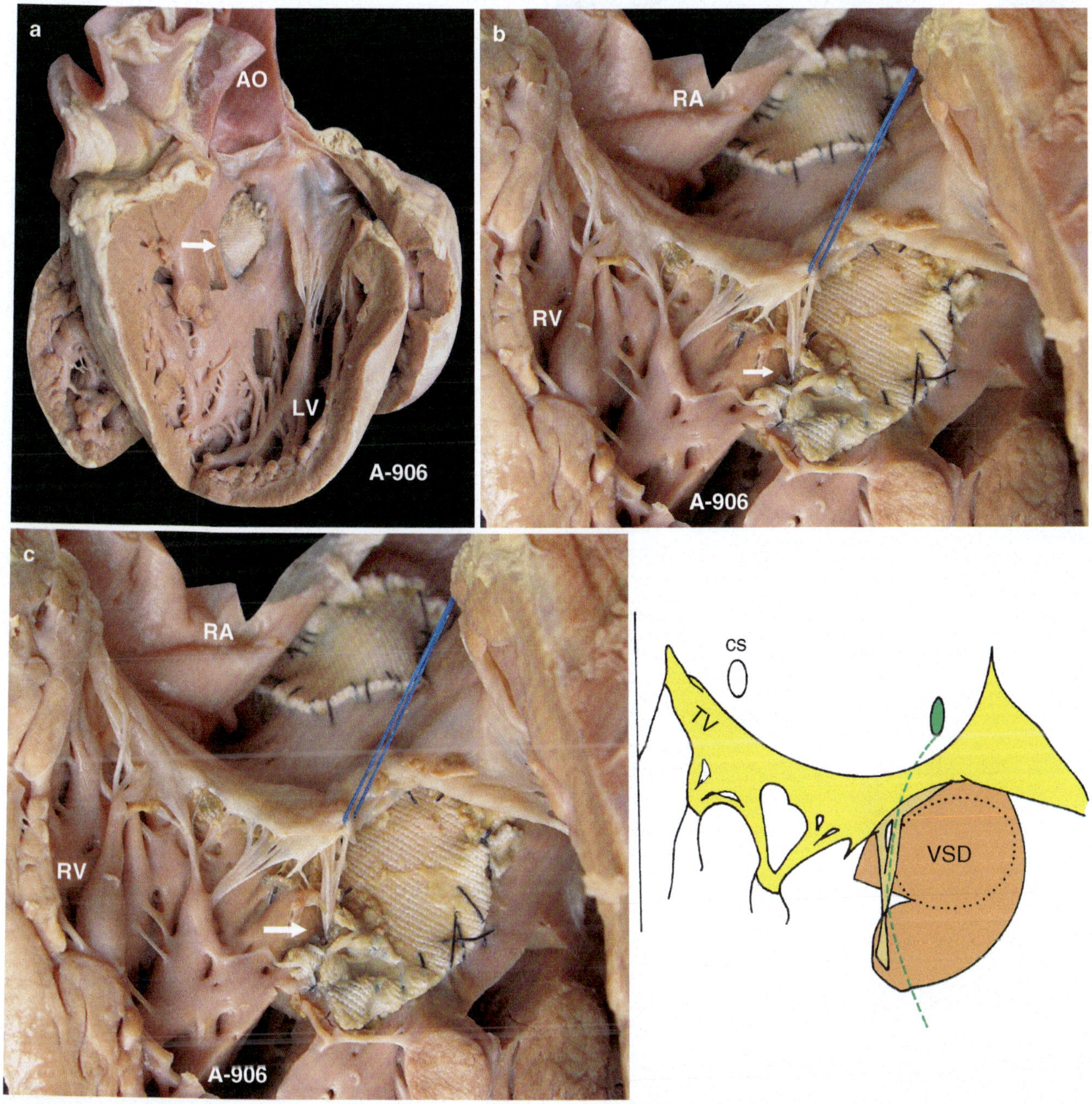

Fig. 13.14 (**a**) Specimen 906. Perimembranous VSD viewed from the left ventricle. MV is normal, and LV outlet tract is spacious. It is different from the AV canal type of VSD. The *arrow* point to the defect. *AO* ascending aorta, *LV* left ventricle. (**b**) Specimen 906. A method of avoiding the conduction bundle in operation. The polyester patch is trimmed with a notch (*arrow*) on the posteroinferior rim. The suture should bypass the conduction bundle. *RA* right atrium, *RV* right ventricle. (**c**) Specimen 906. Diagram of VSD repair. Posterior rim of the patch bypasses the conduction bundle. *RA* right atrium, *RV* right ventricle

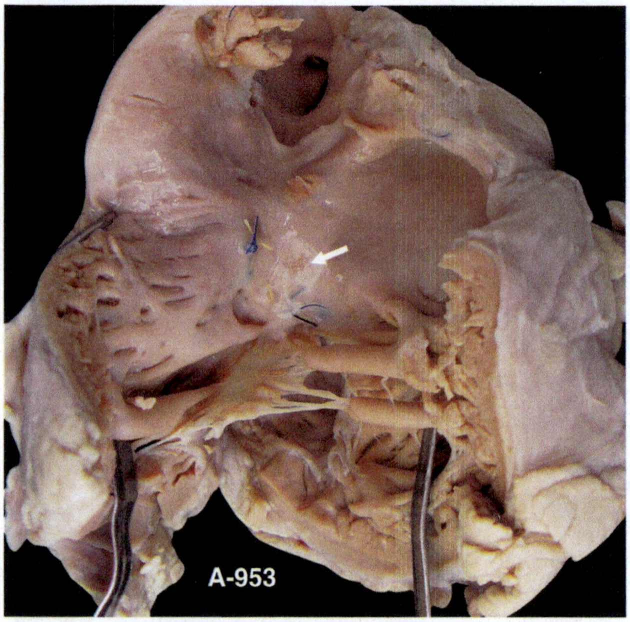

Fig. 13.15 Specimen 953. The patch is covered by endocardium with smooth surface (*arrow*) long term after repair operation

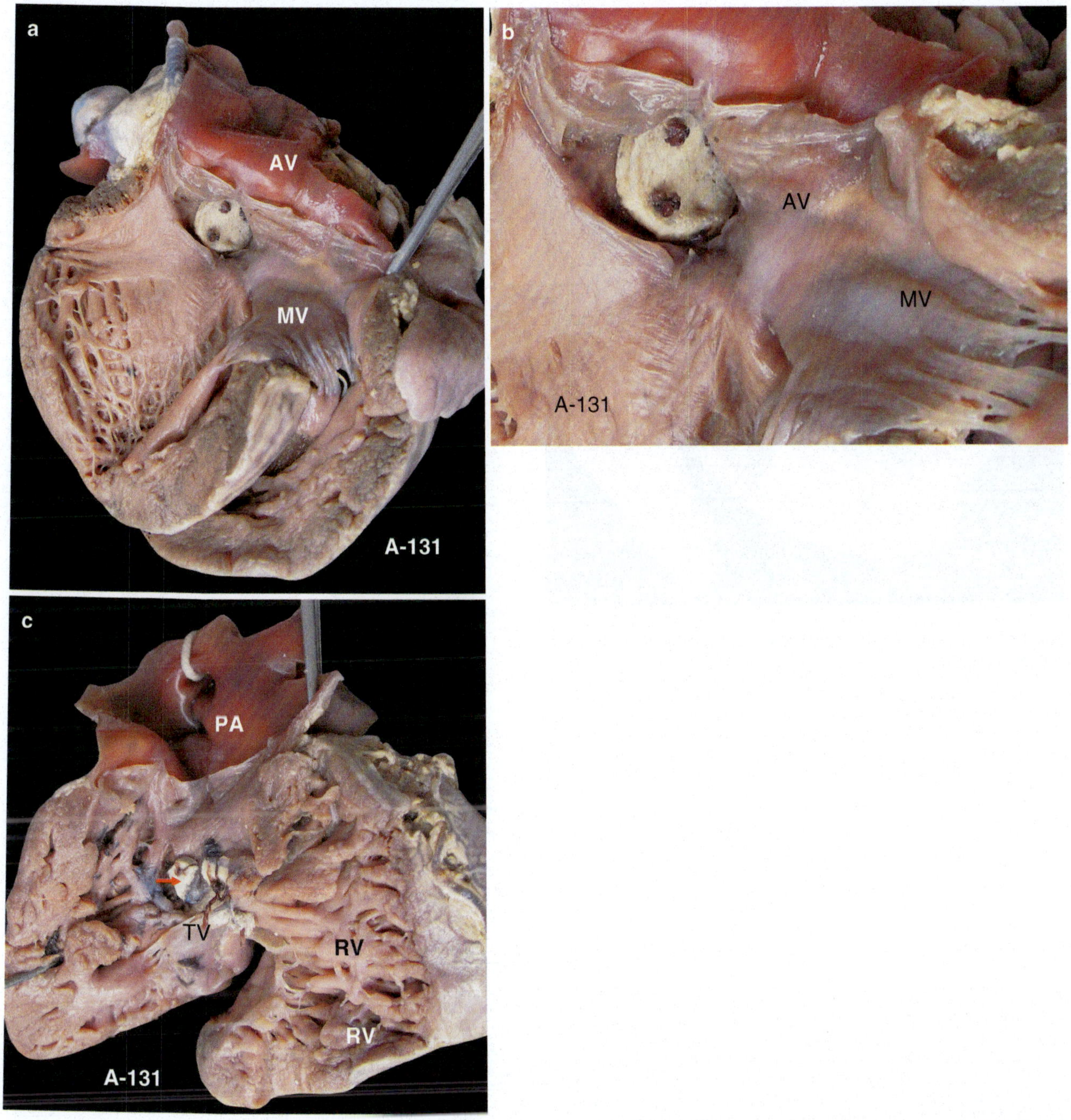

Fig. 13.16 (**a**) Specimen 131. Patch placed on left side. Viewed from the LVOT, a circular thick felt patch is placed below the AV at the left side of the septum. The felt patch is fixed by an interrupted mattress suture knotted on the RV side. This technique is different from the method placing the patch on the right side. VSD occlusion is similar to this method. *AV* aortic valve, *MV* mitral valve. (**b**) Specimen 131. VSD patch viewed from the left side. A residual gap is seen around the patch, indicating the patch is small but is smooth and fixed tightly. *AV* aortic valve, *MV* mitral valve. (**c**) Specimen 131. VSD patch on the left ventricular side with the suture knotted on the right ventricular side. The *red arrow* indicates the knotted suture with pledget. TV is not affected. *PA* pulmonary artery, *RV* right ventricle, *TV* tricuspid valve

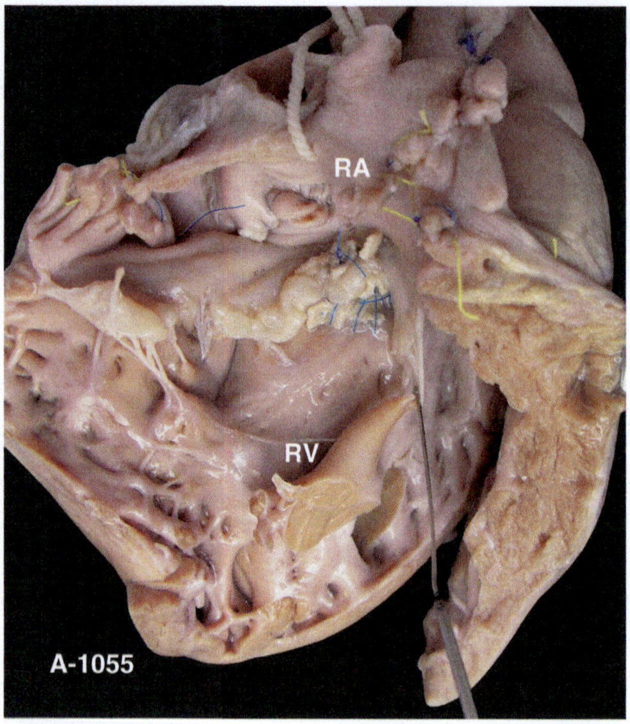

Fig. 13.17 Specimen 1055. Small VSD repaired by direct mattress suture with pledget

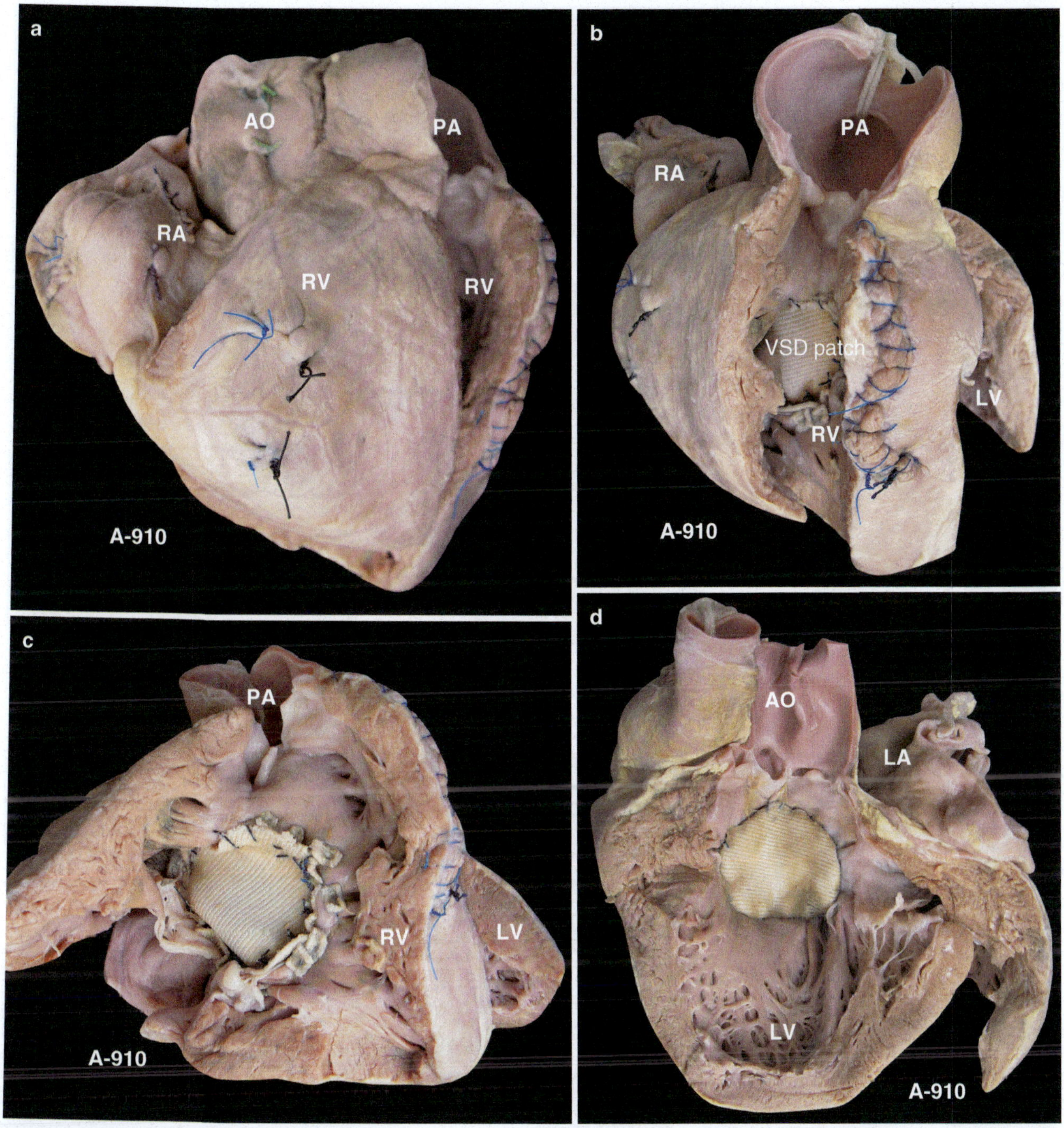

Fig. 13.18 (**a**) Specimen 910. Large perimembranous ventricular septal defect combined with pulmonary arterial hypertension, aortectasia, and right atrial dilatation. *AO* ascending aorta, *PA* pulmonary artery, *RA* right atrium, *RV* right ventricle. (**b**) Specimen 910. Incision on anterior wall of right ventricle. Infracristal VSD can be seen after cutting open the RV. Continuous suture is placed on the anterior and superior rim of the defect and on the posterior and inferior rim. This technique is used frequently. The pulmonary artery is highly dilated. *PA* pulmonary artery, *RA* right atrium, *RV* right ventricle. (**c**) Specimen 910. Anterior wall of right ventricle is raised to expose the whole VSD patch. There is no conduction bundle in the safe area of the rim of the defect, so an interrupted mattress suture with pledgets is used. *PA* pulmonary artery, *LV* left ventricle, *RV* right ventricle. (**d**) Specimen 910. VSD viewed from the left ventricle outlet tract. The defect lies beneath the right coronary cusp and the noncoronary cusp of the AV, involving partial membranous septum and supraventricular crest. The upper margin is close to the aortic annulus, and the AV may be injured in operation. *AO* ascending aorta, *LA* left atrum, *LV* left ventricle. (**e**) Specimen 910. Unfolding left atrium and left ventricle MV is normal, and the inferior edge of patch (*arrow*) can be seen beneath it. *LA* left atrum, *LV* left ventricle. (**f**) Specimen 910. Ventricular septal defect viewed from right atrium. Note the suture placed on the attached edge of the septal leaflet. The bundle of His is along the posterior edge of the defect. *TV* tricuspid valve, *RA* right atrium, *RV* right ventricle, *FO* fossa ovalis

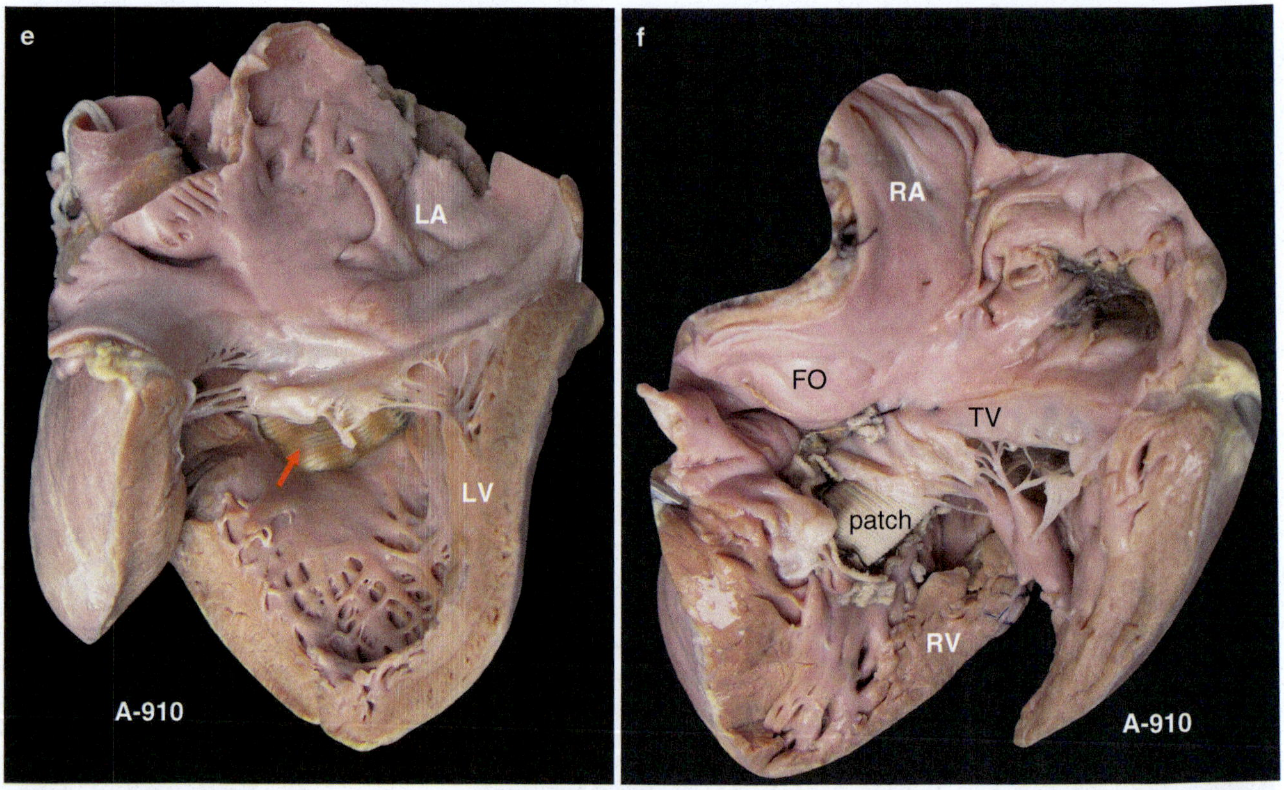

Fig. 13.18 (continued)

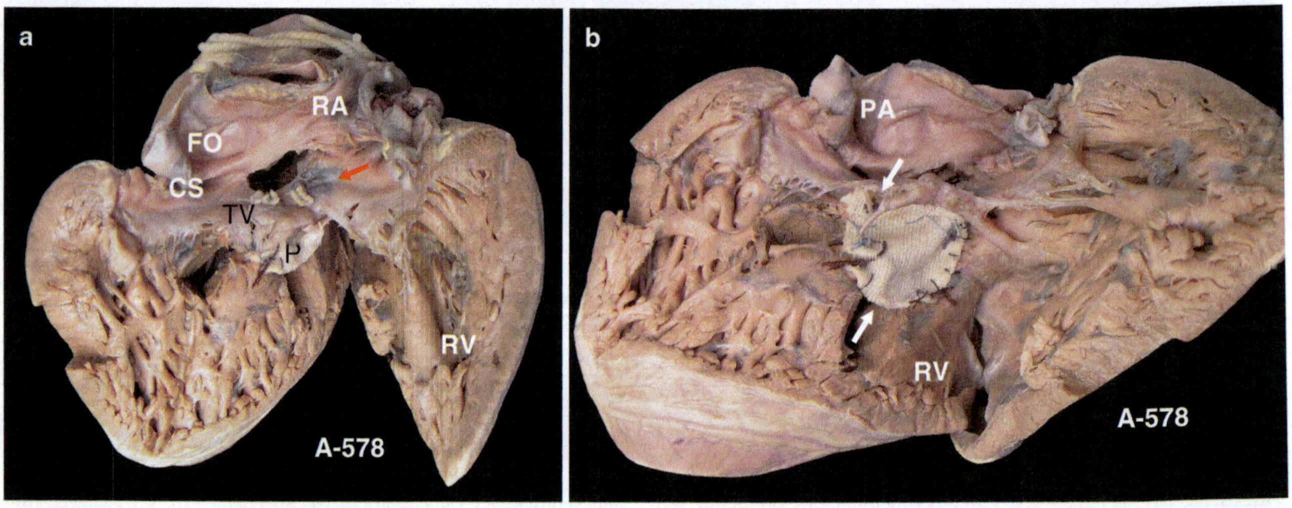

Fig. 13.19 (**a**) Specimen 578. Perimembranous VSD after surgery repair viewed from the right atrium. Subendocardial hematoma (*arrow*) due to the operation can be seen within the AV node area. Pledget reinforcing mattress stitch is used on the atrial side of the TV. Partial patch appears below the septal leaflet. *TV* tricuspid valve, *RA* right atrium, *FO* fossa ovalis, *CS* coronary sinus, *RV* right ventricle. (**b**) Specimen 578. Defect viewed from the right ventricle outlet tract. There is a gap of residual fistula at the edge of the VSD. The chordae of the papillary muscle of the conus artery are cut off. Sparse suture line at the superior and inferior margin leads to residual fistula. *PA* pulmonary artery, *RV* right ventricle

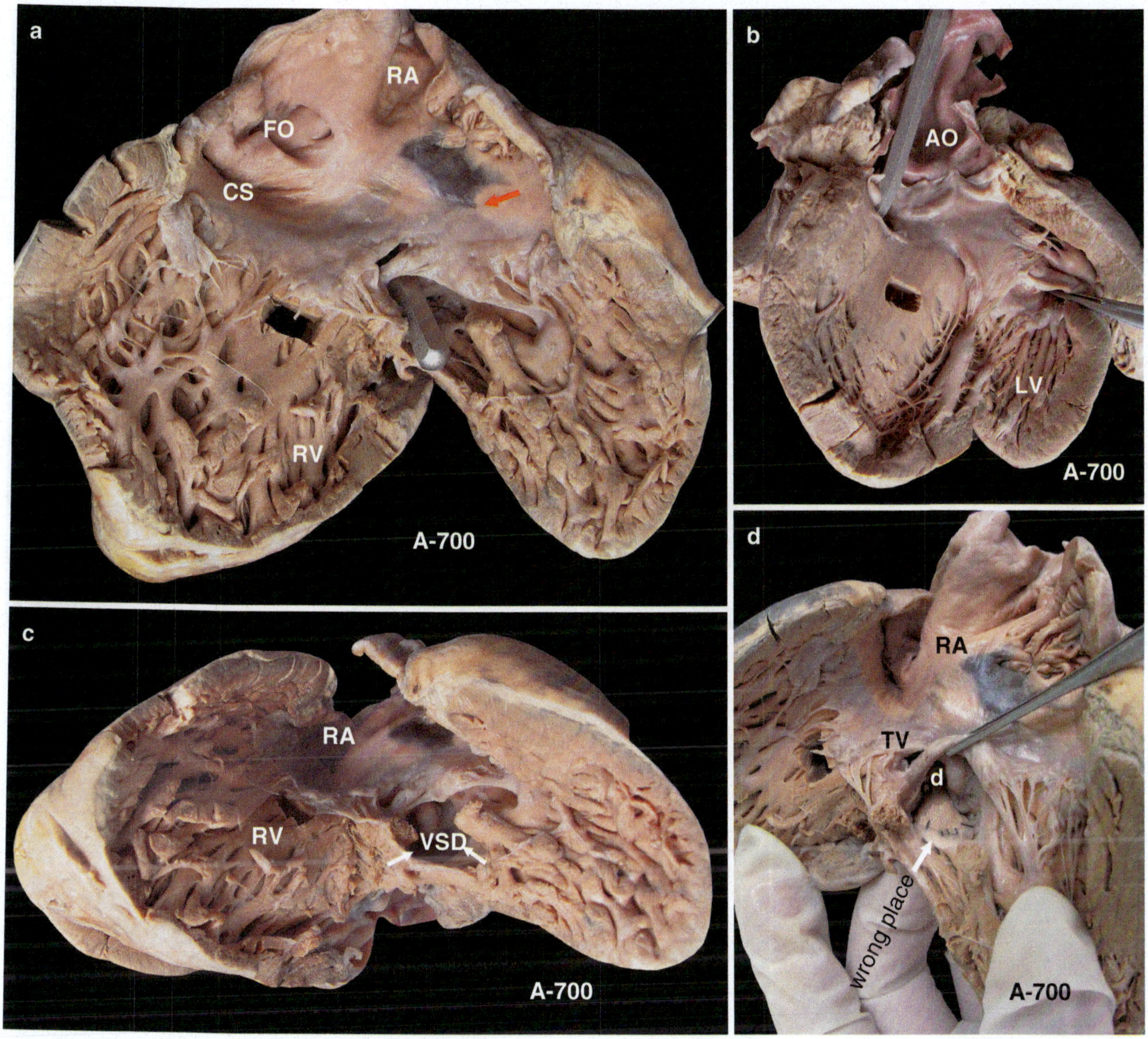

Fig. 13.20 (**a**) Specimen 700. Large perimembranous VSD. Anatomical relationship was not identified properly during surgery, so the patch is placed incorrectly. Probe penetrating the defect demonstrates the location of the VSD. The *red arrow* indicates subendocardial hematoma near the AV node. *RA* right atrium, *FO* fossa ovalis, *CS* coronary sinus, *RV* right ventricle. (**b**) Specimen 700. Left ventricle outlet tract dissected to show VSD. The defect lies beneath the AV. Probe penetrates into the RV through the unpatched defect. *AV* aortic valve, *LV* left ventricle. (**c**) Specimen 700. Defect viewed from the right ventricle. The two *arrows* indicate unpatched VSD through which the AV can be dimly seen. *RA* right atrium, *RV* right ventricle. (**d**) Specimen 700. Right ventricle outlet tract is dissected. The patch is exposed below the PV. The *arrow* indicates that the patch was not placed over the VSD but on the chordae of the TV because the surgeon mistook the gap formed by chordae for the defect. *RA* right atrium, *RV* right ventricle. (**e**) Specimen 700. The patch has not been placed over the VSD; viewed from the right ventricle. *PA* pulmonary artery, *RV* right ventricle, *LV* left ventricle. (**f**) Specimen 700. This defect combines with prolapsed aortic valve (*arrow*). *AV* aortic valve, *LV* left ventricle

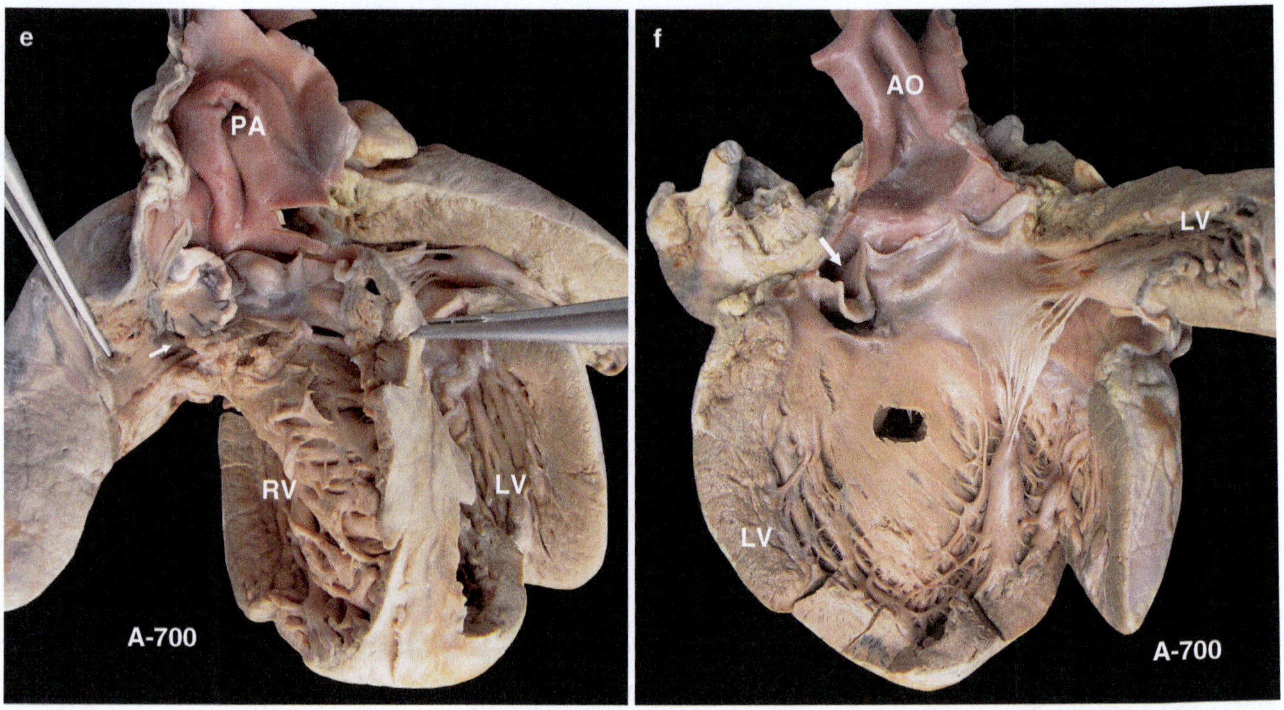

Fig. 13.20 (continued)

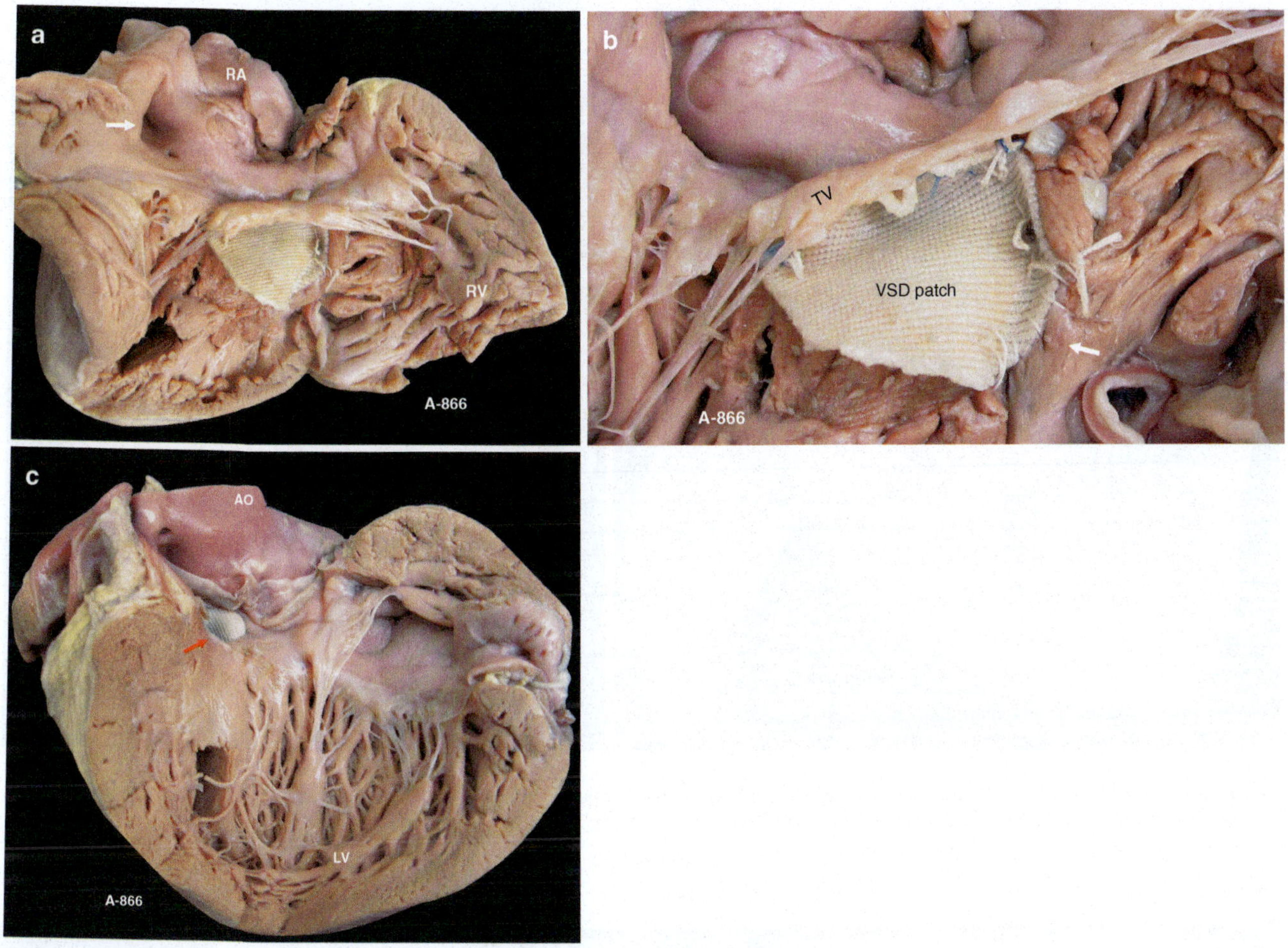

Fig. 13.21 (**a**) Specimen 866. Repair of ventricular septal defect. Continuous suture line placed in rim of defect is avulsed because the edge of the oversized patch is a distance from the rim of the defect, which is a small perimembranous VSD viewed from the left side. Continuous stitch is used along the entire course, with only one mattress stitch reinforced by pledget; as a result, the patch is dislodged after suture avulsion. This defect is combined with the left SVC draining into the coronary sinus (*arrow*), and the coronary sinus is clearly dilated. *SVC* superior vena cava, *RA* right atrium, *RV* right ventricle. (**b**) Specimen 866. The *arrow* indicates avulsed continuous suture line. *TV* tricuspid valve. (**c**) Specimen 866. Residual fistula (*arrow*) viewed from left ventricle, *AV* aortic valve, *LV* left ventricle

13.4.2 Lessons of Repair of Ventricular Septal Defect in Individual Cases

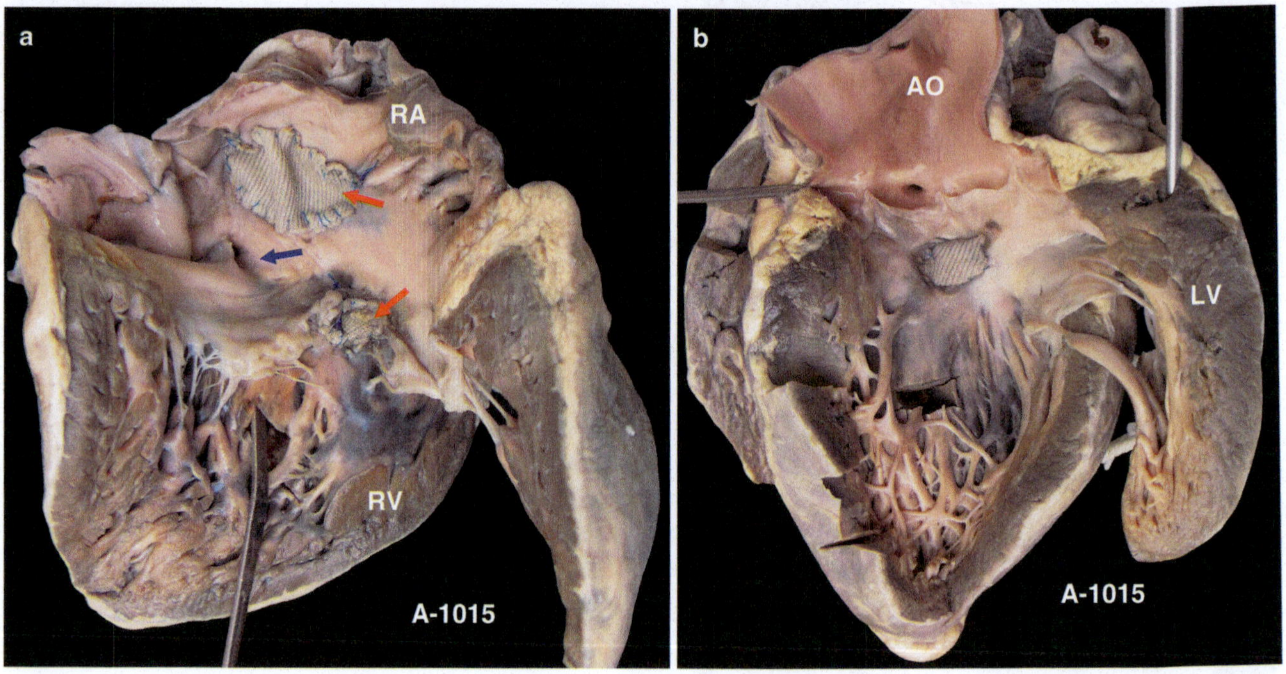

Fig. 13.22 (**a**) Specimen 1015. Perimembranous VSD combined with secundum ASD. *RA* right atrium, *RV* right ventricle. (**b**) Specimen 1015. Tightly repaired VSD viewed from the left ventricle outlet tract. *AO* ascending aortia, *LV* left ventricle

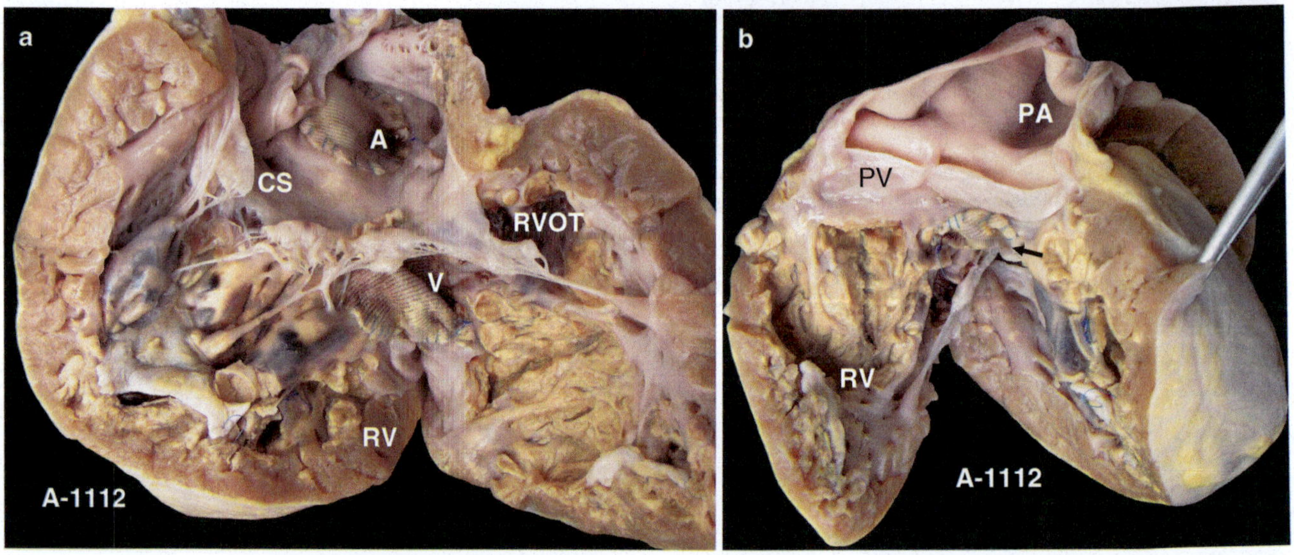

Fig. 13.23 (**a**) Specimen 1112. Secundum ASD combined with perimembranous VSD. The patch is demonstrated through the RA and the ventricular incision. A indicates the ASD patch; the coronary sinus is beneath the defect, so the defect is a secundum ASD. *V* indicates the polyester patch of the VSD beneath the intact TV, so it is not an AV canal-type VSD, *CS* coronary sinus, *RVOT* right ventricle outlet tract, *RV* right ventricle. (**b**) Specimen 1112. Patch of VSD viewed from the right ventricle outlet tract. *PV* pulmonary valve, *PA* pulmonary artery, *RV* right ventricle

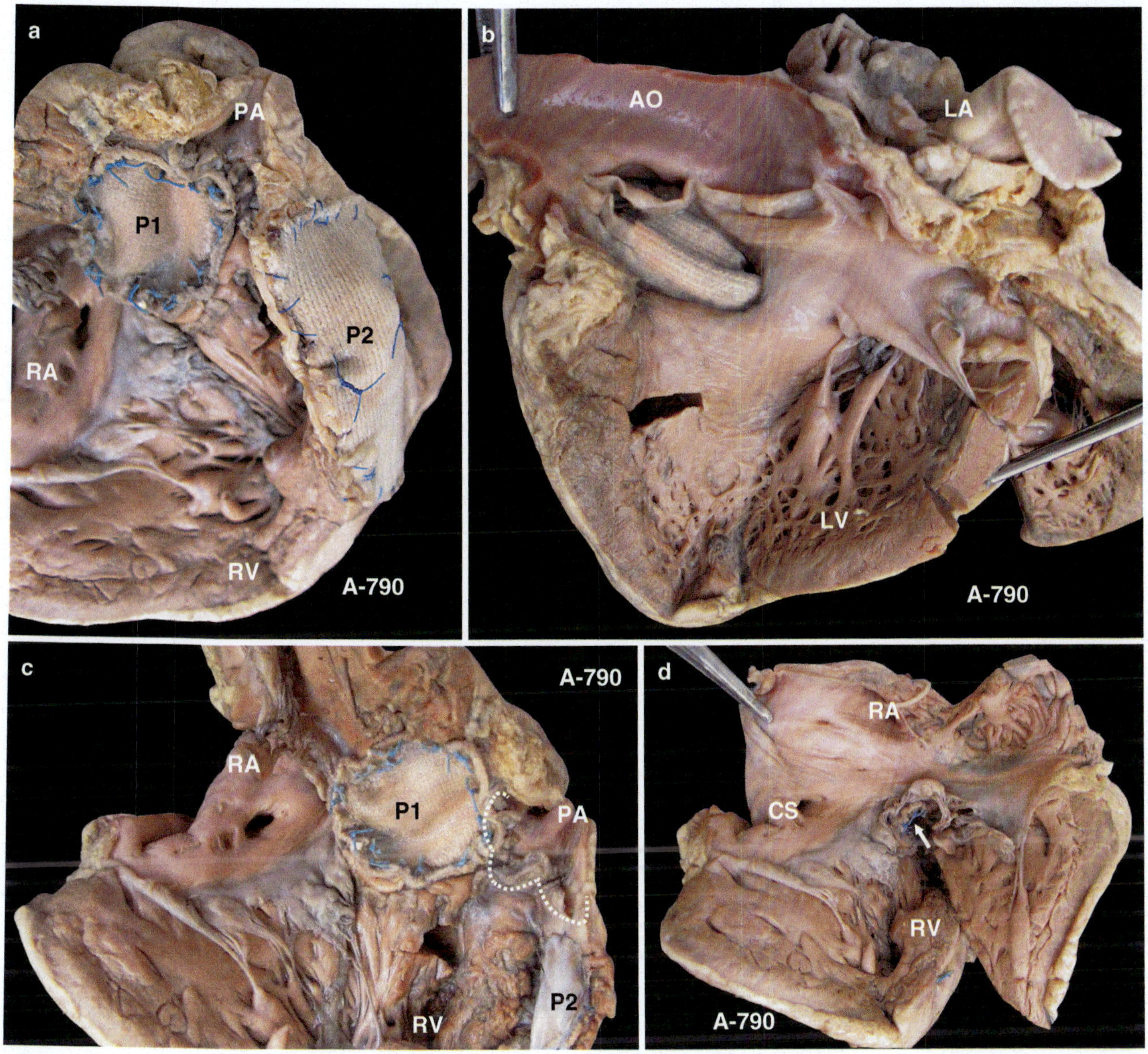

Fig. 13.24 (**a**) Specimen 790. Right ventricle outlet tract obstruction combined with large infundibular VSD. The patch (*P1*) occupies the entire infundibular portion. The RVOT is broadened by the patch (*P2*). *RV* right ventricle, *RA* right atrium, *RV* right ventricle, *PA* pulmonary artery. (**b**) Specimen 790. The left ventricular side of the VSD viewed from left ventricle outlet tract. The patch is located beneath the left and right coronary cusps of the AV and remains a distance from the membranous septum. Dextropositioned aorta is not obvious, and aortic-mitral fibrous continuity exists. It can be mistaken for DORV when observing the AV through a right ventriculotomy and VSD in operation. *LA* left atrium, *LV* left ventricle, *AO* ascending aortia. (**c**) Specimen 790. Patch of VSD (*P1*) and right ventricle outlet tract (*P2*) viewed from right ventricle outlet tract. The *dotted line* between the two patches indicates pulmonary annulus to demonstrate that there is no pulmonary stenosis. *RA* right atrium, *RV* right ventricle. (**d**) Specimen 790. Right atrium and ventricle are dissected. Atrial septum is intact, and partial suture line of the patch can be seen below the *TV*. *TV* tricuspid valve, *RA* right atrium, *RV* right ventricle, *CS* coronary sinus

13.4.3 Repair Technique of Ventricular Septal Defect

A small defect can be repaired with single-needle stitching or mattress suture directly, but a large defect needs a patch. Surgeons usually choose polyester patches or, occasionally, an autogenous pericardium patch. An interrupted mattress suture with pledgets or a continuous suture can be used to fix the patch. Suture avulsion should be avoided in a continuous suture. Injury to the AV, TV, and conduction bundle should be avoided in all operations. AV, aortic valve; TV, tricuspid valve

Fig. 13.25 (**a**) Specimen 2. Large ventricular septal defect combined with right ventricle outlet tract obstruction. Muscles beneath the AV become thick. RA and RV are cut open simultaneously, and the tricuspid annulus is transected. A large patch of the VSD is beneath the unfolded TV. There are several interrupted mattress stitches through the tricuspid annulus with pledget on the atrial side of the TV, and there is a patch as a part of the RV wall. a Patch of VSD, b patch of anterior wall of right ventricle. Surgical suggestions: Oversized patch may lead to obstruction of the RVOT. Stitches through the tricuspid annulus are tight when repairing the upper margin of the defect. The conduction system along the posteroinferior rim of the defect should be protected. An oversized patch of the anterior wall of the RV may affect right ventricular function. *RV* right ventricle, *RA* right atrium, *TV* tricuspid valve. (**b**) Specimen 2. Patch viewed from the front. The main pulmonary artery maintains normal relation to the aorta. There is a big patch on the anterior wall of the RV. The RV and the main pulmonary artery are dissected longitudinally, and the pulmonary annulus is transected. *PA* pulmonary artery. (**c**) Specimen 2. Patch of anterior wall of the right ventricle and patch of VSD (*a*) have covered the supraventricular crest because the defect is large. The parietal band has been removed partly to avoid obstruction of the RV, and another patch, (*b*) is used to broaden the anterior wall of the RV. *PV* pulmonary valve, *RV* right ventricle. (**d**) Specimen 2. Left ventricle outlet tract is unfolded. After the anterior leaflet of the MV is unfolded, the large VSD and the patch (*a*) and subendocardial hematoma (*h*) (black area on left ventricular side of septum) are clear. *LA* left atrium, *LV* left ventricle. (**e**) Specimen 2. Patch of VSD (*a*) and subendocardial hematoma (*h*) are unfolded. *LA* left atrium, *LV* left ventricle. (**f**) Specimen 2. Left ventricle outlet tract cut open. Clamp penetrates the outlet tract from the left ventricular apex to the ascending aorta. *LA* left atrium, *LV* left ventricle, *PA* pulmonary artery, *RV* right ventricle. (**g**) Specimen 2. Left ventricular incision is connected to aortic incision. Great artery relationship is normal. Patch of the VSD and subendocardial hematoma can be seen. *PA* pulmonary artery, *LV* left ventricle. (**h**) Specimen 2. Hypertrophic muscle of the intraventricular septum beneath the AV is exposed after cutting open the LV and aorta. This hypertrophic muscle may lead to mild LVOT stenosis. The *red dotted line* indicates a section of the hypertrophic muscle beneath the AV. Note the fibrous continuity between the AV and the anterior leaflet of the MV: there is no muscular tissue, which indicates no dextropositioned aorta. This is a basic difference between a large VSD and a DORV. The *red dotted line* indicates the section of the hypertrophic muscle beneath the AV

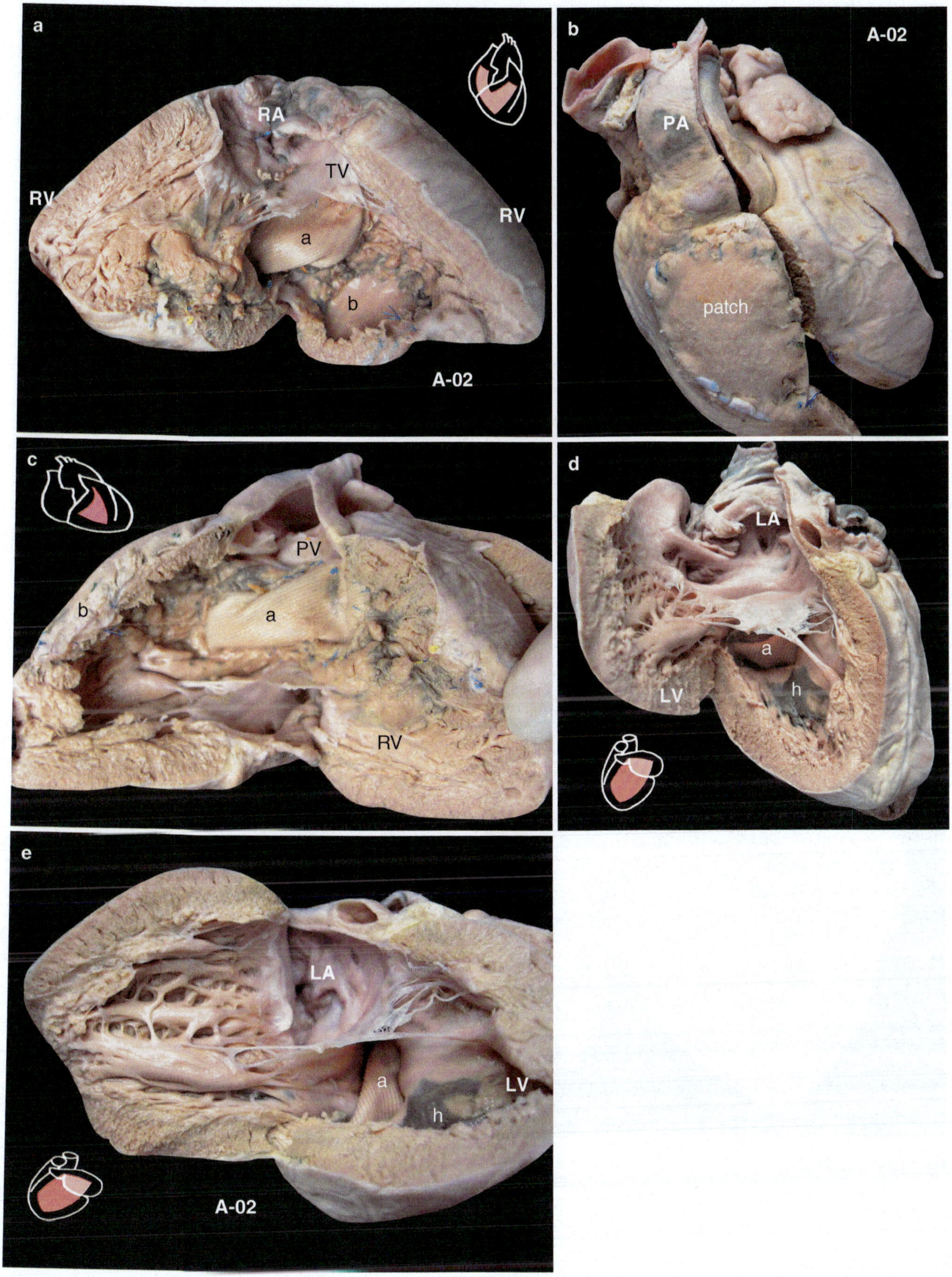
a
RA
TV
RV
RV
a
b
A-02
b
A-02
PA
patch
c
PV
b
a
RV
d
LA
a
LV
h
e
LA
a
LV
h
A-02

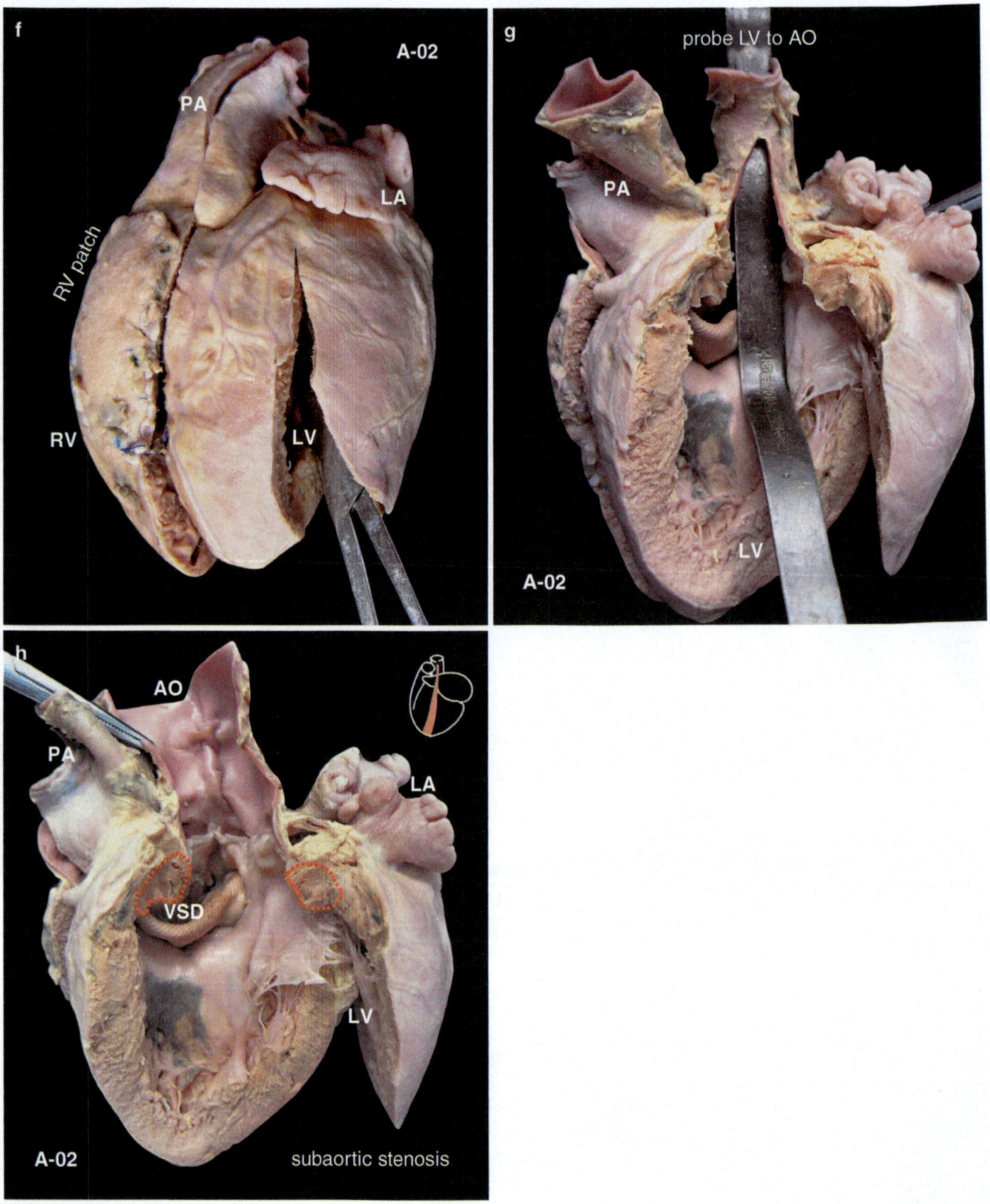

Fig. 13.25 (continued)

14 Anomalous Pulmonary Venous Drainage

14.1 General Considerations

Anomalous pulmonary venous drainage, also known as *anomalous pulmonary venous connection*, refers to the condition in which the pulmonary veins are not connected directly to the LA but are abnormally connected to the systemic venous or RA. If none of the four pulmonary veins drains into the LA, the condition is called *total anomalous pulmonary venous drainage*. When fewer than three pulmonary veins are connected to the LA, the condition is called *partial anomalous pulmonary venous drainage*. Pulmonary venous drainage often occurs concurrently with ASD or other complicated malformations.

14.2 Embryology

Two mechanisms lead to anomalous pulmonary venous drainage: (1) hypoplasia, the common pulmonary vein developed from the LA posterior wall fails to connect with the distal pulmonary vein, and (2) patent and dilated embryonic collateral circulation between the pulmonary venous plexus and venous system.

After birth, the pulmonary venous blood drains into the system vein through the remnants of the branch. An anomalous connection of the pulmonary veins can occur in any part of the normally or abnormally developed venous system. Following is a brief description of the development process of the systemic venous and the pulmonary venous.

14.2.1 Systemic Venous Development

Early in the development of the embryo, the primitive heart tube is divided into the bulbus cordis, the ventricle, the atrium, and the sinus venosus. Three pairs of large veins are connected to the rear of the sinus venosus. The outermost pair is composed of the cardinal veins. The two pairs in the middle are the umbilical veins and the vitelline veins.

The cardinal veins include the common cardinal vein (duct of Cuvier), anterior cardinal vein, and posterior cardinal vein. The left common cardinal vein develops into the right SVC, and the right common cardinal vein develops into the coronary sinus. Both sides of the posterior cardinal veins join to form the azygos vein-half azygos vein (formed by the supracardinal vein) and the middle and distal segments (formed by the subcardinal vein) of the IVC.

The right vitelline vein develops into the proximal segment of the IVC, and the left vitelline vein develops into the portal vein. The umbilical vein is also involved in the formation of the hepatic vein. Between the portal vein and the IVC, in the embryonic period there is a direct channel called the *ductus venosus* that degrades after birth.

The above segments of the vein may be associated with anomalous pulmonary venous connection. Common situations are summarized as follows:

1. Connected to the left anterior cardinal vein and finally draining into the left SVC or innominate vein
2. Connected to the right anterior cardinal vein or right common cardinal vein and finally draining into the right upper vena cava
3. Connected to the left common cardinal vein and finally draining into the coronary sinus
4. Connected to the umbilical vein and vitelline vein and finally draining into the venous duct or proximal IVC

14.2.2 Pulmonary Venous Development

Originating from the venous plexus of the lung bud, the pulmonary veins merge into the main branch of the four pulmonary veins. The pulmonary venous plexus is located at the rear of the atrium and has collateral connection with each pair of the systemic venous veins. At the same time, a cystic protrusion from the left posterior wall of the primitive atrium forms the common trunk of the pulmonary veins and connects with the four pulmonary veins. After the common trunk enlarges into a part of the LA, the four pulmonary veins are drained into the LA.

X. Zhu (ed.), *Surgical Atlas of Cardiac Anatomy*,
DOI 10.1007/978-94-017-9409-1_14, © Springer Science+Business Media Dordrecht and People's Medical Publishing House 2015

14.3 Anomalous Pulmonary Venous Connections

Developmental disorders of the common trunk or developmental abnormalities of the four main branches can cause anomalous pulmonary venous connections.

14.3.1 Partial Anomalous Pulmonary Venous Drainage (PAPVD)

Not all branches (1–3) of the pulmonary veins fail to connect with the LA normally. Some are defined as *partial anomalous pulmonary venous drainage* (PAPVD). According to the number of abnormally connected pulmonary veins, we can divide the PAPVD into the unilateral single-branch type, unilateral two-branch type, and bilateral single-branch type. According to the site of the abnormal connection, a PAPVD can be categorized into supracardiac type, cardiac type, or infracardiac type. The cardiac type is the most common.

The cardiac type of PAPVD usually occurs concurrently with an ASD. With repair of the ASD, the abnormal pulmonary vein can be drained to the LA by using a large patch. In the case of an abnormal pulmonary vein returning to the vena cava, surgery is more complex. Depending on any specific circumstance, different techniques, such as widening of the SVC or other diversion surgery, may be necessary (Figs. 14.1, 14.2, 14.3, and 14.4).

14.3.2 Total Anomalous Pulmonary Venous Drainage

Total anomalous pulmonary venous drainage (TAPVD) is defined as a condition in which all four pulmonary veins fail to connect normally with the LA. The four pulmonary veins can converge together prior to connecting with the LA, or they can connect directly with the RA. In most cases, the four pulmonary veins converge into a horizontal pulmonary trunk behind the atrium and then drain into the left SVA or the IVC through the vertical vein. TAPVD is generally associated with a large ASD, which can compensate for hemodynamic disorders.

According to the site of the abnormal connection, TAPVD can be categorized into supracardiac type, cardiac type, and infracardiac type. The cardiac type is the most common, and the infracardiac type is rare.

Surgical corrections of TAPVD are generally radical. The techniques include anastomosis of the abnormally drained pulmonary veins and LA, closure of the ASD, and ligation of the vertical vein connected with the systemic vein (Figs. 14.5, 14.6, 14.7, 14.8, and 14.9).

Before operating, one should:

1. Check for the pulmonary venous pathway obstruction and the site of obstruction.
2. Identify the location of the pulmonary trunk and determine the approach to anastomose the abnormal drained pulmonary veins with the LA.
3. For patients with the supracardiac or the infracardiac type, in order to facilitate the ligation of the vertical vein, make sure the location of the vertical vein is clear. At the same time, attention should be given to the size of the atrium; during the operation, attention should be paid to the location of the coronary sinus.

If strictures appear at the anastomosis of the pulmonary vein trunk and LA during TAPVD repair, pulmonary congestion may occur after the operation. Hence, the anastomotic stoma should be made as large as possible. The stoma is located at the posterior wall of the heart. The anastomotic stoma bleeding is a common complication.

Supracardiac- and infracardiac-type TAPVD both have a vertical vein that should be ligated. The vessel wall is both very weak and easily torn open during dissection, leading to hemorrhage. Because setting up the extracorporeal circulation may be difficult, many surgeons recommend that the vertical vein be dissected under cardiopulmonary bypass.

14.4 Specimen Demonstrations

14.4.1 Partial Anomalous Pulmonary Venous Drainage

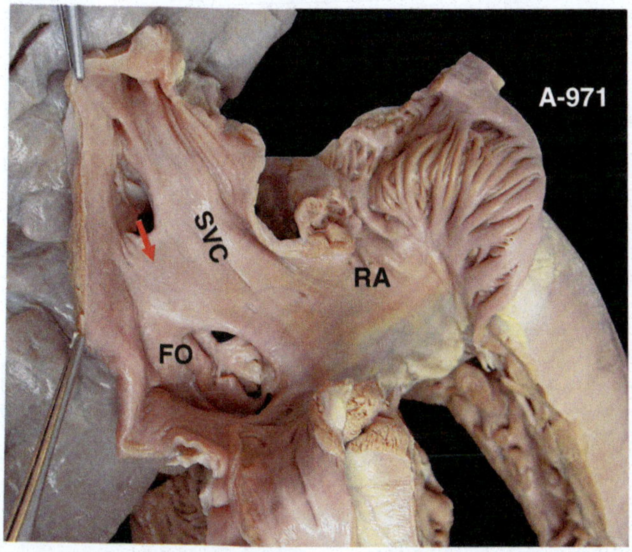

Fig. 14.1 Specimen 971. Right pulmonary vein abnormally drains to the superior vena cava. *Arrow* indicates the right pulmonary vein draining into the proximal SVC. Repair operation of this type of abnormity includes enlarging the foramen ovale, draining the abnormal pulmonary vein to the LA by using a large patch to repair the enlarged foramen ovale, and broadening the SVC (if necessary) in order to avoid stenosis. *FO* foramen ovale, *RA* right atrium, *SVC* superior vena cava

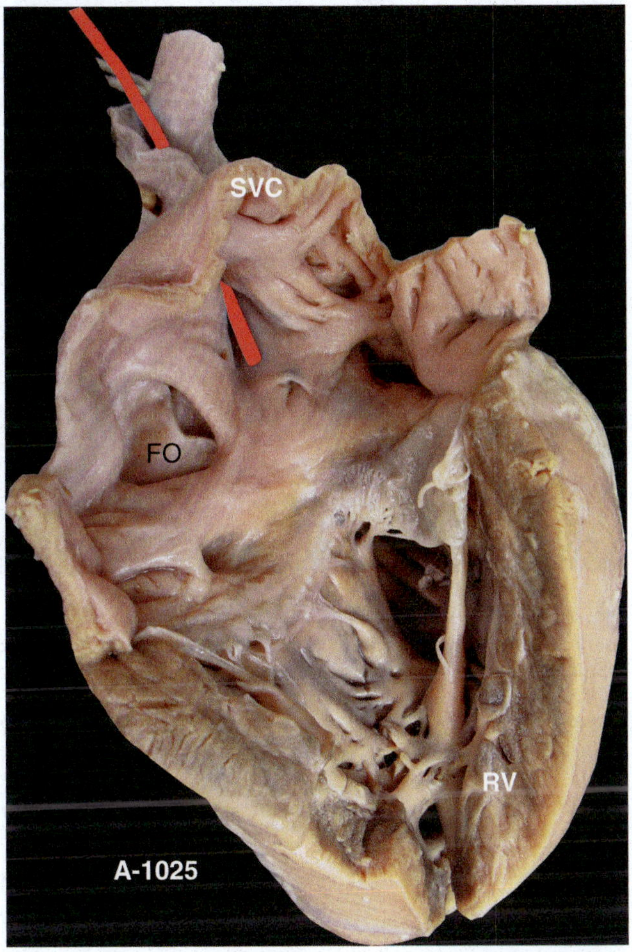

Fig. 14.2 Specimen 1025. Right pulmonary vein abnormally drains to the superior vena cava. *Red probe* indicates the site of the abnormal connection. *RV* right ventricle, *SVC* superior vena cava, *FO* fossa ovalis

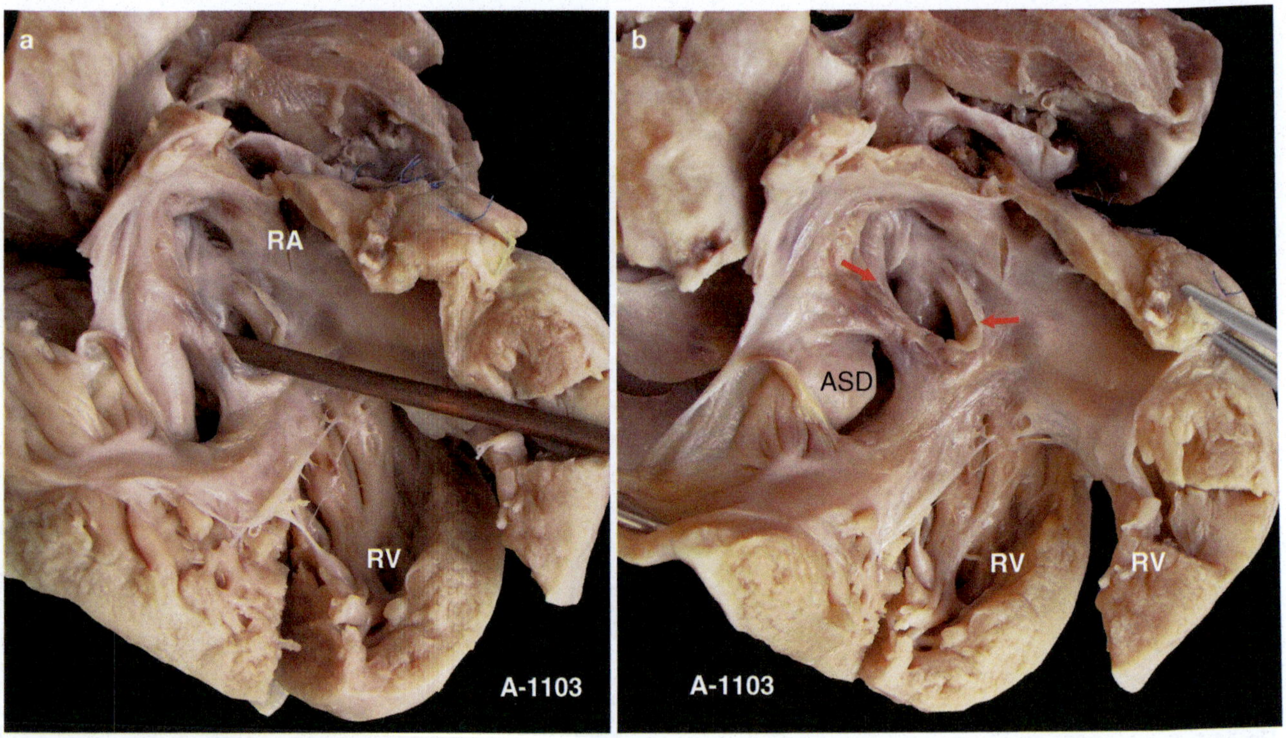

Fig. 14.3 (**a**) Specimen 1103. Right pulmonary vein abnormally drains to the right atrium (probe placed at the inlet of right pulmonary vein). *RA* right atrium, *RV* right ventricle. (**b**) Specimen 1103. Right pulmonary vein abnormally drains to the right atrium. Entrance of the right pulmonary vein (*red arrow*) is located under the entrance of the SVC. *RV* right ventricle, *ASD* atrial septal defect

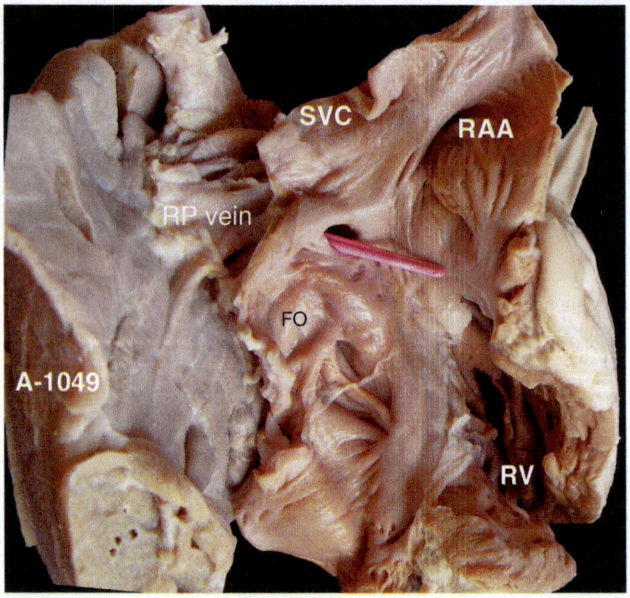

Fig. 14.4 Specimen 1049. Right pulmonary vein abnormally drains to the right atrium. A probe is placed from the right pulmonary vein to the RA. *FO* fossa ovalis, *RAA* auricle of right atrium, *RP vein* right pulmonary vein, *RV* right ventricle, *SVC* superior vena cava

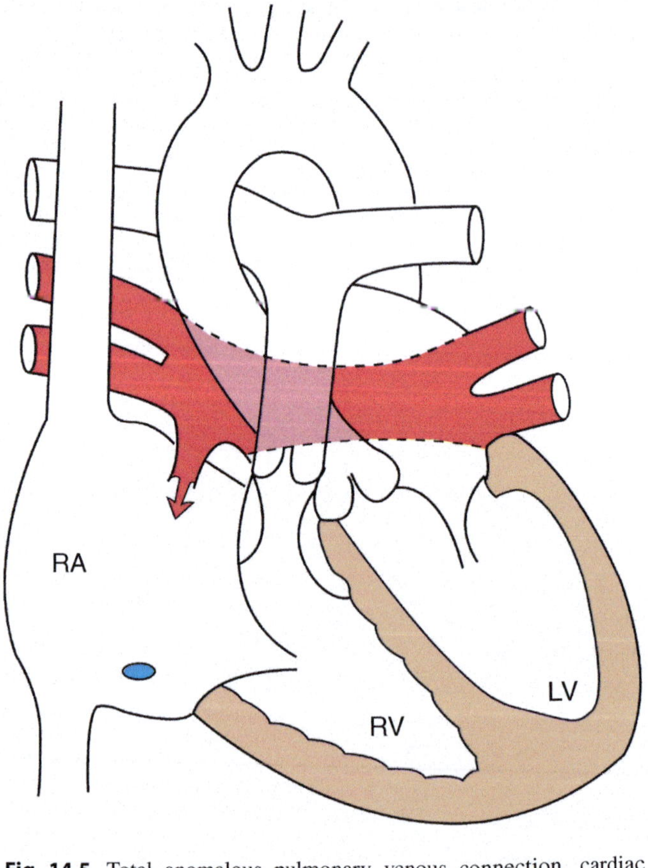

Fig. 14.5 Total anomalous pulmonary venous connection, cardiac type. Oxygenated blood from the lung flows through the pulmonary trunk, four pulmonary veins (*red area*), and the ASD, draining into the RA. *LV* left ventricle, *RV* right ventricle

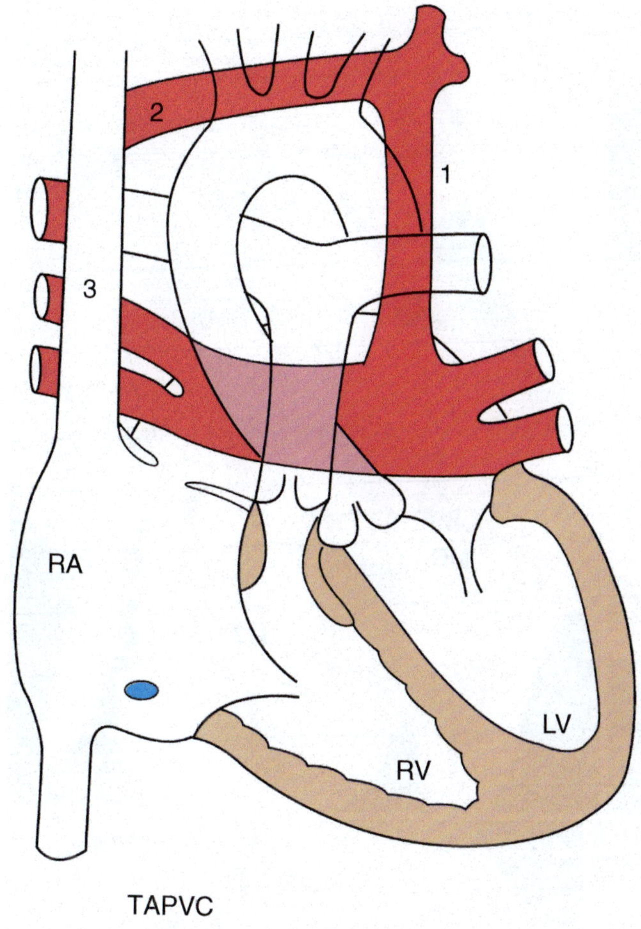

Fig. 14.6 Total anomalous pulmonary venous connection, supracardiac type. Oxygenated blood from the lung flows through the pulmonary trunk, four pulmonary veins (*red area*), vertical vein (*1*), and anonymous veins (*2*), draining into the SVC (*3*). Some blood flows through ASD to the LA, *LV* left ventricle, *RV* right ventricle, *RA* right atrium

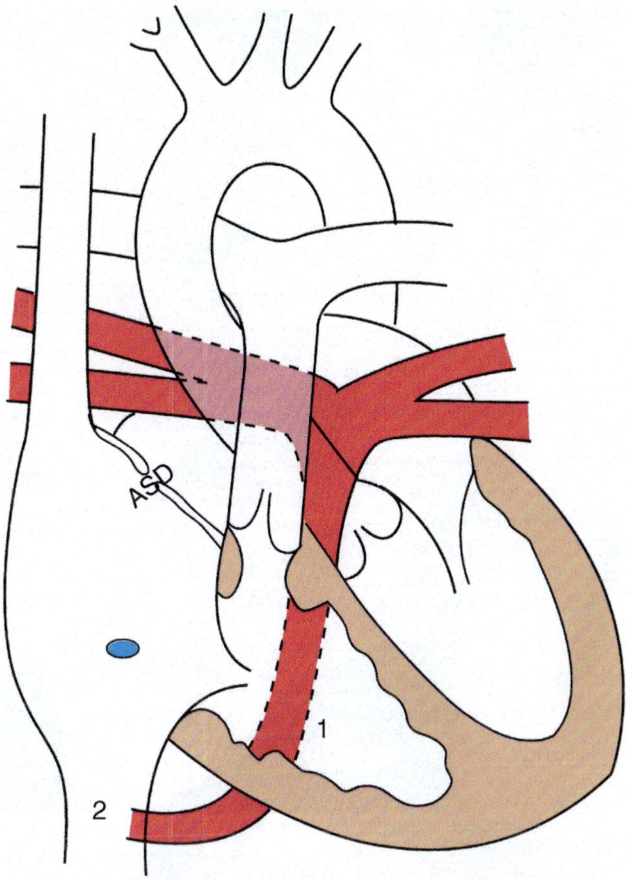

Fig. 14.7 Total anomalous pulmonary venous connection, infracardiac type. Oxygenated blood from the lung flows through the pulmonary trunk, four pulmonary veins (*red area*), and vertical vein under the heart (*1*), draining into the IVC (*2*). Some blood flows through the ASD to the LA

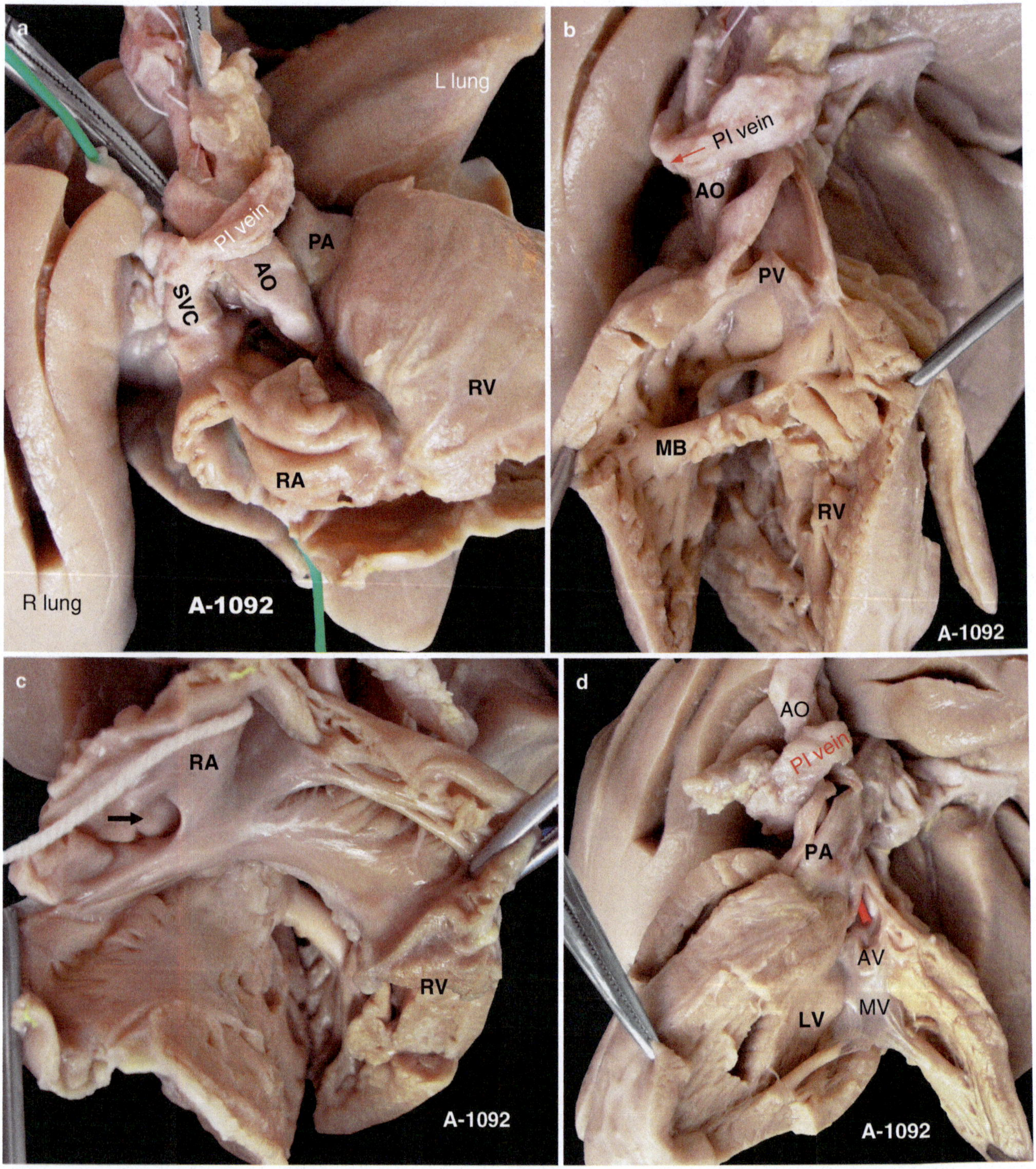

Fig. 14.8 (**a**) Specimen 1092. Total anomalous pulmonary venous drainage, supracardiac type. Front view of the heart shows RA and RV cut open. At the root of the great vessels, the pulmonary vein is connected to anonymous veins and then to the SVC in front of the ascending aorta. The *green probe* is placed in the SVC. *PI vein* means that blood from the pulmonary vein flows through the anonymous vein. *Ao* aorta, *RA* right atrium, *RV* right ventricle, *PA* pulmonary artery, *L lung* left lung, *R lung* right lung. (**b**) Specimen 1092. View from the left to observe the pulmonary vein. PI vein: The pulmonary vein common trunk is connected to the anonymous vein in front of the aorta. This is the normal great artery relationship. The RV has a coarse moderator band. *MB* moderate band, *PV* pulmonary valve, *RV* right ventricle, *AO* ascending aorta. (**c**) Specimen 1092. View from right atrium and right ventricle. The *arrow* indicates the foramen ovale, *RA* right atrium, *RV* right ventricle. (**d**) Specimen 1092. View from left ventricle outlet tract. The left heart is normal. *Red probe* is in the aorta (*Ao*). The pulmonary vein (*PI vein*) originates from the left and right lungs, then passes by the aorta and returns to the SVC. *LV* left ventricle, *MV* mitral valve, *PA* main pulmonary vein, *AV* aortic valve. (**e**) Specimen 1092. Dorsal view of the heart and lung. The left and right pulmonary veins converge, then go forward and return to the SVC. The *red arrow* indicates the pulmonary artery blood flow. (**f**) Schematic diagram (dorsal view). *PI vein* means that blood from the pulmonary vein flows through the anonymous vein. *LA* left atrium, *LV* left ventricle, *AO* ascending aorta, *SVC* superior vena cava and PI vein

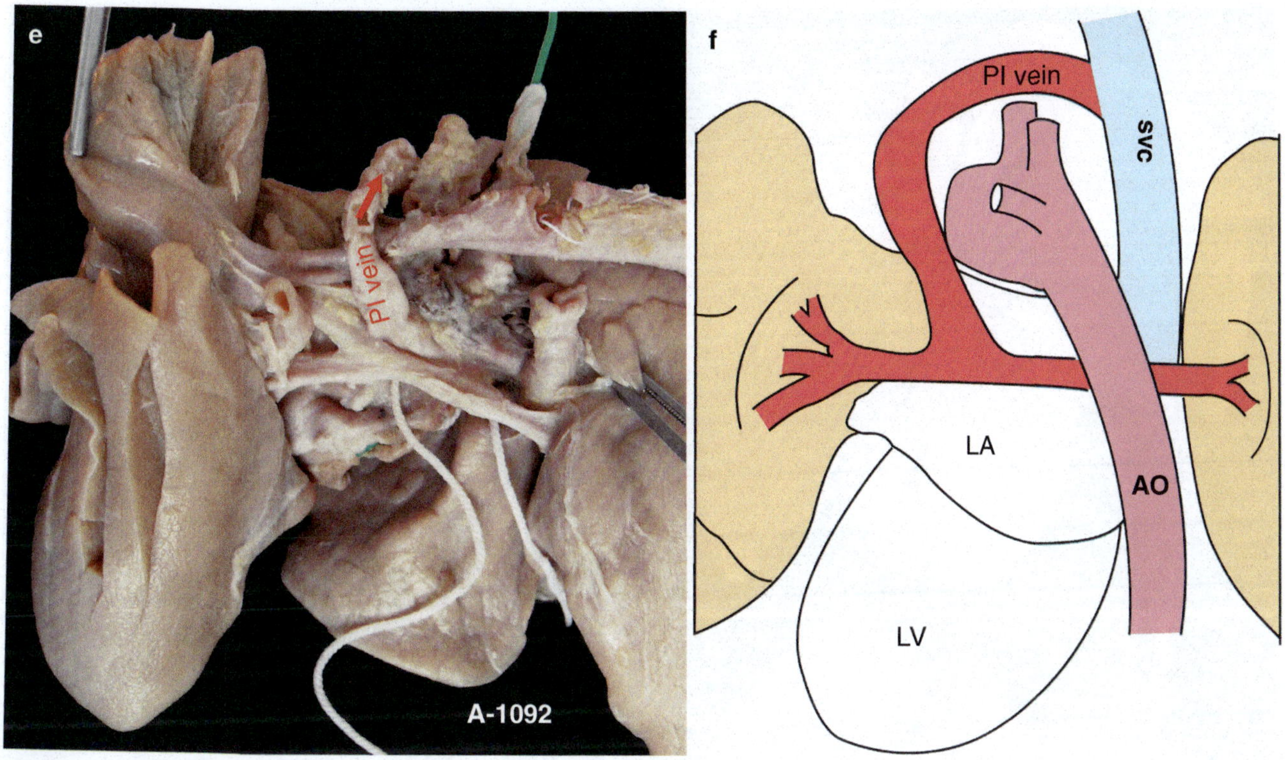

Fig. 14.8 (continued)

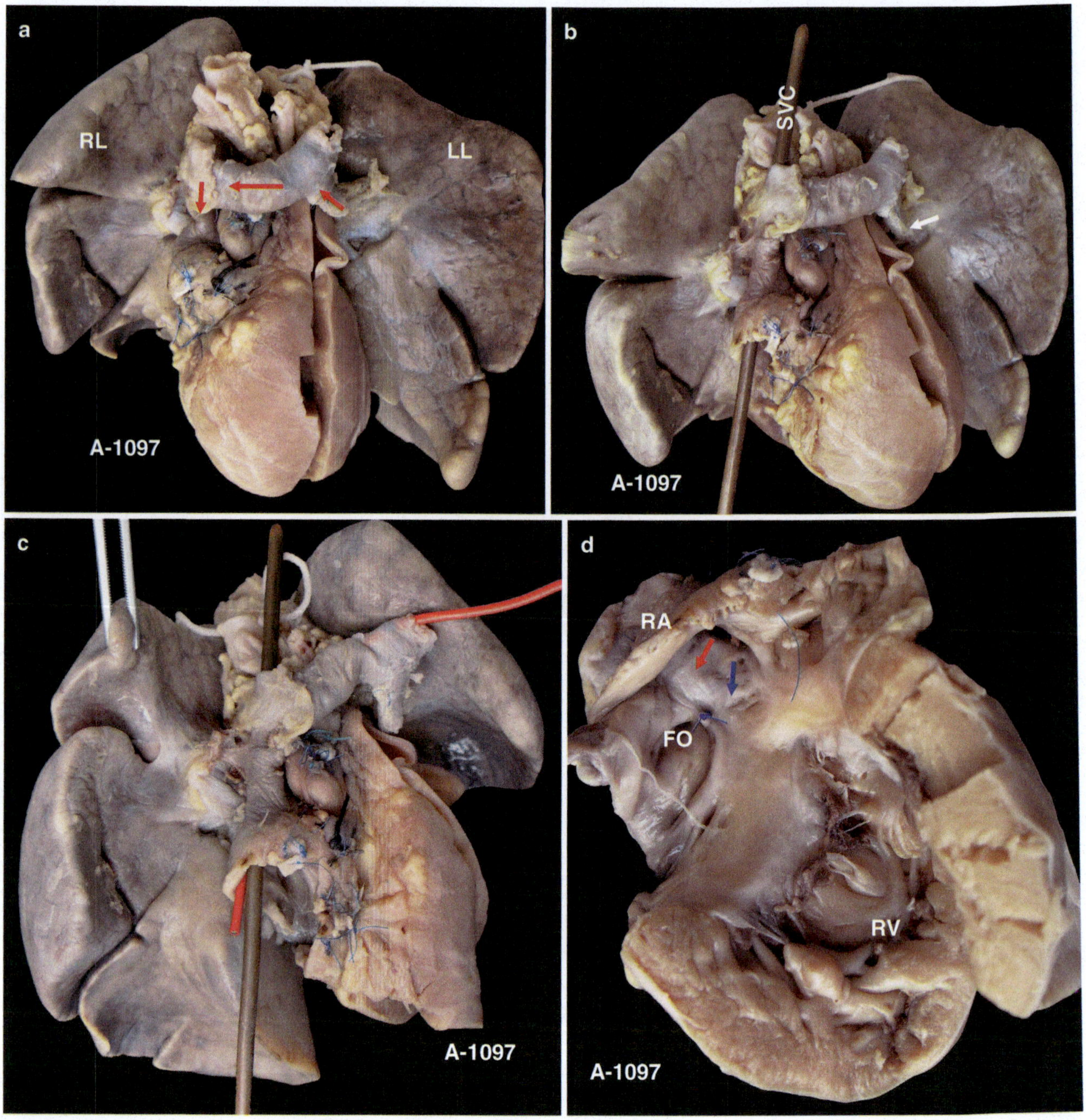

Fig. 14.9 (**a**) Specimen 1097. Postoperative total anomalous pulmonary venous drainage, supracardiac type. View is from the anterior specimen of the heart. The *arrow* indicates the direction of the pulmonary vein blood flow. *LL* left lung, *RL* right lung. (**b**) Specimen 1097. Probe in the superior vena cava connected with horizontal anonymous vein. The *arrow* indicates the vertical vein, which connects with the pulmonary common trunk. Pulmonary vein blood flows through the vertical vein-anonymous vein and returns to the SVC. (**c**) Specimen 1097. View from anterior specimen. The dark probe is in the SVC. The *red probe* is placed through an anonymous vein toward the SVC. The horizontal anonymous vein is in front of the aorta. (**d**) Specimen 1097. View from the right atrium and the right ventricle. Both pulmonary vein blood (*red arrow*) and systemic vein blood (*blue arrow*) return to the entrance of the SVC. *FO* fossa ovalis, *RA* right atrium, *RV* right ventricle. (**e**) Specimen 1097. The *arrow* indicates the ligated vertical vein. The *red probe* is in the gross anonymous vein, *RV* right ventricle, *LV* left ventricle. (**f**) Specimen 1097. View from the dorsal of a specimen of the heart. Incision of the pulmonary vein trunk and the LA (*arrows*) is shown in the dash line frame. The two are parallel, and anastomosis is easy. *LVOT* left ventricular outlet tract. (**g**) Specimen 1097. Incised pulmonary vein trunk and left atrium. Two parallel incisions should be made as large as possible. Anastomosed incision of the pulmonary vein trunk and the LA is shown in the figure, and the anastomosis line is visible. The *red probe* is in the pulmonary vein common trunk. *LA* left atrium, *LV* left ventricle. (**h**) Specimen 1097. View from a short distance from the anastomosis of the left atrium and the pulmonary vein common trunk. The *red probe* is at the auricle of the LA. The anastomotic stoma should be made as large as possible to avoid anastomosis stoma stricture. (**i**) Specimen 1097. Schematic diagram of operation. The LA and the pulmonary vein common trunk are incised and then anastomosed to drain the pulmonary vein blood to the LA. The anastomotic stoma should be made as large as possible to avoid anastomosis stoma stricture, *LV* left ventricle

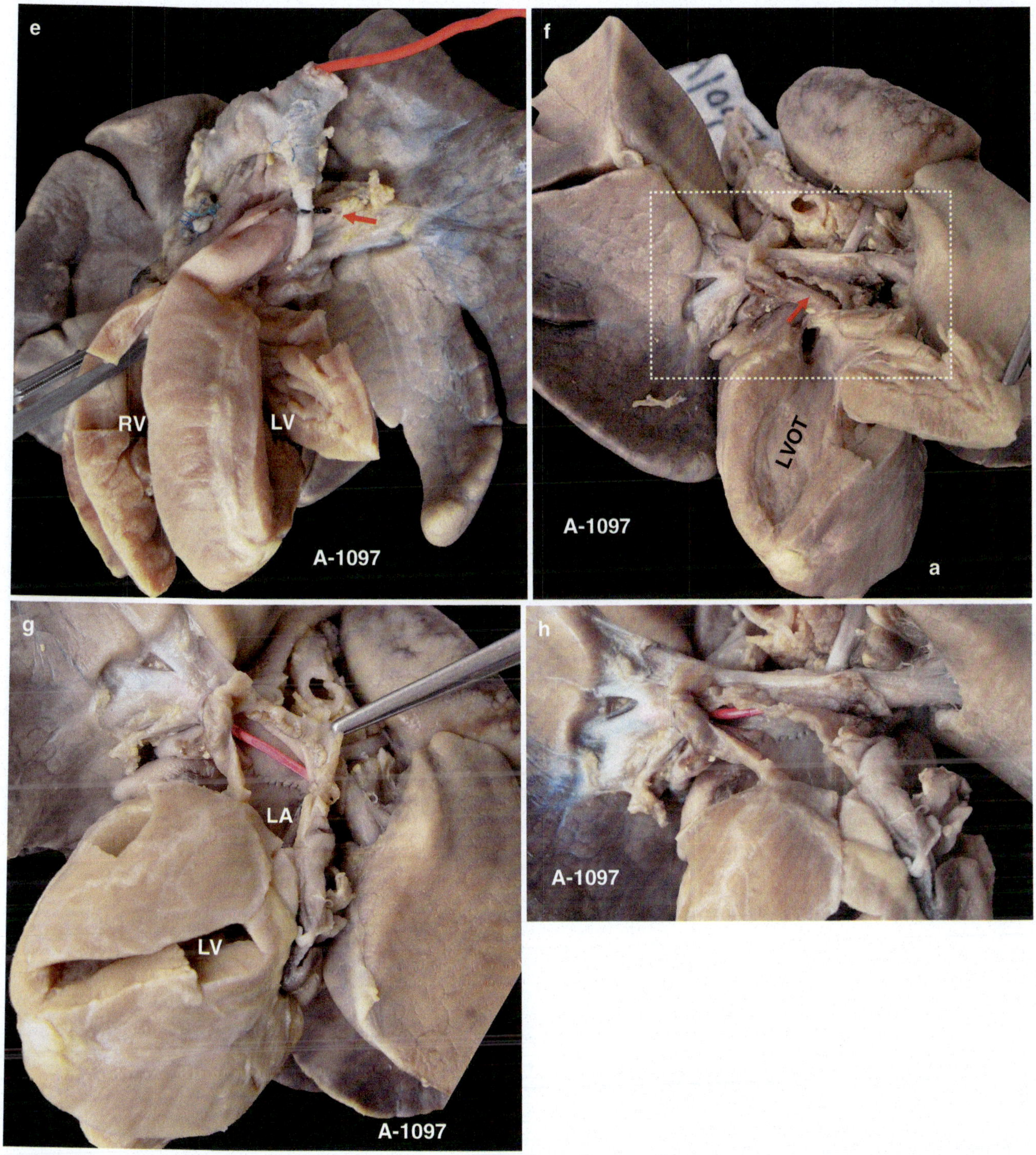

Fig. 14.9 (continued)

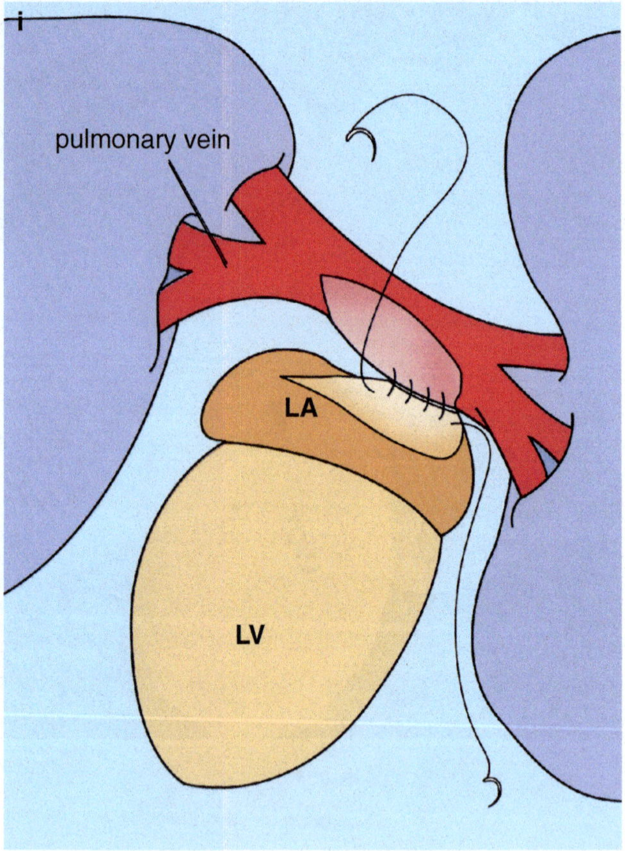

Fig. 14.9 (continued)

15 Aortic Arch Coarctation and Interrupted Aortic Arch

15.1 General Considerations

The proximal aortic arch is the beginning part of the innominate artery; the distal aortic arch is the isthmus where the ligamentum arteriosum is attached and is also the starting point of the descending thoracic aorta. In the embryonic stage, the aortic arch initially is paired and later turns into a single aortic arch (lactation type). Therefore, the region of the aortic arch should include the portion from the beginning part of the innominate artery to the aorta between the first pair of intercostal arteries. Branches of the aortic arch are the innominate artery, the left common carotid artery, the left subclavian artery, and the left patent ductus arteriosus (after birth, it closes and becomes the ligamentum arteriosum). Different regions of the aortic arch are derived from different embryonic components. Aortic arch deformities inevitably include many variations of each branch. Hence, aortic arch anomalies can be divided into many complex types. Aortic arch malformations include the elongation, twists, lumen atresia, interruption, and vascular ring.

15.2 Aortic Arch Coarctation

Congenital narrowing between the innominate artery and the left subclavian artery is called the *aortic arch coarctation*. Most of the aortic arch coarctation is located between the left common carotid artery and the left subclavian artery. The coarctation can be divided into a membrane-like stenosis or a tube-like constriction (Figs. 15.6, 15.1, 15.2, 15.7, and 15.3).

15.3 Interrupted Aortic Arch

Interrupted aortic arch is defined as luminal atresia or discontinuity between the ascending aorta and the descending aorta. Some have a residual aortic wall structure. They are categorized into three types, type A, type B, and type C, according to the site of the interruption. The three types are further classified into several subtypes based on the opening and closure of the ductus arteriosus and the concurrent cardiac malformations.

15.3.1 Type A

In type A, the site of the interruption is located between the left subclavian artery and the patent ductus arteriosus. When a concurrent VSD is present, the descending aorta receives the mixed blood from the pulmonary artery and oxygenated blood from the ascending aorta through collateral circulation. Infants with this condition may survive to childhood, usually with pulmonary hypertension. The goals of surgical correction of type A interrupted aortic arch are (1) to build communication between the ascending aorta and the descending aorta and (2) to cut off the connection between the patent ductus arteriosus and the pulmonary artery. Considering that the traditional method of dissecting the free patent ductus arteriosus and aorta involves a complicated procedure and a high risk of bleeding, we use the pulmonary intracavitary correction method without dissecting free the great vessels, which renders the repair of the VSD easier.

15.3.2 Type B

In the type B, the site of interruption is located between the left subclavian artery and the left common carotid artery.

15.3.3 Type C

The type C involves a site of interruption between the left common carotid artery and the innominate artery (Figs. 15.8, 15.4, 15.9, and 15.5).

X. Zhu (ed.), *Surgical Atlas of Cardiac Anatomy*,
DOI 10.1007/978-94-017-9409-1_15, © Springer Science+Business Media Dordrecht and People's Medical Publishing House 2015

15.4 Specimen Demonstrations

15.4.1 Aortic Arch Coarctation, Conduit Connected with the Aortic Arch and the Descending Aorta

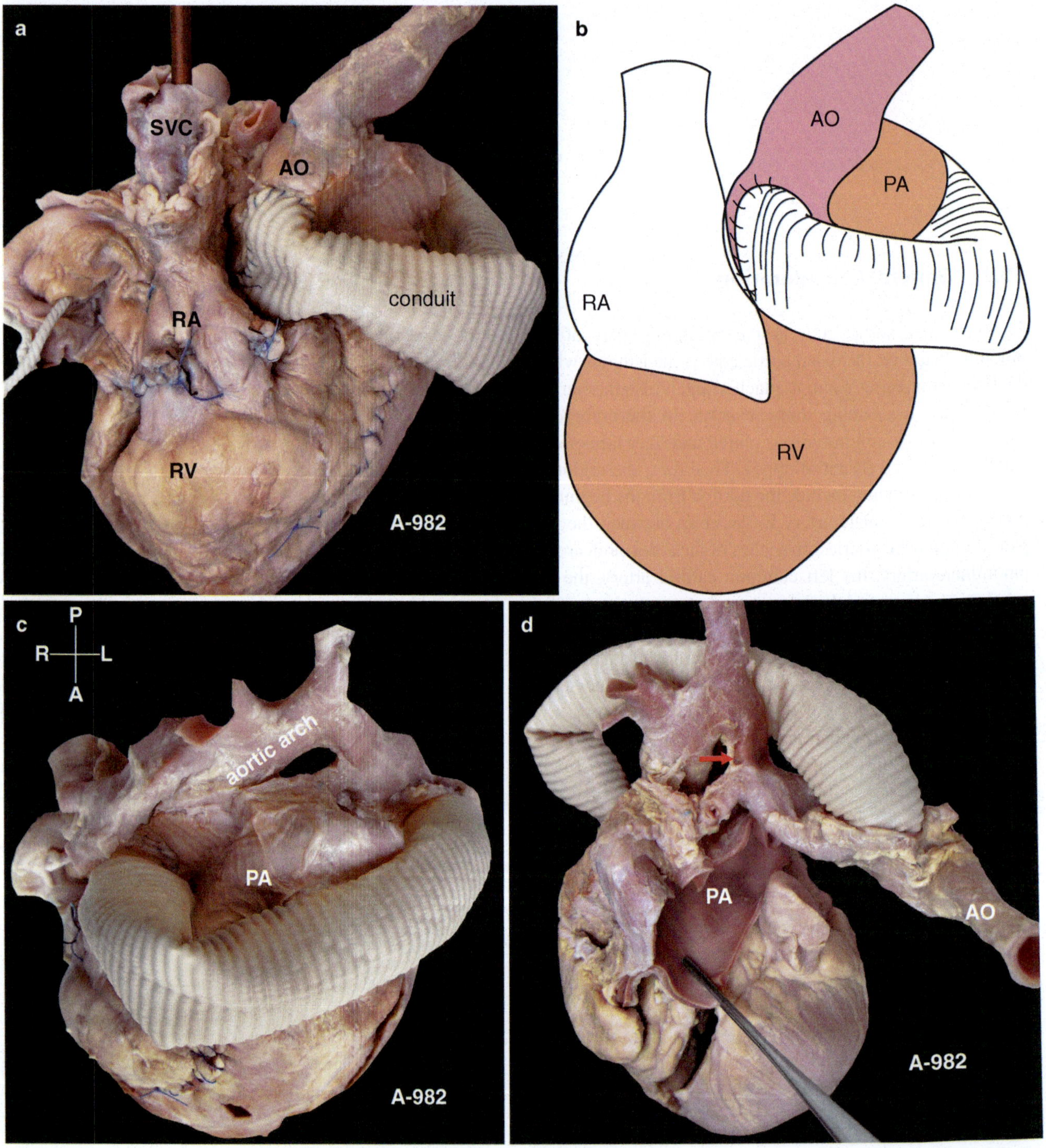

Fig. 15.1 (**a**) Specimen 982. Aortic arch coarctation. The ascending aorta and the descending aorta are connected with the artificial blood vessel. The figure demonstrates the proximal anastomosis of the artificial blood vessel (conduit). *AO* aortic arch, *RA* right atrium, *RV* right ventricle, *SVC* superior vena cava. (**b**) Specimen 982. Schematic diagram of Fig. 15.1a. (**c**) Specimen 982. The ascending aorta and descending aorta viewed from the cephalic position. The aortic arch coarctation (aortic arch), ascending aorta, and descending aorta are connected with the artificial blood vessel. *PA* pulmonary artery. (**d**) Specimen 982. View from the anterior aspect of the heart. The *arrow* indicates the coarctation at the distal part of the left common carotid and the dilated main pulmonary artery (PA). (**e**) Specimen 982. Aortic arch coarctation viewed from another visual angle (*arrow*). (**f**) Specimen 982. View of ventricular septal defect (*arrow*) from the left ventricle. *AV* aortic valve, *LV* left ventricle. (**g**) Specimen 982. Aortic arch coarctation. The VSD patch (D) is viewed from the RV; the distal coarctation of the aortic arch, with concurrent VSD, is shown

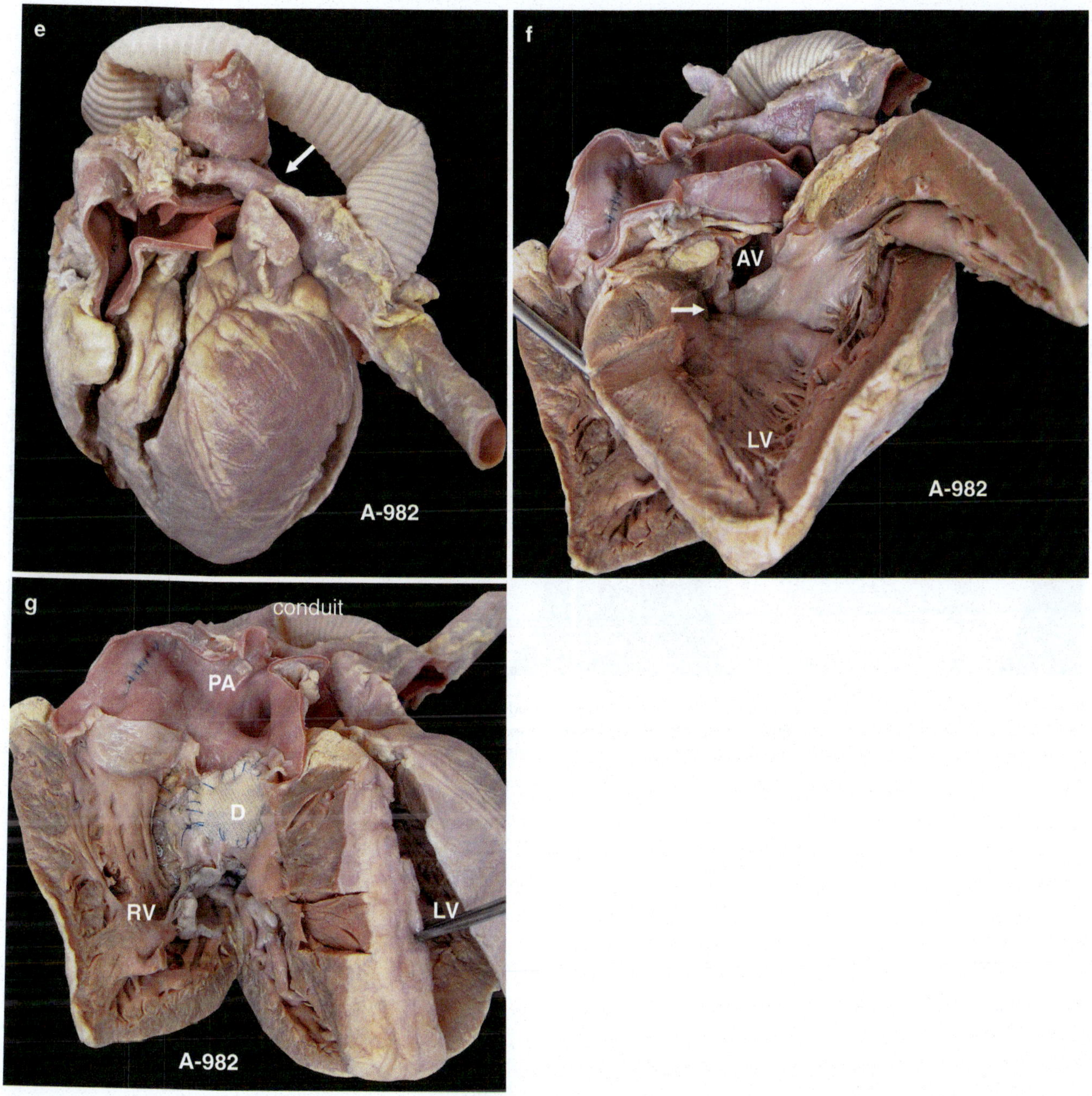

Fig. 15.1 (continued)

15.4.2 Dysplasia of the Aortic Arch and the Ascending Aorta, with Concurrent Ventricular Septal Defect

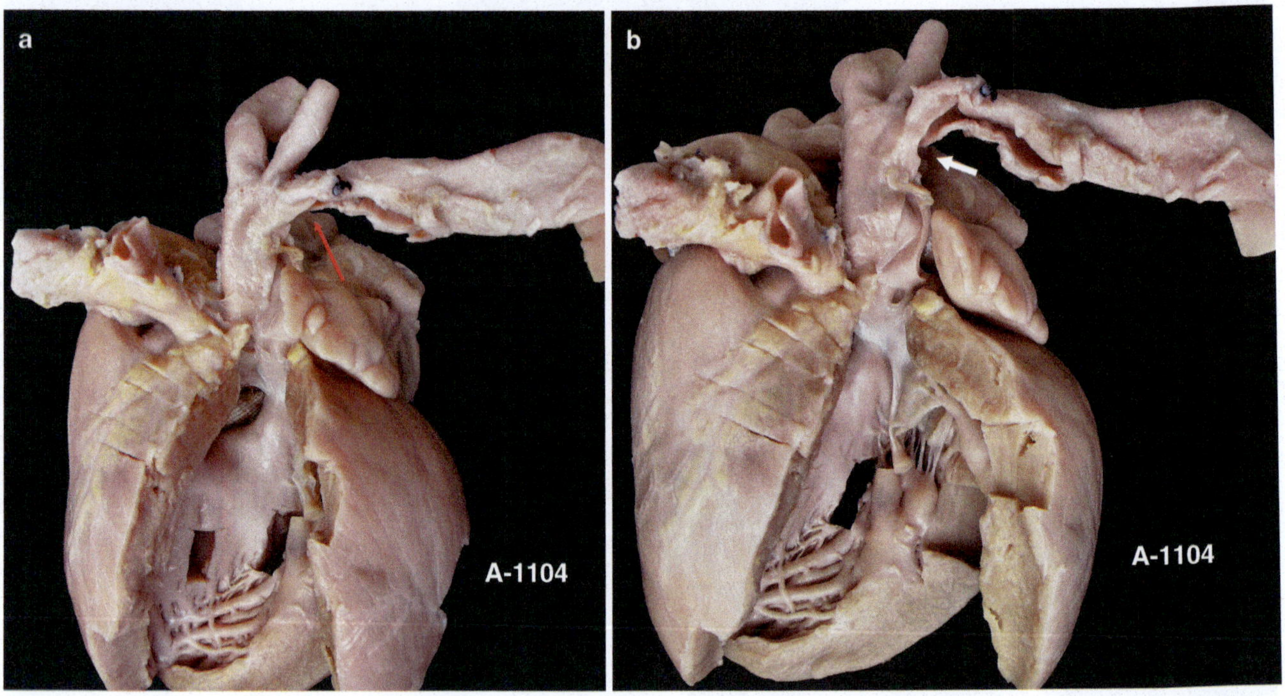

Fig. 15.2 (**a**) Specimen 1104. Aortic arch coarctation at the root of the subclavian artery, with concurrent ventricular septal defect. Viewed from LVOT, the thin ascending aorta, distal coarctation of aortic arch (*arrow*), and repaired VSD are shown. (**b**) Specimen 1104. View from the posterior aspect of the heart. The aortic arch coarctation is seen at the root of the subclavian artery (*arrow*), with no coarctation at the descending aorta

15.4.3 Distal Aortic Arch Coarctation, with Concurrent Ventricular Septal Defect

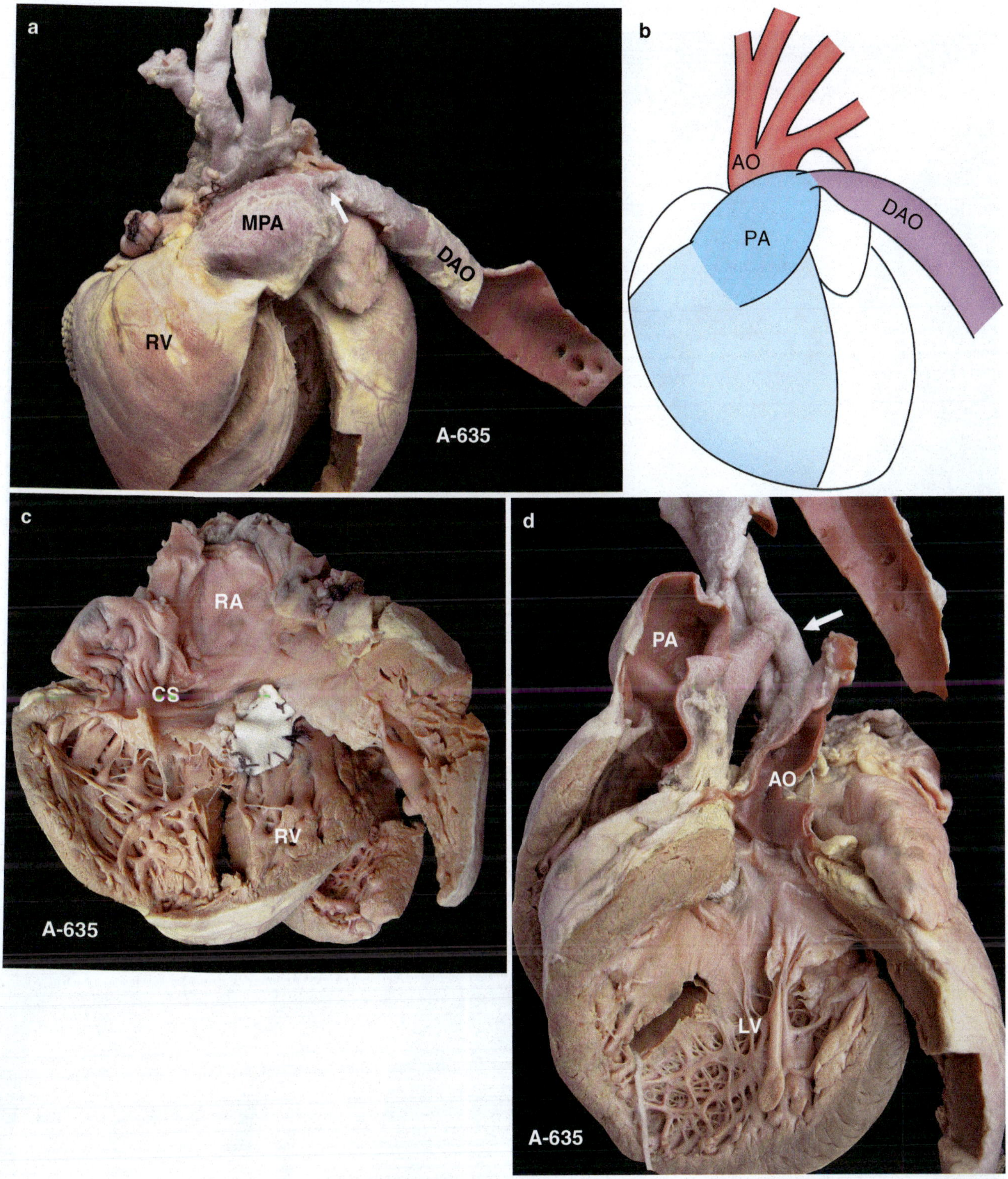

Fig. 15.3 (**a**) Specimen 635. Dysplasia of the aortic arch and the ascending aorta. Site of the interruption of the aorta is at the distal part of the left subclavian artery. The patent ductus arteriosus is indicated by the arrow. Pulmonary artery hypertension is present, and the main pulmonary artery is dilated. *DAO* descending aorta, *MPA* main pulmonary artery, *RV* right ventricle. (**b**) Schematic diagram of SPEC 635. (**c**) Specimen 635. The repaired ventricular septal defect viewed from the right ventricle. *CS* coronary sinus, *RA* right atrium, *RV* right ventricle. (**d**) Specimen 635. Dysplastic ascending aorta and aortic arch. Pulmonary artery hypertension and dilation are present. The *arrow* indicates the site of the coarctation. The distal part is the descending aorta. *AO* ascending aorta, *LV* left ventricle

15.4.4 Aortic Arch Coarctation, with Concurrent Ventricular Septal Defect and Pulmonary Artery Hypertension

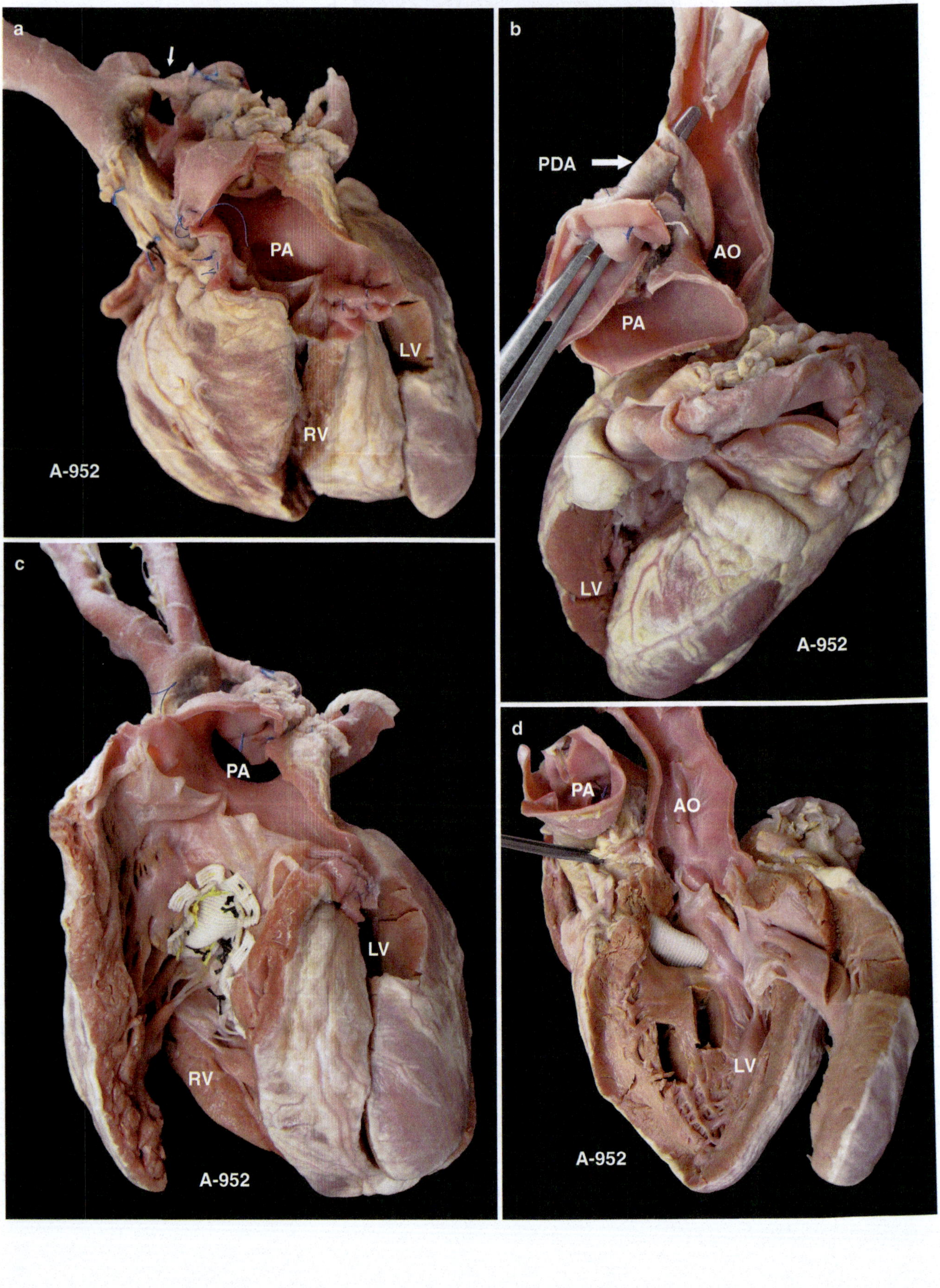

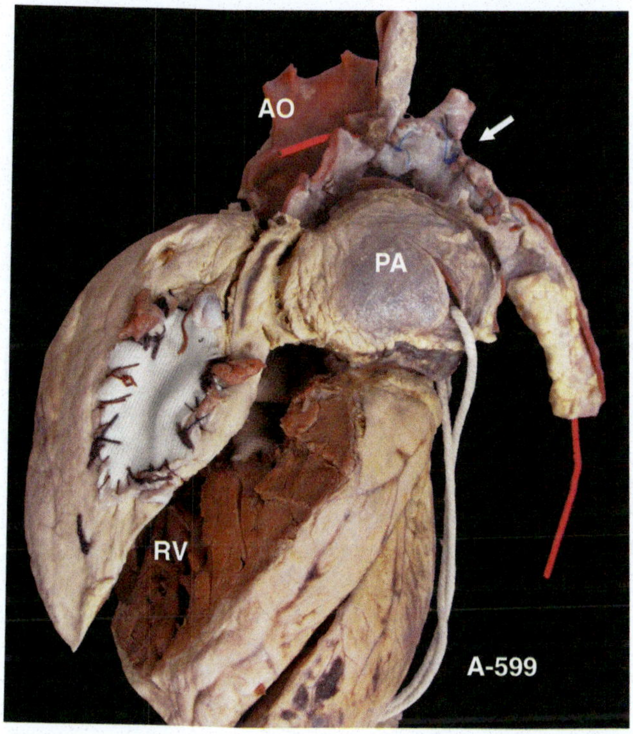

Fig. 15.5 Specimen 599. Aortic arch coarctation. A red probe is placed from the ascending aorta to the descending aorta. An aortic arch coarctation with dilated pulmonary artery and RVOT stenosis are demonstrated. The RVOT is enlarged with a patch

Fig. 15.4 (**a**) Specimen 952. Aortic arch coarctation. The site of the coarctation of the aorta is at the distal part of the left common carotid artery (*arrow*), with a thin patent ductus arteriosus. There is no left common carotid artery; a VSD was present. *LV* left ventricle, *PA* main pulmonary artery, *RV* right ventricle. (**b**) Specimen 952. *Arrow* indicates patent ductus arteriosus. Surgical forceps are placed in the patent ductus arteriosus. The proximal part of the main pulmonary artery has been cut open. The descending aorta is not shown. *AO* aorta, *LV* left ventricle, *PA* main pulmonary artery. (**c**) Specimen 952. Repaired ventricular septal defect viewed from the right ventricle. *LV* left ventricle, *PA* pulmonary artery, *RV* right ventricle. (**d**) Specimen 952. Incised left ventricular outlet tract and aorta (*AO*). The repaired VSD is shown from the LV side. A normal aortic valve can be seen. *PA* pulmonary valve

15.4.5 Interrupted Aortic Arch, Conduit Connected with the Aortic Arch and the Descending Aorta

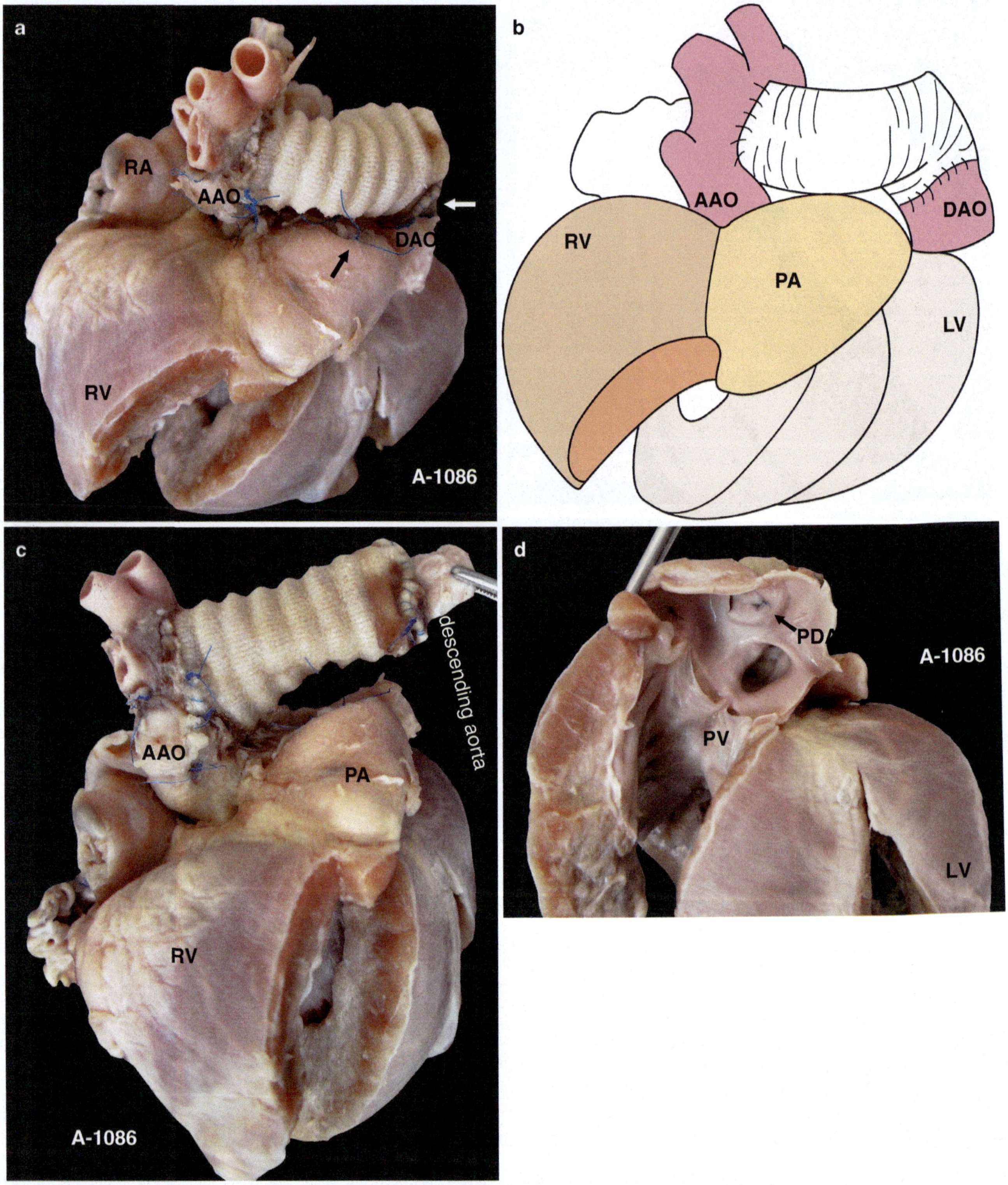

Fig. 15.6 (**a**) Specimen 1086. Interrupted aortic arc viewed from the anterior aspect of heart. The proximal part of the artificial blood vessels is anastomosed with the beginning part of the aortic arch. The distal part of the artificial blood vessels is anastomosed with the descending aorta. The *arrow* indicates artificial blood vessels. *AAO* ascending aorta, *DAO* descending aorta, *RA* right atrium, *RV* right ventricle. (**b**) Specimen 1086. Schematic diagram. (**c**) Specimen 1086. Interrupted aortic arch. The left anterior oblique demonstrates both ends of the anastomosis of the artificial blood vessel. *AAO* ascending aorta, *PA* pulmonary artery. (**d**) Specimen 1086. The opening of the patent ductus arteriosus (PDA) viewed from the aortic lumen (*arrow*). (**e**) Specimen 1086. Dacron patch from left ventricle (LV)

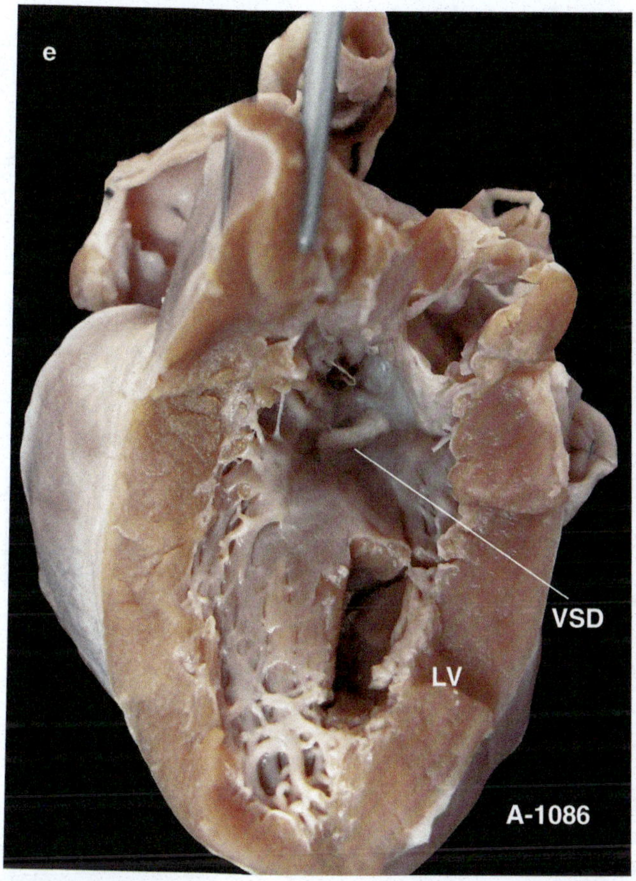

Fig. 15.6 (continued)

15.4.6 Interrupted Aortic Arch. The Ascending Aorta and Descending Aorta Are Directly Anastomosed

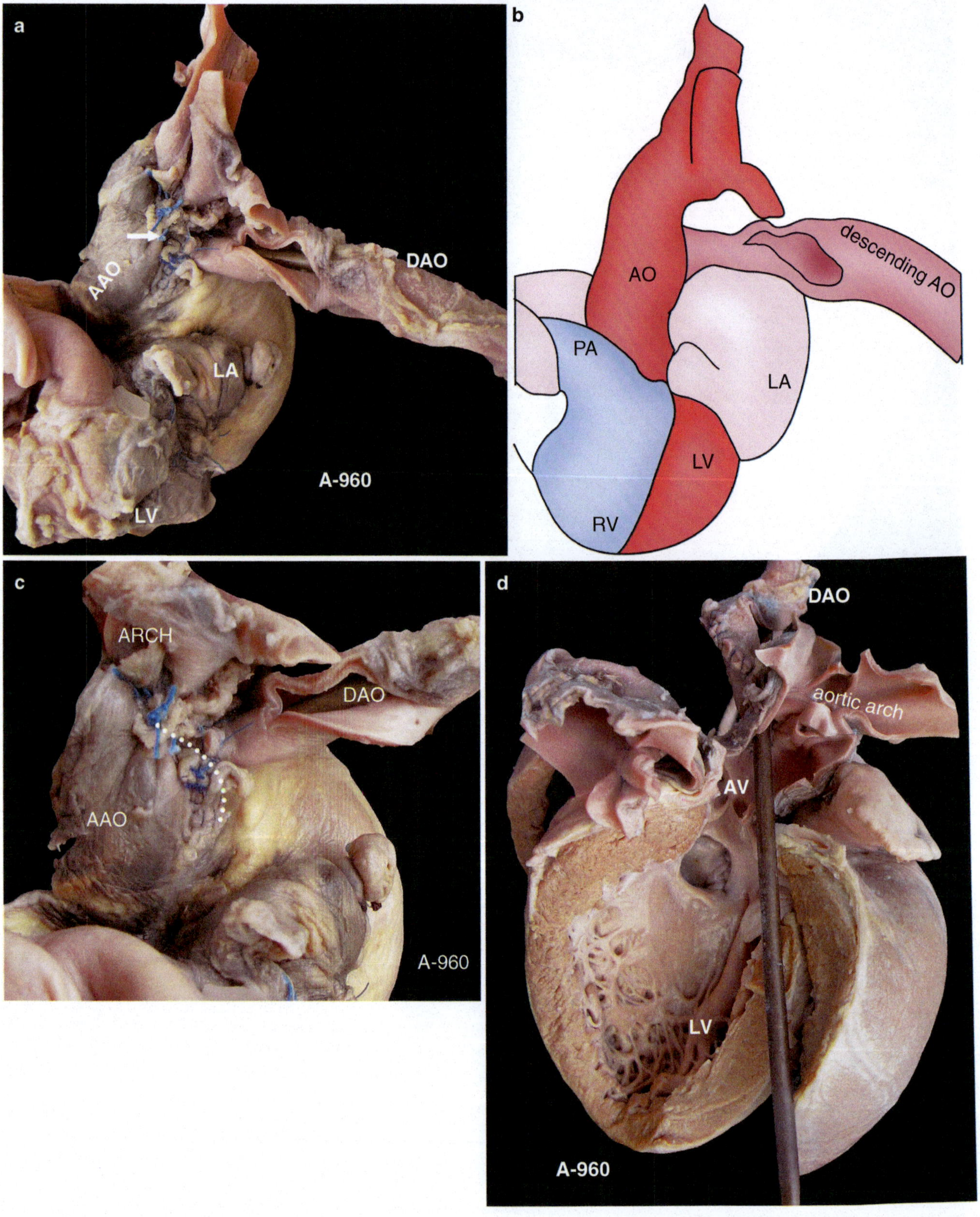

Fig. 15.7 (**a**) Specimen 960. Interrupted aortic arch. The interrupted site of the aorta is located at the distal part of the left subclavian artery. The ascending aorta and the descending aorta are directly anastomosed. The probe is put in the stoma. *AAO* ascending aorta, *DAO* descending aorta. (**b**) Schematic diagram of SPEC 960. (**c**) Specimen 960. Interrupted aortic arch. The anastomosis line (*white dash line*) is shown here. *AAO* ascending aorta, *ARCH* aortic arch, *DAO* descending aorta. (**d**) Specimen 960. The descending aortic stoma viewed from the ascending aorta. The probe is put in the stoma. *AV* aortic valve, *DAO* descending aorta, *LV* left ventricle

15.4.7 Interrupted Aortic Arch. The Ascending Aorta and Descending Aorta Are Directly Anastomosed

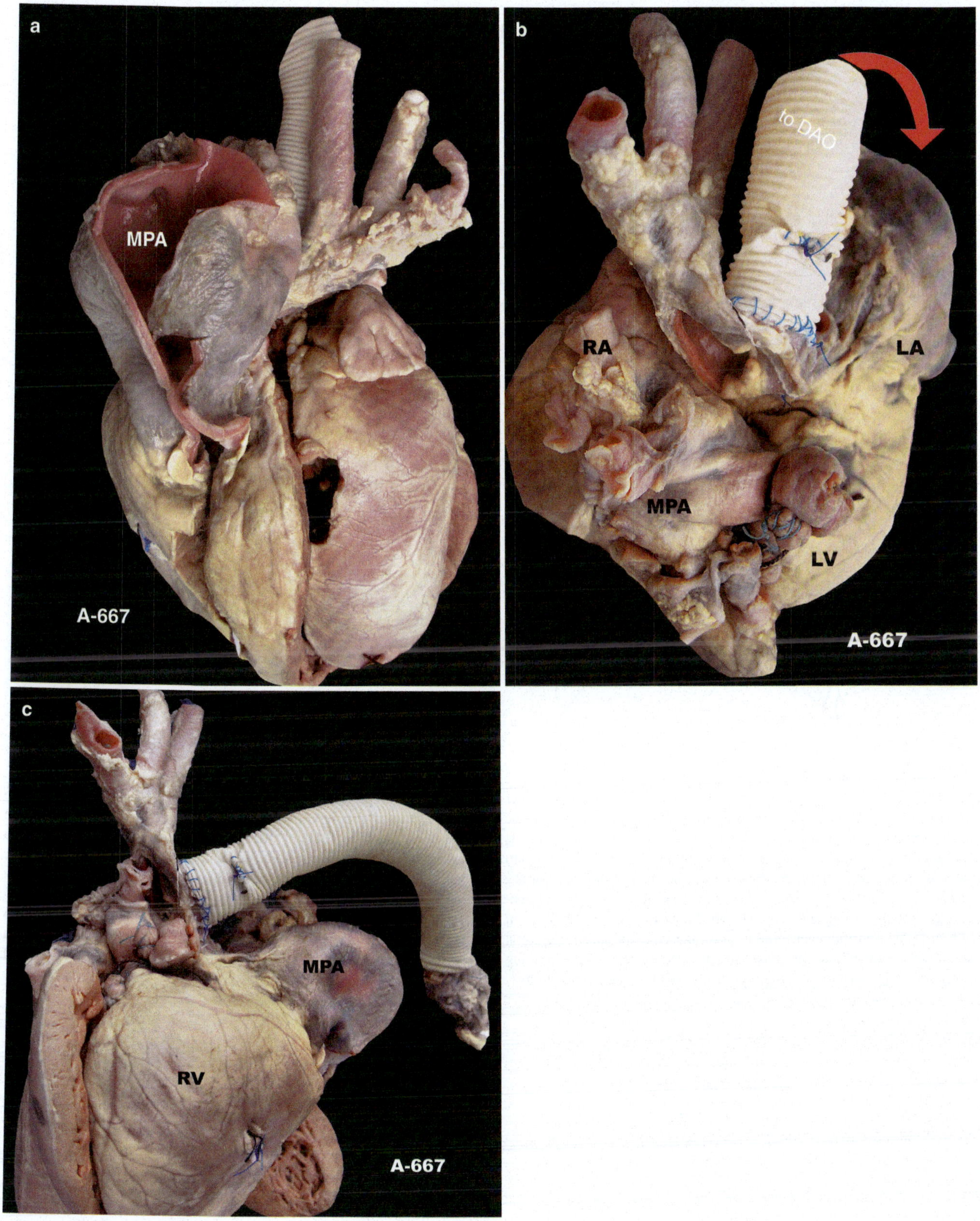

Fig. 15.8 (**a**) Specimen 667. Type A interrupted aortic arch. Severely dilated main pulmonary artery is shown. The site of the interruption of the aorta is at the distal part of the left subclavian artery. *MPA* main pulmonary artery. (**b**) Specimen 667. The ascending aorta and descending aorta directly anastomosed with the artificial blood vessel. The figure shows the proximal stoma. *DAO* descending aorta, *LA* left atrium, *LV* left ventricle, *MPA* main pulmonary artery, *RA* right atrium. (**c**) Specimen 667. The artificial blood vessel between the ascending aorta and the descending aorta. MPA is the dilated main pulmonary artery

15.4.8 Interrupted Aortic Arch, with Concurrent Ventricular Septal Defect and Patent Ductus Arteriosus Connecting the Ascending Aorta to the Descending Aorta

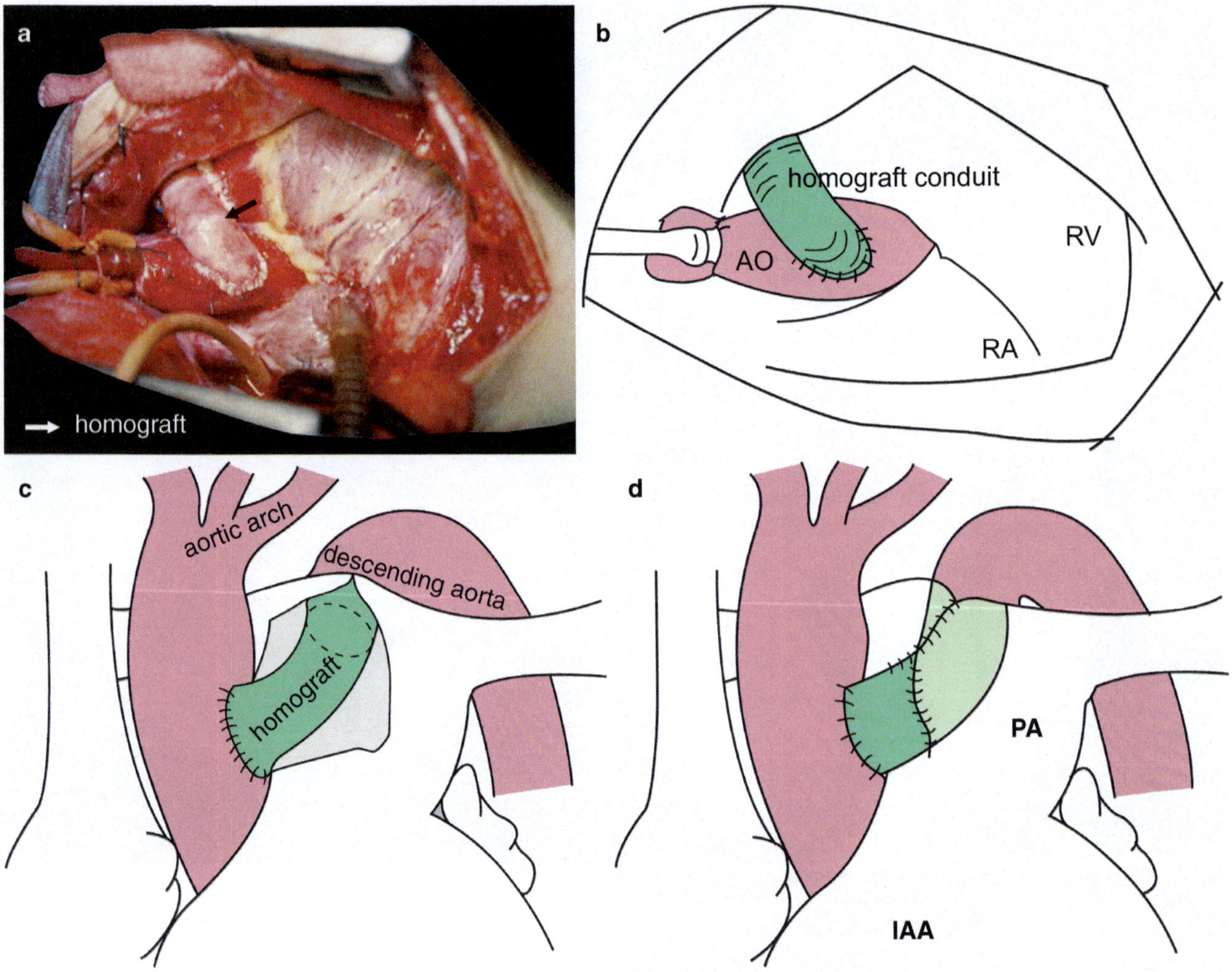

Fig. 15.9 (**a**) Operative figure of type A interrupted aortic arch. Under hypothermic cardiopulmonary bypass, low-flow perfusion, and aortic occlusion, the main pulmonary artery is incised. The allogeneic graft is end to end anastomosed to the descending aorta. Its other side is end to side anastomosed to the ascending aorta. The main pulmonary artery incision is finally closed. This figure shows anastomosis of the allogeneic graft to ascending aorta (*arrow*). (**b**) Simplified schematic diagram of Fig. 15.5a. The allogeneic graft is anastomosed to the ascending aorta, and its other side is anastomosed to the descending aorta. *AO* ascending aorta, *RA* right atrium, *RV* right ventricle. (**c**) Operating procedure1. After the ascending aorta and the descending aorta are occluded, the severely dilated main pulmonary artery is cut open to identify the entrance of the patent ductus arteriosus, which is the entrance of the descending aorta, and to anastomose it end to end to the allogeneic graft. A lateral incision is made on the ascending aorta to anastomose it to the other side of the allogeneic graft. At last, the cut edge of the pulmonary artery is sutured to the wall of allogeneic graft and the occlusion clamp is removed. The ascending aorta now can communicate with the descending aorta. (**d**) Operating procedure 1 of Fig. 15.5c. The distal part of the allogeneic graft is connected with the descending aorta

16 Congenital Anomalies of the Tricuspid Valve

16.1 General Considerations

16.1.1 Tricuspid Stenosis and Tricuspid Atresia

Tricuspid stenosis covers a spectrum of diseases, including tricuspid dysplasia (leaflets dysplasia and small annulus), supravalvular stenosis caused by the fiber membrane in the atrial side of the tricuspid, and double-orifice TV caused by the fiber cord dividing the tricuspid orifice. Tricuspid atresia is defined as the TV being completely undeveloped. The equivalent area of the tricuspid can refer to the umbilication.

Tricuspid atresia or severe stenosis is often associated with other complex malformations, mainly hypoplasia of the RV, VSD, single ventricle, and pulmonary stenosis or atresia. Principles of surgical treatment are to drain the systemic venous blood from the vena cava or RV to the pulmonary artery without depending on the function of the RV, aimed to increase substantially the arterial oxygen saturation and to alleviate the ventricular volume load. If the RV has enough volume and contractility and is without pulmonary stenosis, right ventricular function should be retained as much as possible. Clinically, we can choose the following surgical bypass approaches:

16.1.1.1 Superior Vena Cava-Pulmonary Artery Anastomosis

The classic method of repair, also known as the *Glenn operation*, is to anastomose the SVC and the right pulmonary artery. The SVC is disconnected at the entrance of the atrium, and the beginning part of the right pulmonary artery is cut off. The two ends are anastomosed. However, this procedure will result in uneven distribution of the blood flow and has been replaced by a modified technique called *bidirectional cavopulmonary anastomosis*. Without disconnecting the right pulmonary artery, the SVC blood is discharged into the bilateral pulmonary artery. For ill children with interruption of the IVC, blood from the IVC flows through the azygos vein and the SVC into the bilateral pulmonary artery.

16.1.1.2 Right Atrium-to-Right Ventricular Bypass

In patients with tricuspid atresia, if the ventricle (RV) from which the pulmonary artery originates develops in a certain degree, the RA and the RV can be connected with an extracardiac conduit, and then the blood from the RA can drain into the pulmonary artery through the RV. If no pulmonary stenosis is present, right ventricular involvement may help with the pump function. The RV participating in the pumping function might bring some disadvantages. RV-to-RA regurgitation is a common occurrence, unless the extracardiac conduit is an implanted valve. However, the implanted valve will increase the ejection resistance of the RA, which one should consider when planning to apply this procedure (Figs. 16.1 and 16.4).

16.1.1.3 Right Atrium-Pulmonary Artery Anastomosis, Also Called Fontan Operation

This surgical approach has been widely used, with increasing indications. Initially, a prosthetic valve was implanted at the opening of the IVC; the general consensus now is that the valve is not necessary. The RA can be seen as a valve-free chamber with systolic function, playing a pumping function between the vena cava and the pulmonary artery. With a well-developed RA, sinus rhythm, and low pulmonary resistance, this operation can result in a good hemodynamic state. On the other hand, an RA with a thin wall will directly affect the operative outcomes. Therefore, patient-selection criteria are sinus rhythm, normal vena cava, RA with normal capacity, mean pulmonary artery pressure <15 mmHg, normal left atrioventricular valve function, and ventricular ejection fraction >0.6 (Fig. 16.3).

X. Zhu (ed.), *Surgical Atlas of Cardiac Anatomy*,
DOI 10.1007/978-94-017-9409-1_16, © Springer Science+Business Media Dordrecht and People's Medical Publishing House 2015

16.1.1.4 Total Cavopulmonary Connection

The SVC is anastomosed to the pulmonary artery. The IVC is connected to the pulmonary artery through the conduit in the RA, which allows blood to flow more smoothly and results in more reasonable hemodynamics and simplified technique.

16.1.2 Congenital Tricuspid Insufficiency

Congenital tricuspid insufficiency often is caused by leaflets, chordae, or papillary muscle deformity or absence or developmental disorders of the endocardial cushion. The specimens in this chapter demonstrate congenital tricuspid papillary muscle absence. Valvuloplasty or replacement can be selected according to the lesion's characteristics (Fig. 16.2).

16.2 Specimen Demonstrations

16.2.1 Tricuspid Atresia, Right Atrium-Right Ventricle Artificial Vessel Bypass

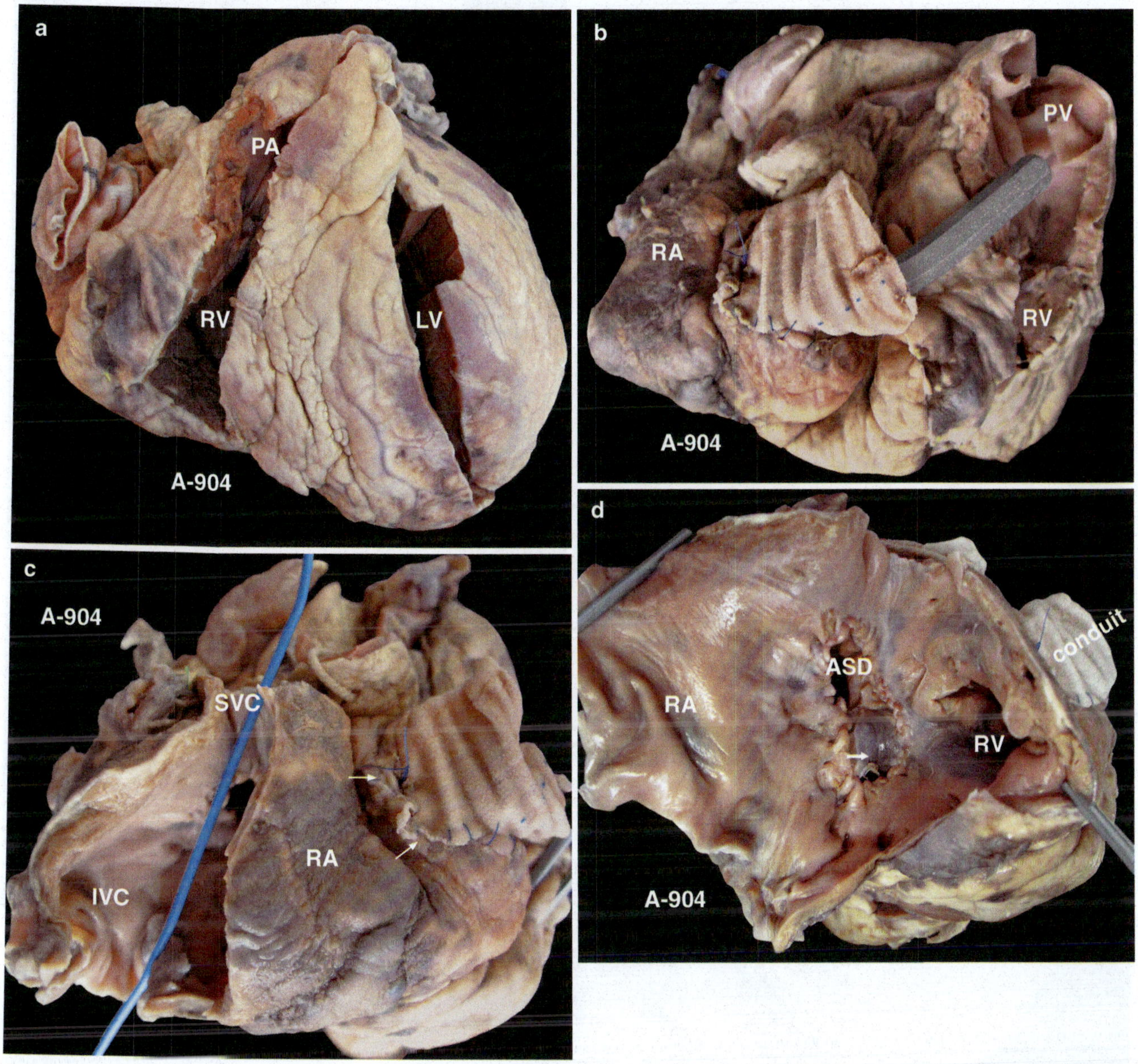

Fig. 16.1 (**a**) Specimen 904. Tricuspid atresia, right atrium-pulmonary artery extracardiac conduit connection, anterior view. The left and right ventricles are incised. An end of the artificial vessel can be seen. *LV* left ventricle, *PA* pulmonary artery, *RV* right ventricle. (**b**) Specimen 904. The anastomosis line of the right atrium and the artificial vessel in the right anterior oblique position. The probe is put into the artificial vessel through the RA. The pulmonary valve (PV) generally is normal, *RA* right atrium, *RV* right ventricle, *PV* pulmonary valve. (**c**) Specimen 904. Tricuspid atresia. The RA is cut open. The blue probe indicates the position of the superior and inferior venae cavae. The RA is connected to the artificial vessel. The *arrow* indicates the anastomosis line. *IVC* inferior vena cava, *RA* right atrium, *SVC* superior vena cava. (**d**) Specimen 904. Tricuspid atresia. The RA is dissected, illustrating the ASD, the orifice of the artificial vessel, and the umbilication formed by the tricuspid atresia (*arrow*), *RA* right atrium, *RV* right ventricle. (**e**) Simplified schematic diagram of Fig. 16.1d, *RA* right atrium, *RV* right ventricle, *ASD* atrial septal defect. (**f**) Specimen 904. The ventricular septal defect (D) seen from the incised right ventricle (RV). The pulmonary valve (PV) is well developed, the RV is small, and no atrioventricular valve tissue is present in the chamber, *C* conduit, *RV* right ventricle, *LV* left ventricle. (**g**) Specimen 904. Ventricular septal defect (D), mitral valve (MV), and aorta (AO) viewed from left ventricle, *LV* left ventricle. (**h**) Specimen 904. Incised left ventricular left outlet. An ASD (*arrow*) is viewed from the LA. The MV is normal. *LV* left ventricle, *MV* mitral valve. (**i**) Specimen 904. Operative schematic diagram of tricuspid atresia. Extracardiac conduit of the RA and the RV is shown. In the specimen, the atrial end of the conduit is kept and right ventricular end is removed

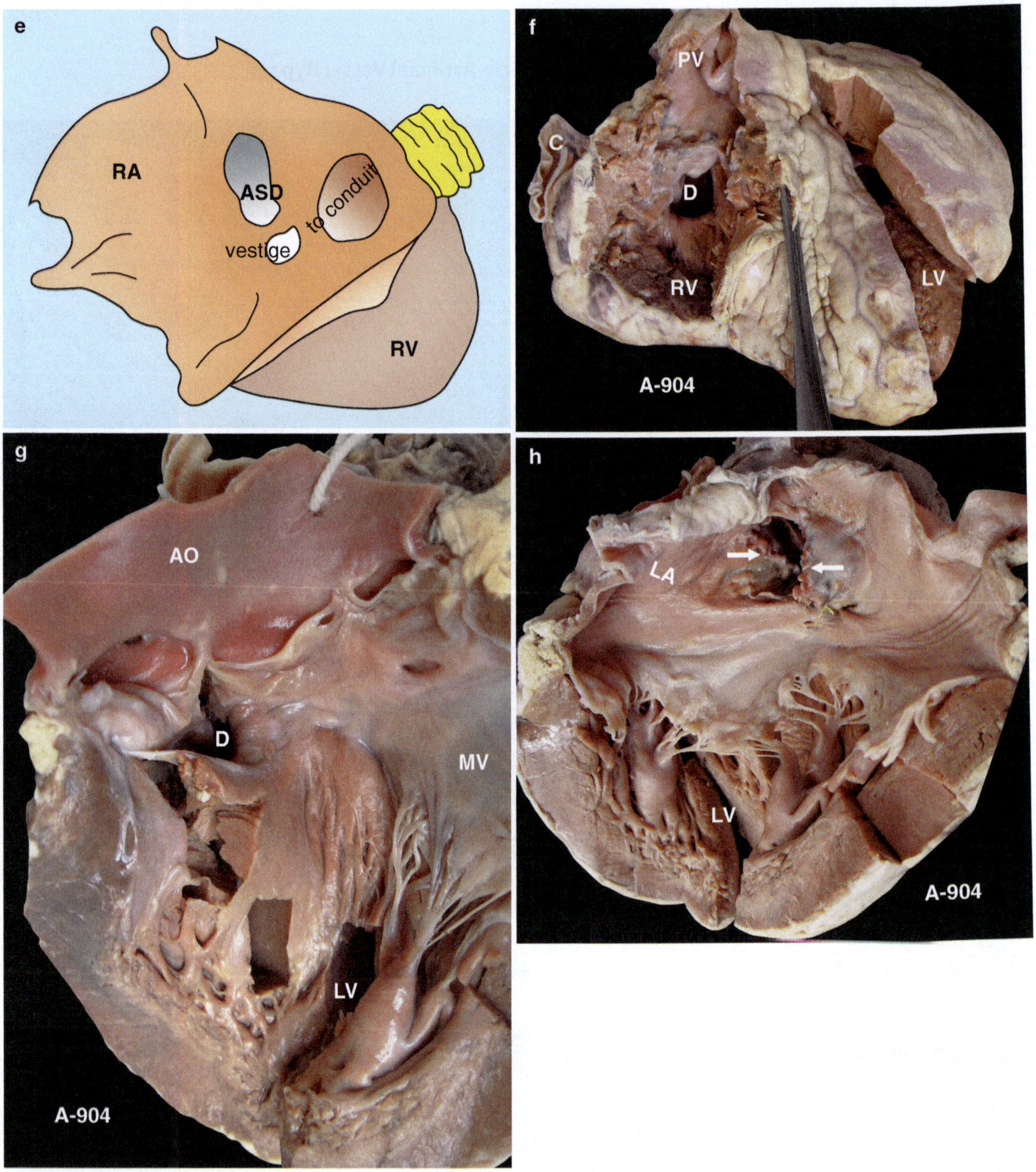

Fig. 16.1 (continued)

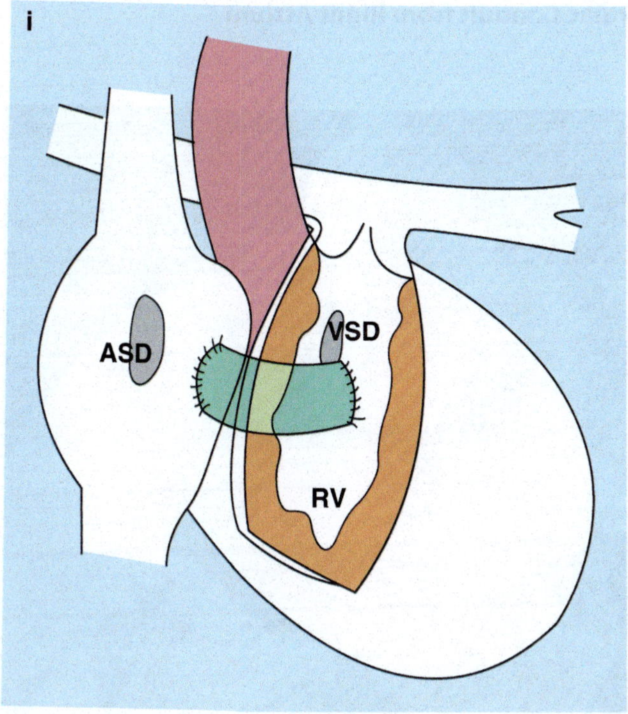

Fig. 16.1 (continued)

16.2.2 Tricuspid Atresia. Bovine Pericardial Extracardiac Conduit from Right Atrium to Right Ventricle

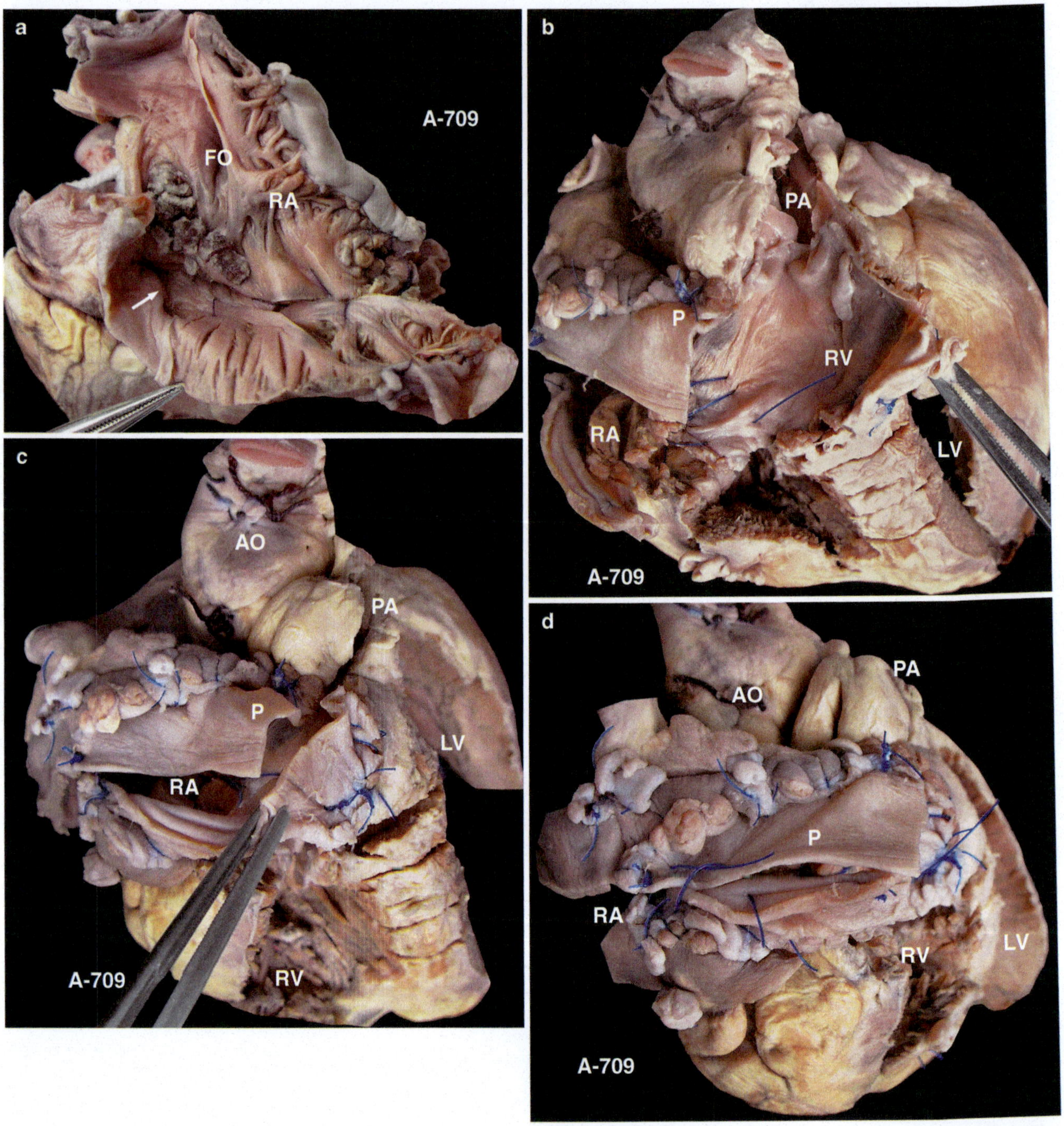

Fig. 16.2 (**a**) Specimen 709. Tricuspid atresia. The TV cannot be seen from the open RA. The *arrow* indicates the umbilication formed by the tricuspid atresia. *FO* fossa ovalis, *RA* right atrium. (**b**) Specimen 709. The right atrium, right ventricle, and pulmonary artery are incised simultaneously. The dysplastic RV, the normal development of the main pulmonary artery, and the PV are illustrated. The blue surgical suture can be seen at the area between the atria and ventricles, and the area constitutes the bottom of the channel of the RA to the RV. A bovine pericardial patch (*P*) is sutured to the top of the extracardiac conduit of the RA-RV-main pulmonary artery, which results in complete extracardiac conduit. (**c**) Specimen 709. The bovine pericardium (P) is seen as the top cover of the extracardiac conduit of the RA-RV-main pulmonary artery. (**d**) Specimen 709. Surgical formation of an extracardiac conduit. P indicates the bovine pericardial extracardiac conduit connecting the RA to the RV. (**e**) Specimen 709. Red probe going through the ventricular septal defect. (**f**) Specimen 709. Normal mitral valve viewed from left ventricle, with small ventricular septal defect (*arrow*). *LV* left ventricle

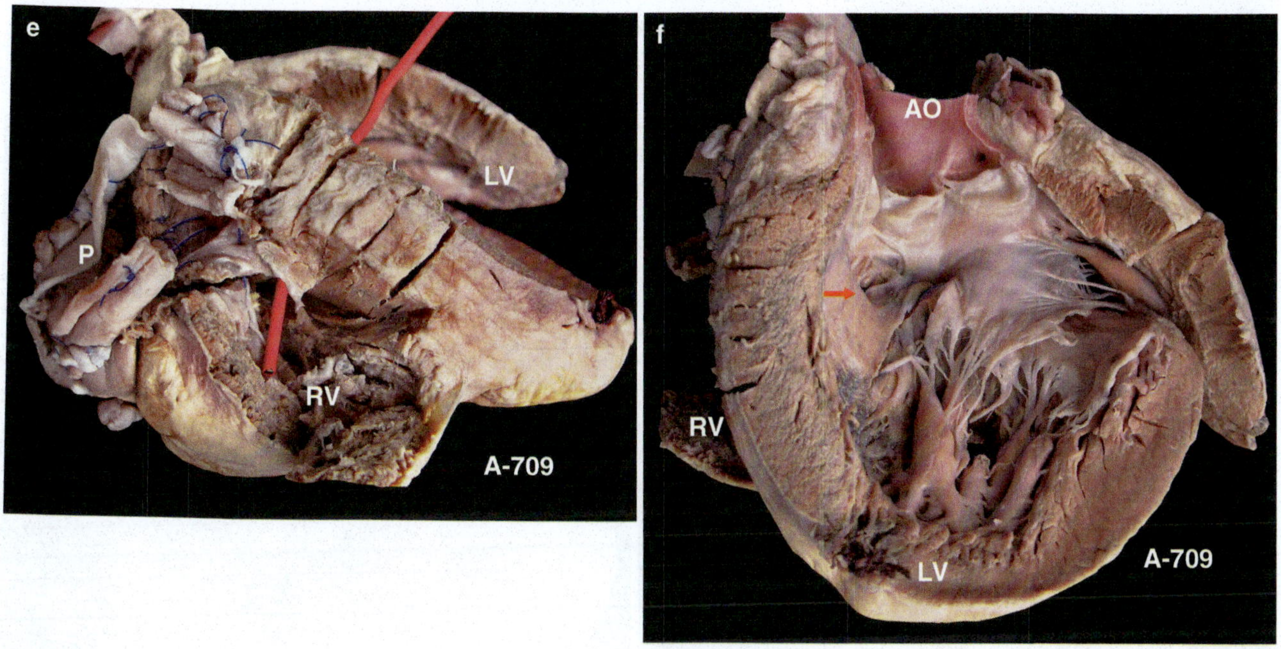

Fig. 16.2 (continued)

16.2.3 Tricuspid Atresia. Autologous Pericardial Extracardiac Conduit from the Right Atrium to the Pulmonary Artery

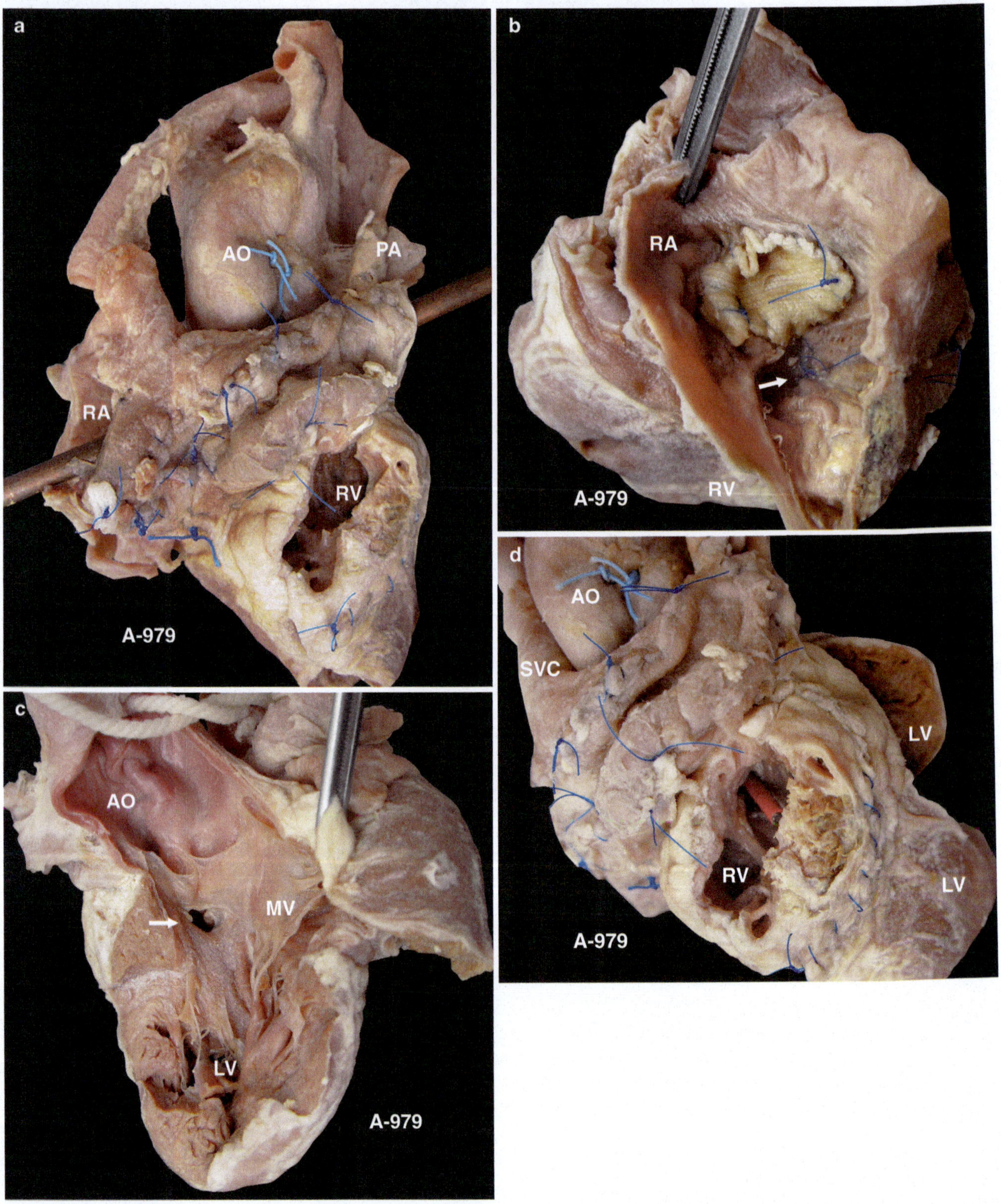

Fig. 16.3 (**a**) Specimen 979. Extracardiac conduit bypass from the right atrium to the pulmonary valve in a patient with tricuspid atresia. A probe is placed from the RA through the extracardiac conduit to the pulmonary artery (PA). *AO* ascending aorta. (**b**) Specimen 979. Top view of an open right atrium. An ASD has been repaired with a Dacron patch (*P*). The *arrow* indicates the umbilication formed by tricuspid atresia. (**c**) Specimen 979. Tricuspid atresia with ventricular septal defect. A small VSD (*arrow*) is viewed from the LV. *LV* left ventricle, *MV* mitral valve. (**d**) Specimen 979. Red probe placed from left ventricle through ventricular septal defect to the right ventricle, *RA* right atrium, *RV* right ventricle, *AO* ascending aorta, *SVC* superior vena cava

16.2.4 Absent Tricuspid Papillary Muscle. Tricuspid Valvuloplasty

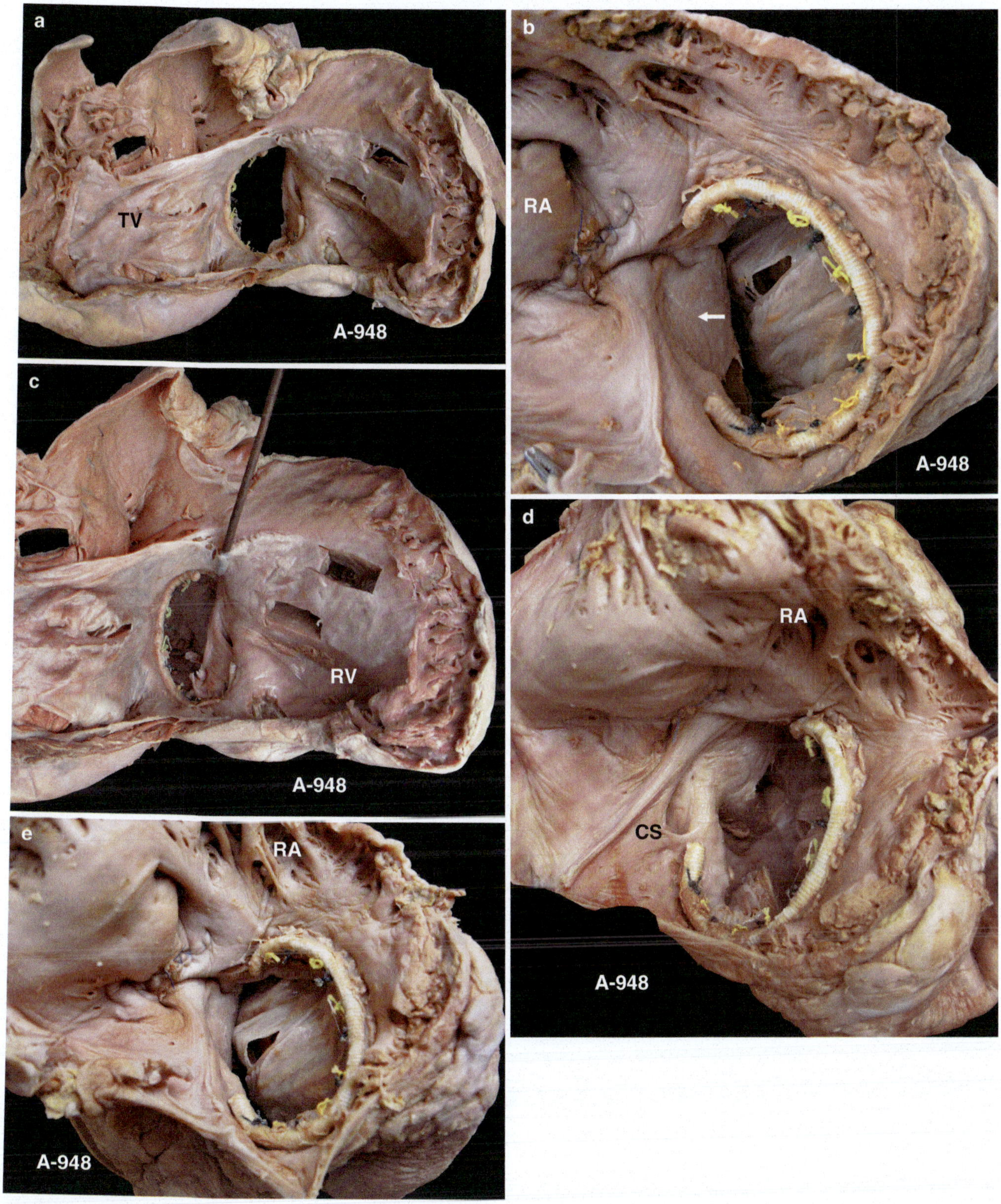

Fig. 16.4 (**a**) Specimen 948. Absent tricuspid papillary muscle. After tricuspid valvuloplasty, the orifice of the TV is viewed from the apex. The TV is underdeveloped, and there is no papillary muscle. The suture line on the tricuspid annulus can be seen. *TV* tricuspid valve. (**b**) Specimen 948. The C-shaped tricuspid annuloplasty ring viewed from the right atrium. A shaping metal ring is sutured to the tricuspid annulus of the anterior and posterior leaflets. The site of the atrioventricular bundle is kept free. The *arrow* indicates the absent septal leaflet. The TV still regurgitates. *RA* right atrium. (**c**) Specimen 948. Tricuspid valve viewed from the apex. Underdeveloped chordae and papillary muscle are shown. A thin chorda is raised by a probe. *RV* right ventricle. (**d**) Specimen 948. Tricuspid valve and right atrium viewed from right atrium. No atrialized RV is present, and there is no downward displacement of the tricuspid annulus. *CS* coronary sinus, *RA* right atrium. (**e**) Specimen 948. Metal annuloplasty rings that may be too large. The valve leaflets are torn. The TV still regurgitates. *RA* right atrium

17 Right Ventricular Outflow Tract Obstruction

17.1 General Considerations

In this chapter, we will discuss congenital RV infundibulum muscular hypertrophy stenosis, including the muscle bundle of the muscular ventricular septum, combined with the main pulmonary stenosis, ASD or VSD. There is no RV or TV aplasia or dextropositioned aorta such as TOF, pulmonary atresia, or transposition of the great arteries (TGA) (Fig. 17.1a–g).

17.2 Embryology

Right ventricular outflow also means the *right ventricular infundibulum*, which originates from the circular cone of the primitive heart tube. The circular cone resides on the left and right ventricles, and the conical septum divides the circular cone into the pulmonary circular cone (right ventricular infundibulum) and the aortic circular cone (left ventricular outflow). If the course does not proceed successfully, the pulmonary circular cone will be smaller. The last growth stage of the circular cone is a rotation forward of the pulmonary circular cone and the absorption of the aortic circular cone. If muscular hypertrophy occurs under the pulmonary circular cone, right ventricular infundibulum stenosis or obstruction will occur (Fig. 17.2). This chapter discusses mainly abnormal infundibulum pachynsis, not rotation or lack of absorption (Fig. 17.1d).

The right ventricular infundibulum is a circular muscular structure composed of muscular fiber and located between the pulmonary and the tricuspid valves. Its outer layer is circuitous muscle, and the inner layer is composed of circuitous and oblique muscle. The septal band passes through the ventricular septum, the parietal band forms the anterior wall of the RVOT, and the oblique band is included in the diagonal muscle. The oblique band originates from the ventricular septum and passes across the entire RV and adheres to the root of the anterior leaflet of the TV, which is called the *moderator band*. The congenital proliferation of the muscle band leads to outflow tract obstruction. The abnormal right ventricular band is defined as an abnormal moderator band that obstructs between inflow and outflow, which separates the RV into two cavities. It is very important to distinguish the abnormal band because we regard the narrow outflow tract as a VSD at times (Fig. 17.3).

17.3 Types of Infundibulum Obstructions

17.3.1 Entrance Obstruction (Fig. 17.4)

The entrance is a fibrous muscular ring that lies on the papillary muscle of the conus arteriosus. The infundibulum is normal, and the hollow can be seen on the surface of the wall; the hole is limited, followed by secondary pachynsis of the proximal muscle, but a normal or larger cavity is seen in the far end.

17.3.2 Tube-Like Obstruction (Fig. 17.5)

A tube-like obstruction involves mainly the muscular pachynsis all around, but with no expansion after the main pulmonary artery in most cases.

17.3.3 Limited Stenosis Under the Pulmonary Valve (Fig. 17.6)

This obstruction involves a fibrous muscular ring or occasionally only a fibrous membrane. The obstruction could be exposed through an incision of the pulmonary artery. The hemodynamics of obstruction under the PV and simple pulmonary obstruction are nearly the same, so there is expansion after the main pulmonary artery.

X. Zhu (ed.), *Surgical Atlas of Cardiac Anatomy*,
DOI 10.1007/978-94-017-9409-1_17, © Springer Science+Business Media Dordrecht and People's Medical Publishing House 2015

17.4 Specimen Demonstrations

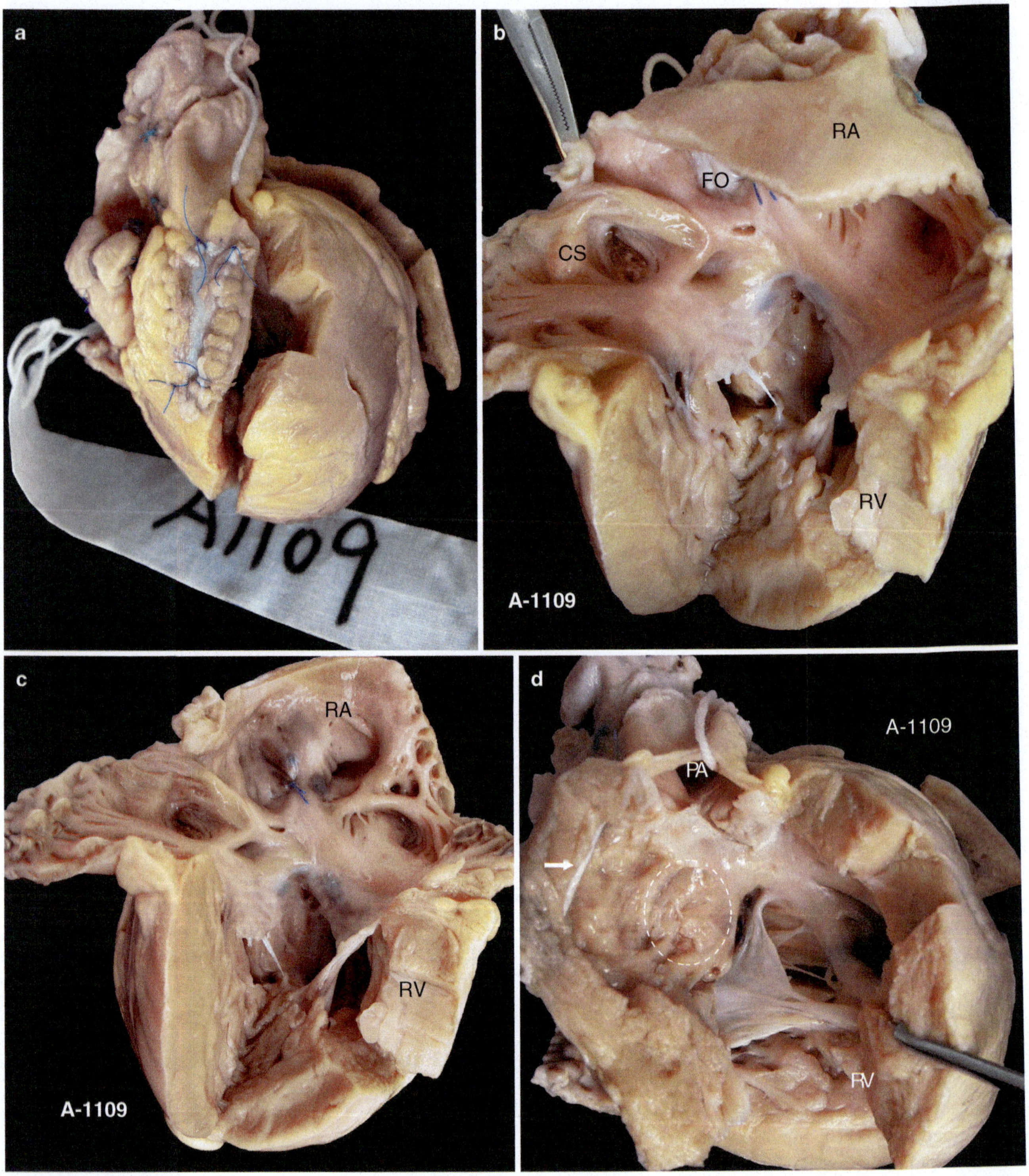

Fig. 17.1 (**a**) Specimen 1109. A single right ventricular infundibulum muscular hypertrophy stenosis, no pulmonary stenosis, a patch-widened operation. (**b**) Specimen 1109. Figure shows RA expansion, coronary sinus expansion, and blood clot. *FO* fossa ovalis, *RA* right atrium. (**c**) Specimen 1109. Figure shows the right ventricular wall thickening, complete atria, and ventricular septum. (**d**) Specimen 1109. Bulky parietal band from right ventricular outflow tract (the *dashed line*). The band has been cut off, the outflow tract has been widened, and there is no ventricular septum or pulmonary or aortic stenosis. (**e**) Specimen 1109. A patch-widened operation on the anterior wall of the right ventricle. The patch is not over the pulmonary ring, and there is no pulmonary stenosis. *P* patch, *PV* pulmonary valve. (**f**) Specimen 1109. After patch-widened operation on the right ventricular outflow tract. (**g**) Specimen 1109. Figure shows normal LV, cutting off of the LVOT, and a complete ventricular septum. *AV* aortic valve, *LV* left ventricle, *MV* mitral valve

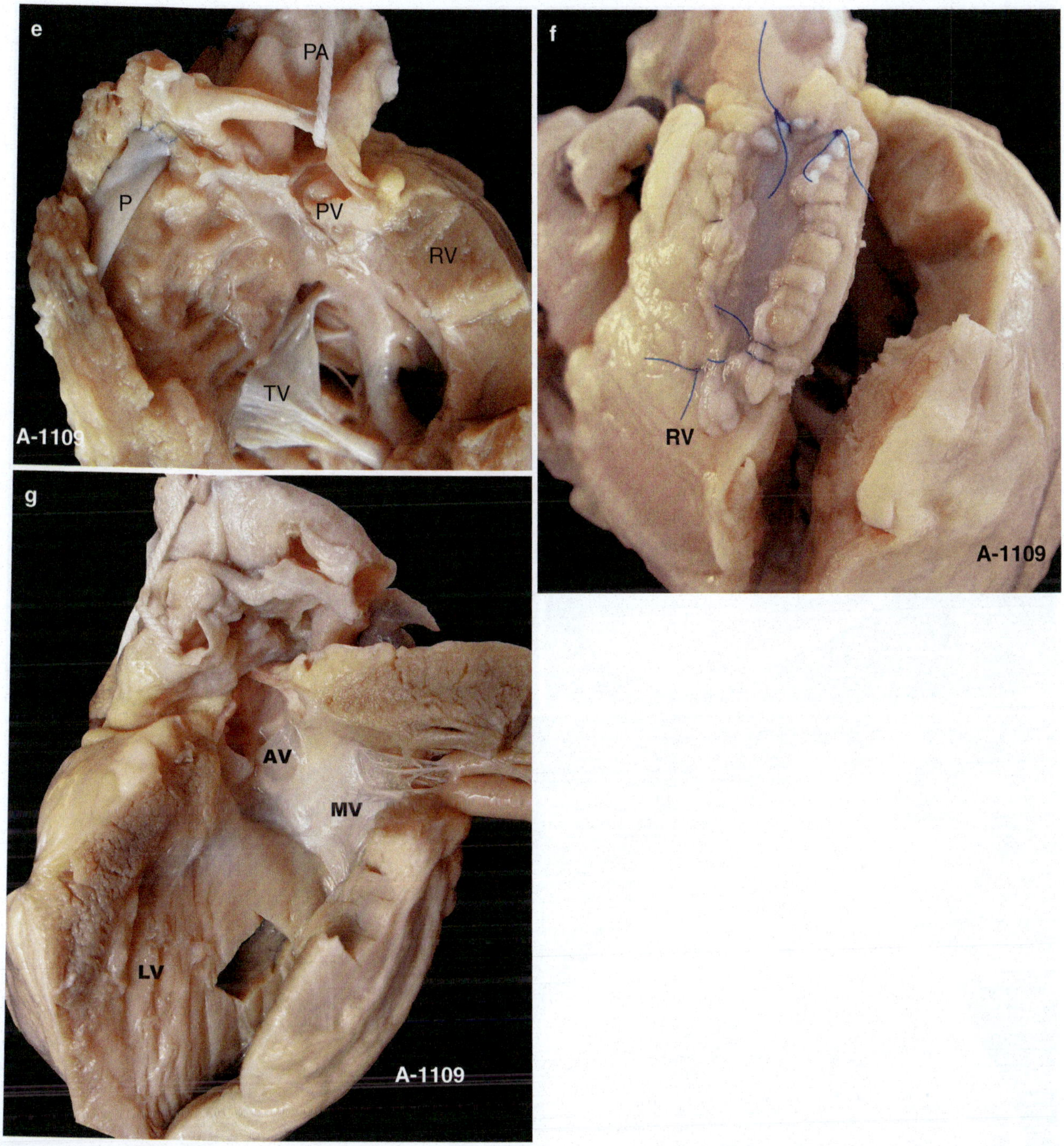

Fig. 17.1 (continued)

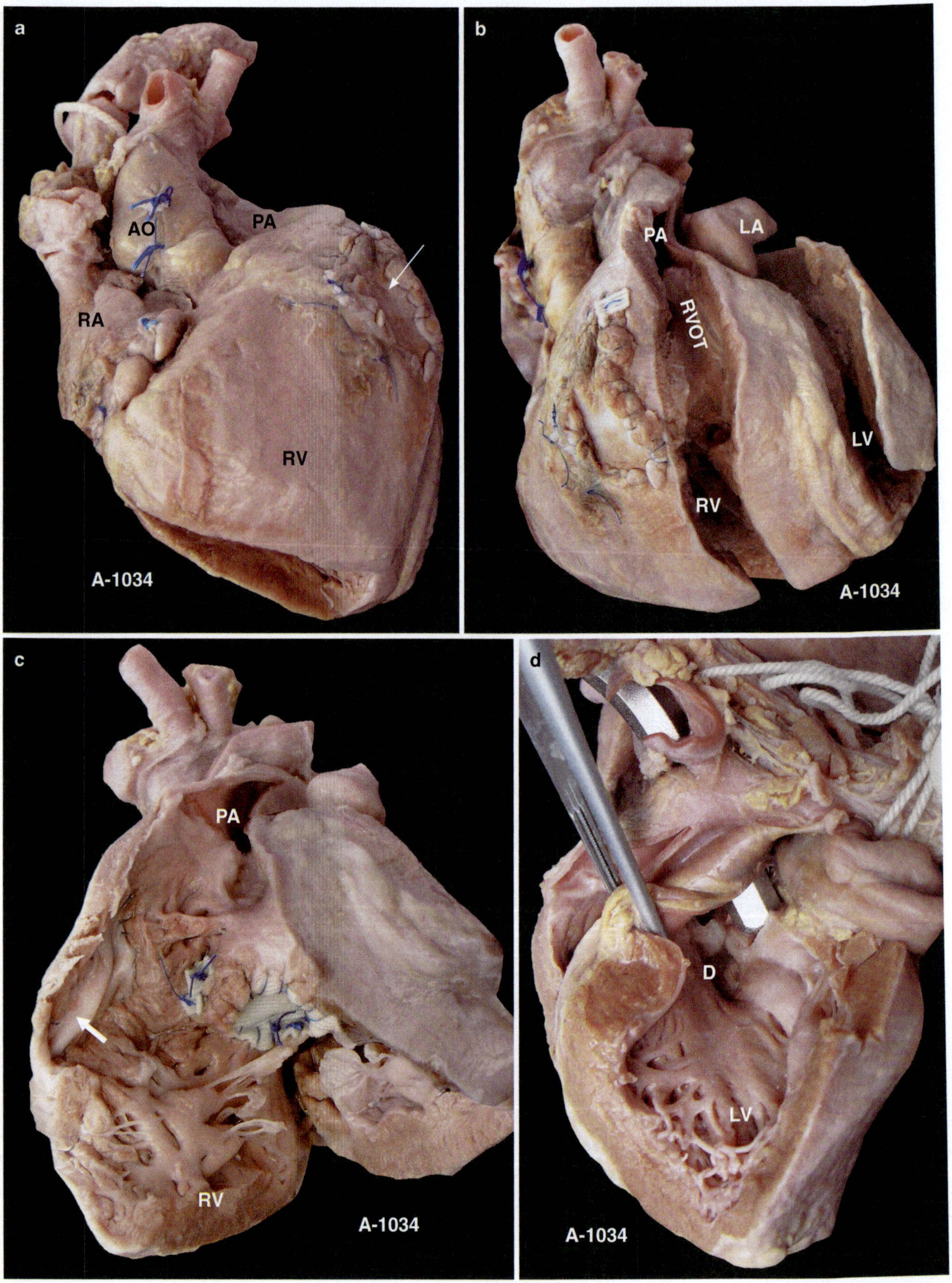
a
AO
PA
RA
RV
A-1034
b
PA
LA
RVOT
LV
RV
A-1034
c
PA
RV
A-1034
d
D
LV
A-1034

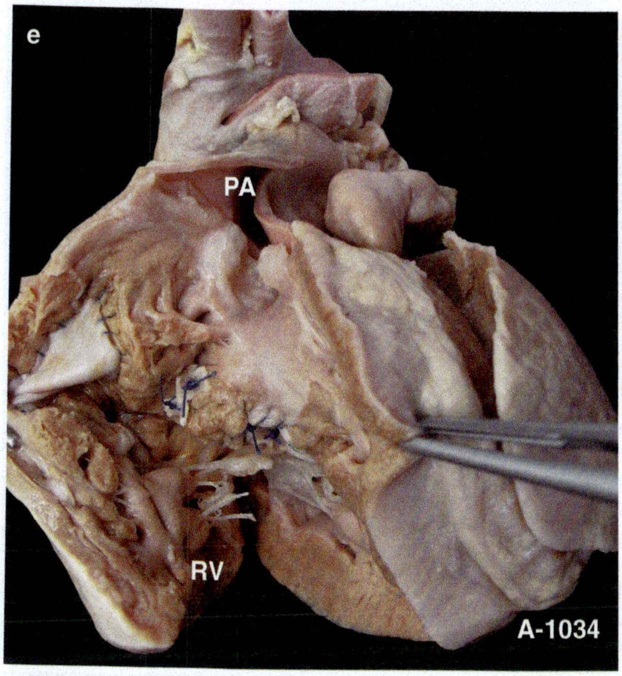

Fig. 17.2 (continued)

Fig. 17.2 (**a**) Specimen 1034. Right ventricular outflow tract obstruction and ventricular septal defect in the front. The pulmonary artery is normal, and the RVOT has been widened by a patch. (**b**) Specimen 1034. The patch in the anterior wall of the right ventricle. Figure shows splitting of the RVOT from the left side of the patch and supraventricular crest hypertrophy. The pulmonary artery is thinner than the AO, and the PV is normal. (**c**) Specimen 1034. Normal pulmonary valve; infundibulum muscle has been cut off. This figure shows that a section of the parietal band has been cut off; the *arrow* indicates the patch used in widening the RVOT. (**d**) Specimen 1034. This figure shows that the LA and LV are normal; the probe is inserted through AO to LV. D, VSD. (**e**) Specimen 1034. This specimen shows the position and size of the patch when the RVOT needs to be widened. The patch is not over the pulmonary ring but at the same level as the parietal band and relieves the obstruction. It will not influence the function of the RV. The parietal band is very clear

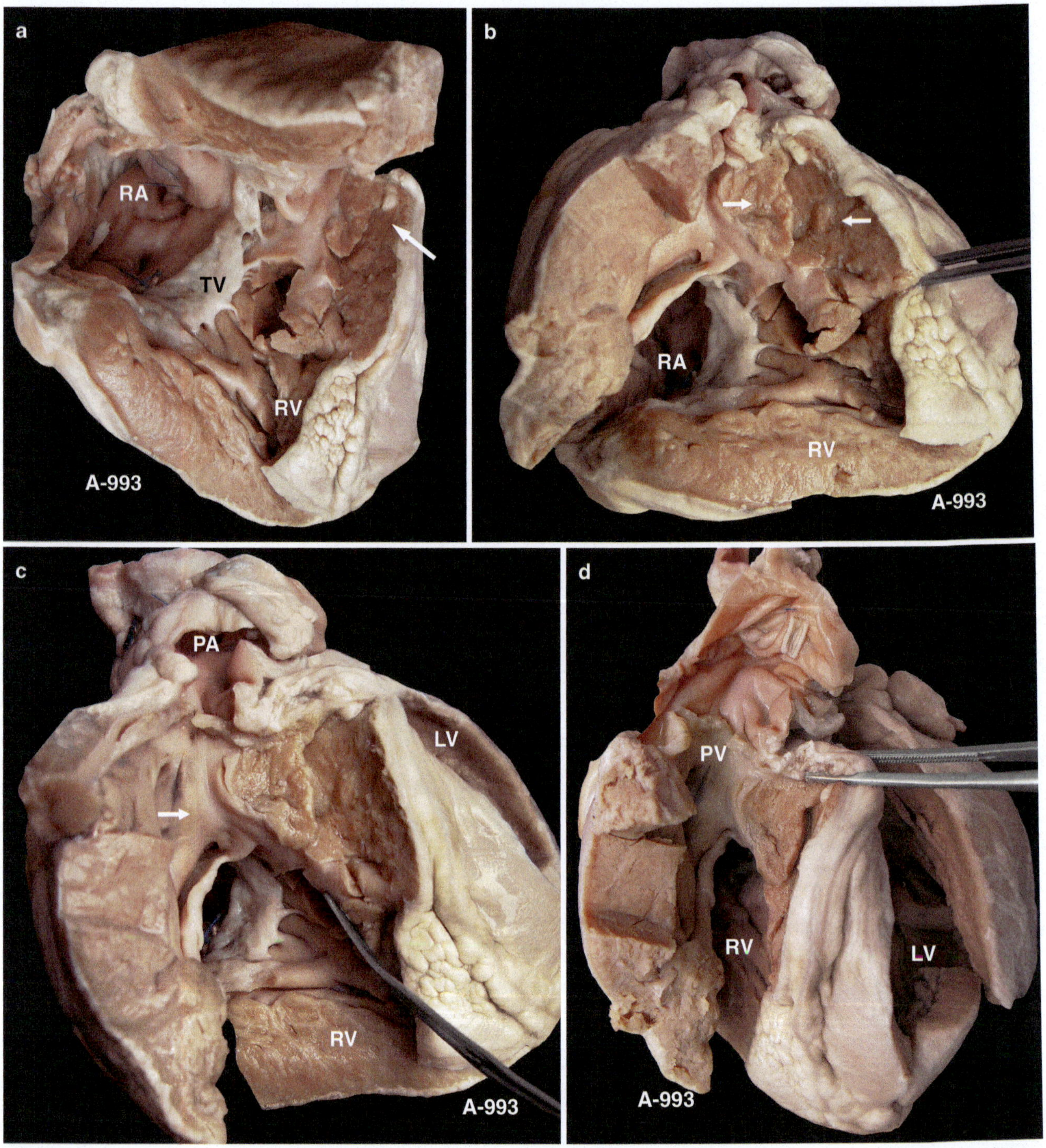

Fig. 17.3 (**a**) Specimen 993. The figure shows single right ventricle infundibulum muscular hypertrophy obstruction, complete atria septum, and RV thickening. *Arrow*, RVOT. (**b**) Specimen 993. Severe right ventricle infundibulum muscular hypertrophy stenosis. The band has obstructed the RVOT; the specimen shows thickening of the TV and the wall of the RV and complete ventricular septum. (**c**) Specimen 993. Bulky supraventricular crest (*arrow*), the section of the parietal band. (**d**) Specimen 993. The PV has three well-grown leaflets, the edge is thickened, the ring is normal, and muscular hypertrophy stenosis is under the PV. (**e**) Specimen 993. The LV and RV are shown at the same time. The RV is thicker than the LV, and the MV is normal. (**f**) Specimen 993. LVOT is shown here with a complete ventricular septum and normal contiguity of the atrial and mitral valves

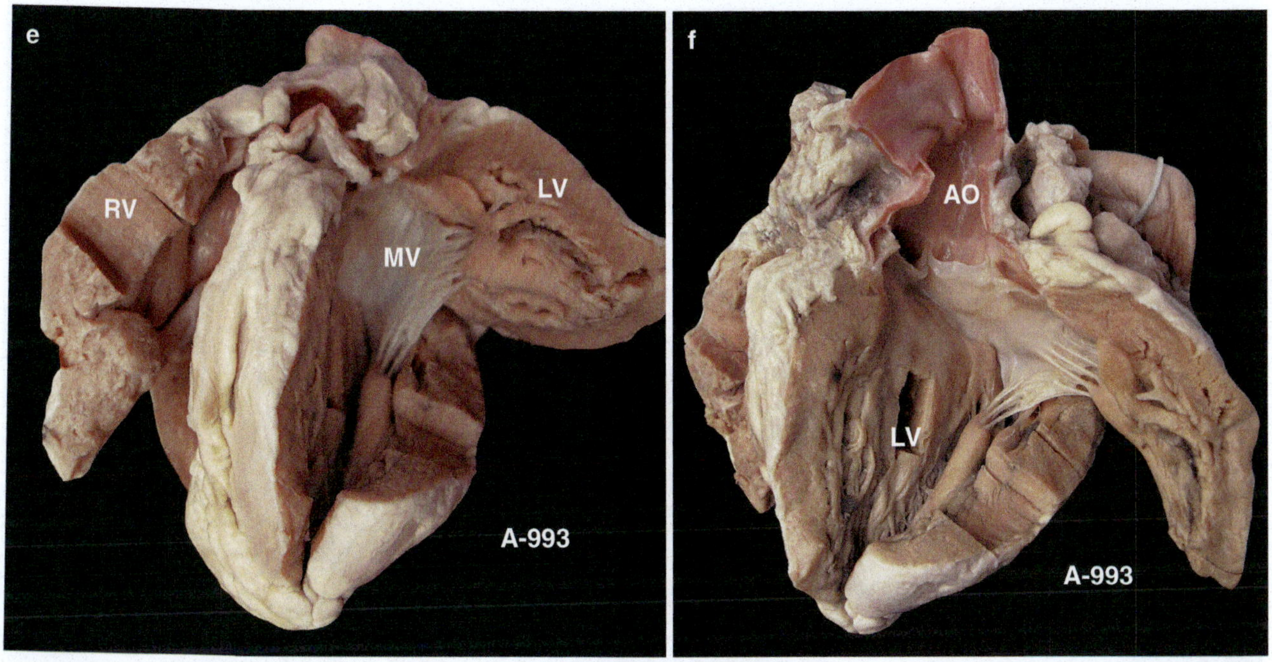

Fig. 17.3 (continued)

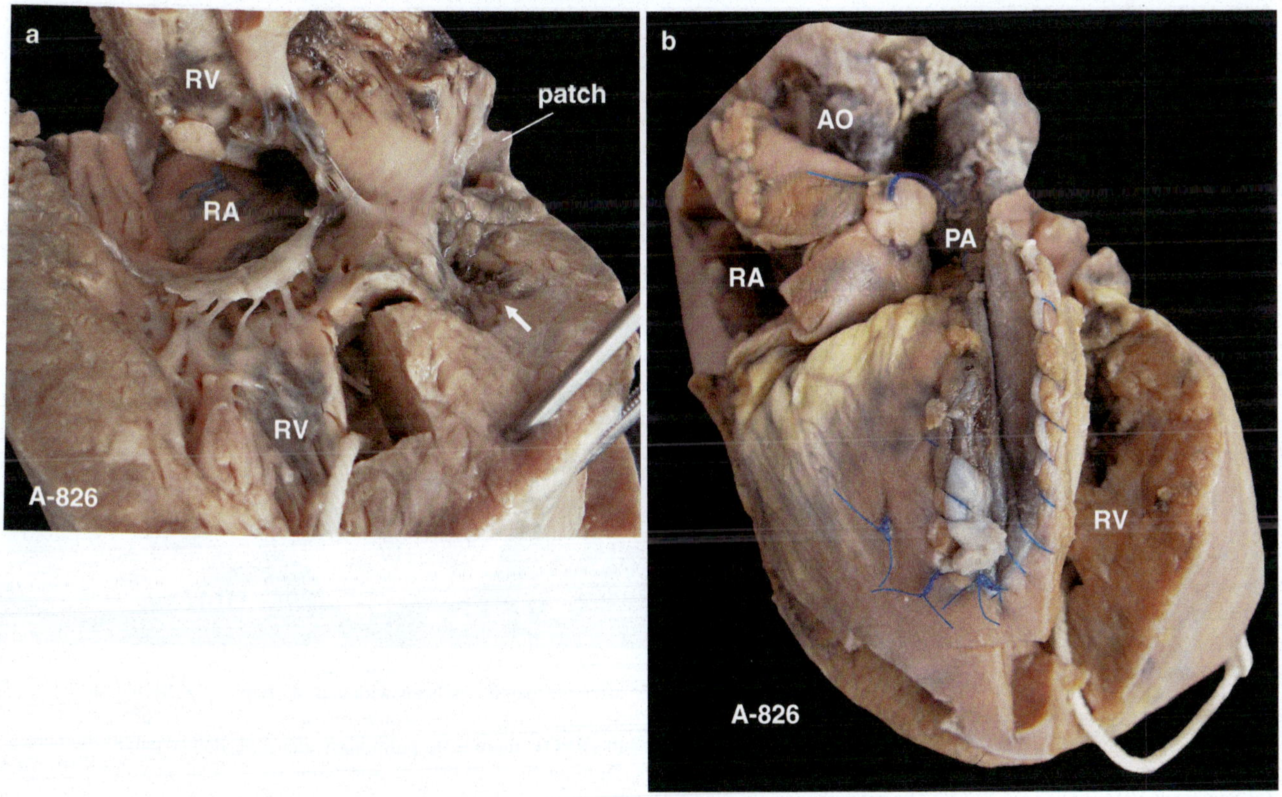

Fig. 17.4 (**a**) Specimen 826. The anterior wall of the RV is lifted to show muscular hypertrophy obstruction of the RVOT (*arrow*). The RVOT has been widened by a pericardium patch and is over the pulmonary ring. (**b**) Specimen 826. A patch-widened right ventricular outflow tract. The pulmonary arterial stenosis is not obvious

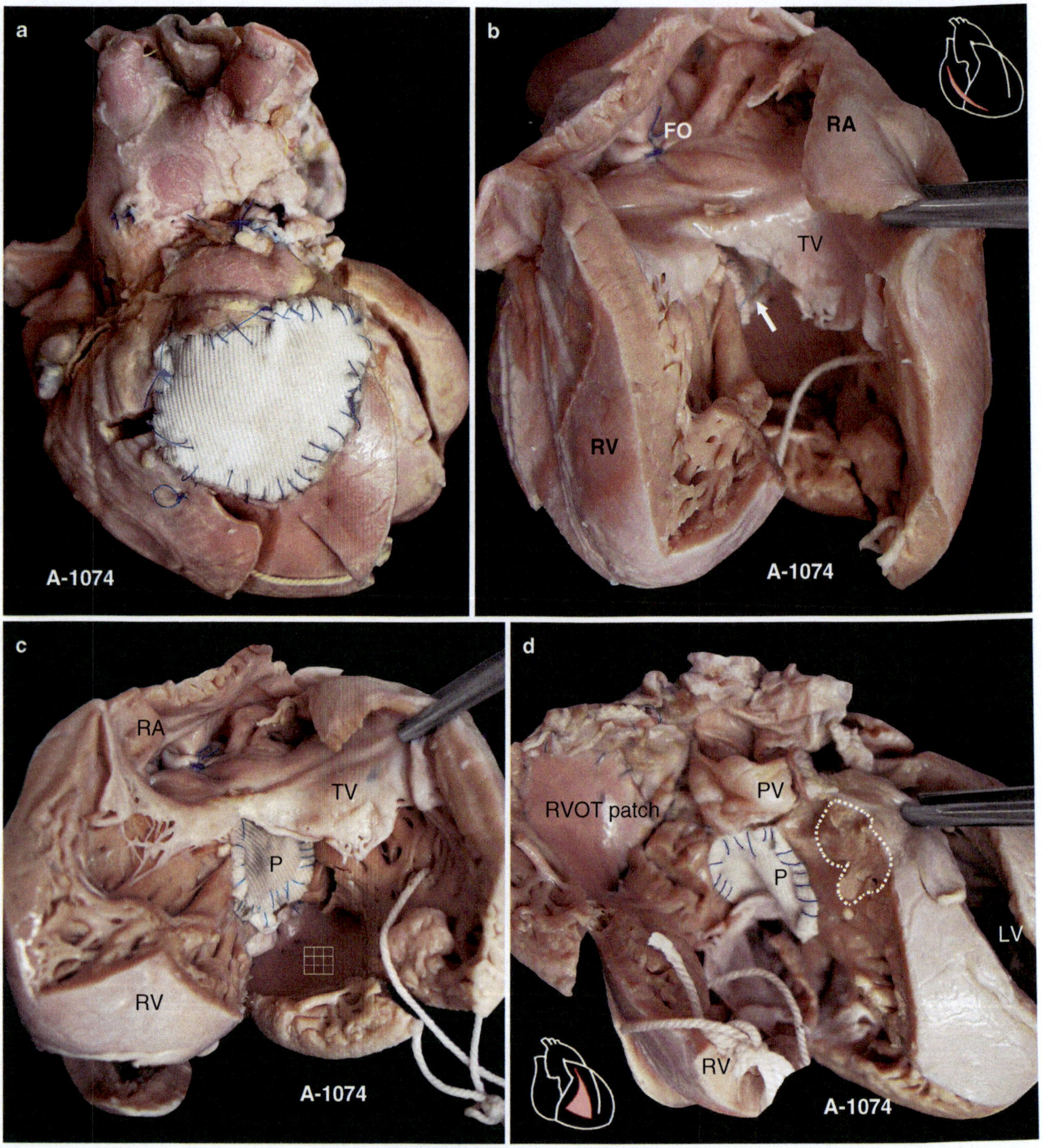

Fig. 17.5 (**a**) Specimen 1074. View from the front. Figure shows RVOT stenosis, patch-widened, and no pulmonary stenosis, with VSD. (**b**) Specimen 1074. The RA is a little larger, and the FO has been closed; the lower limb of the VSD patch (*arrow*) can be seen. (**c**) Specimen 1074. Expanded RVOT. The net is the intracardiac side after the patch has been widened. *P*, patch, *TV* tricuspid valve. (**d**) Specimen 1074. The *dashed line* indicates the hypertrophy band of the RVOT; there is no pulmonary stenosis. RVOT patch: the pericardium patch of RVOT. (**e**) Specimen 1074. VSD is viewed from left ventricular side

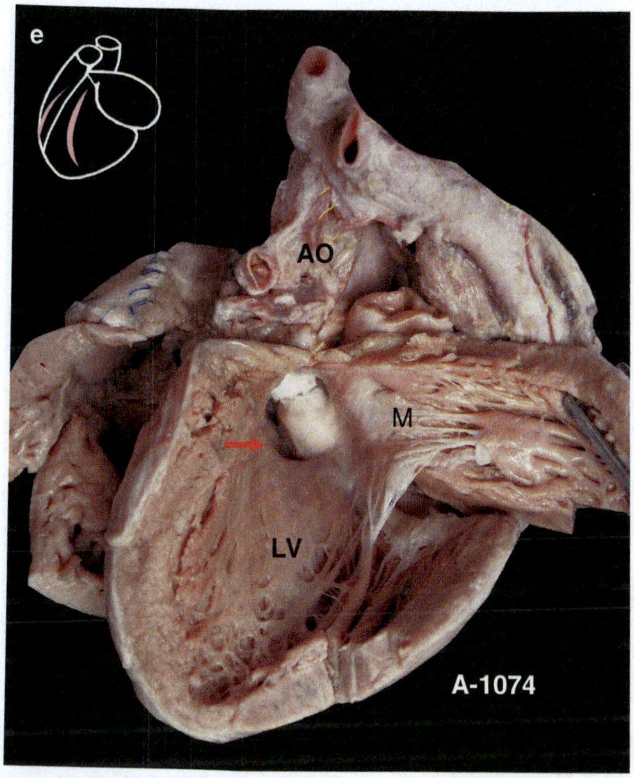

Fig. 17.5 (continued)

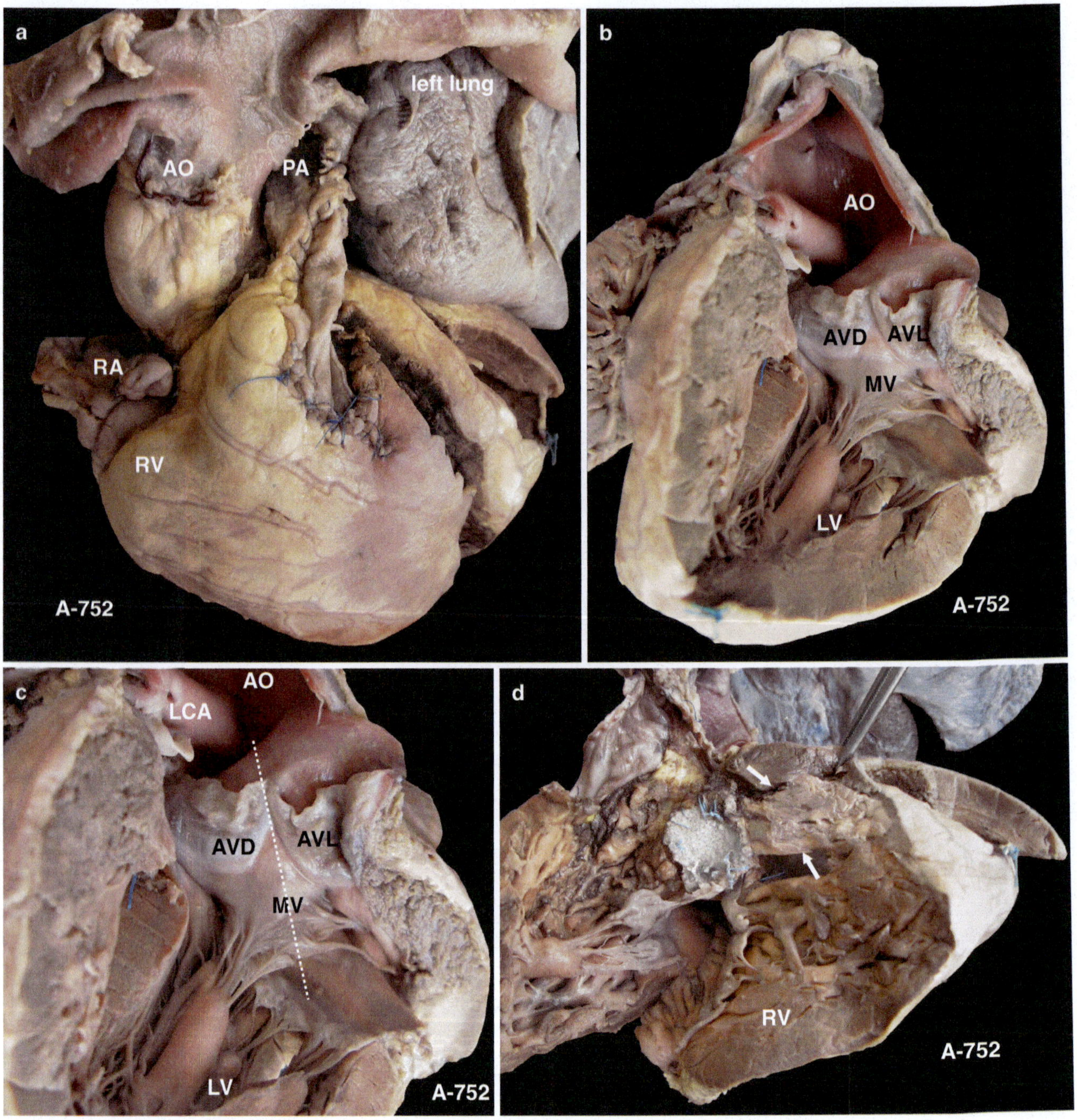

Fig. 17.6 (**a**) Specimen 752. Right ventricle outflow tract stenosis, pulmonary stenosis, and VSD. This specimen shows the front of the heart, with a thin pulmonary artery, widened RVOT, and the main pulmonary artery with a patch over the pulmonary ring. (**b**) Specimen 752. The LVOT and AV are shown. The back valve (AVD) and the left valve (AVL) are totally in the left ventricular side. The contiguity of the two aortic leaflets and the anterior leaflet of the MV is normal, indicating there is no bestriding of the aorta. The lower limb of the VSD patch can be seen. (**c**) Specimen 752. Identification with tetralogy of Fallot. The *dashed line* is the mark of the middle point of the anterior leaflet of the MV, and it heads on the triangle of the AV (between the left valve and the back valve). It indicates that there is no rotation or shifting of the AV. There is no bestriding of the aorta. The VSD is small, and there is no evidence of TOF. (**d**) Specimen 752. The *arrow* indicates the bulky parietal band of the RVOT. The VSD is small and has been repaired. (**e**) Specimen 752. The figure shows the patch over the pulmonary ring from the RVOT to the main pulmonary artery. The patch (P) is autologous pericardium, there is no artificial semilunar valve, and the patch has reached to the furcation of the pulmonary artery

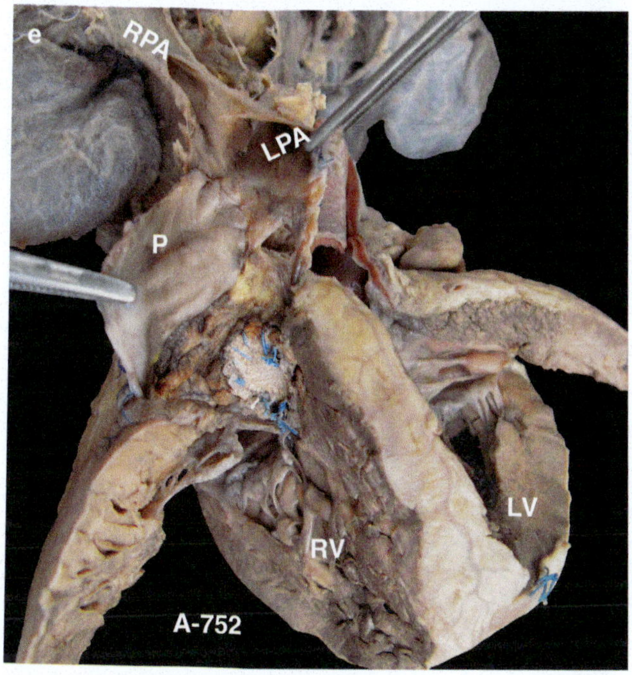

Fig. 17.6 (continued)

Mitral Atresia and Hypoplastic Left Ventricle

18

18.1 General Considerations

Hypoplastic left ventricle, also termed *hypoplastic left heart syndrome*, is often associated with a variety of deformities, such as AV and aortic arch anomalies. The specimens presented in this chapter are limited to show mitral atresia and mild hypoplastic LV.

18.2 Mitral Atresia

Muscular atresia and membranous atresia are two morphological subtypes of mitral atresia. The former may result from the directional abnormity of the development of the ventricular septum, and the muscular tissue occludes the atrioventricular orifice. The latter may result from the over fusion of the endocardial cushion, which results in the MV leaflets being fused together. Mitral atresia inevitably combines defects of the atrial or ventricular septum, patent ductus arteriosus, and hypoplastic LV.

18.3 Hypoplastic Left Ventricle

The specimens shown in this chapter are mainly those of mild hypoplasia of the LV. The LV is relatively small and thin walled, with mitral or aortic valve malformation. Specimen NO. 1025 (Fig. 18.2) shows dysplasia of the LV, MV, and aorta. It could be referred to as *hypoplastic left heart syndrome*. The RA and RV may have varying degrees of dilation.

X. Zhu (ed.), *Surgical Atlas of Cardiac Anatomy*,
DOI 10.1007/978-94-017-9409-1_18, © Springer Science+Business Media Dordrecht and People's Medical Publishing House 2015

18.4 Specimen Demonstrations

18.4.1 Mitral Atresia Combined with Pulmonary Atresia

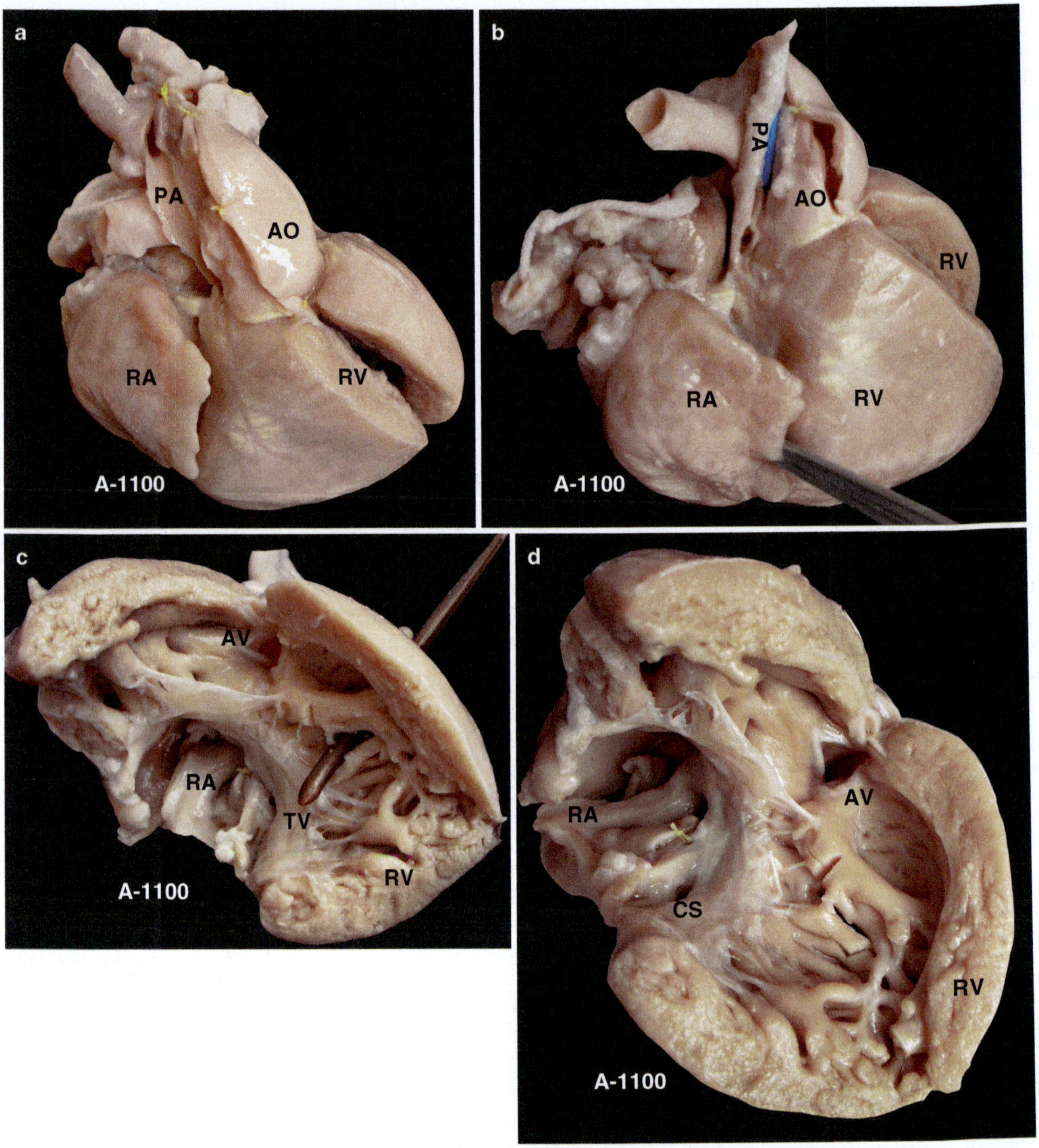

Fig. 18.1 (**a**) Specimen 1100. The frontal view of mitral atresia combined with pulmonary atresia shows the fine main pulmonary artery at the back of the aorta. (**b**) Specimen 1100. The blue bougie indicates the fine main pulmonary artery. The blind-end root confirms the pulmonary atresia. *RA* hypertrophic and dilated right atrium, *RV* right ventricle, the only ventricular chamber of this specimen. (**c**) Specimen 1100. Ventricular anterior wall opened to expose the right atrium and outflow tract of the right ventricle. The structure of the atrioventricular valve conforms to the characteristics of the TV. The multiple muscle trabeculae, conical muscle-like structures between the aortic and tricuspid valves, could confirm that the chamber is morphologically the RV. The RVOT links with the AV. The PV does not exist in the chamber. The features confirm that the left side does not have a chamber, which is equivalent to a single ventricle with right ventricular morphology. *AV* aortic valve, *RA* right atrium, *RV* morphological right ventricle, *RVOT* right ventricular outflow tract, *TV* tricuspid valve. (**d**) Specimen 1100. Exposure of the aortic valve orifice and subvalvular myocardial infundibulum. *AV* aortic valve, *CS* coronary sinus, *RA* right atrium, *RV* right ventricle. (**e**) Specimen 1100. Exposure of the left atrium without the mitral valve. *a*, atrial septal defect; *b*, left atrial appendage; *c*, rupture of the coronary sinus; *LA* left atrium, *RA* right atrium. (**f**) Specimen 1100. Mitral atresia. No MV is found in the LA. The bougie indicates the residual recess. The ventricular wall under the atrium belongs to the morphological RV (RV wall). *AO* aorta. (**g**) Schematic diagram of specimen 1100. The ventricle has structures characteristic of the RV. The RA has inlets of the coronary sinus and the vena cava. These features prove the atrium is in situ solitus. The mitral atresia remains a recess without a LV chamber

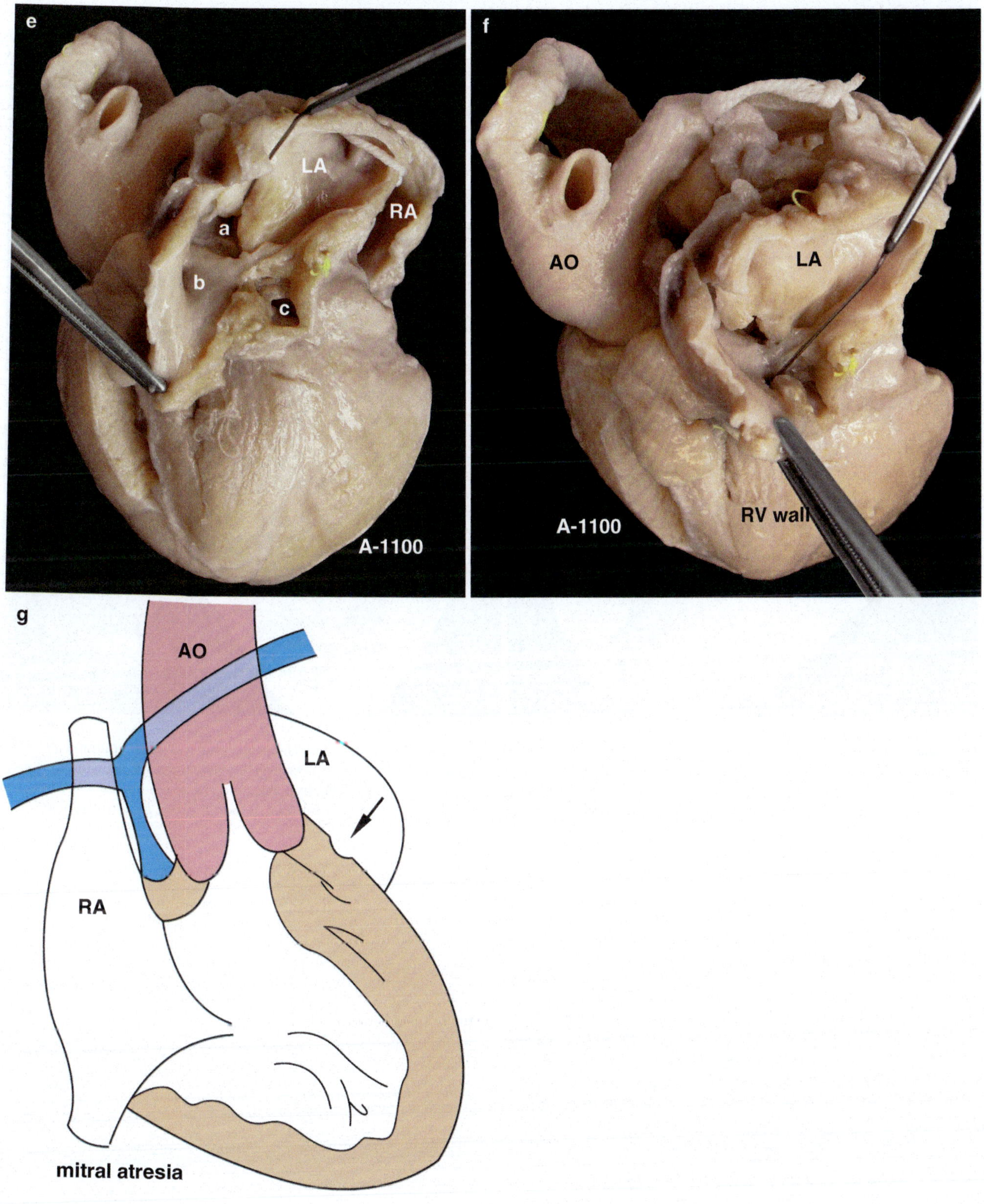

Fig. 18.1 (continued)

18.4.2 Hypoplastic Left Ventricle + Mitral Valve Malformation + Fine Aorta

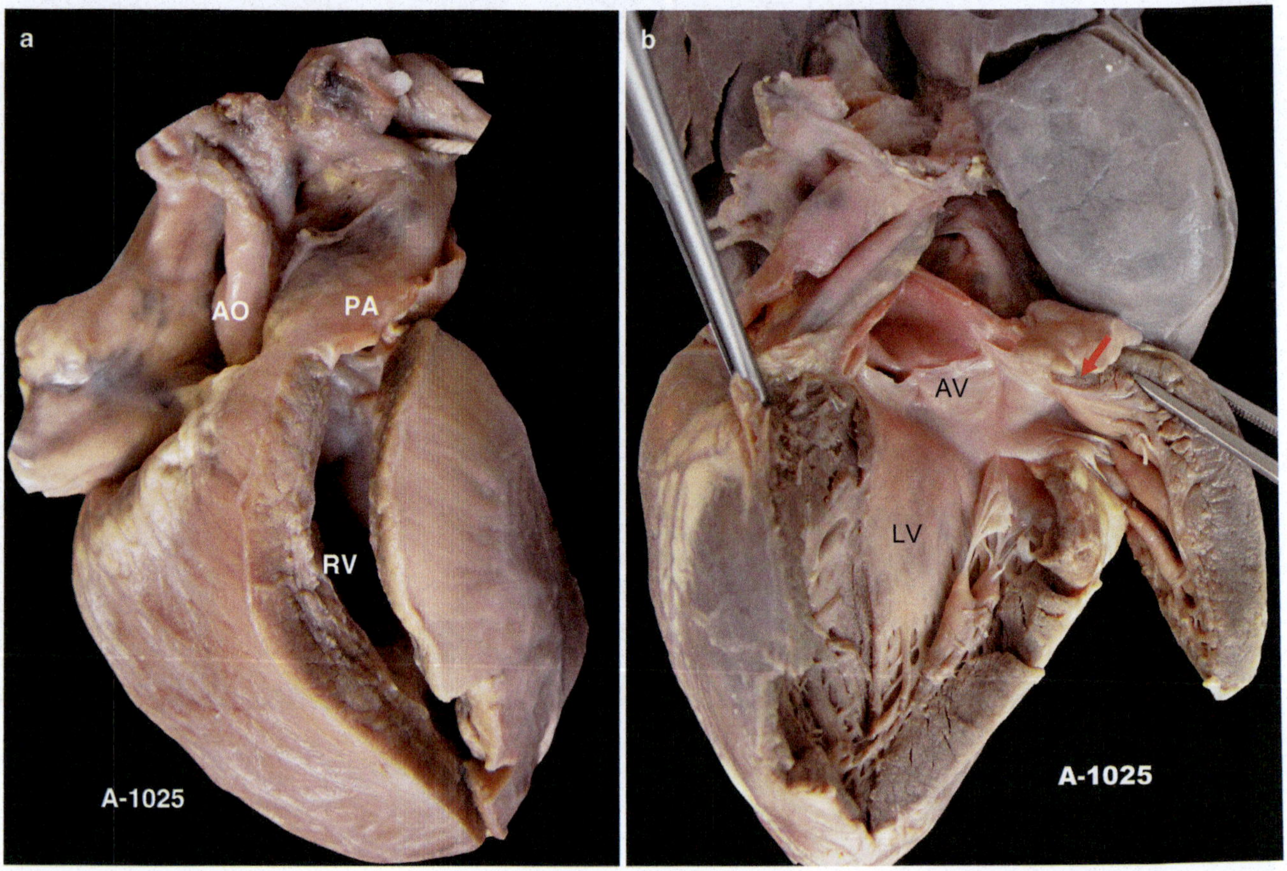

Fig. 18.2 (**a**) Specimen 1025. The frontal view of mild hypoplasia of the left ventricle and aorta. The aorta is small, and the main pulmonary is developed fairly well. (**b**) Specimen 1025. The ascending aorta is thin, whereas the AV is developed fairly well and keeps a fibrous connection with the MV. The shape and size of the pulmonary artery remain normal. The RV is enlarged, whereas the LV is underdeveloped. The *arrow* indicates the hypoplastic chordae tendineae of the MV and the shortened papillary muscle. *AV* aortic valve, *LV* left ventricle

18.4.3 Hypoplastic Left Ventricle Combined with Mitral Valve Malformation

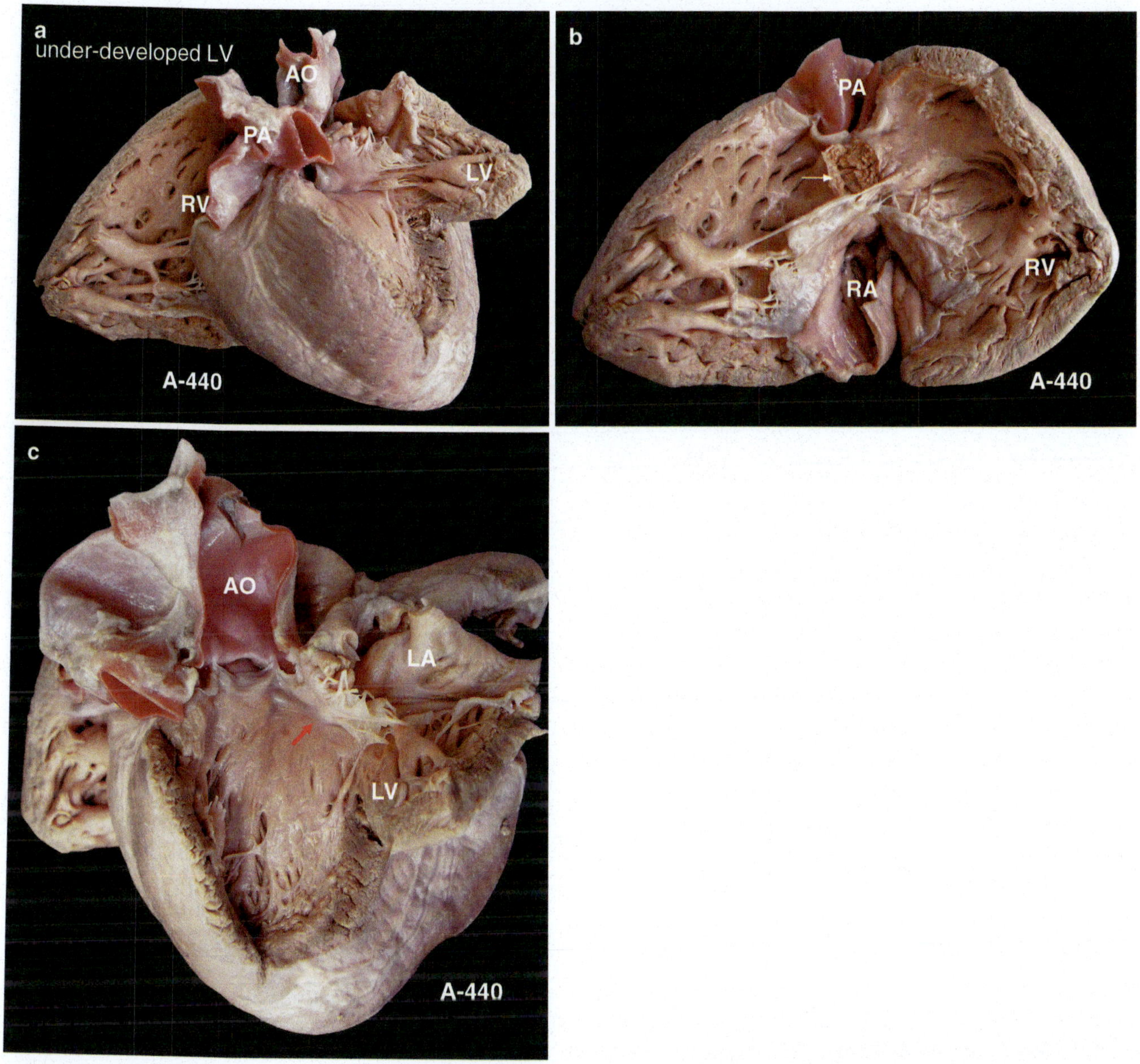

Fig. 18.3 (**a**) Specimen 440. Hypoplastic left ventricle, the left anterior oblique section of the left and right ventricles. The RV is developed well, whereas merely 30 % of the LV has developed. The ventricular septum is not hypertrophic. The LV is small and thin walled, with malformations of the MV and the chordae tendineae. The AV remains fibrously connected with the MV. (**b**) Specimen 440. The RV is developed well, and its chamber is mildly enlarged. The supraventricular crest is hypertrophic. The *arrow* indicates the section of the supraventricular crest myocardium. (**c**) Specimen 440. Section of the left ventricle. The *arrow* indicates the hypoplastic but unobstructed MV. The LV chamber is obviously small, and the aorta is mildly thin. The chordae tendineae and papillary muscle of the MV are hypoplastic

19 Single Ventricle

19.1 General Considerations

Single ventricle is a congenital cardiac malformation in which both atria are connected to only one ventricle chamber by either two separate atrioventricular valves or a common AV. Usually, there are two ventricles. One is the large main chamber, whereas the other is always incomplete (rudimentary) and hypoplastic. However, there is only one chamber. The two great arteries originate from the single ventricle, and the interrelationship can be normal or transpositioned. Therefore, the single ventricle is supposed to be a group of integrated deformities involving the ventricle, atrioventricular valves, and great arteries. The single ventricle due to hypoplasia of ventricular septum is also known as the *common ventricle*.

19.2 Embryology

The embryonic mechanism of the single ventricle is the developmental abnormality of the ventricular segment of the primitive heart tube. The primitive heart tube is right loop shaped and bent normally. Lying in the bottom section of the heart tube is the primitive ventricle. Its right end develops into the sinus part of the RV and is linked to the atrioventricular canal. The muscular ventricular septum exists between the left and right sinus parts. The conus part rotates and straddles the two ventricles, forming the ventricular outflow tract. Thus, whether it is hypoplasia of the right end (sinus of right ventricle) or the left end (sinus of left ventricle) of the primitive ventricle, or muscular ventricular septum, it results in a single ventricle.

19.3 Typing

There are a variety of typing methods for single ventricle. From the practical point of surgery, the following characteristics should be emphasized:

1. The main chamber of the single ventricle is equivalent to the left or right ventricle.
2. Either a common AV valve or two separate atrioventricular valves.
3. Conus structure under the valve of the great arteries or not.
4. Rudimentary ventricular septum or not.
5. Positional relationship between the great arteries.
6. Pulmonary stenosis or atresia or not.
7. Other combined deformities.

19.4 Hemodynamics

The hemodynamics can be analyzed using the "imaginary blood flow direction." If the blood flows from the LA to the AV through the arterial atrioventricular valve, it would be assumed to be the arterial blood line; then, the blood flow from the RA to the pulmonary valve through the venous atrioventricular valve would be the venous blood line. If these two lines are parallel to each other, the two blood flows could be separated by an artificial ventricular septum. However, if these two lines cross in the heart, the malformation could not be repaired by artificial ventricular septum.

19.5 Surgical Management

Palliative surgery is the main therapy for the single ventricle. The shunt operation, such as the vena cava-pulmonary artery shunt or the bidirectional Glenn operation, is indicated for less pulmonary blood. The pulmonary banding is indicated for more pulmonary blood. The ideal separation operation is to divide the single ventricle into left and right ventricles; however, that approach cannot be implemented in most cases because of the swing of the artificial ventricular septum and the abnormality of transduction bundles.

X. Zhu (ed.), *Surgical Atlas of Cardiac Anatomy*,
DOI 10.1007/978-94-017-9409-1_19, © Springer Science+Business Media Dordrecht and People's Medical Publishing House 2015

19.6 Specimen Demonstrations

19.6.1 Single Ventricle, Double-Inlet Ventricle

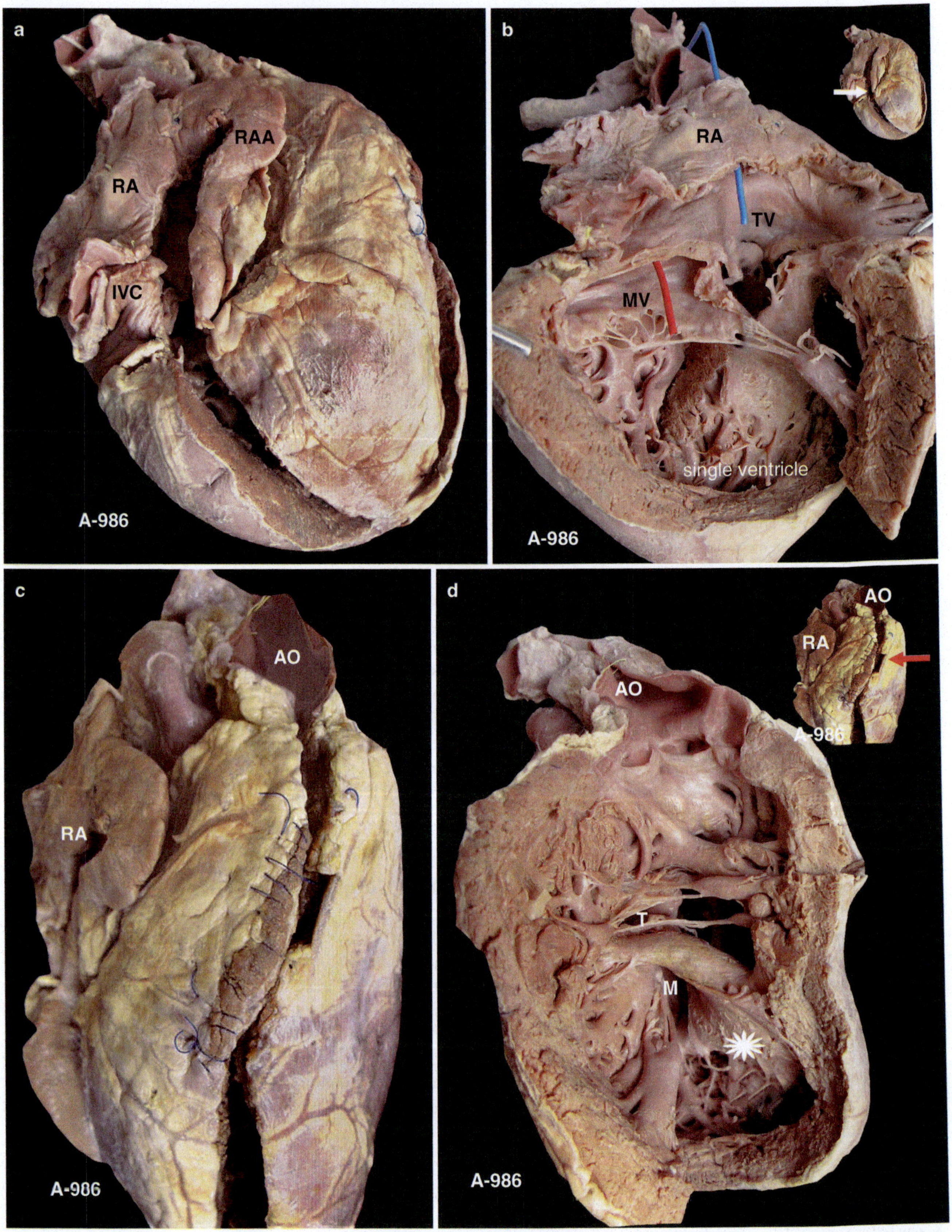

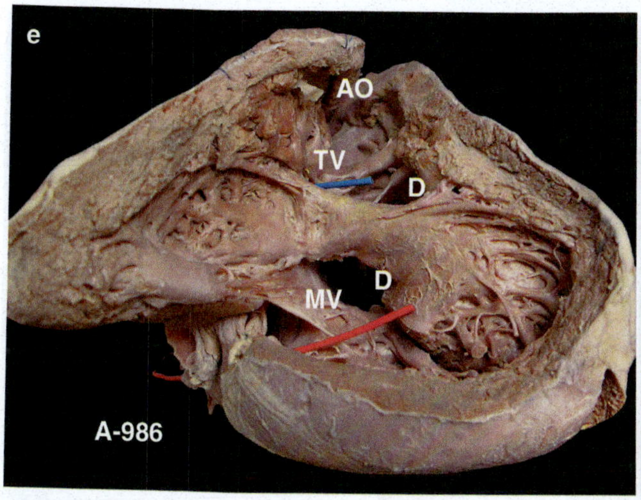

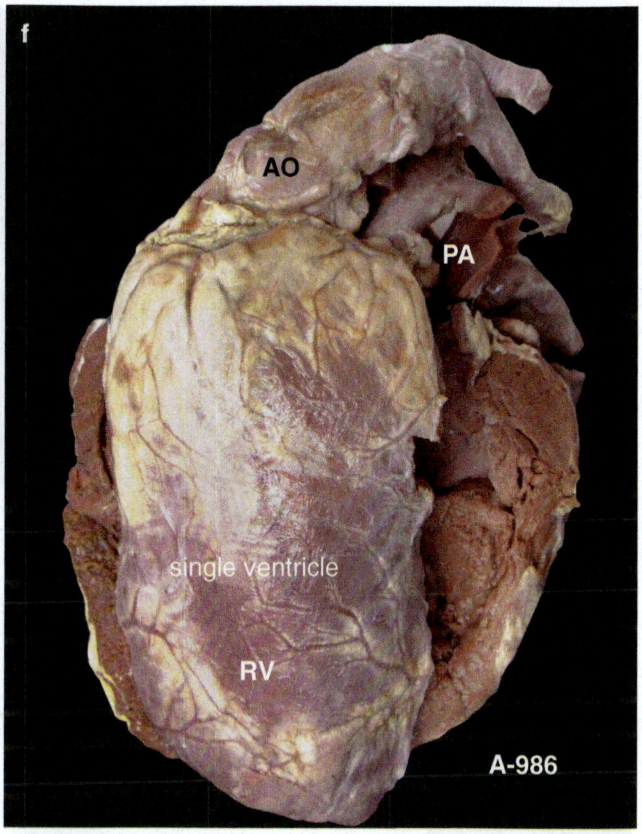

Fig. 19.1 (continued)

Fig. 19.1 (**a**) Specimen 986. The right-sided view of the double-inlet ventricle and the conus under the aortic valve. The RA, RAA, and IVC all are on the right. The right-sided ventricle is the subaortic valve chamber, which can be interpreted as the main chamber of the single ventricle. *IVC* inferior vena cava, *RA* right atrium, *RAA* right atrial appendage. (**b**) Specimen 986. Inflow tract of right-sided ventricle. The *blue* and *red* indicate the blood flow of the right or left atria, respectively. The MV and TV are open to the same huge ventricular chamber, which is the main chamber of the single ventricle. The *arrow* in the upper right corner indicates the incision site. *MV* mitral valve, *TV* tricuspid valve. (**c**) Specimen 986. Subaortic ventricular incision. The sutures of the operative incision are shown. The anterior aorta and inferior PV indicate the abnormal position relationship of the great arteries. *AO* aorta. (**d**) Specimen 986. Sufficient exposure of the ventricular chamber under the aortic valve, which is the main chamber of a single ventricle. Trabecular area (*asterisk*) is the free wall of the ventricle but not of the ventricular septum. The TV and AV are separated by the conus muscle. The ventricular septum nearly disappears; remaining are the subaortic conus muscle, the bundles of thick muscle, and the apical trabecular muscle. *AV* aortic valve, *TV* tricuspid valve. (**e**) Specimen 986. The thick muscle bundles and abundant trabecular muscle in accord with morphological right ventricle structure. "D" indicates the huge VSD. The *blue* and *red* indicate the blood of both the TV and MV inflow to the subaortic ventricular chamber. *MV* mitral valve, *TV* tricuspid valve, *VSD* ventricular septal defect. (**f**) Specimen 986. Left-side view of the positional relationship of the great arteries. The aorta and main pulmonary artery originate from the same ventricle. The former is located anteriorly and the latter inferiorly. This positional relationship belongs to transposition of the great arteries. (**g**) Specimen 986. A small ventricle under the pulmonary valve, which could be interpreted as a subpulmonary chamber. Actually, it belongs to the same ventricular chamber as does the subaortic chamber. No ventricular septum exists between them, and only bundles of free muscle are found. Continuous fiber exists between the pulmonary valve and the TV. The blue was inserted from the TV into the subpulmonary chamber. The inferior wall is the free wall of the contralateral main ventricular chamber. *TV* tricuspid valve. (**h**) Specimen 986. Range of ventricular chamber. Above the dotted line is the subpulmonary chamber, which is hypoplastic. Below the line is the right side of the subaortic chamber. (**i**) Specimen 986. The comparison of two ventricular chambers. The *white* and *yellow dotted lines* indicate the subaortic chamber and the subpulmonary chamber, respectively. *PA* pulmonary artery, *PV* pulmonary valve. (**j**) Specimen 986. Schematic diagram of double-inlet ventricle. LV indicates the residual cavity of the morphological LV, whereas RV indicates the morphological RV. There is no ventricular septum. The *blue* and *red lines* indicate the tricuspid and mitral valves, respectively. *AO* aorta, *LA* left atrium, *LV* left ventricle, *PA* pulmonary artery, *RA* right atrium, *RV* right ventricle

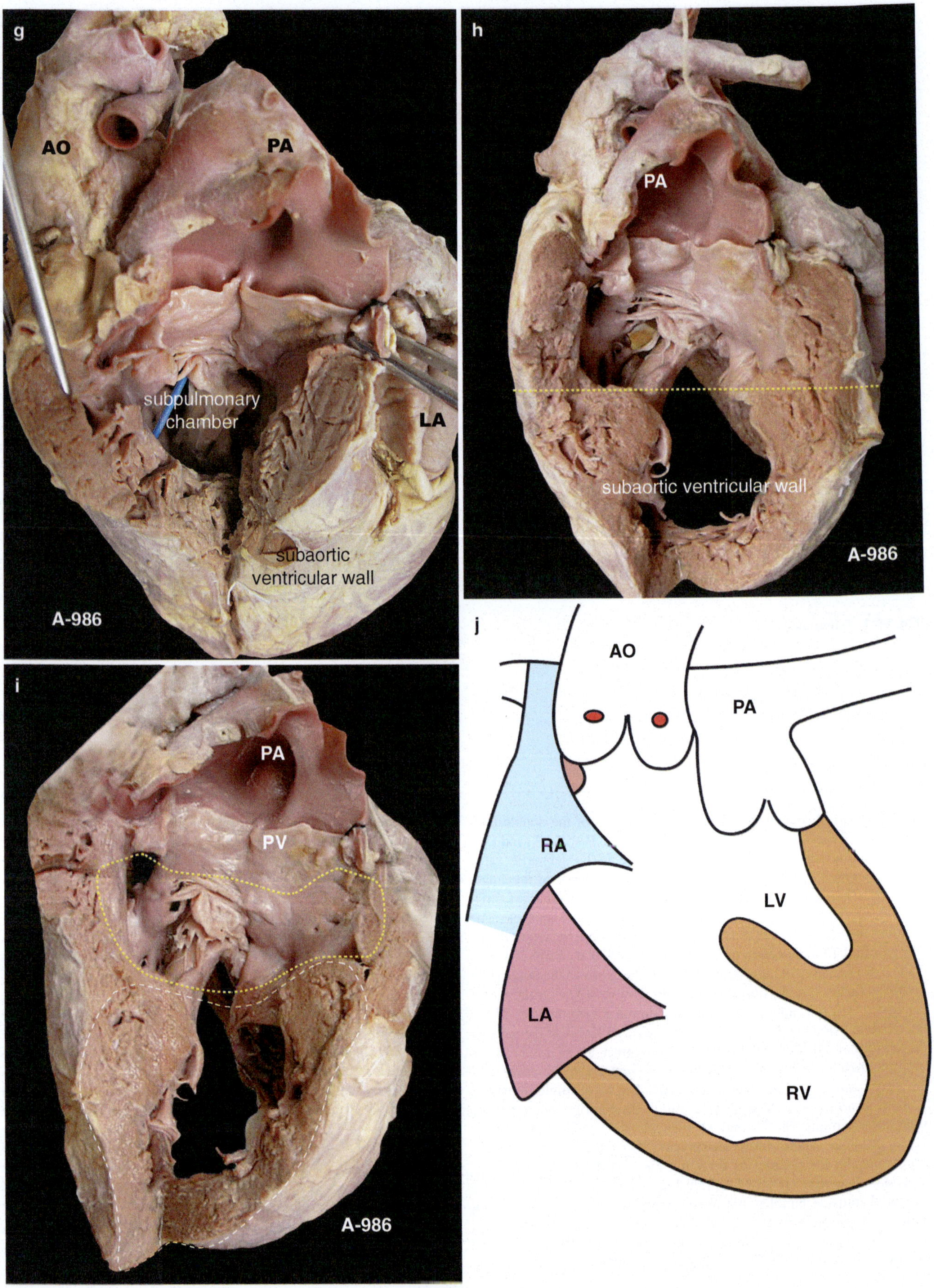

Fig. 19.1 (continued)

19.6.2 Single Ventricle, Common Atrium

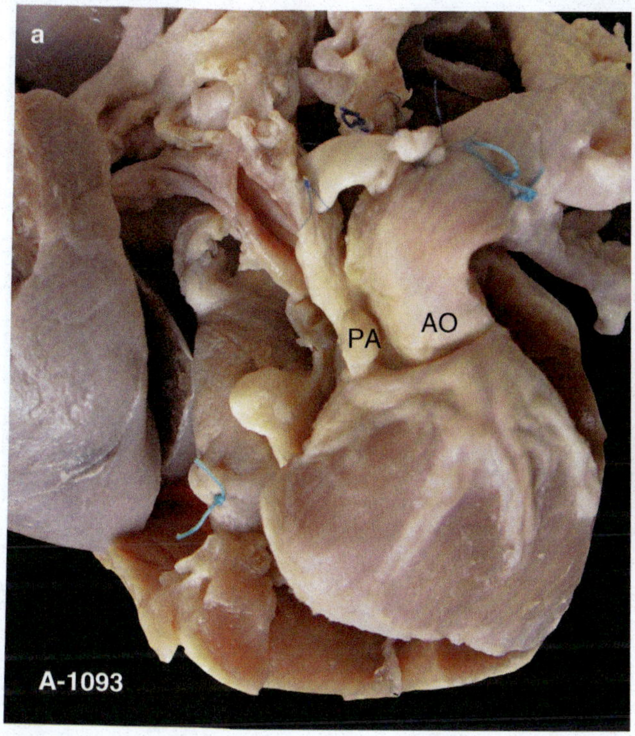

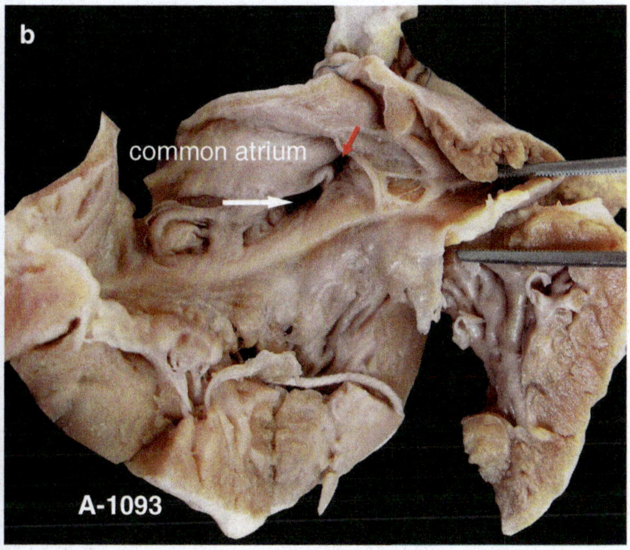

Fig. 19.2 (**a**) Specimen 1093. Single ventricle, common atrium, common atrioventricular valve, and Waterston shunt for pulmonary stenosis. The frontal view presents the small main pulmonary artery and the large ascending aorta. The aorta is on the left of the pulmonary artery. (**b**) Specimen 1093. The atrium without septum. The back wall has the inlets of the pulmonary veins (*red arrow*). The *white arrow* indicates the left appendage. There is only one atrioventricular ring between the atrium and ventricle. (**c**) Specimen 1093. Pulmonary veins are linked to the common atrium. The *arrow* indicates the right pulmonary vein connecting to the back wall of the atrium. The apex is a single ventricle. (**d**) Specimen 1093. Only one ventricle chamber is seen after opening the anterior wall. Its inner wall has large muscular bundles but no ventricular septum or small infundibulum chamber. The atrioventricular valve is on the right side, and the two great arteries are on the left side. The red probe indicates the AV; the blue one is in the PV. (**e**) Specimen 1093. Aortic valve. Large muscular bundles are seen under the AV after opening the ventricular anterior wall. The aortic and atrioventricular valves are separated by muscular tissue. *AO* aorta, *M* the wall muscular bundles under the aortic valve. (**f**) Specimen 1093. Waterston shunt operation (shunt between the ascending aorta and right pulmonary artery). The PV is on the right side of the AV. The diameter of the pulmonary artery is only one fourth that of the aorta. There is pulmonary stenosis and muscular tissue under the PV. There is an artificial graft between the main pulmonary artery and ascending aorta to guarantee the shunt to the pulmonary artery. *AO* aorta, *PA* pulmonary artery. (**g**) Specimen 1093. The Waterston shunt operation presents the artificial graft between the aorta and the pulmonary artery. (**h**) Schematic diagram of specimen 1093. *A* common atrium, *V* single ventricle

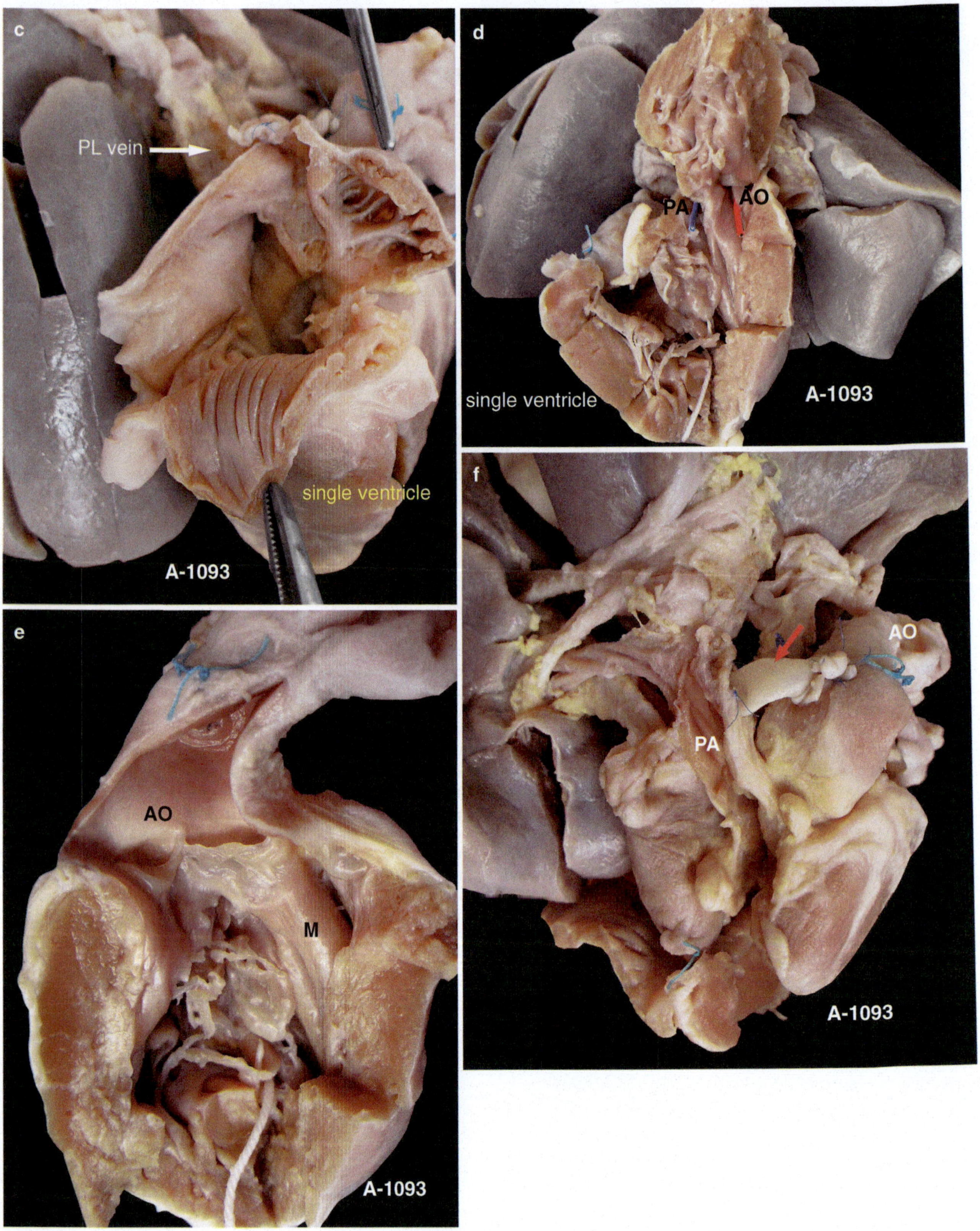

Fig. 19.2 (continued)

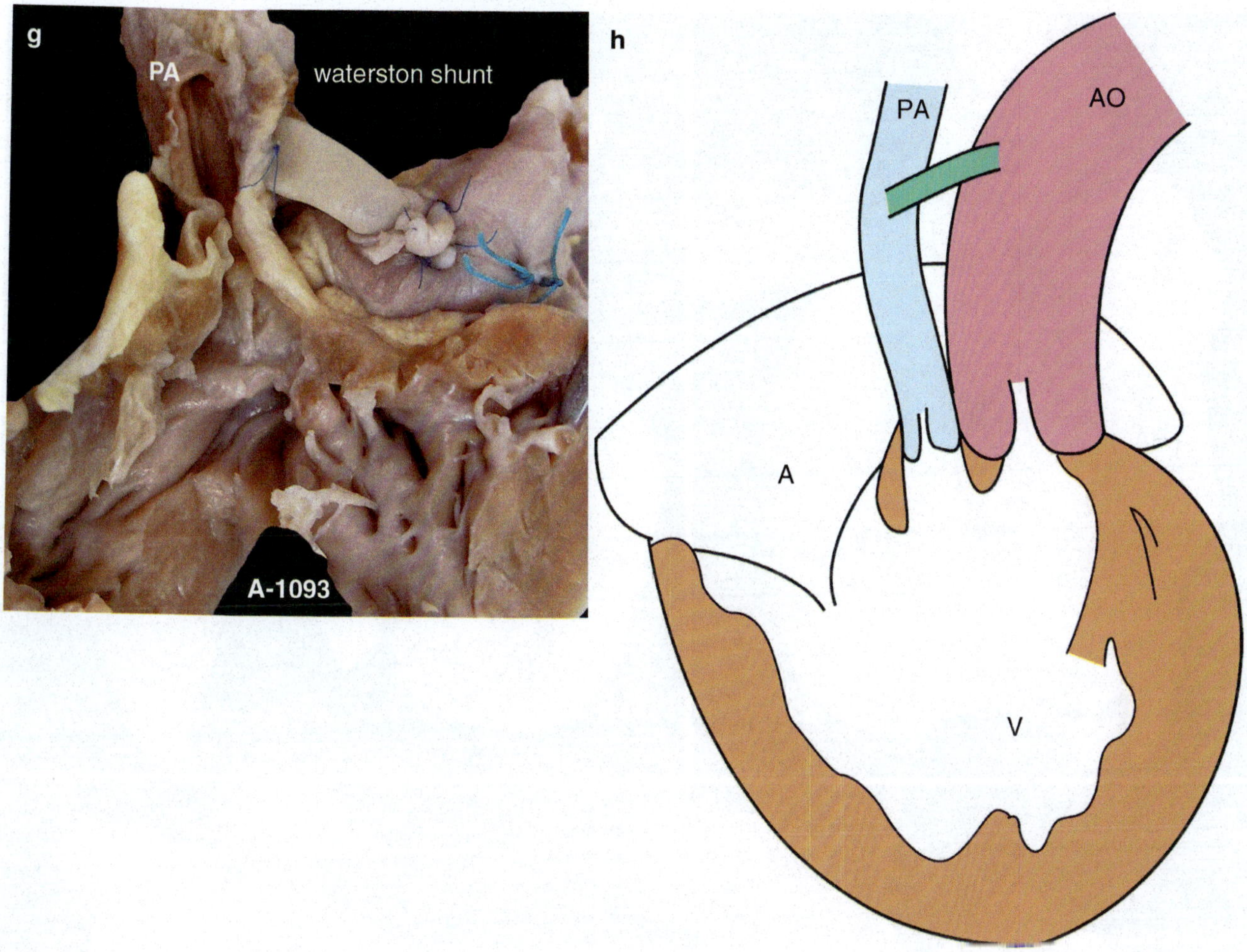

Fig. 19.2 (continued)

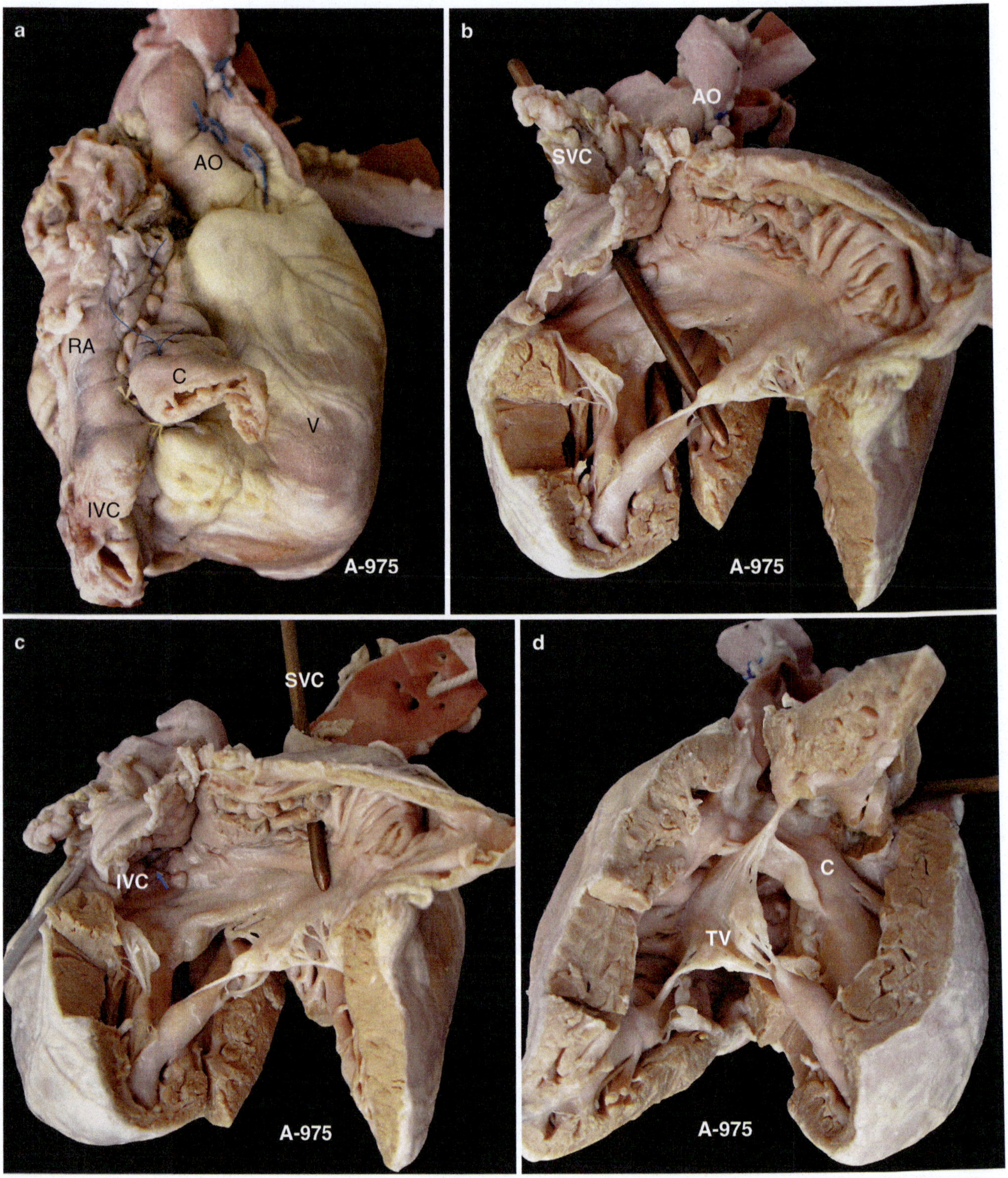

Fig. 19.3 (**a**) Specimen 975. The frontal view shows a single atrium and a single ventricle combined with pulmonary stenosis. *AO* aorta, *C* stub end of the extracardiac conduit between the RA and pulmonary artery, *IVC* inferior vena cava, *RA* right atrium, *V* ventricular free wall. (**b**) Specimen 975. Opening of the atrium and ventricle. The specimen is in the SVC. Its front end is in the ventricular chamber. One huge papillary muscle of the atrioventricular valve is similar to the TV. (**c**) Specimen 975. Presentation of single atrioventricular valve. The *arrow* indicates the anastomoses of the conduit between the RA and the pulmonary artery. *IVC* inferior vena cava. (**d**) Specimen 975. Opening of the anterior wall of the ventricle to expose the single ventricle. *C* the muscular tissue of supraventricular crest indicates the morphological right ventricle; *TV* tricuspid valve. (**e**) Specimen 975. Close view of pulmonary valve. The thick muscle encompasses the PV. The PV and the subvalvular tissue are narrow, and the AV is on its right front. (**f**) Specimen 975. Exposure of oversized aorta, three valve leaflets, and coronary artery orifices. The PV is sheltered by a huge muscular bundle. *AO* aorta, *AV* aortic valve. (**g**) Specimen 975. Presentation of single ventricle with the pulmonary valve stenosis and normal aortic valve. *LA* left atrium. (**h**) Schematic diagram of specimen 975. Single ventricle, pulmonary stenosis and subvalvular pulmonary stenosis

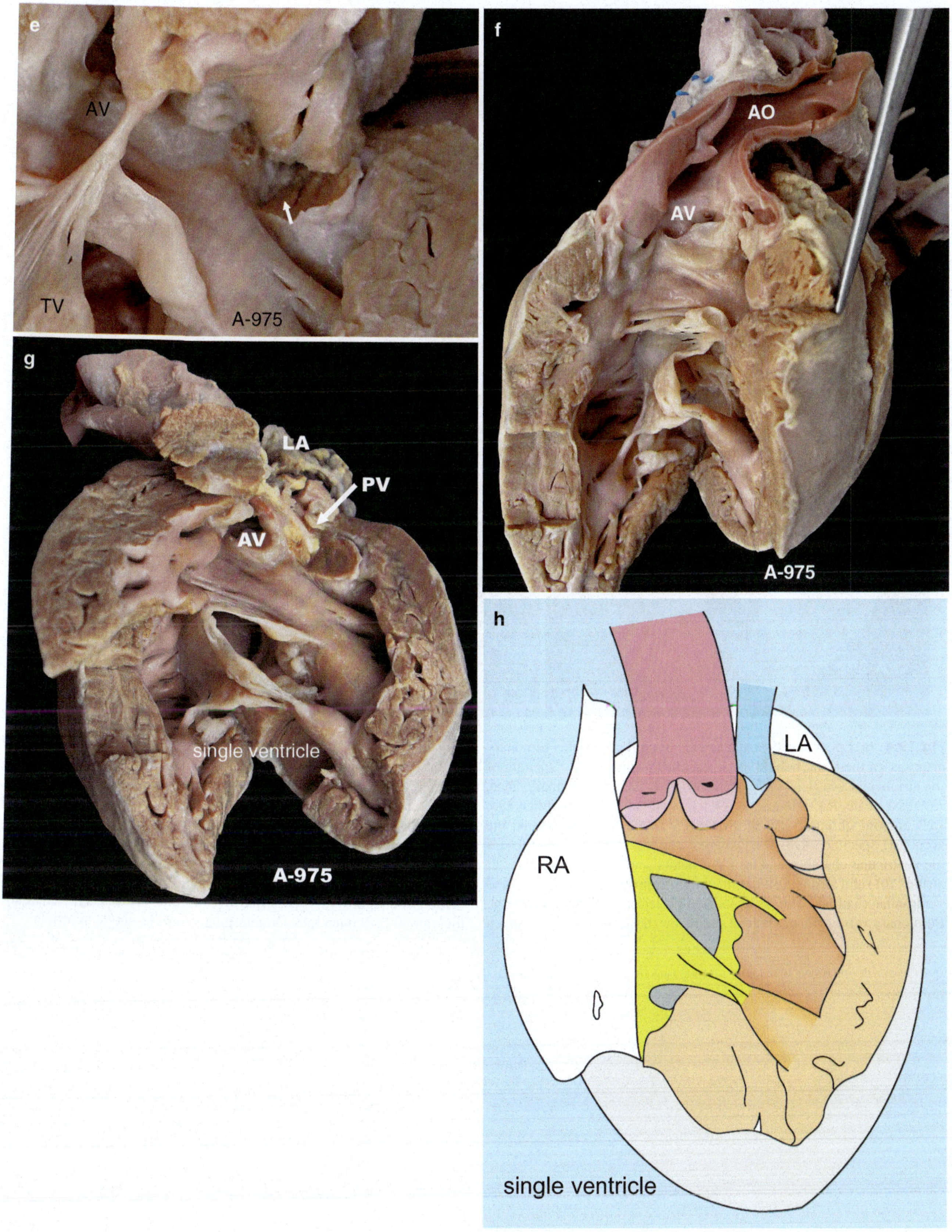

Fig. 19.3 (continued)

19.6.3 Double Inlets of Ventricle

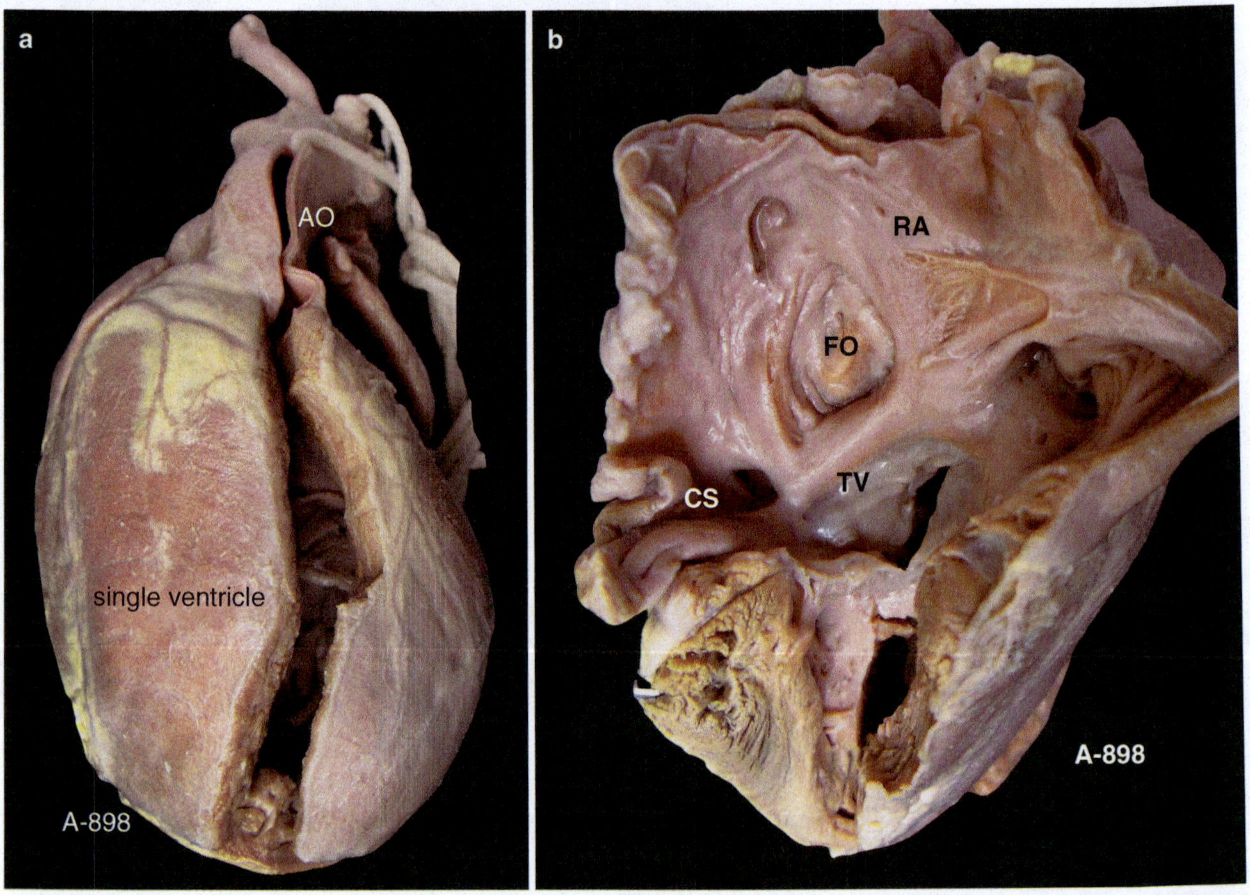

Fig. 19.4 (**a**) Specimen 898. Appearance of single ventricle Pulmonary artery is on the right, behind the aorta. (**b**) Specimen 898. Opening of the atrium and ventricle. Only one ventricle is seen. *CS* coronary sinus, *FO* fossa ovale, *RA* right atrium, *TV* tricuspid valve. (**c**) Specimen 898. Left anterior oblique view shows the PV is on the right, behind the aorta. (**d**) Specimen 898. The two great arteries originate from the single ventricular chamber, and the PV is behind and below the AV. *LA* left atrium, *RA* right atrium. (**e**) Specimen 898. Exposure of aortic valve and ventricular chamber. The PV is located posteriorly. (**f**) Specimen 898. Pulmonary valve orifice is behind and below the aortic valve. The green probe is in the pulmonary artery; the blue probe indicates the RA blood inflows to the ventricular chamber. (**g**) Specimen 898. Two atrioventricular valves are connected to the same single ventricular chamber, which can confirm the double-inlet ventricle. The red probe is in the MV orifice (LA), the blue probe is in the TV orifice (RA), and the green probe is in the PV (PA). (**h**) Specimen 898. Presentation of the double-inlet ventricle again. The blue probe is in the RA, and the red probe is in the LA. *LA* left atrium. (**i**) Specimen 898. Presentation of the left (*red arrow*) and right (*blue arrow*) atrioventricular orifices. (**j**) Schematic diagram of Specimen 898. Double-inlet ventricle, single ventricle

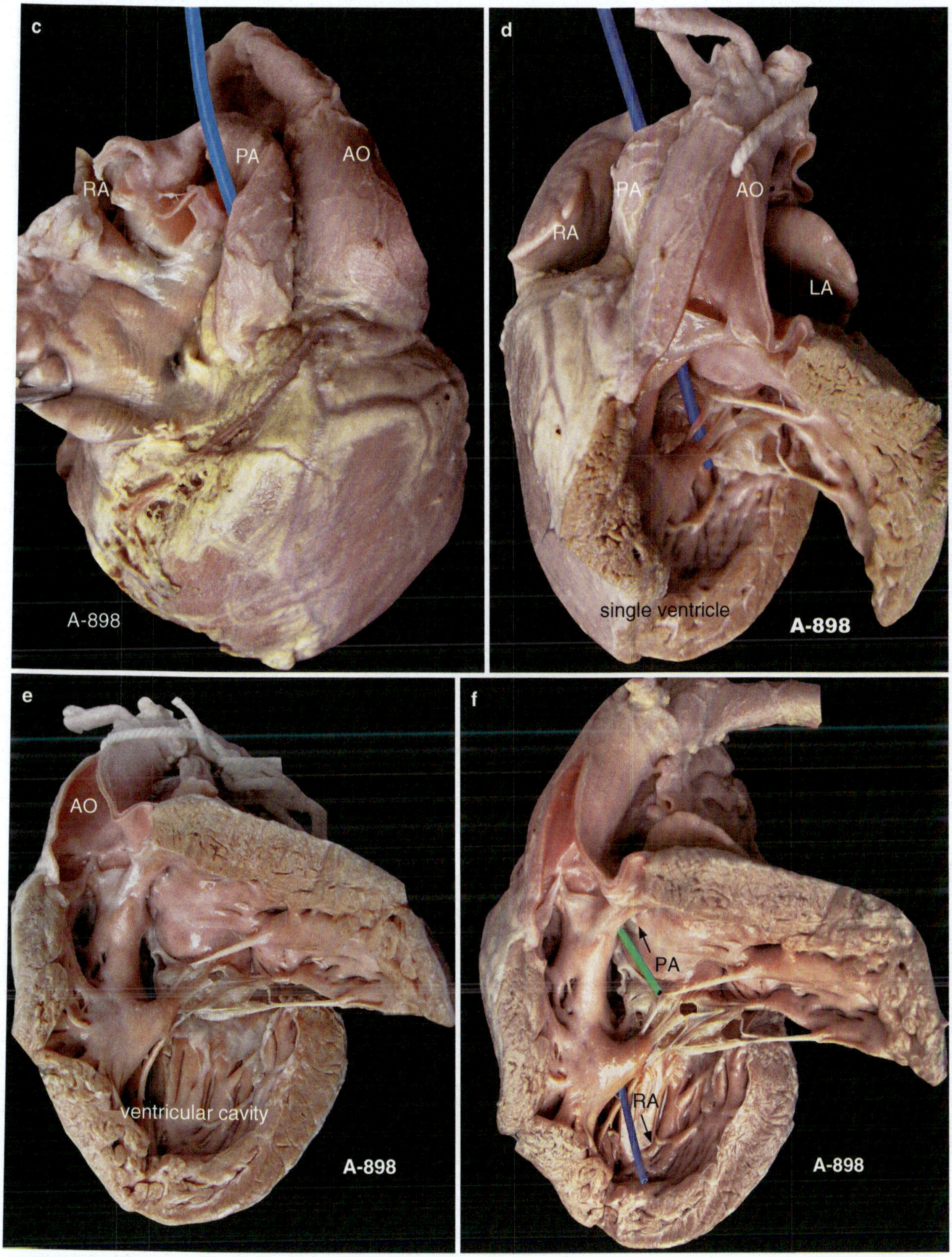

Fig. 19.4 (continued)

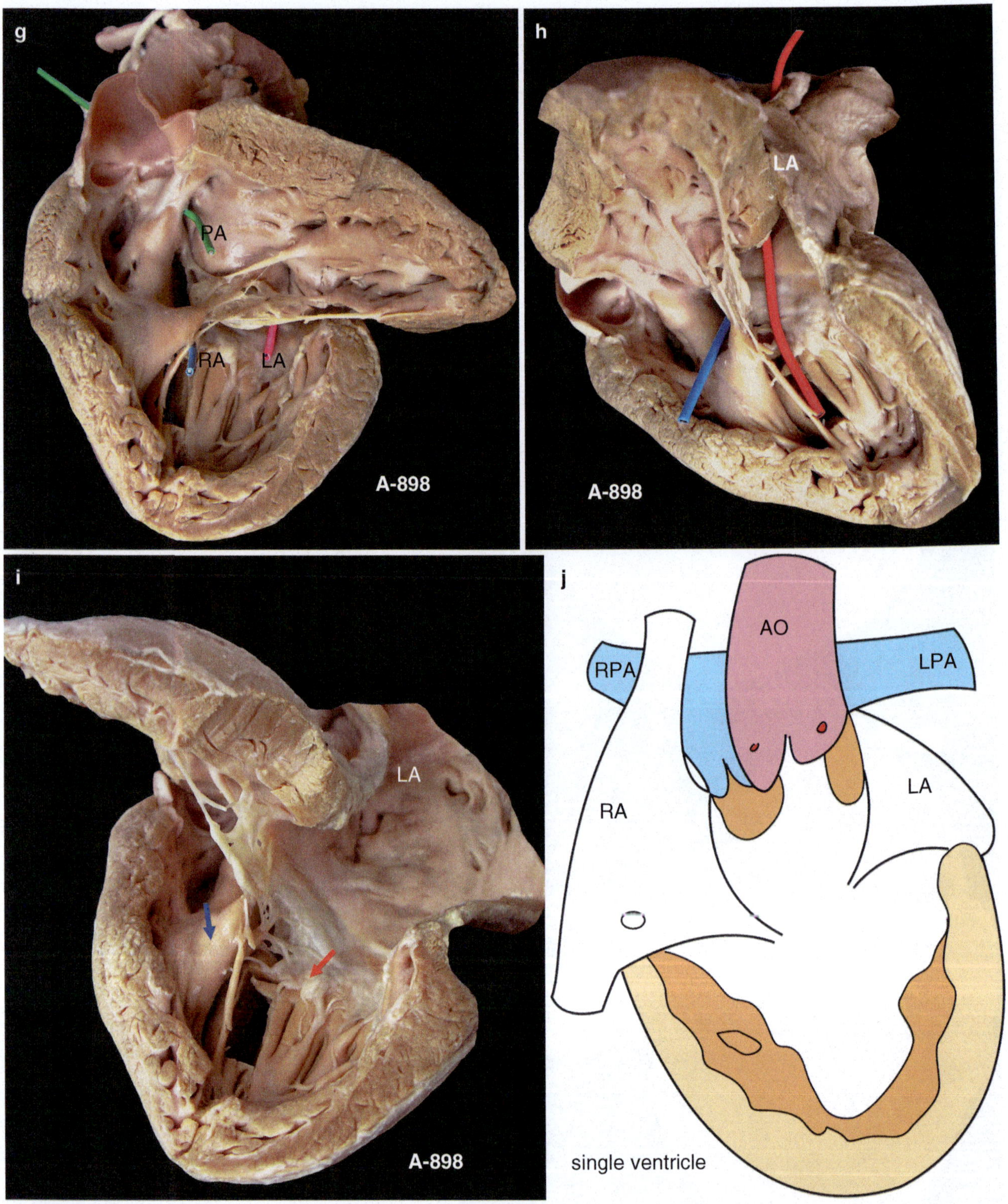

Fig. 19.4 (continued)

19.6.4 Single Ventricle Combined with Main Pulmonary Artery Stenosis

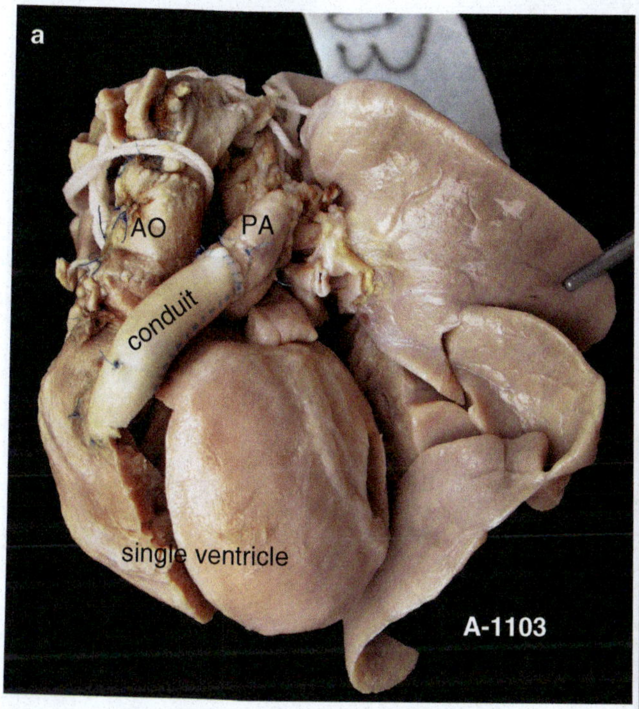

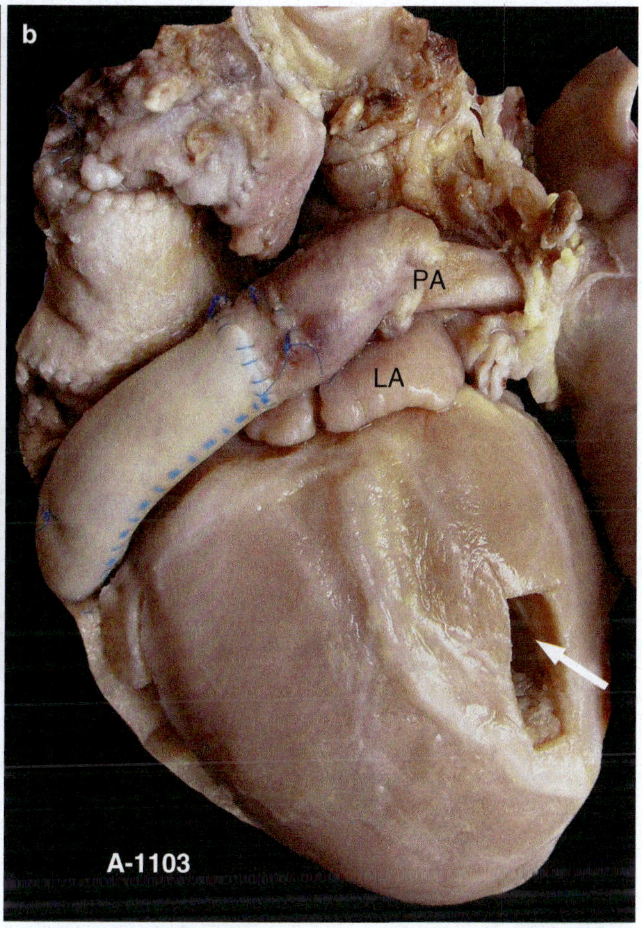

Fig. 19.5 (**a**) Specimen 1103. Single ventricle combined with main pulmonary artery stenosis. Development of the left and right pulmonary arteries is still good. There is an artificial conduit between the ventricle and the main pulmonary artery. The AO is covered by pericardium; the conduit is the artificial extracardiac conduit. *AO* ascending aorta. (**b**) Specimen 1103. Left anterior oblique view shows the left free wall of the single ventricle. *LA* left atrium, *PA* pulmonary artery. (**c**) Specimen 1103. Presentation of the combined anomalous pulmonary venous drainage. The RA is open; the ventricular chamber is the RV, and there is an ASD. The *red arrow* indicates the anomalous pulmonary venous drainage into the RA. *RA* right atrium, *RV* right ventricle. (**d**) Specimen 1103. The ventricular wall has been cut from the apex. This view confirms that there is only one ventricle and one atrioventricular valve. The cone muscle between the AV and the atrioventricular valve indicates the chamber is the RV. There is an ASD between the left and right atria. *TV* tricuspid valve. (**e**) Specimen 1103. The left free wall confirms that there is only one ventricular chamber. The anterior wall of the ventricle is the anterior wall of the single ventricle. The right view confirms there is only one atrioventricular valve with a structure that conforms to the TV. The blood of the LA flows with that of the RA into the single ventricle through the ASD. *LA* left atrium, *RA* right atrium, *TV* tricuspid valve. (**f**) Specimen 1103. Close view of the aortic valve orifice. On the top is the ventricular end of the artificial conduit. (**g**) Schematic diagram of Specimen 1103

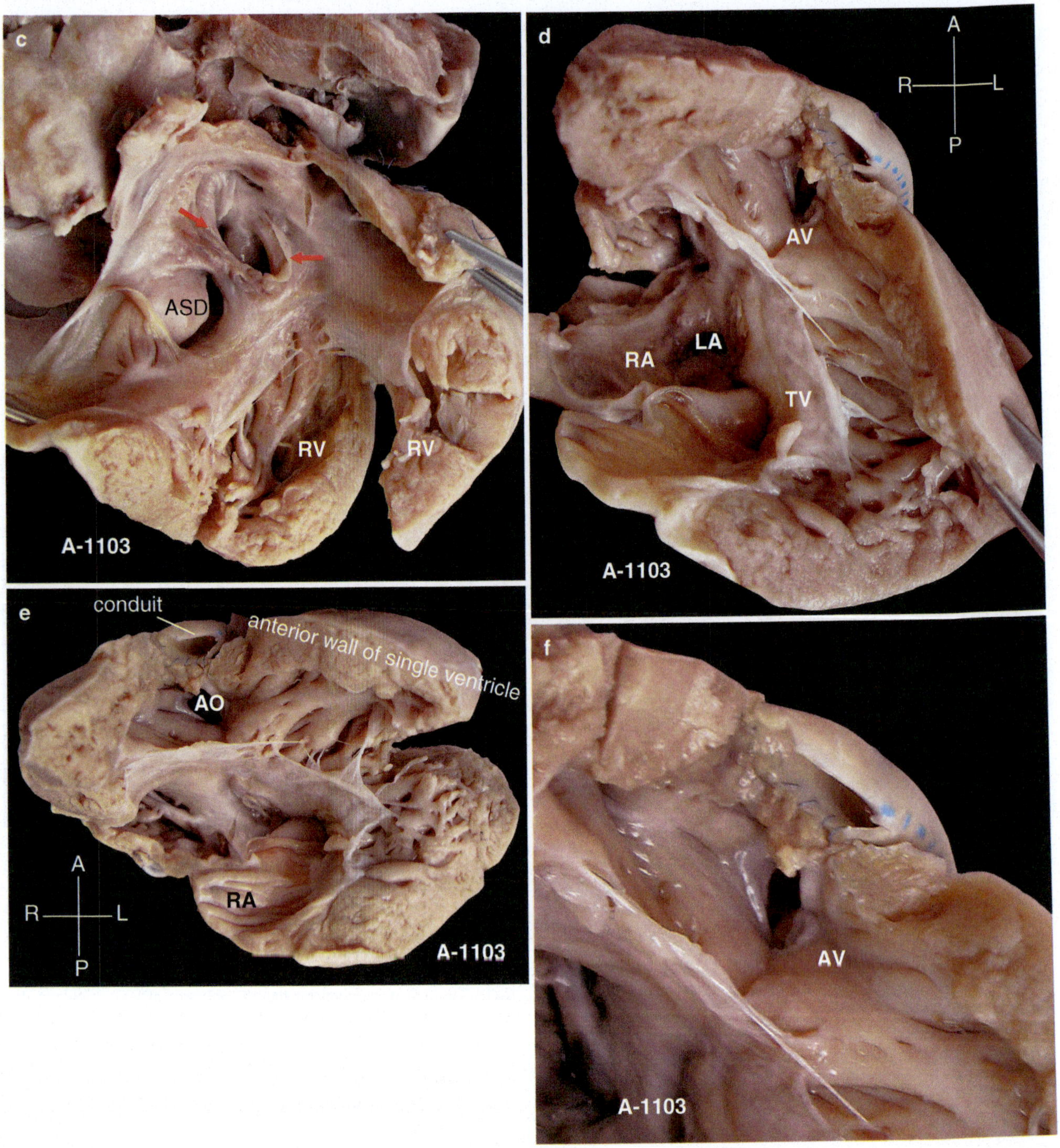

Fig. 19.5 (continued)

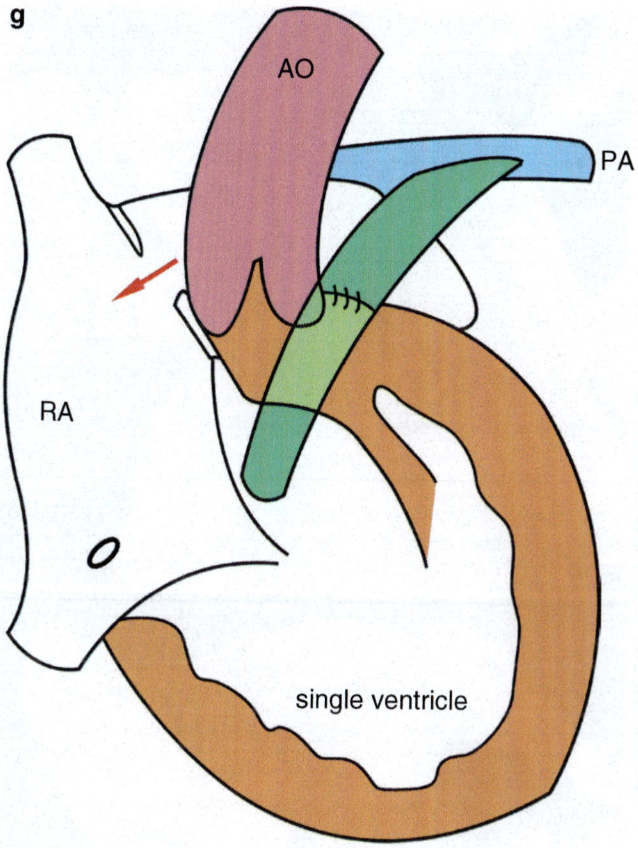

Fig. 19.5 (continued)

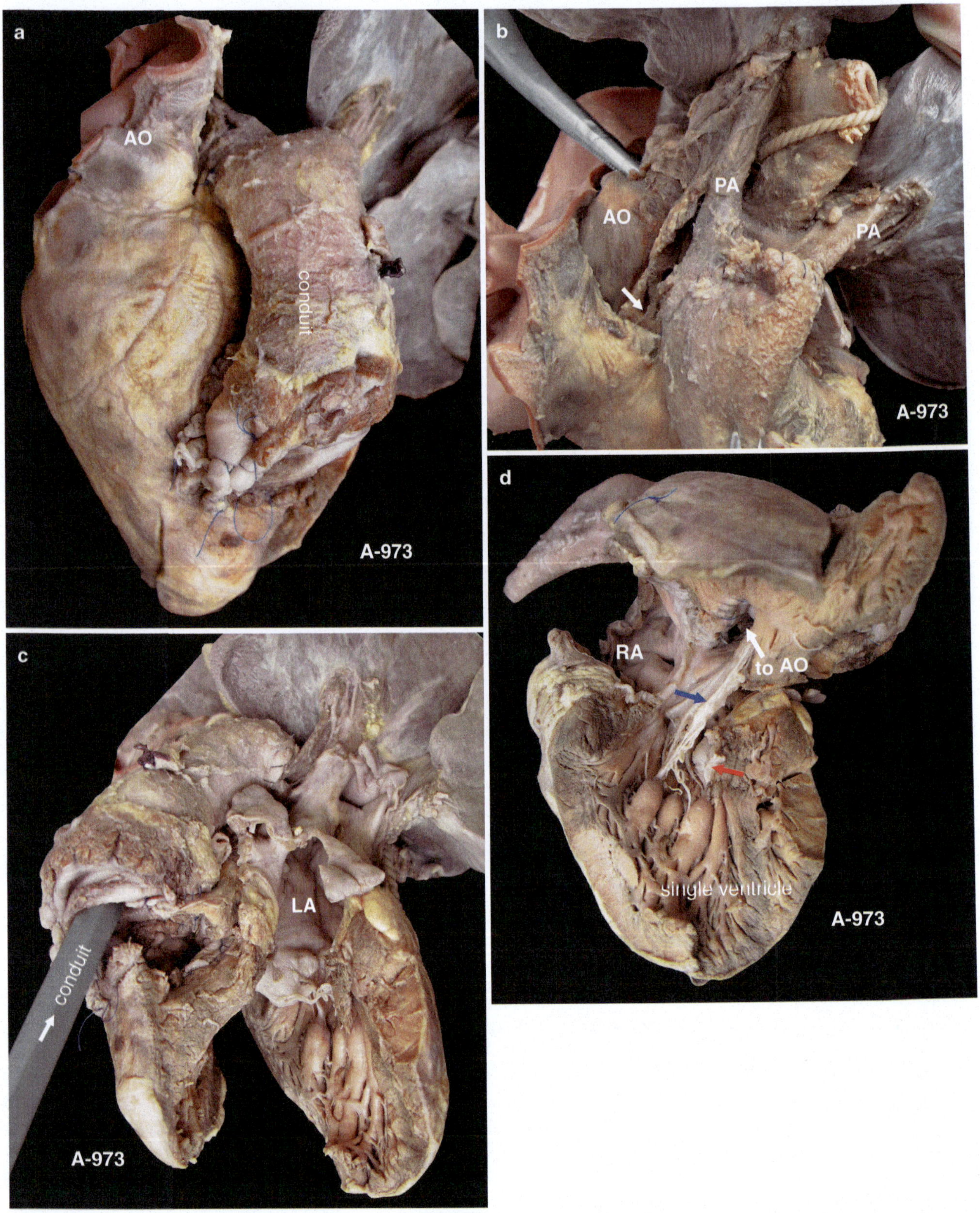

Fig. 19.6 (**a**) Specimen 973. Double-inlet ventricle, single ventricle, pulmonary artery stenosis, extracardiac conduit between the ventricle and pulmonary artery. (**b**) Specimen 973. Close view of the anastomoses between the pulmonary artery and the conduit. The *arrow* indicates the pulmonary stenosis. (**c**) Specimen 973. The ventricle is cut from the LA; only one ventricular chamber is left. The proximal end of the conduit can be seen from the ventricle anterior wall. The valve leaflets under the LA are connected with the single ventricular chamber. *LA* left atrium. (**d**) Specimen 973. The single ventricle is cut from the RA. The right atrioventricular valve is connected with the single ventricular chamber. The *blue*, *red*, and *white arrows* indicate the right, left atrioventricular valve, and subaortic tissue, respectively. *RA* right atrium. (**e**) Specimen 973. The aortic valve orifice (probe in it) is under the right atrioventricular valve. No fibrous connections exist between the right atrioventricular valve leaflets and the AV. *AV* aortic valve, *RA* right atrium, *TV* tricuspid valve. (**f**) Specimen 973. Left-side view of the atrioventricular valve. The blood of the LA flows into the single ventricle through the MV. The conduit to the pulmonary artery is at the top. (**g**) Schematic diagram of Specimen 973

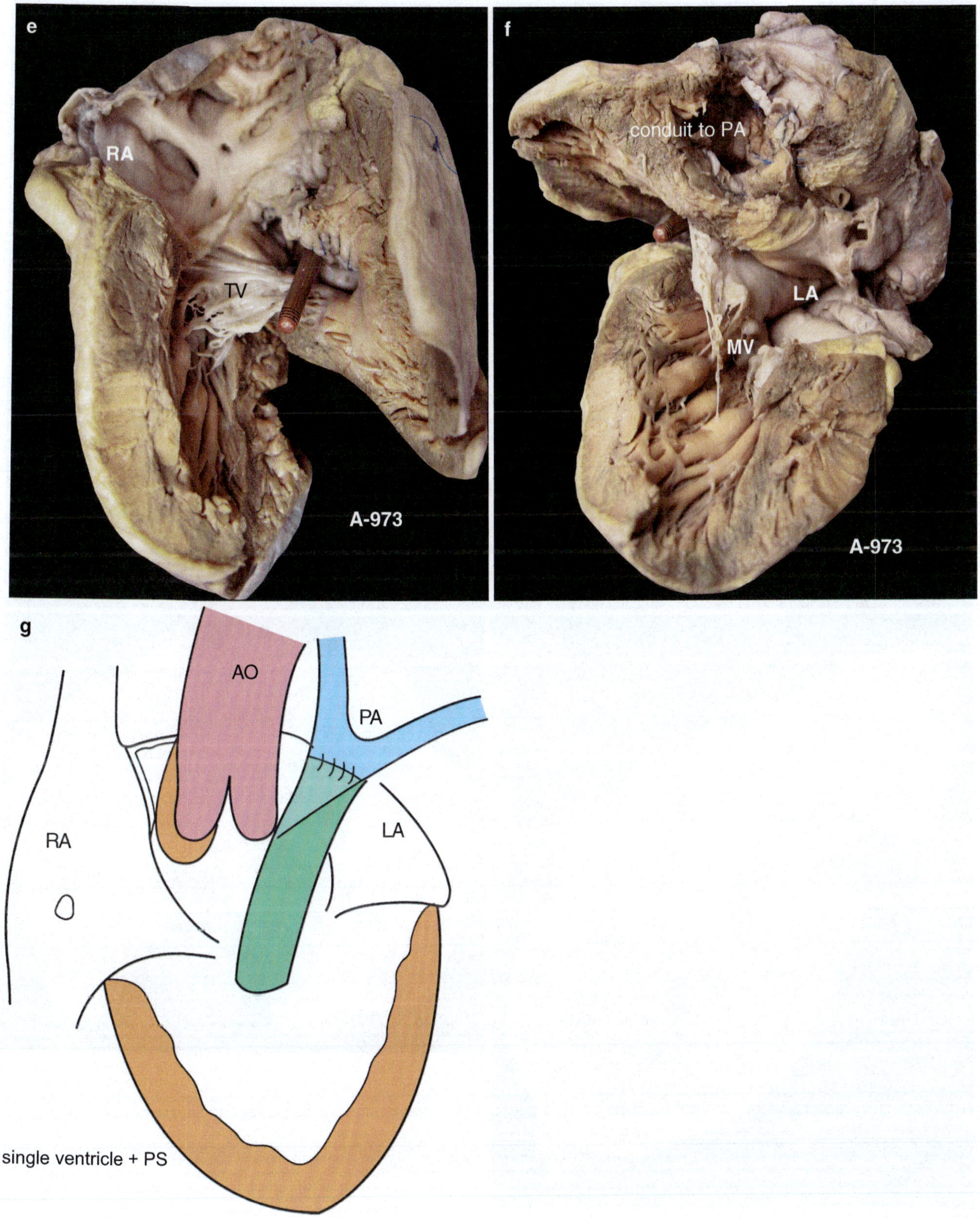

Fig. 19.6 (continued)

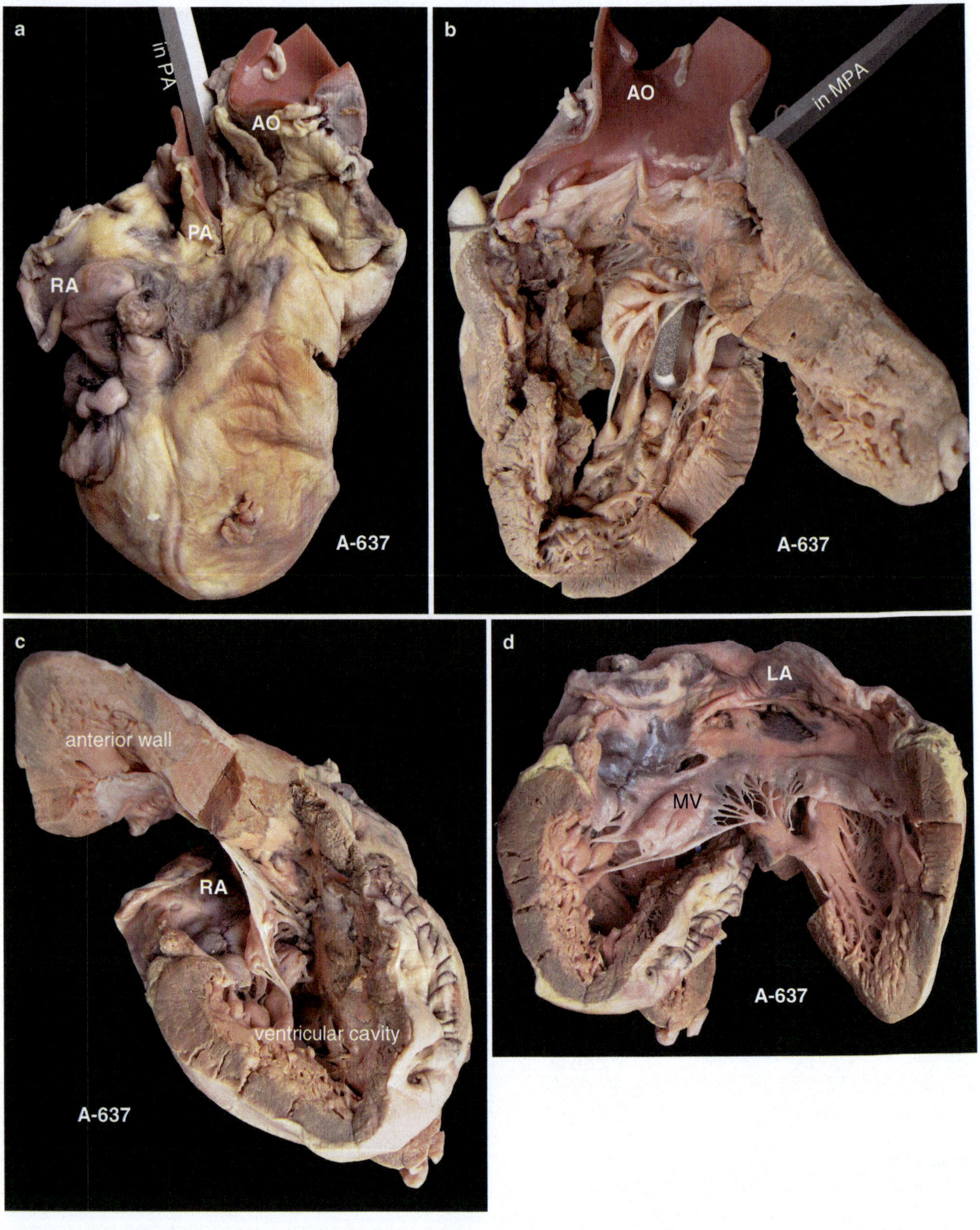

Fig. 19.7 (**a**) Specimen 637. The front view. Double-inlet single ventricle; the cone is under the two arterial valves; the small pulmonary artery is on the right behind the aorta; the probe is inserted into the ventricular chamber through the PV. (**b**) Specimen 637. The probe is inserted into the ventricular chamber through the PV. The PV orifice is on the right, behind the AV; there is no conal muscle between these two valves. (**c**) Specimen 637. The anterior wall is open. Presentation of right atrioventricular valve; the single ventricular cavity is under the valve. (**d**) Specimen 637. Presentation of the mitral valve and single ventricular chamber through the left atrium. The single ventricle conforms to the LV structure. *LV* left ventricle, *MV* mitral valve. (**e**) Specimen 637. Double-inlet ventricle. The *red* and *blue arrows* indicate the blood of the left and right atria, respectively. Blood from both atria flows into the single ventricle

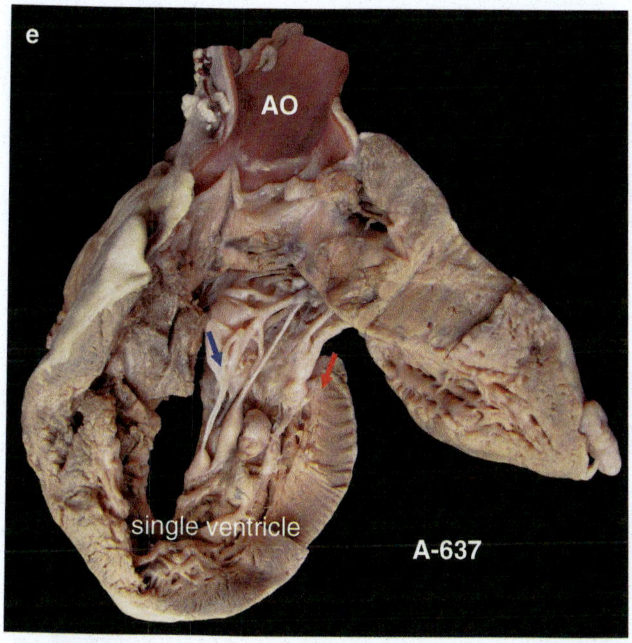

Fig. 19.7 (continued)

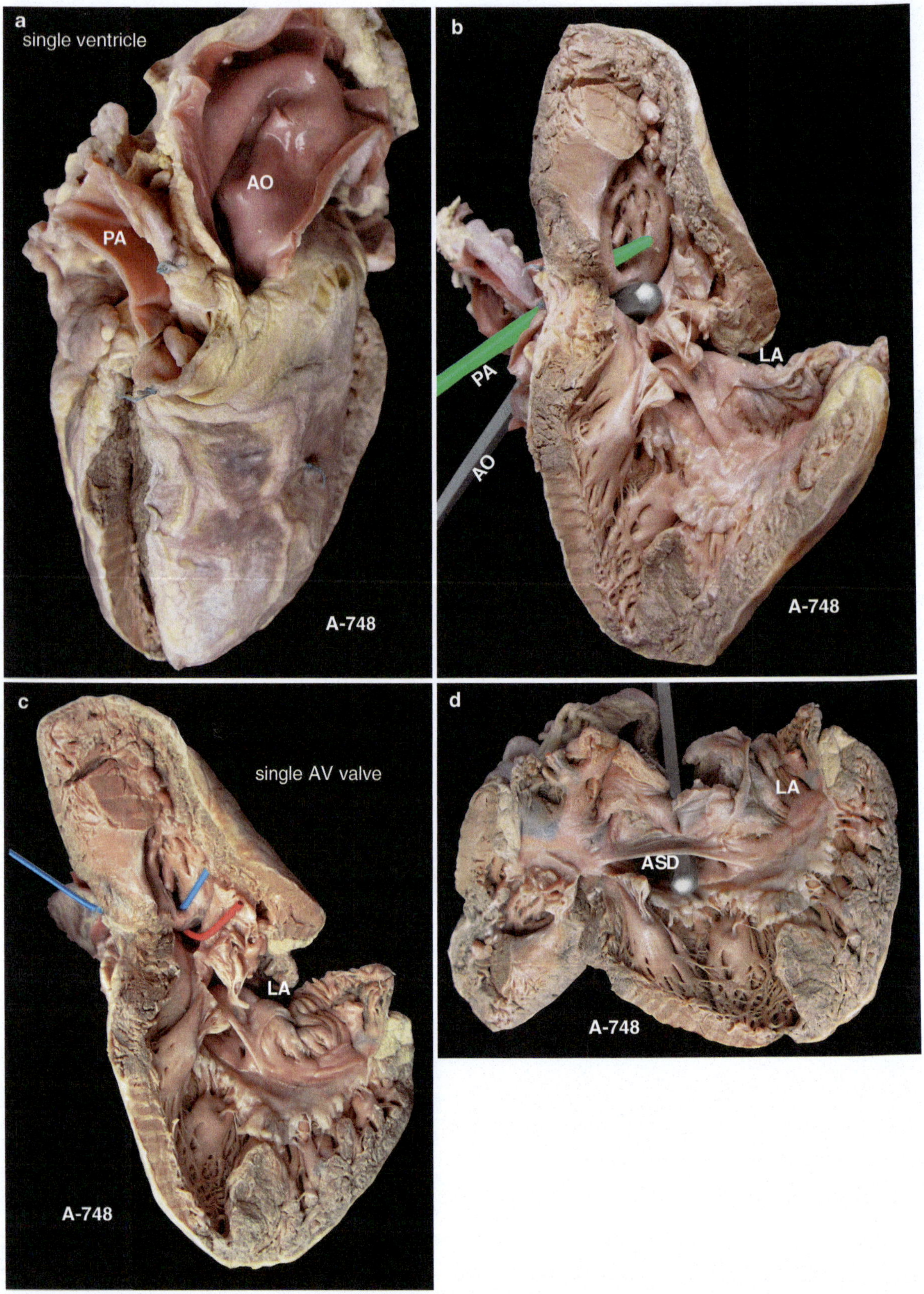

Fig. 19.8 (**a**) Specimen 748. Single ventricle, single atrioventricular valve, left ventricular structure, left anterior oblique. The aorta is on the left, behind the pulmonary artery; the semilunar valves are developed well. (**b**) Specimen 748. The green and blue probes are inserted into the PV and the AV, respectively. The anterior wall is opened to show there is only one ventricular chamber. Two semilunar valves are surrounded by muscular tissue, which confirms the double-cone structure. (**c**) Specimen 748. Observation from the left atrium. There is a muscle beam high in the ventricular chamber; however, it is not the ventricular septum because there are no independent ventricular structures bilaterally. These features could be diagnosed as single ventricle. (**d**) Specimen 748. Presentation of the atrial septal defect from the left atrium. The RA communicates with the LA through the ASD. The RA does not connect with the ventricle. There is only one atrioventricular valve. *LA* left atrium, *RA* right atrium

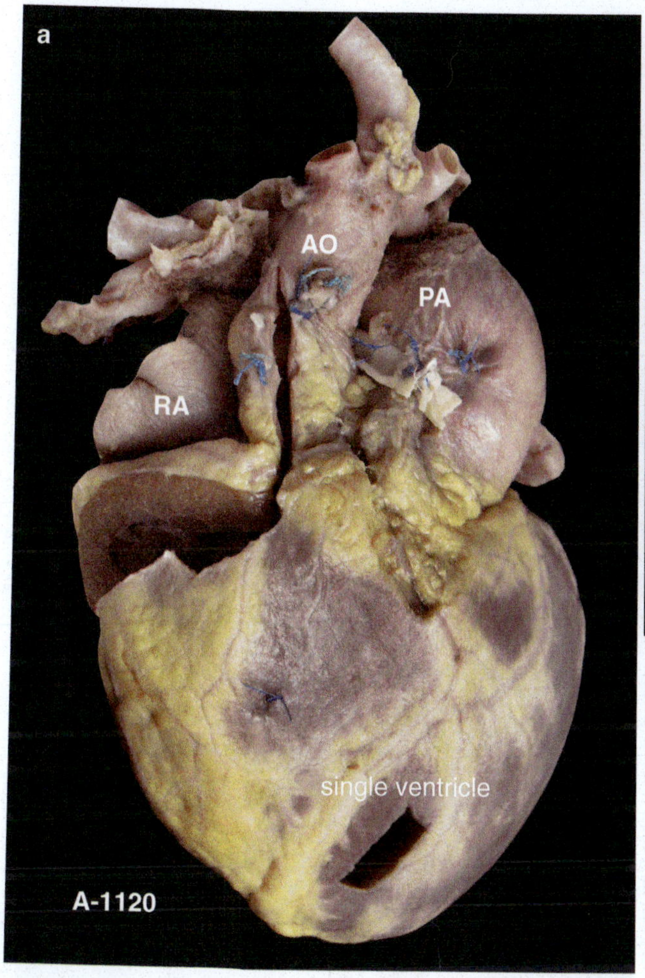

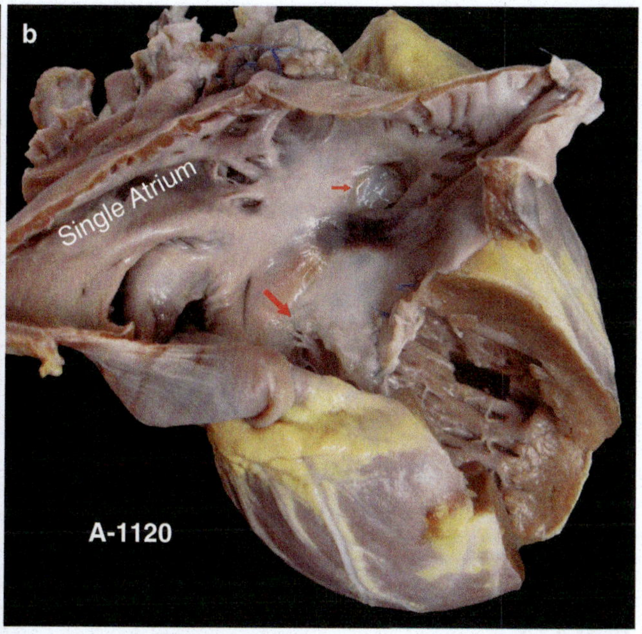

Fig. 19.9 (**a**) Specimen 1120. Single ventricle, single atrium, single atrioventricular valve. The aorta and pulmonary arteries are paratactic. The apex is the anterior wall of the single ventricle. (**b**) Specimen 1120. Opened atrium from the right side to present the two atrioventricular valve orifices. The *smaller arrow* indicates the remaining trace of the undeveloped atrioventricular valve, which has no normal leaflets; the *larger arrow* indicates the intact atrioventricular valve, which is similar to the MV. There is no atrial septum in the atrium. The left and right atrial appendages communicate with the single atrium. *MV* mitral valve. (**c**) Specimen 1120. The red probe is inserted into the undeveloped atrioventricular valve; it could reach the ventricular residual cavity under the AV. *LAA* left atrial appendage, *RAA* right atrial appendage. (**d**) Specimen 1120. Presentation of the undeveloped atrioventricular valve from the ventricle. The red probe is inserted from the atrium into the infundibulum muscle through the white thin membrane. (**e**) Specimen 1120. Presentation of the main ventricular chamber and the rudimentary infundibular chamber. There is white residual trace of the atrioventricular valve in the muscular rudimentary infundibular chamber under the AV. It is surrounded by a thin membrane with a diameter of 8 mm, and no normal chordae tendineae or papillary muscle exists. The *dotted line* indicates the blood direction from the single ventricle to the pulmonary artery. *AV* aortic valve. (**f**) Specimen 1120. The main ventricle is opened to expose the single atrioventricular valve. The *arrow* indicates the fibrous connection between the pulmonary and the atrioventricular valves. It could be diagnosed as single ventricle combined with single atrioventricular valve. (**g**) Specimen 1120. The probe is inserted into the ventricular chamber through the pulmonary artery. The rudimentary chamber is under the AV. *AV* aortic valve, *C* rudimentary cavity, *PV* pulmonary valve. (**h**) Schematic diagram of specimen 1120

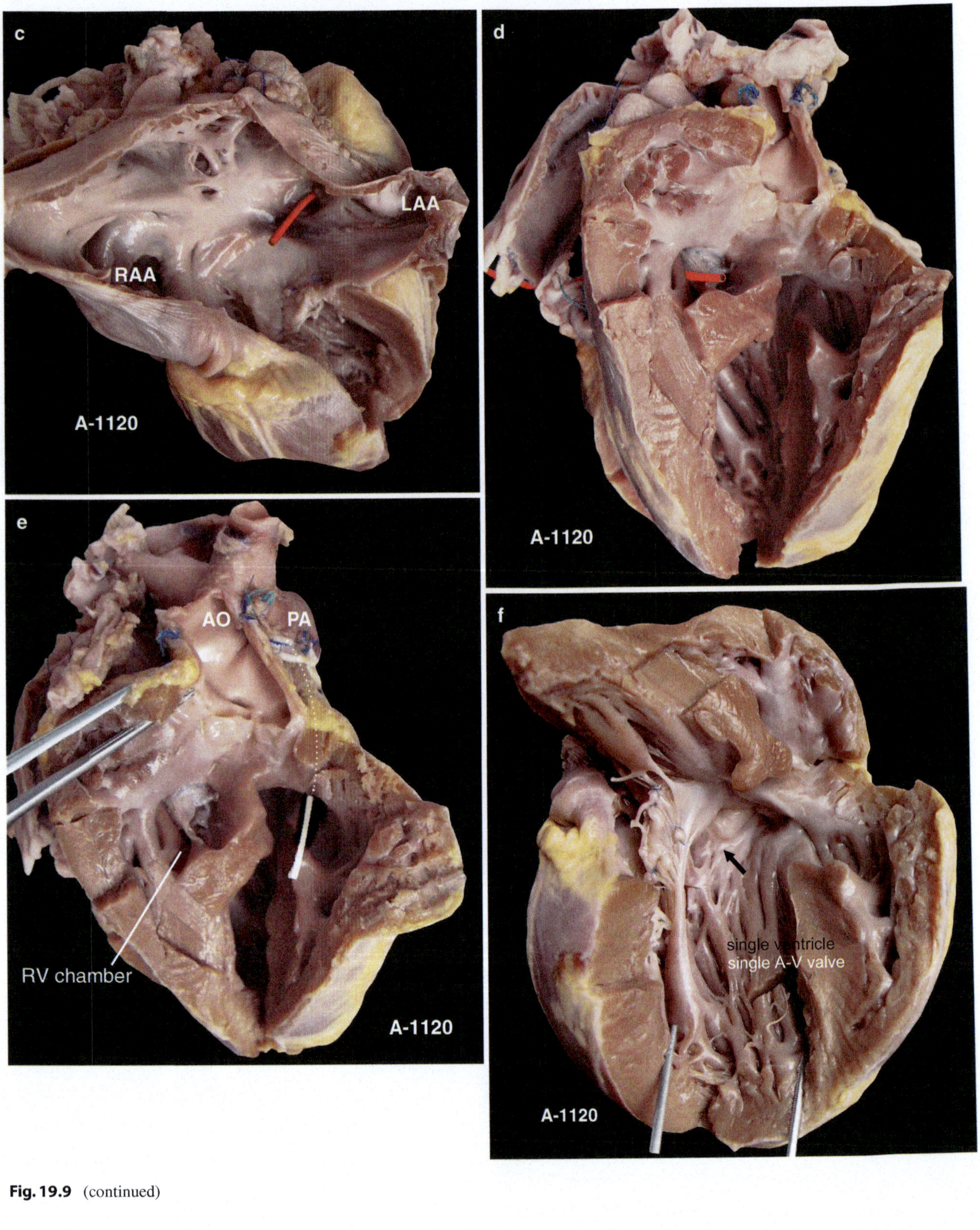

Fig. 19.9 (continued)

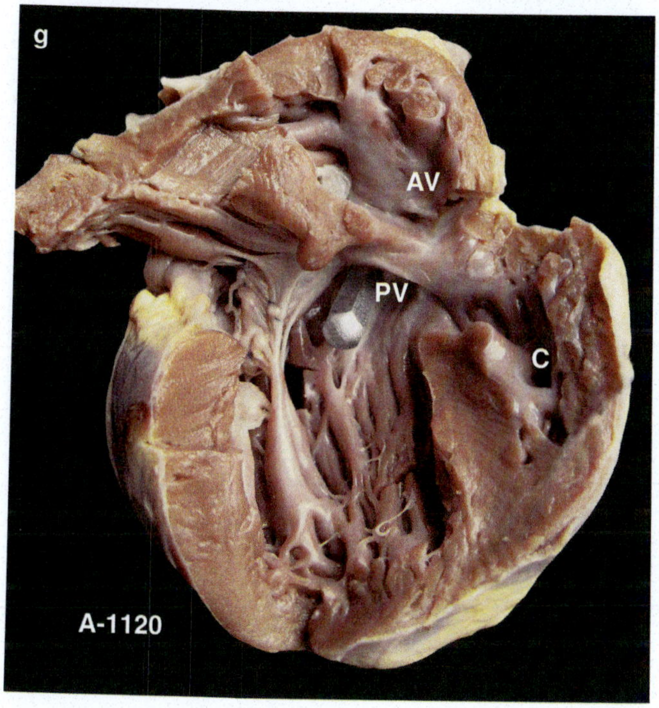

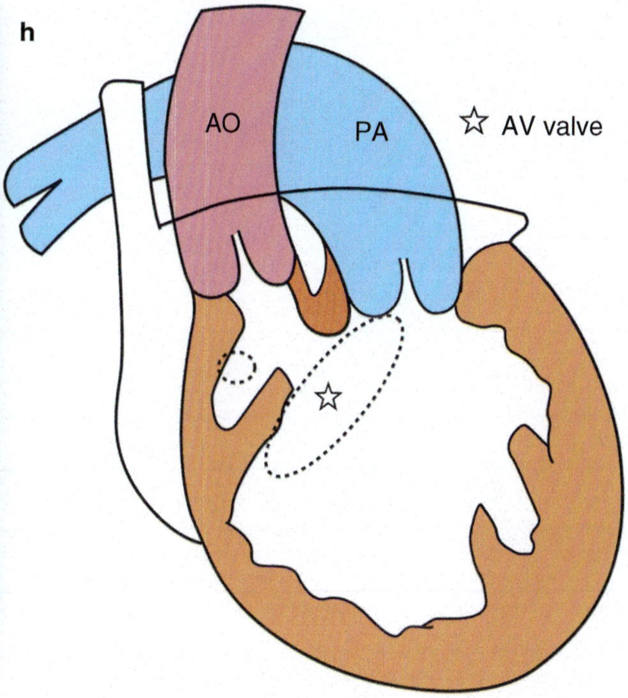

Fig. 19.9 (continued)

20 Persistent Truncus Arteriosus

20.1 General Considerations

A single trunk, with just one group of semilunar valves, arises from the base of the heart. All coronary arteries and the pulmonary artery originate from the trunk. This semilunar organ can be separated into three leaflets (Fig. 20.2f) (similar to normal AVs) or into two or four leaflets, which is the common arterial trunk, also known as *persistent truncus arteriosus*. A VSD is usually associated with this disorder (Fig. 20.2e).

In this condition, the embryological structure known as *the truncus arteriosus* fails to divide properly into the pulmonary trunk and aorta, resulting in a series of anatomical changes, including VSD and semilunar dysplasia. This great vessel usually has one large valve that usually has three leaflets. The aortic trunk customarily sits over both ventricles, but occasionally it can originate from only one of them. The upper portion of the wall between these two chambers is missing, resulting in what is known as a *subarterial ventricular septal defect*. If the artery cone remains its proximal residue, then the VSD is known as an *intracristal* type. In this condition, the heart conduction bundle is a far distance from the border of the septum defect.

Four anatomical types of truncus arteriosus (Fig. 20.1) are recognized on the basis of the anatomical origin of the pulmonary arteries:

Type I, a short pulmonary trunk arises from the truncus arteriosus, giving rise to both pulmonary arteries (Fig. 20.2a).

Type II, each pulmonary artery arises separately from the truncus, without the pulmonary trunk.

Type III, the unilateral pulmonary artery arises from the truncus; the other collateral vessels rising from the aortic arch or the descending aorta can complete the similar function of blood supply to the lung, just as can aberrant pulmonary arteries.

Type IV, bilateral aberrant pulmonary arteries rise from the aortic arch or the descending aorta, with nothing essentially different from the pulmonary atresia.

X. Zhu (ed.), *Surgical Atlas of Cardiac Anatomy*,
DOI 10.1007/978-94-017-9409-1_20, © Springer Science+Business Media Dordrecht and People's Medical Publishing House 2015

20.2 Specimen Demonstrations

Fig. 20.1 Specimen 1063. Persistent truncus arteriosus includes four basic types. Type I: a short pulmonary trunk arises from the truncus arteriosus, giving rise to both pulmonary arteries. Type II: each pulmonary artery arises separately from the truncus. Type III: unilateral pulmonary artery arises from the truncus. Type IV: bilateral aberrant pulmonary arteries rising from the aorta and its branches

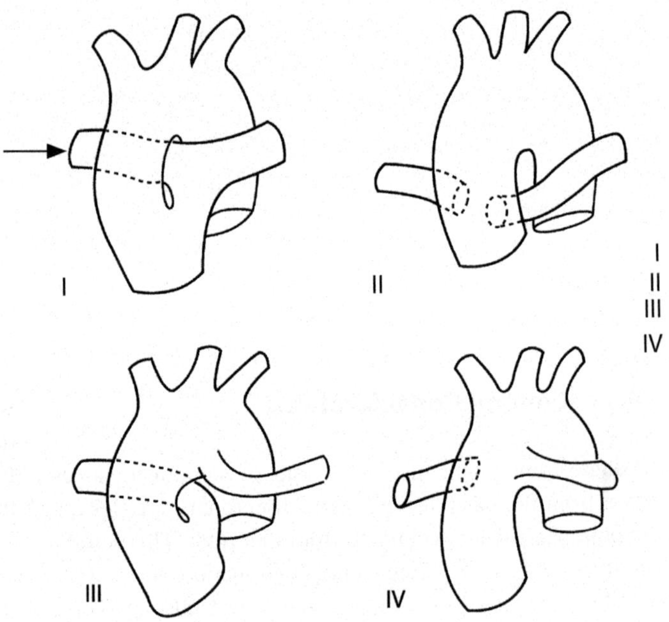

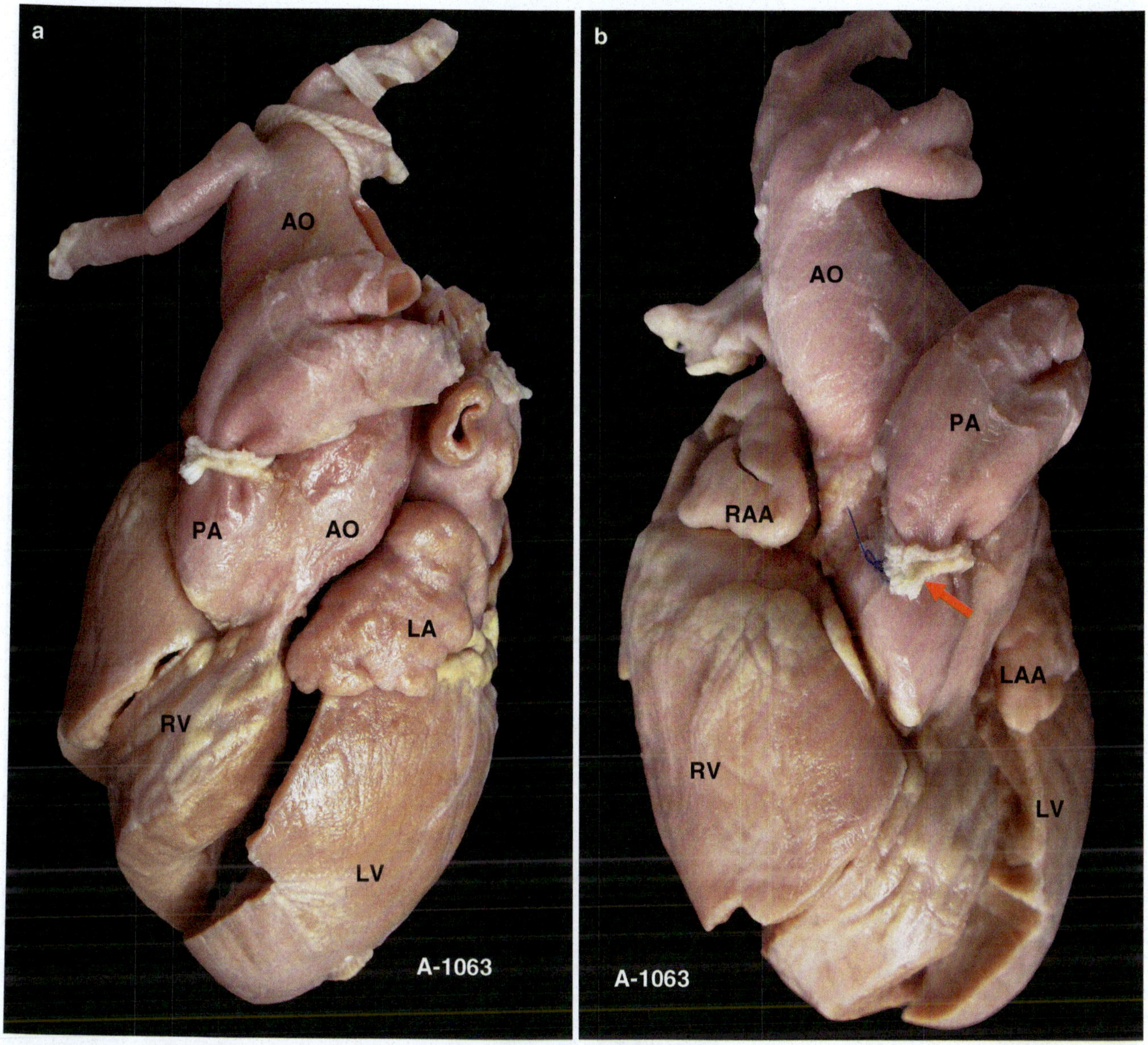

Fig. 20.2 (**a**) Specimen 1063. Persistent truncus arteriosus (type I), pulmonary hypertension, banding surgery. The pulmonary trunk arises from anterior right aspect of the ascending truncus, without a PV but with a VSD. The pulmonary trunk is banding with tape. (**b**) Specimen 1063. View from the right aspect. Arrow is the banding tape used to narrow the root of the pulmonary trunk. The pulmonary artery arises from the anterior of the AO. *RAA* right auricle, *LAA* left auricle. (**c**) Specimen 1063. View from the right aspect. The red probe is from SVC to IVC by RA. The *arrow* indicates the banding tape. (**d**) Specimen 1063. View from the aspect of the right atrium and the right ventricle. FO and coronary sinus are in the RA, right ventricular hypertrophy. TV is generally normal. (**e**) Specimen 1063. The anterior wall of the right ventricle is open. No origin of trunks is seen in the RVOT; only a VSD is seen as an outlet. (**f**) Specimen 1063. Viewed from the section of the left ventricle outflow and the AO. A VSD is under the AV, which has three leaflets. The *red arrow* indicates the origin of the pulmonary trunk without a PV. (**g**) Specimen 1063. The red probe shows that the pulmonary trunk rises from the anterior right aspect of the ascending aorta. (**h**) Specimen 1063. The probe is from the RV to AO by the VSD; the RV does not run through the PA directly. (**i**) Specimen 1063. The LA and MV are generally normal. The *arrow* indicates the origin of the PV. (**j**) Specimen 1063. Schematic diagram

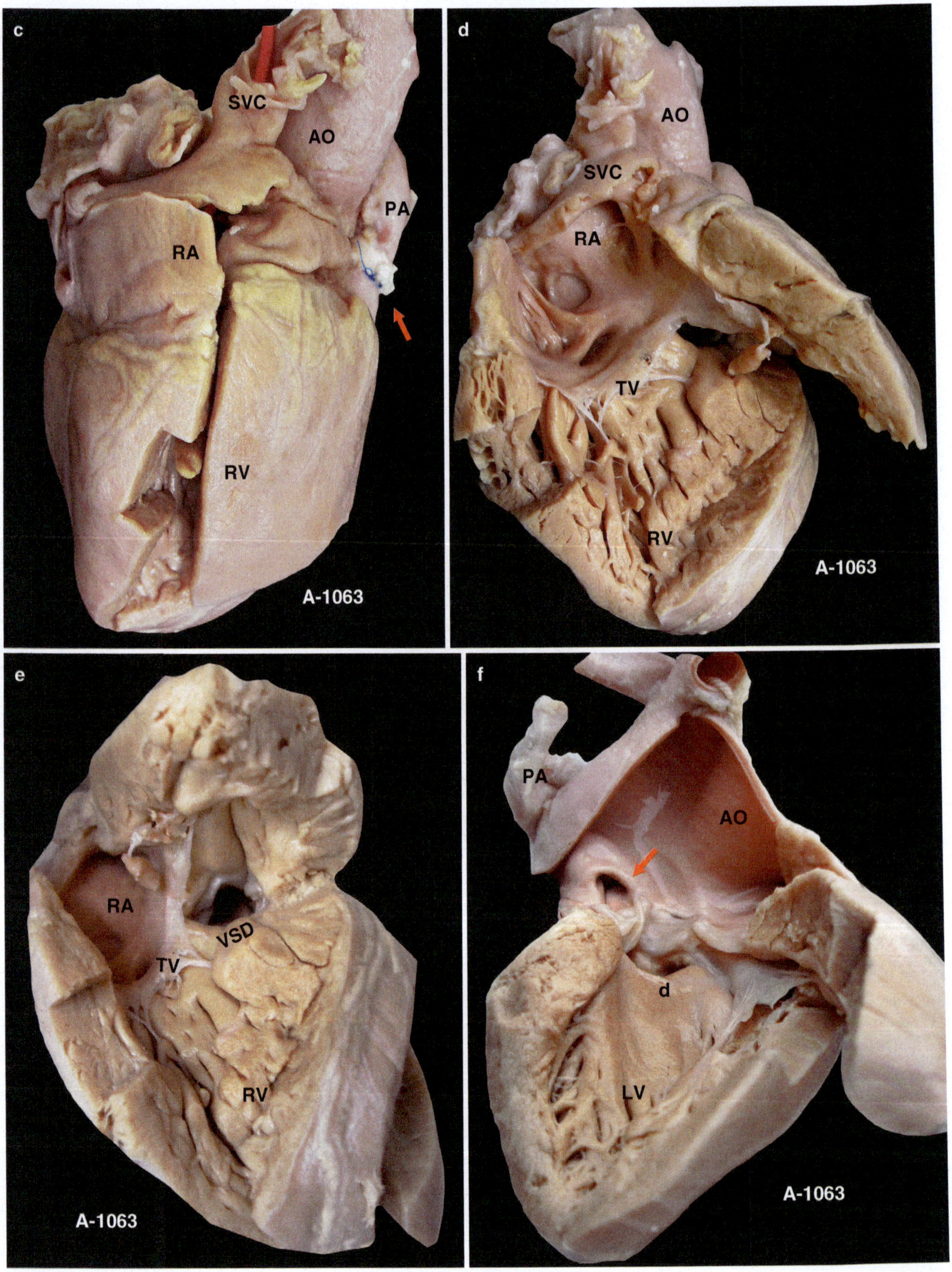

Fig. 20.2 (continued)

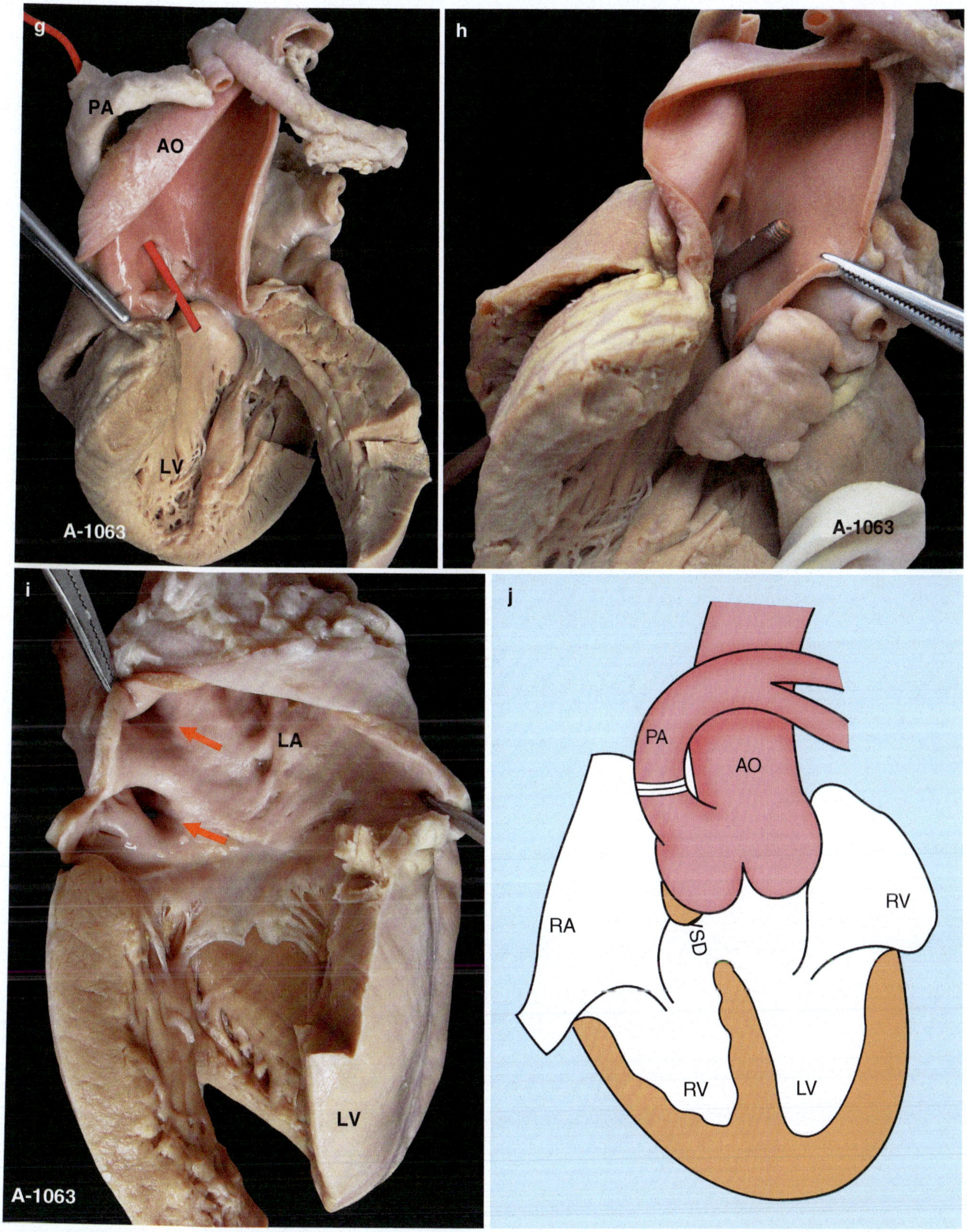

Fig. 20.2 (continued)

21 Aortopulmonary Window

21.1 General Considerations

Aortopulmonary window is also called *aortopulmonary septum defect*. Aortic and pulmonary valves are developed well. At the root of the AV and PV, above the position of the semilunar valve, is a septum for blood injection into the lumen. The difference between an aortopulmonary window and a persistent truncus arteriosus is the development of a semilunar valve (Fig. 21.1e–g). The proximal aortic sac suffers a developmental disorder in a patient with an aortopulmonary window, leading to septum dysplasia. Hence, this chapter describes it independently from the persistent truncus arteriosus.

21.2 Classification

The clinical classification includes:

Type I: defects occur in the proximal part of the aortopulmonary septum, just above the sinus of Valsalva, forming a window or a short tube (Fig. 21.1f).

Type II: defects occur in the distal part of the aortopulmonary septum, which are more cephalad on the ascending aorta, and the septum just above the sinus of Valsalva is intact.

Type III: defect is a combination of types I and II, and both the left and right pulmonary arteries rise from the ascending aorta.

During the operative repair, the important matters to consider include the size of the defect, the position of the aortopulmonary window, and the malformation. In cases of distal septum defect, an aortic cross-clamping will provide worse exposure to some surgeons during the operation.

X. Zhu (ed.), *Surgical Atlas of Cardiac Anatomy*,
DOI 10.1007/978-94-017-9409-1_21,

21.3 Specimen Demonstrations

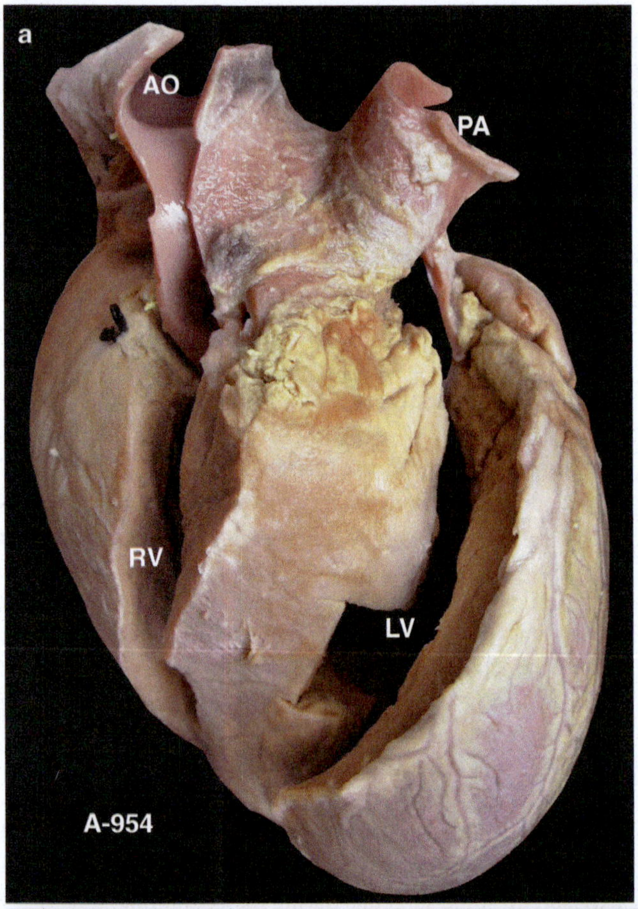

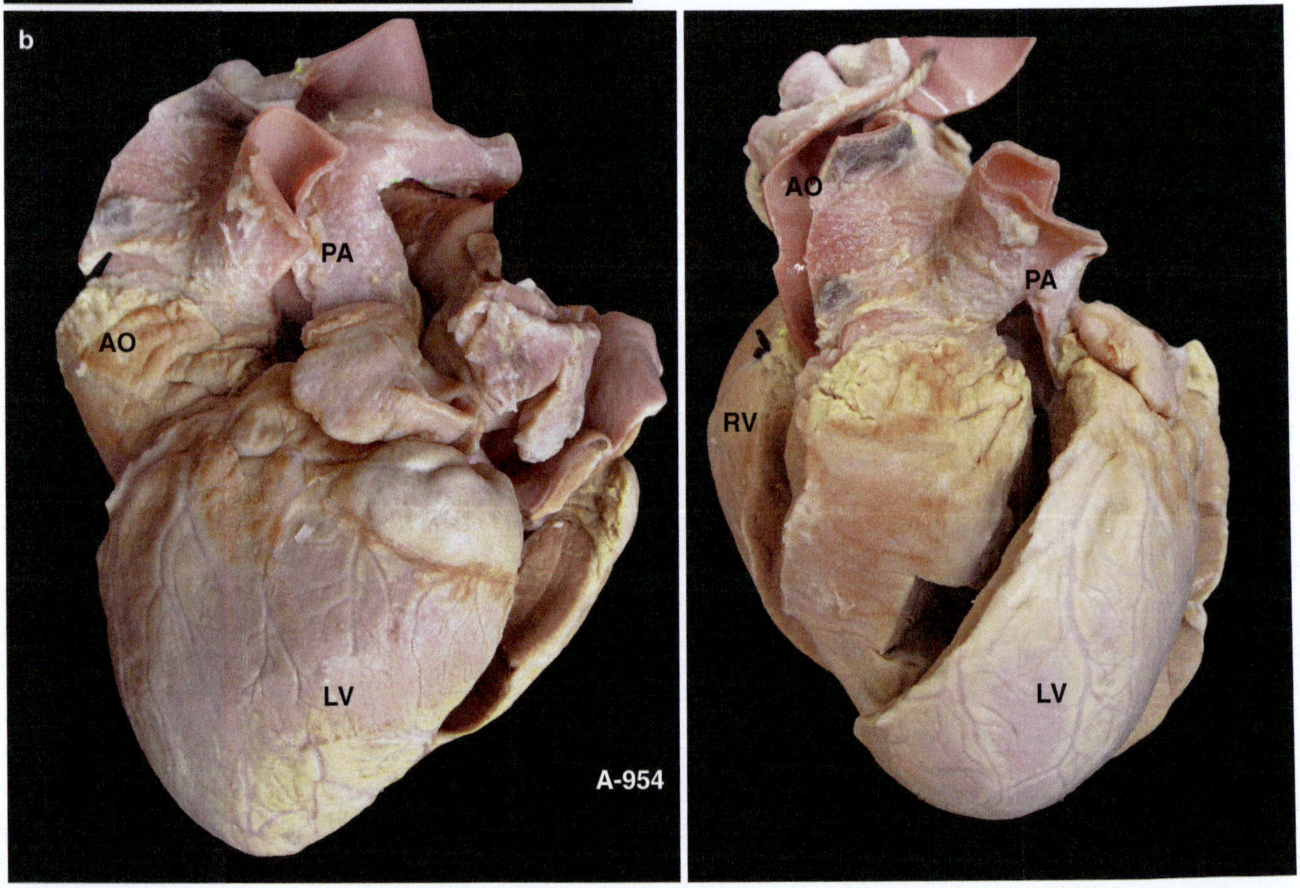

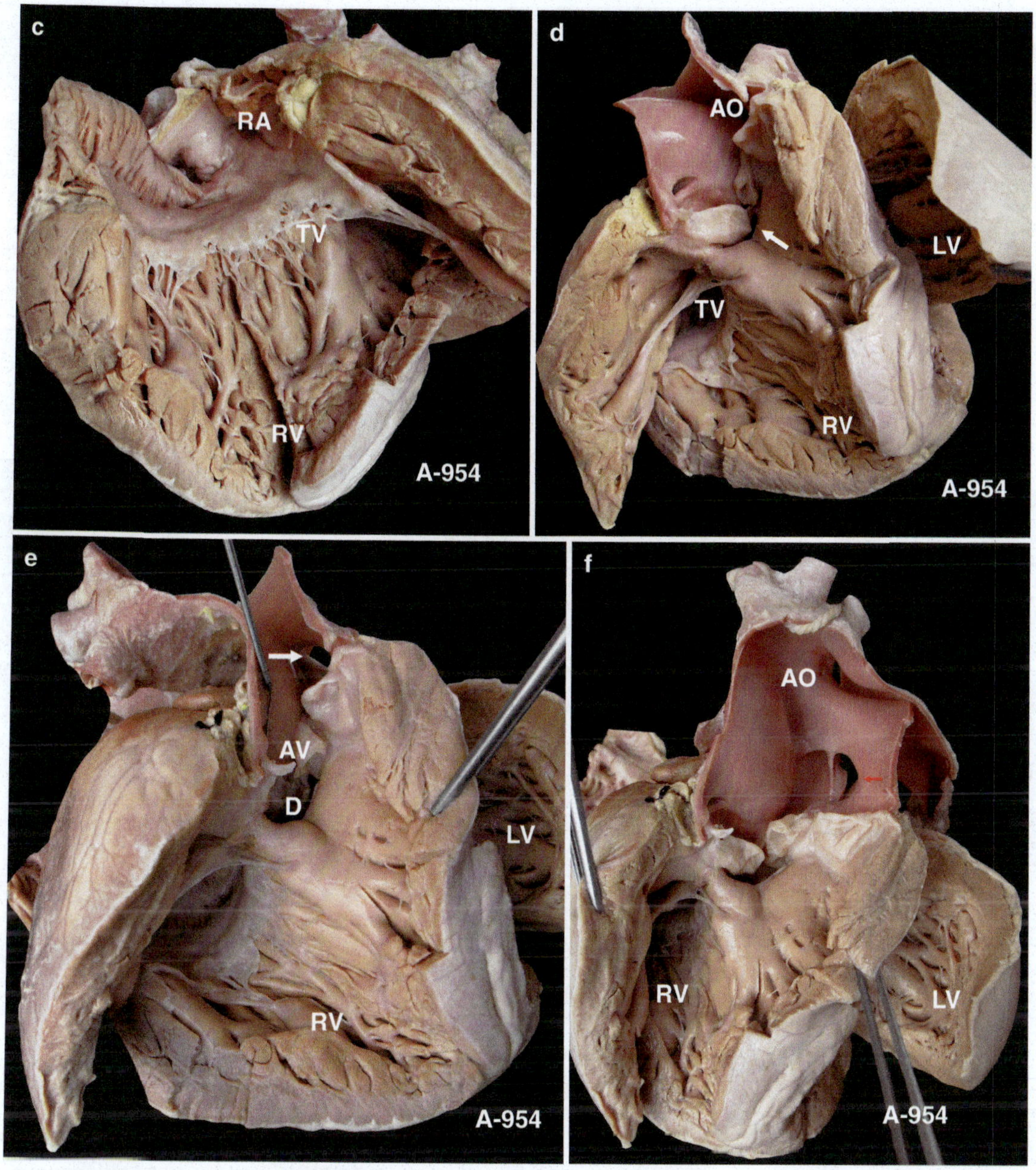

Fig. 21.1 (continued)

Fig. 21.1 (**a**) Specimen 954. Aortopulmonary window, complete transposition of great arteries, ventricular septum defect. The root of the aorta and pulmonary artery is a huge aortic trunk in appearance, but each of the arteries has its own isolated opening; meanwhile, the AO is connected to the RV, and the PA is connected to the LV, named as a TGA with VSD. (**b**) Specimen 954. Different aspects of the abnormal connection of the aorta and the ventricle, just as complete transposition of great arteries. Aorta and pulmonary artery have a septum defect, but they also have their isolated valves. (**c**) Specimen 954. Illustration of inflow tract of right ventricle. TV is generally normal. (**d**) Specimen 954. Section of outflow tract of the right ventricle. The aorta and the AV rise from the RV completely. Between the AV and the TV is the infundibular muscle. The *arrow* indicates the subaortic VSD. (**e**) Specimen 954. A subaortic ventricular septal defect (*D*) from an aspect of outflow tract of the right ventricle. The *arrow* indicates a lateral opening in the wall of AO, which is the window between the AO and the PA. (**f**) Specimen 954. The *red arrow* indicates the aortopulmonary window from the section of the aortic lumen. (**g**) Specimen 954. The left ventricle section showing that the origin of the pulmonary artery is completely from the left ventricle. The pulmonary and mitral valves have a fibrous connection. The *arrow* indicates the VSD. (**h**) Specimen 954. The probe shows a subpulmonary VSD from the section of the LVOT. (**i**) Specimen 954. The fibrous connection of pulmonary and mitral valves. The aorta and the pulmonary artery have their own valves independently, which is the essential difference from persistent truncus arteriosus. (**j**) Specimen 954. MV is shown as normal. (**k**) Specimen 954. Scheme. The *red arrow* point to A-P window

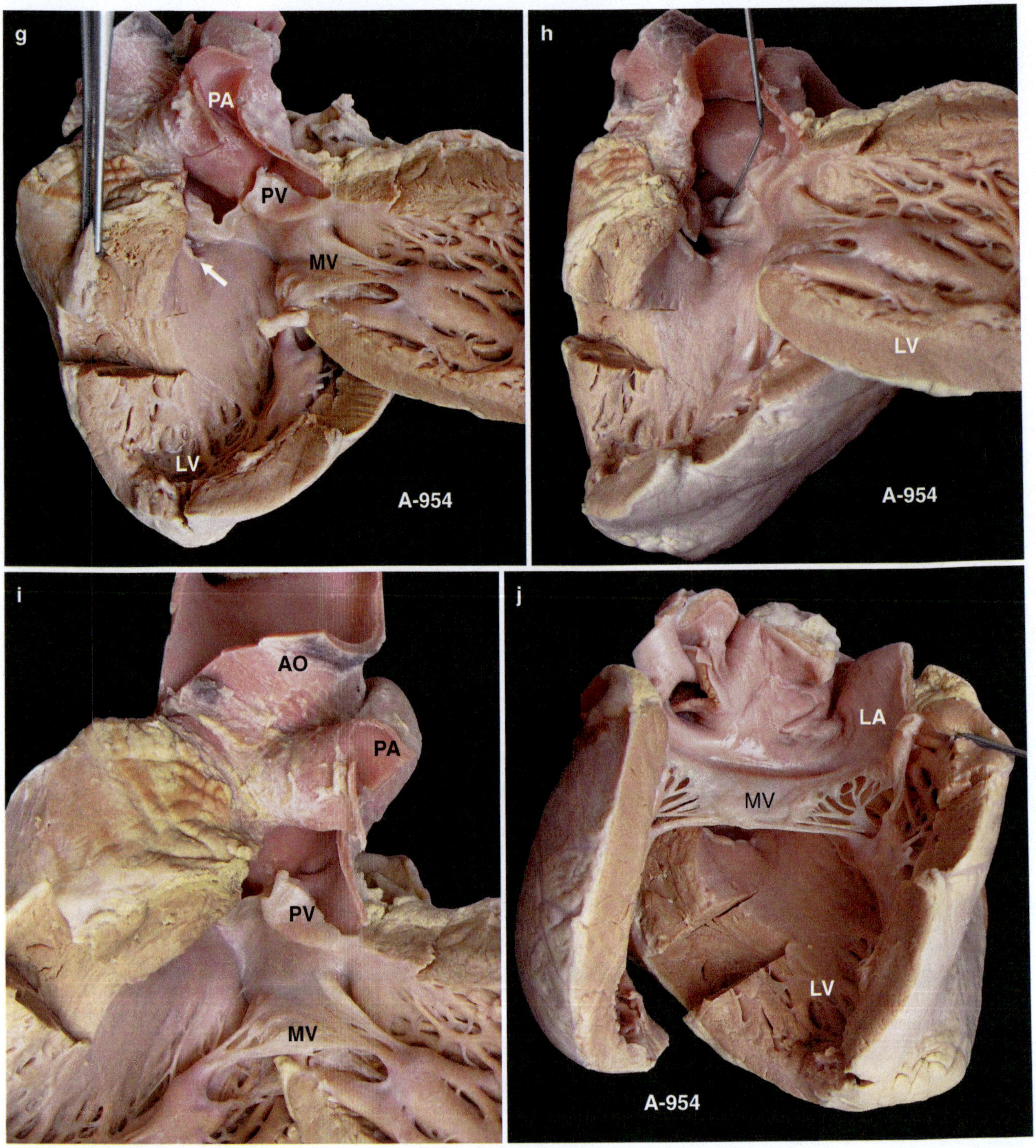

Fig. 21.1 (continued)

k

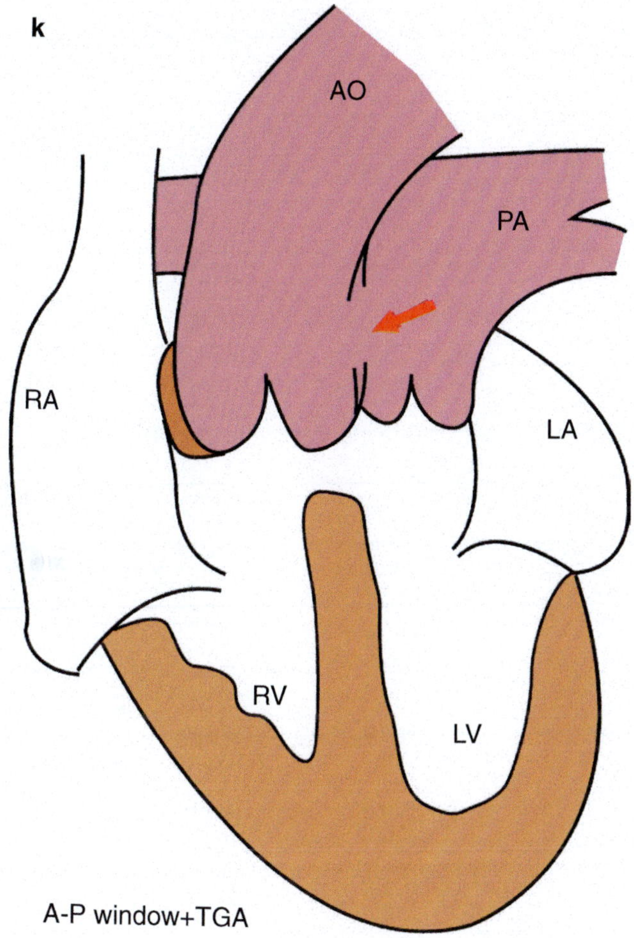

Fig. 21.1 (continued)

Aneurysm of the Sinus of Valsalva

22

22.1 General Considerations

The sinus of Valsalva is a structure that connects the aortic annulus and the aortic wall. Aortic cusps join with the aortic wall at the sinotubular junction. The aortic sinuses (sinuses of Valsalva) are dilated pockets of the aortic root located between the sinotubular junction and the aortic annulus. The walls of the sinuses are considerably thinner, or in some cases their supporting muscular tissues are underdeveloped, with the result that the walls of the sinuses are inclined to dilation and forming of aneurysms under the pressure of blood flow. The dilation and aneurysm is termed *sinus of Valsalva aneurysm*. The wall of the aneurysm continuously becomes thinner under blood pressure and may eventually rupture if factors such as infection and displacement are involved, which is seen most commonly in the right coronary sinus (Fig. 22.1a–d).

With the rupture of the sinus of Valsalva aneurysm, blood drains from the aorta into adjacent heart chambers, leading to hemodynamic disorders. Different rupture locations result in different drainages of the heart chamber (Figs. 22.2 and 22.3). Unruptured aneurysms usually are asymptomatic, unless they block the outflow portion of the RV or fall into the LV, causing aortic insufficiency. A large amount of left-to-right blood flow or severe aortic insufficiency inevitably leads to pulmonary hypertension and heart failure.

22.2 Classifications

Several different classifications of sinus of Valsalva aneurysm, based on clinical practices, have been accepted and applied:

22.2.1. Based on the heart chamber into which the aneurysm ruptures.

22.2.2. Based on the severity of the aortic insufficiency. This is an important factor in surgical considerations. If aortic insufficiency is present and left untreated or even aggravated during surgery, left heart failure and difficulty in heart beat recovery may be induced, in which case aortic valvuloplasty or valve replacement is then needed.

22.2.3. Based on the complication of VSD. Aortic insufficiency presents with a higher rate in aneurysms complicated with VSD. The risk of injuring the AV during surgery to repair the aneurysm is greater if the VSD is complicated, and the surgery approach is different.

In this chapter, one specimen of an aneurysm of the right sinus of Valsalva after surgical repair is shown. The diameter of the rupture is approximately 1.5 cm. The aneurysm has a very thin wall, complicated with a crescent-shaped subarterial VSD, and its rupture has been repaired with a patch on the right ventricular side. An incision of the right ventricular outflow portion is usually made to obtain an adequate exposure of the aneurysm for surgical repair, and sometimes, incision of the ascending aorta is also necessary.

X. Zhu (ed.), *Surgical Atlas of Cardiac Anatomy*,
DOI 10.1007/978-94-017-9409-1_22, © Springer Science+Business Media Dordrecht and People's Medical Publishing House 2015

22.3 Specimen Demonstrations

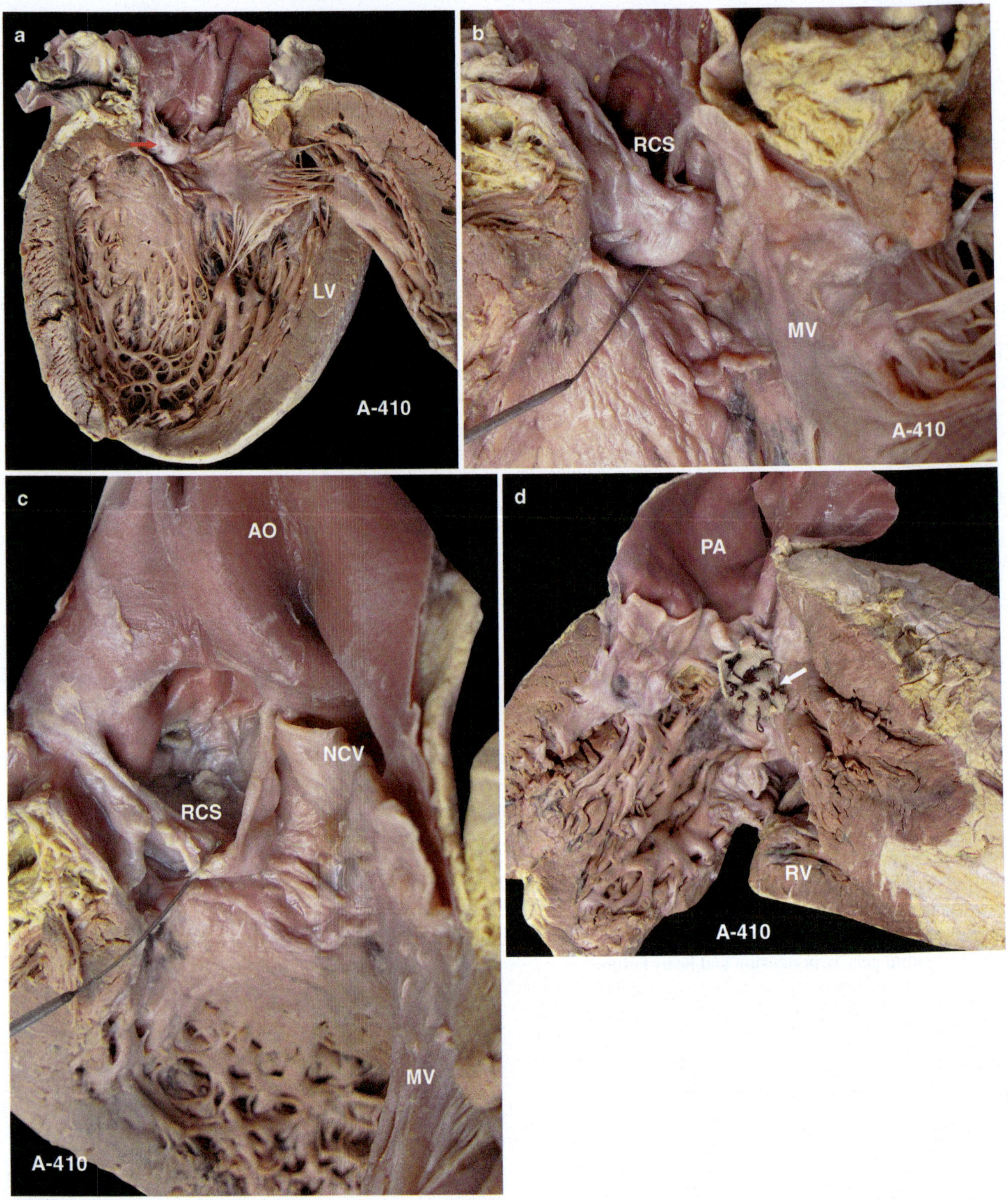

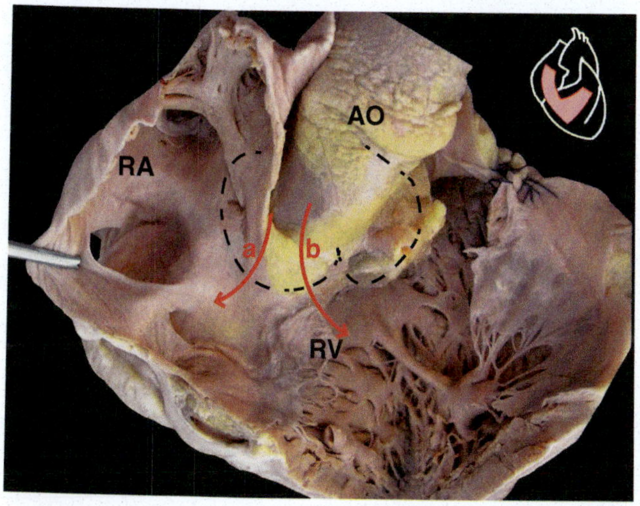

Fig. 22.2 Relationship between the normal sinus of Valsalva and the right chambers. The *dashed lines* show the contour of the sinus of Valsalva. *Red arrows* indicate the directions of the perforation. *a* Rupture into the right atrium, *b* rupture into the right ventricle

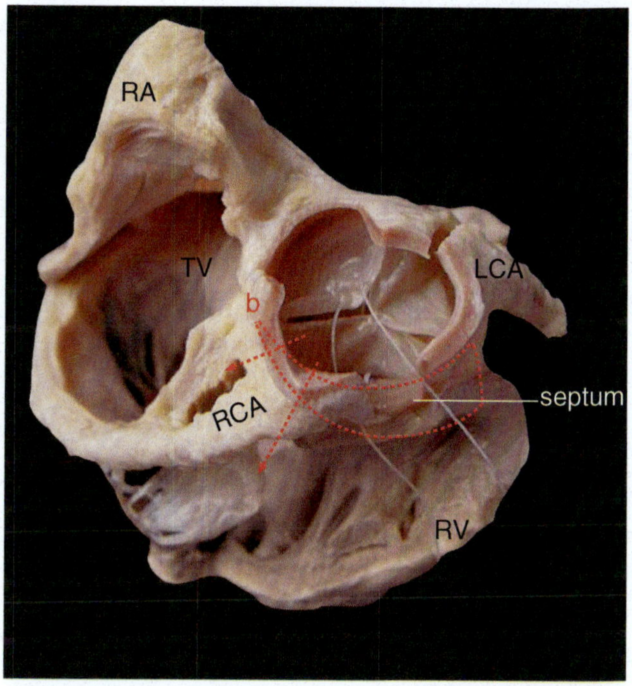

Fig. 22.3 Another normal heart specimen. The *dashed arrows* show the directions in which the right sinus of Valsalva ruptures into the RA or the RV. *b* membranous septum, *LCA* left coronary artery, *RA* right atrium, *RCA* right coronary artery, *RV* right ventricle; septum, ventricular septum

Fig. 22.1 (**a**) Specimen 410. Right coronary sinus of Valsalva aneurysm. The right coronary cusp enlarges and prolapses, with a VSD underlying the leaflet, which is a crescent fissure and repaired from the right ventricular side. The rupture point of the sinus aneurysm has been sewn from the interior of the sinus (which is not shown in this figure). *Arrow* indicates the right coronary cusp. (**b**) Specimen 410. The inferior part of the right coronary sinus (RCS) of Valsalva enlarges and partly protrudes into the VSD. The probe points at the fissure of VSD, *MV* Mitral valve. (**c**) Specimen 410. The suture in the right coronary sinus View from inside the aortic sinus shows the operation technique. The RVOT and the aortic root may be opened at the same time during the operation. The perforation is sewn from the sinus using interrupted pledgeted mattress sutures and covered with a polyester patch from the right ventricular side in order to prevent residual shunt caused by postoperative rupture due to the thin-walled sinus. *NCV* noncoronary valve, *RCS* right coronary sinus (right sinus). (**d**) Specimen 410. Sinus of Valsalva aneurysm ruptured into the right ventricle (postoperative). The perforation of the sinus of Valsalva aneurysm into the RV is above the crista supraventricularis, which is repaired with a polyester patch. The arrow points to the interrupted mattress sutures

Tetralogy of Fallot

23

23.1 General Considerations

23.1.1 Embryology and Pathology

The diagnosis criteria, with regard to pathology and anatomy, of tetralogy of Fallot (TOF) include (1) the aorta overrides on both the left and the right chambers, when the level of the AV orifice is heaved higher (Fig. 23.11a–c); (2) subaortic VSD is present; (3) right ventricular infundibulum stenosis or complication with PV stenosis is present; (4) fibrous connections remain between the aortic and mitral valves, when the aortic and tricuspid valves are joined with fiber or separated by muscle; and (5) right ventricular hypertrophy is present.

The embryological basis of TOF is malformation of the conotruncus:

1. The rotation of the conotruncus is not adequate. The aorta is not completely connected to the LV. It is biventricular in origin and overrides on the ventricular septum, which is a result of the AV not adequately moving to the left, posterior, and inferior of the PV from the position at the early embryonic stage, when both valves are at the same level as the ventricles (AV is on the right, and PV is on the left).
2. The conotruncus does not divide equally, and the pulmonary artery is smaller than the aorta (the same size at normal condition).
3. The conal septum does not join with the membranous ventricular septum and the muscular ventricular septum to close the interventricular foramen, and a subaortic VSD is in position. The subpulmonary infundibular hypoplasia, the right ventricular infundibulum shortening and stenosis, and the inadequate rotation of the conotruncus lead to the subaortic VSD and the aorta overriding. During the fetus period, the pressures of the right and left ventricles are equal, and the right ventricular hypertrophy is acquired.

23.2 Key Points in the Anatomy

In the normal condition, a longitudinal line through the midpoint of the anterior mitral leaflet, termed *midline of mitral valve*, always goes across the commissure of the left coronary aortic cusp and noncoronary cusp. In TOF, the commissure displaces rightward and leaves the midline when the aorta rotates rightward and anteriorly. With regard to the relationship between the membranous septum and AV, the membranous septum is immediately below the commissure of the right coronary aortic cusp and the noncoronary cusp. In the case of TOF, the commissure rotates rightward and anteriorly, and the septum is below the noncoronary cusp.

The aortic annulus is lifted higher: the VSD is subaortic and equal to an intracristal VSD, the inferior edge of which is formed by a muscular septum and a septal band. The aortic annulus overrides rightward and forms the superior edge of the VSD when the anterior edge is the anteriorly displaced parietal band. Hence, the VSD in TOF might be considered an infundibular VSD.

23.2.1 **Features of the VSD** (Figs. 23.12a–j, 23.13a–d, and 23.7a–d)

The superior edge of the VSD is the connection portion of the aortic annulus and the front wall of the RV, which is equal to the inferior of the right coronary aortic cusp or the noncoronary cusp or both. In all the specimens, this edge is composed of muscular tissue, and the anterior end of the muscular edge extends to the parietal band when the posterior end is near the TV annulus.

The anterior edge is the anteriorly displaced parietal end of the crista supraventricularis, which arises from the front portion and extends to the free wall of the RV to form the posterior border of the RVOT. This edge is composed of muscular tissue in all the specimens.

X. Zhu (ed.), *Surgical Atlas of Cardiac Anatomy*,
DOI 10.1007/978-94-017-9409-1_23, © Springer Science+Business Media Dordrecht and People's Medical Publishing House 2015

The posterior edge descends from the aortic annulus, then crosses the front of the TV annulus and reaches to the posterior end of the ventricular septum. It is actually part of the RV free wall in front of the TV annulus. The muscle bundle (or fiber bundle) of the edge is equal to the posterior limb of the parietal band. Intracristal VSDs occur commonly in TOF, but occasionally, the VSD extends to the membranous septum.

From the surgical point, the posterior edge of the VSD is very important because it is near the atrioventricular bundle. Based on the strategy of the operation, VSDs in TOF might be divided into two types:

1. There is no muscular bundle between the posterior edge of the VSD and the TV annulus, and the posterior edge is composed of the fibrous connections between the tricuspid and the aortic annuli. With the malformation of the membranous septum, the atrioventricular bundle passes along the border of the membranous septum, which should be avoided during the repair of the VSD.
2. There are muscular bundles between the posterior edge of the VSD and the TV annulus, and muscles surround the VSD completely; therefore, the aorta is separated from the tricuspid annulus. A well-developed membranous septum is located behind the muscular bundle, in which case there is no conduction bundle, and it is safe to sew the rim of the muscles directly during repair. The others belong to the intermediate type.

23.2.2 Features of Pulmonary Artery Stenosis

The infundibular stenosis is sometimes associated with PV stenosis and stenosis of the pulmonary artery trunk or the branches (Fig. 23.8a–d). The stenosis of the valve might be a result of bicuspid or commissural fusion. Because of the infundibular stenosis, dilation of the pulmonary artery after the stenosis is rarely seen in clinic, and the diameter of the pulmonary trunk is smaller than that of the ascending aorta. Because different surgical strategies are selected based on the type and extent of the infundibular stenosis, it is important that these factors be estimated preoperatively:

1. The stenosis is confined and located at the proximal part of the infundibulum, forming a "third ventricle." With a well-developed pulmonary annulus, the RVOT obstruction can be relieved simply by excising the hypertrophic crista supraventricularis (Fig. 23.5a–c).
2. The stenosis is diffuse, and a patch in the RVOT is always necessary to enlarge the pulmonary artery during the surgery (Figs. 23.6a–f, 23.14a–b, 23.15a–d, and 23.2a–c).
3. In the malformation or the absence of the infundibulum, a valve conduit is commonly needed to link the RV to the pulmonary trunk to correct the pulmonary valvular stenosis or the pseudotruncus arteriosus due to pulmonary valve atresia (Fig. 23.24a–f). As the burden increases after birth, the patients' RV becomes increasingly hypertrophied, which adds to the RVOT obstruction. As the coronary arteries on the surface of the RV become larger, tortuous, and dilated, making an incision into the RV safely becomes more difficult.

23.3 Technical Risks with Repair of TOF

The basic principles of repair of TOF are to relieve the ROVT obstruction and the PV stenosis and to repair the VSD (Figs. 23.19a–f, 23.9a–e, 23.16a–f, 23.21a–f, 23.3a–e, 23.4, and 23.1). The most common risks include (Fig. 23.17a–e):

1. There are often many aortopulmonary collateral arteries (APCAs) in adult patients. If this condition is associated with patent ductus arteriosus, a large amount of blood flow can drain into the pulmonary artery, even after an aorta clamp is placed, which not only affects the view of the surgery field but also causes destruction of the blood.
2. Large coronary artery branches on the surface of the RV can be damaged when the right ventriculotomy is made, which might cause localized myocardial infarction (Fig. 23.23a–e).
3. Residual RVOT obstruction may occur. One reason is that the RVOT muscular obstruction is inadequately relieved, and the other is that repairing the subaortic VSD might induce RVOT obstruction, requiring that the RVOT be widened by using a patch.
4. Residual pulmonary artery and RVOT obstructions may occur, usually the result of not being adequately widened by using a patch (Fig. 23.8a–d).
5. PV regurgitation and RV dysfunction may result. With hypoplasia PV, enlarging a pulmonary annulus might cause severe regurgitation, so the PV homograft is recommended.
6. Residual shunt of the VSD may occur, usually at the transitional portion of the joint between the TV annulus and the ventricular septum, where the sutures on the patch often are unsteady and continuous suture may tear the muscular tissue in infants (Figs. 23.18a–f and 23.22a–e).
7. Damage to the AV may occur. The AV might be injured during resection of the RVOT muscles and repair of the VSD, resulting in severe regurgitation.
8. Complete atrioventricular block (Fig. 23.20a–b).

23.4 Specimen Demonstrations

23.4.1 Typical Postoperative Specimen of TOF Showing the Override of the Aortic Valve

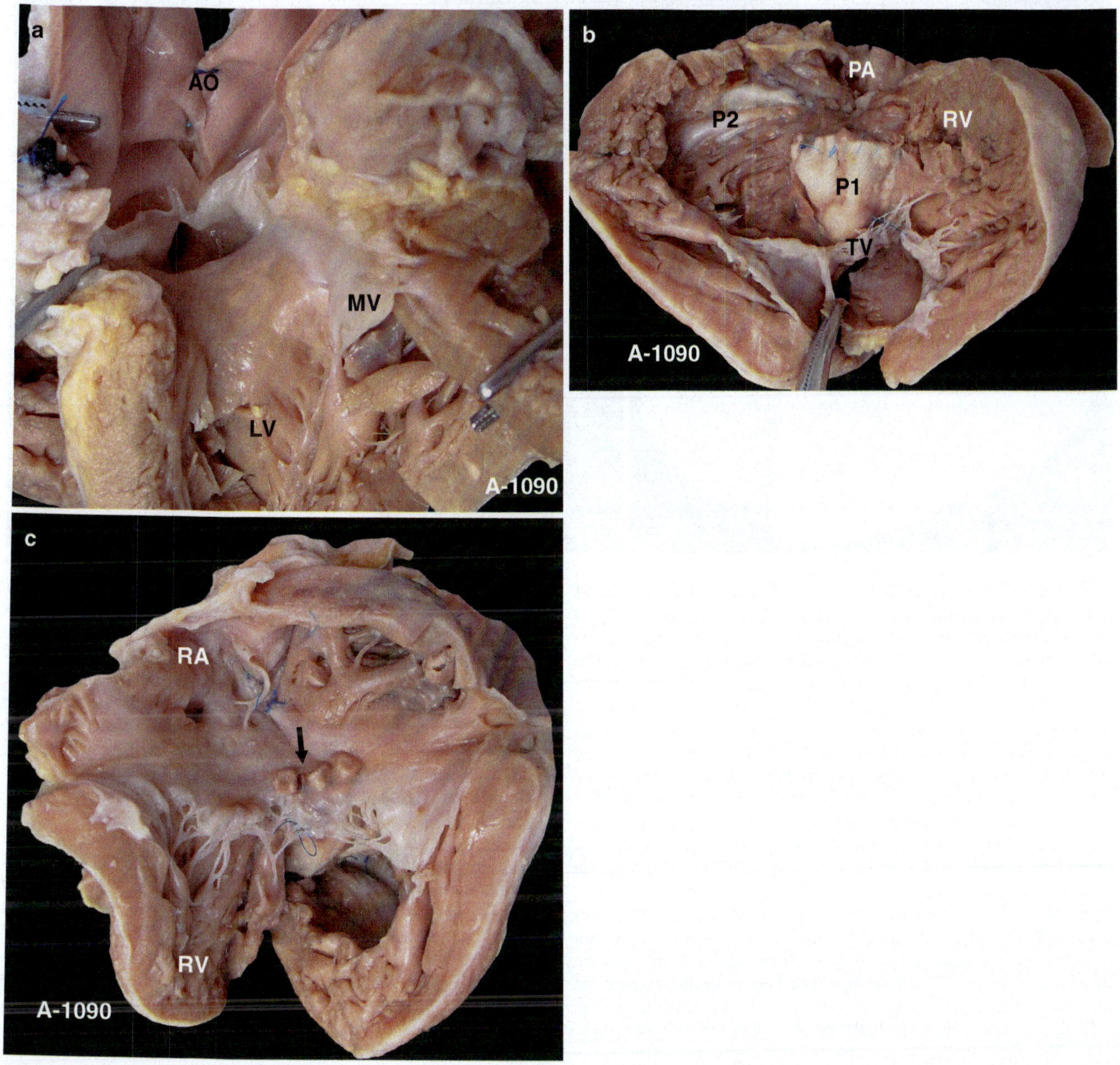

Fig. 23.1 (**a**) Specimen 1090. TOF. Viewed from the LV, two cusps of the AV can be seen to override across the VSD, which infers a severe 70 % override. However, the AV retains fibrous continuity with the MV, which distinguishes this case from the DORV. *AO* aorta, *LV* left ventricle, *MV* mitral valve. (**b**) Specimen 1090. Viewed from the front side of the heart, the large VSD patch (*P1*) is seen adjacent to the area above of the crista supraventricularis. The "P2" marks the RVOT expanding patch. The muscular bands of the RVOT are transected. *RV* right ventricle, *TV* tricuspid valve. (**c**) Specimen 1090. The *arrow* shows the correct position of the reinforcement pledgets used to patch the VSD via the RA

23.4.2 TOF, the VSD Patch Removed to Show the Stitched Edge of the AV and the VSD

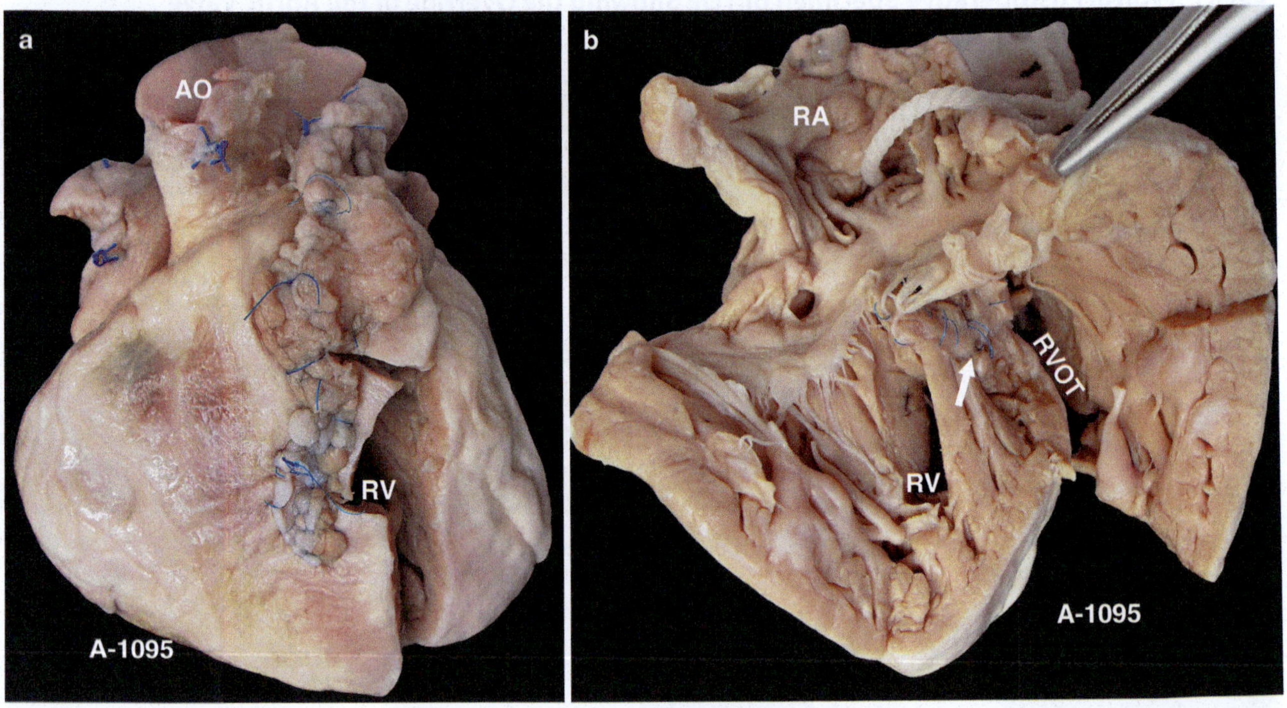

Fig. 23.2 (**a**) Specimen 1095. External view of the heart after repair of TOF. The RVOT is enlarged by a transannular patch, and the RVOT and the PV annulus are exposed through an incision carried along the ventricular septum. *AO* aorta, *RV* right ventricle. (**b**) Specimen 1095. The RA (RA), RV (RV), and TV can be seen simultaneously through the RVOT. The VSD patch is removed, but the sutures of the inferior rim of the VSD remain. The *arrow* points at the blue sutures, among which one entangles the chordae tendineae of the TV. The "RVOT" shows the narrowing of the RVOT. (**c**) Specimen 1095. The right ventricle and the left ventricle cut open along both sides of the ventricular septum. The transannular patch crossing from the RVOT to the PA is totally demonstrated. Meanwhile, the aorta (*AO*) enlargement and the pulmonary artery (*PA*) stenosis also can be seen. (**d**) Specimen 1095. The *dash line* shows the border of the patch from the RV to the pulmonary artery, and the *arrow* indicates the transected muscles of the parietal band. (**e**) Specimen 1095. The relationship between the aortic valve and the rim of the ventricular septal defect. Both the front wall of the RV and the ascending aorta are incised and then raised above to fully expose the override of the AV and the exact position of the stitched edge of the VSD patch. The *red arrow* indicates the upper edge of the patch. Residual polyester patch remains in front of the *left red arrow*, which indicates that the AV may be wounded here by the suture. The *white arrows* indicate the lower edge of the VSD patch. (**f**) Specimen 1095. Diagram of Fig. 23.12e. The *dashed line* indicates the incision. *PB* parietal band, *SB* septal band. (**g**) Specimen 1095. Coronal plane view of the heart to expose the override of the aortic valve and the stitched edge of the ventricle septal defect patch. The *red arrow* denotes that the AV might be wounded here easily by the suture. *a* orifice of coronary artery, *b* residual patch of the VSD, *c* residual sutures on the lower edge of the VSD, *d* VSD. (**h**) Specimen 1095. The image shows that the RVOT, especially the muscles of parietal band (*arrow*), lies adjacent to the AV leaflets, which can be injured during resection of the hypertrophied muscles of the RVOT. (**i**) Specimen 1095. Demonstration of the override of the aorta and direction of the right-to-left shunt. The *blue dashed line* shows that the blood flow is drained from the RV into the aorta, and the red bougie represents the LVOT. (**j**) Specimen 1095. Although two leaflets of the AV override onto the RV, there are fibrous connections between the aortic and mitral valves. Hence, the case is diagnosed as TOF rather than DORV

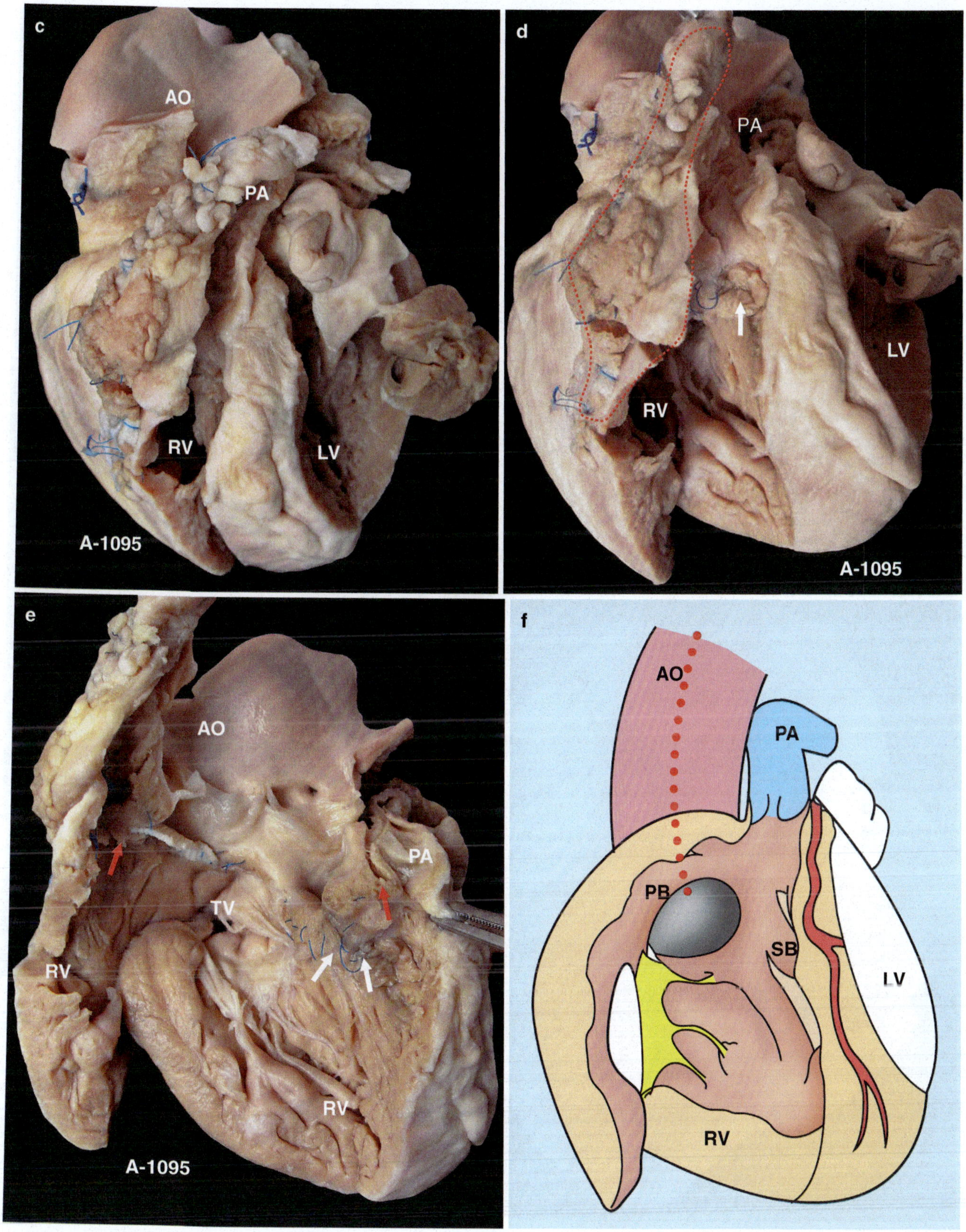

Fig. 23.2 (continued)

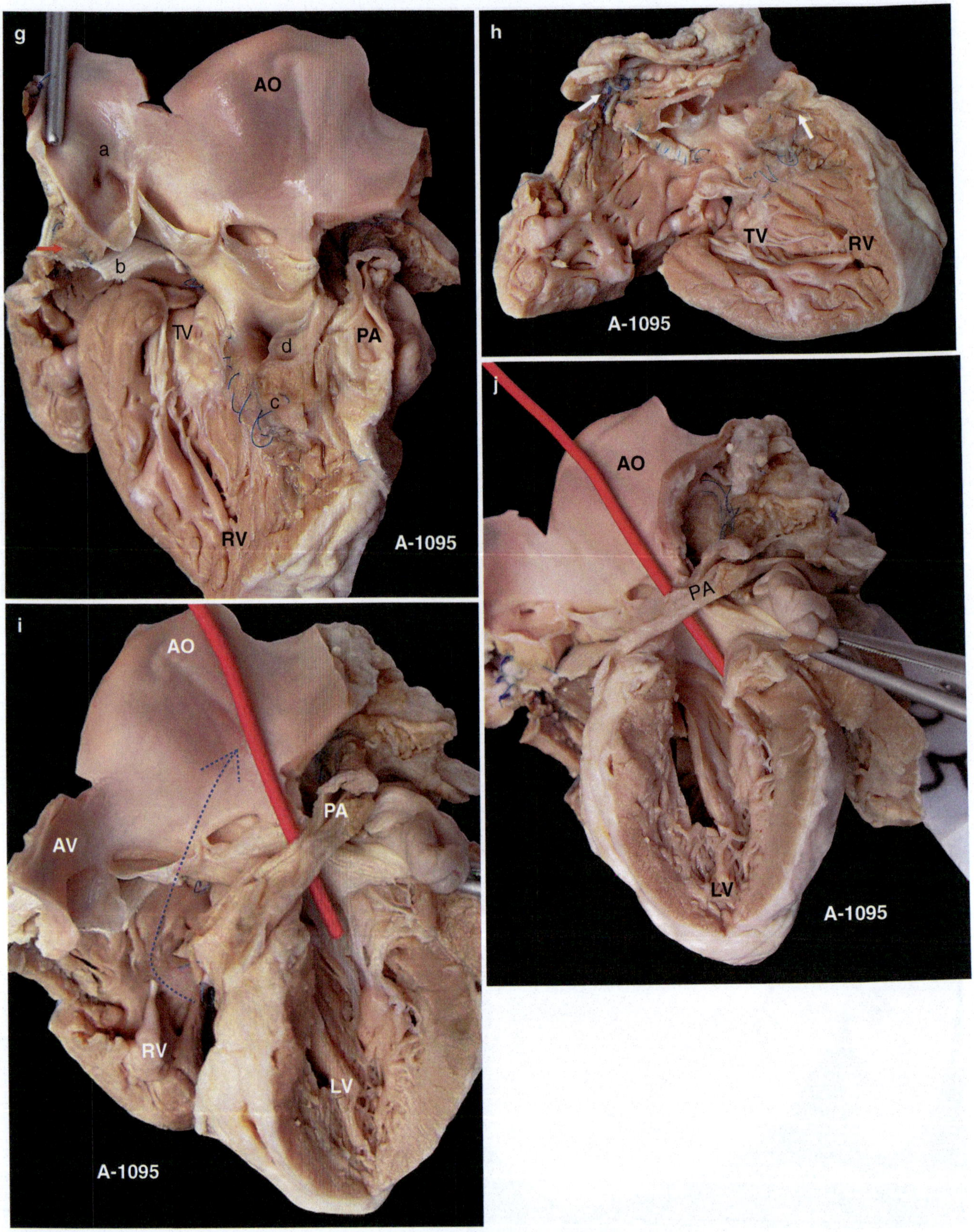

Fig. 23.2 (continued)

23.4.3 View of Intact VSD in TOF

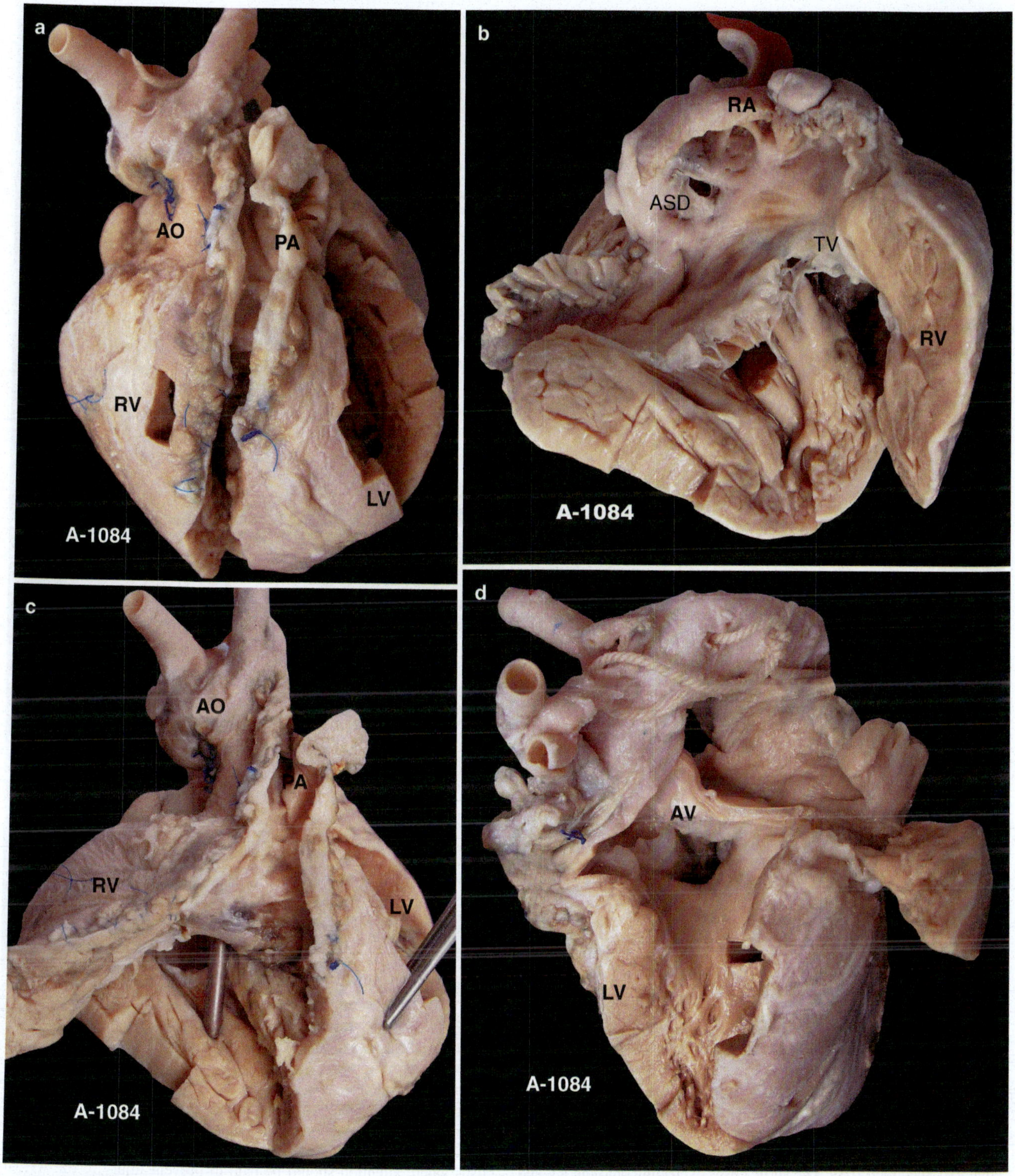

Fig. 23.3 (**a**) Specimen 1084. Specimen of postoperative TOF. The VSD patch is removed during the autopsy to show an overview of the VSD. This is a front view of the specimen. *AO* aorta, *LF* left ventricle, *PA* pulmonary artery, *RV* right ventricle. (**b**) Specimen 1084. An ASD is accompanied. The RV is obviously hypertrophied. *RA* right atrium, *TV* tricuspid valve. (**c**) Specimen 1084. The patch expanding the front wall of the RV is cut open to reveal the VSD below the crista supraventricularis. Remnant polyester pledgets remain on the edge of the VSD, but the patch is removed. The bougie is conducted across the VSD into the RV from the ascending aorta. (**d**) Specimen 1084. The VSD and the overriding aortic valve viewed from the left ventricle side. There is no trace indicating a repair on the left side of the VSD

23.4.4 Repair of TOF, Cut Across the VSD Patch

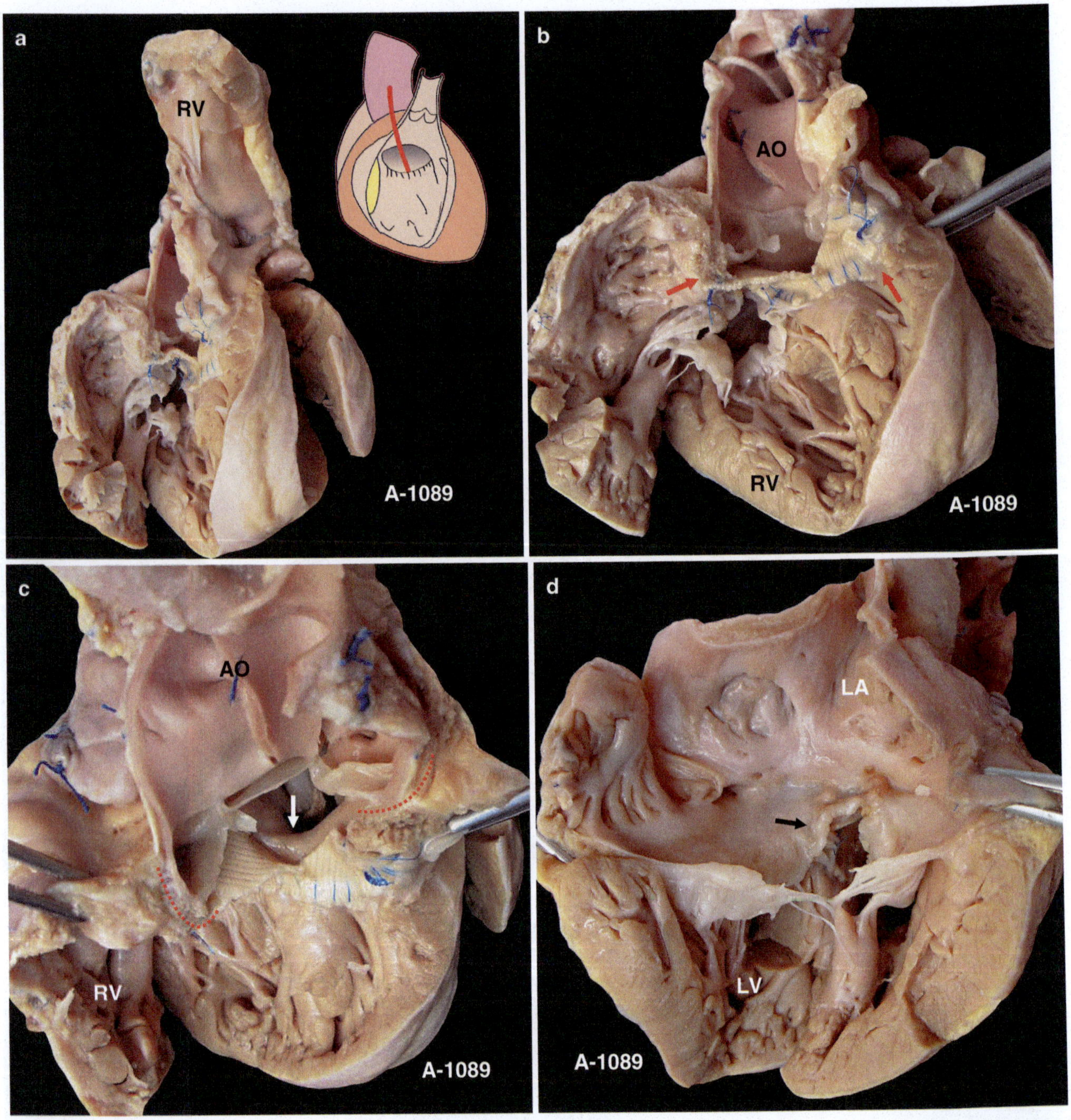

Fig. 23.4 (**a**) Specimen 1089. View from the opened right ventricle with an incision carried across the front of the ascending aorta and the ventriculoinfundibular fold. The front wall of the RV is lifted up by cutting apart the aortic annulus and across the VSD patch from the upper edge. (**b**) Specimen 1089. The aortic cusps overriding onto the RV and the trace of repair on the inferior edge of the VSD patch are demonstrated by retracting the rim of the incision of the aortic annulus and the crista supraventricularis. The areas to which the *red arrows* point are far from the conduction bundle and are safe for suturing. However, they are close to the aortic cusps, so care should be taken to avoid injuring the aortic cusps when relieving the RVOT obstruction. (**c**) Specimen 1089. The *white arrow* indicates the lower edge of the VSD and position of the patch. The *red dash lines* show the location of the overriding aortic cusps where attention should be paid to prevent injury to the AV. (**d**) Specimen 1089. The thickened and curled leaflet near the posterior commissure (*arrow*) infers that this case of TOF is accompanied by an acquired lesion of the MV

23.4.5 TOF with Distal Pulmonary Artery Stenosis and Patent Ductus Arteriosus (PDA)

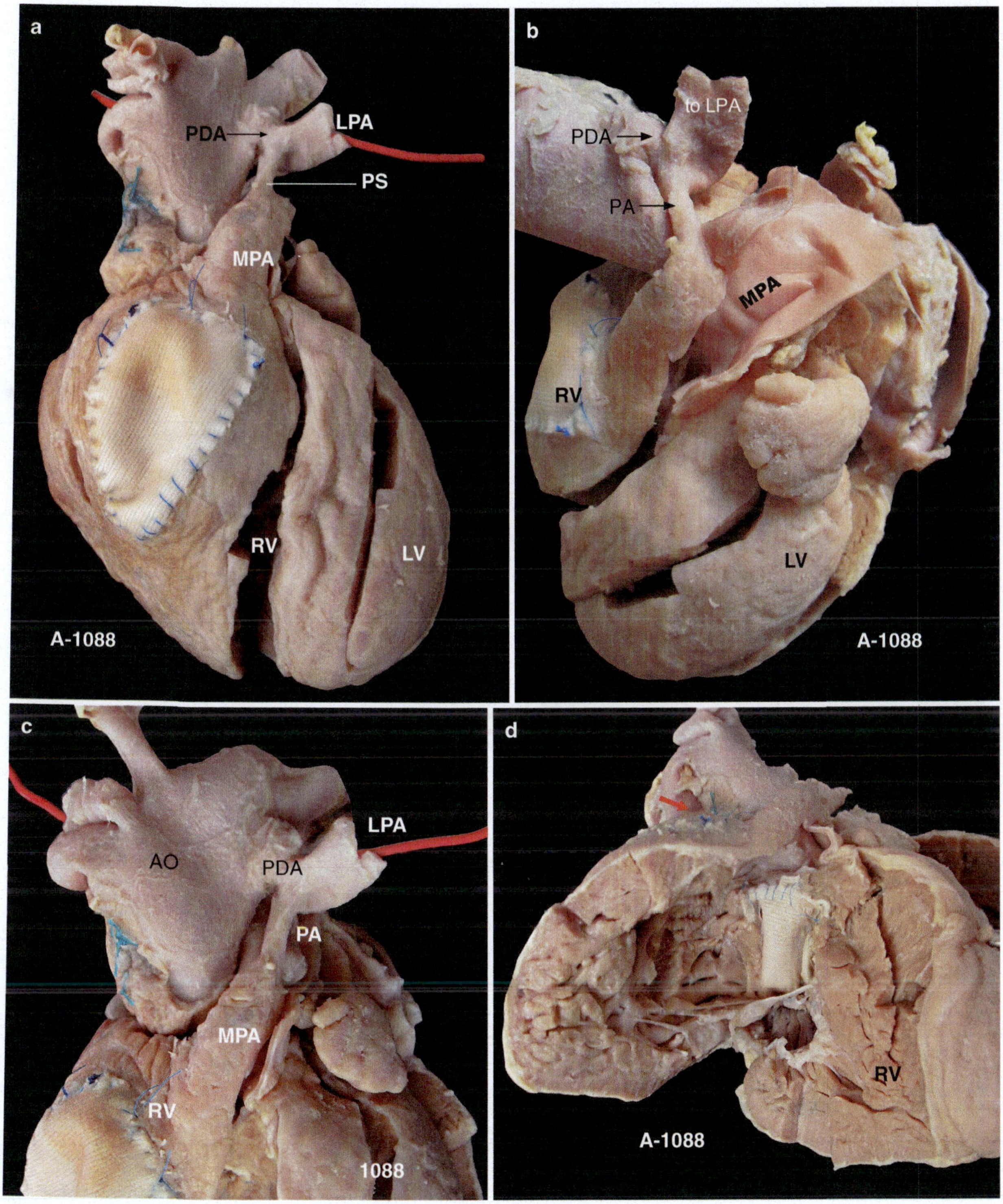

Fig. 23.5 (**a**) Specimen 1088. TOF, PDA, right pulmonary artery absence. A patch expanding the RVOT can be seen in front of the RV. The diameter of the original part of the main pulmonary artery (MPA) is approximately half that of the ascending aorta, but the vessel is obviously narrow between the MPA and the left pulmonary artery (*LPA*), which has a diameter of only 8 mm. The right pulmonary artery (RPA) is absent. The red bougie goes from the LPA into the ascending aorta through the PDA. (**b**) Specimen 1088. The right pulmonary artery is absent. The ascending aorta is drained into the LPA by the PDA. The proximal part of the MPA is free from stenosis as it is extremely narrow at the distal part, which should be expanded during the operation. (**c**) Specimen 1088. Anteriorly exposed relationship between the aorta (*AO*) and the left pulmonary artery (*LPA*). *MPA* main pulmonary artery trunk without stenosis, *PA* the stenotic distal part of the main pulmonary artery (the stenosis is not relieved), *PDA* patent ductus arteriosus. (**d**) Specimen 1088. The RVOT is repaired without a transannular patch, and the infundibular muscular bands below the pulmonary valve are inadequately relieved. The *arrow* indicates that the patch of the RVOT is too far away from the PV

23.4.6 Repair of TOF Without Expansion of RVOT

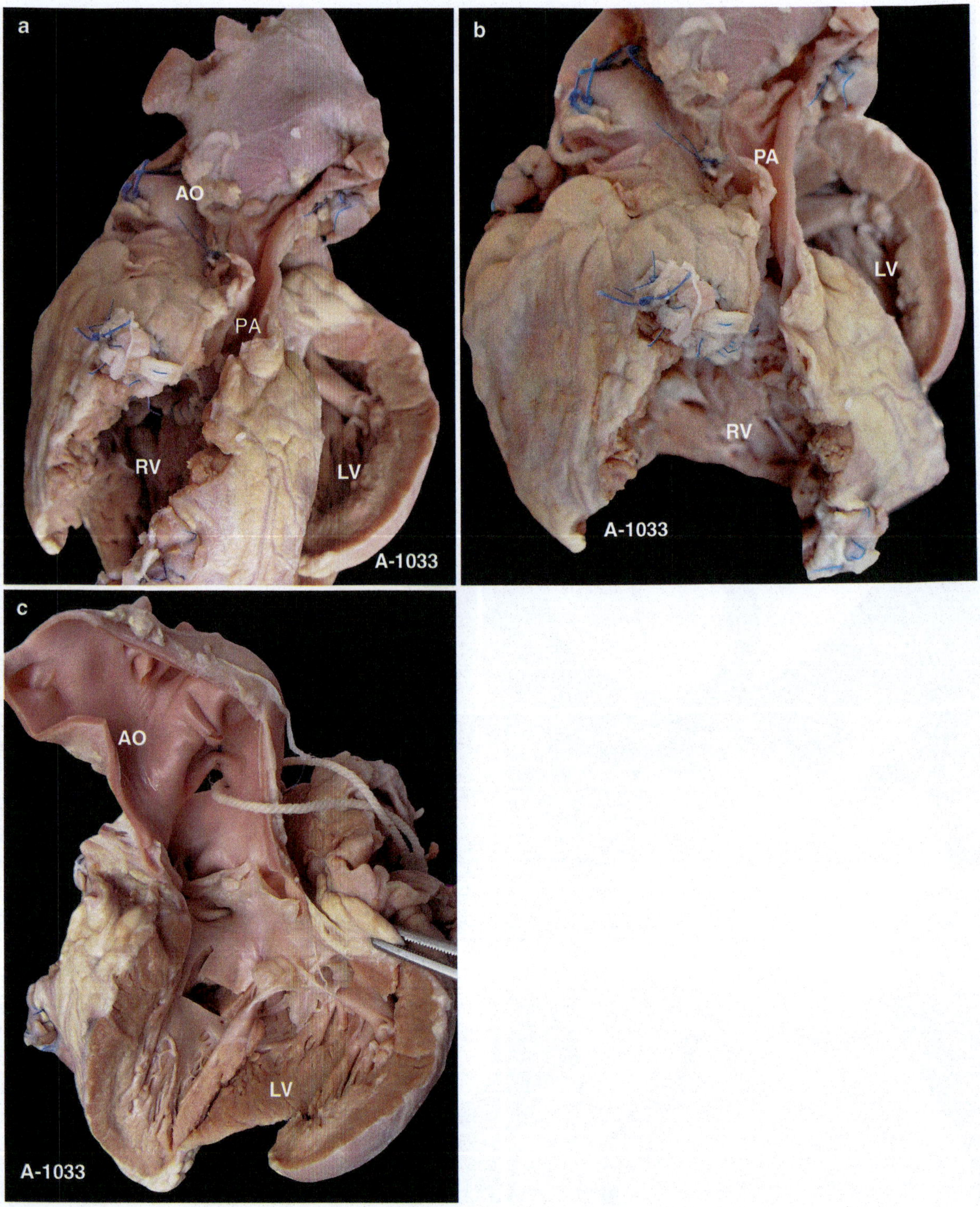

Fig. 23.6 (**a**) Specimen 1033. Left anterior oblique view of TOF. The RV and LV are opened. The aorta (*AO*) is obviously dilated. The pulmonary artery (*PA*) is small, but the leaflets exist. (**b**) Specimen 1033. The relief of the RVOT is adequate, so no patch is needed. Many pledgeted sutures are added to prevent the right ventricular incision from bleeding. The PV has moderate regurgitation, with nearly normal leaflets. (**c**) Specimen 1033. The VSD exposed via the LVOT. The AV overrides more than 50 %. There are fibrous connections between the AV and the MV

23.4.7 Oversized Transannular Patch in TOF

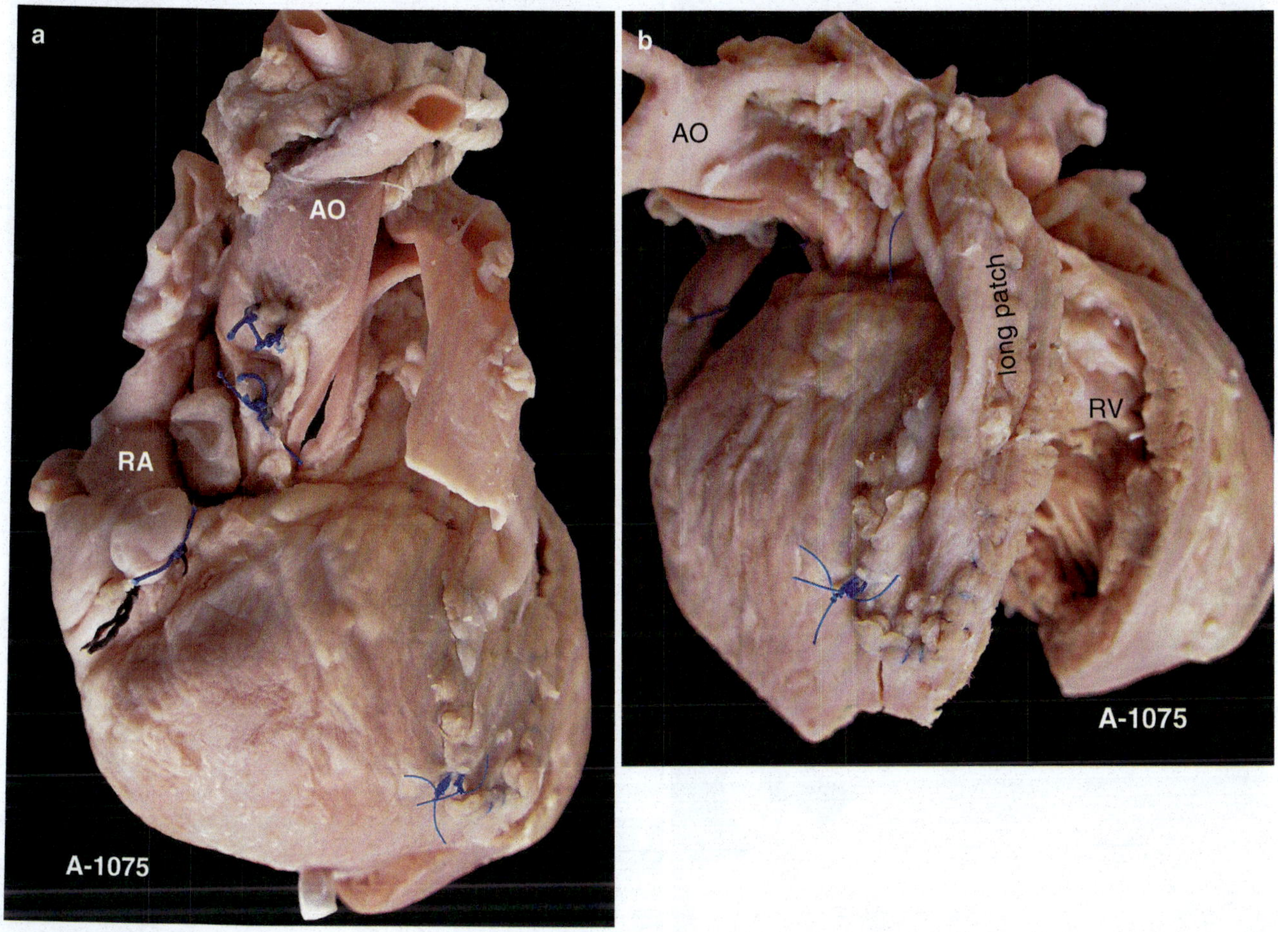

Fig. 23.7 (**a**) Specimen 1075. External view of specimen after repair of TOF. Viewed from the front to observe the patch of the RVOT and the enlarged ascending aorta. The pulmonary artery is sheltered. (**b**) Specimen 1075. The RV and the pulmonary artery are dissected. The RVOT and the PV annulus are expanded by pericardial patch, the lower end and upper end of which reach, respectively, the apex and the pulmonary artery bifurcation. Oversized patch at the front wall of the RV may affect the function of the RV. (**c**) Specimen 1075. A large patch used to expand the RVOT and the pulmonary artery. The image shows the small pulmonary artery and the hypoplasia PV, which have been adequately widened with a large patch. An overbroad transannular patch might cause regurgitation of the PV. *PV* pulmonary valve. (**d**) Specimen 1075. A closer view shows that the PV annulus has been expanded sufficiently, but postoperative severe PV regurgitation might occur due to the hypoplasia of the pulmonary valve (*PV*). (**e**) Specimen 1075. View of left ventricular outflow tract (LVOT). The VSD is below the AV, and two AV leaflets override above the VSD, which means the aorta arises more than 50 % from the RV. (**f**) Specimen 1075. Large ventricular septal defect patch and RVOT patch. The front wall of the RV is lifted in this specimen to show the polyester patch repairing the VSD, beside which are the septal band and the parietal band that have been partly resected. The transannular patch of the RVOT should not be oversized

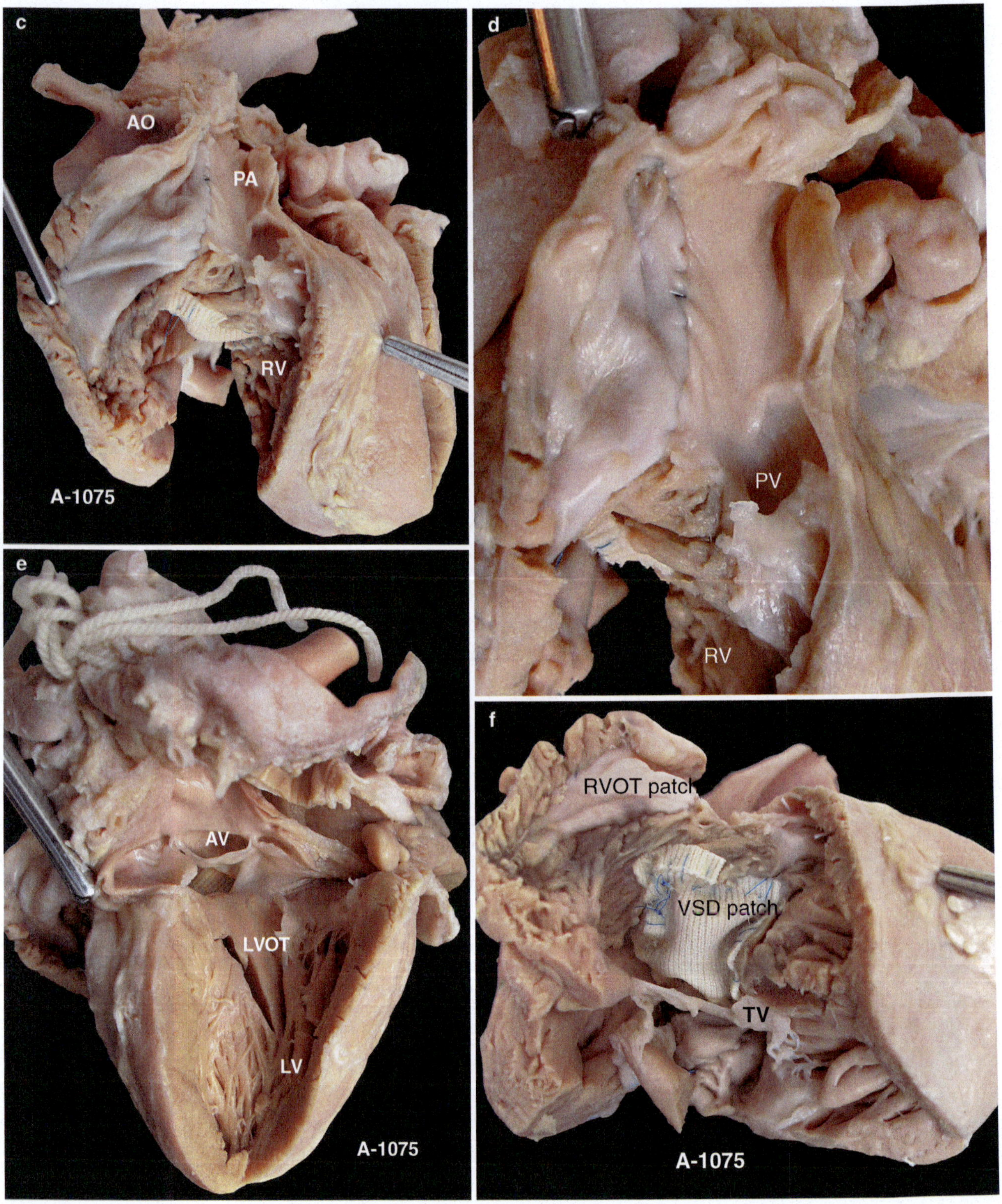

Fig. 23.7 (continued)

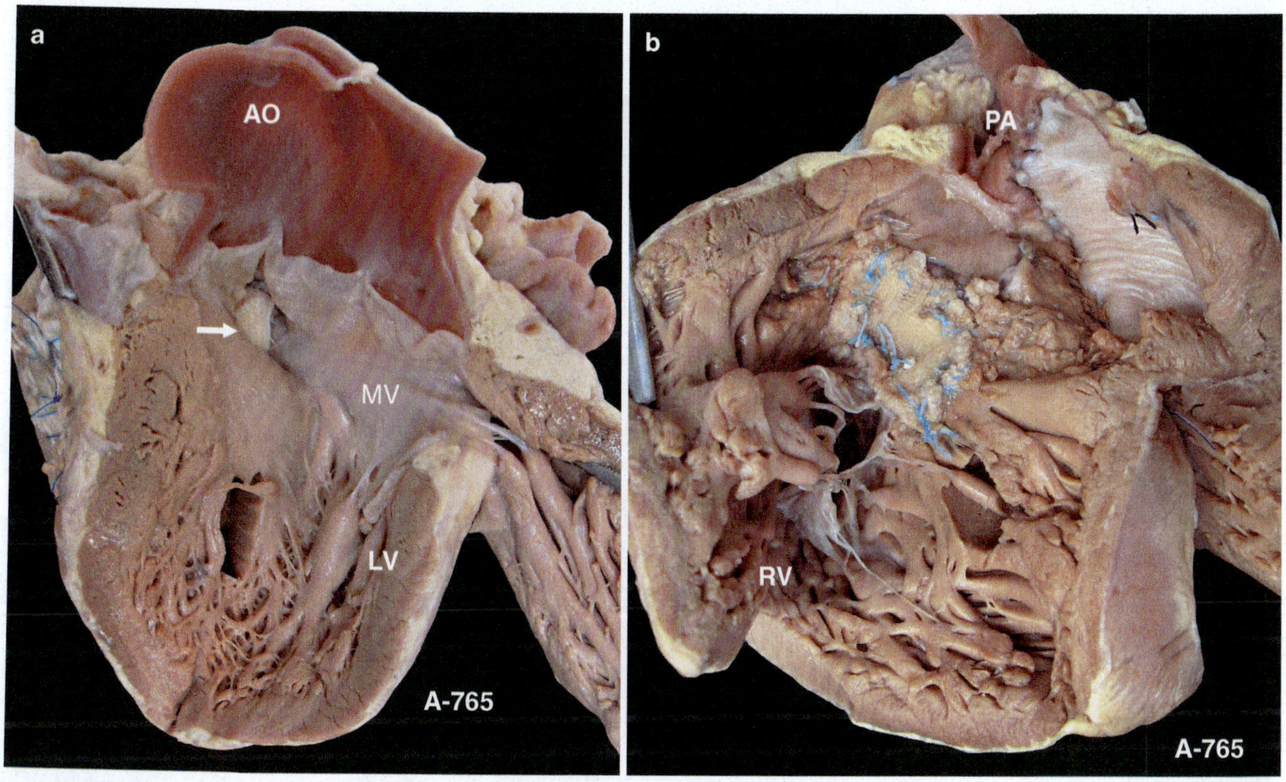

Fig. 23.8 (**a**) Specimen 765. View from the RVOT to observe the VSD and the polyester patch on the RV side (*arrow*) after repair of TOF. *AO* aorta, *LV* left ventricle, *MV* mitral valve. (**b**) Specimen 765. Exposure of the polyester patch repairing the VSD and the artificial vessel patch widening the pulmonary artery via the right ventricle. The hypertrophied muscular bands of the RVOT have been removed, and in order to avoid influencing the RV function, there is no need to use a large patch

23.4.8 Enlarged RV After Expansion of RVOT

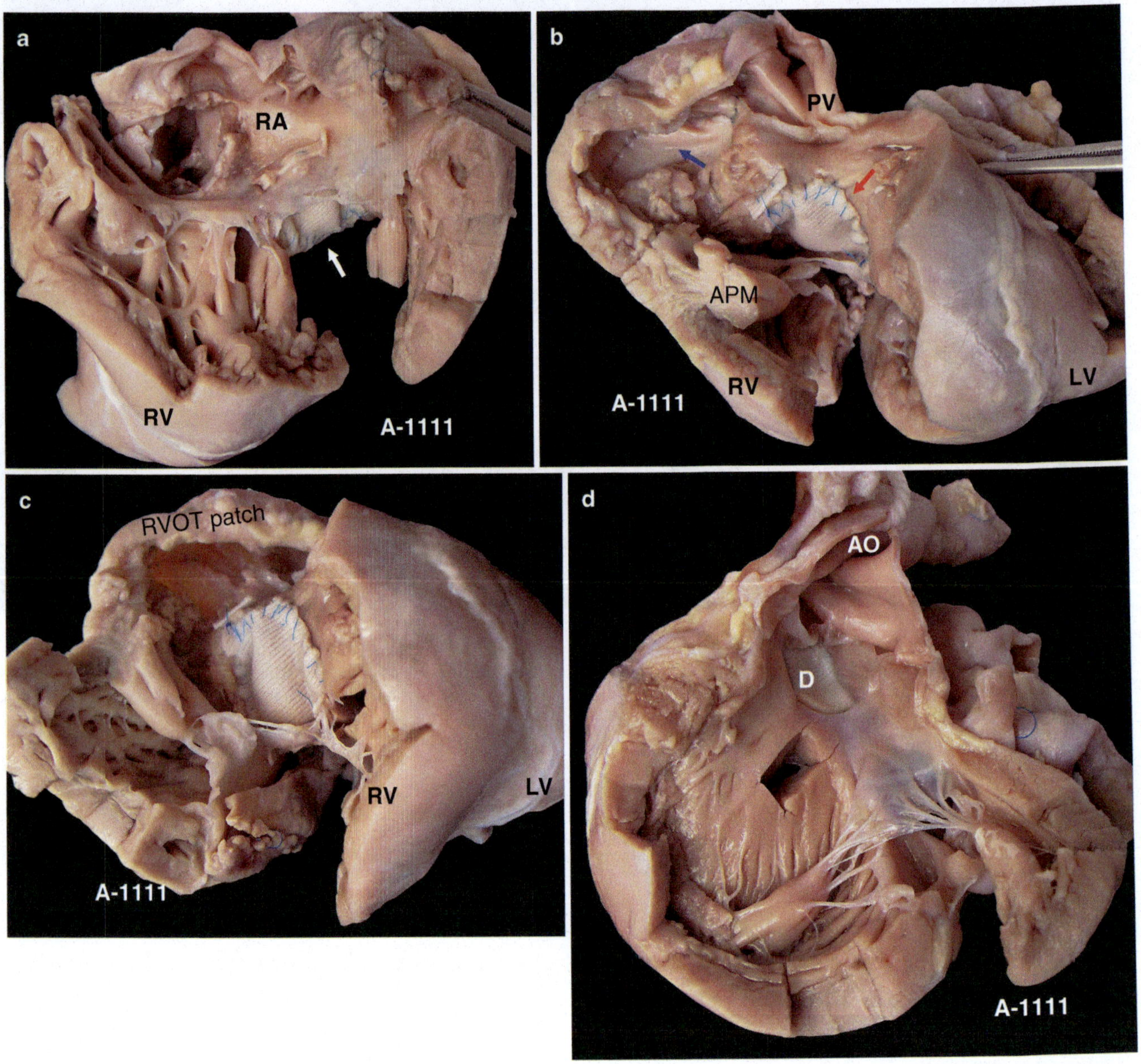

Fig. 23.9 (**a**) Specimen 1111. Specimen of postoperative TOF. The RA and the RV are incised to expose the TV and the sutures (*arrow*) on the posteroinferior edge of the VSD patch. (**b**) Specimen 1111. The RVOT is demonstrated through opened RV and pulmonary artery. Development of the pulmonary valve leaflets is fine, but there is obvious narrowing of the RVOT. The *blue arrow* shows the patch expanding the RVOT, below which is the transverse section of the resected muscles of the thick parietal band. The *red arrow* indicates the anterior edge of the VSD patch. The "APM" marks the anterior papillary muscle of the TV. Special attention should be paid to ensure that the patch area of the RVOT cannot contract or make a contrary pulse. (**c**) Specimen 1111. The figure mainly demonstrates the RVOT patch, which, after removal of so many muscular tissues of the parietal band, can induce an enlarged RV that might influence the heart function. (**d**) Specimen 1111. The left ventricular outflow tract view. Fibrous connections exist between the aortic and the mitral valves. The aorta overrides about 50 %. The VSD patch (*D*) is very smooth

23.4.9 Oversized RVOT Patch

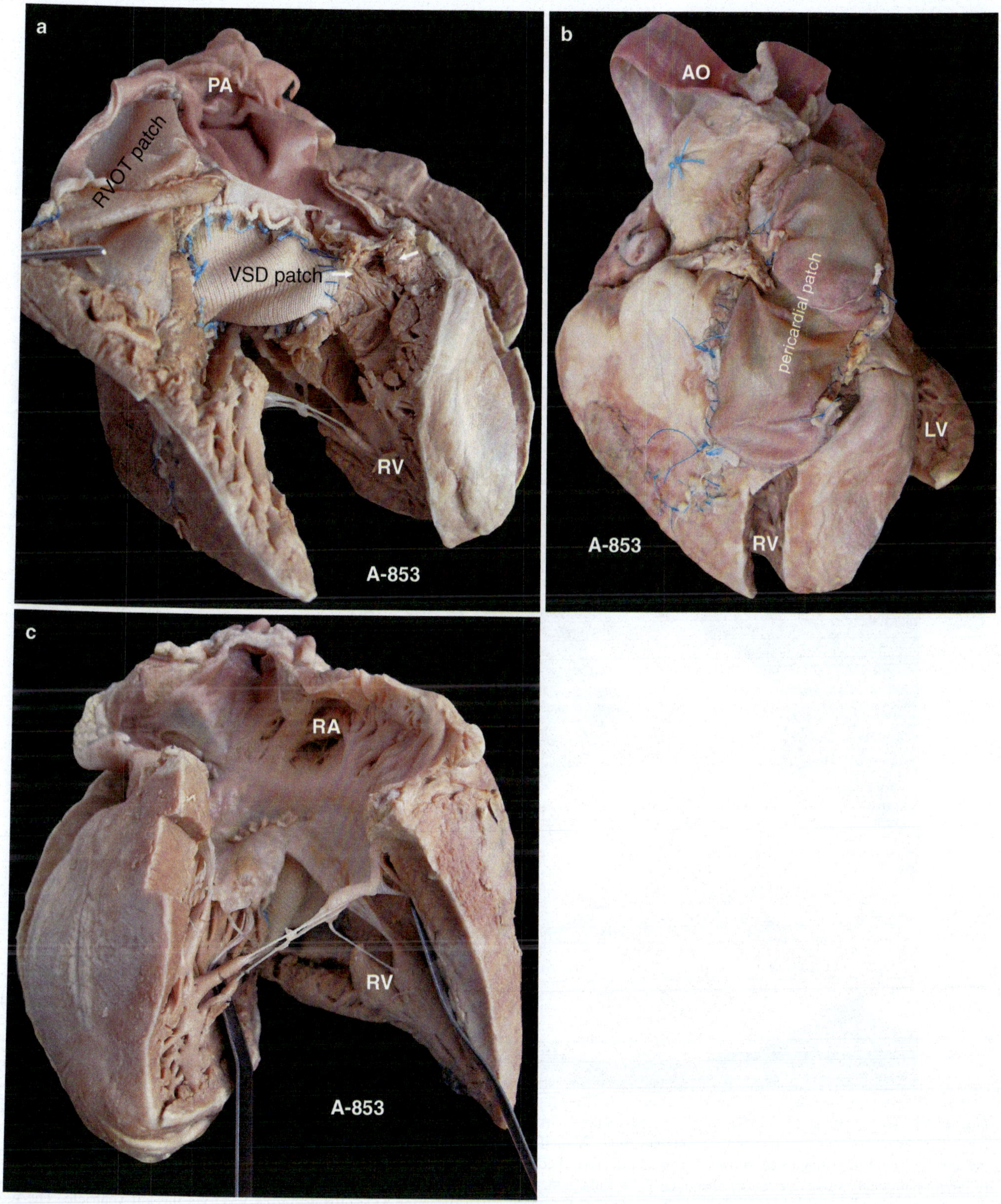

Fig. 23.10 (**a**) Specimen 853. TOF, hypoplasia pulmonary valve. The RVOT patch is oversized and causes severe postoperative regurgitation of the pulmonary valve. The *arrow* marks the stenotic RVOT, which is nearly atresic, and this case is equal to that of the TOF with pseudotruncus arteriosus. The VSD is precisely closed (VSD patch). (**b**) Specimen 853. Front view of the same specimen. Oversized pericardial patch of the RV is present. (**c**) Specimen 853. The pledgeted mattress sutures above the TV annulus are applied to reinforce the VSD patch

23.4.10 TOF with Pseudotruncus Arteriosus

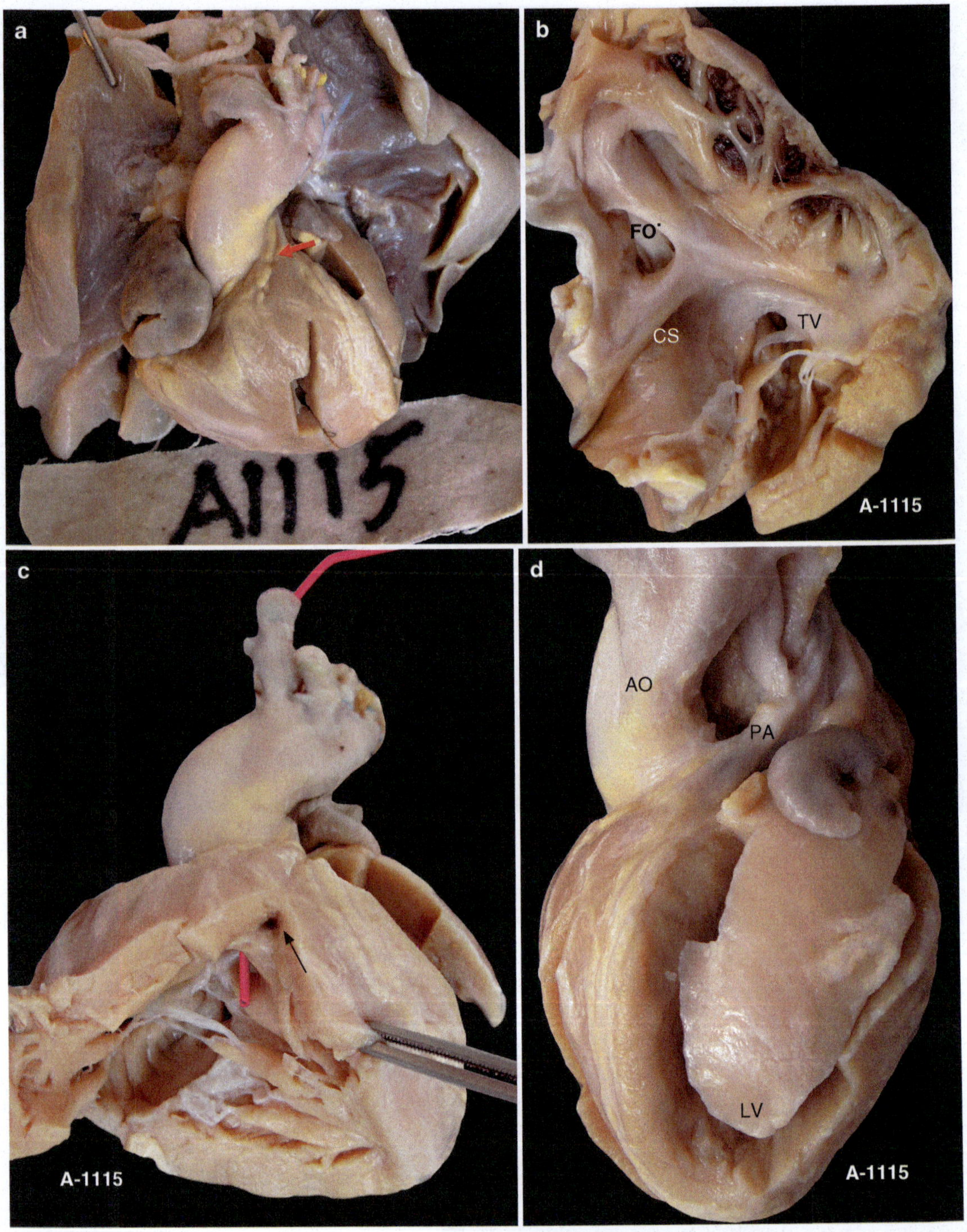

Fig. 23.11 (**a**) Specimen 1115. TOF anterior view of the heart and lung. The *red arrow* indicates the small pulmonary trunk. Behind it is the enlarged aorta. It is diagnosed as *pseudotruncus arteriosus* because of the subpulmonary valve atresia. (**b**) Specimen 1115. Enlargement of the RA and thickening of the TV leaflets are demonstrated. Coronary sinus is below the FO. *FO* fossa ovalis, *CS* coronary sinus. (**c**) Specimen 1115. The RVOT is cut open without transannular incision. The blood flows directly into the aorta from the RV due to the aorta overriding onto the RV. The red bougie is the RV and the ascending aorta. The concave part (*black arrow*) in the anterior end of the RVOT is a stenotic subpulmonary artery orifice, which tends to be a caecum. It should be diagnosed as *pseudotruncus arteriosus*. (**d**) Specimen 1115. Left ventricle (*LV*) incision. The side wall of the LV is lifted to expose the LV inflow and outflow tracts. *AO* aorta, *LV* left ventricle, *PA* pulmonary artery. (**e**) Specimen 1115. The LVOT, especially the MV, is exposed. The cavity in the wall of the LV is made for a histological examination. (**f**) Specimen 1115. Exposing the VSD via the left ventricle. The red bougie shows the connection between the LV and the aorta, as the situation of an overriding aorta. *AO* aorta, *LAA* left atrial appendage, *LV* left ventricle, *PA* pulmonary artery. (**g**) Specimen 1115. Fibrous connections between the aortic valve and the mitral valve. The *white arrow* indicates the VSD. The *red arrow* identifies the overriding aorta. The *black arrow* indicates that there are still fibrous connections between the AV leaflet and the anterior MV leaflet. *MV* mitral valve

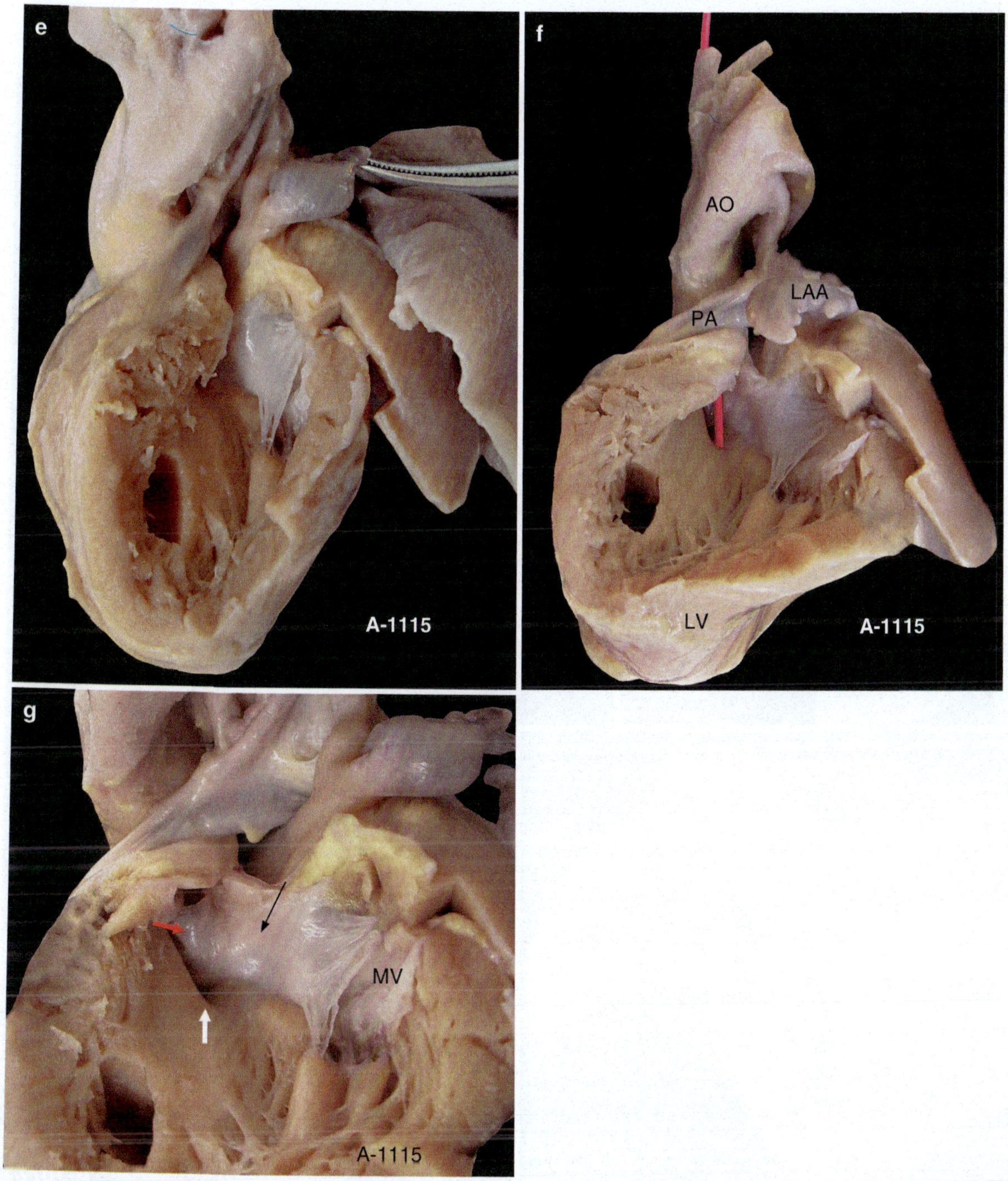

Fig. 23.11 (continued)

23.4.11 Typical Repair of TOF

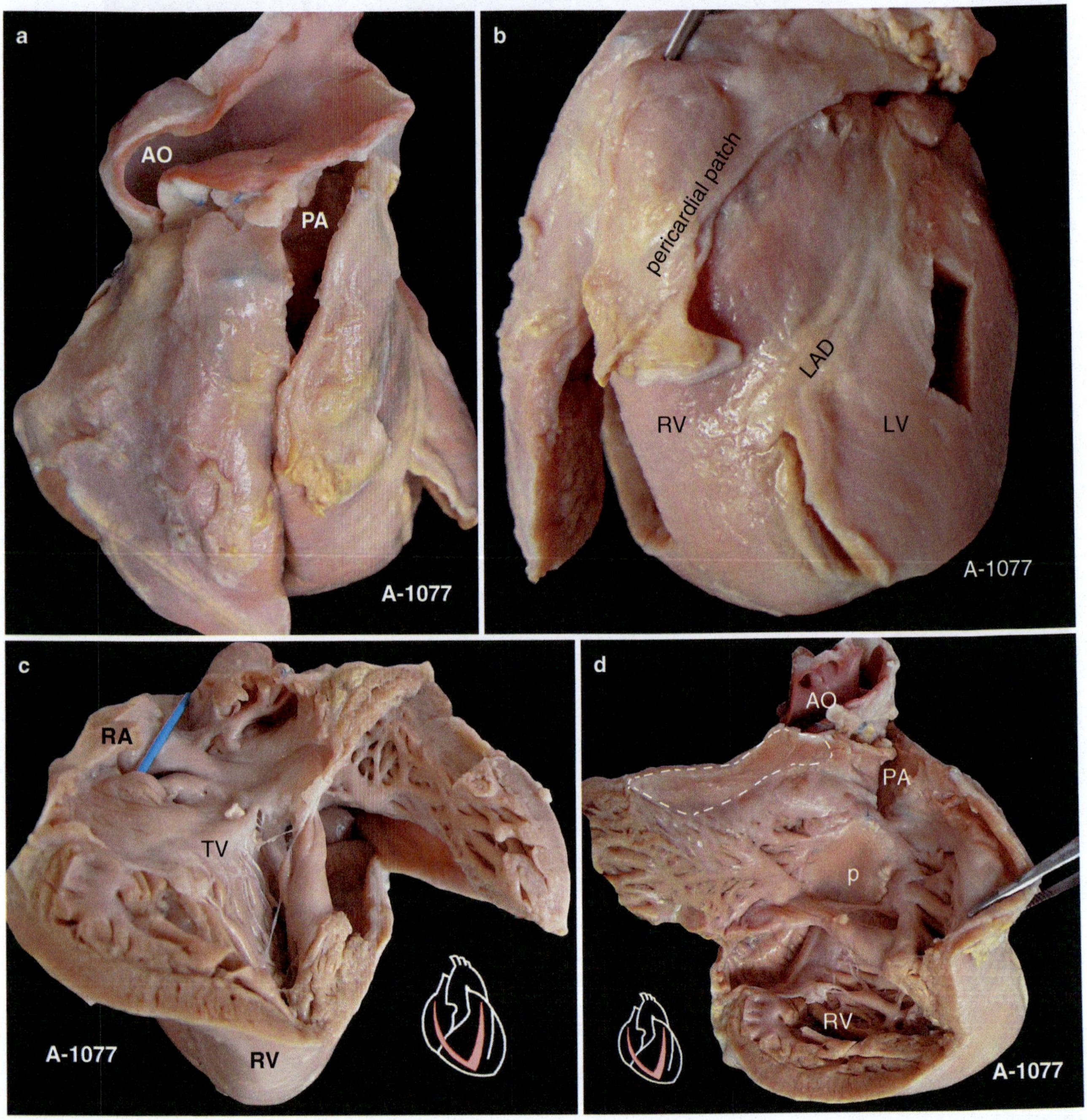

Fig. 23.12 (**a**) Specimen 1077. View of postoperative TOF. Patch-widened RVOT can be seen from the anterior view of the heart, and the diameter of the ascending aorta is two times bigger than that of the pulmonary trunk. Transannular widening of the PV is present. *AO* aorta, *PA* pulmonary artery. (**b**) Specimen 1077. Demonstration of the pericardial patch used to enlarge the RVOT. The coronary artery on the left of the patch is the LAD artery. *LV* left ventricle, *RV* right ventricle. (**c**) Specimen 1077. Dissected right atrium and right ventricle. The blue bougie passes through the superior and the IVC. A polyester mattress stitch, which indicates the place of a reinforcement suture placed during the patching of the VSD, can be seen at the bottom of the septal TV leaflet. Hypertrophied RV wall and moderator band also exist. (**d**) Specimen 1077. A patch (*P*) that is sewn precisely and neatly is clearly shown after the RV and the pulmonary artery are cut open. Enlarged right ventricular trabeculae are exposed, and the *dash line* marks the patch that widens the RVOT. (**e**) Specimen 1077. The LVOT is dissected to show that the VSD has been repaired from the RV side. The AV, which still fibrously joins to the MV, partly overrides the septum. The "M" refers to the papillary muscle of the MV. (**f**) Specimen 1077. View from the left side of the VSD. The *arrows* denote the inferior rim of the defect, where the conduction bundle goes below the left ventricular endocardium. This image clearly shows the fibrous connections between the aortic and mitral valves

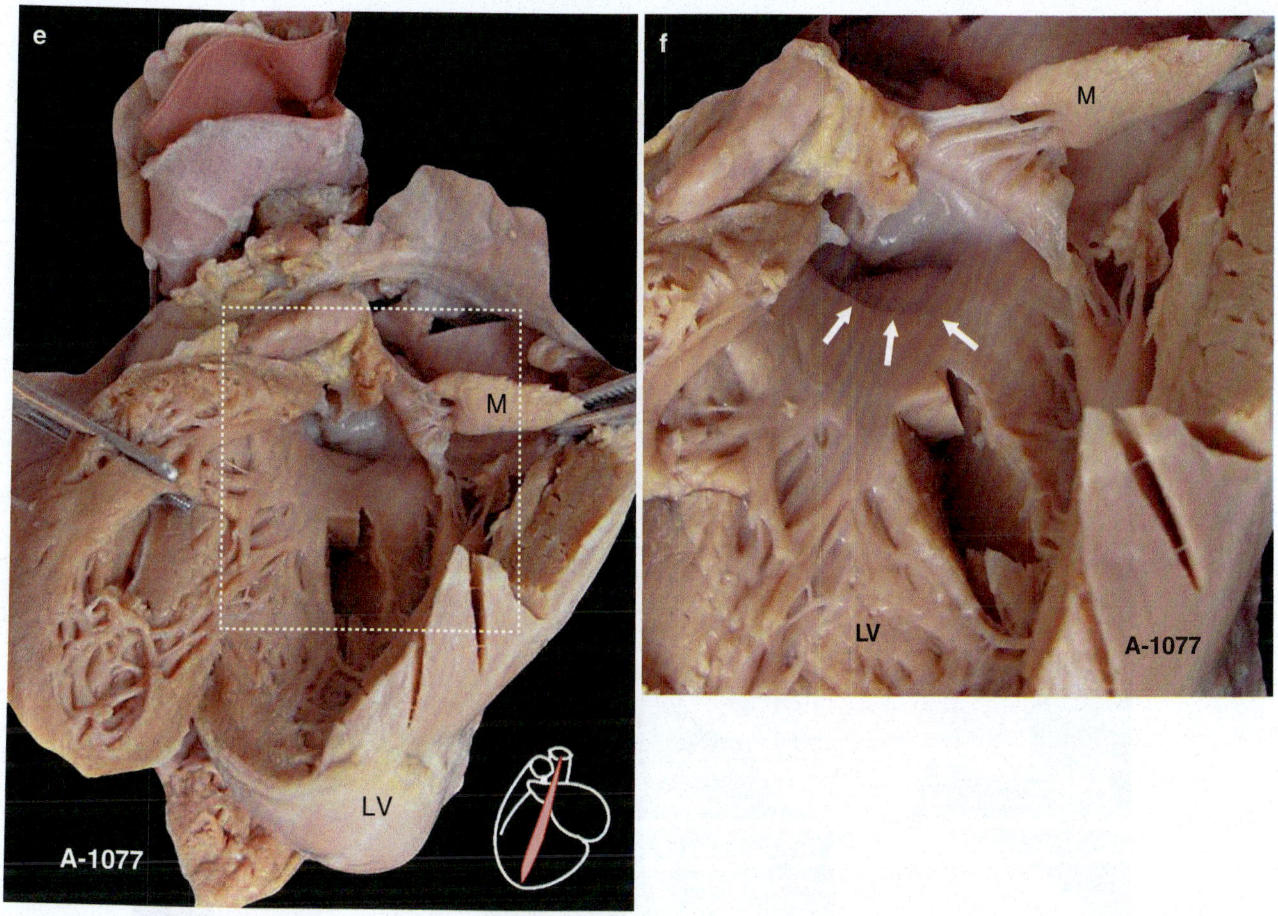

Fig. 23.12 (continued)

23.4.12 Fresh Specimen After Repair of TOF

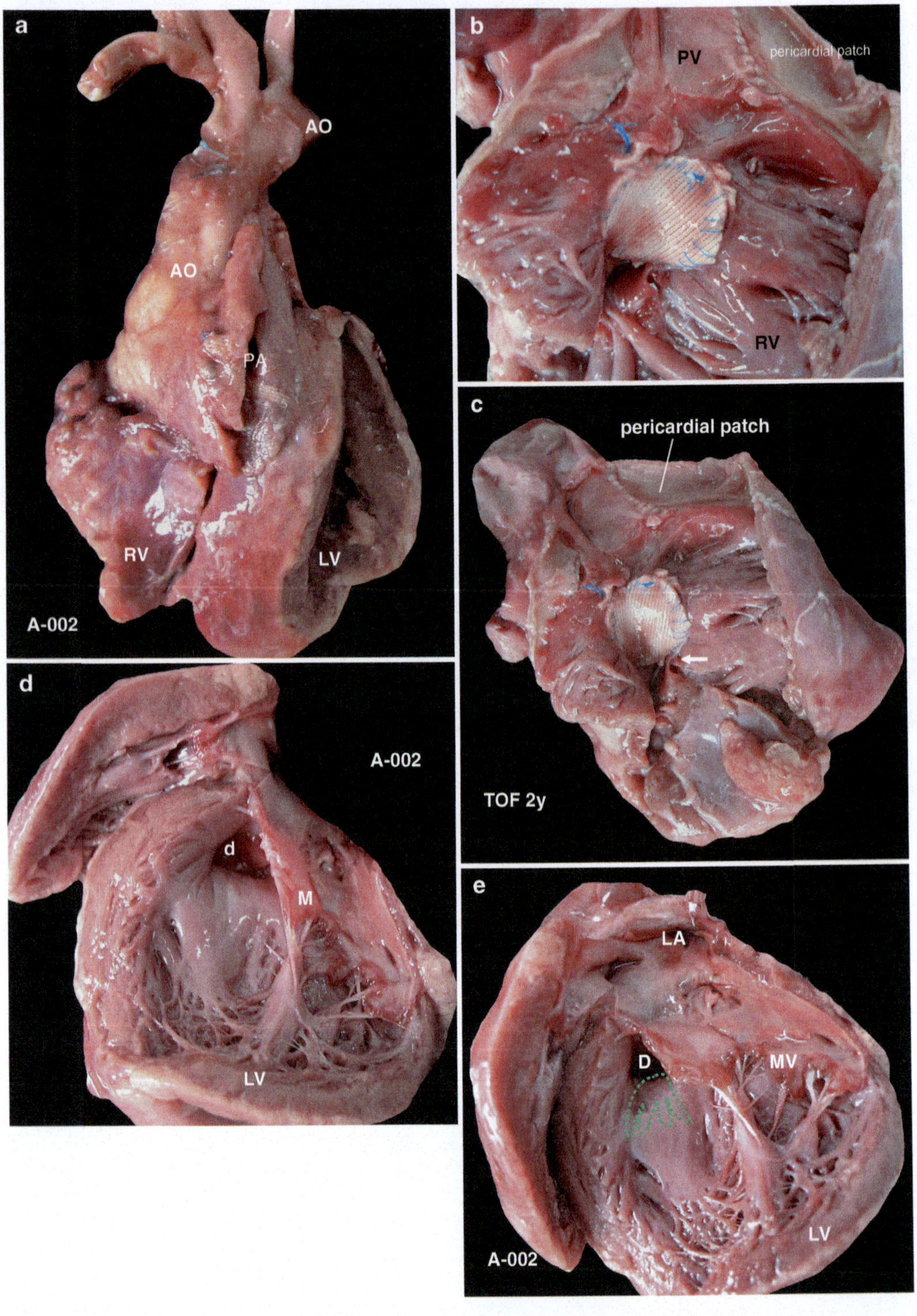

Fig. 23.13 (**a**) Specimen 002. Postoperative specimen of a 2-year-old child after the repair of TOF. An enlarged ascending aorta and a small pulmonary artery can be seen from an anterior view. Both the LV and the RV have been cut open. (**b**) Specimen 002. A right ventriculotomy is made to show the patch of an infundibular VSD. The septal band and the parietal band stand on each side of the precisely sewn Dacron patch, the upper edge of which lies adjacent to the pulmonary valve annulus. A pericardial patch might be seen sutured to the anterior wall of the RV to expand the RVOT and the pulmonary artery. The two pulmonary valve leaflets are approximately normal. (**c**) Specimen 002. The conal papillary muscle anterior, behind, and below the ventricular septal defect. The right bundle branch goes across the root of the conal papillary muscle and then courses into the muscles of the moderator band. So it is safe to sew at the edge of the VSD, which is in front of the conal papillary muscle. (**d**) Specimen 002. View from the left ventricle side to observe the ventricular septal defect (*d*). Fibrous connections are seen between the aortic and mitral valves (M). (**e**) Specimen 002. Both the left atrium and left ventricle exposed at the same time. The "D" indicates the VSD, and the *green dash line* shows the left bundle branch under the left ventricular endocardium

23.4.13 Typical Repair of VSD in TOF

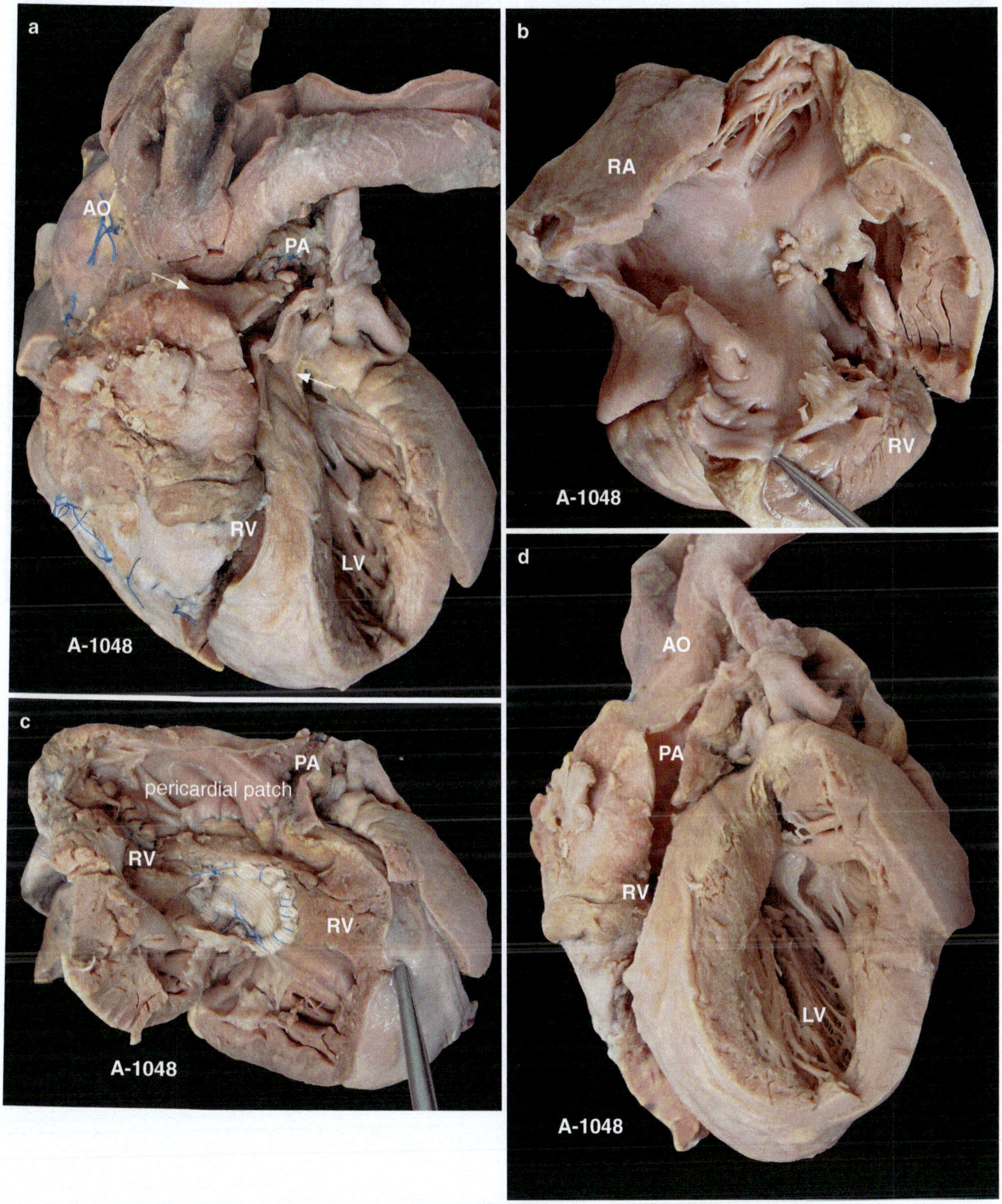

Fig. 23.14 (**a**) Specimen 1048. Front view of the specimen. Figure shows the dilated ascending aorta, small pulmonary artery, hypertrophied RV, and boot-shaped heart. The *arrows* show the stenosis of right ventricular outflow tract. (**b**) Specimen 1048. Very large patch expanding the RVOT. Several polyester pledgets can be seen at the root of the septal TV leaflet and are used to fasten the edge of the VSD patch. (**c**) Specimen 1048. View through opened right ventricle. A precisely sewn VSD patch can be seen, the edge of which is partly sutured to the root of the septal TV leaflet. The pericardial patch widening the RVOT is smooth and viewed from the RV side. (**d**) Specimen 1048. The pulmonary artery (*PA*) is adequately expanded by the transannular patch of the RVOT. (**e**) Specimen 1048. Overriding aortic valve. The VSD is fully revealed through the opened LV, upon which the AV overrides. The rectangular defect above the muscular septum is artificial due to histological examination. (**f**) Specimen 1048. The red bougie is conducted from the LV into the aorta through the VSD, which indicates a mild override of the aorta

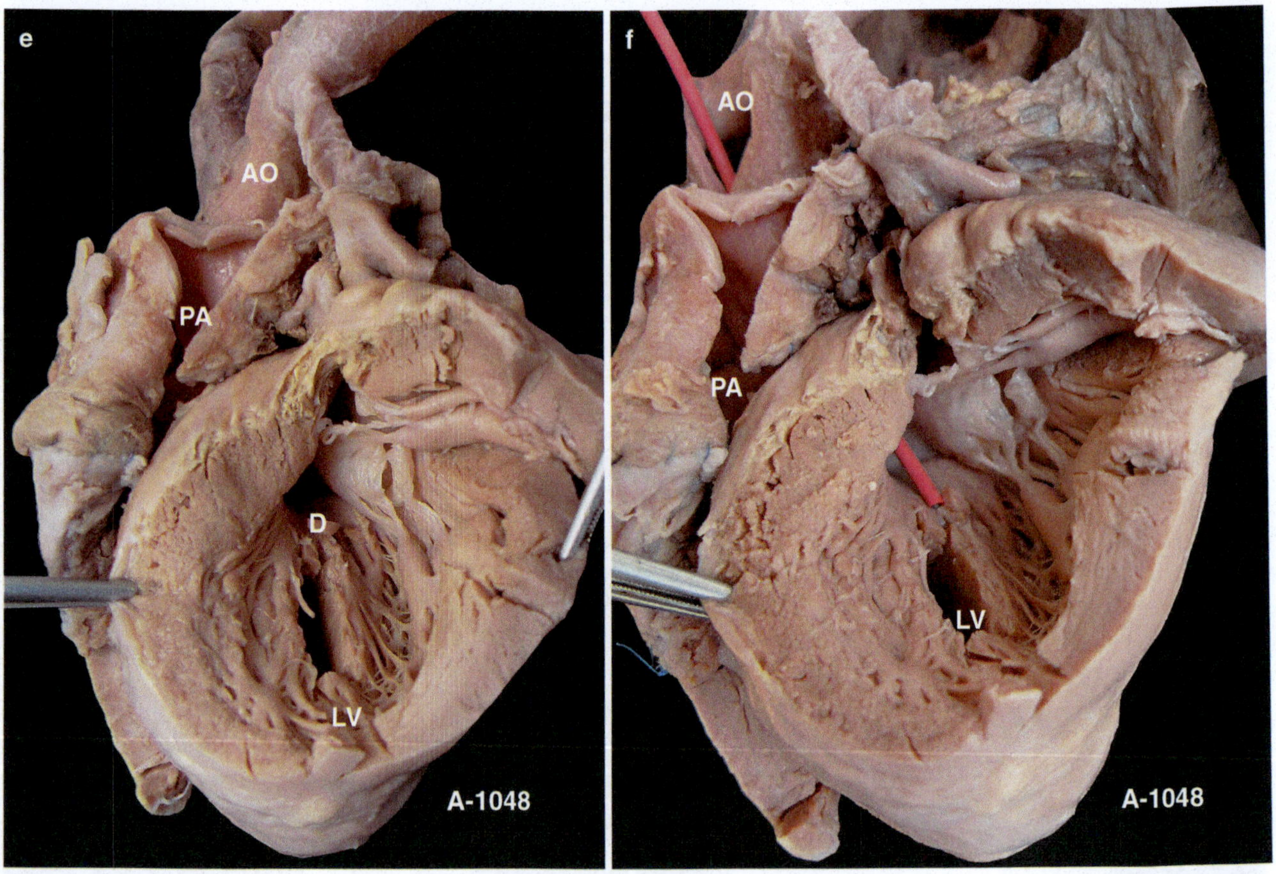

Fig. 23.14 (continued)

23.4.14 TOF with Atrial Septal Defect

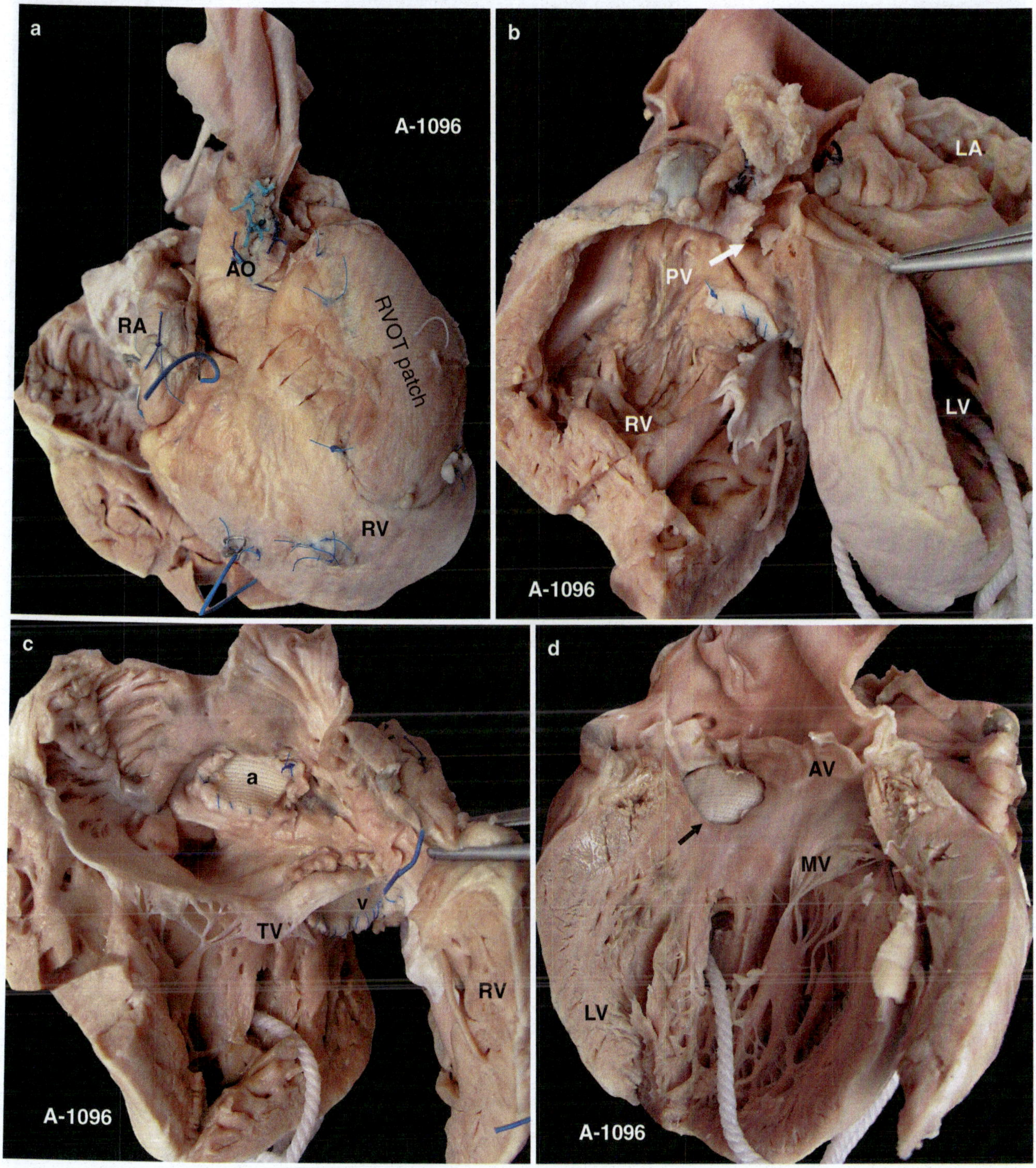

Fig. 23.15 (**a**) Specimen 1096. Front view of the heart. Figure shows TOF after operation. The aorta is enlarged. An artificial patch expanding the RVOT is present, and pacemaker wires can be seen on the surface of the RA and the RV. (**b**) Specimen 1096. The *arrow* shows the small hypoplasia pulmonary valve. The RV is hypertrophied. (**c**) Specimen 1096. The ASD patch (*a*) and the VSD patch (*v*) are revealed. There are polyester pledgets firming the VSD patch at the bottom of the septal TV leaflet (*TV*). This case undergoes third-degree atrioventricular block after operation, probably resulting from damage to the conduction bundle during the repair of the VSD. (**d**) Specimen 1096. View from the left ventricular outflow tract to show the left ventricle side of the VSD (*arrow*). The AV overrides about 50 % and fibrously connects to the MV. (**e**) Specimen 1096. The VSD (*d*) and the mitral valve viewed from the LVOT. (**f**) Specimen 1096. The left atrial side of atrial septal defect patch (*arrow*) viewed from the left ventricular inflow tract

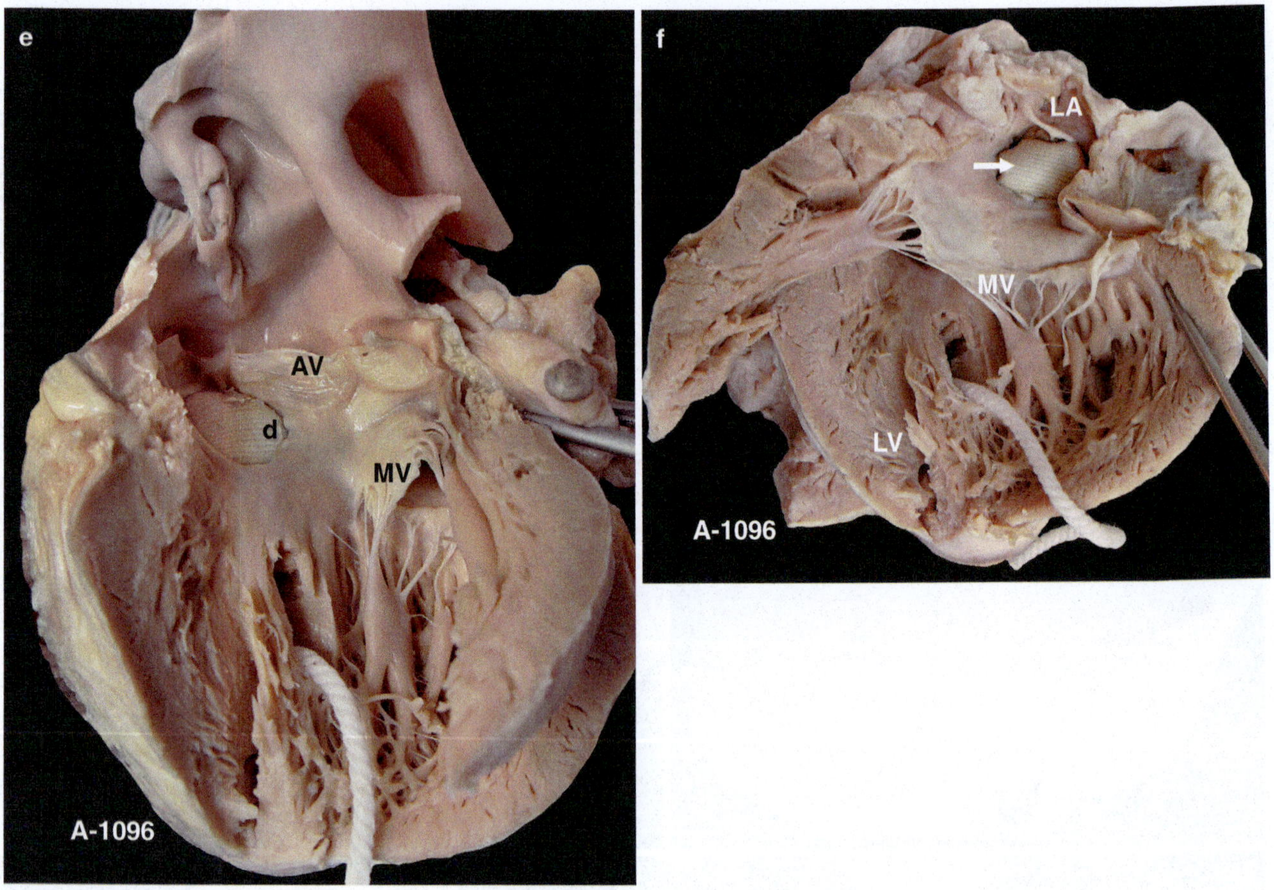

Fig. 23.15 (continued)

23.4.15 Postoperative TOF, Subendocardial Hematoma, Thick Muscular Band in ROVT

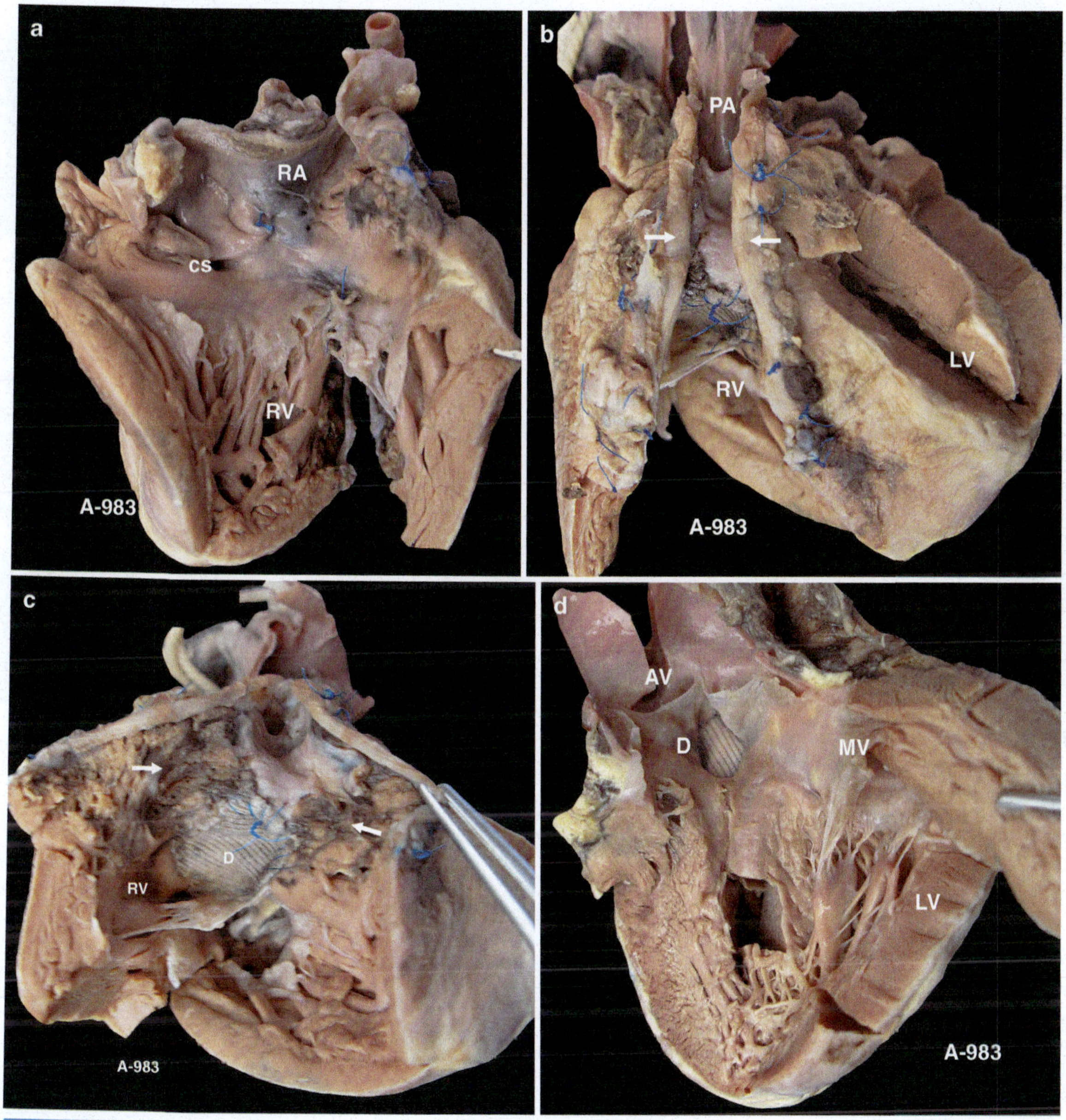

e

Interpretation of specimen 983.

1. Typical TOF. The aorta overrides approximately 50%.
2. The pulmonary valve develops well, and the obstruction is localized mainly at the RVOT.
3. The expanding patch of RVOT is decent.

Fig. 23.16 (**a**) Specimen 983. Typical specimen of TOF after operation. An operative suture can be seen on the FO through the incisions of the RA and the RV. There is also a polyester pledget at the anterior area of the septal TV leaflet, where the atrioventricular node lies. Mind the blue area of the subendocardial hematoma near the pledget. *CS* coronary sinus. (**b**) Specimen 983. A patch is seen inserting into the anterior wall of RV to expand the RVOT. A vertical incision on the RV patch extending to the main pulmonary artery is made. The pulmonary artery is small, but the valve grows well. The hypertrophied crista supraventricularis is seen below the PV. The *arrow* points at the patch expanding the RVOT, which is adequate in length. (**c**) Specimen 983. The VSD patch (*D*). The *arrows* indicate the transverse sections after resecting muscles of the parietal band. (**d**) Specimen 983. The VSD (*D*) and the mitral valve viewed from the LV. A smooth patch precisely sutured on the RV side can be seen through the VSD. Fibrous connections remain between the aortic and mitral valves. The aorta overrides approximately 50 %. The rectangular defect at the septum is made because of pathological examination. (**e**) Specimen 983. Interpretation of specimen 983

23.4.16 Typical External View of Postoperative Specimen of TOF

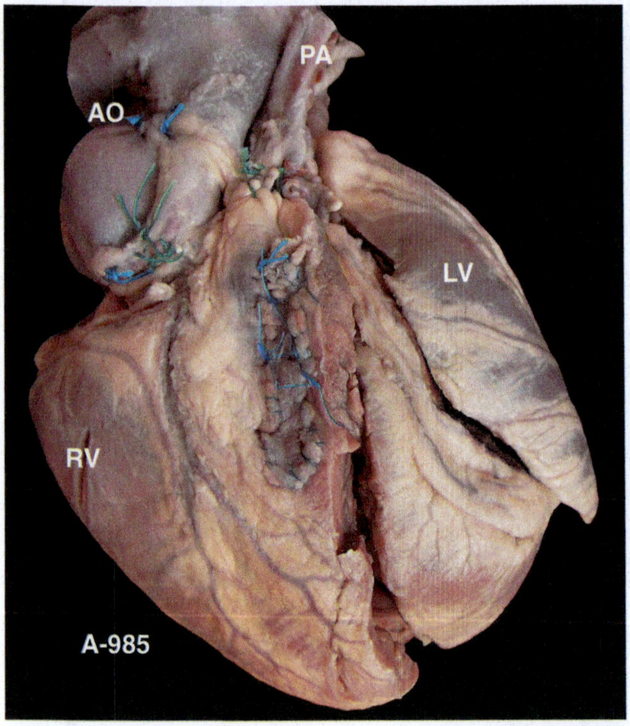

Fig. 23.17 Specimen 985. Typical external view of specimen of TOF. The dilated ascending aorta (*AO*), small pulmonary artery (*PA*), hypertrophied right ventricle (*RV*), RVOT expanded by a patch, relatively small left ventricle (*LV*), and boot-shaped heart are viewed from the front

23.4.17 ASD is Closed During the Repair of TOF

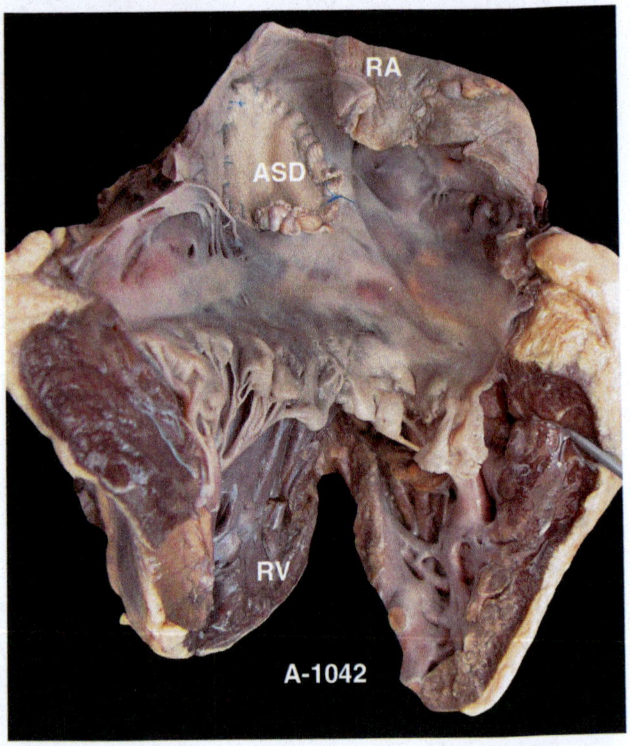

Fig. 23.18 Specimen 1042. TOF with ASD. The ASD is closed correctly. The RV is highly hypertrophied, and the VSD is sheltered by the TV

Fig. 23.19 (**a**) Specimen 1110. Mild TOF. The diameter of the pulmonary artery (*PA*) is about half that of the aorta (*AO*). A patch is applied to correct the infundibular stenosis. (**b**) Specimen 1110. A patch on the FO can be seen through the RA. RV is mildly hypertrophied. The polyester pledgets used to fasten the VSD patch can be seen at the commissure of the septal leaflet and anterior leaflet of the TV. The *arrows* show the torn muscles and leaflet tissues related to the suture of the posterior edge of the VSD near the medial papillary muscle, along which the conduction bundles go. So, special attention should be paid. *TV* tricuspid valve. (**c**) Specimen 1110. The LVOT and the VSD (*D*) are revealed through opened LV. *AO* aorta, *PA* pulmonary artery. (**d**) Specimen 1110. The extent of the override of the AV is shown. Two leaflets of the AV totally override onto the RV. The polyester patch repairing the VSD is folded, which might result in residual shunt. (**e**) Specimen 1110. The left atrium and the left ventricular inflow tract. The MV is normal. Features of the specimen 1110: (1) Repair of TOF with ASD. (2) The sutures of the lower edge of the VSD patch injure the TV. (3) The aorta overrides about 50 %. (4) The VSD patch is carefully sutured, but not so smooth

23.4.18 Wounded Tricuspid Valve After Repair of TOF

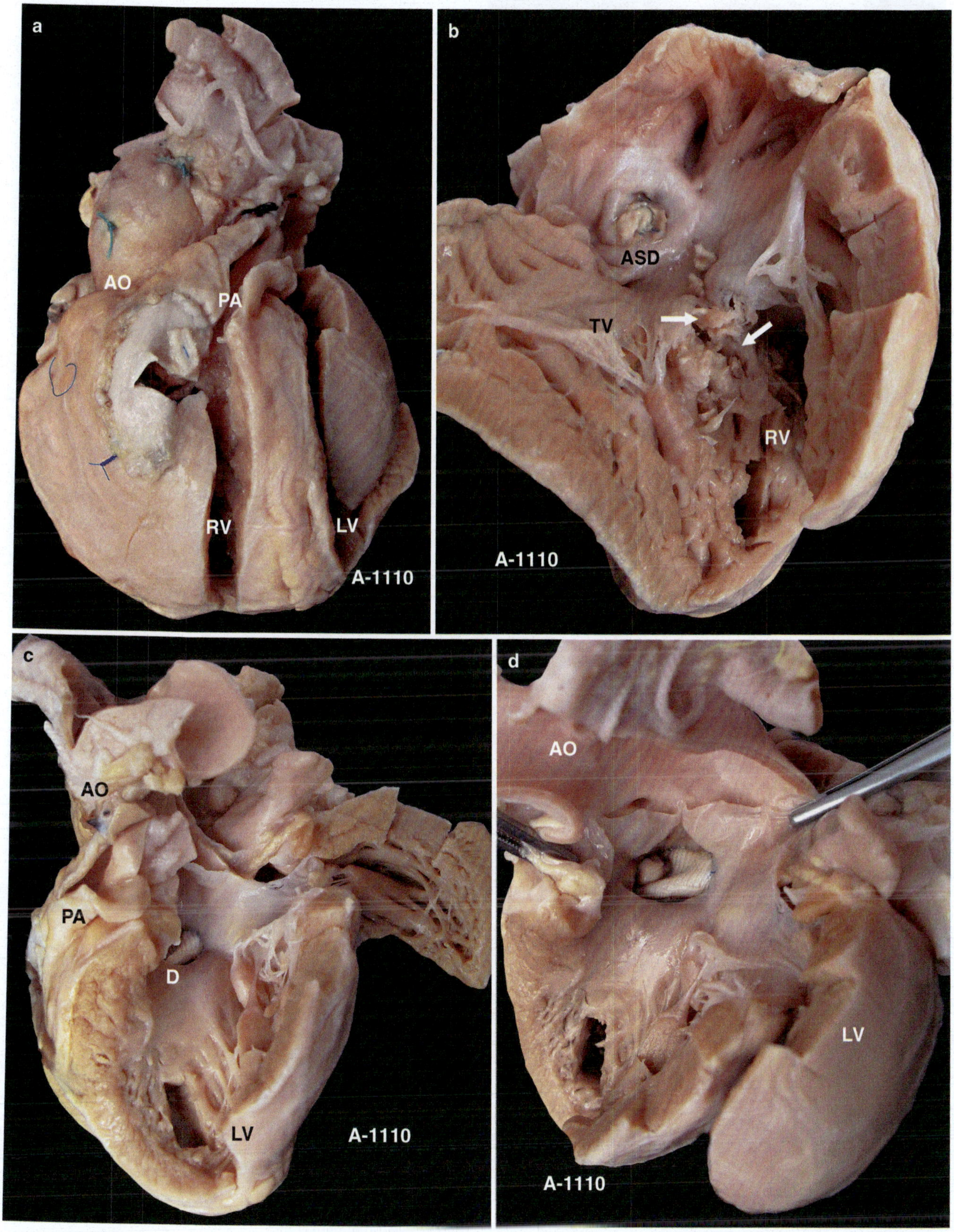

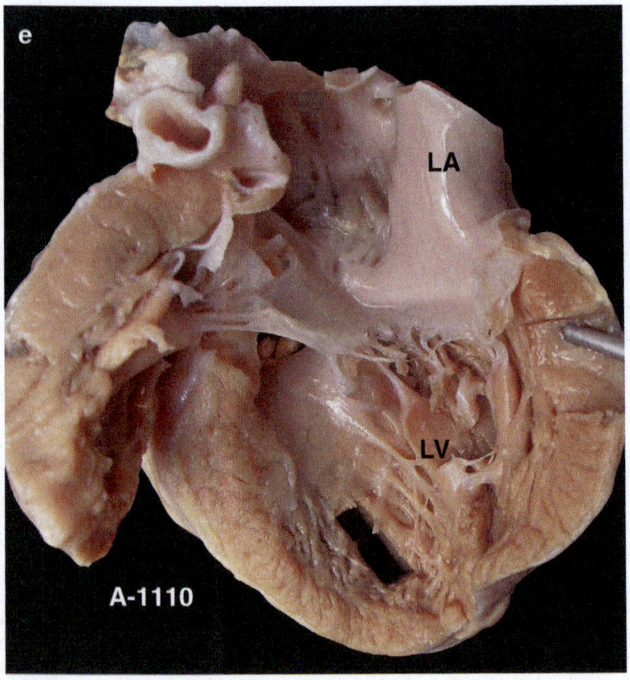

Fig. 23.19 (continued)

23.4.19 Incision of the RV Is Too Low

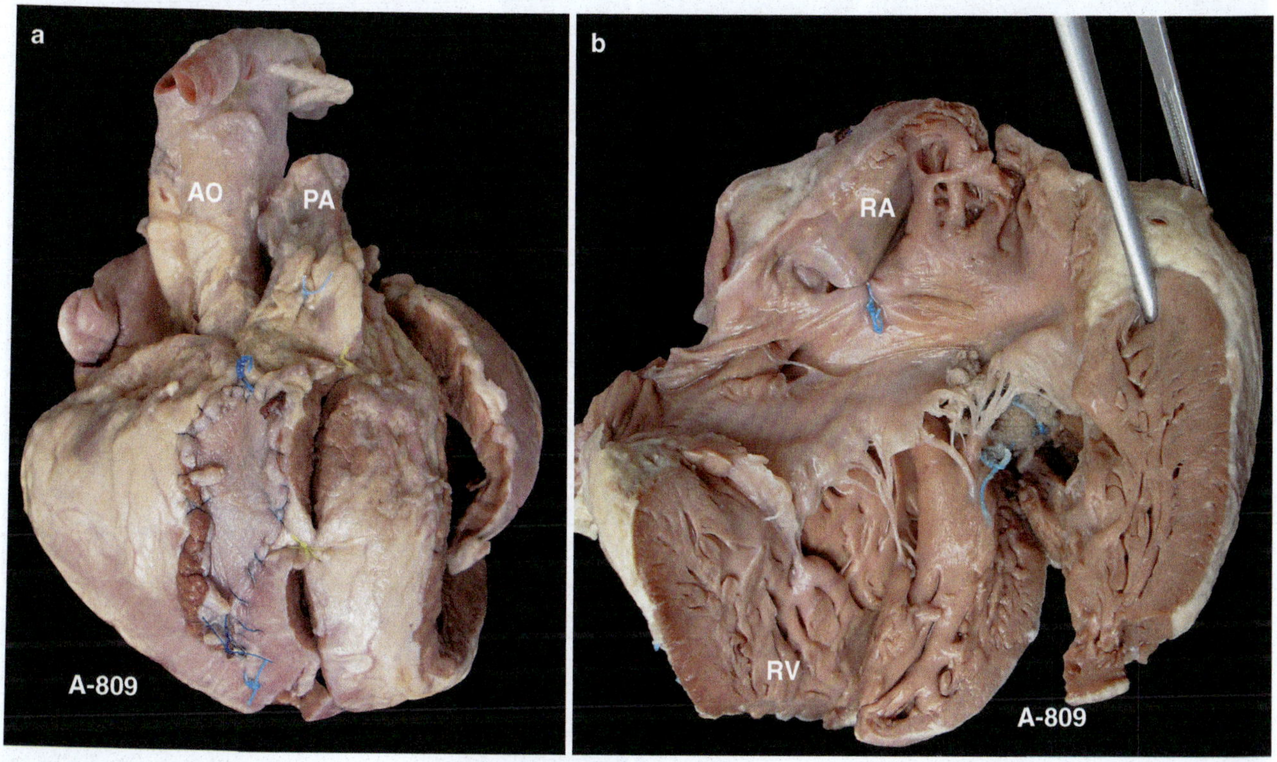

Fig. 23.20 (**a**) Specimen 809. The RVOT is expanded by an oversized polyester patch, the lower end of which reaches to the apex. The right ventricular incision carried to the apex not only does no good to exposure but also severely affects the right ventricular function. The RV is obviously hypertrophied, and the LV is relatively small. (**b**) Specimen 809. The RA and RV are opened. A small fissure with a suture is seen on the FO. The lower half of the VSD patch can be seen below the TV. (**c**) Specimen 809. Several muscular bands are anterior to the VSD, so injury to the conduction bundle does not occur easily. The *arrow* shows the large transverse sections of the parietal band. (**d**) Specimen 809. The VSD (*arrow*) observed from the left ventricle. The aorta overrides mildly. *AV* aortic valve, *LV* left ventricle, *MV* mitral valve. (**e**) Specimen 809. The LA and the MV are nearly normal

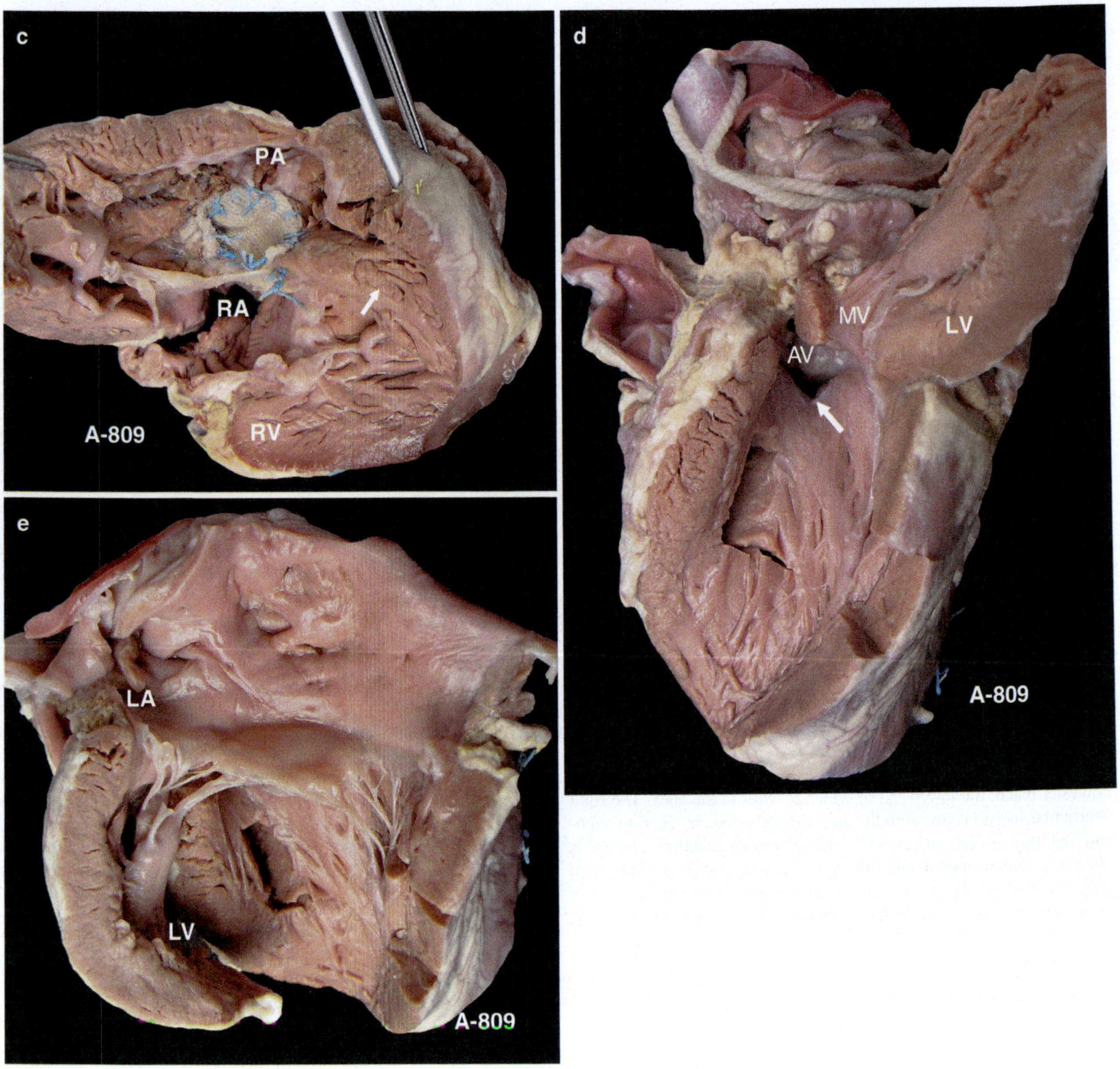

Fig. 23.20 (continued)

23.4.20 Residual Shunt of VSD After Repair of TOF

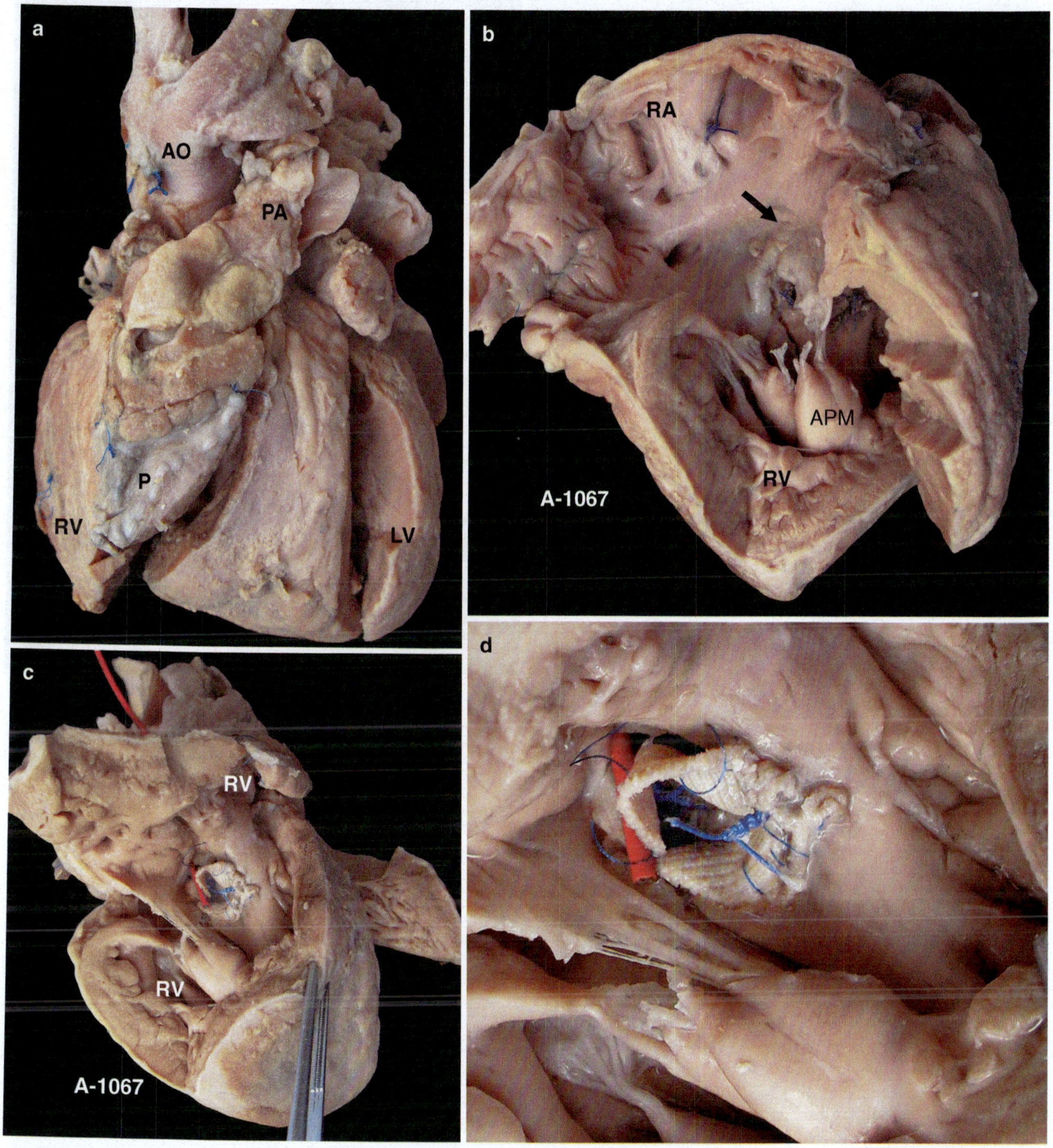

Fig. 23.21 (**a**) Specimen 1067. Typical external view of specimen of TOF. The LV and the RV are cut open. A patch expands the RVOT in the front wall of the RV, which is a little lower. *AO* aorta, *LV* left ventricle, *PA* pulmonary artery, *RV* right ventricle. (**b**) Specimen 1067. The right atrium and right ventricle opened to reveal the tricuspid valve. The "APM" indicates the anterior papillary muscle of the TV. Polyester pledget fastening the posterior edge of the VSD patch can be seen at the root of the septal TV leaflet. The *arrow* denotes the torn area of the VSD patch (concealed by the tricuspid valve). (**c**) Specimen 1067. The VSD patch viewed from the RVOT. The posterosuperior stitched edge of the patch is split apart, from which the red bougie can be conducted through the defect. The defect is below the crista supraventricularis, which is the most common position of a residual shunt and is adjacent to the AV. (**d**) Specimen 1067. Close-up view of the torn area of the VSD patch. The fissure formed by the avulsion due to the disconnection of the continuous suture lies near the posterosuperior edge of the VSD, where discontinued pledgeted mattress sutures can be applied to reinforce the patch. (**e**) Specimen 1067. The relationship of the aortic valve (*AV*) and the VSD is mainly demonstrated. The red bougie denotes that the AV cusps might be cut or sutured during operation. The *black arrow* indicates the TV. (**f**) Specimen 1067. The VSD viewed from the LVOT. The red bougie is inserted into the LV from the RV through the fissure of the VSD patch, which shows the existence of the residual shunt

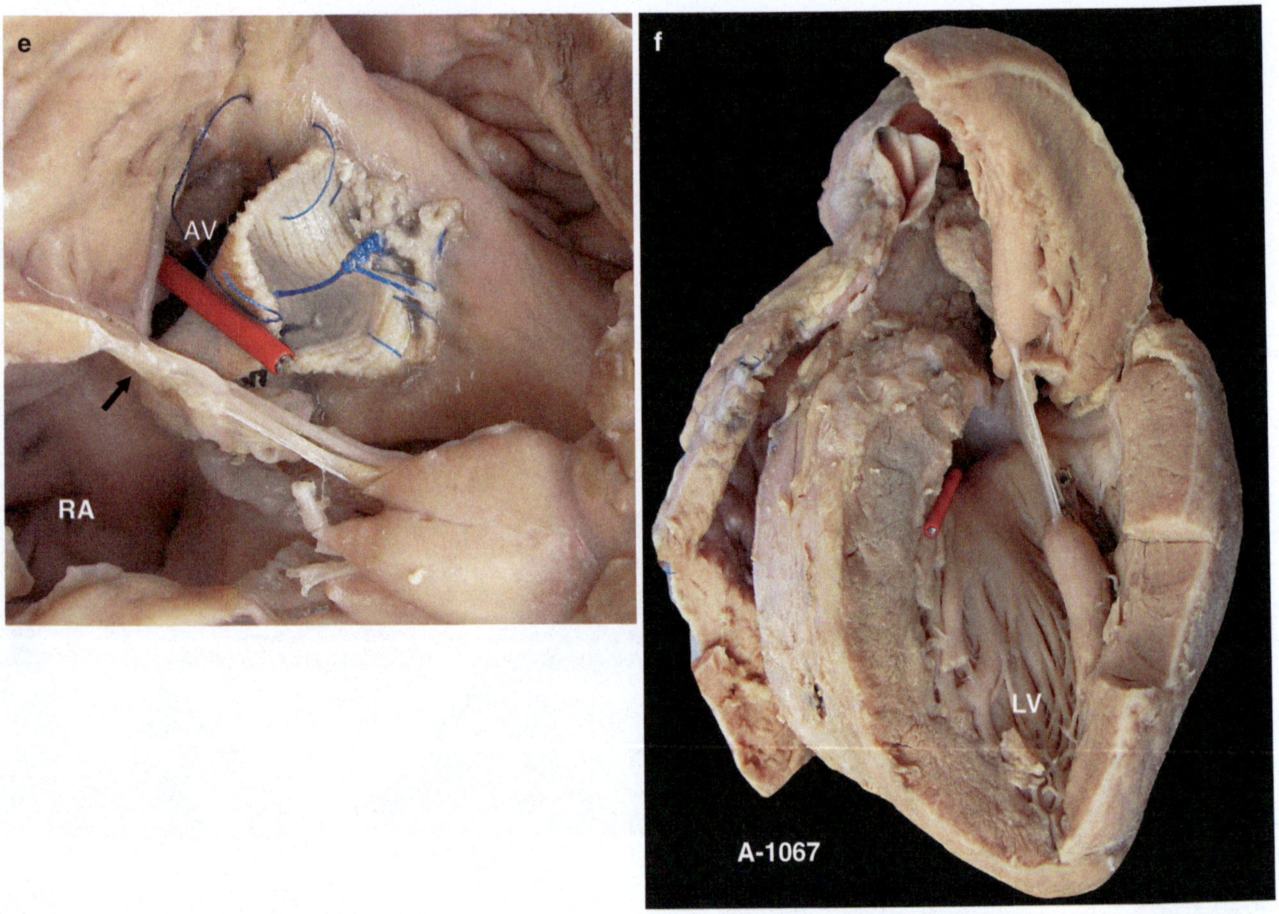

Fig. 23.21 (continued)

23.4.21 Residual Shunt of VSD After Repair of TOF

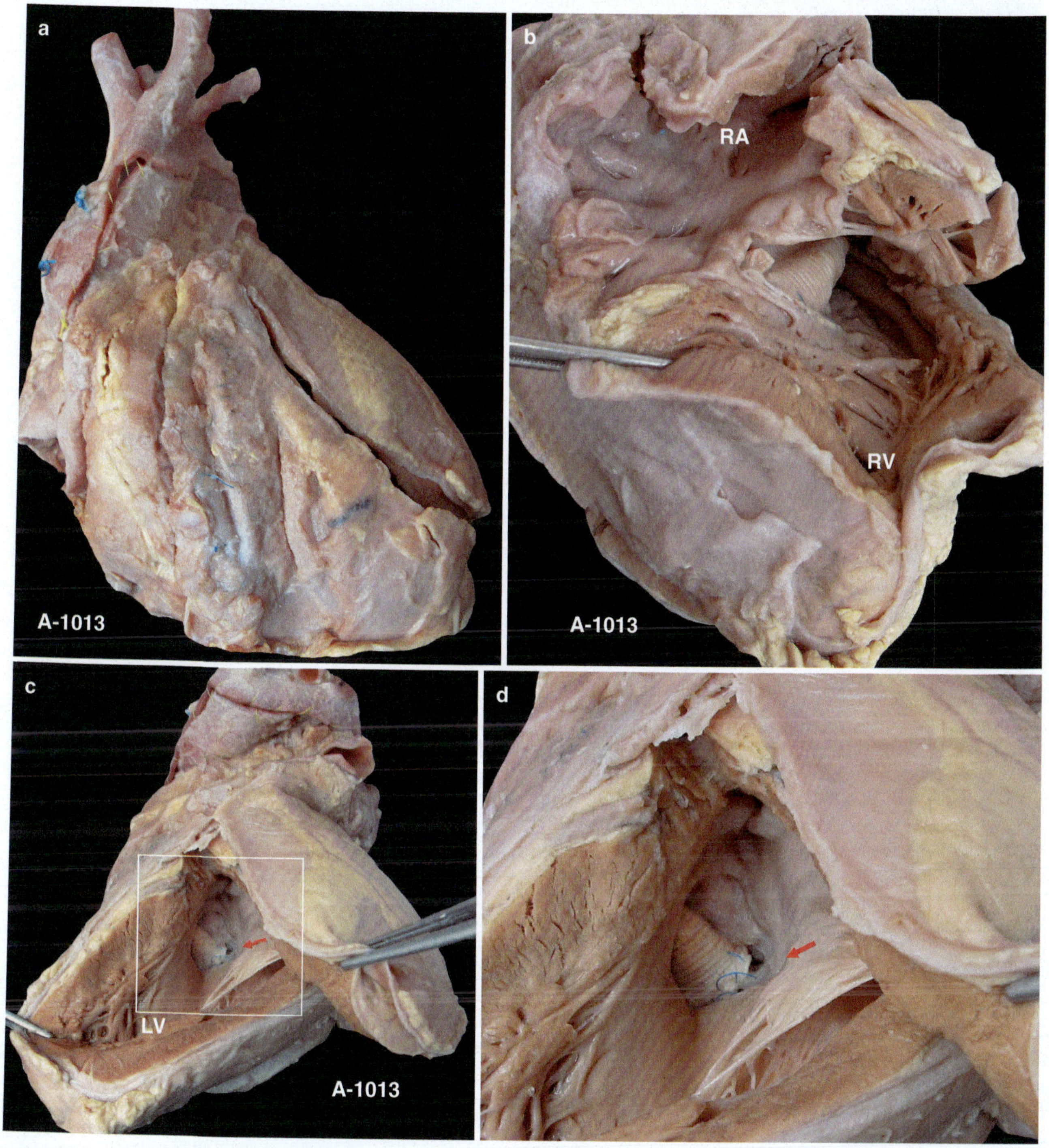

Fig. 23.22 (**a**) Specimen 1013. TOF, boot-shaped heart. (**b**) Specimen 1013. View from the right atrium. No reinforcement pledget is used in suturing the anterior edge of the VSD, which might be associated with the residual shunt of the VSD. *RA* right atrium, *RV* right ventricle. (**c**) Specimen 1013. The fissure (*arrow*) of the VSD patch is seen through the LV side. The *dash line* area marks the VSD patch. (**d**) Specimen 1013. Close-up view of the fissure (*arrow*) of the VSD patch. (**e**) Interpretation of the specimen 1013. The VSD patch below the TV is observed from the RV. The *arrow* shows that there is no reinforcement suture, which results in the residual shunt. The fissure near the postero-inferior edge of the VSD patch is observed from the LV

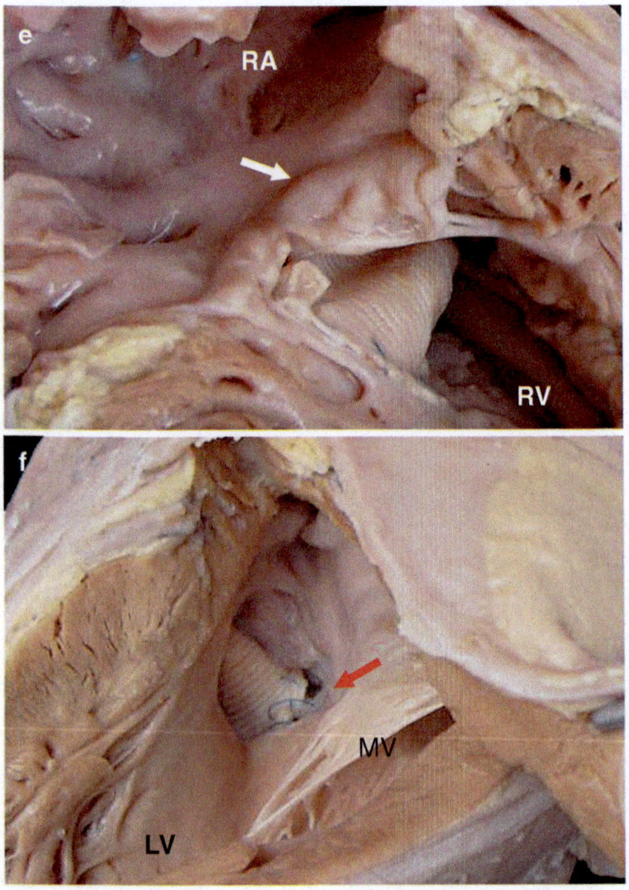

Fig. 23.22 (continued)

23.4.22 Repair of TOF and the Conduction Bundle

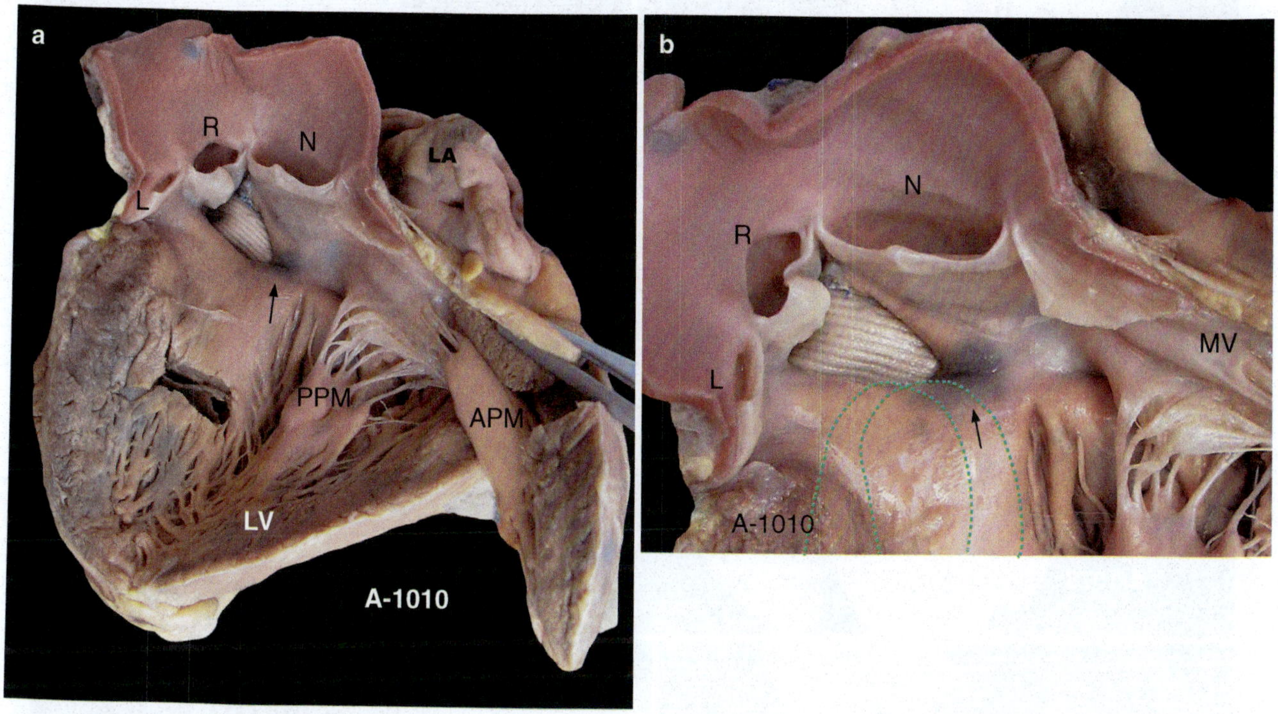

Fig. 23.23 (**a**) Specimen 1010. View of VSD in TOF from the left ventricle. This is the most common type of TOF. The image shows that the VSD is right below the right coronary cusp and the noncoronary cusp. The *arrow* points to the subendocardial hematoma near the posterosuperior edge of the VSD. *APM* anterior papillary muscle of the MV, *L* left coronary cusp, *N* noncoronary cusp, *PPM* posterior papillary muscle of the MV, *R* right coronary cusp. (**b**) Specimen 1010. Close-up view of the left ventricular subendocardial hematoma after repairing the VSD. The *green dash lines* show the distribution area of the left anterior bundle branch and the left posterior bundle branch

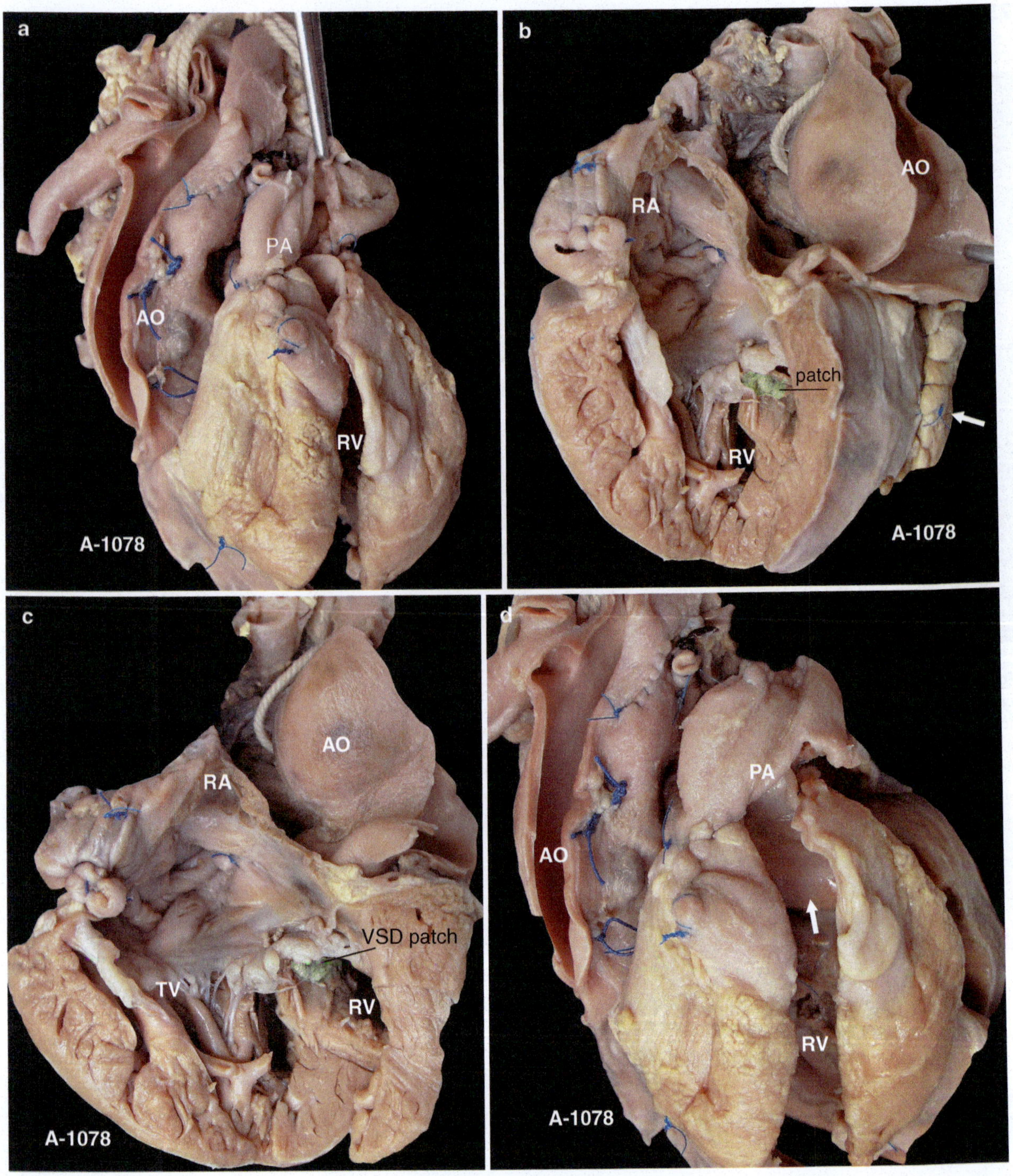

Fig. 23.24 (**a**) Specimen 1078. Front view of postoperative specimen of TOF with aortic valve stenosis. The ascending aorta is dilated, and the pulmonary artery is small. The patch of the RVOT is present. *AO* aorta, *PA* pulmonary artery, *RV* right ventricle. (**b**) Specimen 1078. The ascending aorta is severely dilated without similar dilation at the level of the AV annulus, which indicates that it is an aortic dilation following the stenosis. The TV and the posteroinferior edge of the VSD patch are seen through the opened RA and RV. The *arrow* indicates the patch enlarging the RVOT. (**c**) Specimen 1078. The RA and the RV are highly hypertrophied. There are sutures on the FO. The polyester pledgets at the root of the septal cusp of the TV can reinforce the stitched edge to avoid residual shunt of the VSD and rarely injure the conduction bundle. The ascending aorta is dilated after the stenosis. (**d**) Specimen 1078. Left anterior oblique view of the right ventricular incision. The *arrow* indicates thickened muscles of the crista supraventricularis, which induce the RVOT stenosis. The pulmonary annulus is well developed, and the transannular patch is not necessary in expanding the RVOT. (**e**) Specimen 1078. The lateral wall of the left ventricle removed to show the mitral valve and the left ventricle hypertrophy (because of associated aortic valve stenosis). The *red arrows* indicate that the stitched edge of the expansion patch of the RVOT already reaches to the left side of the ventricular septum, where the left anterior descending and diagonal branches might be affected. (**f**) Specimen 1078. Localized dilation of the ascending aorta. (**g**) Specimen 1078. The stenotic aortic valve annulus and the accompanied dilated aortic lumen. The aortic valve is tricuspid

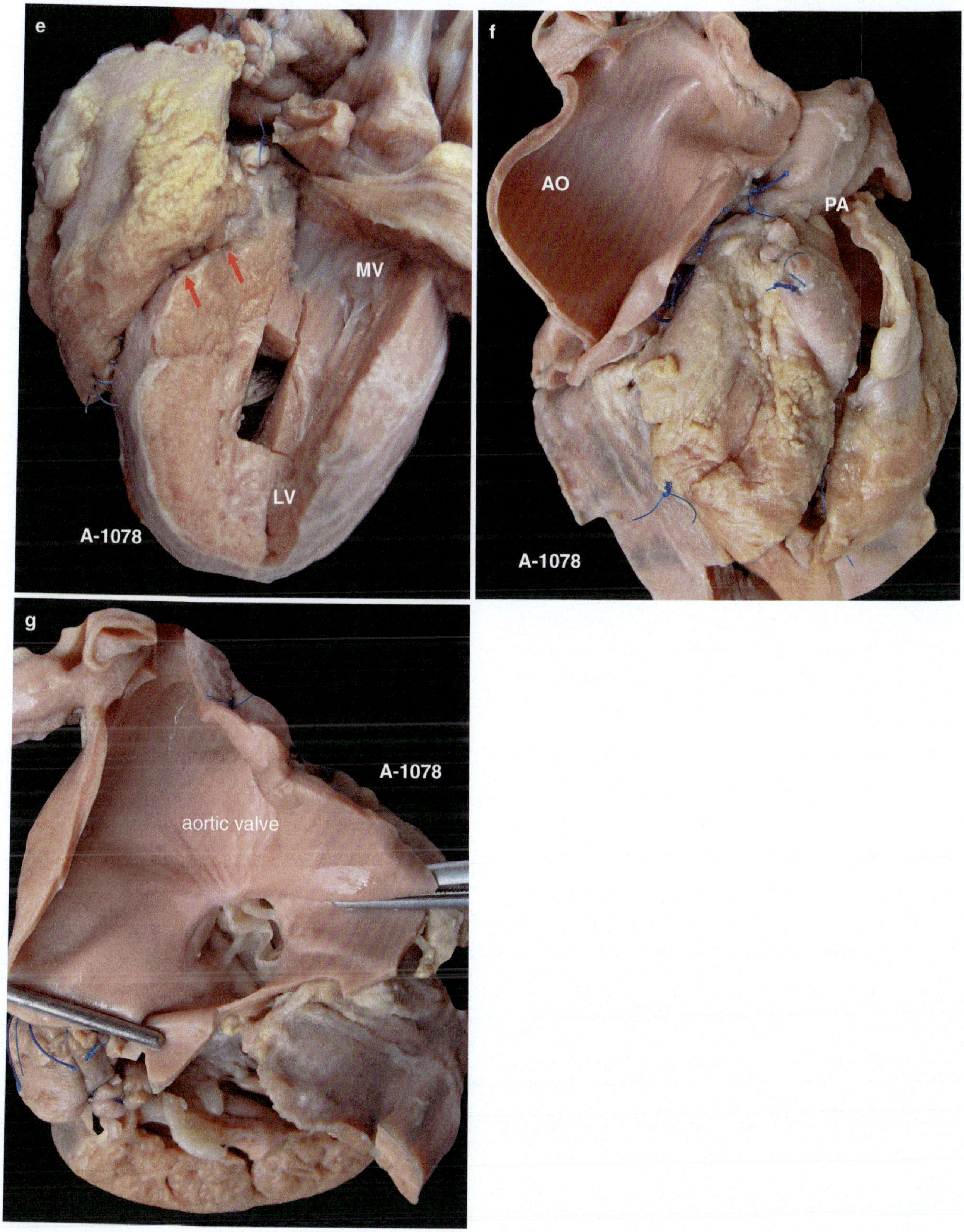

Fig. 23.24 (continued)

24 Double-Outlet Right Ventricle

24.1 General Considerations

24.1.1 Definition

The definition of double-outlet right ventricle (DORV) remains controversial. The consensus is that it is a type of ventriculoarterial connection in which both great vessels arise either or predominantly from the RV. Strictly speaking, the standard is either a semilunar valve that was separated by the cone muscle but is not a fibrous connection or some cone muscle exists between the two semilunar valves if the AV arises more than 90 % from the RV.

24.1.2 Embryological Basis

DORV is a kind of congenital truncal defect characterized by its abnormal absorption and rotation. It could be understood as a transition between the TOF and the TGA.

The two great arteries all arise from the RV in the embryonic stage. With the formation of left and right ventricles and the rotation and absorption of the cone, the PV connects to the RV and the AV connects to the LV. Otherwise, the two arteries would stay at the primary status and develop into a DORV. The cone structure that remains under the two arteries in most of the patients is called the *bilateral cone*. The two semilunar valves are located side by side, the AV on the right and PV on the left.

24.1.3 Categorization

DORV is always classified according to the following characteristics:

1. Location of the VSD
2. With/without pulmonary stenosis
3. With/without pulmonary hypertension
4. Associated anomalies

24.1.4 Surgical Application

The LV can connect only with the artery through the VSD because there is no normal outflow tract. As far as the surgical treatment is concerned, the fibrous connection between the semilunar valve and the atrioventricular valve is beneficial to confirm the anomaly. The surgical technique depends on the size of the VSD and the development of pulmonary artery.

Palliative Operation: The pulmonary artery banding can be used in patients with DORV and pulmonary hypertension. A systemic-to-pulmonary artery shunt can be used for patients with pulmonary stenosis.

Radical Operation:

1. Subaortic VSD with adequate distance between the TV and the PV is treated by simple intraventricular baffling that directs LV blood through the VSD to the aorta.
2. Subpulmonary VSD without pulmonary stenosis is treated with an arterial switch procedure and a simple VSD closure.
3. Blood flow in doubly committed VSDs usually is baffled using a more complex patch from the LV to the aorta.
4. Noncommitted VSD may be treated by baffling of the blood through the VSD to the aorta associated with a myocardial flap reinsertion of the straddling TV or by a Fontan type of procedure.

X. Zhu (ed.), *Surgical Atlas of Cardiac Anatomy*,
DOI 10.1007/978-94-017-9409-1_24, © Springer Science+Business Media Dordrecht and People's Medical Publishing House 2015

24.2 Specimen Demonstrations

24.2.1 DORV, AVSD, L-Malposition

Fig. 24.1 Schematic diagram of common types of DORV. *RA* right atrium, *RV* right ventricle, *LA* left atrium, *LV* left ventricle, *VSD* ventricular septal defect, *ASD* atrial septal defect

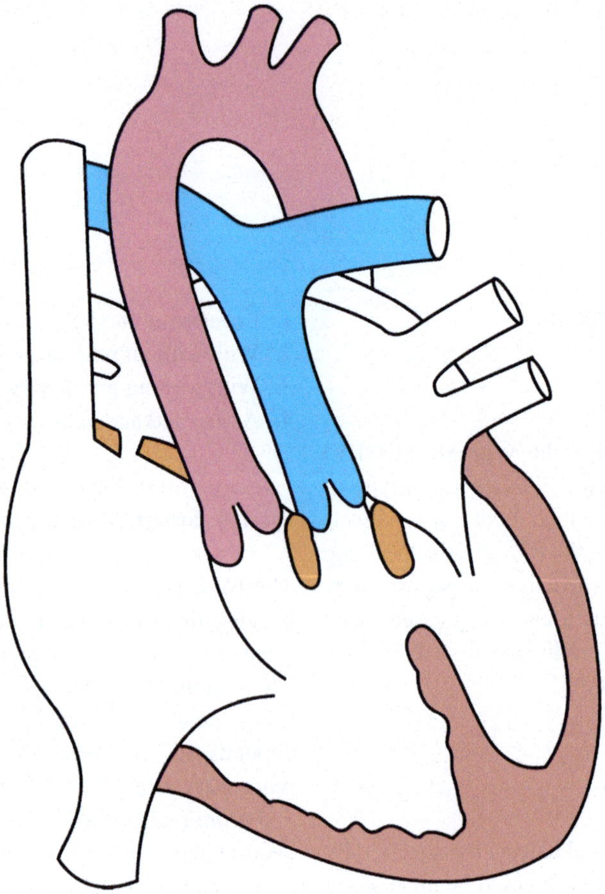

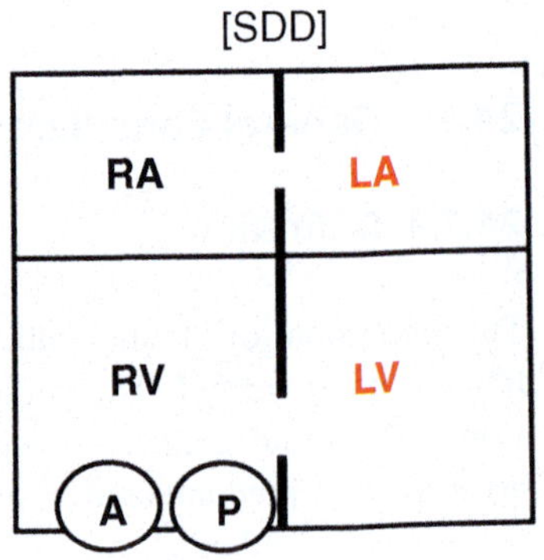

DORV+VSD+ASD

Fig. 24.2 (**a**) Specimen NO. 1079. DORV with AVSD and L-malposition. Morphological RV is on the left. (**b**) Specimen NO. 1079. Exposure of the left ventricle to demonstrate the outflow tract and leaflets. The PV is located in the upper right of the AV. (**c**) Specimen NO. 1079. Demonstration of the pulmonary valve, the thick cone muscle under the valve, and the morphological right ventricle. (**d**) Specimen NO. 1079. The bougie passing from the aortic valve to the ventricle, and the two muscle bundles under the valve. These characteristics confirmed that the aortic and pulmonary valves arise from the same ventricle. (**e**) Specimen NO. 1079. Outflow tract of morphological right ventricle. The *red line* indicates the blood flow in the aorta; the *arrow* indicates the pulmonary valve. (**f**) Specimen NO. 1079. Structure of the morphological right ventricle. (**g**) Specimen NO. 1079. The *red arrow* indicates the moderator band; the *blue arrow* indicates the APM. (**h**) Specimen NO. 1079. Cut of the right atrium and ventricle. An undeveloped right atrioventricular valve forms a common atrioventricular valve with the left one. (**i**) Specimen NO. 1079. Left-side view of the atrioventricular tract. (**j**) Schematic diagram of specimen 1079, it's a heart of complete atrioventricular septal defect. It's situs sulitus and L-loop. Atrium at right side is right atrium because of it's connection to vena cava. Atrium at left side is left atrium because of it's connection to pulmonary vein. *PA* pulmonary artery, *AO* ascending aorta, *RAA* right atrium auricle, *LAA* left atrium auricle, *PV* pulmonary valve, *RV* right ventricle, *LA* left atrium, *AVSD* atrioventricular septal defect, *RA* right atrium, *LV* left ventricle, *D* ventricular septal defect

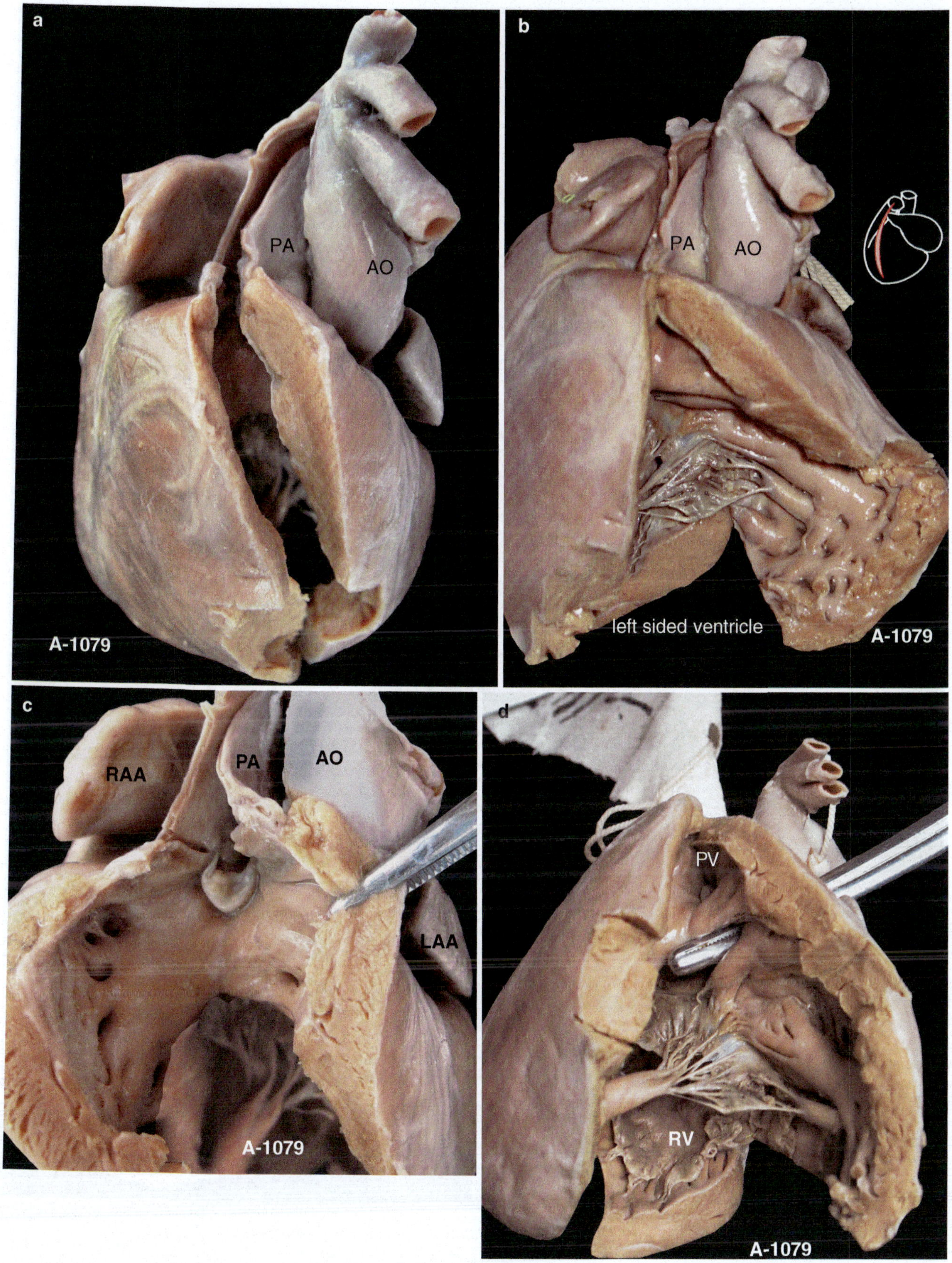
a
PA
AO
A-1079
b
PA
AO
left sided ventricle
A-1079
c
RAA
PA
AO
LAA
A-1079
d
PV
RV
A-1079

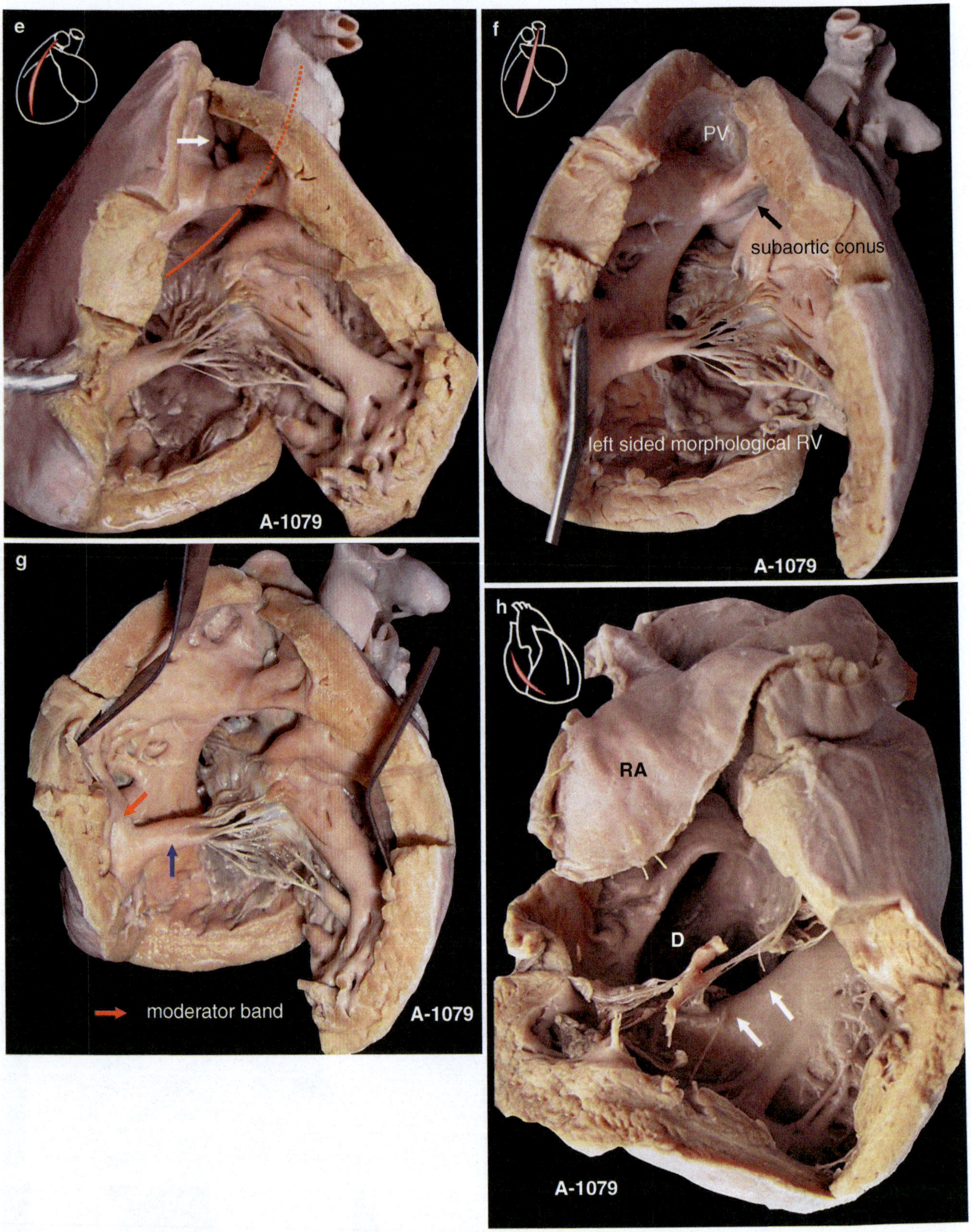

Fig. 24.2 (continued)

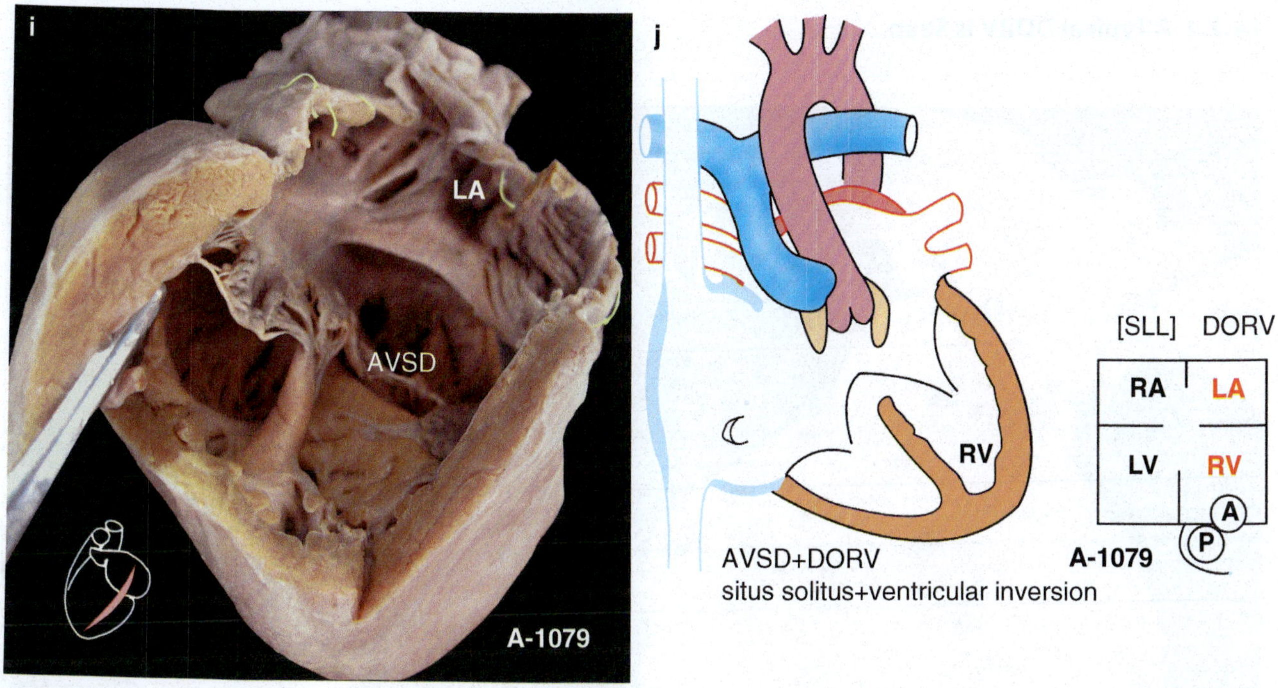

Fig. 24.2 (continued)

24.2.2 A Typical DORV Is Seen

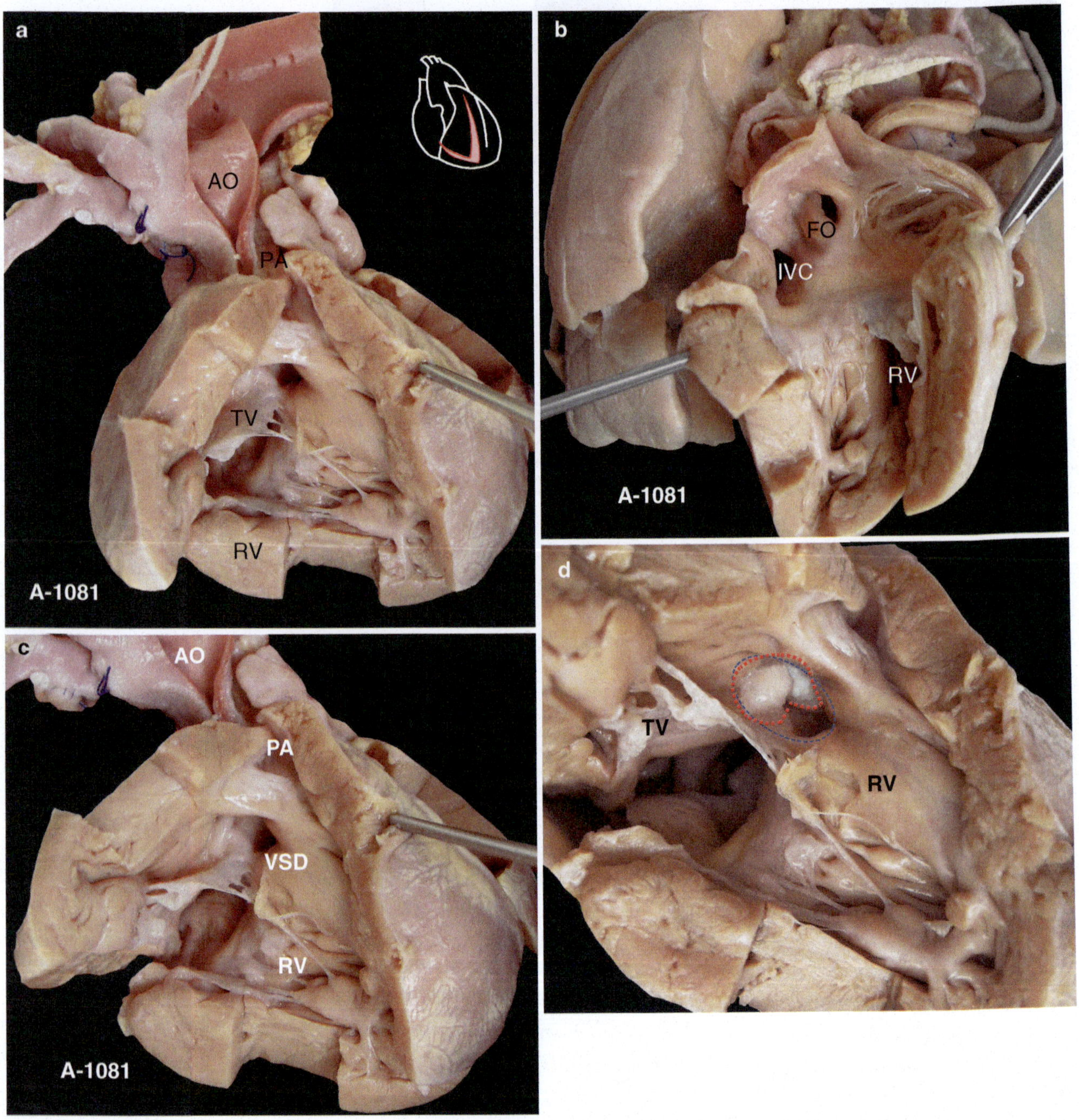

Fig. 24.3 (**a**) Specimen NO. 1081. Specimen of DORV. (**b**) Specimen NO. 1081. Exposure of the right atrium and ventricle to demonstrate its inflow tract. (**c**) Specimen NO. 1081. Demonstration of the outflow tract of the right ventricle. (**d**) Specimen NO. 1081. Exposure of the VSD from the outflow tract of the right ventricle. The AV overlies the VSD. (**e**) Specimen NO. 1081. Left ventricle cut to demonstrate its outflow tract. (**f**) Specimen NO. 1081. Left atrium and ventricle cut to demonstrate the left atrium, the mitral valve, and the inflow tract. *AO* ascending aorta, *PA* pulmonary artery, *TV* tricuspid valve, *RV* right ventricle, *FO* fossa ovalis, *IVC* inferior vena cava, *VSD* ventricular septal defect, *M* mitral valve, *LV* left ventricle, *LA* left atrium

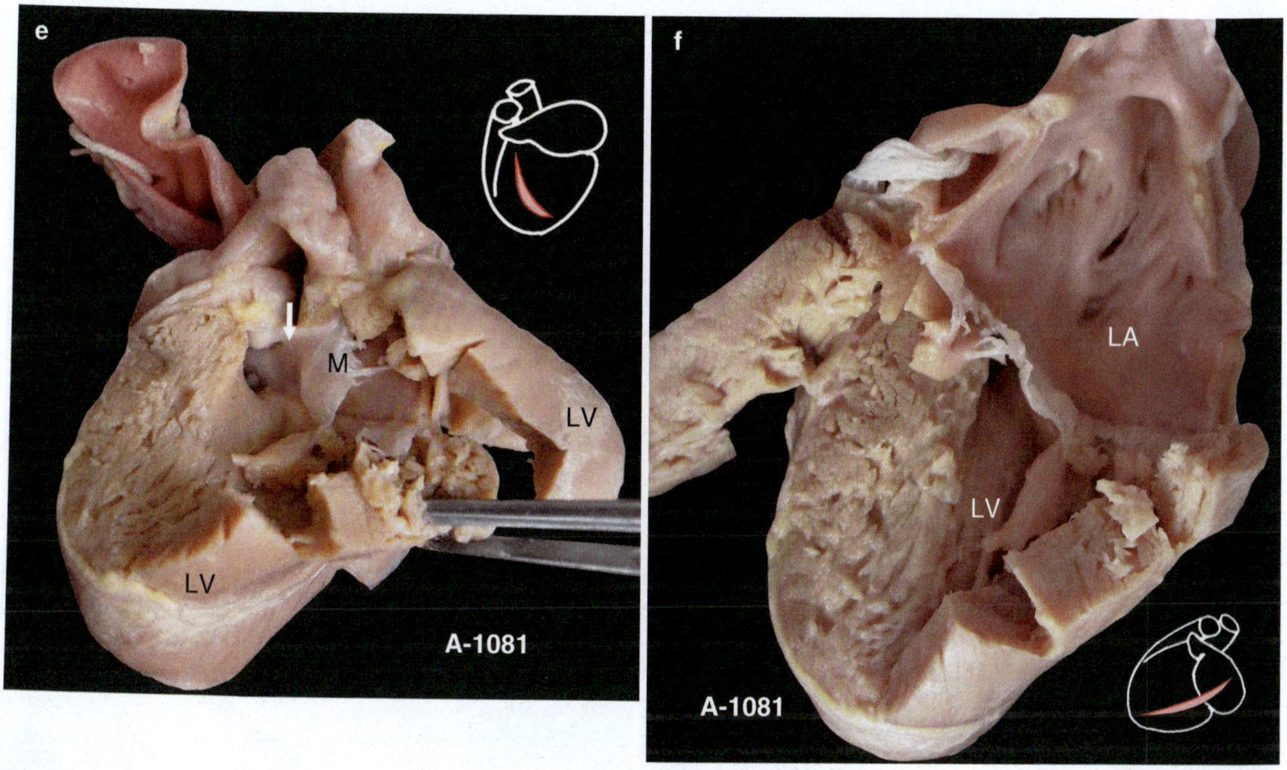

Fig. 24.3 (continued)

24.2.3 DORV, Dextrocardia, and Exterior Channel Surgery are Shown

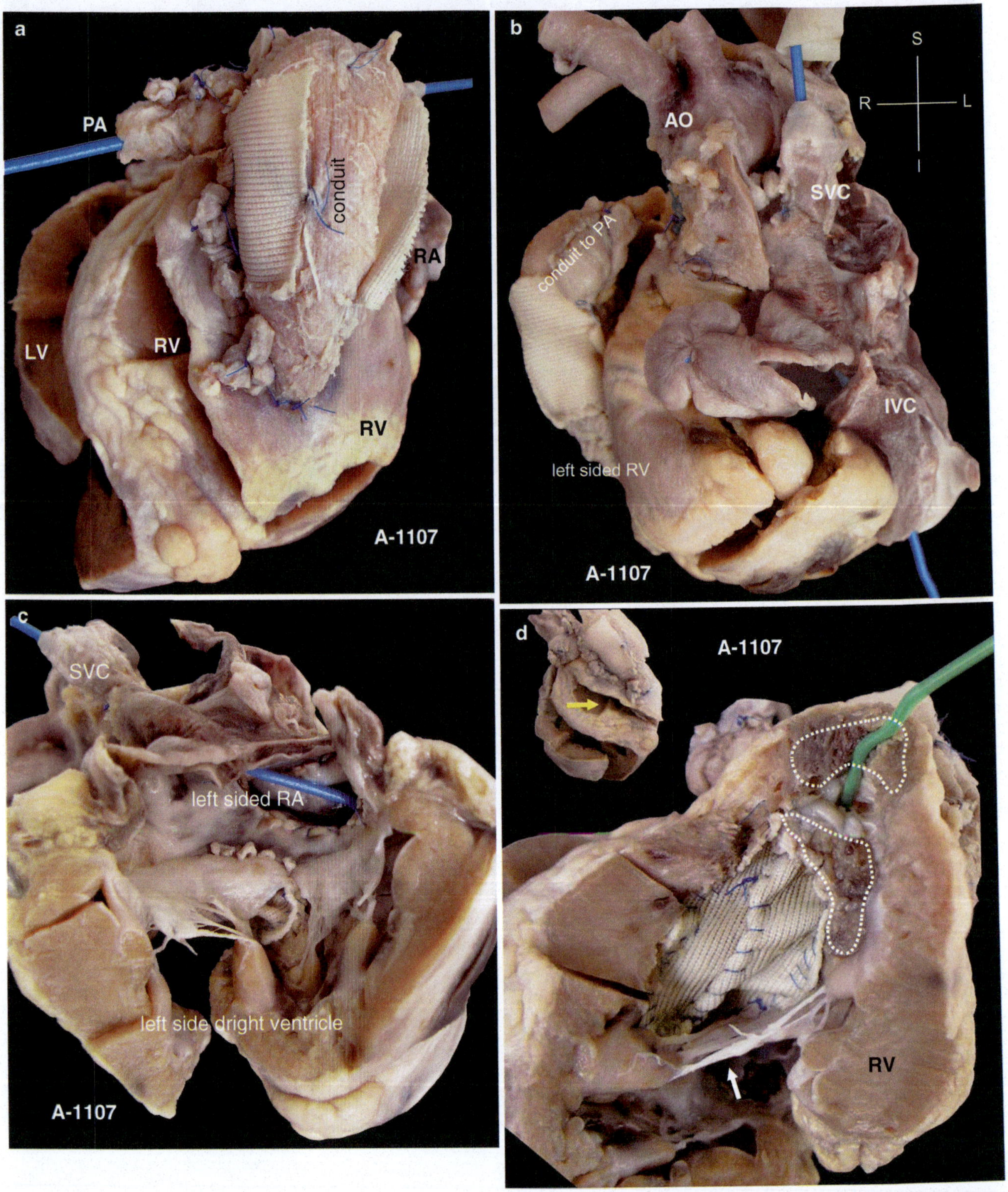

Fig. 24.4 (**a**) Specimen NO. 1107. DORV and dextrocardia, pulmonary stenosis. RV connects with the pulmonary artery with graft. (**b**) Specimen NO. 1107. Left anterior oblique view. RA and RV are on the left; bougie is in the vena cava. (**c**) Specimen NO. 1107. Left-side view. RA and RV are on the left; the VSD patch is in the TV. (**d**) Specimen NO. 1107. Opened outflow tract of the right ventricle. The aorta and PV arise from the RV. (**e**) Specimen NO. 1107. Clearer demonstration of the structure under the pulmonary valve. (**f**) Specimen NO. 1107. *Dotted line* indicates the connection of RV and the exterior channel. (**g**) Specimen NO. 1107. The MV is away from the AV, there is no fibrous connection, and the AV arises from opposite RV. The *arrow* indicates the VSD patch, which forms a part of LVOT. (**h**) Specimen NO. 1107. The same figure with 24-4 g with closer observation. (**i**) Specimen NO. 1107. Exposure of the left atrium and ventricle on the right. (**j**) Specimen NO. 1107. Schematic diagram of specimen 1107. *PA* pulmonary artery, *PV* pulmonary valve, *RV* right ventricle, *LA* left atrium, *AO* ascending aorta, *RA* right atrium, *LV* left ventricle, *TV* tricuspid valve, *IVC* inferior vena cava, *SVC* superior vena cava, *MV* mitral valve, *PV* pulmonary vein

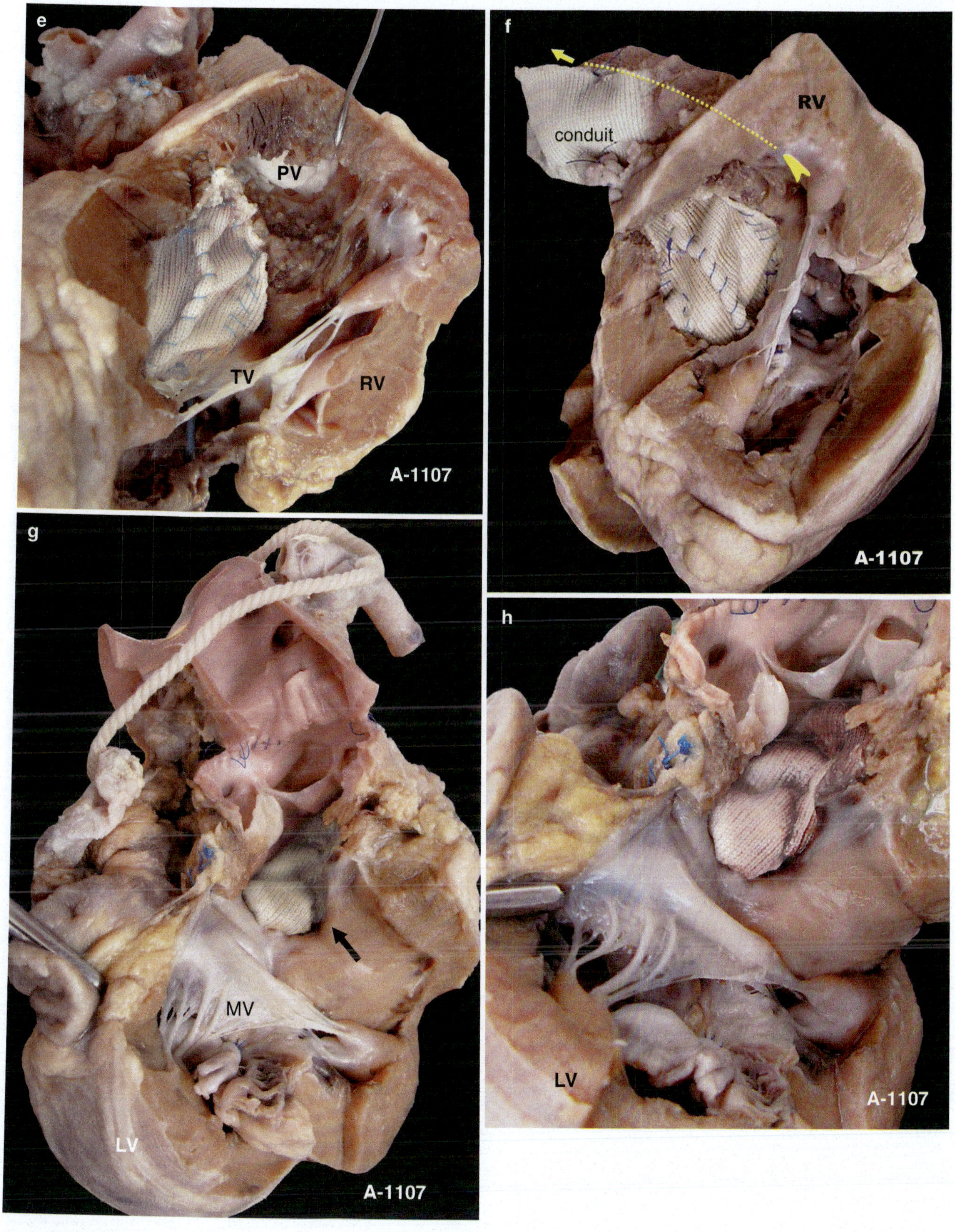

Fig. 24.4 (continued)

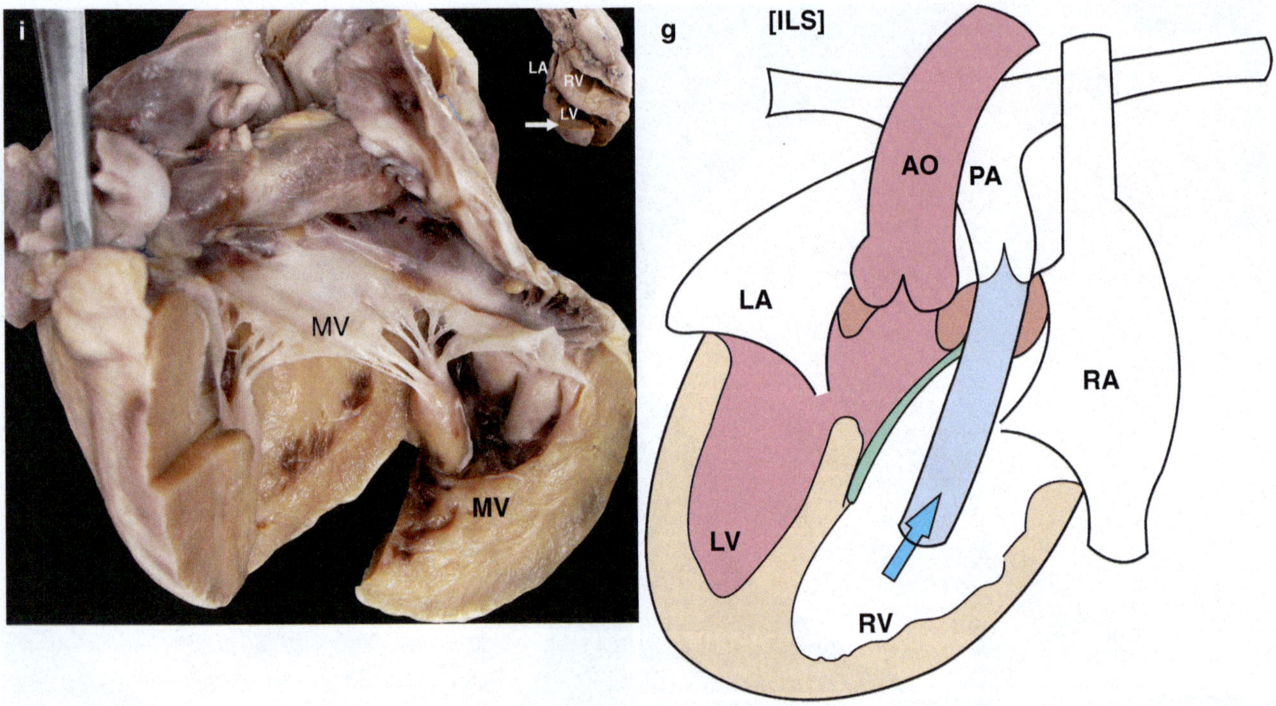

Fig. 24.4 (continued)

24.2.4 DORV, Hypoplasia of Pulmonary Artery

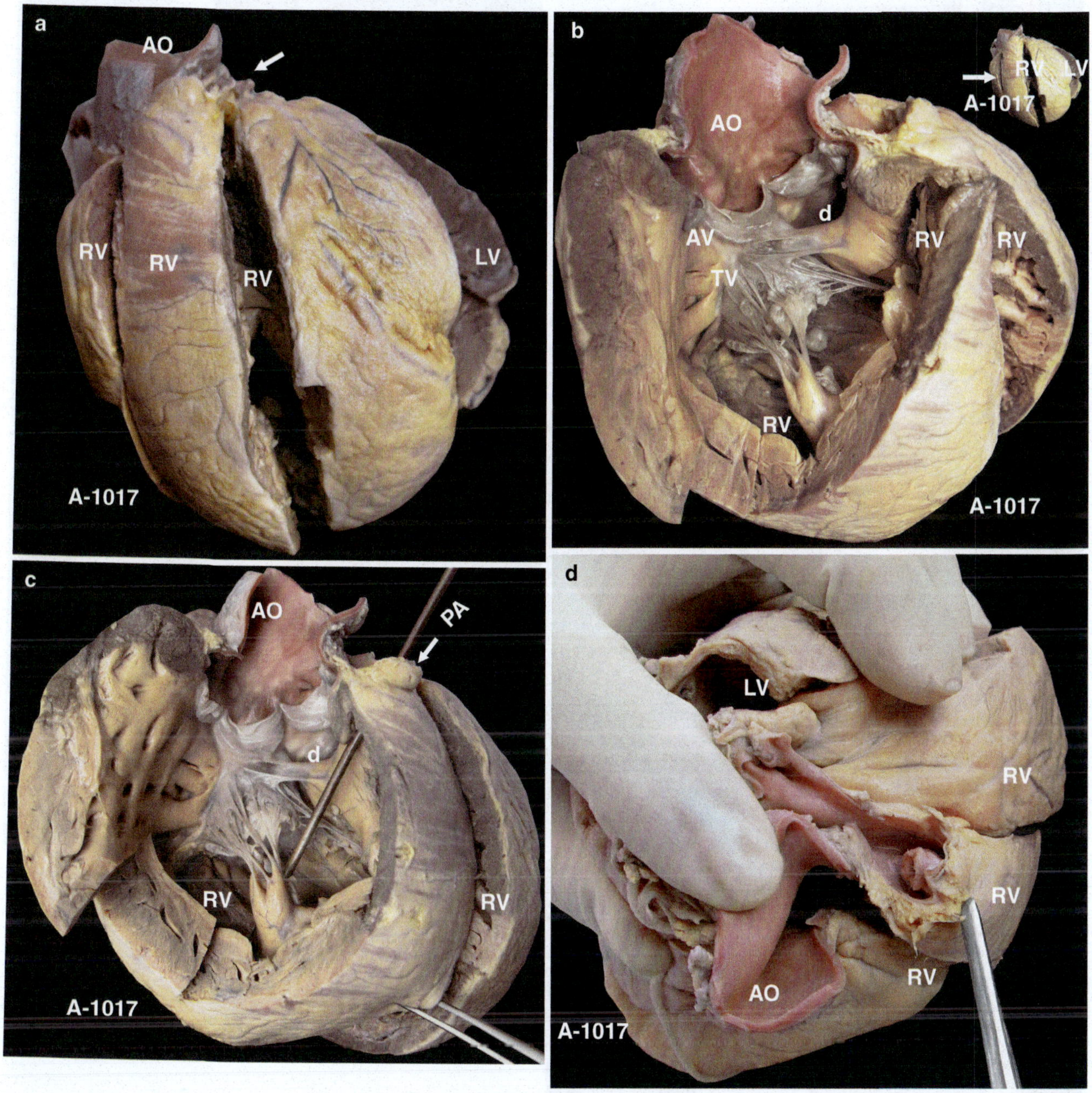

Fig. 24.5 (**a**) Specimen NO. 1017. Two incisions show the chamber of the right ventricle. (**b**) Specimen NO. 1017. The AV arises from the RV and with fibrous connection with the TV. The VSD locates in the infundibulum of the RV. (**c**) Specimen NO. 1017. Exposure of the chamber of the right ventricle. (**d**) Specimen NO. 1017. Top view of the pulmonary valve. (**e**) Specimen NO. 1017. The left front view. The bougie passes from the PV to the RV. (**f**) Specimen NO. 1017. The bougie passes through the PV. (**g**) Specimen NO. 1017. The LV is much smaller than the RV. (**h**) Specimen NO. 1017. Exposure of the inflow tract of right atrium and ventricle. (**i**) Specimen NO. 1017. Schematic diagram of specimen 1017. The aorta arises from the RV; the pulmonary is hypoplastic but not atresia. *PA* pulmonary artery, *AO* ascending aorta, *RV* right ventricle, *LA* left atrium, *RA* right atrium, *LV* left ventricle, *D* ventricular septal defect, *TV* tricuspid valve, *MV* mitral valve, *AV* aortic valve

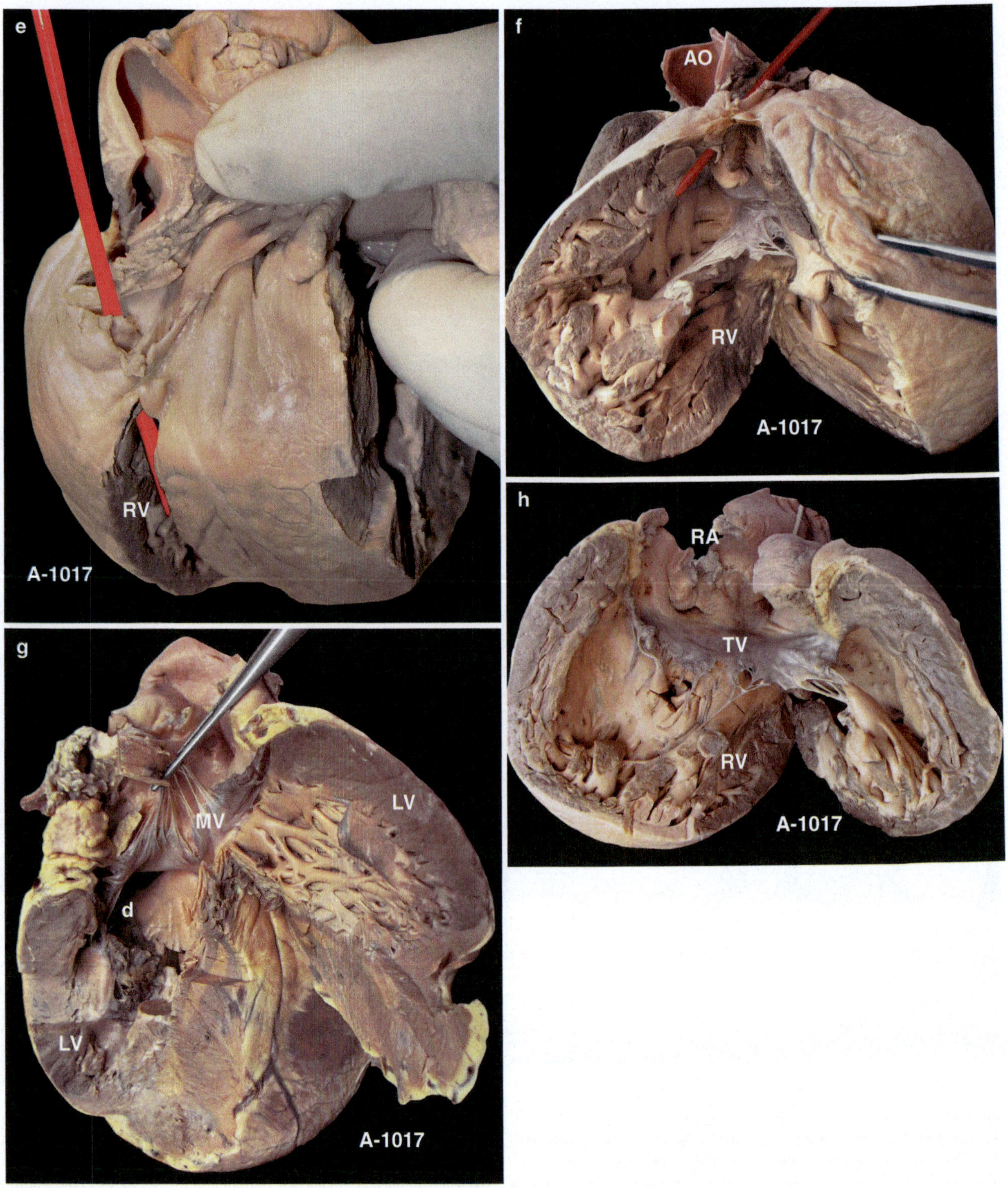

Fig. 24.5 (continued)

i

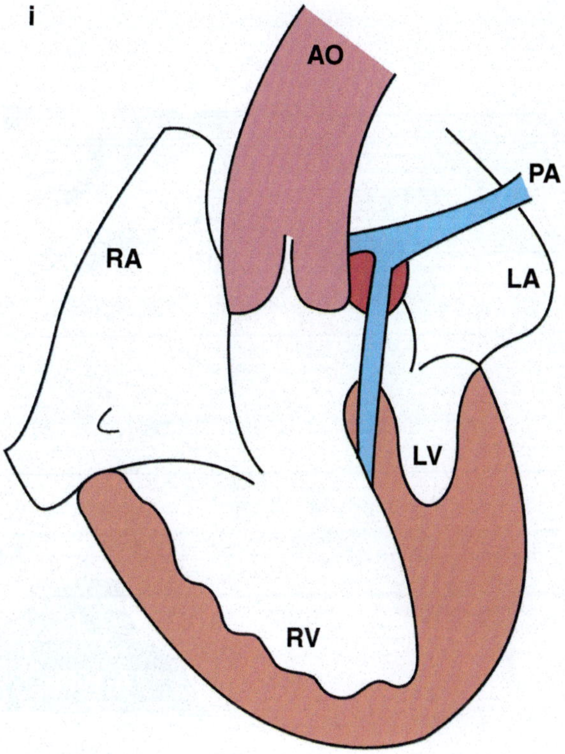

Fig. 24.5 (continued)

24.2.5 DORV, Subaortic VSD

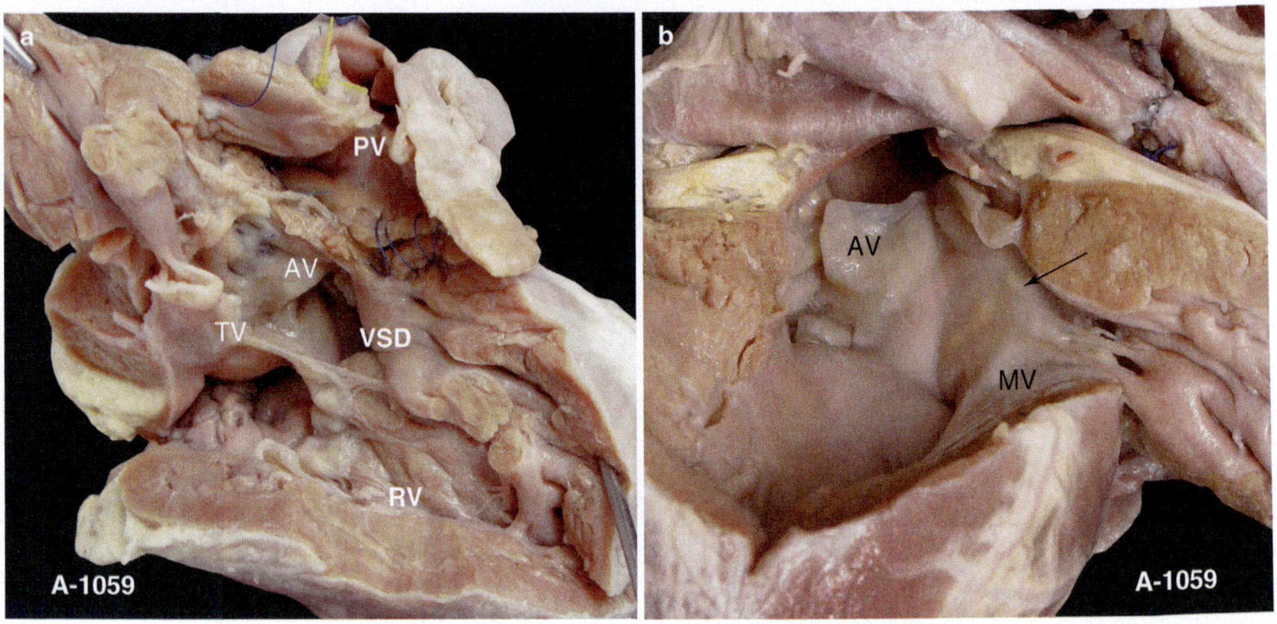

Fig. 24.6 (**a**) Specimen NO. 1059 DORV. The AV overlies the RV. (**b**) Specimen NO. 1059. The *arrow* indicates that the AV and THE PV are separated by muscle. *PV* pulmonary valve, *AV* aortic valve, *TV* tricuspid valve, *VSD* ventricular septal defect, *RV* right ventricle, *MV* mitral valve

24.2.6 DORV, Subpulmonary VSD

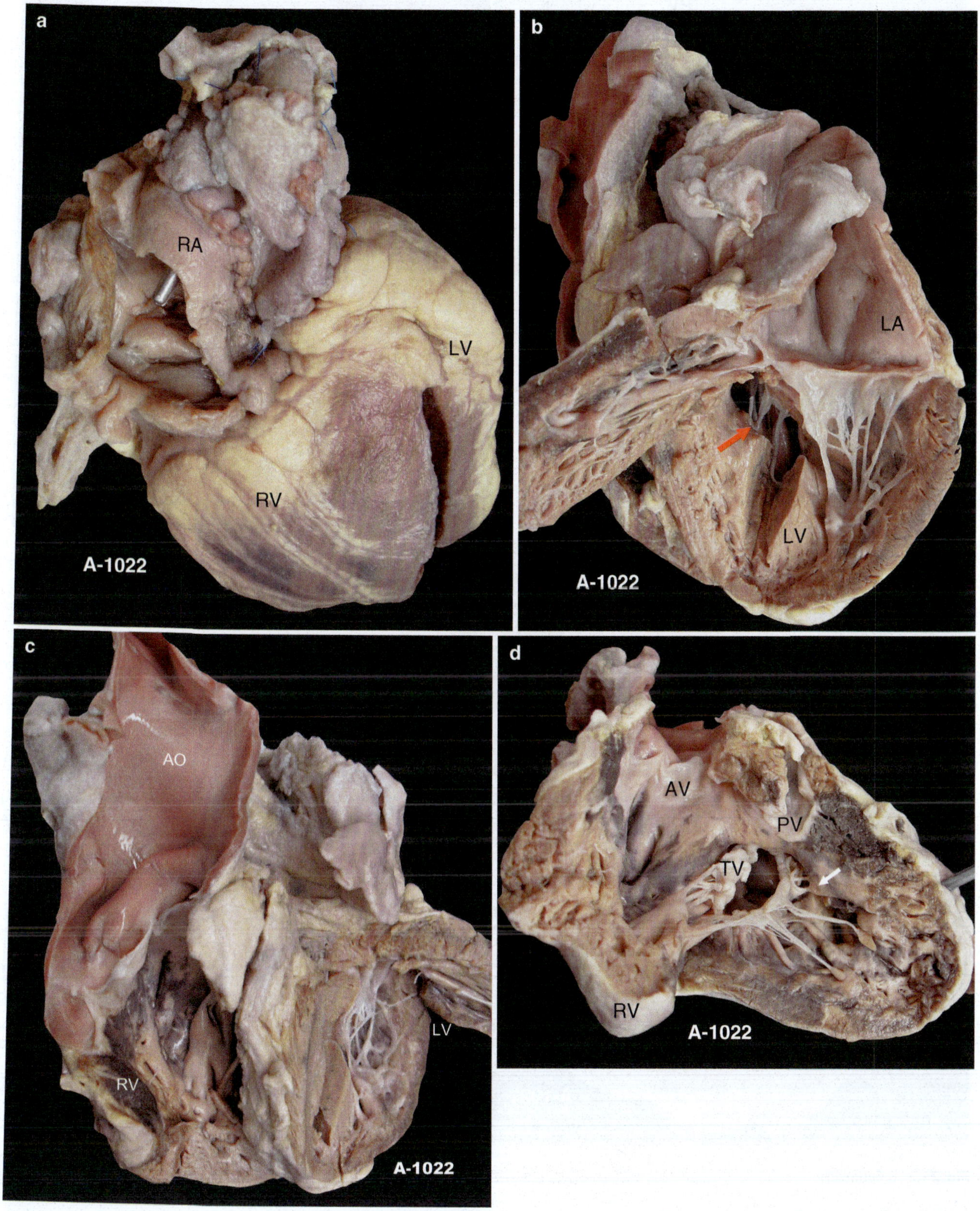

Fig. 24.7 (**a**) Specimen NO. 1022. Anterior view of DORV. (**b**) Specimen NO. 1022. Observation of the mitral valve from the left atrium and ventricle. No common atrioventricular valve is present. (**c**) Specimen NO. 1022. The left and right ventricles are cut. The aorta arises from the RV. (**d**) Specimen NO. 1022. Observation from the outflow tract of the right ventricle. The AV and PV arise from the RV. There is cone muscle tissue between the aortic and tricuspid valves. The VSD is under the PV. (**e**) Specimen NO. 1022. Closer observation of the subpulmonary ventricular septal defect. (**f**) Specimen NO. 1022. The PV arises from the RV. (**g**) Specimen NO. 1022. The *red arrow* indicates the lower border of the VSD. (**h**) Specimen NO. 1022. Schematic diagram showing DORV, double-cone, noncommitted VSD. *AO* ascending aorta, *PV* pulmonary valve, *PA* pulmonary artery, *RV* right ventricle, *LA* left atrium, *RA* right atrium, *LV* left ventricle, *TV* tricuspid valve, *VSD* ventricular septal defect, *MV* mitral valve, *AV* aortic valve, *PV* pulmonary vein

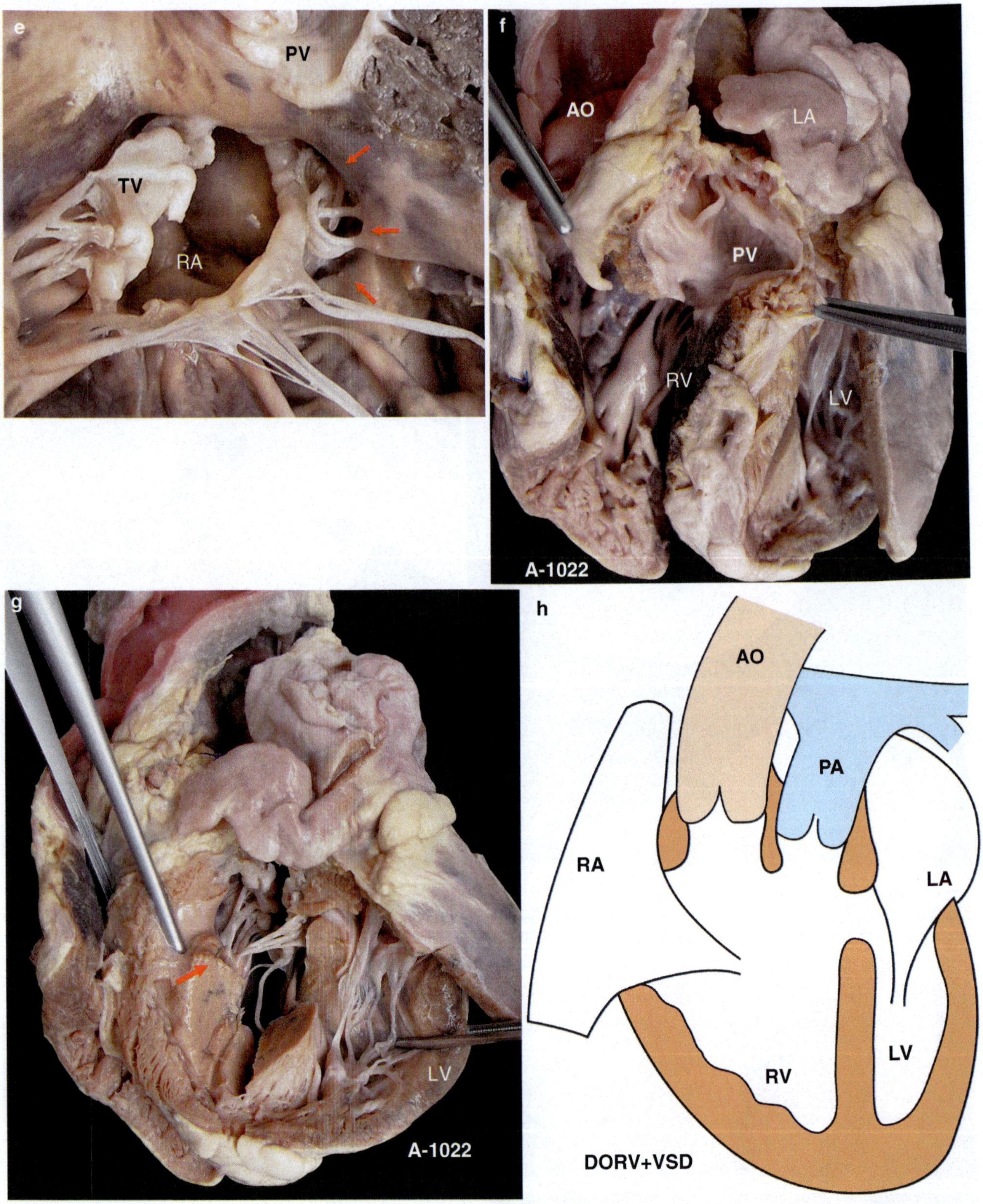

Fig. 24.7 (continued)

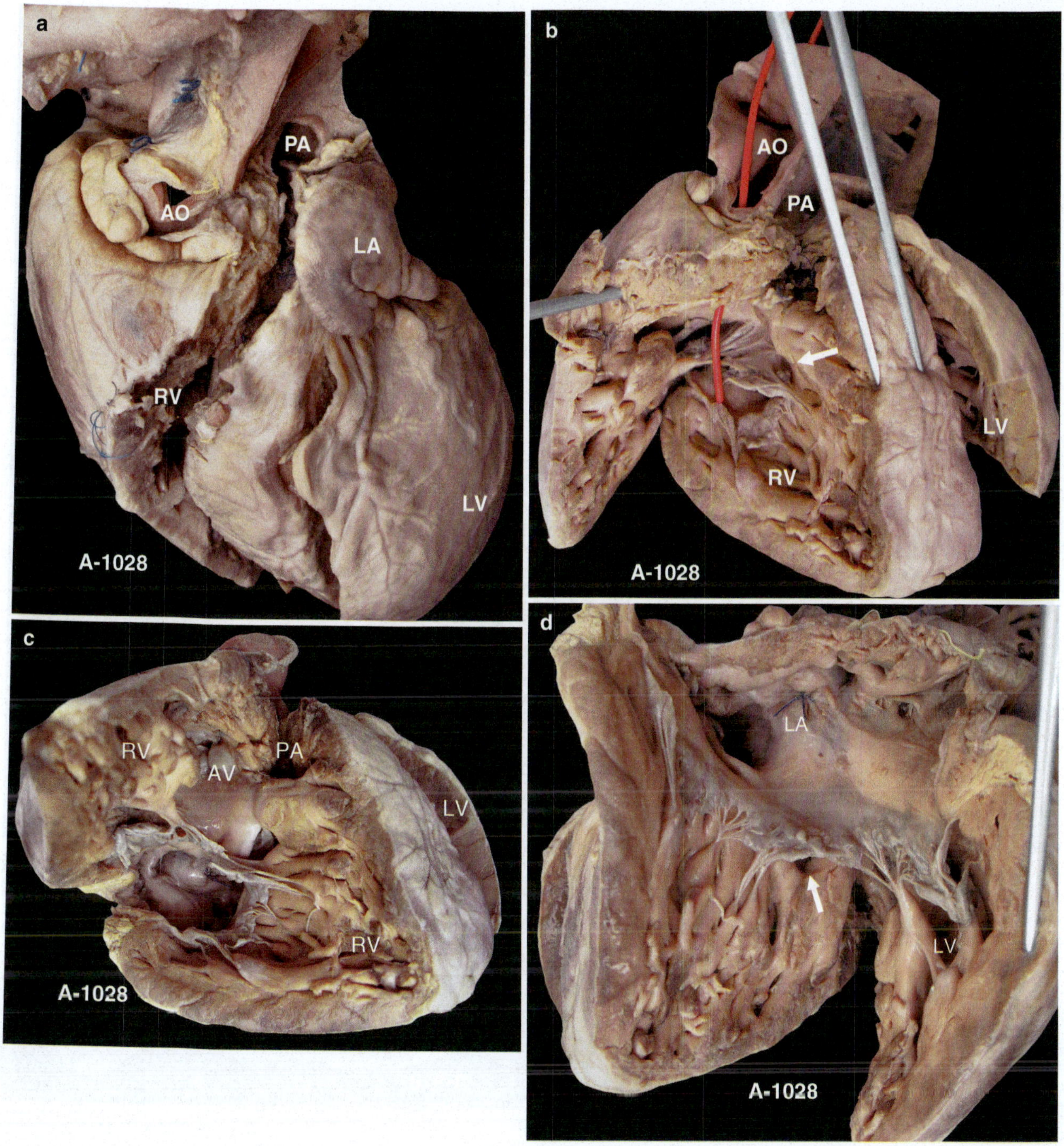

Fig. 24.8 (**a**) Specimen NO. 1028. Anterior view of DORV. (**b**) Specimen NO. 1028. The aorta and pulmonary artery arise from the RV. (**c**) Specimen NO. 1028. Demonstration of aortic valve, pulmonary valve, and ventricular septal defect from ROVT. (**d**) Specimen NO. 1028. Observation from the left atrium. *PA* pulmonary artery, *AO* ascending aorta, *RV* right ventricle, *LA* left atrium, *LV* left ventricle, *AV* aortic valve

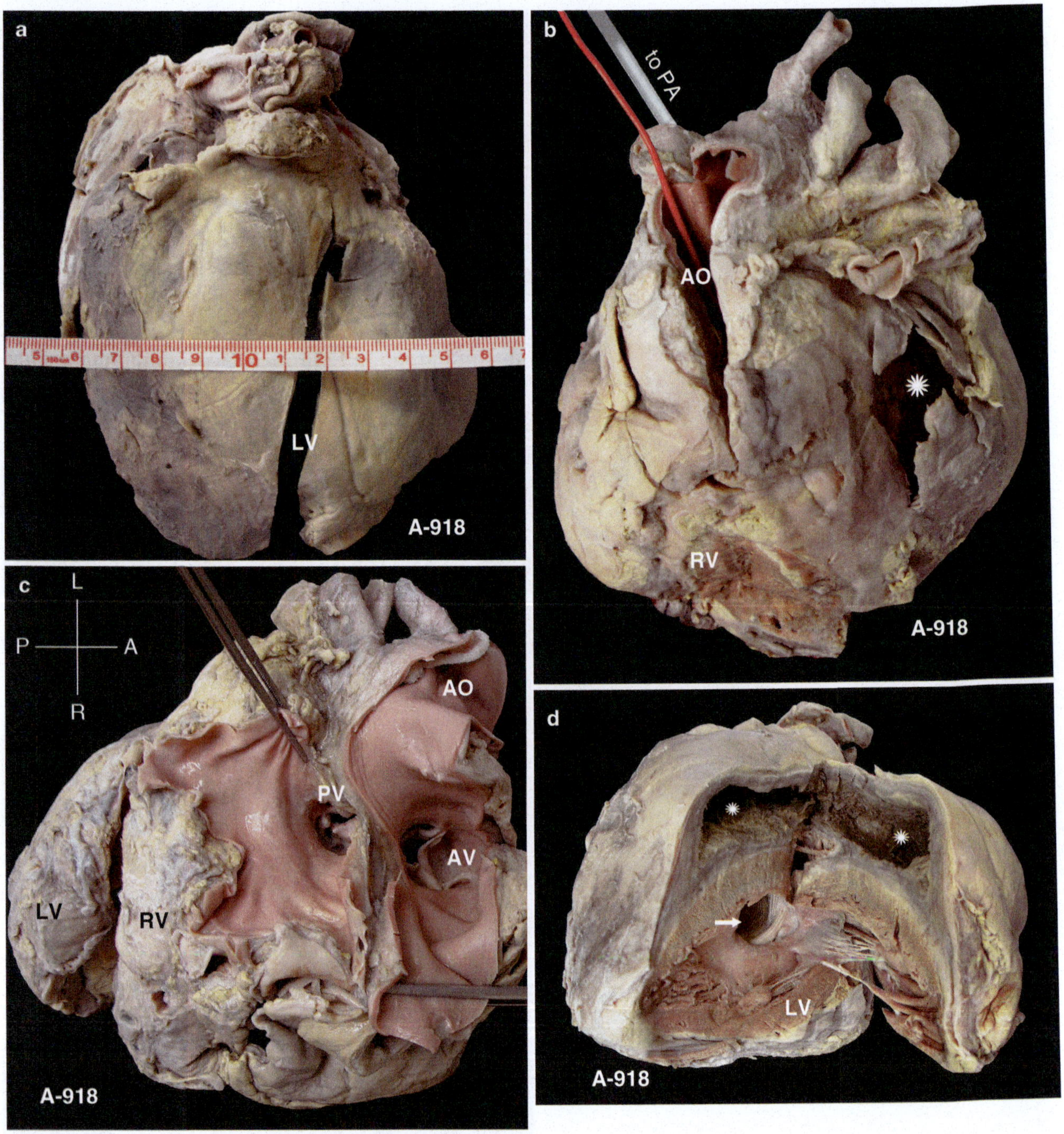

Fig. 24.9 (**a**) Specimen NO. 918. DORV. (**b**) Specimen NO. 918. DORV. (**c**) Specimen NO. 918. DORV top view from the right. (**d**) Specimen NO. 918. The LV and the aneurysm cavity are cut. (**e**) Specimen NO. 918. Demonstration of the interior channel from the outflow tract. (**g**) Specimen NO. 918. Normal inflow tract of the left atrium and ventricle. (**f**) Specimen NO. 918. Demonstration of the interior channel from the inflow tract. (**h**) Specimen NO. 918. Schematic diagram. *PA* pulmonary artery, *AO* ascending aorta, *PV* pulmonary valve, *RV* right ventricle, *LA* left atrium, *RA* right atrium, *LV* left ventricle, *TV* tricuspid valve, *VSD* ventricular septal defect, *AV* aortic valve, *PV* pulmonary vein

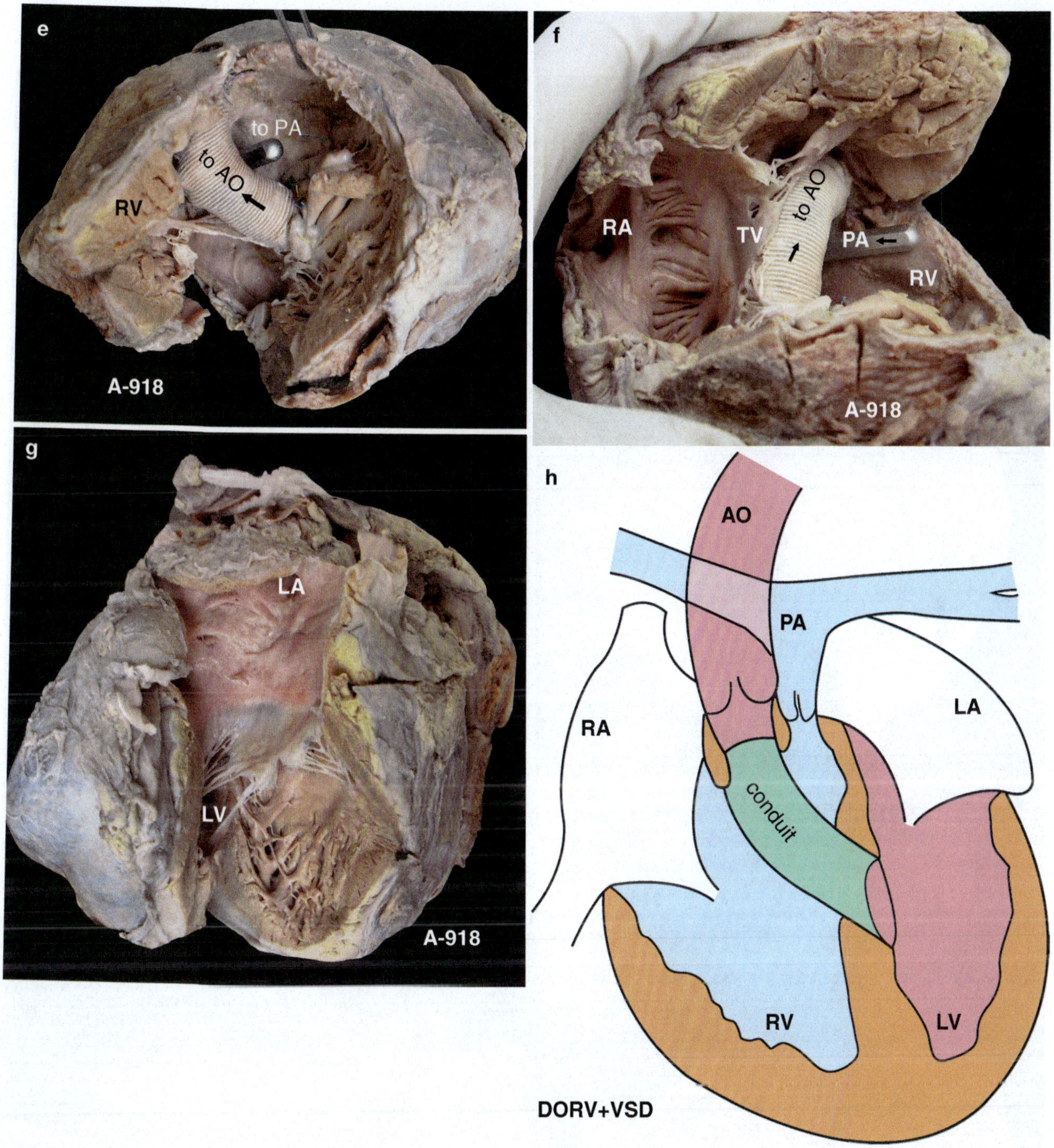

Fig. 24.9 (continued)

24.2.7 DORV, REV Surgery

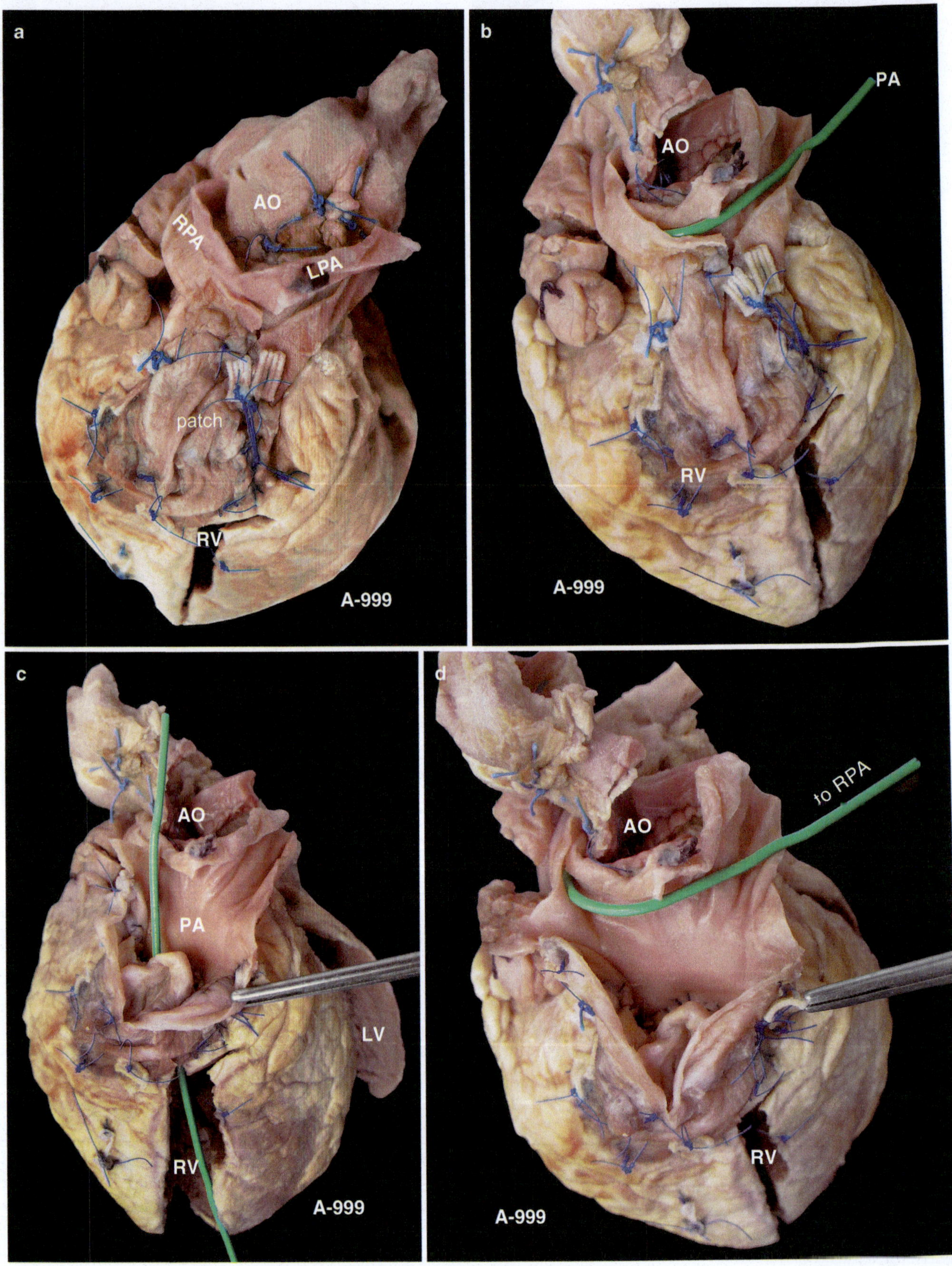

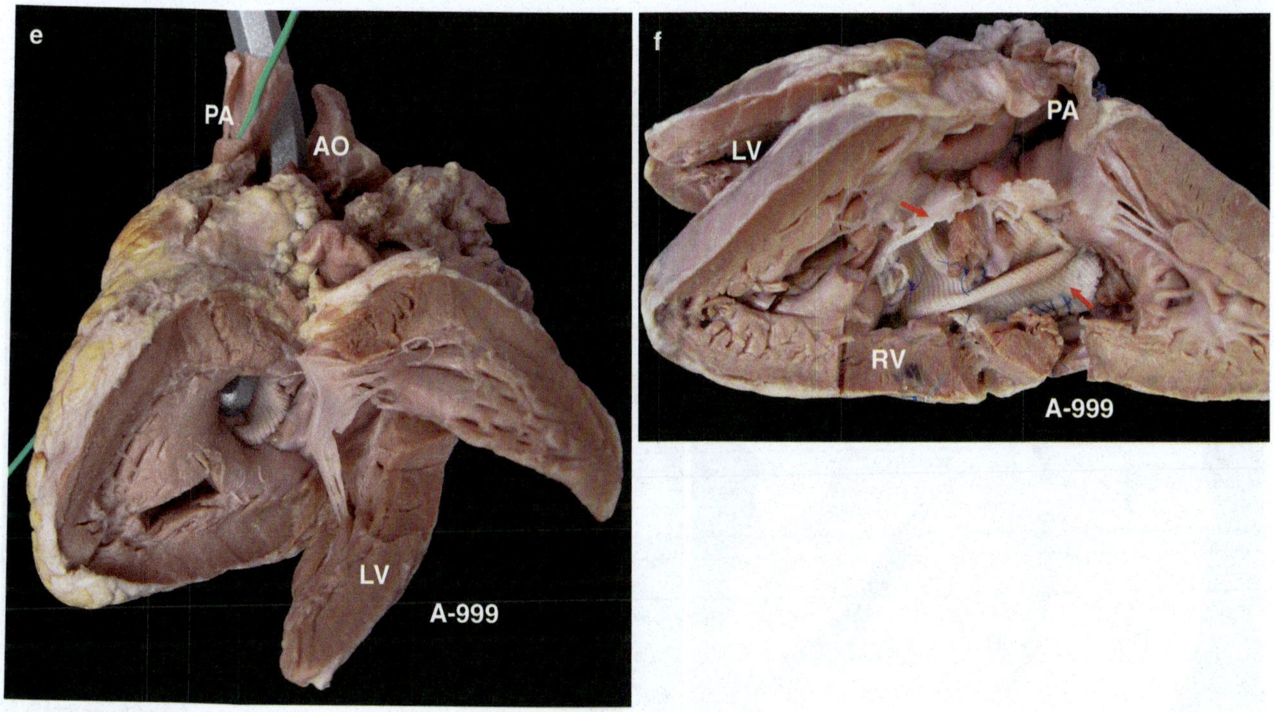

Fig. 24.10 (continued)

Fig. 24.10 (**a**) Specimen NO. 999. DORV. VSD patch separates the AV to LV. (**b**) Specimen NO. 999. The ascending aorta is cut, and the left and right pulmonary arteries are in front of the aorta. (**c**) Specimen NO. 999. *Green bougie* indicates the RVOT and the main pulmonary artery. (**d**) Specimen NO. 999. Left, right pulmonary arteries and right ventricle outflow tract. (**e**) Specimen NO. 999. The LV is cut to show the VSD patch. (**f**) Specimen NO. 999. The pulmonary and aortic valves arise from the RV; a huge VSD is seen apart from them. *PA* pulmonary artery, *AO* ascending aorta, *RV* right ventricle, *LV* left ventricle, *RPA* right pulmonary artery, *LPA* left pulmonary artery

24.2.8 DORV, Graft

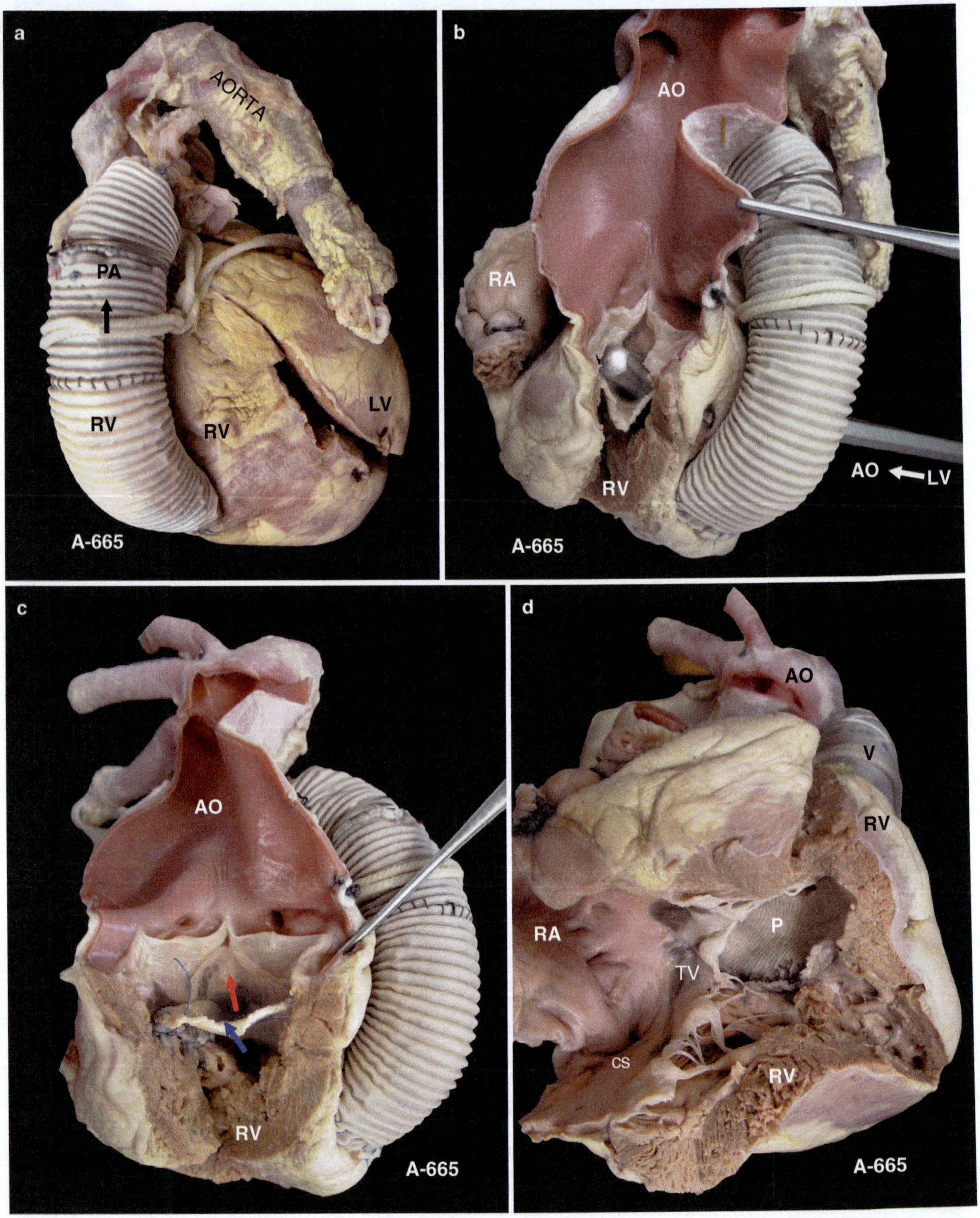

Fig. 24.11 (**a**) Specimen NO. 665. DORV+VSD. Muscle tissue is seen between the VSD and the great valves. (**b**) Specimen NO. 665. Overhead view. The PV connects with the graft. (**c**) Specimen NO. 665. Overhead view of the subaortic valve structure. (**d**) Specimen NO. 665. The right atrium and ventricle are cut. (**e**) Specimen NO. 665. The blood flow direction after the correction of DORV. (**f**) Specimen NO. 665. Cut the left ventricle to show the VSD. (**g**) Specimen NO. 665. Normal left atrium and mitral valve. (**h**) Specimen NO. 665. Schematic diagram before the correction is made. (**i**) Specimen NO. 665. Schematic diagram after the correction is made. *PA* pulmonary artery, *AO* ascending aorta, *RV* right ventricle, *LA* left atrium, *RA* right atrium, *LV* left ventricle, *TV* tricuspid valve, *CS* coronary sinus

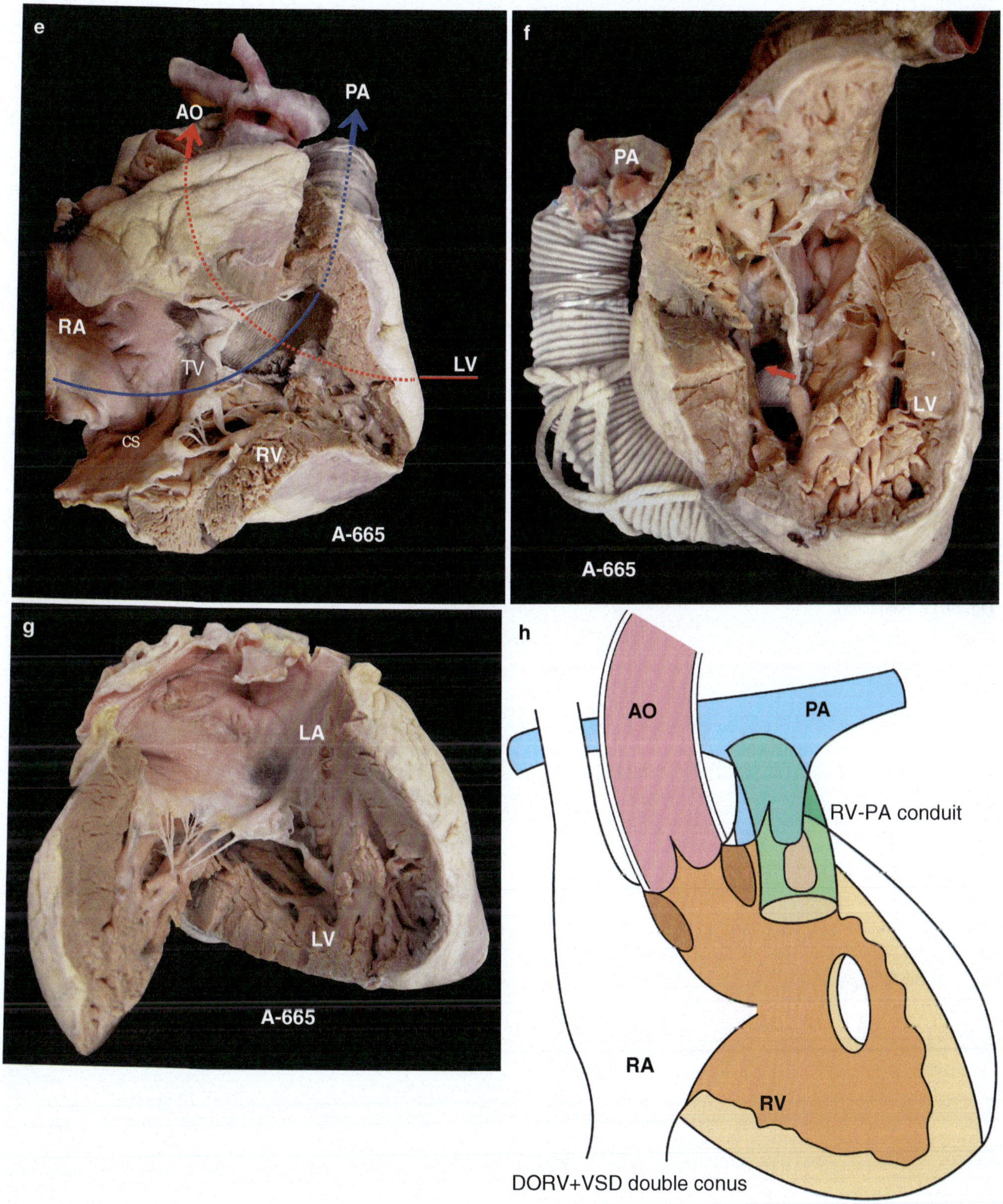

Fig. 24.11 (continued)

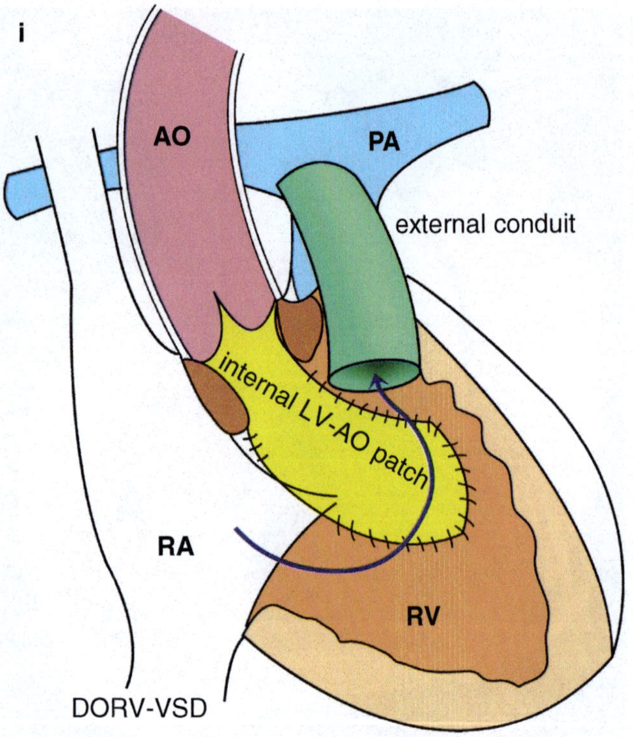

Fig. 24.11 (continued)

Fig. 24.12 (**a**) Specimen NO. 981. Anterior view of DORV with pulmonary stenosis. (**b**) Specimen NO. 981. The *blue bougie* is in the RV and pulmonary artery. (**c**) Specimen NO. 981. The *blue bougie* passes through the pulmonary artery from the RVOT, and the VSD patch separates the AV to the LV. (**d**) Specimen NO. 981. The AV arises from the RV, with no fibrous connection between the AV and the MV. (**e**) Specimen NO. 981. The *red bougie* passes through the VSD from the LV to the aorta. (**f**) Specimen NO. 981. Intact atrium septum. *PA* pulmonary artery, *AO* ascending aorta, *RV* right ventricle, *LA* left atrium, *RA* right atrium, *LV* left ventricle, *D* ventricular septal defect, *MV* mitral valve

24.2.9 DORV, Pulmonary Stenosis

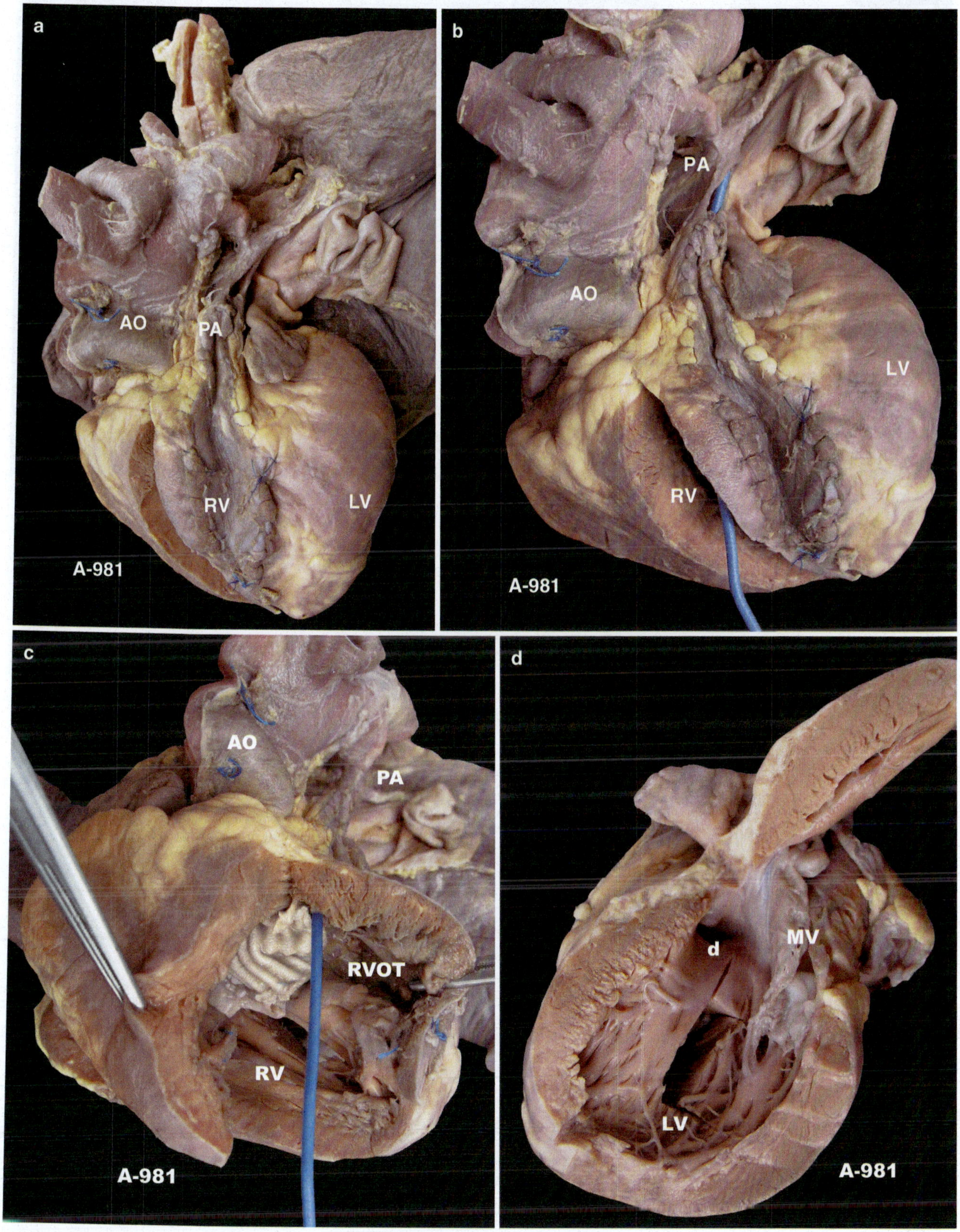

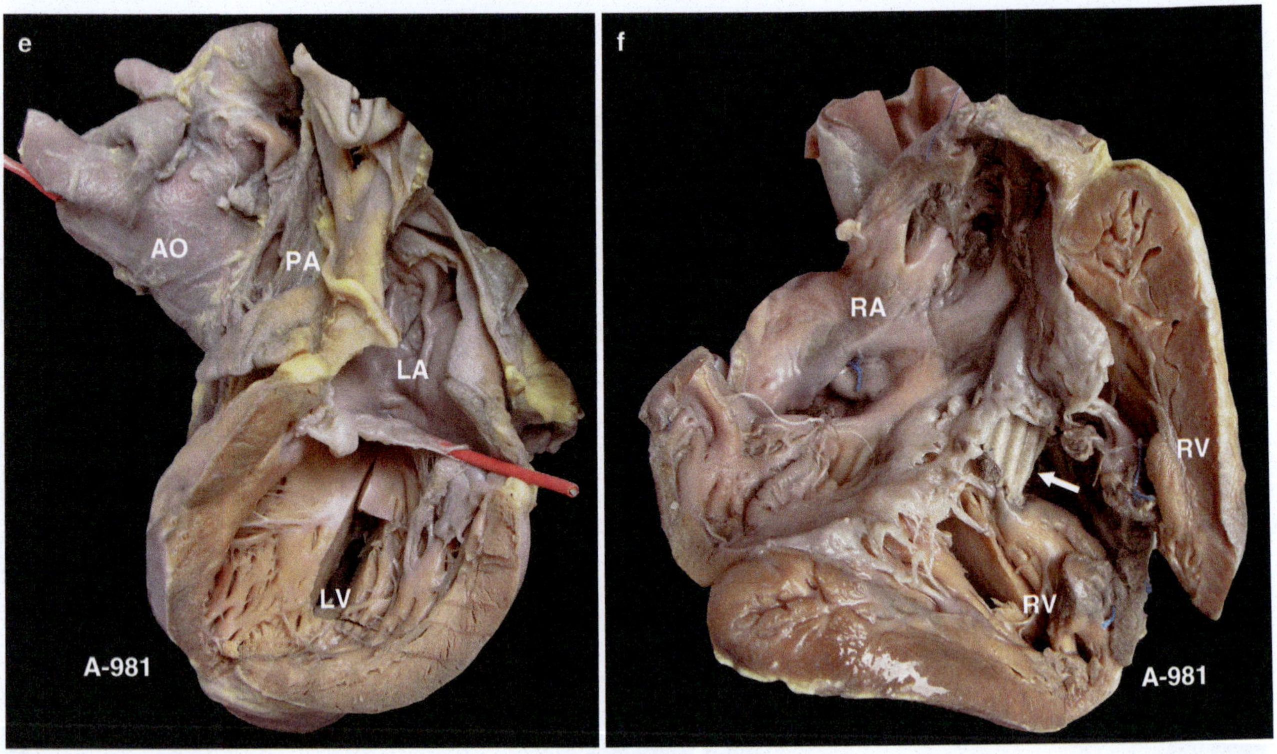

Fig. 24.12 (continued)

Fig. 24.13 (**a**) Specimen NO. 915. DORV, exterior conduit between the right ventricle and main pulmonary artery. (**b**) Specimen NO. 915. The LV is apart from the aorta; exterior conduit of homograft is seen. (**c**) Specimen NO. 915. Demonstration of the anastomosis of the exterior conduit with pulmonary artery. (**d**) Specimen NO. 915. Top view of the relationship of the aortic valve, right ventricle, pulmonary valve, and exterior conduit. (**e**) Specimen NO. 915. The LV is cut to demonstrate that the aorta arises from the RV. (**f**) Specimen NO. 915. Schematic diagram. *RA* right atrium, *AO* ascending aorta, *PA* pulmonary artery, *RV* right ventricle, *AV* aortic valve, *LV* left ventricle, *D* ventricular septal defect

24.2.10 DORV, Exterior Conduit

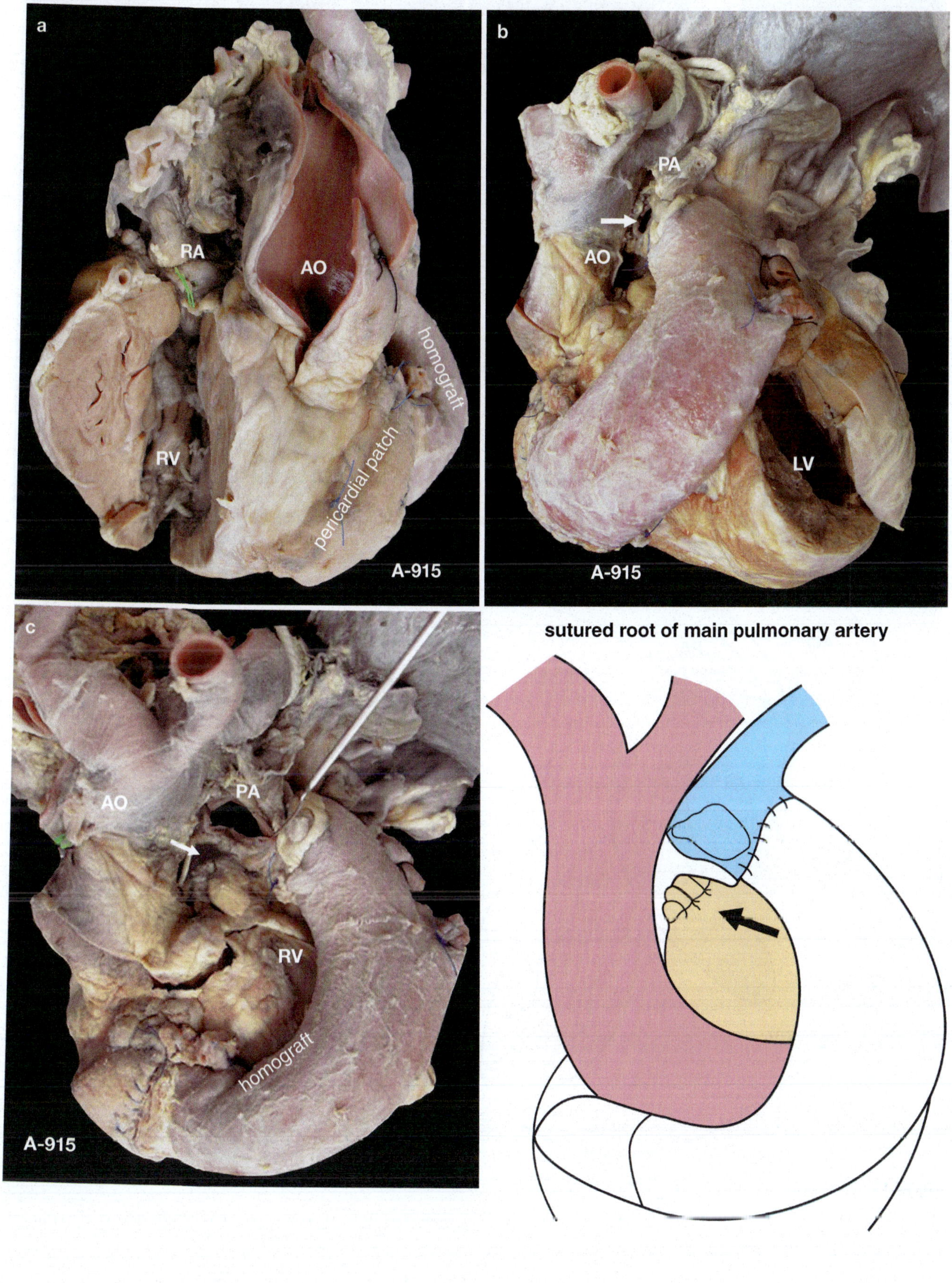

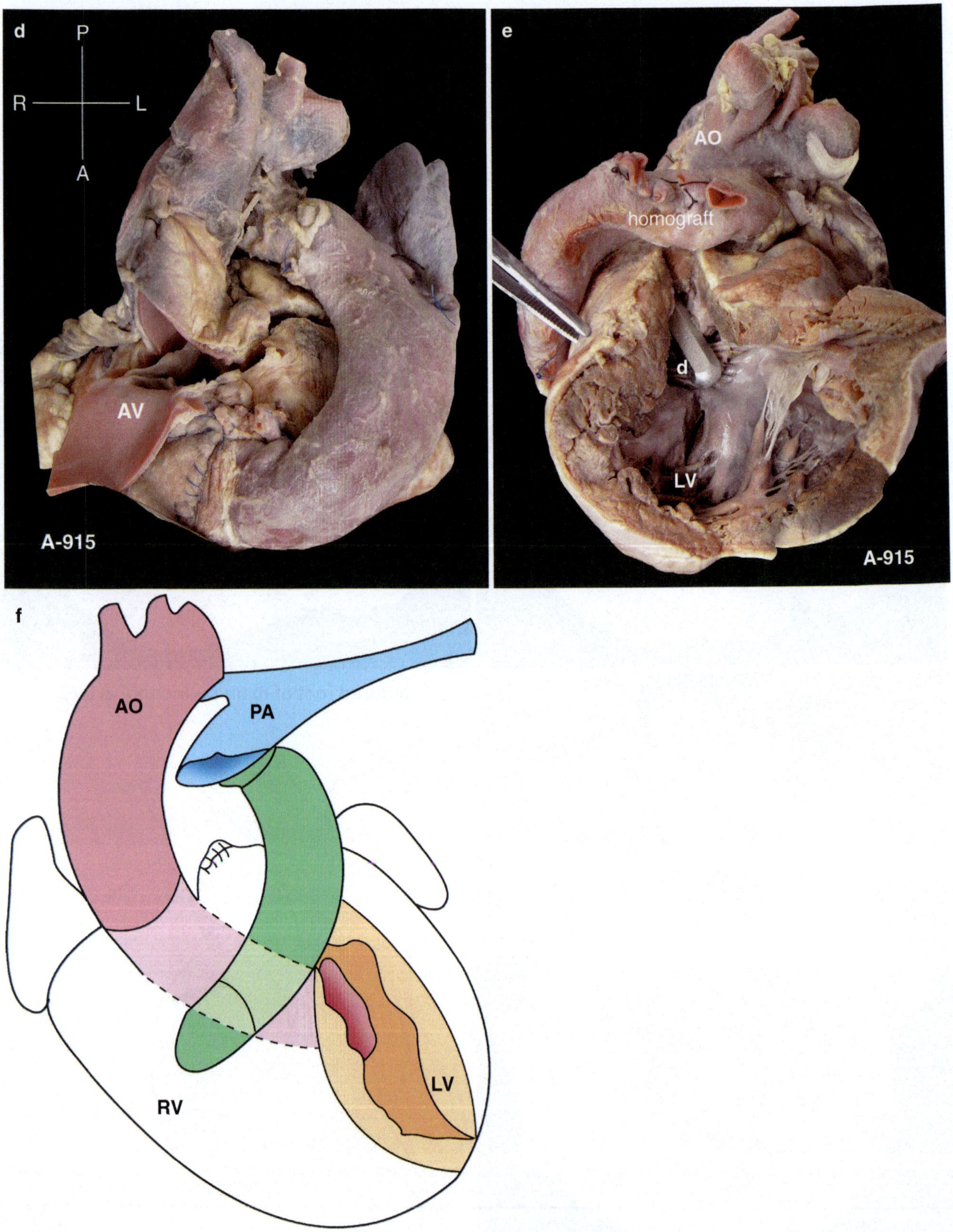

Fig. 24.13 (continued)

24.2.11 Surgery

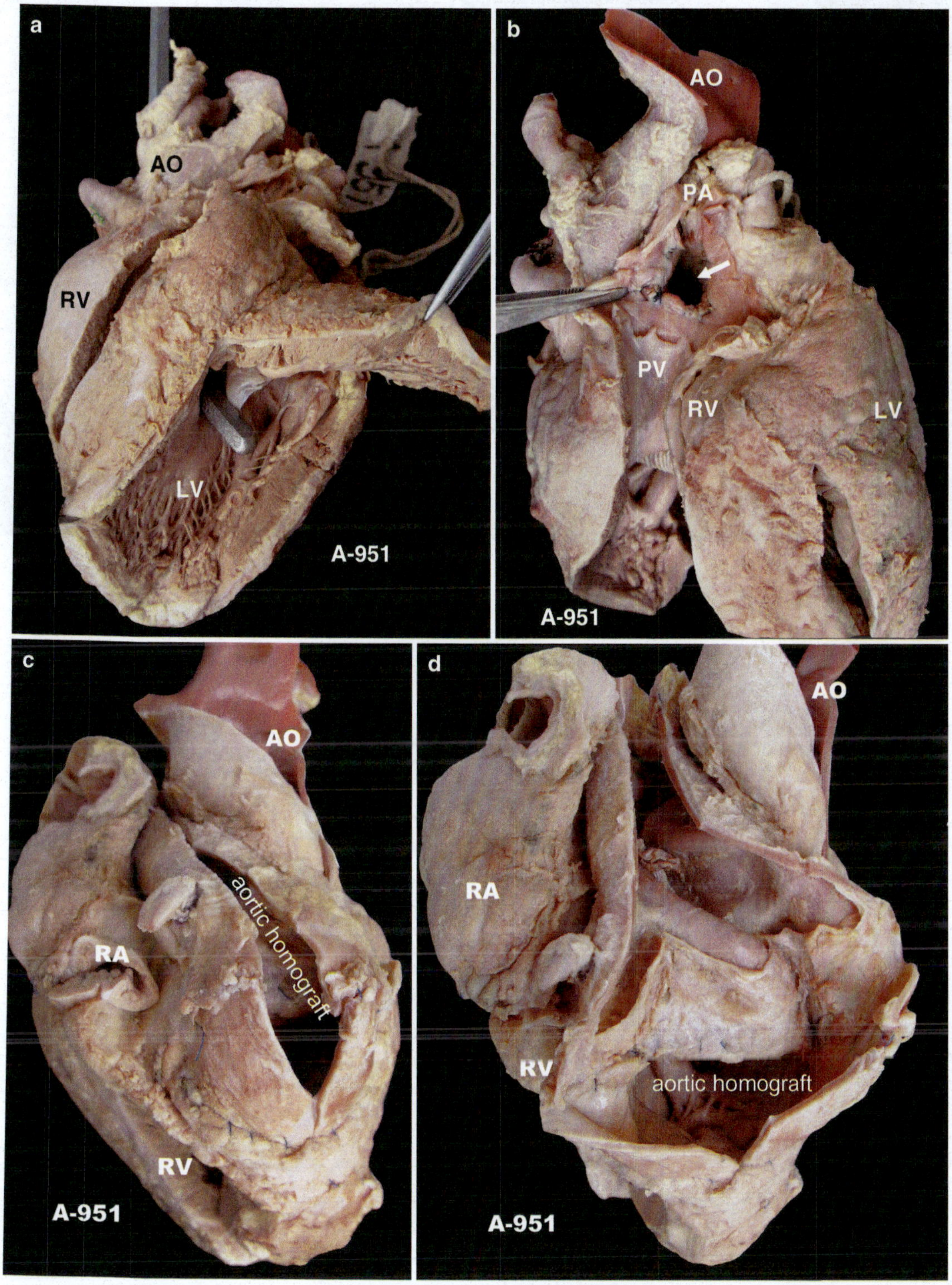

Fig. 24.14 (**a**) Specimen NO. 951. DORV view from the left ventricle. (**b**) Specimen NO. 951. The ROVT is cut. (**c**) Specimen NO. 951. The exterior conduit is cut between the RV and the pulmonary artery. (**d**) Specimen NO. 951. The exterior conduit is cut. (**e**) Specimen NO. 951. Exterior conduit, the view from the right ventricle. (**f**) Specimen NO. 951. Left ventricle inflow tract (LVIT). (**g**) Specimen NO. 951. Schematic diagram. *PA* pulmonary artery, *AO* ascending aorta, *PV* pulmonary valve, *RV* right ventricle, *LA* left atrium, *RA* right atrium, *LV* left ventricle, *VSD* ventricular septal defect, *PV* pulmonary vein

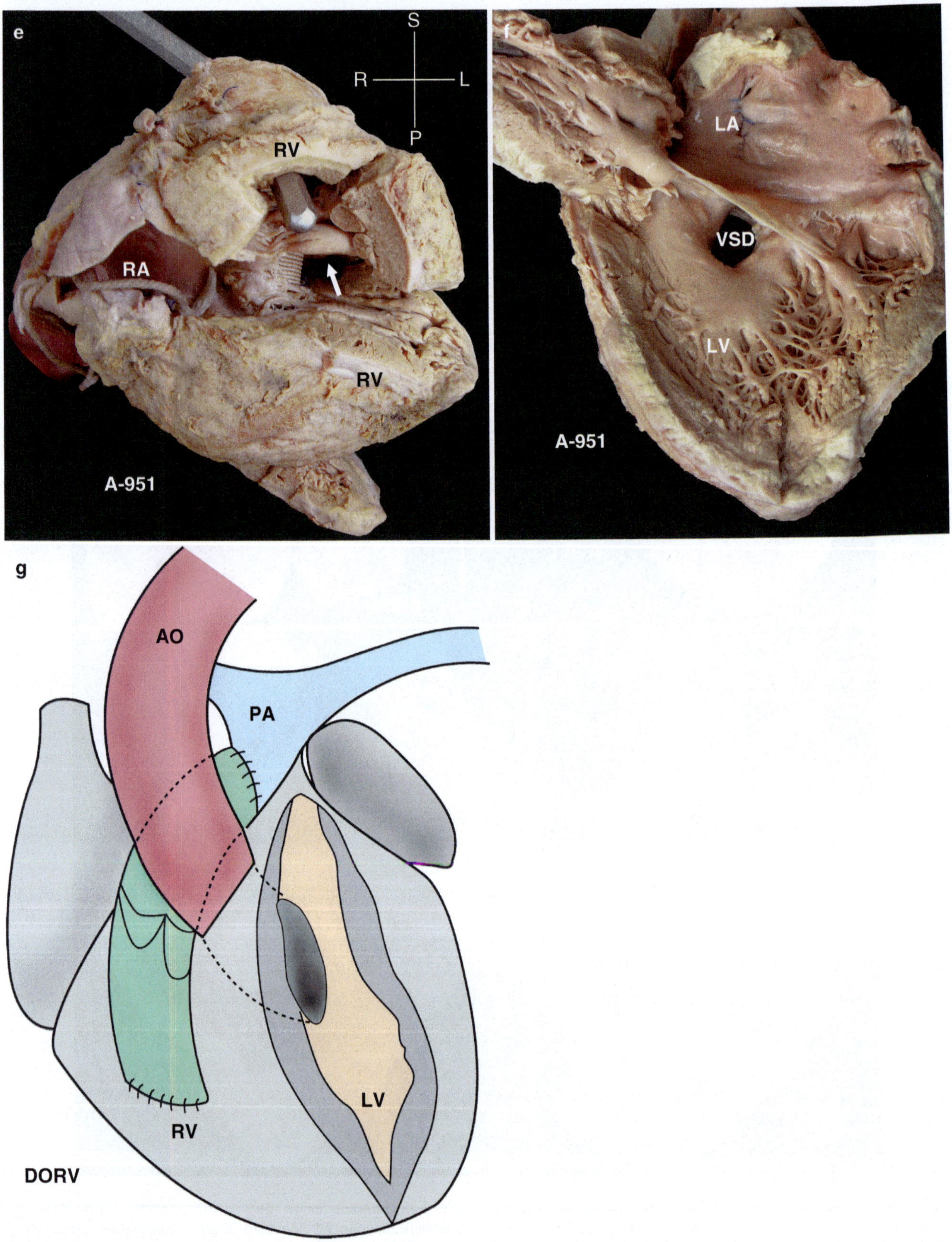

Fig. 24.14 (continued)

24.2.12 DORV, Subpulmonary VSD

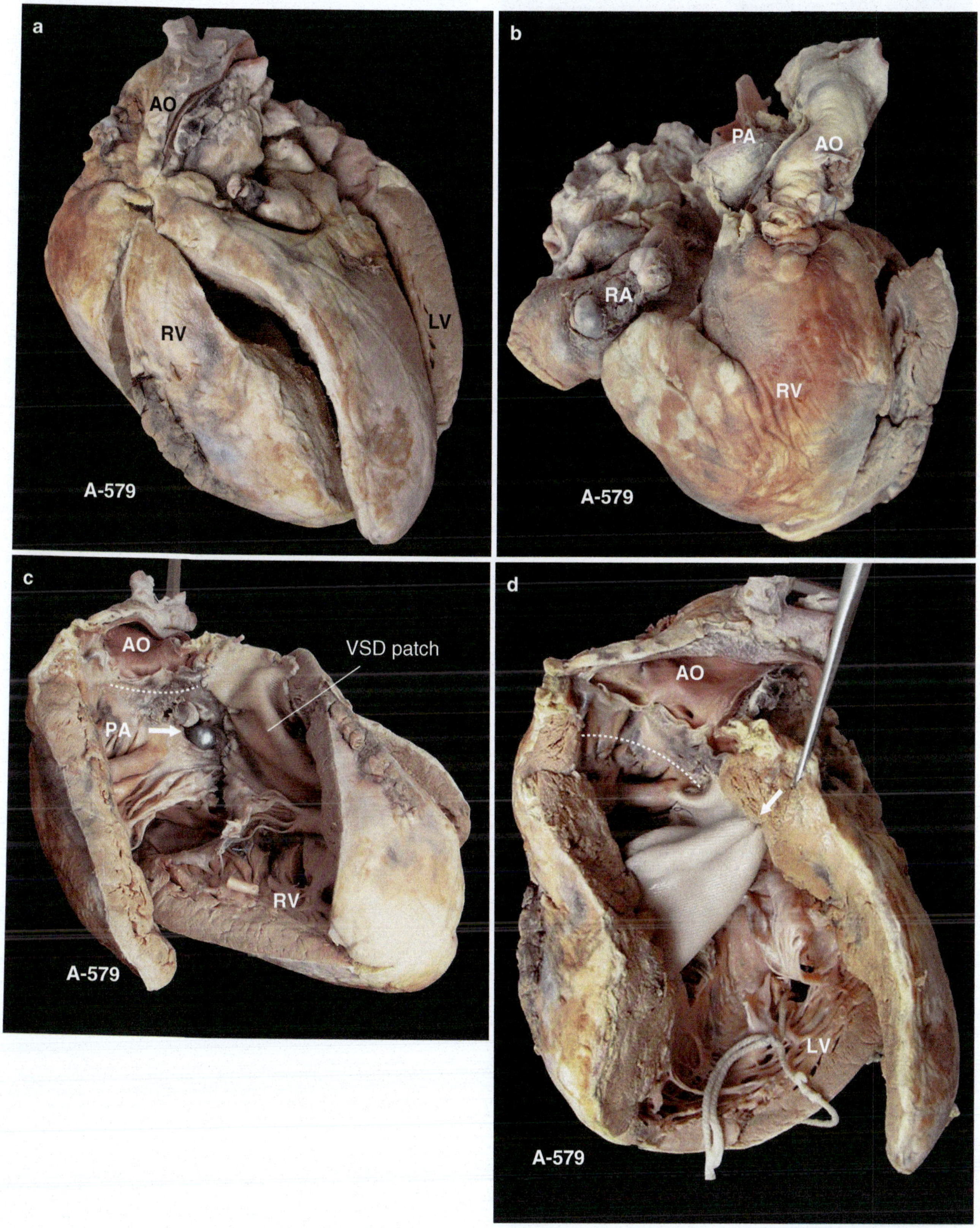

Fig. 24.15 (**a**) Specimen NO. 579. The front view of DORV. (**b**) Specimen NO. 579. The main pulmonary artery is behind the ascending aorta. (**c**) Specimen NO. 579. The RVOT is cut, and the PV is under the AV. (**d**) Specimen NO. 57.9. DORV, the view from the left ventricle. (**e**) Specimen NO. 57.9. The LVIT is cut to show the normal LA. *PA* pulmonary artery, *AO* ascending aorta, *RV* right ventricle, *LA* left atrium, *RA* right atrium, *LV* left ventricle

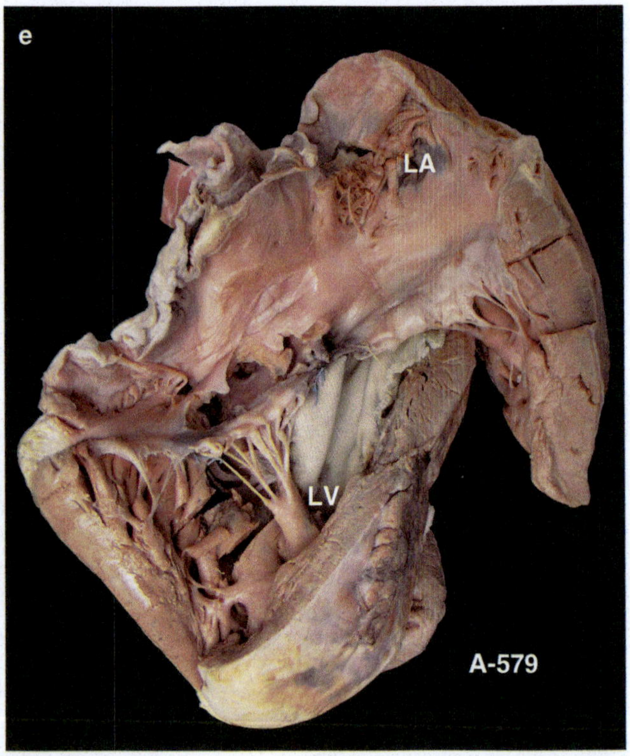

Fig. 24.15 (continued)

24.2.13 DORV, Subaortic VSD

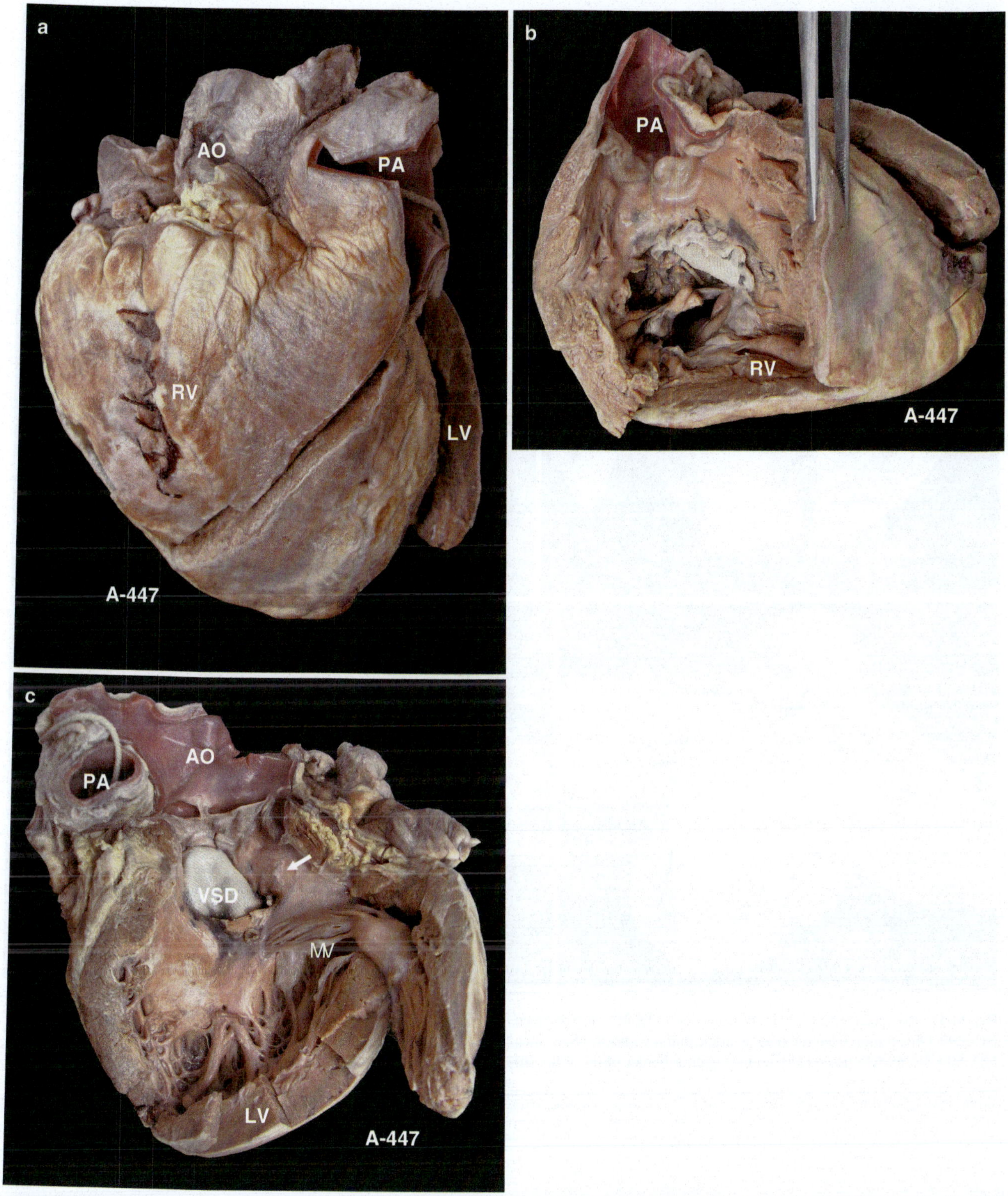

Fig. 24.16 (**a**) Specimen NO. 447. DORV. (**b**) Specimen NO. 447. Right ventricle outflow tract and a normal pulmonary valve. (**c**) Specimen NO. 447. Demonstration of the VSD under aortic valve from the left ventricle outflow tract. *PA* pulmonary artery, *AO* ascending aorta, *RV* right ventricle, *LV* left ventricle, *LA* left atrium, *VSD* ventricular septal defect, *MV* mitral valve

24.2.14 DORV, Doubly Committed VSD

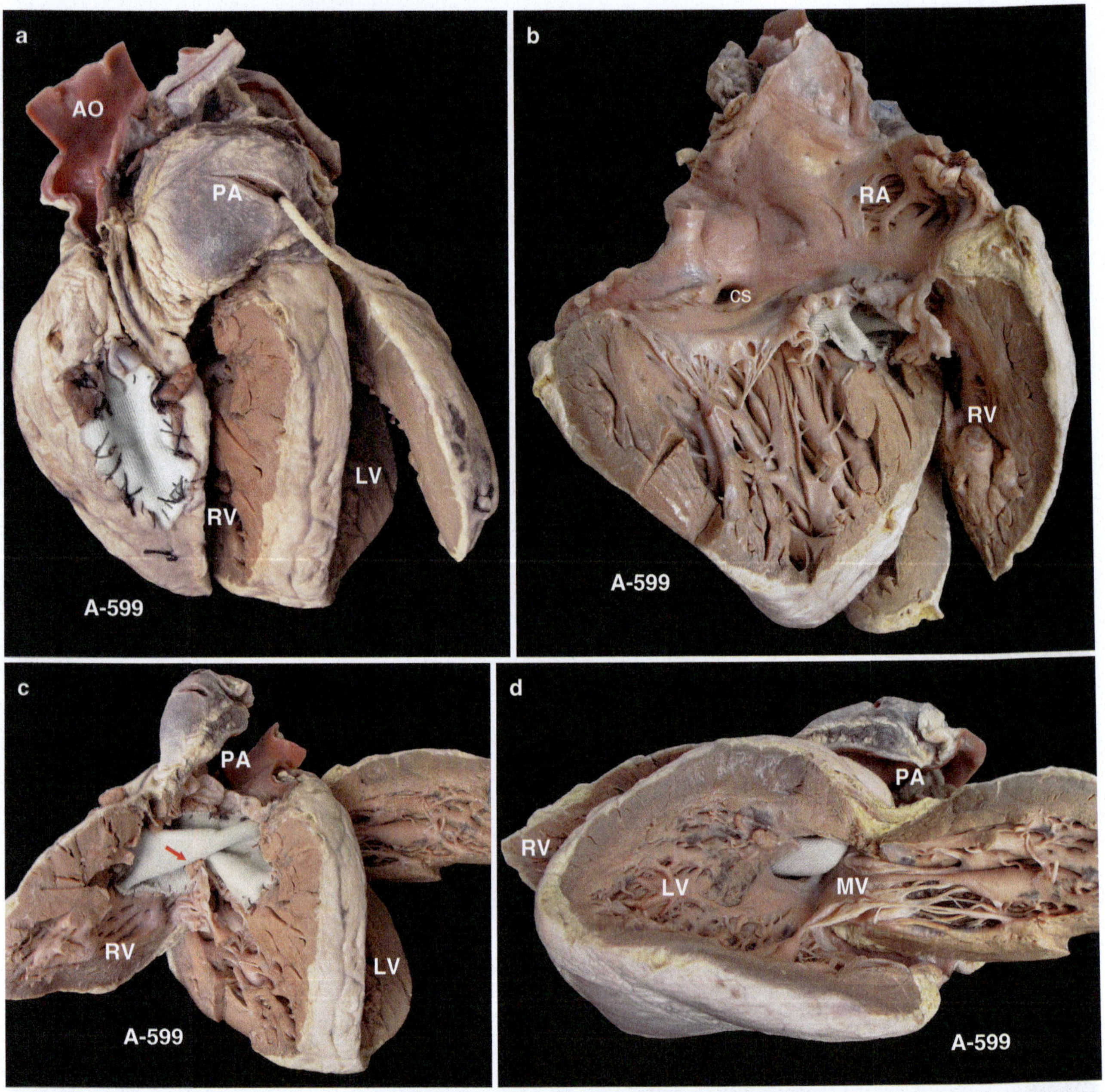

Fig. 24.17 (**a**) Specimen NO. 599. The front view of DORV. (**b**) Specimen NO. 599. Observation from the right ventricle outflow tract. (**c**) Specimen NO. 599. No muscle tissue exists in the superior border of the VSD. (**d**) Specimen NO. 599. No fibrous connection exists between the AV and the MV. *PA* pulmonary artery, *AO* ascending aorta, *RV* right ventricle, *RA* right atrium, *LV* left ventricle, *MV* mitral valve, *CS* coronary sinus

Fig. 24.18 (**a**) Specimen NO. 653. Front view of the DORV. (**b**) Specimen NO. 653. The right atrium and ventricle are cut to show the border of the VSD patch. (**c**) Specimen NO. 653. The outflow tract and main pulmonary artery are cut to show the normal PV. (**d**) Specimen NO. 653. The RVOT and the pulmonary artery are cut to show the huge VSD patch. (**e**) Specimen NO. 653. The bougie passes through the VSD from the aorta. (**f**) Specimen NO. 653. The red wire indicates the blood tract from the VSD to the ascending aorta. *PA* pulmonary artery, *AO* ascending aorta, *RV* right ventricle, *RA* right atrium, *CS* coronary sinus, *LV* left ventricle

24.2.15 DORV, Cut the VSD Patch

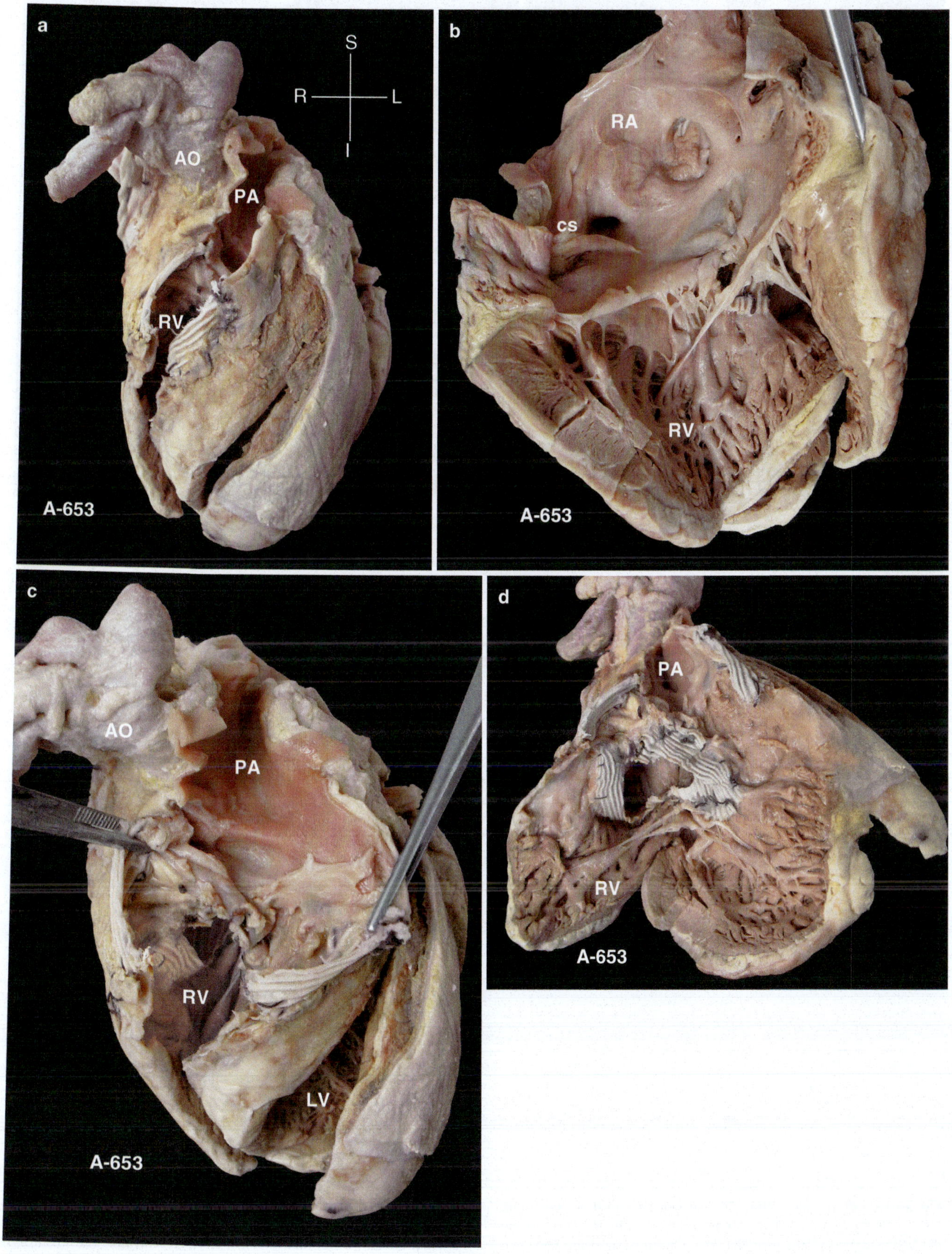

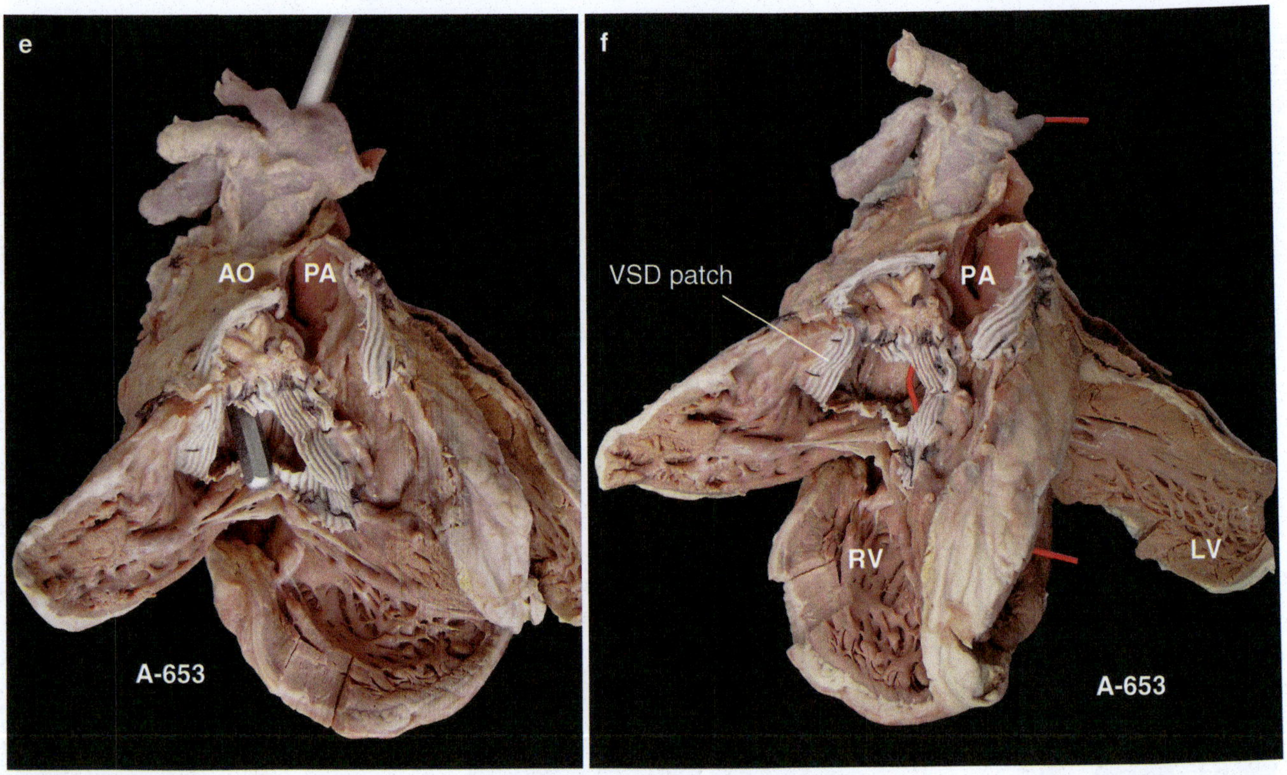

Fig. 24.18 (continued)

Fig. 24.19 (**a**) Specimen NO. 743. Exterior conduit from right ventricle to pulmonary artery. (**b**) Specimen NO. 743. The RV is cut to show the VSD patch. (**c**) Specimen NO. 743. Observation of the VSD patch from the left ventricle. (**d**) Specimen NO. 743. The graft connects the RV and the pulmonary artery. (**e**) Specimen NO. 743. The anastomosis of the graft and pulmonary artery. (**f**) Specimen NO. 743. Schematic diagram. *PA* pulmonary artery, *AO* ascending aorta, *RV* right ventricle, *RA* right atrium, *CS* coronary sinus, *LV* left ventricle, *LA* left atrium, *MV* mitral valve

24.2.16 DORV + PS + VSD Correction, Exterior Conduit from Right Ventricle to Pulmonary Artery

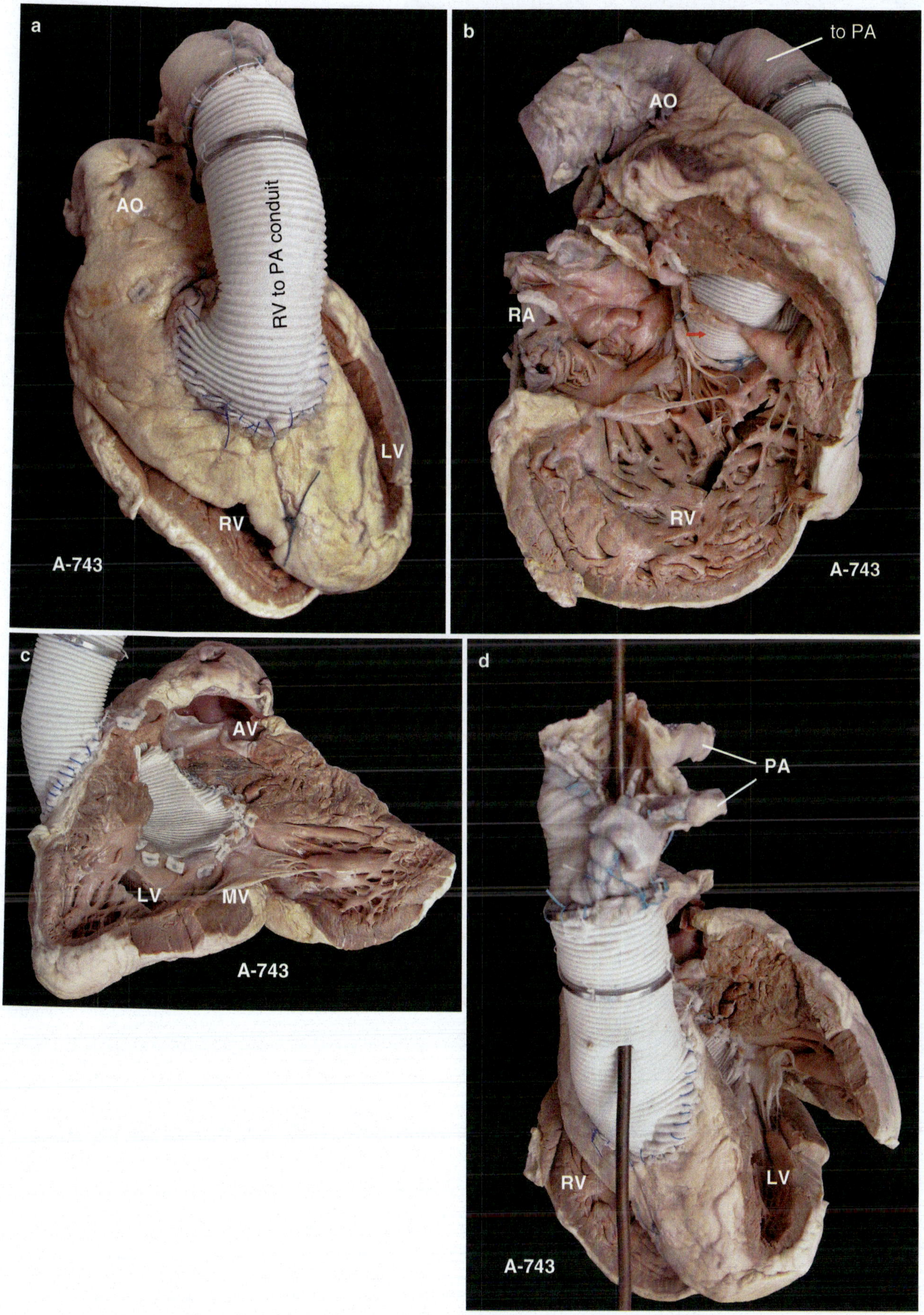

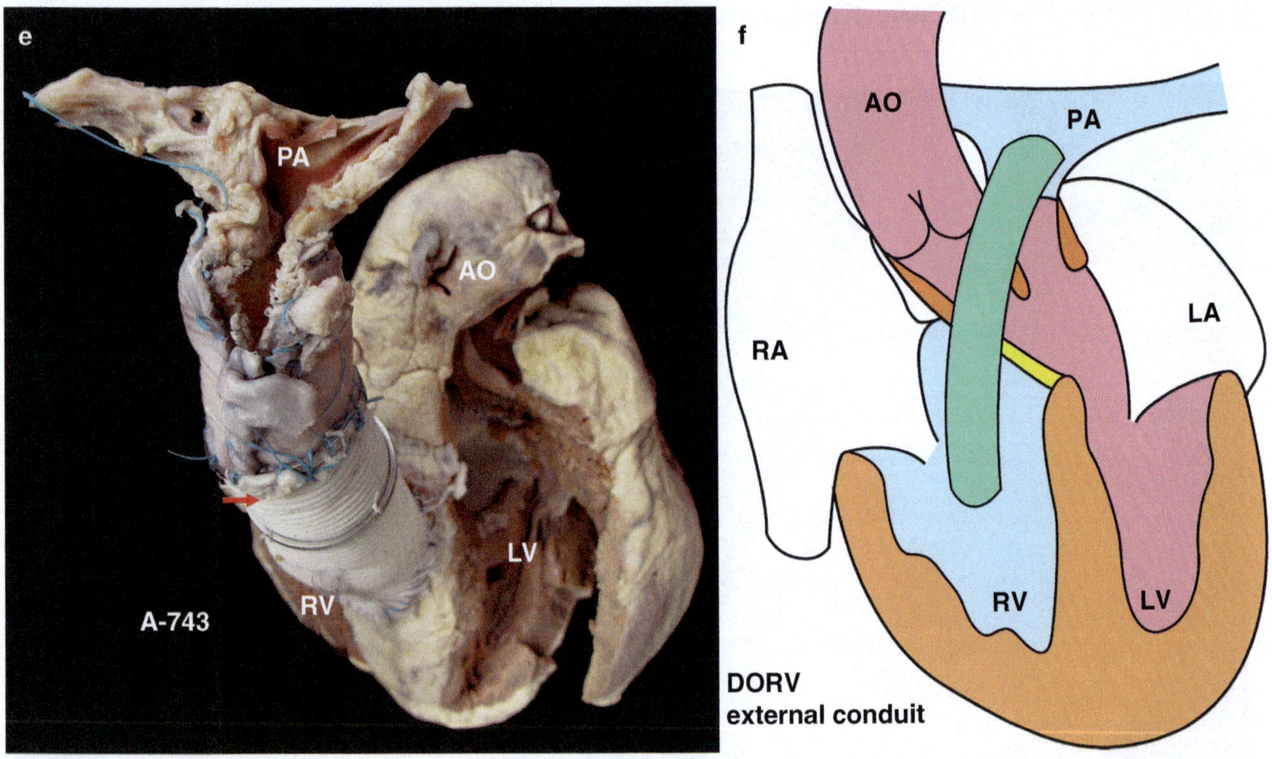

Fig. 24.19 (continued)

24.2.17 DORV, Mirror Dextrocardia

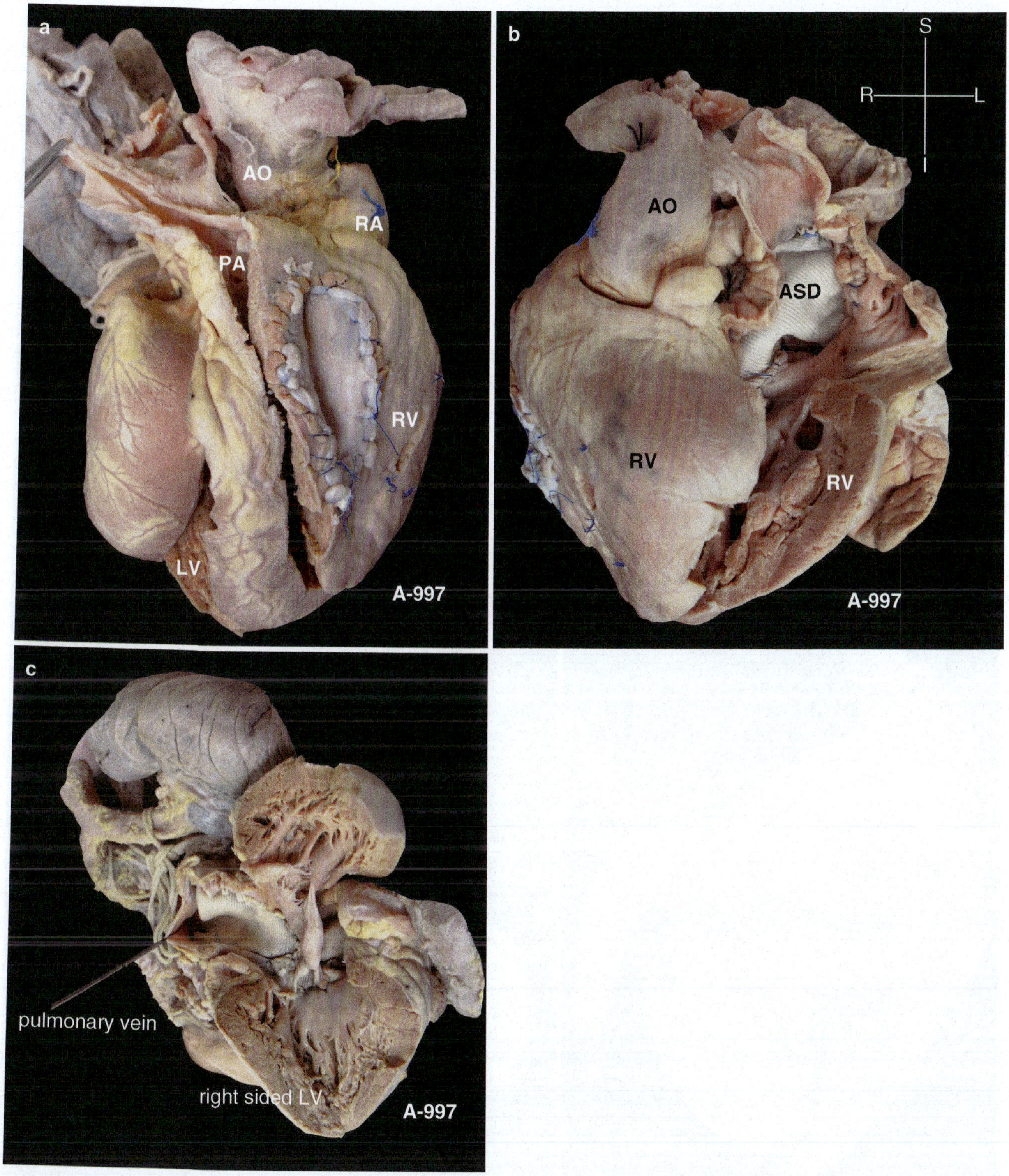

Fig. 24.20 (**a**) Specimen NO. 997. DORV, mirror dextrocardia. (**b**) Specimen NO. 997. Right atrium and ventricle are located on the left side. (**c**) Specimen NO. 997. The LA is located on the right side. *PA* pulmonary artery, *AO* ascending aorta, *RV* right ventricle, *RA* right atrium, *LV* left ventricle, *ASD* atrial septal defect

24.2.18 DORV, Subpulmonary VSD, Intracardiac Channel

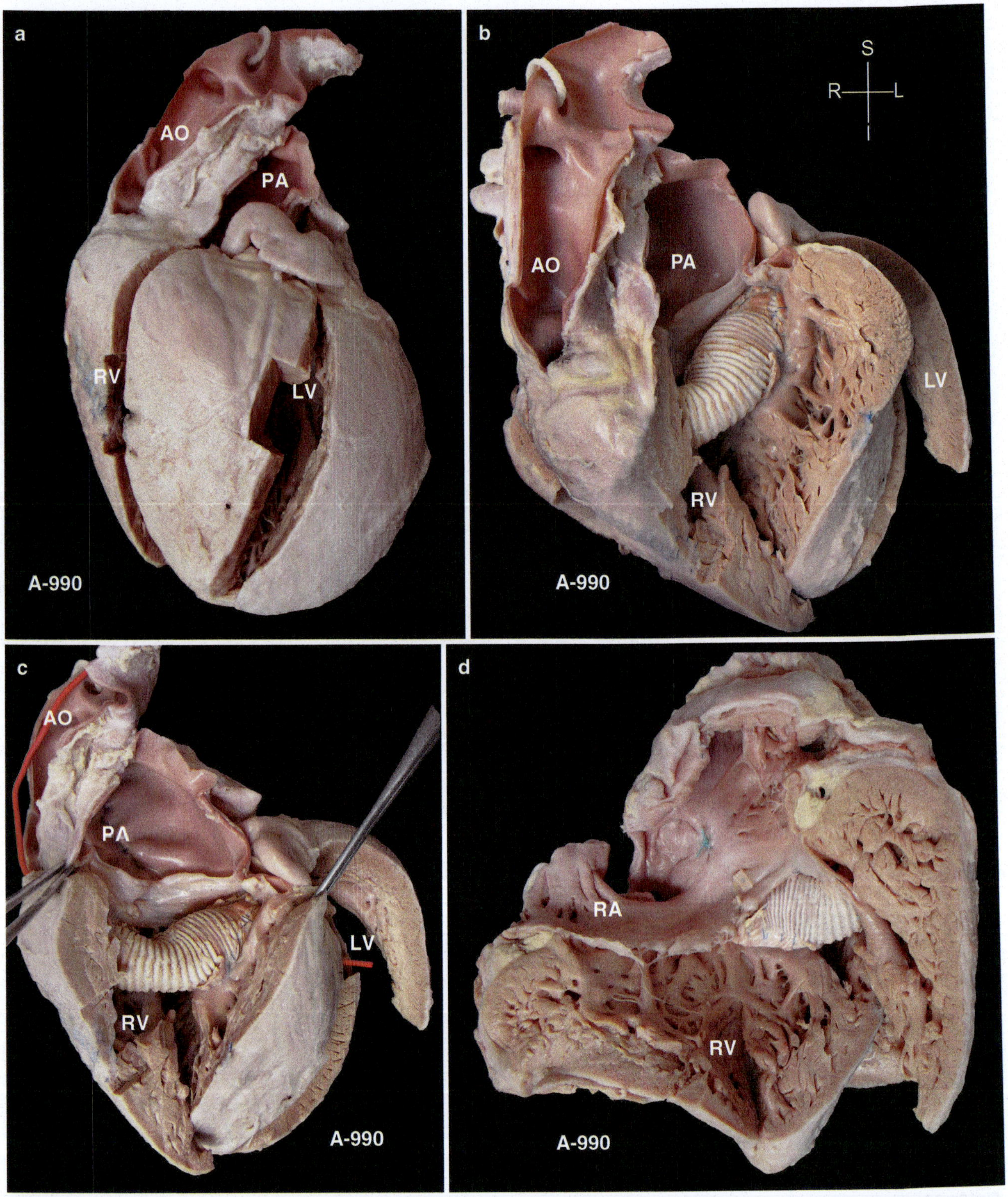

Fig. 24.21 (**a**) Specimen NO. 990. Front view of DORV. (**b**) Specimen NO. 990. The aorta and pulmonary artery arise from the RV. (**c**) Specimen NO. 990. A graft connects the VSD and the AV. (**d**) Specimen NO. 990. The RV is cut to show the intracardiac channel. (**e**) Specimen NO. 990. Schematic diagram. *PA* pulmonary artery, *AO* ascending aorta, *RV* right ventricle, *RA* right atrium, *LV* left ventricle, *LA* left atrium

e

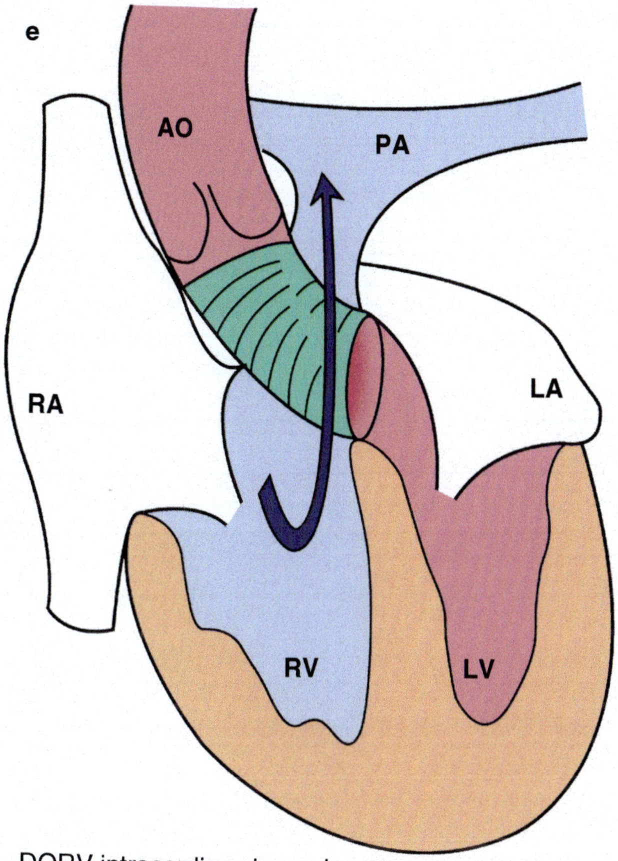

DORV intracardiac channel

Fig. 24.21 (continued)

25 Complete Transposition of the Great Arteries

25.1 General Considerations

TGA, also referred to as *transposition of the great arteries*, is a case in which the primary arteries (the aorta and the pulmonary artery) are transposed rather than being in an abnormal position. TGA is a congenital cardiac malformation characterized by atrioventricular concordance and ventriculoarterial discordance (Fig. 25.2).

In the patient with TGA, the aorta is in front of the pulmonary artery and the plate line of the AV ring is higher than the pulmonary artery. If the aorta is right anterior of the pulmonary artery, the condition is called *dextro-transposition TGA* (D-TGA); if the aorta is left anterior of the pulmonary artery, it is called *levo-transposition TGA* (L-TGA). The aorta and the pulmonary artery rise from different ventricles separately, rather than from a common chamber. If the connection of the valves of the arteries and ventricles is normal, the condition is called *malposition of the great arteries*, in which case, the position of the arteries could be abnormal.

In the case of TGA, the arterial cone suffers dysplasia during the embryonic period, and the mechanism includes abnormal rotation and abnormal absorption of the arterial cone.

In clinic, we can classify patients with TGA by these features:

1. VSD or not
2. Pulmonary stenosis
3. Pulmonary hypertension
4. Developmental condition of ventricles
5. Disposition of the atria and ventricle
6. With other malformations

In the case of TGA, the aorta receives venous blood from the right atria, arterial oxygen saturation is reduced, and the patient turns blue.

The classic TGA belongs to SDD (situs solitus, dextroloop, D-TGA) type in clinic: the formations of the atria and ventricle are similar to the normal ones, but the ascending aorta rises from the right outlet of the RV, the left and right coronary arteries rise from the anterior part of the root of the aorta, and the major pulmonary artery is pushed to a position left posterior of the aorta (Fig. 25.1). The position of the AV is high and far away from the atrioventricular valve; the position of the PV is low, buried deeply between the MV and the TV. Some fibrous connection exists between the pulmonary and mitral valves. In the RV, the TV and a supraventricular crest can be seen. The AV and TV are separated by muscle of the supraventricular crest. If there is a huge VSD, the PV can be seen through the VSD (Fig. 25.10c). A fibrous connection exists between the pulmonary and mitral valves.

Surgical therapy for TGA includes:

1. Palliative surgery, such as atrial septostomy and banding
2. Atrial bypass surgery, such as Mustard surgery and Senning surgery (Fig. 25.3)
3. Switch surgery (Fig. 25.5)
4. Surgery for complications

X. Zhu (ed.), *Surgical Atlas of Cardiac Anatomy*,
DOI 10.1007/978-94-017-9409-1_25, © Springer Science+Business Media Dordrecht and People's Medical Publishing House 2015

25.2 Specimen Demonstrations

25.2.1 Typical TGA, Without Pulmonary Malformation

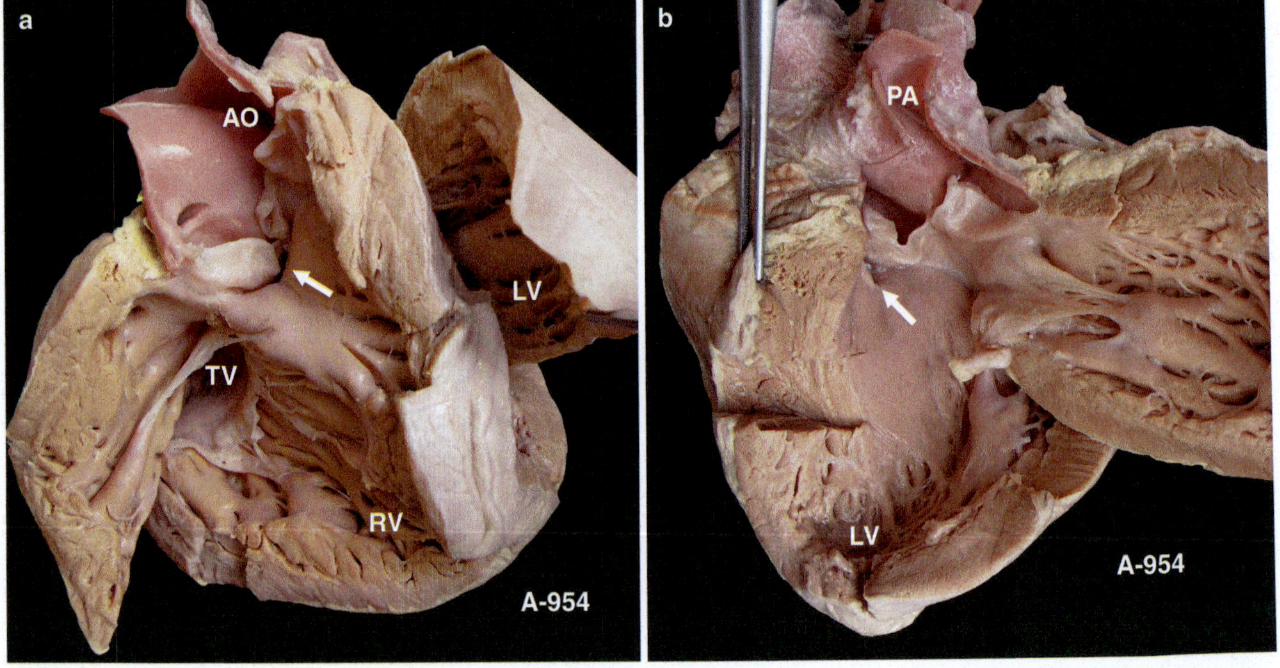

Fig. 25.1 (**a**) Specimen 954. TGA+VSD aspect of the right ventricle. The aorta rises from the RV. The supraventricular crest is between the aorta and RV. The wall of the RV is very thick. The *arrow* shows a subarterial VSD. Above all is a classic anatomy case of TGA+VSD. *TV* tricuspid valve, *LV* left ventricle, *AO* ascending aorta. (**b**) Specimen 954. An aspect of the left ventricle. The pulmonary artery rises from the LV, and a fibrous connection is found between the pulmonary and mitral valves. The *arrow* shows the VSD under the pulmonary vein. A membranous ventricular septum is anterosuperior of the MV. *PA* pulmonary artery, *LV* left ventricle

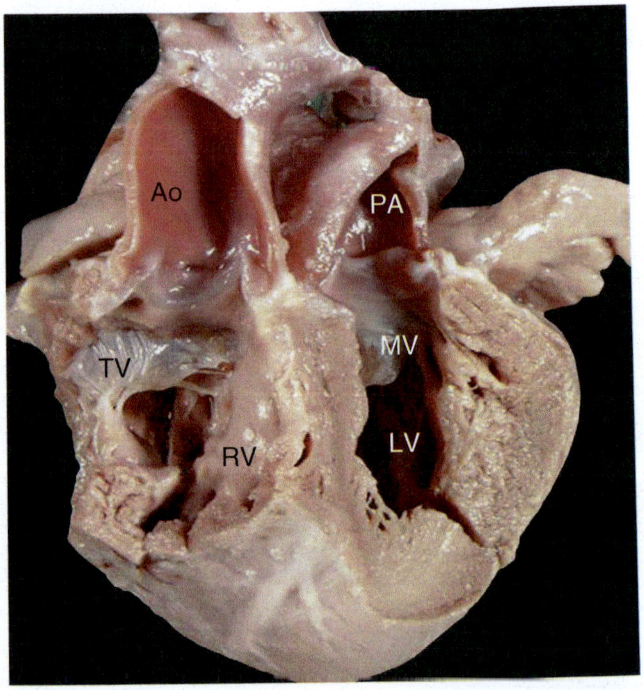

Fig. 25.2 Frontal view of TGA. A long-axis, four-chamber heart aspect shows the anatomical relationship of the TGA: the aorta and pulmonary artery have complete transposition. The supraventricular crest is between the aorta and the TV, and a fibrous connection is between the pulmonary artery and the MV. *AO* ascending aorta, *RV* right ventricle, *LV* left ventricle, *PA* pulmonary artery

25.2.2 TGA with VSD, Postoperation with RV-PA Bypass

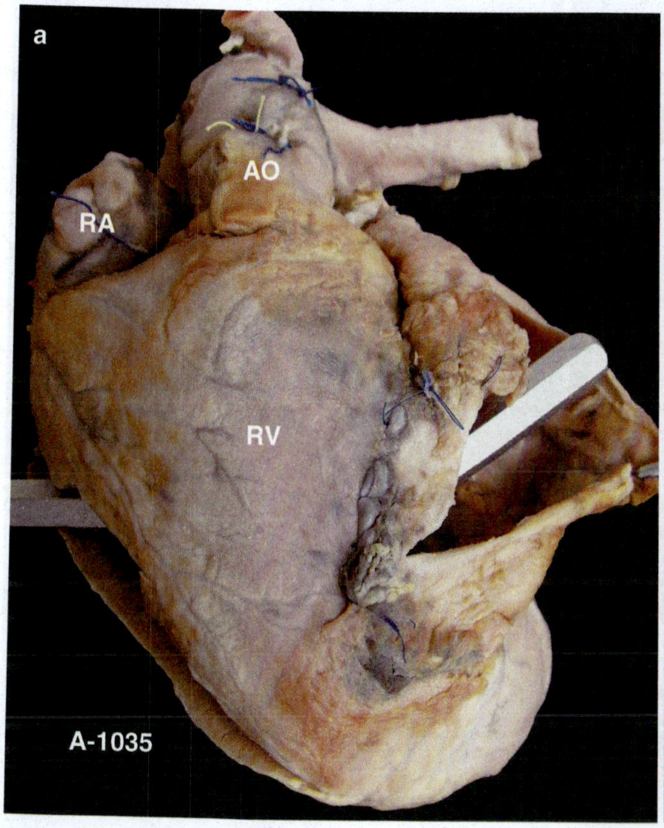

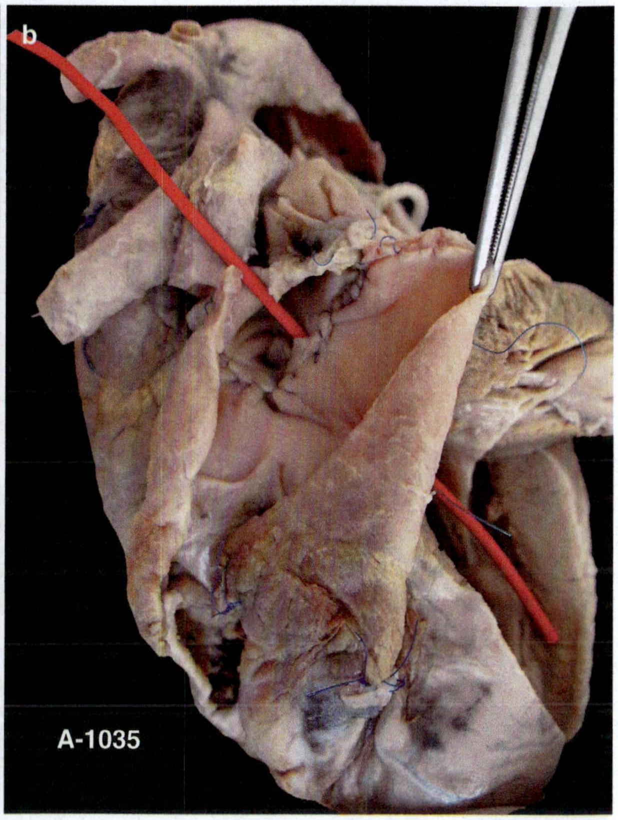

Fig. 25.3 (**a**) Specimen 1035. Front view of TGA. The aorta rises from the RV; the anatomical position of the atrium and the ventricle is correct, demonstrating a TGA type. The probe is from the RV to the external conduit. *RA* right atrium, *AO* ascending aorta. (**b**) Specimen 1035. Left view of TGA and VSD. The pulmonary artery has a passage to the LV. The *red probe* is from the PV to the LV, so this case has pulmonary stenosis. An external conduit made of a homograft valved artery is at a junction with the left/right pulmonary artery. (**c**) Specimen 1035. From right atrium view, the septal leaflet of the tricuspid valve. A revision patch is under the AV and connects the aorta to the LV. Meanwhile, the throughway can be seen from the outlet of the RV to the artificial external conduit. An incision in the front wall of the RV will be selected for the VSD revision and then connected to the external conduit as an entrance. Therefore, a part of the revision patch cannot be seen in this figure. *RA* right atrium, *AO* ascending aorta, *TV* tricuspid valve, *RV* right ventricle, *CS* coronary sinus. (**d**) Specimen 1035. From the right view. The anatomical relationship between the external conduit and RV is shown; the silver probe is inserted into the RV through the external conduit. *RA* right atrium, *TV* tricuspid valve, *RV* right ventricle, *CS* coronary sinus. (**e**) Specimen 1035. From the left posterior view. Figure shows the left and right pulmonary arteries, which are well developed, without stenosis. The connection to the external conduit is satisfied. *AO* ascending aorta, *LPA* left pulmonary artery, *RPA* right pulmonary artery, *LV* left ventricle. (**f**) Specimen 1035. From left anterior view, showing the unbounded external conduit. The probe crosses through the gap posterior to the external conduit, which means that the conduit is entirely outside the RV, the remote of the external conduit anastomoses to the pulmonary bifurcation. "Homograft" is a homograft valved artery. *AO* ascending aorta, *PA* pulmonary artery. (**g**) Specimen 1035. From front view. Figure shows the external conduit from the RV to the pulmonary artery. The upper part of the conduit is an artificial patch, and the lower part is a homograft valved artery. The left anterior descending artery is next to the conduit. *AO* ascending aorta, *RV* right ventricle, *LV* left ventricle. (**h**) Specimen 1035. From cutaway view of the external conduit. Figure shows the homograft valves; the opening of the PV is closed, and the external conduit is directly connected to the pulmonary artery. The blood in the RV flows into the pulmonary artery through the external conduit, and the blood in the LV flows into the aorta. "RV" indicates the anastomosis between the RV and the external conduit. *AO* ascending aorta. (**i**) Specimen 1035. From left upper view. Figure shows the relationship of the pulmonary artery opening and the RV. The *red probe* is inserted into the LV through a gap of the pulmonary artery, which excludes the diagnosis of DORV. (**j**) Specimen 1035. Cutaway into the aorta and left ventricle. The valves of the aorta rise from the RV, and a VSD patch blocks the aorta to the LV, which becomes a part of the LV outlet. The patch also can be thought of as an inner channel of isolated blood flow from the LV to the aorta. *AO* ascending aorta, *AV* aortic valve. (**k**) Specimen 1035. From left lateral view, the structure of the left atrium and ventricle. A defect (*D*) is on the top of the ventricular septum, and the MV has a little cleft in the anterior leaflet, the border of which is fibrous and thick. *LA* left atrium, *LV* left ventricle. (**l**) Specimen 1035. Spatial structure diagram of postoperation. A VSD patch connects the aorta to the LV, and an external conduit connects the pulmonary artery to the RV; the pulmonary valve opening is closed. *AO* ascending aorta, *PA* pulmonary artery, *LA* left atrium, *RA* right atrium

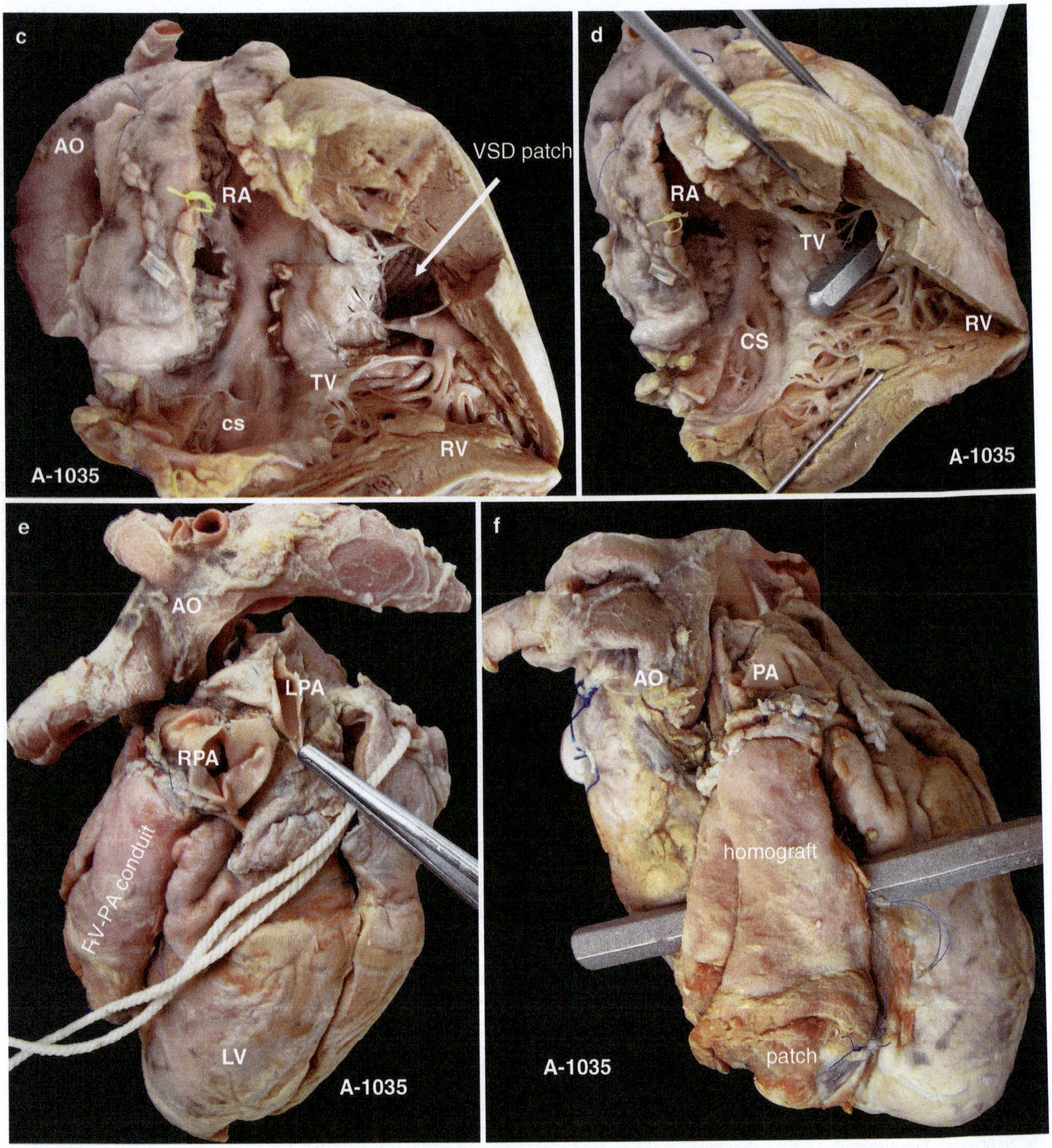

Fig. 25.3 (continued)

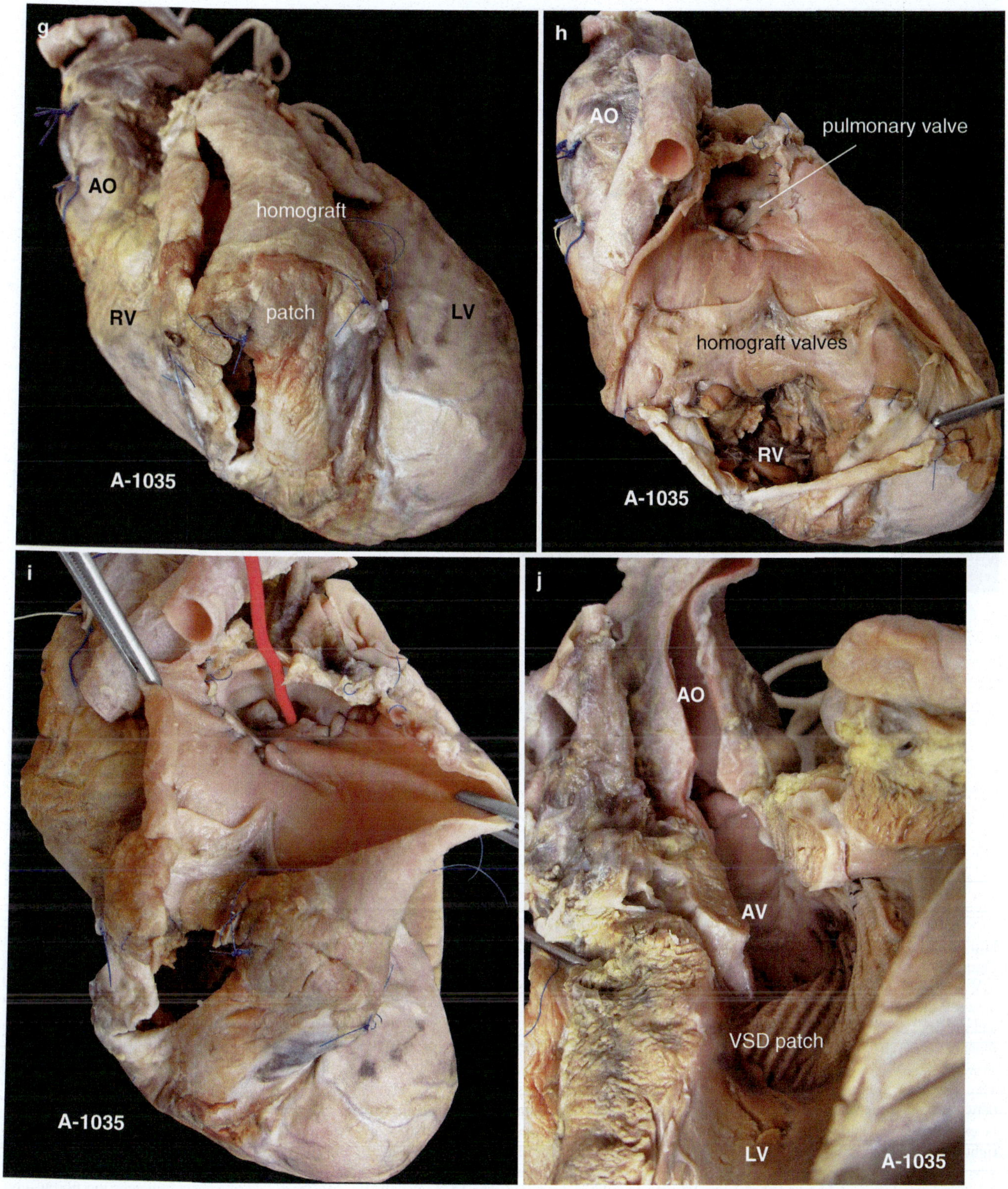

Fig. 25.3 (continued)

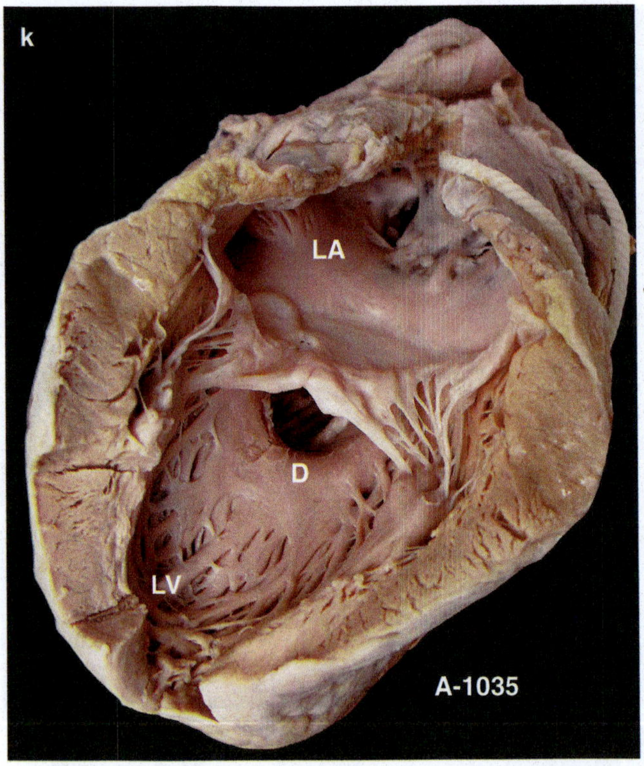

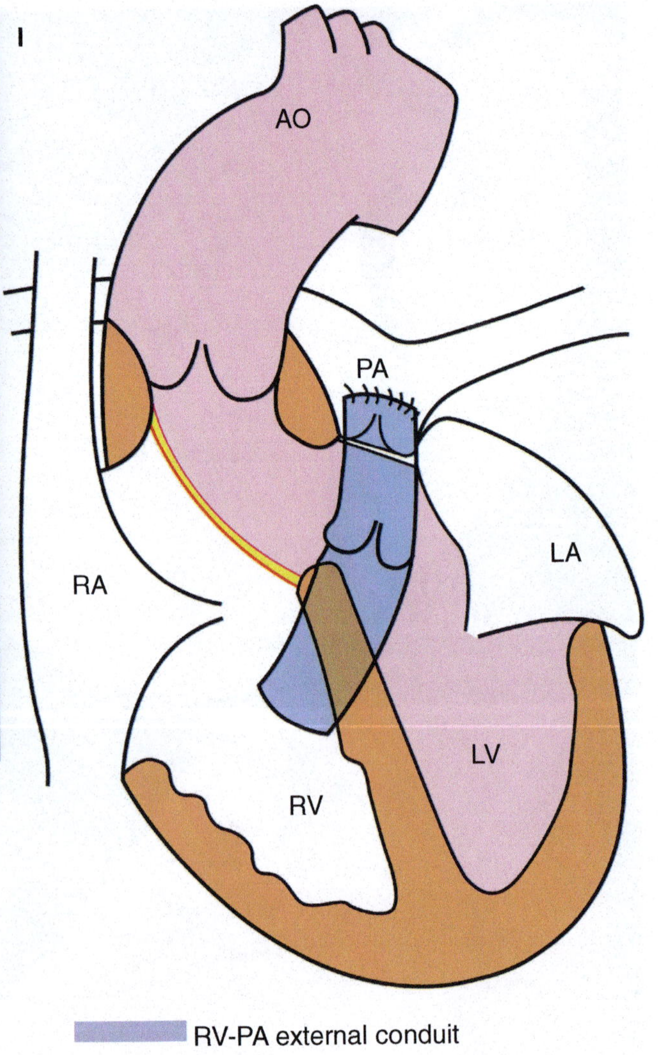

Fig. 25.3 (continued)

Fig. 25.4 (**a**) Specimen 930. TGA, VSD, PS, TS, MV overriding. Figure shows a cutaway of the anterior wall of the RV; the tendon of the MV is overriding the RV; the aorta rises from the RV. Fibrous tissue is between the pulmonary and mitral valves. The surgical suture is in the anterior wall of the RV. *AO* ascending aorta, *RV* right ventricle, *PA* pulmonary artery. (**b**) Specimen 930. Cutup of the anterior wall of the RV, showing the MV (*thin probe*) and the TV (*thick probe*) simultaneously. The cone muscle is between the AV and the TV, and no crack exists between the MV and the TV, so the atrioventricular defect channel can be excluded. The tendon of the MV is anchored to the papillary muscle, right to the ventricular septum. The TV has mild stenosis, the leaflets of which have a fibrous uneven thick border and a very smooth surface as a result of secondary reactive hyperplasia caused by the blood flow impact. *AO* ascending aorta. (**c**) Specimen 930. View looking down. A homograft external conduit overrides the bottom of the heart, showing that the MV is overriding. "*P*" indicates the papillary muscle of the RV, and the two probes show the opening of two atrioventricular valves. A major part of the MV junctions to the LV, and there is a fibrous connection between the mitral and pulmonary valves, which is different from DORV. The wall of the RV is thick. *TV* tricuspid valve, *PV* pulmonary valve. (**d**) Specimen 930. Demonstration of right atrium, right auricle, and homograft valved external conduit. The external conduit is a homograft artery root with AVs. The conduit ascends posterior to the aorta. The relationship of the aorta and the outlet of the RV can be found. *AO* ascending aorta, *RA* right atrium, *RAA* right atrium auricle. (**e**) Specimen 930. Opening of the tricuspid valve from right atrium view. The leaflets are thickened, and the TV opening has stenosis. *RA* right atrium, *RV* right ventricle, *TV* tricuspid valve. (**f**) Specimen 930. The tweezers clamp the wall of the RA, part of which is connected to the external conduit. The *arrow* shows the valves of the homograft conduit. *AO* ascending aorta. (**g**) Specimen 930. From left anterior oblique aspect of the pulmonary artery and a homograft anastomosis (*arrow*). Pulmonary and mitral valves have a fibrous connection; the ascending aorta is right anterior to the pulmonary artery. *AO* ascending aorta, *PV* pulmonary vein, *MV* mitral valve, *VSD* ventricular septal defect, *P* papillay muscle. (**h**) Specimen 930. Left and right atrioventricular valves have a connection but without crack. The *red arrow* indicates the blood flow passage from the LV to the RV by a VSD; the *white arrow* indicates the fibrous connection between the pulmonary and mitral valves. *PA* pulmonary artery, *PV* pulmonary valve, *LAA* left atrium auricle, *TV* tricuspid valve, *MV* mitral valve. (**i**) Specimen 930. Figure shows the inlet of the LV; the LA is normal. The MV is completely connected to the LV. The inner wall of the LV is smooth, the chamber of the LV is similar to that of the RV, and the tendon of the MV is overriding to the RV. *D* ventricular septal defect. (**j**) Specimen 930. Diagram TGA + VSD + PS + TS with the MV overriding. The RA is connected to the pulmonary artery with a homograft valved conduit. The pulmonary artery receives blood from the RA, but other malformations have not been corrected. *PS* pulmonary stenosis, *AO* ascending aorta, *PA* pulmonary artery, *LA* left atrium, *RA* right atrium, *RV* right ventricle, *LV* left ventricle

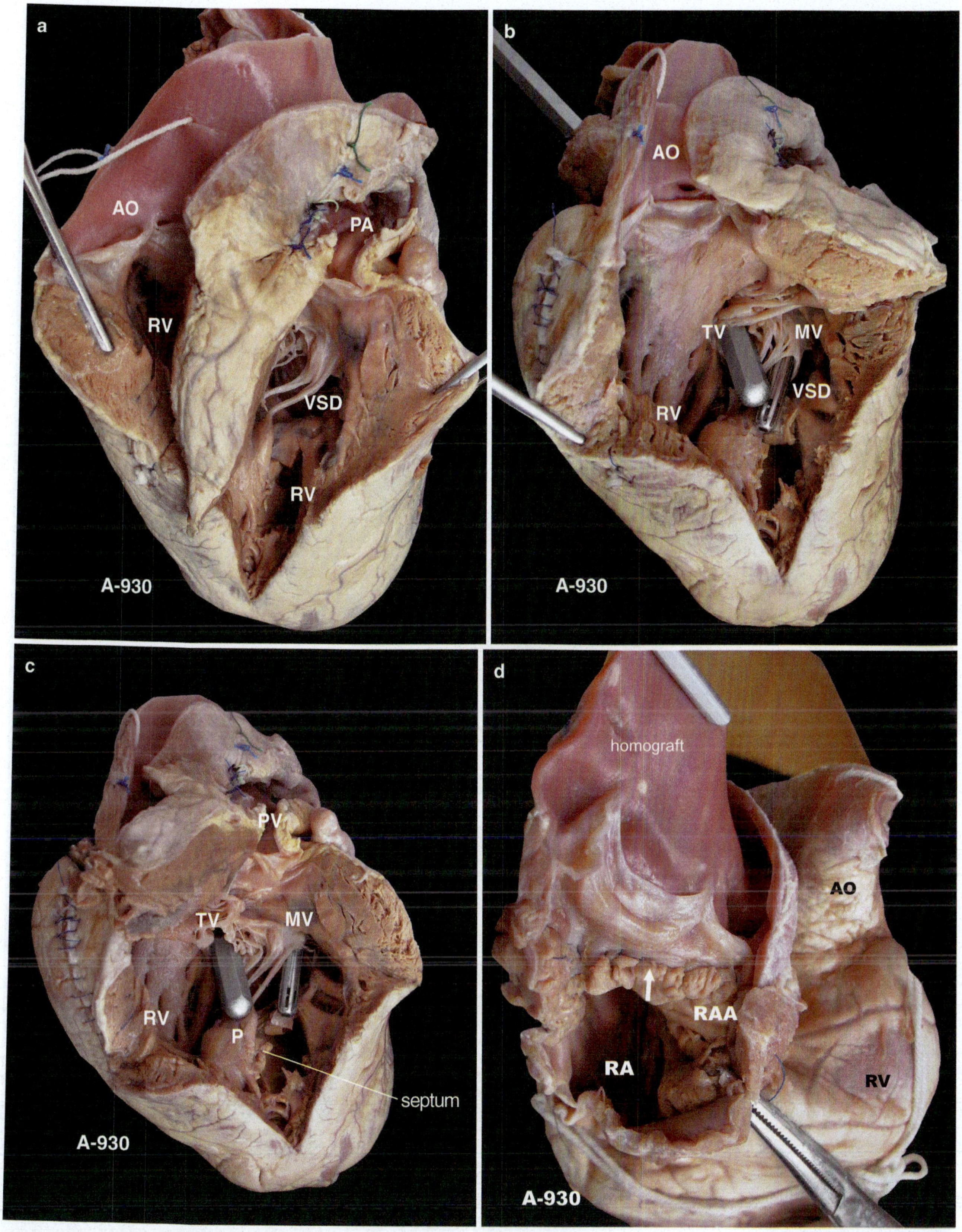
a
AO
PA
RV
VSD
RV
A-930
b
AO
TV
MV
VSD
RV
A-930
c
PV
TV
MV
RV
P
septum
A-930
d
homograft
AO
RAA
RA
RV
A-930

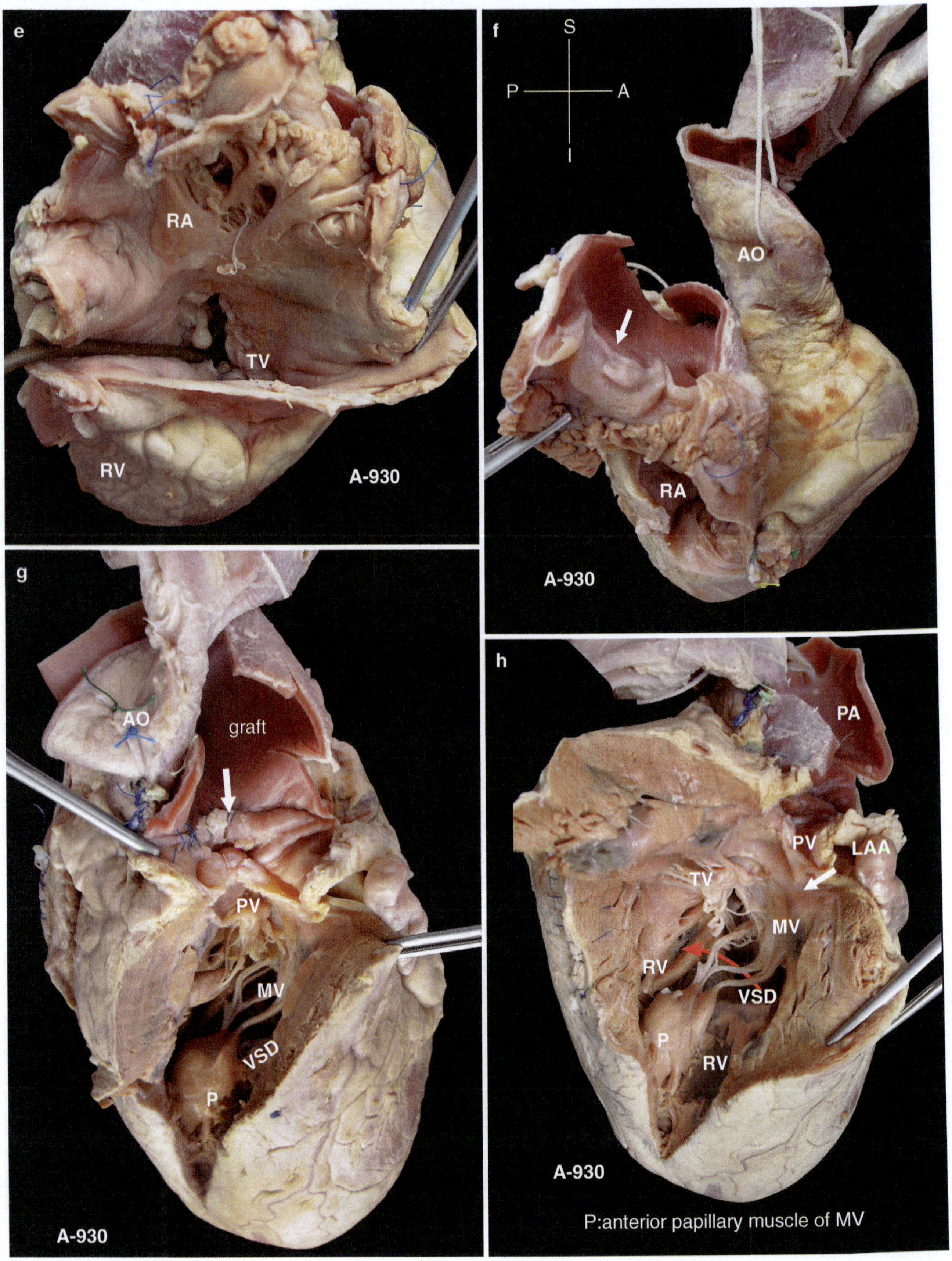

Fig. 25.4 (continued)

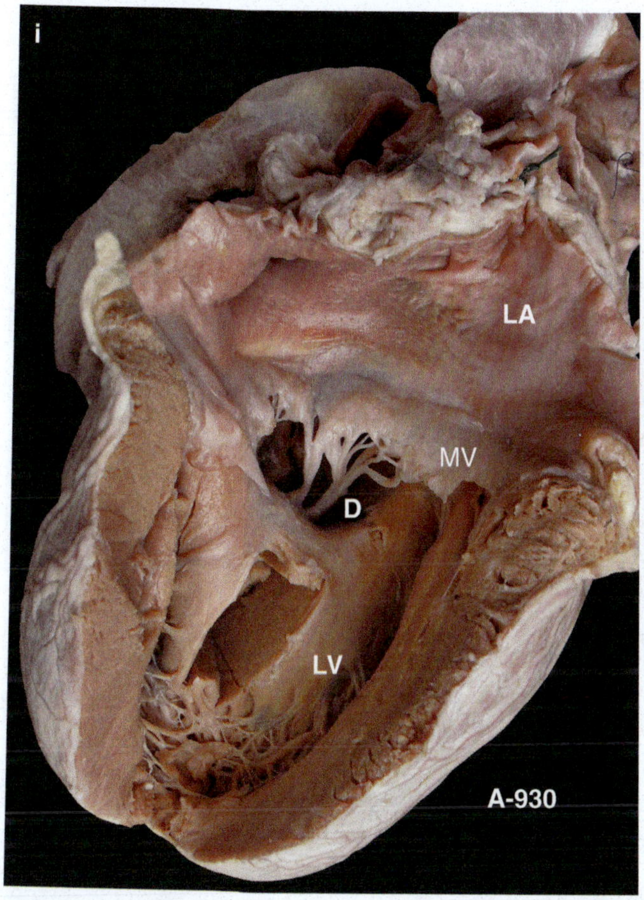

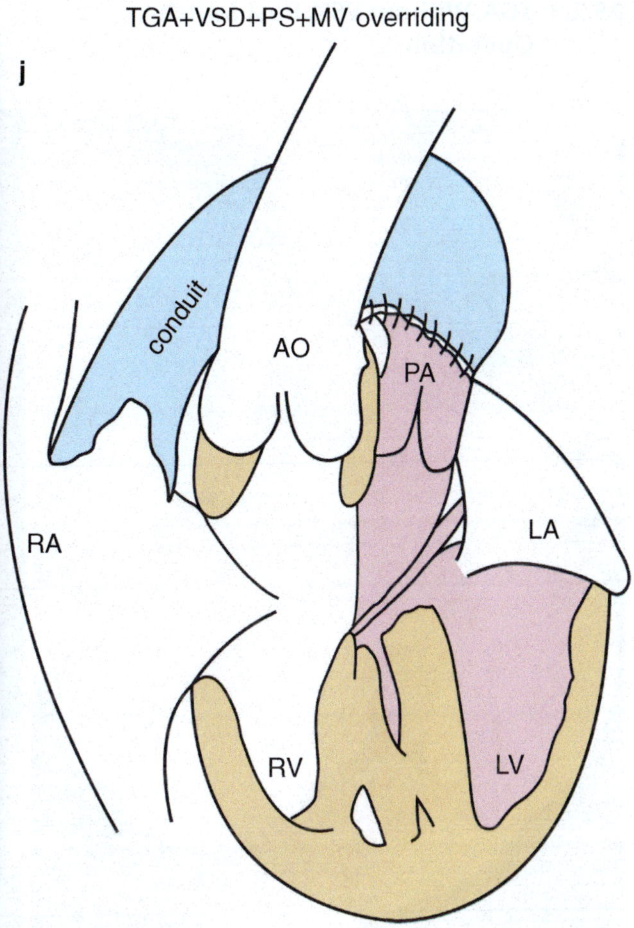

Fig. 25.4 (continued)

25.2.3 TGA, Without VSD, Post Switch Operation

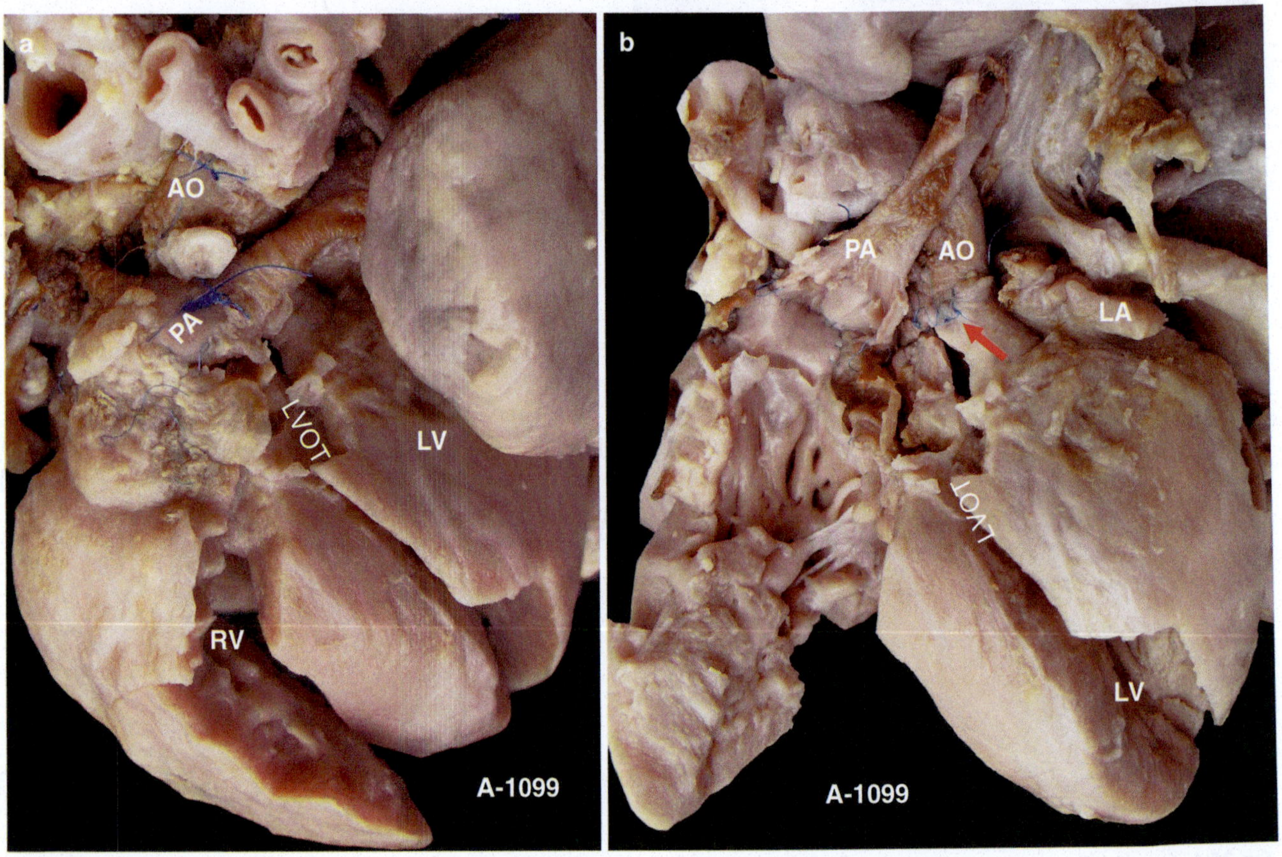

Fig. 25.5 (**a**) Specimen 1099. Post switch operation. The RV is the right ventricle, the pulmonary artery has been moved in front to connect to the RV, and the original ascending aorta was cut off from the RV and connected to the LV back of the pulmonary artery. LVOT represents the left ventricular outlet tract; the ventricular septum is intact. *AO* ascending aorta, *PA* pulmonary artery. (**b**) Specimen 1099. Post switch operation. The aorta is connected to the LV, and the pulmonary artery is moved in front of the aorta and connected to the RV. The *arrow* shows the anastomotic position of the new aorta. *PA* pulmonary artery, *AO* ascending aorta, *LA* left atrium. (**c**) Specimen 1099. View from the aspect of the LVOT. The root of the aorta can be seen after the operation; the opening of the coronary artery also has been transpositioned well. The *arrow* shows the anastomosis of the aorta and the LV. *AO* ascending aorta, *MV* mitral valve. (**d**) Specimen 1099. Close observation of the new aorta. The specimen shows the detailed new aorta root after the switch operation. The opening of the coronary artery is also moved to the right position. The *arrow* is the postoperation opening of the coronary artery. *MV* mitral valve

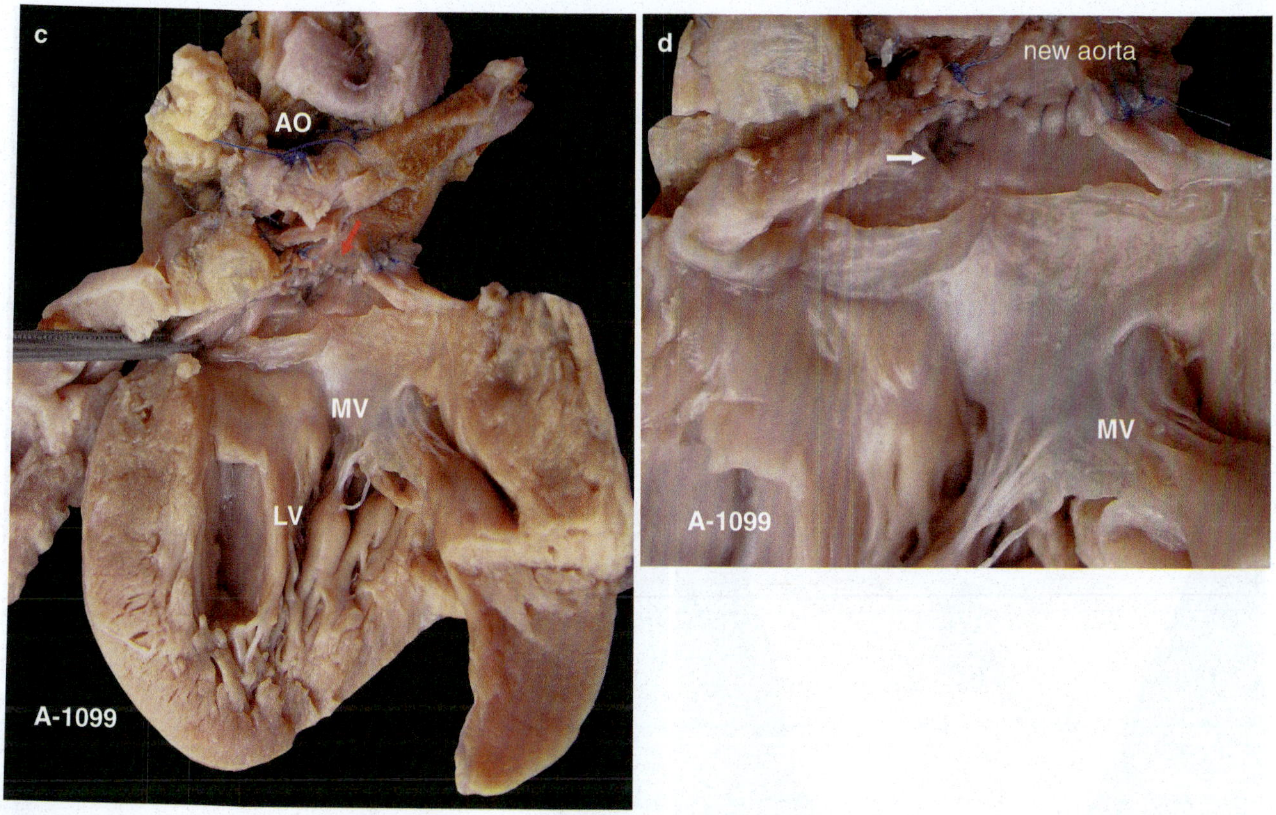

Fig. 25.5 (continued)

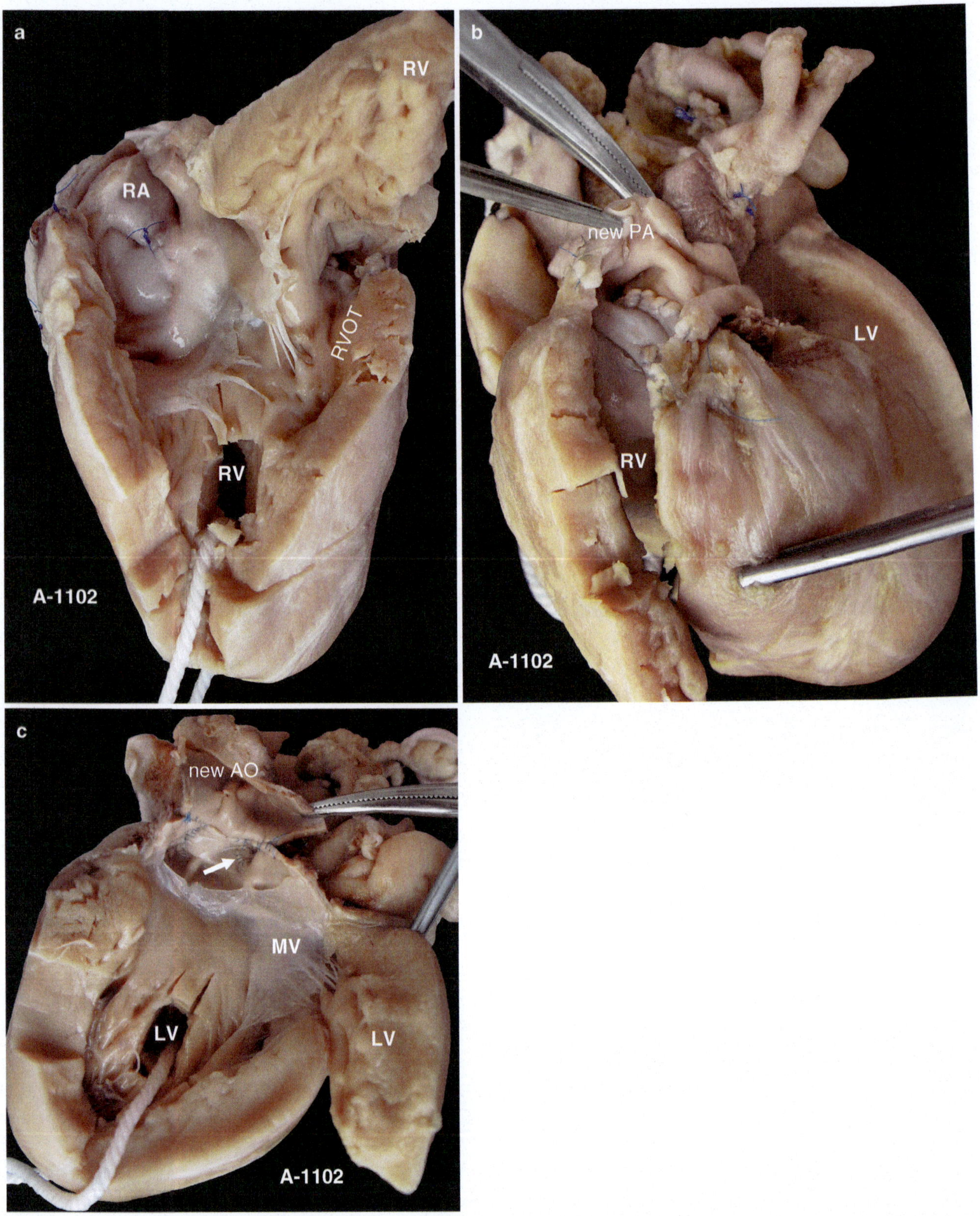

Fig. 25.6 (**a**) Specimen 1102. From right anterior oblique view, TGA shows. After the transposition operation for an intact ventricular septum case is completed, the outlet and inflow tracts are unobstructed. *RA* right atrium, *ROVT* right outflow of the ventricular tract, *RV* right ventricle. (**b**) Specimen 1102. Post switch operation of TGA. Figure shows the new pulmonary artery is connected to the RV at the position of the original aortic valves. *LV* left ventricle. (**c**) Specimen 1102. New aorta. Figure shows the anastomosis of the aorta and the LVOT; the ventricular septum is intact. The new aorta is the postoperation aorta. *LV* left ventricle, *MV* mitral valve

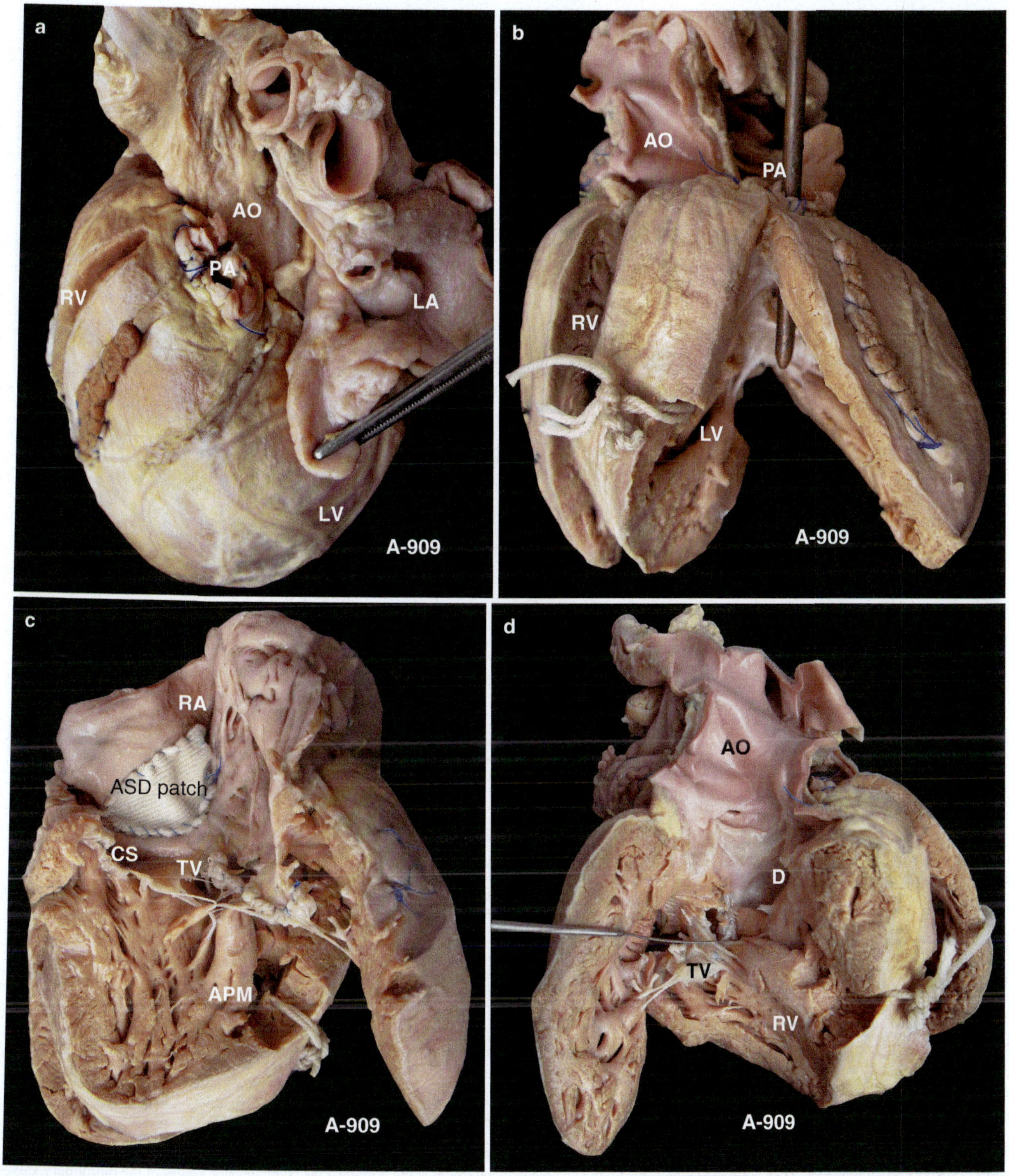

Fig. 25.7 (**a**) Specimen 909. Anterior view. TGA with subpulmonary stenosis (equated to a LVOT obstruction), VSD and secondary ASD. *LV* left ventricle, *AO* ascending aorta, *PA* pulmonary artery, *LA* left atrium, *RV* right ventricle. (**b**) Specimen 909. The pulmonary artery with an *iron probe* is left anterior to the ascending aorta, which is from the LV; there is a surgical incision in the wall of the LV. *LV* left ventricle, *AO* ascending aorta, *PA* pulmonary artery, *RV* right ventricle. (**c**) Specimen 909. The atrium is normal, and the orifice of the CS and the TV is shown. The wall of the ventricle is thickened, and the anterior papillary muscle is bulky. A huge ASD patch can be found in the interatrial septum as a secondary ASD. *RA* right atrium, *APM* anterior papillary muscle. (**d**) Specimen 909. Cutaway of the RVOT. The AV rises from the RV. A drag hook shows some muscle connection between the AV and TV, the latter of which has been torn. "*D*" indicates the VSD, *AO* ascending aorta. (**e**) Specimen 909. Muscle obstruction of the LVOT. After lifting up the LVOT, the subpulmonary stenosis (equated to LVOT obstruction) is shown (indicated by the *white arrow*); the red probe is in the VSD. *APM* anterior papillary muscle, *LV* left ventricle. (**f**) Specimen 909. The MV structure is unfolded. The *white arrow* indicates a subpulmonary obstruction of the LVOT. *D* ventricular septal defect, *LV* left ventricle. (**g**) Specimen 909. Left atrium and inflow tract of the left ventricle. The ASD has been closed with a patch; a surgical suture is on the right side of the interatrial septum. *LV* left ventricle, *LA* left atrium. (**h**) Specimen 909. Diagram. The aorta rises from the RV, the pulmonary artery rises from the LV, the VSD is on the top of ventricular septum, and the muscle structure of the LVOT is thickened, meaning there is an obstruction of the LVOT. *RA* right atrium, *AO* ascending aorta, *PA* pulmonary artery, *LA* left atrium

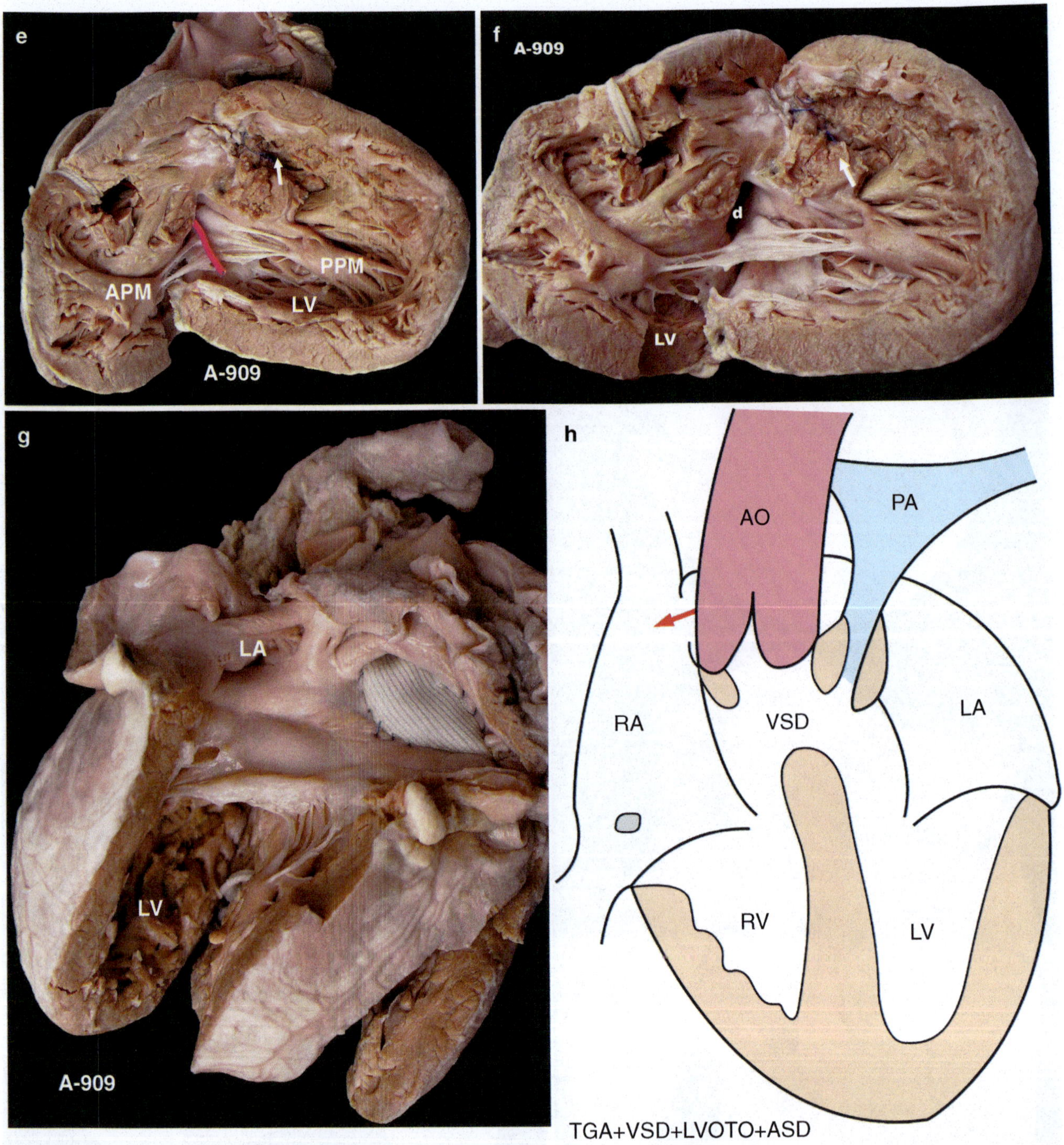

Fig. 25.7 (continued)

Fig. 25.8 (**a**) Specimen 854. TGA with ASD, anterior view. The aorta rises from the RV, and the pulmonary artery rises from the LV, both of which are in parallel positions. *AV* aortic valve, *PA* pulmonary artery, *AO* ascending aorta. (**b**) Specimen 854. The free ventricular wall of the LV and the RV lifted up to expose the communication of the ventricle and great artery. Under the AV is a large bond of muscle. The morphological RV is on the right. There is no fibrous connection between the TV and the AV, and the aorta rises from the RV. The morphological LV is on the left. The pulmonary artery rises from the LV, and there is some fibrous connection between the pulmonary and mitral valves. *PV* pulmonary valve, *MV* mitral valve. (**c**) Specimen 854. Relationship of aorta and TV. The TV and AV are separated by a huge cone muscle. *RV* right ventricle, *AO* ascending aorta, *PA* pulmonary artery. (**d**) Specimen 854. Relationship of pulmonary and mitral valves. The pulmonary and mitral valves have a fibrous connection instead of smooth muscle. A valvular perforation can be found at the border of the right anterior leaflet of the pulmonary valve; due to the special overlapping effect when the pulmonary valve is closed, it will not lead to regurgitation. MV mitral valve, AO ascending aorta, *PA* pulmonary artery, *LV* left ventricle, *RV* right ventricle. (**e**) Specimen 854. Close-up view of fibrous connections between the pulmonary and mitral valves. The ventricular septum is intact. *MV* mitral valve, *PV* pulmonary valve, *PA* pulmonary artery. (**f**) Specimen 854. Secondary ASD. There is a muscle crest between the lower edge of the defect and the atrioventricular valves. CS, coronary sinus. *RA* right atrium, *RV* right ventricle. (**g**) Specimen 854. Left heart view of ASD. The muscle crest is between the lower edge of the ASD and the MV, proving that this ASD is a secondary type. *LV* left ventricle, *LA* left atrium. (**h**) Specimen 854. Right view showing the opening of the LAD (*arrow*) and bulky muscle bundle in the right ventricle. *LV* left ventricle, *AO* ascending aorta, *PA* pulmonary artery, *RV* right ventricle

25.2.4 TGA with Secondary ASD

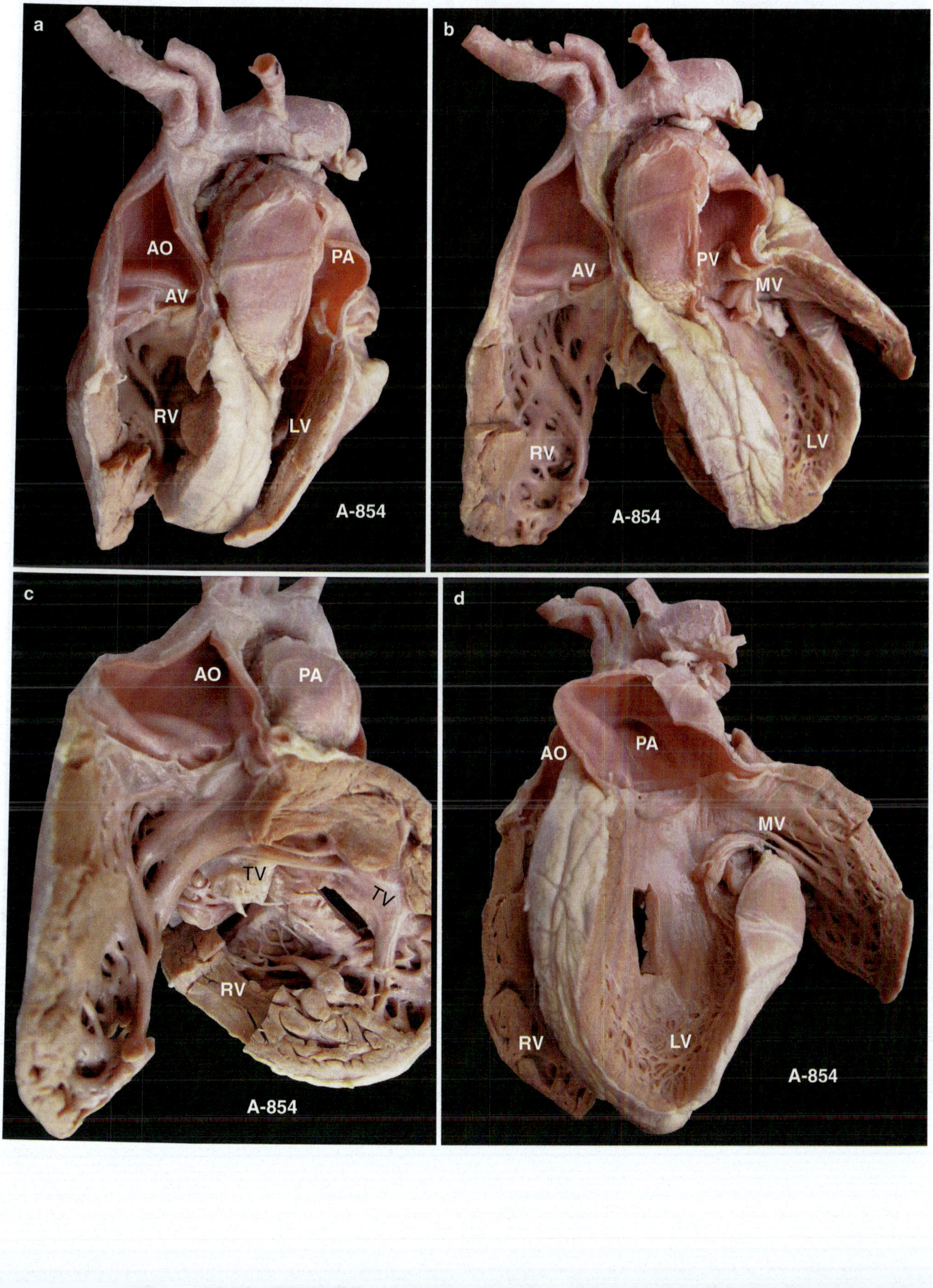

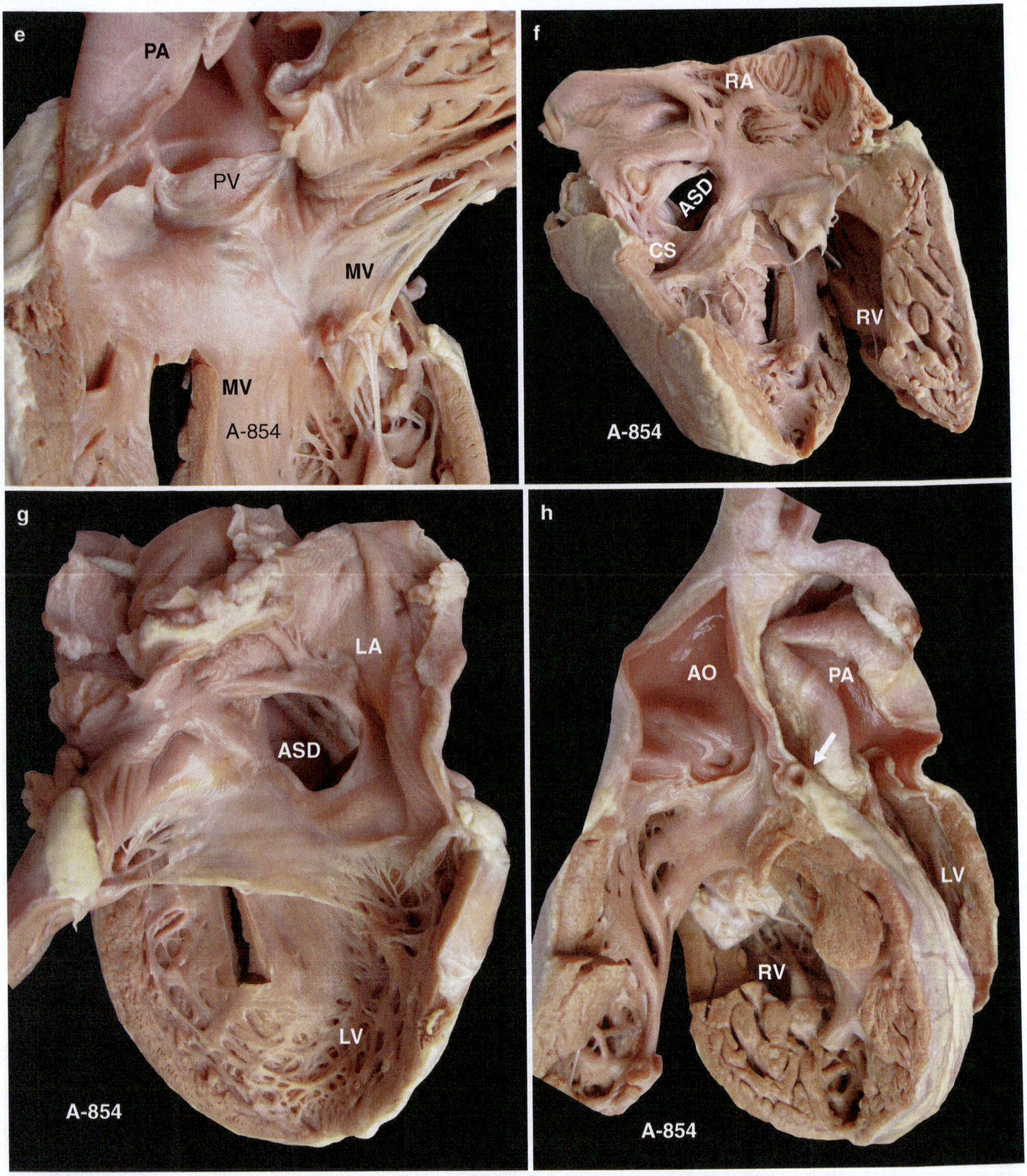

Fig. 25.8 (continued)

25.2.5 TGA with External Conduit

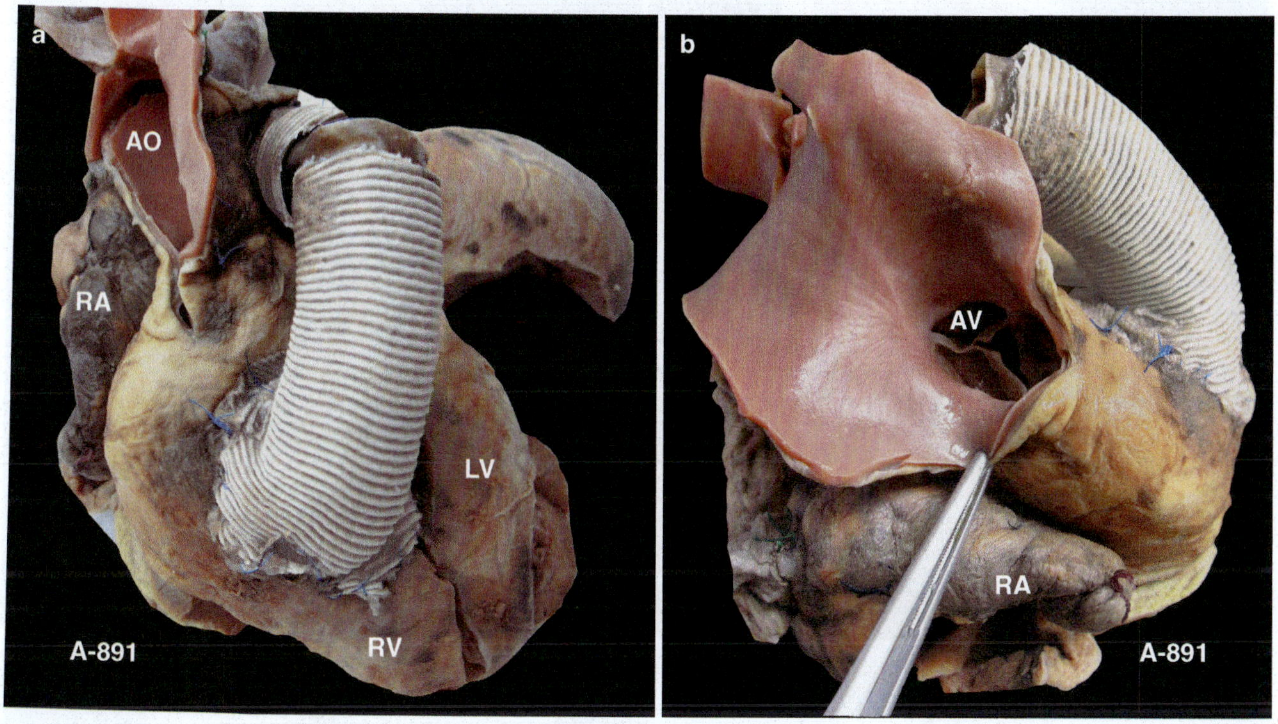

Fig. 25.9 (**a**) Specimen 891. TGA with Rastelli operation. The artificial conduit is from the RV to the pulmonary artery, and the aorta is from the RV. *AO* ascending aorta, *RA* right atrium, *LV* left ventricle. (**b**) Specimen 891. Up-down view. The aorta is right anterior, and the pulmonary artery is left posterior (behind the aorta); the external conduit is artificial material with an inner wall of bovine pericardium. *AV* aortic valve, *RA* right atrium. (**c**) Specimen 891. Cutaway anterior wall of the right ventricle showing the entrance of the artificial external conduit (*arrow*). The RV outlet is completely closed at the VSD position by a pericardium patch. The AV is left of the LV, and blood in the LV flows into the aorta by the VSD. The pulmonary artery connects to the artificial external conduit. (**d**) Specimen 891. The surgical clamp in the external conduit shuttles to the RV, pointing to the VSD patch; blood in the RV has to flow into the pulmonary artery through the external conduit. *LV* left ventricle. (**e**) Diagram of Fig. 25.9d. The outlet of the RV is completely closed by the pericardium patch (*green* patch in RV); the other *green* patch is the ASD patch. *RA* right atrium, *LV* left ventricle. (**f**) Specimen 891. View of VSD in the left ventricle. The pulmonary valve is in the LV and has a fibrous connection with the MV, establishing a diagnosis of TGA. The LV is connected to the aorta through the VSD (probe is in VSD). *PV* pulmonary valve, *MV* mitral valve, *LV* left ventricle. (**g**) Specimen 891. The anastomosis of the artificial external conduit and the pulmonary artery. The PV has been closed, which makes blood in the external conduit flow into the RV. *LV* left ventricle, *PA* pulmonary artery, *PV* pulmonary valve, *LA* left atrium. (**h**) Specimen 891. Right anterior view of the right atrium and valve. The wall of the RV is thick, a huge secondary ASD is closed with a patch, and there is a residual shunt at the top of the patch, with an opening of pulmonary vein. *RA* right atrium, *RV* right ventricle. (**i**) Specimen 891. View of the left atrium and valve. The MV is normal, and the ASD is closed with a patch but still has a residual shunt. *LA* left atrium, *LV* left ventricle. (**j**) Specimen 891. Comprehensive schematic diagram (Notes: 1, the PV is closed; 2, the outlet of the RV is closed by a patch; 3, the green region shows the external conduit, the pulmonary artery has to receive the blood from the RV; and 4, the VSD is not large enough, which may explain the failure of surgery; the VSD should be enlarged in this case). *PA* pulmonary artery, *AO* ascending aorta, *LA* left atrium, *RA* right atrium, *RV* right ventricle, *LV* left ventricle

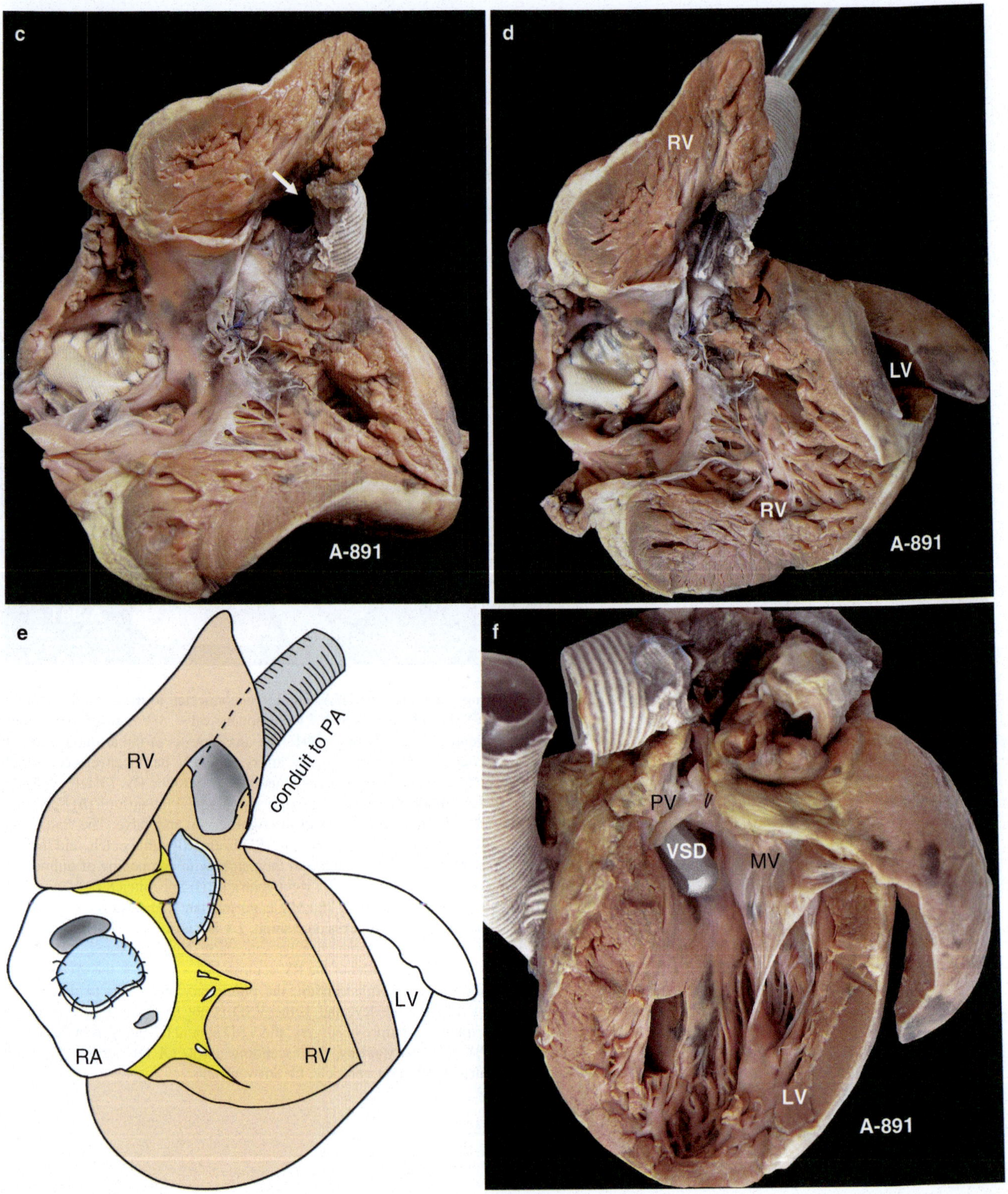

Fig. 25.9 (continued)

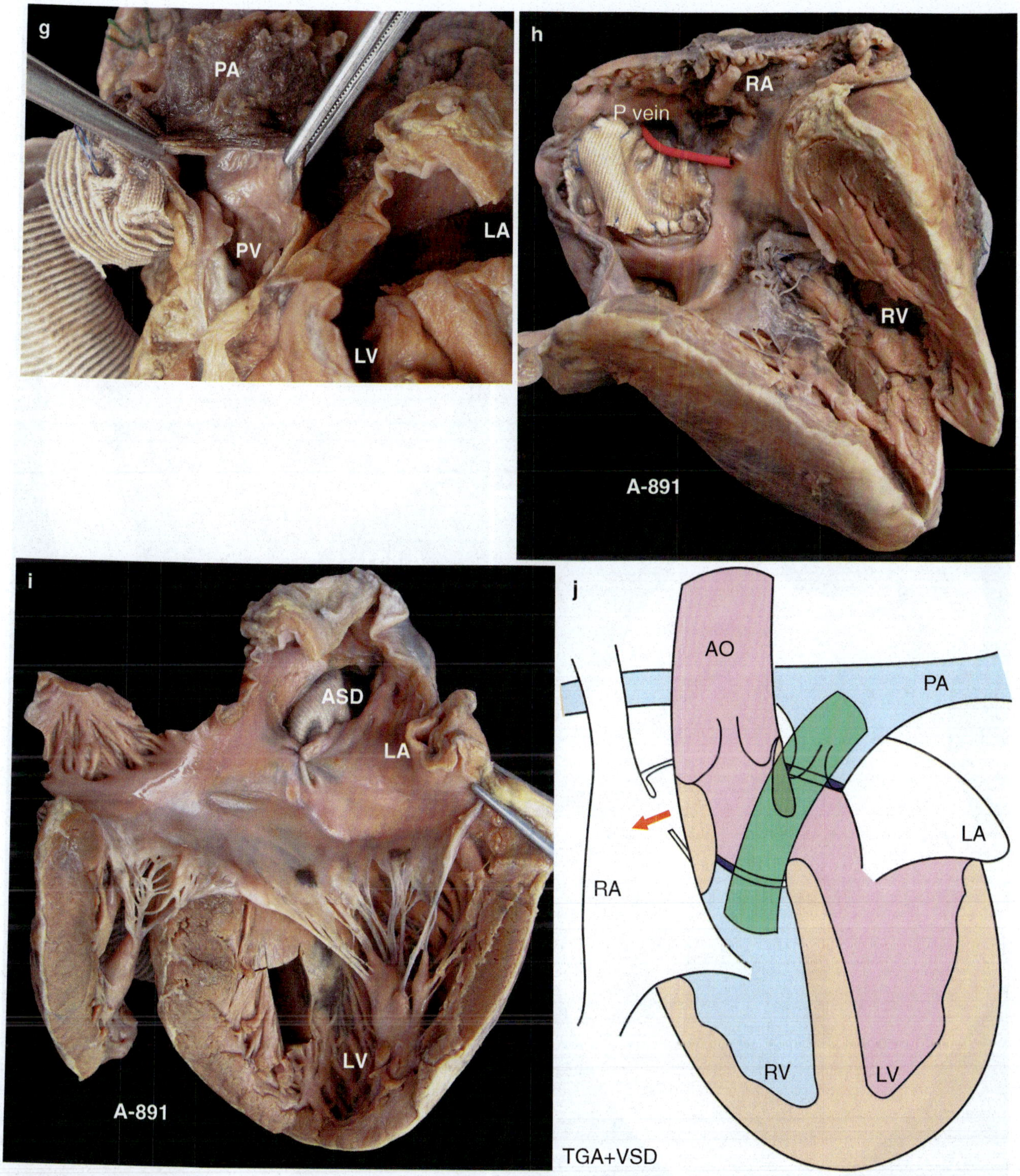

Fig. 25.9 (continued)

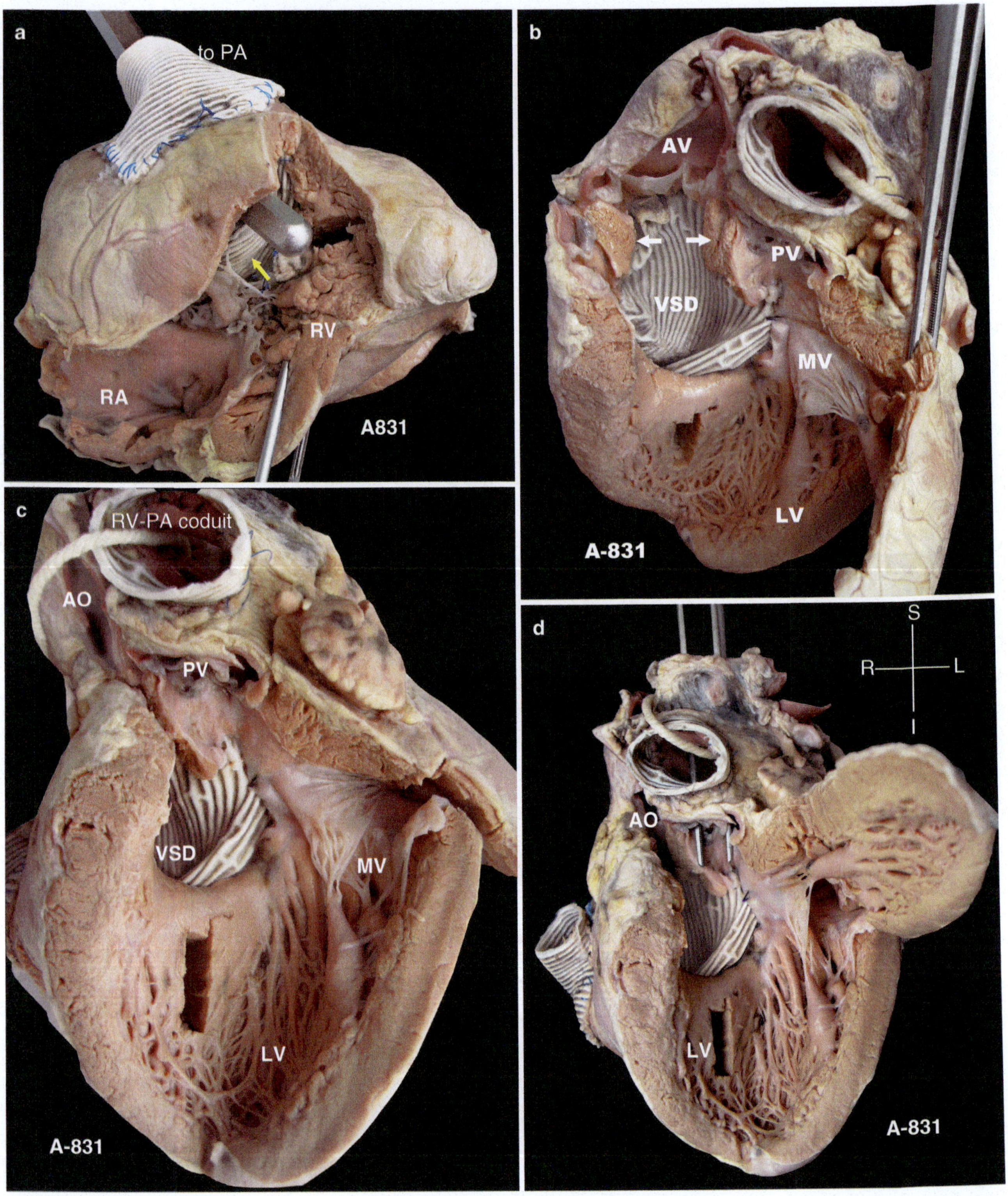
a
to PA
RV
RA
A831
b
AV
PV
VSD
MV
LV
A-831
c
RV-PA coduit
AO
PV
VSD
MV
LV
A-831
d
S
R
L
I
AO
LV
A-831

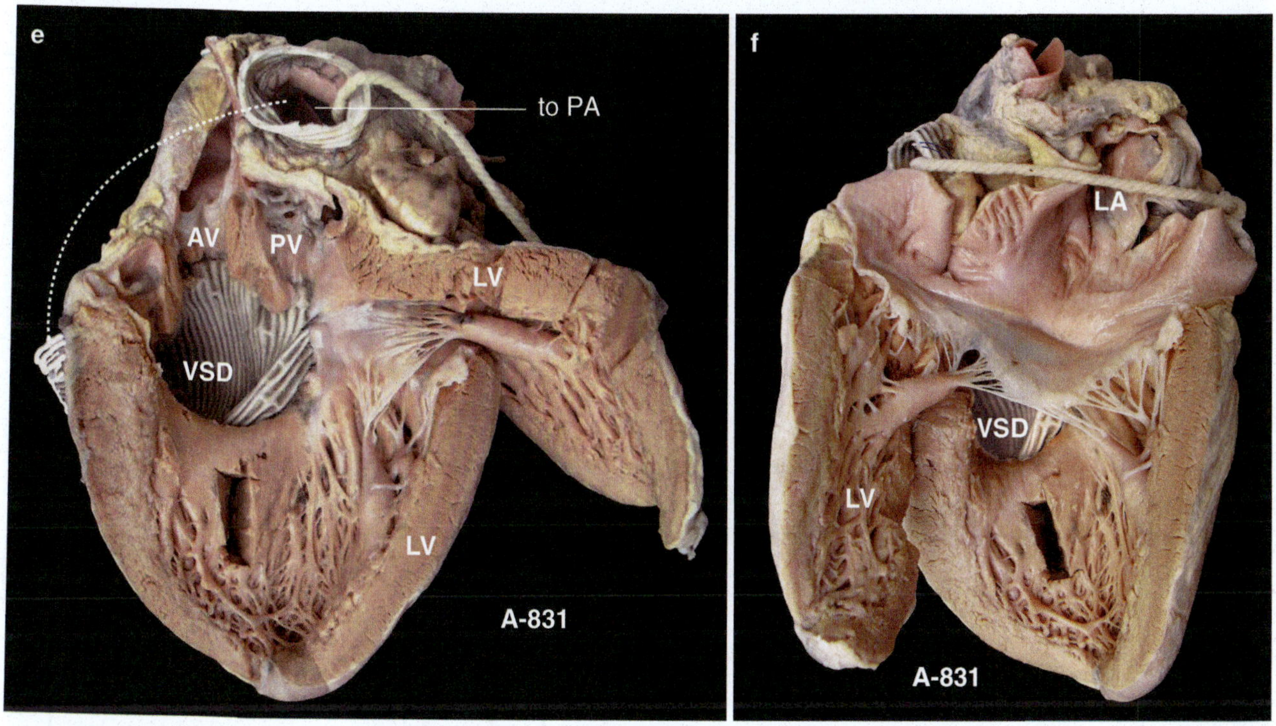

Fig. 25.10 (continued)

Fig. 25.10 (**a**) Specimen 831. A TGA case with VSD and double arterial cones under the aortic valve. The VSD is closed by an artificial patch (*yellow arrow*), which makes the LV connected to the aorta, and then an artificial external conduit completes the track from the RV to the pulmonary artery. An iron probe inserts into the RV through the external conduit. *RA* right atrium. (**b**) Specimen 831. Cut into the left ventricle. The PV rises from the LV completely and has been closed to a little gap. The AV rises completely from the RV, and the VSD is far away from that two arterial valves. Cutting off the ceiling of the VSD reveals the section, with *arrows*. The MV has no fibrous connection to the two groups of semilunar valves, which confirms that this case is a remote type of VSD. (**c**) Specimen 831. In a similar view, the sections of the VSD return to the original positions, proving that the aorta rises completely from RV; a section of the artificial conduit shows the anastomotic stoma. *AO* ascending aorta, *PV* pulmonary valve, *MV* mitral valve, *LV* left ventricle. (**d**) Specimen 831. Tweezers inserted into the residual gap of the closed PV confirm that the PV rises completely from the LV; an artificial conduit can be found at the right border of the specimen. *AO* ascending aorta. (**e**) Specimen 831. Posterior-lateral view of the left ventricle. The external conduit has been cut off, and a huge muscular VSD can be found at the top of the ventricular septum. There is no fibrous connection between the pulmonary and mitral valves, and the VSD is far away from the arteries. In this case, the PV rises from the LV but has no fibrous connection to the MV, which is very rare. The *dotted line* shows the direction of the blood flow in the conduit. *AV* aortic valve, *LV* left ventricle, *PV* pulmonary valve, *PA* pulmonary artery. (**f**) Specimen 831. Left posterior view showing the structure of the left atrium, the mitral valve, and the left ventricle. There is a huge muscular VSD at the upper part of the ventricular septum, which is under the anterior leaflet of the MV. *LA* left atrium, *LV* left ventricle. (**g**) Specimen 831. Right view of the heart. Figure shows the relationship of the RA and the inlet of the RV. The TV is a septum leaflet, the base of which is free. A VSD patch (P) divides some part of the subarterial septum to the LV. The upper hole of the patch is an opening of the external conduit. (**h**) Schematic diagram of specimen 831. The PV has been closed, and the blood flows from the PV into the external conduit through the RV. The lower part of the ventricular septum and the outlet of the RV have been closed, which forms a new outlet of the LV: it is a Rastelli surgery for TGA. *PA* pulmonary artery, *AO* ascending aorta

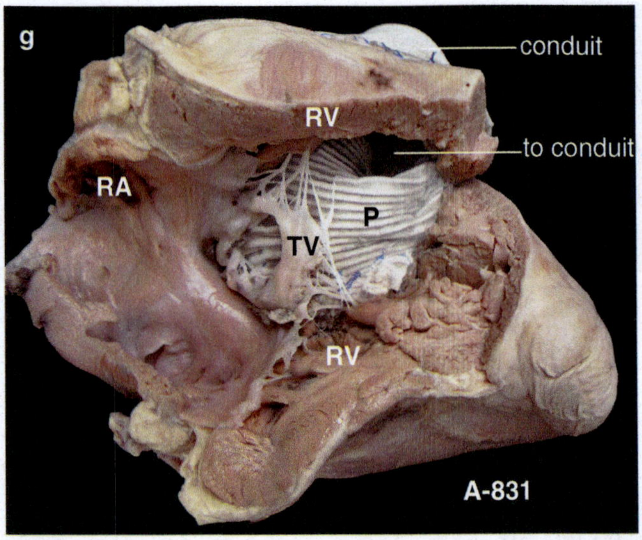

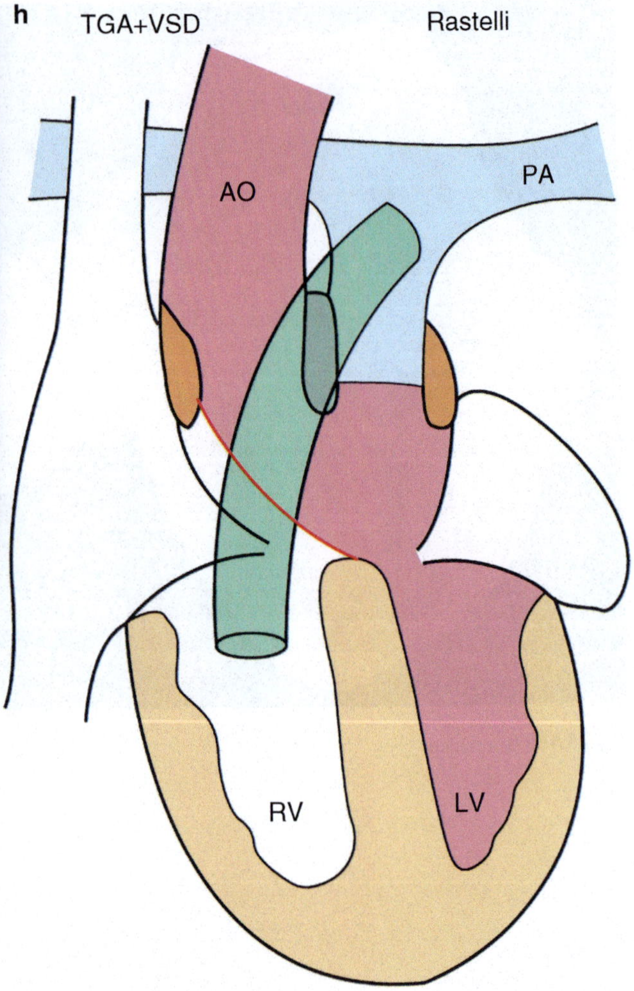

Fig. 25.10 (continued)

Fig. 25.11 (**a**) Specimen 1119. TGA. The RA and the RV both are left posterior, but the LA and the LV both are right anterior; this case suffers from an ASD and a VSD. We completed the switch operation and repaired the ASD and the VSD. *PA* pulmonary artery, *AO* ascending aorta. (**b**) Specimen 1119. Cut into the outlet of LV and aorta shows sutures for switch operation, proving transposition of the AV and aorta. It can be considered as a new aorta connected to the LA by transplantation from the RV. *MV* mitral valve. (**c**) Specimen 1119. Detailed view of the left ventricle, mitral valve, and aortic valve. There is no muscular connection. Surgical sutures can be seen in the conduit (*arrow*). *AV* aortic valve, *MV* mitral valve, *LV* left ventricle. (**d**) Specimen 1119. The right atrium and the tricuspid valve. An ASD patch can be seen, as can some subintimal hematoma that may constrict the conduction bundle. *RA* right atrium, *RV* right ventricle. (**e**) Specimen 1119. Figure shows the outlet of the RV, which is connected to the new transplanted pulmonary artery, and there are surgical sutures at the stoma. The VSD is so large that the surgeon had to use two patches. This diagram also shows the hypoplasia morphological RV. *TV* tricuspid valve, *RA* right atrium. (**f**) Specimen 1119. The RV and the LV can be seen at the same time by retracting their lateral walls. The morphological LV is very thick, and the morphological RV has hypoplasia. The *black arrows* shows the new pulmonary artery stoma. *TV* tricuspid valve

25.2.6 Arterial Ventricular Transposition, TGA + ASD + VSD

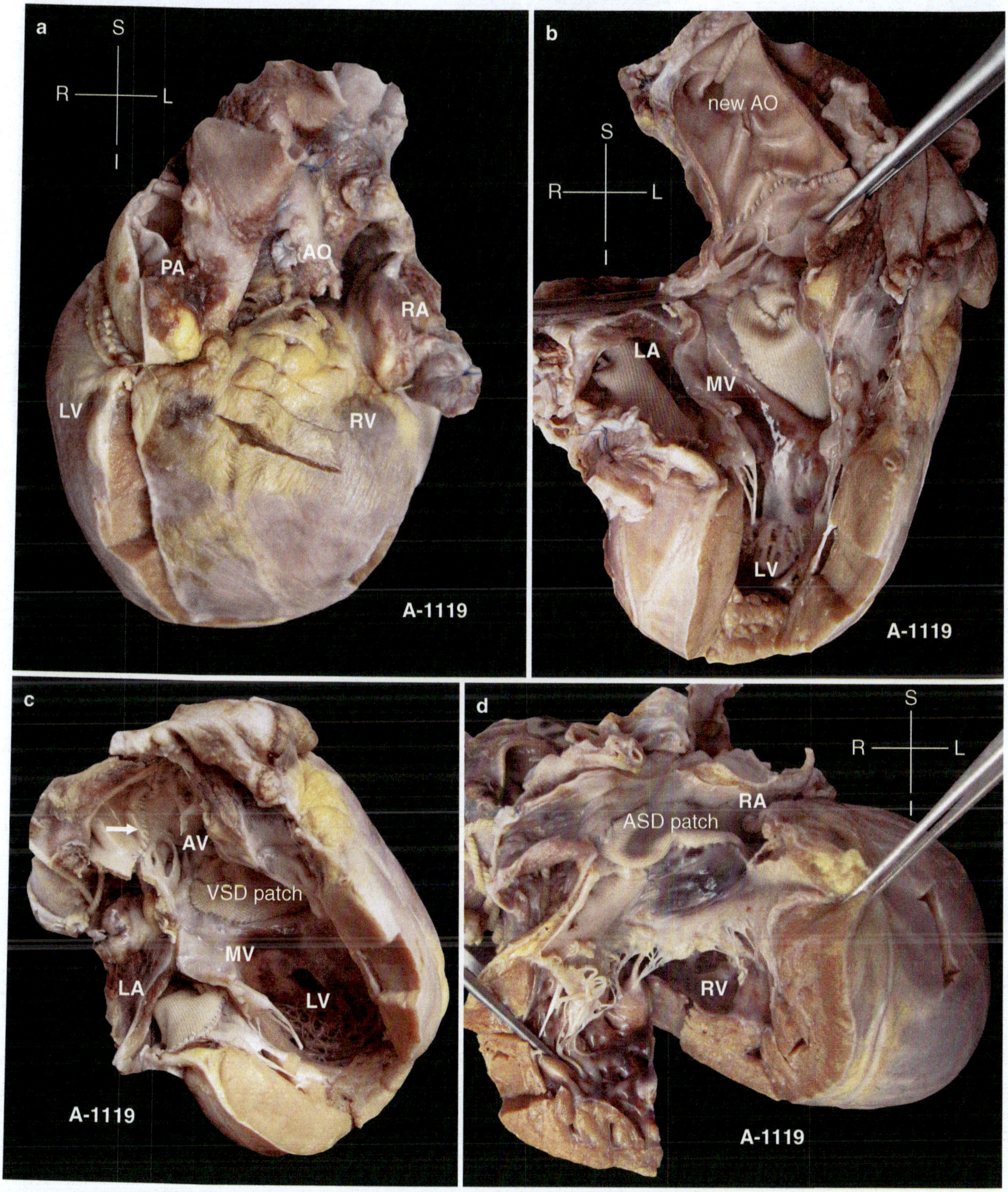

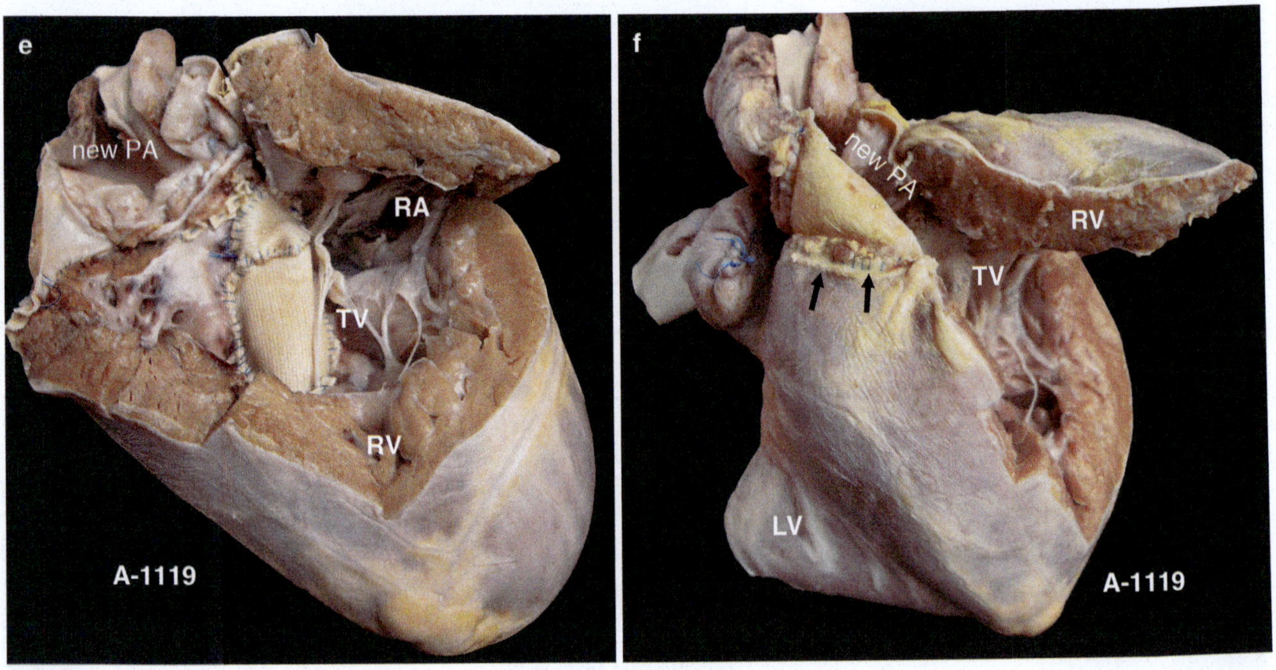

Fig. 25.11 (continued)

26 Corrected Transposition of the Great Arteries

26.1 General Considerations

26.1.1 Definition

Corrected transposition of the great arteries (CTGA) is a condition in which the connections of the great artery valves with the ventricles are inverted, as are the atrium-ventricle connections (Fig. 26.1a, b). Oxygenated blood of the morphological LA can still flow into the morphological RV and be pumped into the aorta; therefore, the blood flow is functionally corrected and clinical cyanosis will not occur.

26.1.2 Classification

Classification of complete TGA and CTGA (connection of great arteries and ventricles are completely inverted, excluding double outlet of the ventricle) are as below:

CTGA, because it has been functionally corrected and the aorta still receives pulmonary vein blood and the pulmonary artery still receives blood from the systemic veins, is clinically asymptomatic only until tricuspid insufficiency occurs. The insufficiency occurs when the patient has grown into childhood or adulthood, and the TV no longer is able to withstand the long-term load of the systemic circulation. If it coexists with another intracardiac abnormality and requires surgery, the anatomical characteristics should be given thorough consideration.

26.2 Surgical Considerations

26.2.1 Cardiac Appearance and Course of the Coronary Artery

The left and right atria are situs, but the left and right ventricles are bilaterally side by side, and the ventricular septum

Usual classifications of complete anomalous arterial ventricular connection (not including partial anomalous connection)

Complete transposition of the great arteries with normal atrioventricular connection					Functionally corrected transposition of the great arteries with atrioventricular discordance			
Situs	Solitus		Inversus		Solitus		Inversus	
Ventricle	Dextro-		Levo-		Levo-		Dextro-	
Transposition of the great arteries	RA	LA	LA	RA	RA	LA	LA	RA
	RV	LV	LV	RV	LV	RV	RV	LV
		P	P		P		P	
	A			A		A		A
	TGA		TGA		CTGA		CTGA	

A aorta; *LA* left atrium; *LV* left ventricle; *P* pulmonary valve orifice; *RA* right atrium; *RV* right ventricle

X. Zhu (ed.), *Surgical Atlas of Cardiac Anatomy*,
DOI 10.1007/978-94-017-9409-1_26, © Springer Science+Business Media Dordrecht and People's Medical Publishing House 2015

is nearly sagittal (Fig. 26.1a). Thus, the anterior interventricular groove is closer to the median line than in the normal heart. The ascending aorta lies to the left anteriorly, and the pulmonary artery lies to the right and posteriorly. The coronary arteries originate from the left and right aortic sinus, and the noncoronary sinus (posterior sinus) is moved to the anterior. The anterior descending branch originates from the main trunk of the RCA.

26.2.2 Atrioventricular Ring and the Membranous Septum

The atrium of the (situs solitus,L-loop, L- transposition) SLL type of CTGA is situs, but the ventricular chamber of the right side is a MV ventricle (Fig. 26.1c), with the adherent fringe of the MV at the ventricular septum still higher than the contralateral adherent fringe of the tricuspid septal leaflet. The membranous septum lies inferior to the MV ring. Therefore, in a dissected cardiac chamber, the membranous septum can be seen to lie inferior to the right atrioventricular ring (MV ring), which is contrary to the right ventricle view in the normal heart.

26.2.3 Complete Muscular Cone Inferior to the Aortic Valve

The AV and the left TV ring are separated by muscular tissue. Contrariwise, the PV is connected by fibrous tissue with the MV to the right. A supraventricular crista-like bundle can be seen from the left side of ventricular septum (Fig. 26.1d).

26.2.4 Anatomical Connection of the Pulmonary Artery Outflow

The pulmonary artery outflow of the CTGA is an oblique canal that lies deeply between the inverted MV and TV ring (Fig. 26.1e). The PV ring is conjuncted [*connected?*] with the posterior central fibrous body. It is fibrously connected with the MV ring to the right, and its left side is the membranous septum (Fig. 26.1f). It lies anteriorly next to the AV ring with the conical ventricular septum. The PV ring rides just over the muscular ventricular septum.

26.2.5 Course of the Conducting System

The conducting system of a CTGA originates from the anterior atrioventricular node, which is located at the junction point of the pulmonary and mitral valves and also where the atrial septum connects with the right auricle (Fig. 26.1g). The conducting bundle travels from the anterior border of the PV ring to the superior border of the muscular ventricular septum; it then divides into the left and right bundles. The left bundle is located in the anatomical LV, whereas the right bundle travels downward into the anatomical RV. If there is a membranous VSD, the right bundle will travel at the anterior border of the defect, whereas in the normal heart, the conduction system travels along the posterior and inferior border of the defect. When suturing a VSD in CTGA, the anterior and superior border of the defect is a dangerous zone where the conducting bundle is easily injured and can cause atrioventricular block.

Fig. 26.1 (**a**) Specimen 850. Anterior view of CTGA. The pulmonary artery is located right and posterior to the aorta, whereas the RA remains at the right side, which belongs to the situs solitus. *AO* ascending aorta; *LV* left ventricle; *PA* main pulmonary artery; *RA* right atrium; *RV* right ventricle. (**b**) Specimen 850. The RA is connected with the morphological LV or anatomical LV. *IVC* inferior vena cava; *RA* right atrium; *SVC* superior vena cava. (**c**) Specimen 850. Atrium and ventricle of the right side are dissected. The superior is the RA, and the atrioventricular valve is the MV. The inferior is a morphological LV (or an anatomical LV). (**d**) Specimen 850. Ventricular septal defect and the anterior border of the defect. The conducting bundle travels underneath the endocardium from the anterior atrioventricular node (*green line*) and along the anterior and superior border of the VSD. *RA* right atrium. (**e**) Specimen 850. The outflow of the morphological LV (*ALV*) extends directly into the pulmonary artery (*blue probe*). (**f**) Specimen 850. A thick probe is inserted into the morphological LV from the pulmonary artery. A VSD can be detected at the same time. *MV* mitral valve, *PA* pulmonary artery, *AO* ascending aorta. (**g**) Specimen 850. Figure demonstrates the fibrous connection of mitral and pulmonary valves (*arrow*). *MV* mitral valve, *PV* pulmonary valve. (**h**) Specimen 850. Observing the LA and left-side ventricle. The left-side ventricle is the morphological RV, which connects with the AV. *AO* ascending aorta, *LA* left atrium. (**i**) Specimen 850. Showing the left-side atrium and ventricle. The atrium (*LA*) has been dissected, showing the unsmooth surface of the atrium septum. The LA is connected with the morphological RV. (**j**) Specimen 850. The ventricular wall has been lifted, and the ventricular inflow can be seen. The TV has been completely unfolded, and a VSD can be seen subvalvularly. The ARV is the morphological RV. *LA* left atrium. (**k**) Specimen 850. Figure demonstrates that the aorta (*AO*) originates directly from the left ventricular outflow. There is a muscular septum between the aorta and TV, the structure of which equals the supraventricular crista, beneath which is the VSD (**d**). *LA* left atrium, *AO* ascending aorta, *D* ventricular septal defect, *TV* tricuspid valve. (**l**) Specimen 850. Figure demonstrates the connection between the AV and the TV. The TV has three leaflets with many small pores at the border of the leaflets, which is a variance of the leaflet development. Because the pores lie above the closing line, they seldom cause insufficiency. The aortic and the tricuspid valves are separated by the supraventricular crista, beneath which is the VSD, with its border adhered by chordae tendineae (*arrow*). (**m**) Schematic diagram of specimen 850. Atria are solitus, whereas the ventricles and atrioventricular valves are inversus. The aorta originates from the morphological RV, and the pulmonary artery originates from the morphological LV, with a muscular-type VSD. *PA* pulmonary artery, *AO* ascending aorta, *LA* left atrium

26.3 Specimen Demonstration

26.3.1 CTGA + VSD, Typical Type

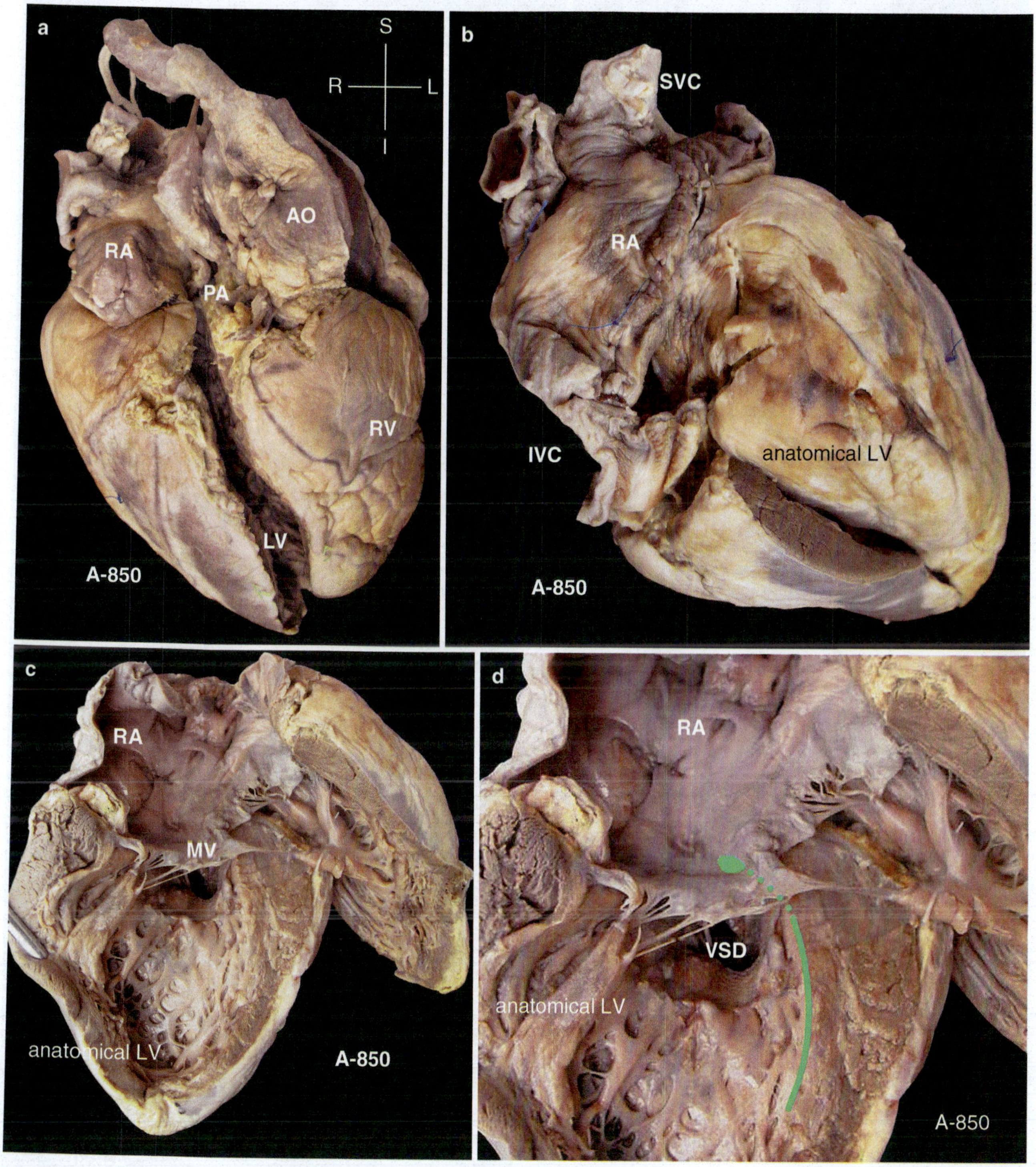

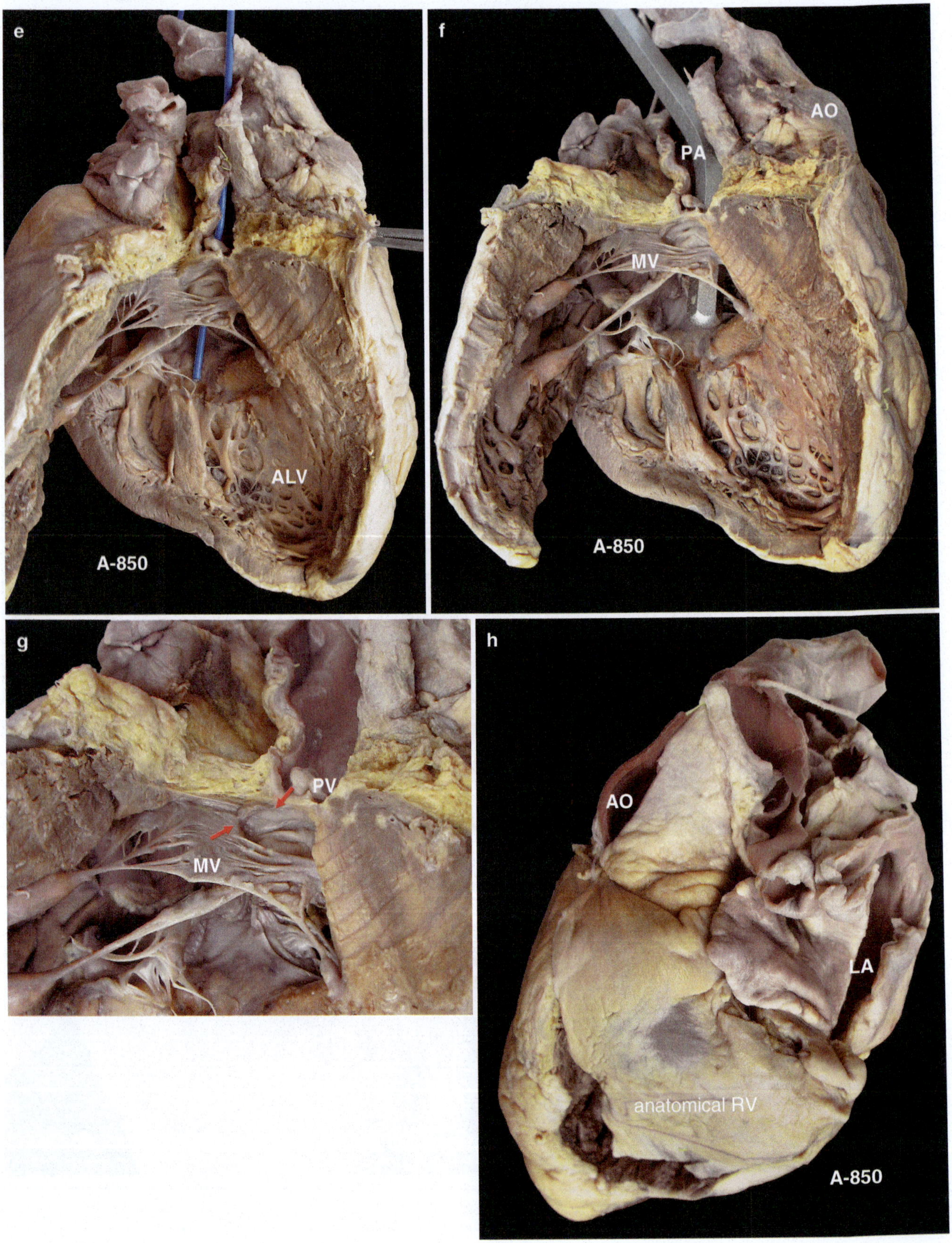

Fig. 26.1 (continued)

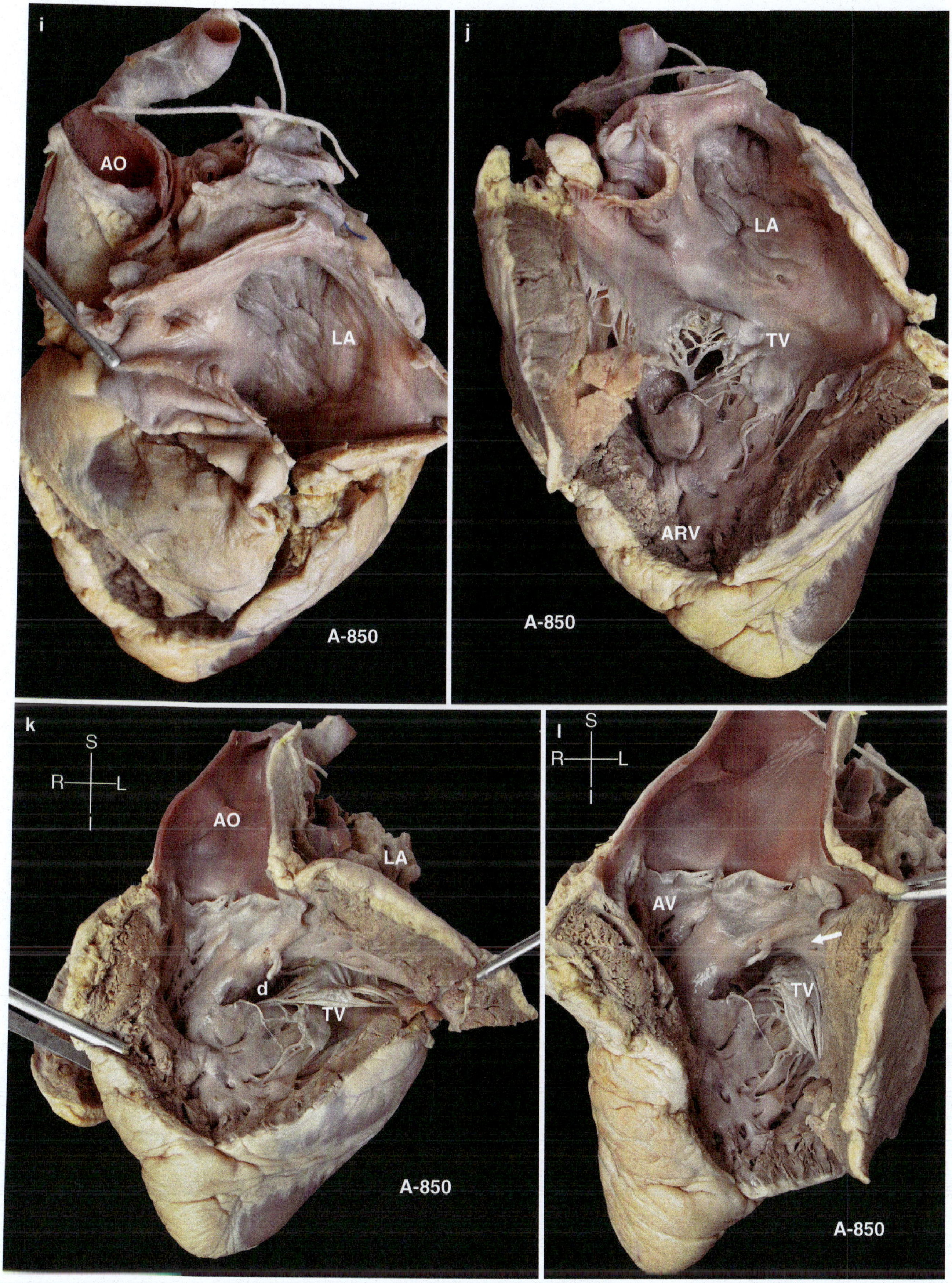

Fig. 26.1 (continued)

m

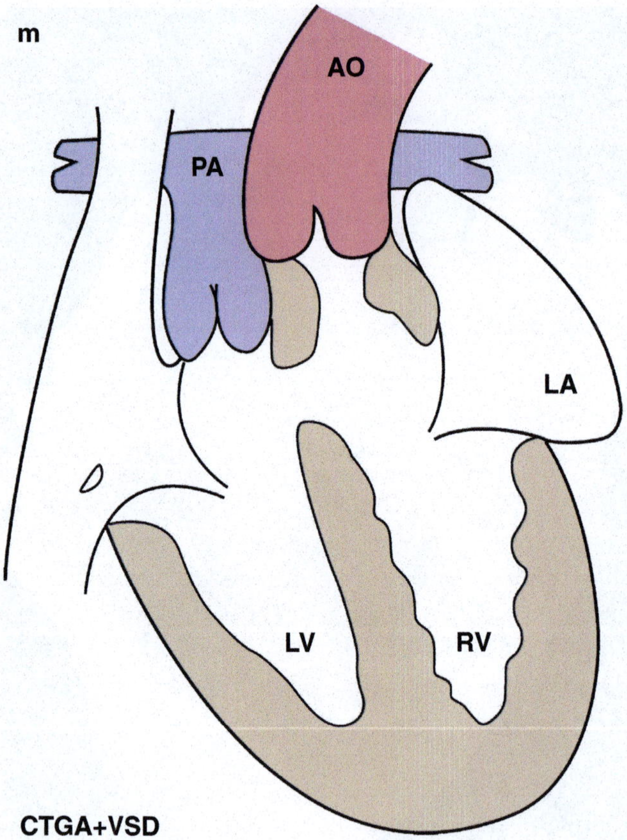

CTGA+VSD

Fig. 26.1 (continued)

26.3.2 Fresh Specimen of CTGA

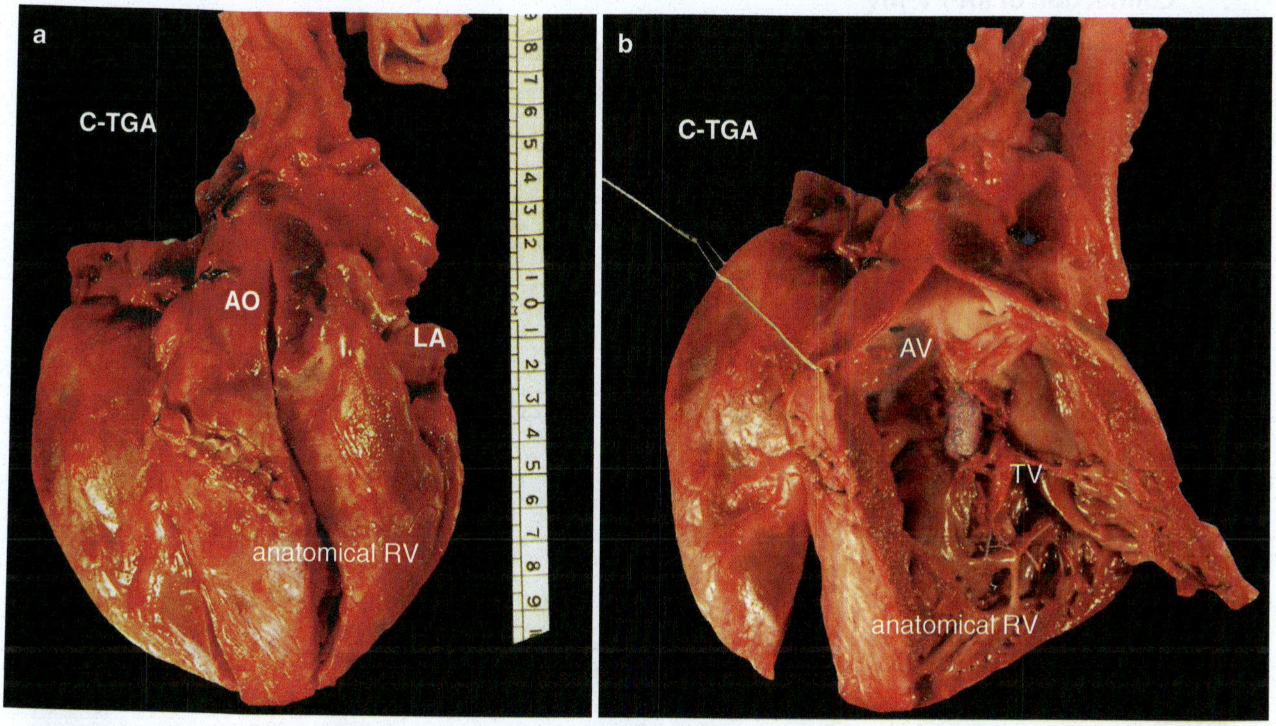

Fig. 26.2 (**a**) Specimen 001. Anterior view of postoperational CTGA. The left-side ventricle is a right ventricular structure, which is the morphological RV (anatomical RV). The aorta originates from the morphological RV, and the pulmonary artery originates from the morphological LV. *AO* ascending aorta, *LA* left atrium. (**b**) Specimen 001. Left view of postoperational CTGA. The left-side ventricle is a right ventricular structure, which is the morphological RV. The aorta originates from the morphological RV. *AV* aortic valve; *TV* tricuspid valve

26.3.3 CTGA + VSD: Analyzing the Adjacent Connection of RA-PV-MV

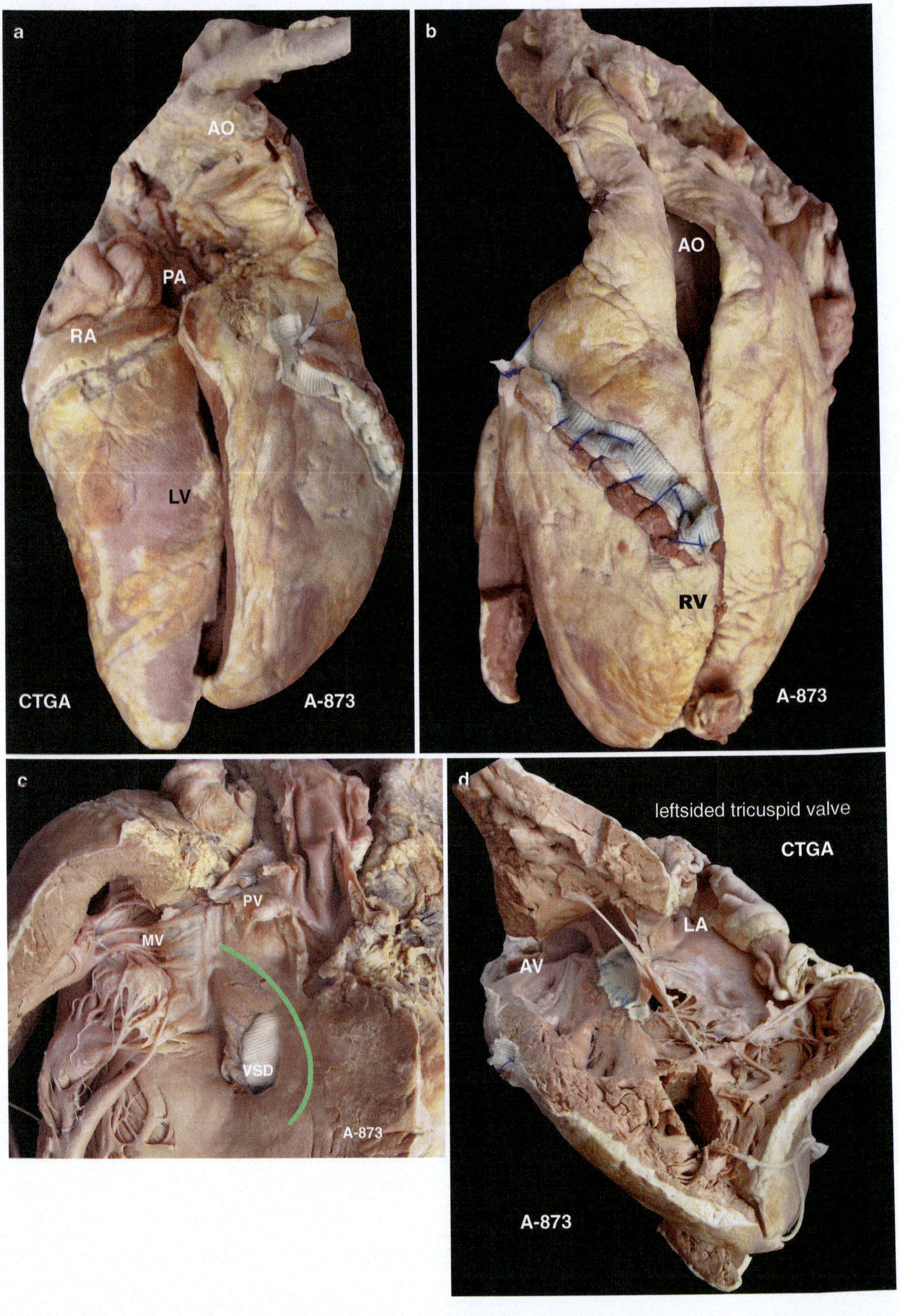

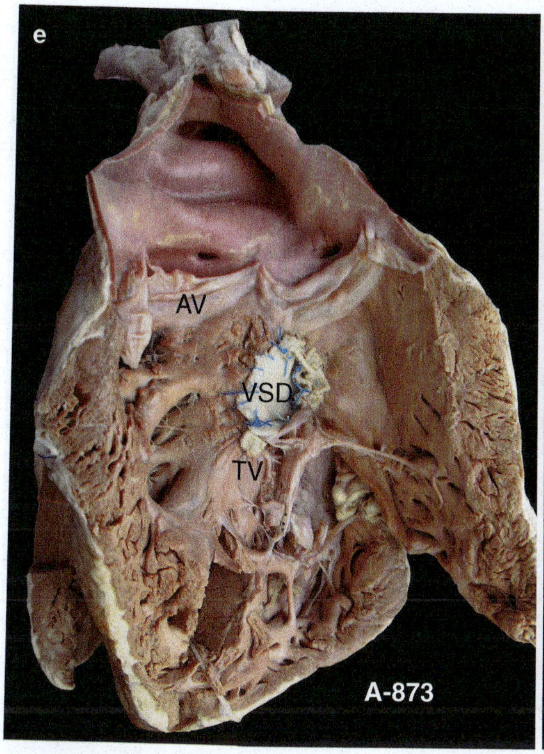

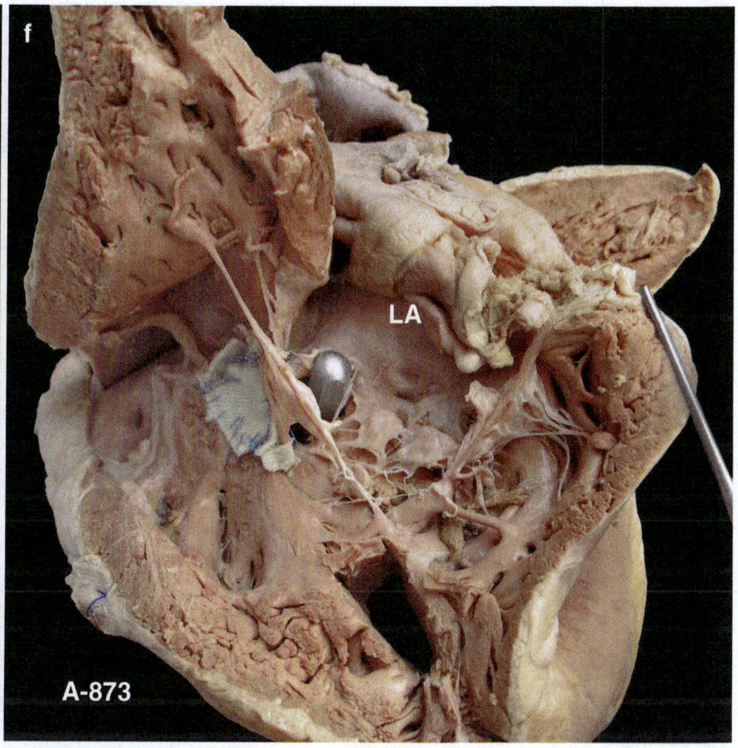

Fig. 26.3 (continued)

Fig. 26.3 (**a**) Specimen 873. Outside view of CTGA. PA is the main pulmonary artery, which lies right and posteriorly to the ascending aorta. The morphological LV on the right side has been dissected. The surgical incision with stitch is located at the anterior wall of the morphological RV. *AO* is aorta; *RA* right atrium. (**b**) Specimen 873. Left-side view of CTGA. This view is just toward the left side of the ascending aorta, with the pulmonary artery blocked. The aorta and ventricle have been completely dissected. The ventricle beneath the AV is the morphological RV, the anterior wall of which has the surgical incision through which the VSD was repaired. (**c**) Specimen 873. The right-sided ventricle has been dissected (morphological LV), showing the VSD and the course of the conducting bundle (*green line*). *MV* mitral valve, *PV* pulmonary valve. (**d**) Specimen 873. The left-sided ventricle has been dissected (morphological RV) to show the TV and the VSD stitching patch. Repairing of the VSD was by incision on the morphological RV, which will influence the pumping function of the systemic circulation. *CTGA* corrected transposition of the great arteries, *AV* aortic valve, *LA* left atrium. (**e**) Specimen 873. Figure demonstrates the AV and the supravalvular cone, proving that the ventricle beneath the AV is the RV. *TV* tricuspid valve. (**f**) Specimen 873. The metallic probe can pass through the VSD patch, proving that there is a remnant shunt after surgery. *LA* left atrium. (**g**) Specimen 873. View showing the tricuspid valve. The arrow points out that the posterior commissure has been banded. *LA* left atrium, *TV* tricuspid valve. (**h**) Specimen 873. The structure shown within the white frame demonstrates the relationship among the RA wall, the pulmonary artery, and the MV. In CTGA, the relationship among them corresponds with the relationship among the LA wall, the AV, and the MV in the normal heart. In those with PV stenosis that requires broadening of the main pulmonary artery, an incision can be made in this area. The meeting border of the MV, the pulmonary artery, and the posterior wall of the RA in CTGA corresponds with the meeting border of the MV, the AV, and the posterior wall of the LA in the normal heart. (**i**) Specimen 873. The relationship among the pulmonary artery valve, the mitral valve, and the posterior wall of right atrium in CTGA. In the schematic diagram on the right, there is a common joining line of the RA, the PV, and the MV, where pulmonary artery valve dilation can be performed. The relationship among the RA, the PV, and the MV corresponds with that among the MV, the AV, and the posterior wall of the LA in the normal heart. The *green line* show the course of the conducting bundle. *PA* pulmonary artery, *VSD* ventricular septal defect, *LV* left ventricle. (**j**) Specimen 873. In contrast with (**i**), showing the relationship among the mitral valve, the aortic valve, and the posterior wall of left atrium in the normal heart. The *arrow* shows the joining point. *AO* ascending aorta, *LA* left atrium, *LV* left ventricle, *ROVT* right outflow of the ventricular tract. (**k**) Specimen 873. Schematic diagram of the relationship among peripheral structures of the pulmonary artery valve. This figure shows one of the approaches by which surgical dilation of the pulmonary artery root can be performed in PV stenosis. Widening the pulmonary artery at the joining point of the PV and the posterior wall of the RA will not injure the anterior atrioventricular node. *PA* pulmonary artery, *AO* ascending aorta, *MV* mitral valve

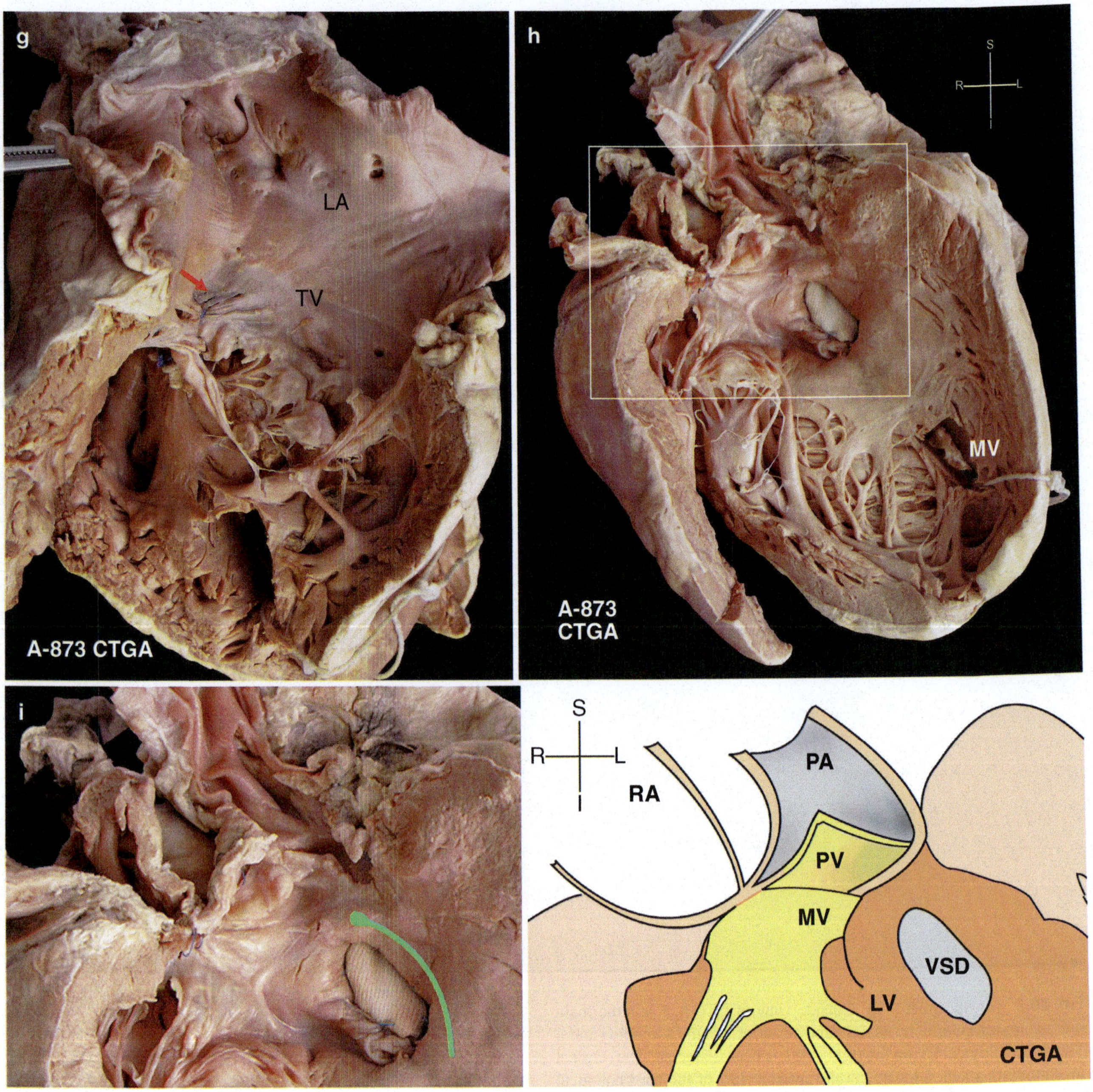

Fig. 26.3 (continued)

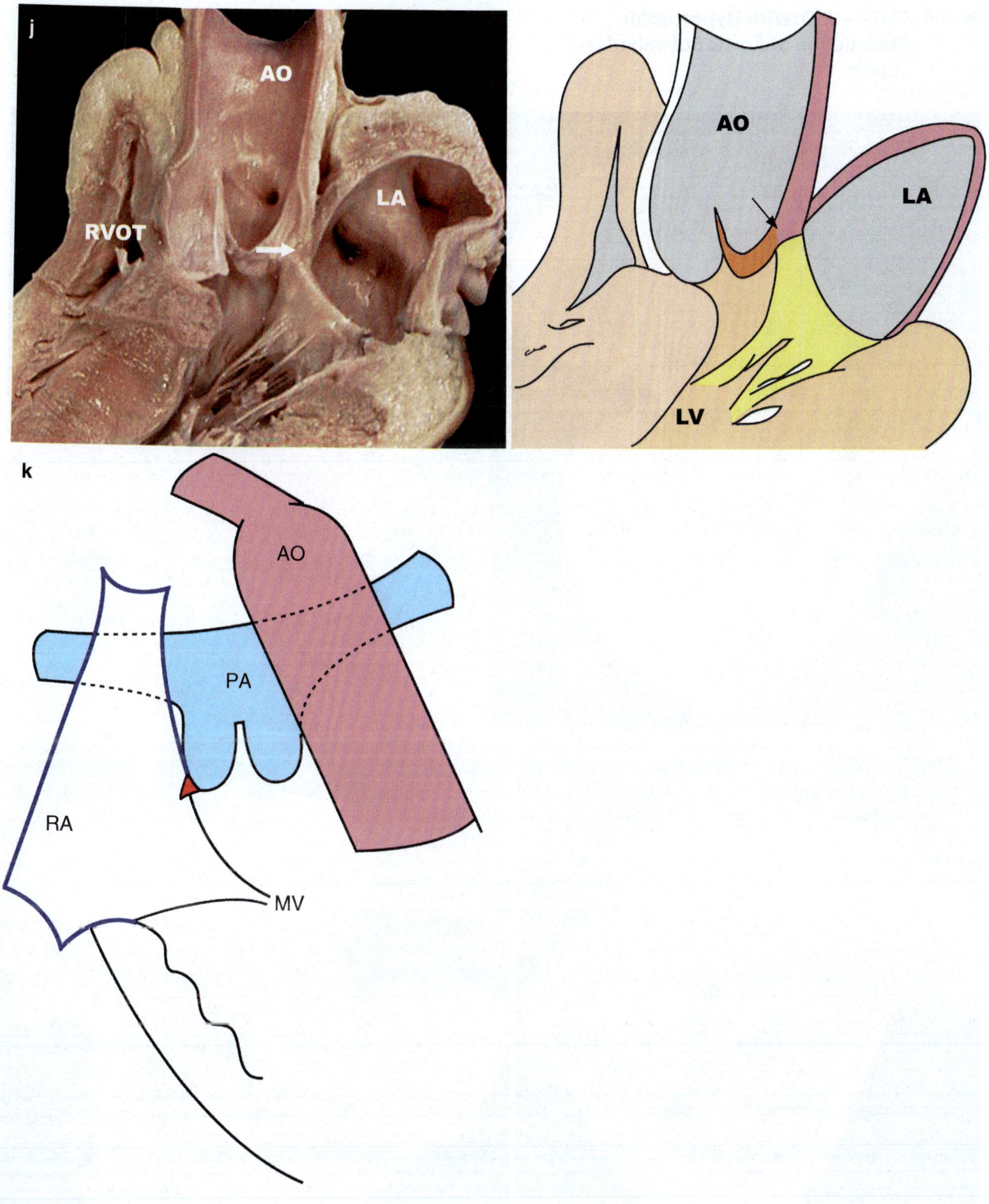

Fig. 26.3 (continued)

26.3.4 CTGA + VSD: With Hypotrophic Obstruction of Aortic Subvalvular Muscle

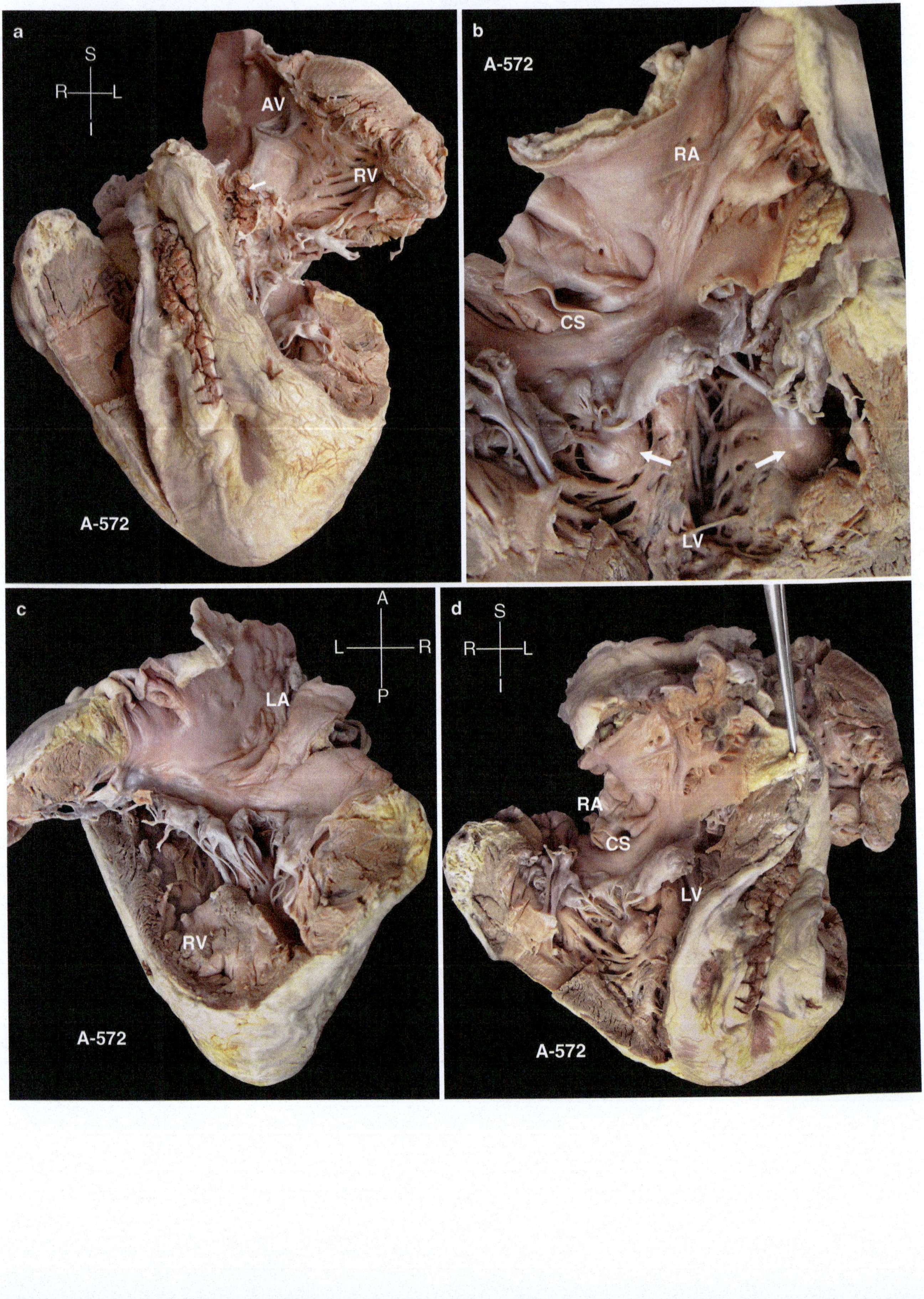

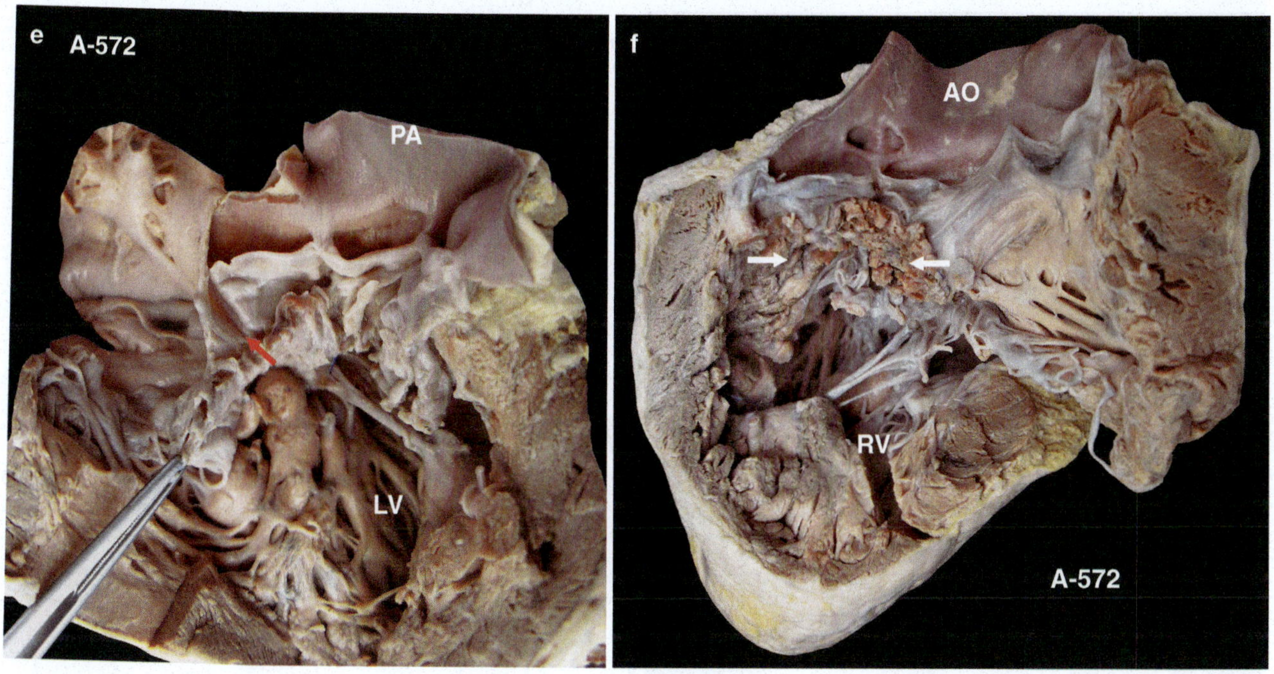

Fig. 26.4 (continued)

Fig. 26.4 (**a**) Specimen 572, CTGA. On the left is the morphological RV connecting with the AV, with its anterior wall lifted. There is an incision on the left ventricular wall (*arrow*). (**b**) Specimen 572. The RA is dissected. The atrioventricular valve is a two-leaflet structure with fibrous thickening of the leaflets and a smooth surface. The *arrow* points to the papillary muscle. CS indicates the orifice of the coronary sinus. *LV* left ventricle. (**c**) Specimen 572. The LA has been dissected, showing that the atrioventricular valve is the TV. *RV* right ventricle. (**d**) Specimen 572. The RA and the ventricular structure have been dissected. The atrioventricular valve is the MV, with major papillary muscle that can be seen. CS indicates the orifice of the CS. There is a sutured incision on the left ventricular wall. *LV* left ventricle. (**e**) Specimen 572. The pulmonary artery originating from the left ventricle, with a VSD. Notice the fibrous connection between the pulmonary and atrioventricular valve leaflets (*arrow*), proving that the PV originates from the morphological LV. Below is the VSD. *LV* left ventricle. (**f**) Specimen 572. Aortic valve originating from the morphological right ventricle on the right. There are large thick muscular bundles between the AV and the atrioventricular valve, without fibrous connection. The muscular bundle beneath the AV (outflow of the RV) causes an obstruction. The *arrow* points to the cross section of the bundles cut in operation. There is a giant VSD beneath the TV. *AO* ascending aorta. (**g**) Schematic diagram of specimen 572. There is a muscular septum between the aortic and atrioventricular valves, whereas there is a fibrous connection between the pulmonary and atrioventricular valves, showing that on the left is a morphological RV and on the right is a morphological LV. There is a giant muscular defect on the ventricular septum. Muscular hypotrophic obstruction can be seen beneath the AV. *LA* left atrium, *AO* ascending aorta, *VSD* ventricular septal defect. (**h**) Schematic diagram of specimen 572

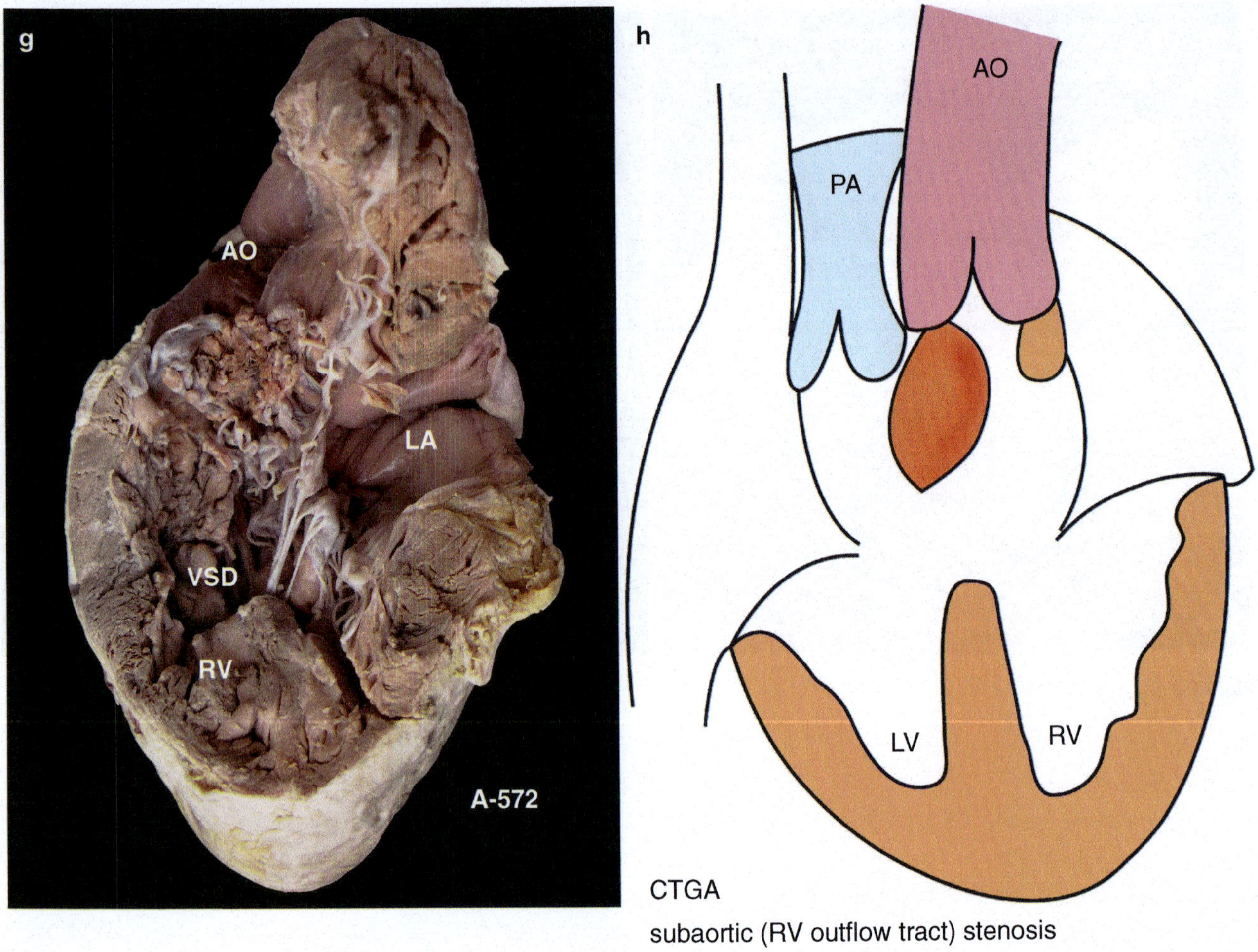

Fig. 26.4 (continued)

Fig. 26.5 (**a**) Specimen 672. CTGA. The AV lies left and anteriorly, with two semilunar valves at the same level. Two probes are inserted into the ventricles through the atrioventricular valves. *Blue* indicates the RA, and the *red* indicates the LA. *PA* pulmonary artery, *AO* ascending aorta, *LV* left ventricle, *RV* right ventricle. (**b**) Specimen 672. The fibrous connection between the pulmonary and mitral valves (*blue arrow*). Beneath the AV and the atrioventricular valve is a muscular structure (*red arrow*). Two probes pass through the atrioventricular pores, respectively. *PA* pulmonary artery, *AO* ascending aorta, *LV* left ventricle, *RV* right ventricle. (**c**) Specimen 672. A close view of the cone muscular structure beneath the aortic valve (*arrow*). *AO* ascending aorta, *RV* right ventricle. (**d**) Specimen 672. The atrioventricular valve beneath the PV is the MV, proving that the structure on the right is a morphological LV. d stands for the giant VSD. The *dashed line* shows the course of the conducting bundles. *PA* pulmonary artery, *AO* ascending aorta. (**e**) Specimen 672. The mitral valve and two major papillary muscles. *APM* anterior papillary muscle; *PPM* posterior papillary muscle. *RA* right atrium. (**f**) Specimen 672. The left atrium dissected to show the tricuspid valve and the giant VSD. *AO* aorta; *LA* left atrium; *TV* tricuspid valve. (**g**) Schematic diagram of Specimen 672. CTGA: The left atrioventricular valve is the TV, and the ventricle is the morphological RV. The right atrioventricular valve is the MV, with a morphological LV. There is a giant defect on the ventricular septum. *PA* pulmonary artery, *AO* ascending aorta

26.3.5 CTGA + VSD: Typical SLL

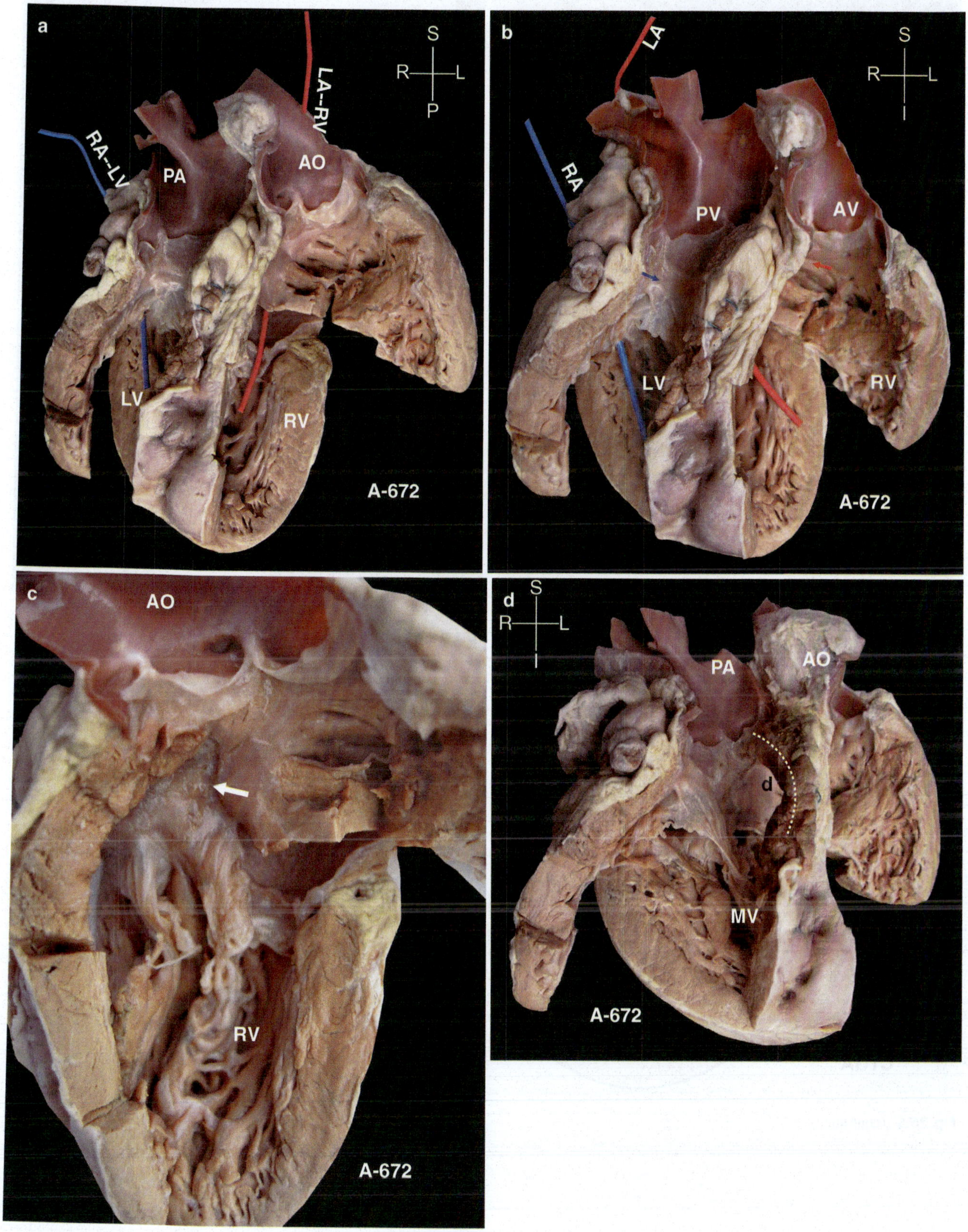

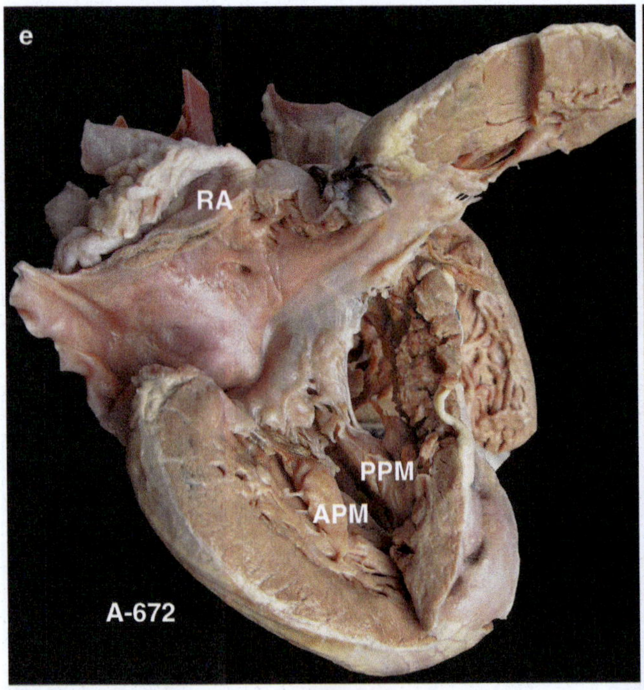

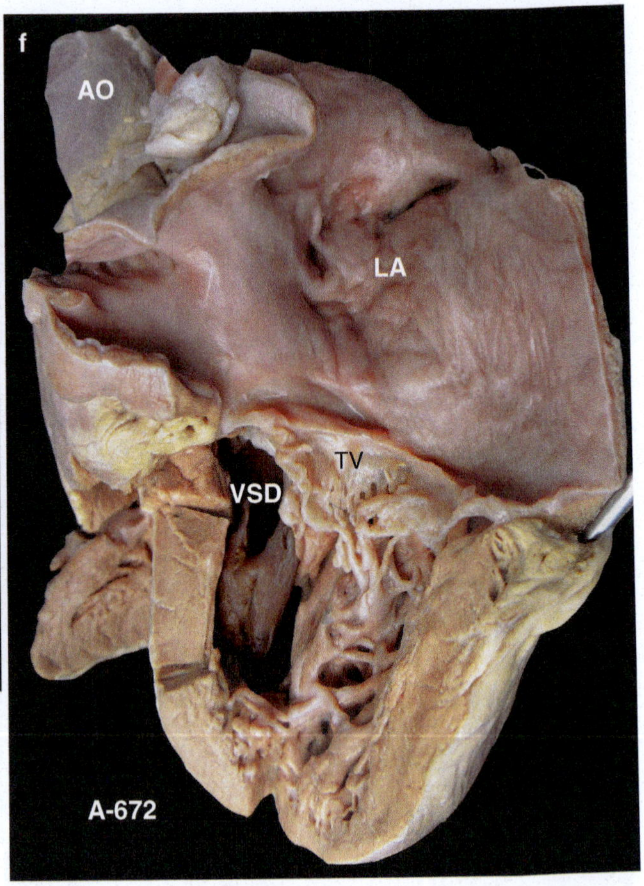

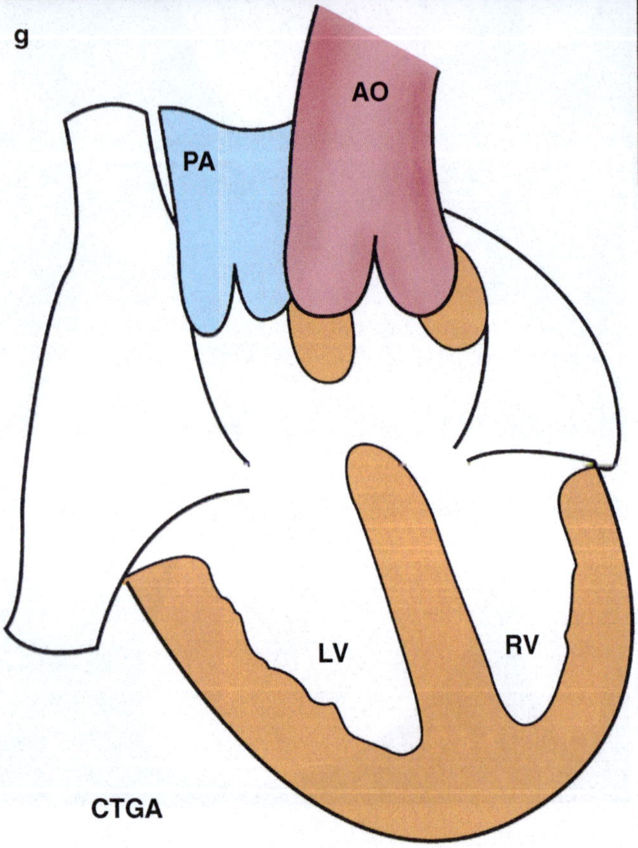

Fig. 26.5 (continued)

26.3.6 CTGA + VSD + PS Morphological Left Ventricle-Pulmonary Artery Exterior Canal

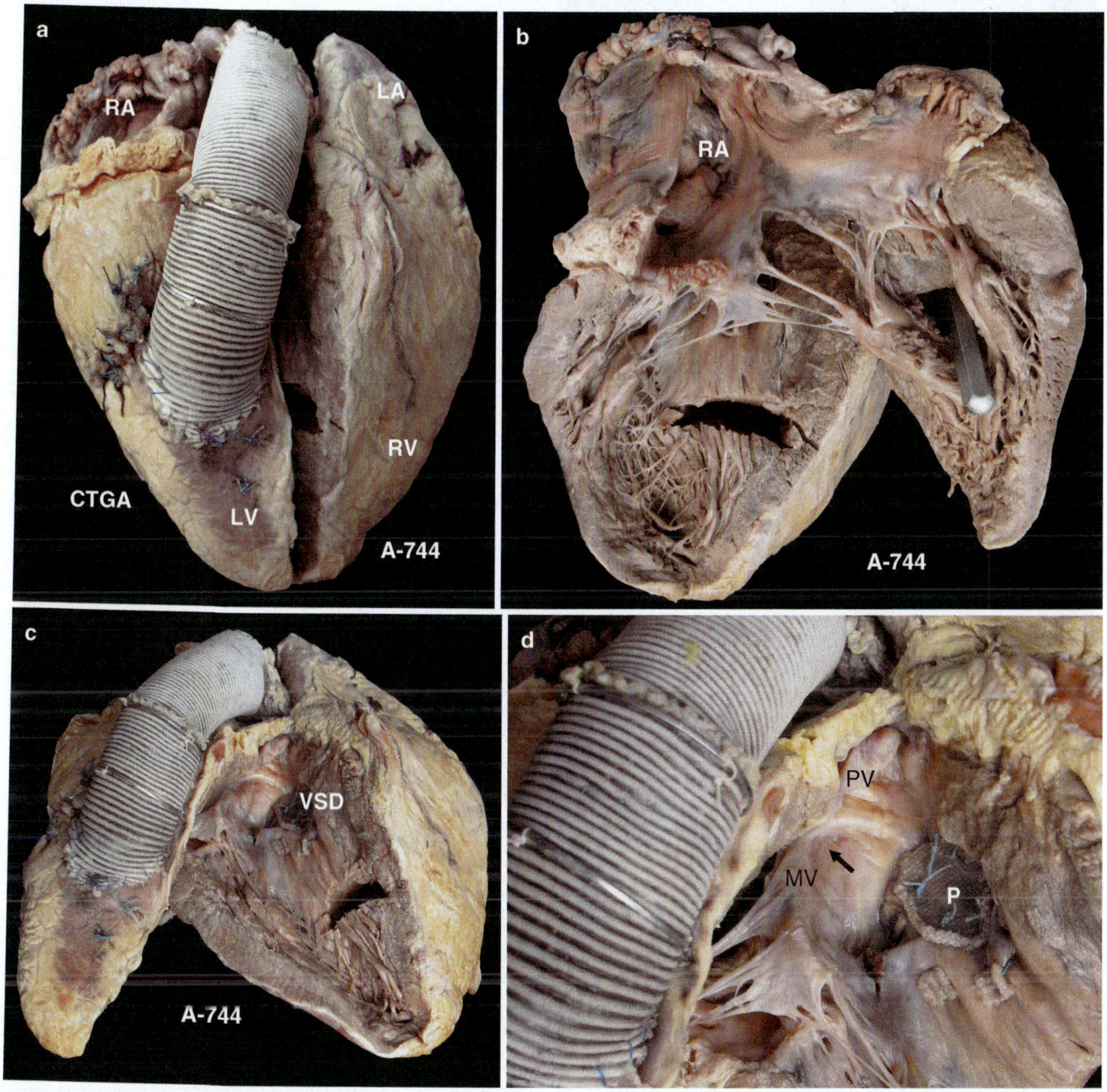

Fig. 26.6 (**a**) Specimen 744. A case with CTGA, pulmonary stenosis, and VSD. A VSD repair and artificial blood vessel replacement have been performed to connect the morphological LV and the main pulmonary artery. *LA* left atrium; *LV* morphological left ventricle; *RA* right atrium; *RV* morphological right ventricle. (**b**) Specimen 744. The RV has been lifted (morphological RV). The metallic probe shows the proximal opening of the artificial blood vessel. (**c**) Specimen 744. Showing the VSD has been repaired by a Dacron patch. (**d**) Specimen 744. A close view of the fibrous connection (*arrow*) between the mitral valve (*MV*) and the pulmonary valve (*PV*). *P* patch of the VSD. (**e**) Specimen 744. A view of the aortic valve and the morphological right ventricle beneath it. The main basis is the muscular structure beneath the AV. There is no fibrous connection between the AV and the atrioventricular valve, proving the left-sided ventricle is a morphological RV. *AO* ascending aorta

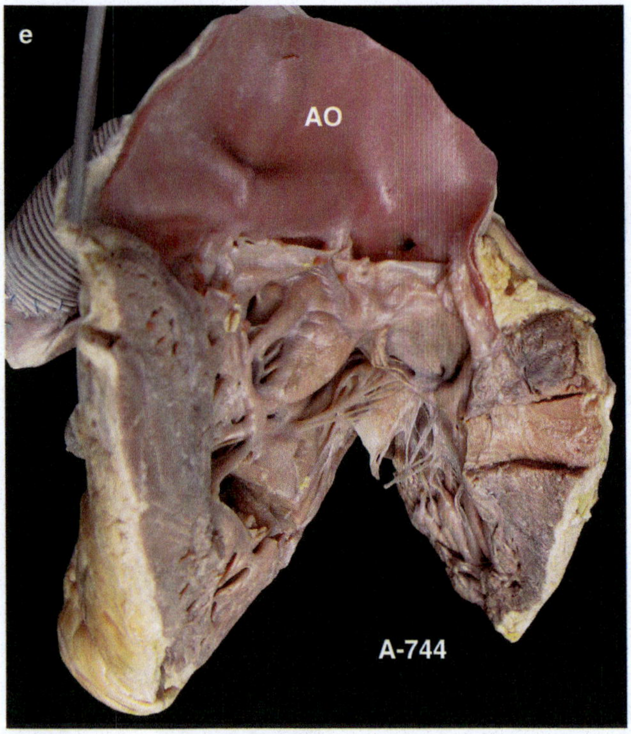

Fig. 26.6 (continued)

Fig. 26.7 (**a**) Specimen 867. A postoperational case with CTGA, AVSD, and pulmonary stenosis and a left superior vena cava drained into the left atrium. The *green probe* is placed in the LV-pulmonary artery exterior canal. On the right is the morphological LV, and on the left is the morphological RV. Cone muscular tissue beneath the AV can be seen. *AV* aortic valve. (**b**) Comprehensive schematic diagram of the lesion. The left-sided ventricle is the morphological RV, from which the aorta originates. The morphological LV is linked with the pulmonary artery through an exterior canal. *AO* ascending aorta. (**c**) Specimen 867. Showing the morphological right ventricle from the left side. The tricuspid (*TV*) and the aortic (*AV*) valves are separated by a muscular bundle, and the ASD is covered by the TV. Conduit means that the exterior canal is made from bovine pericardium. *AO* ascending aorta, *RV* right ventricle. (**d**) Specimen 867. Observing from the cross section of the right-side atrium and the ventricle. The giant defect patch can be seen. There is almost no atrial septum remaining, and it has been repaired with the autologous pericardium. The VSD is small and has been repaired with a semicircle patch. The MV leaflet has a fissure on it, which belongs to the atrioventricular septal defect. The *blue arrow* points to the left superior vena cava, which is now drained into the RA

26.3.7 CTGA + AVSD + PS Morphological Left Ventricle-Pulmonary Artery Exterior Canal

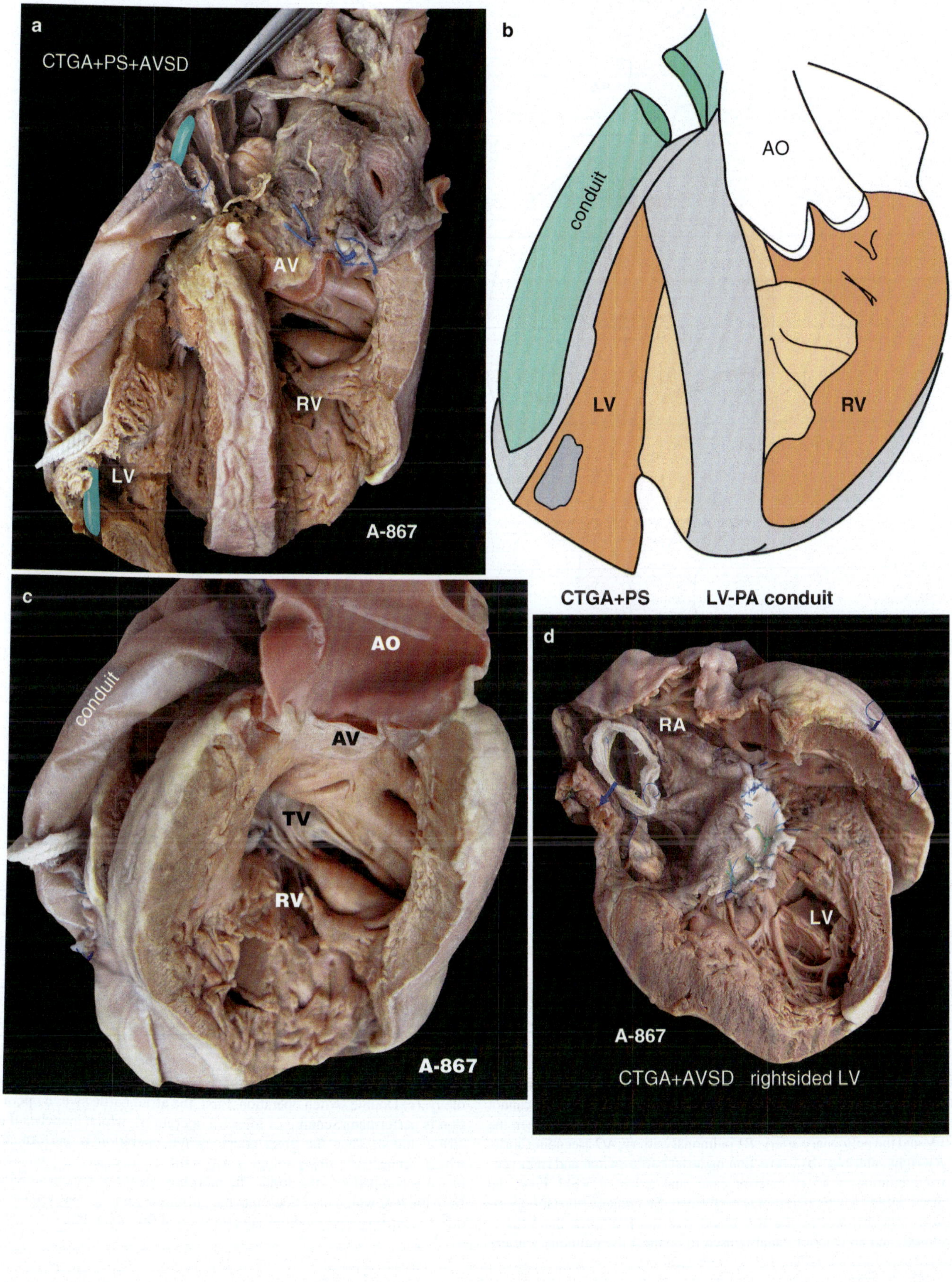

26.3.8 TGA + VSD + PS: Several Surgical Correcting Approaches

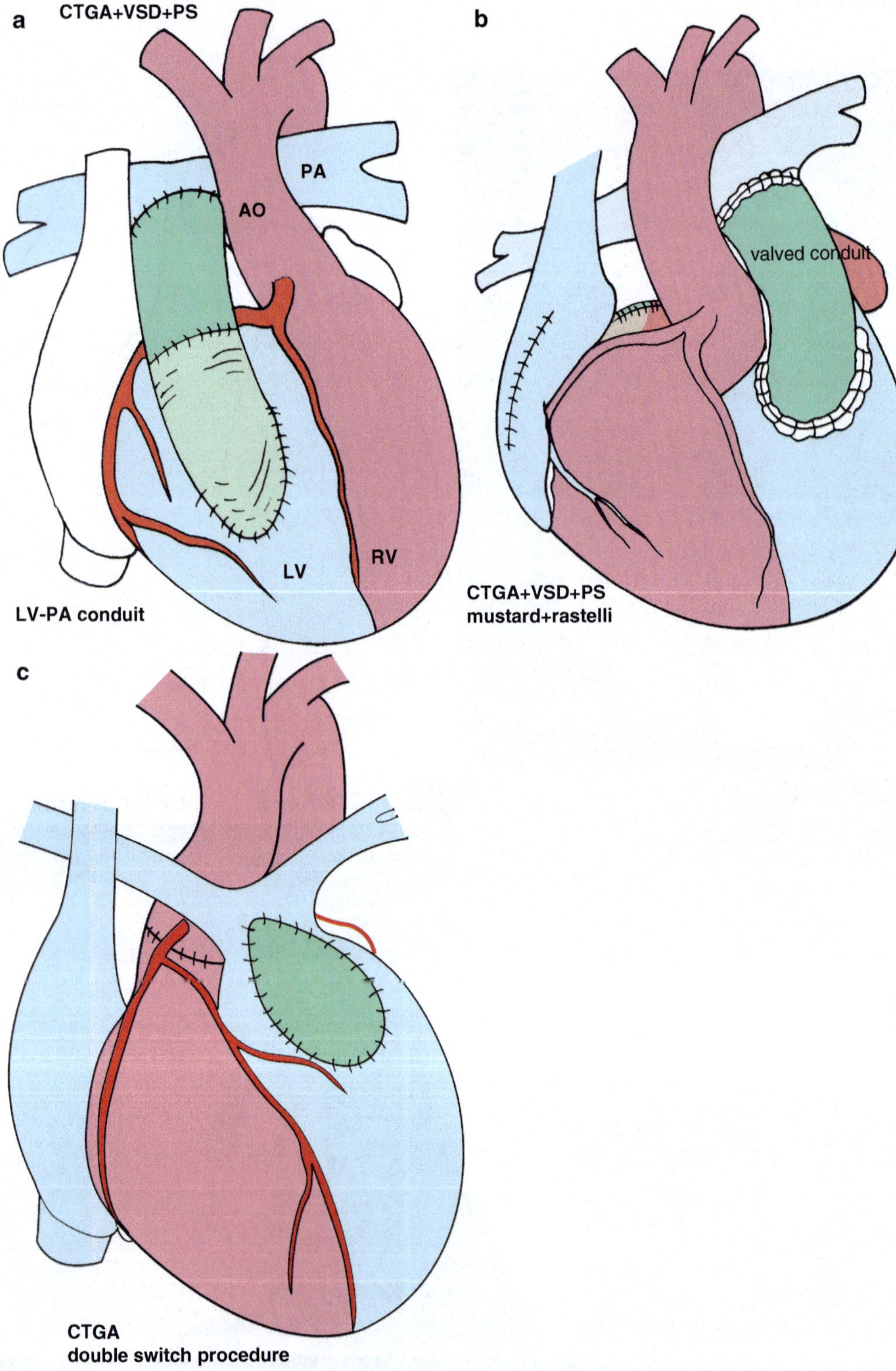

Fig. 26.8 (**a**) Traditional method for CTGA + VSD + PS correction. First, the VSD is repaired. If the pulmonary subvalvular stenosis cannot be deoppilated, an exterior valvular conduit can be placed between the LV and the pulmonary artery. *PA* pulmonary artery, *AO* ascending aorta, *RV* right ventricle. (**b**) Correction by atrial baffle switch and right ventricle-pulmonary artery exterior canal and repair of VSD. First, the atrial baffle switch operation is performed (Senning or Mustard operation), which transfers the RA blood into the RV. Then the VSD is closed, and an exterior canal is used to connect the pulmonary artery and morphological RV, which is more in accordance with normal physiology. (**c**) Double switch operation. First, the atrial baffle switch operation is performed (Senning or Mustard operation), which switches the atrium blood. Then the great artery switch operation is performed, which connects the morphological LV with the LA, and oxygenated blood is pumped into the aorta. The morphological RV receives blood from the RA, which then flows into the pulmonary artery. The VSD and other coexisting abnormalities are corrected at the same time

27 Pulmonary Atresia

27.1 General Considerations

Pulmonary atresia (PA) is a form of congenital pulmonary dysplasia in which the PV does not form properly; the duct of the RV and the pulmonary artery have an interruption in which there may be a residual ligament, or the artery may be totally separated. The range of pulmonary atresia may be limited or diffuse, and the portion of the atresia may be single or multiple. Generally, PA is classified according to two types: 1) PA with intact ventricular septum (PA-IVS) and 2) PA with VSD (PA-VSD) (Fig. 27.1c). The IVS type involves the arterial duct, accompanied by hypoplasia of the RV and the TV and may even include deformity of the coronary artery. The VSD type is characterized by underdevelopment of the RVOT accompanied by a large VSD and overriding of the aorta.

27.2 Embryology and Pathology

The PA-VSD shares the same basis of embryology with that of TOF, but with a different extent of stenosis of the pulmonary artery (Fig. 27.1e). The aortic and pulmonary arteries originate from the arterial stem, which comes from the far end of the bulbus cordis, and then the conotruncus rotates and becomes the aortic artery and the main pulmonary artery. However, if the conotruncus fails to rotate completely and the conotruncal septum fails to connect the ventricular septum, such that the ventricular septum with membranous septum and muscular septum are not closed simultaneously, a VSD results (Fig. 27.1f). However, if the muscular septum grows well, the whole septum should be intact. The forward malposition of the conotruncus may cause right ventricular infundibulum stenosis or atresia.

The PA-IVS is usually caused by a mild form of dysontogenesis of the main pulmonary artery, which may be accompanied by dysontogenesis of the RV.

In the case of the PA-VSD, the left and right pulmonary arteries mostly connect with each other and extend to the atretic main pulmonary artery. In some mild cases, there is some lumen in the root atresia, and the branches of the pulmonary artery grow well, so the arterial duct affords the blood supply. In severe cases, the pulmonary artery is long and narrow or agenesis, followed by stenosis of the left and right pulmonary arteries (Fig. 27.2g).

In addition to the arterial duct, the lateral branch between the aortic artery and the pulmonary artery is another source of the blood supply for the lung and often originates from the proximal end of the descending aorta; the branches may join with the artery of the lung lobe or the pulmonary arterioles. The lateral branch has some of the characteristics of muscular arteries, but the stenosis caused by the pressure from the descending aorta should not be ignored.

According to the levels of growth of the PA, with or without VSD, the source of the blood supply to the lung provides a means to classify the PAs. From the clinical perspective, we should emphasize four points: (1) whether there is remaining lumen in the far end of the root atresia or not; (2) the growth and connection of the left and right pulmonary arteries; (3) the main blood supply to the lung, the quantity and position of the atrial branch between the aortic artery and the pulmonary artery, and whether the blood comes totally from the lateral branch; and (4) the relationship between the VSD and the AV and the PV (Fig. 27.3).

Because of the separation between the pulmonary artery and the ventricle, the author separates the pulmonary atresia from the deformity of the connection between the ventricle and the artery (e.g., TGA).

X. Zhu (ed.), *Surgical Atlas of Cardiac Anatomy*,
DOI 10.1007/978-94-017-9409-1_27, © Springer Science+Business Media Dordrecht and People's Medical Publishing House 2015

27.3 Policy of Surgical Operation

27.3.1 Palliative Operation (Fig. 27.4)

When the blood supply is diminished, we preferentially increase the blood supply using procedures such as the improved Blalock-Taussig bypass or the Waterston bypass:

1. *Improved Blalock-Taussig bypass*. A tunnel should be built from the right subclavian artery to the right pulmonary artery with an artificial shunt. According to the age and weight of the child, we choose different calibers of polytetrafluoroethylene (PTFE), usually 4–5 mm in diameter for an infant and 6 mm in diameter for children (Fig. 27.1a).
2. *Waterston anastomosis*. If the pulmonary artery is wide enough, a side-to-side anastomosis can be executed between the ascending aorta and the main pulmonary artery.
3. *Bidirectional Glenn operation*. The SVC is disconnected in the root, the proximal end to the heart is closed, and the far end to the right pulmonary artery is sutured. After 1–3 years, a second thoracotomy for shutting the VSD can be considered; if the case is accompanied with obstruction of the RVOT, reconstruction, including widening the outflow tract with a patch or replacing the old shunt from the RV to the pulmonary artery, can be considered.

27.3.2 One-Stage Radical Operation

The corrective surgery for PA-VSD involves building a tunnel from the RV to the pulmonary artery and repairing the VSD. In the case of main pulmonary atresia, one can adopt the pulmonary fusion, which involves building an outer tunnel between the RV and the lumen composed of the combination of the lateral branch between the aortic artery, the pulmonary artery, and the left and right pulmonary arteries. One-stage pulmonary fusion has become an important way to treat pulmonary atresia (Fig. 27.5).

A one-stage radical operation adopts deep hypothermic circulatory arrest (DHCA) or extracorporeal circulation. One should ligature the arterial duct before blocking the ascending aorta. The left and right pulmonary arteries should be dissociated; if the main pulmonary artery degenerates, the dissociation should be continued to the crotch. The left and right pulmonary arteries and lateral branches should be joined together, and a new lumen should be made by combining the RV through the outer tunnel. The VSD is repaired at last.

Fig. 27.1 (**a**) Specimen 1069. Appearance of the heart and lung in pulmonary atresia. The inflow and outflow tracts of the RV have been divided, the ascending aorta is thick and large but without the main pulmonary artery, and a shunt is connected with the left pulmonary artery in the left side of the ascending aorta, through which is a lumen. This case has accepted pulmonary fusion. *AO* ascending aorta. (**b**) Specimen 1069. Outflow tract of right ventricle. Figure shows that the AV (*a*) originates mostly from the R, and the VSD (*d*) is under the AV, without a PV. There is a thick parietal band hypertrophic supraventricular crest, and the wall of the RV is obviously thickening. *Dotted line*, thick parietal band; *c* artificial shunt. (**c**) Specimen 1069. VSD. The *red probe* is in the VSD; the AV mostly bestrides on the RV and extends to the TV. *AO* ascending aorta. (**d**) Specimen 1069. Left atrium slightly enlarged. The *green arrow* indicates the entrance of the pulmonary vein; the MV bestrides on the edge of VSD. *LA* left atrium, *LV* left ventricle. (**e**) Specimen 1069. The left atrium and the left ventricle. The VSD (d) can be seen through the left outflow tract under the MV; the depth of the wall of RV and LV is comparable. *LA* left atrium. (**f**) Specimen 1069. View of the left side. The *red probe* is through the VSD. A leaflet of the AV remains in the side of the LV and extends to the MV. There is no trace of PV in the left outflow tract. *AO* ascending aorta. (**g**) Specimen 1069. The fibrous connection of the aortic and mitral valves. Figure shows the VSD bestriding the AV. The *red probe* is through the VSD, and a fibrous connection of the AV and MV is seen. *RV* right ventricle, *LV* left ventricle. (**h**) Specimen 1069. The relationship of the AV, the MV, and the TV. The ventricular septum is cut up in the edge of the VSD in order to show the MV and the TV. This picture shows the fibrous connection of the leaflet of the AV and the MV and the TV. The *dotted line* indicates the point for observing. *RV* right ventricle, *LV* left ventricle. (**i**) Specimen 1069. The structure of Fig. 27.1h and the *dotted line*. The doorway of the AV is mostly in the RV, with only a leaflet of AV that bestrides the deficient ventricular septum; there is a fibrous connection of the AV and the MV and the TV. (**j**) Specimen 1069. The aorta-to-pulmonary artery bypass. The procedure divides the left and right pulmonary arteries (*yellow arrow*) and makes a tunnel with a piece of pericardium (*blue arrow*) so as to enlarge the pipeline of the pulmonary artery; the artificial shunt (*red shunt*) is sutured to the side wall of the aorta, and the artificial shunt is connected with the pericardial tunnel to finish the bypass. (**k**) Specimen 1069. Same as Fig. 27.1j, the figure shows the details of the bypass. *a*, the artificial shunt; *b*, pericardial tunnel; *c*, the left pulmonary artery. *AO* ascending aorta. (**l**) Specimen 1069. Sketch map of the operation. Attention should be given to the shunt from the aorta to the pulmonary artery; there is a fibrous connection of the AV and the MV and the TV

27.4 Specimen Demonstrations

27.4.1 Pulmonary Atresia, Ascending Aorta-Pulmonary Bypass

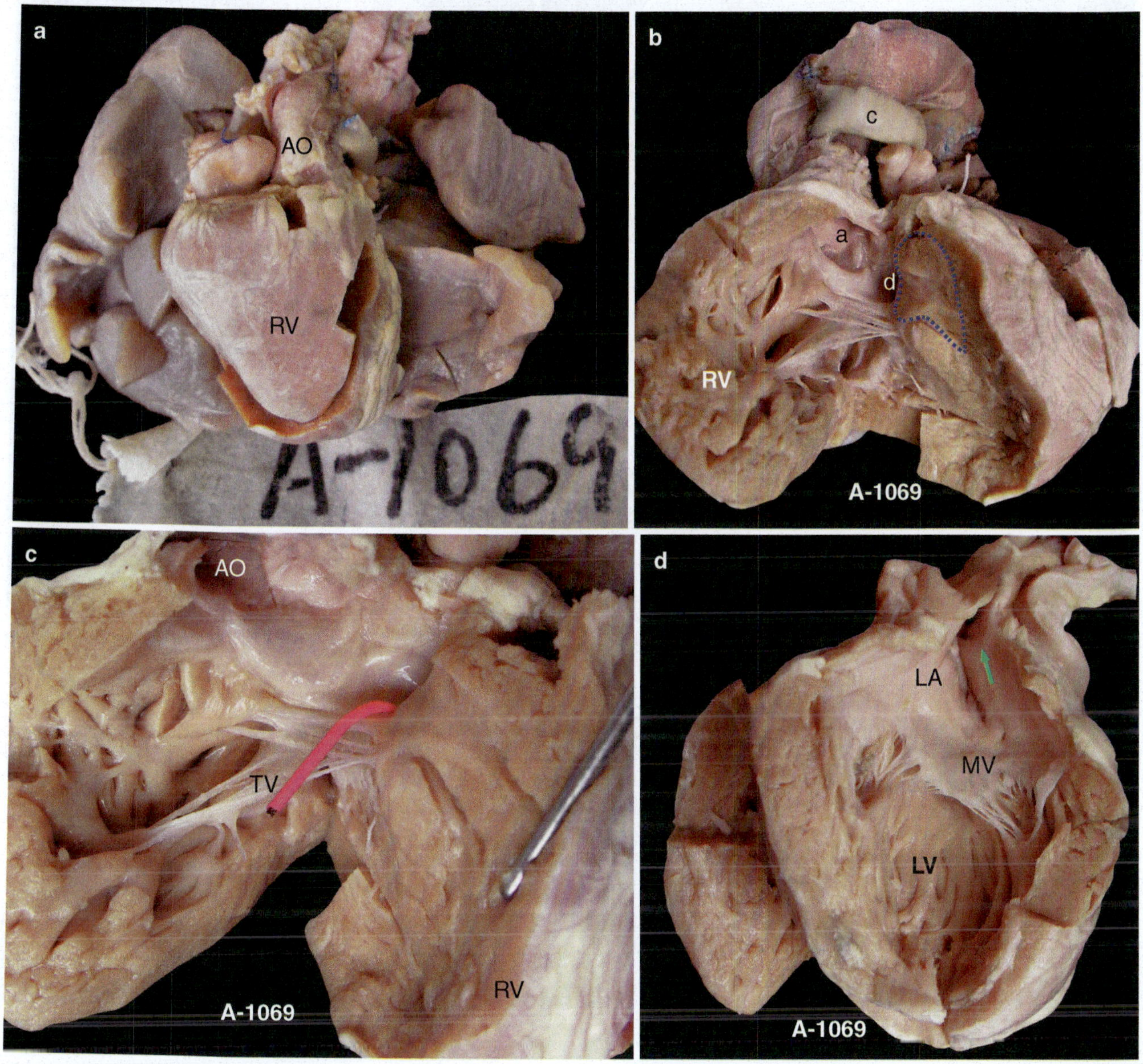

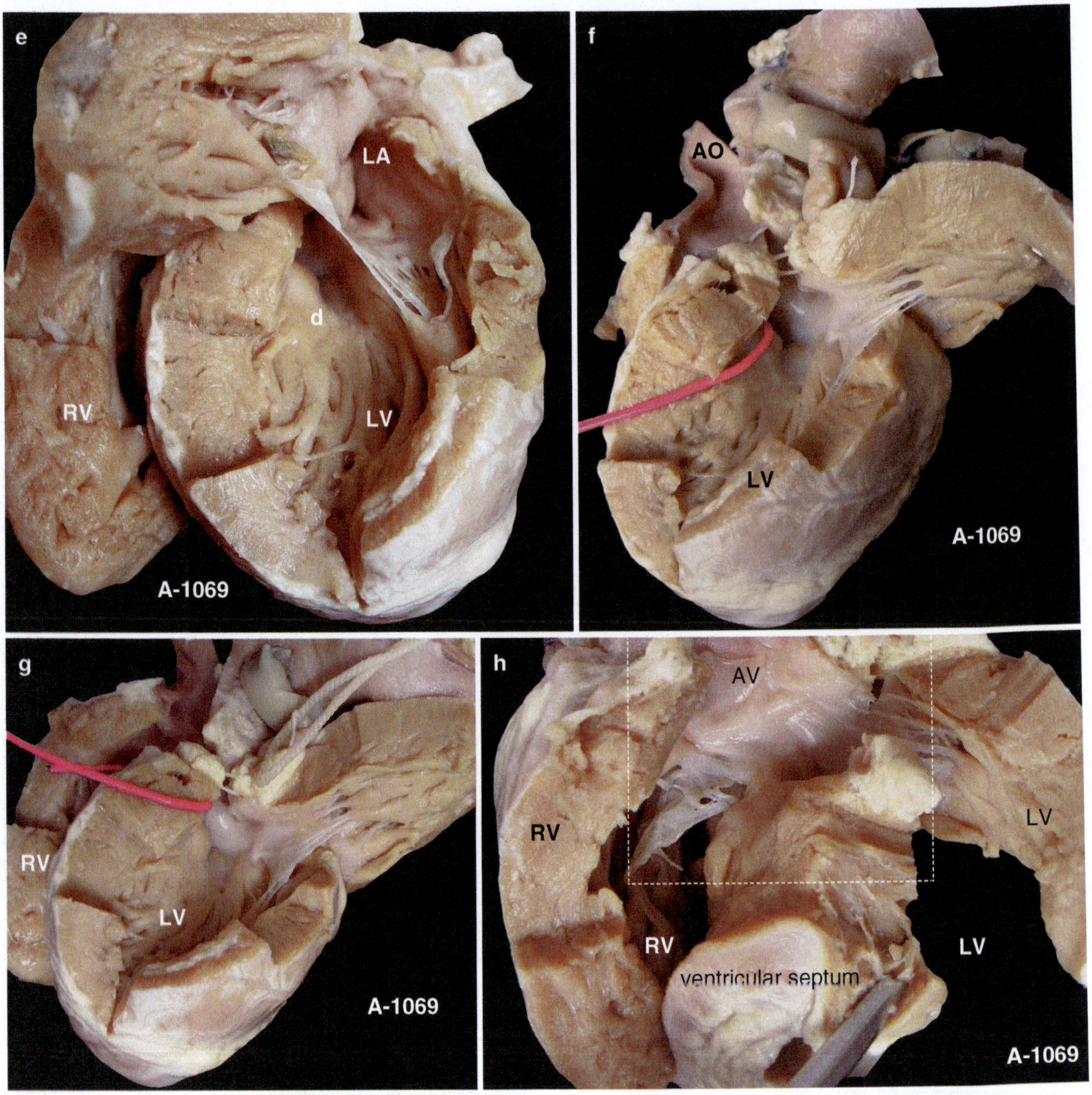

Fig.27.1 (continued)

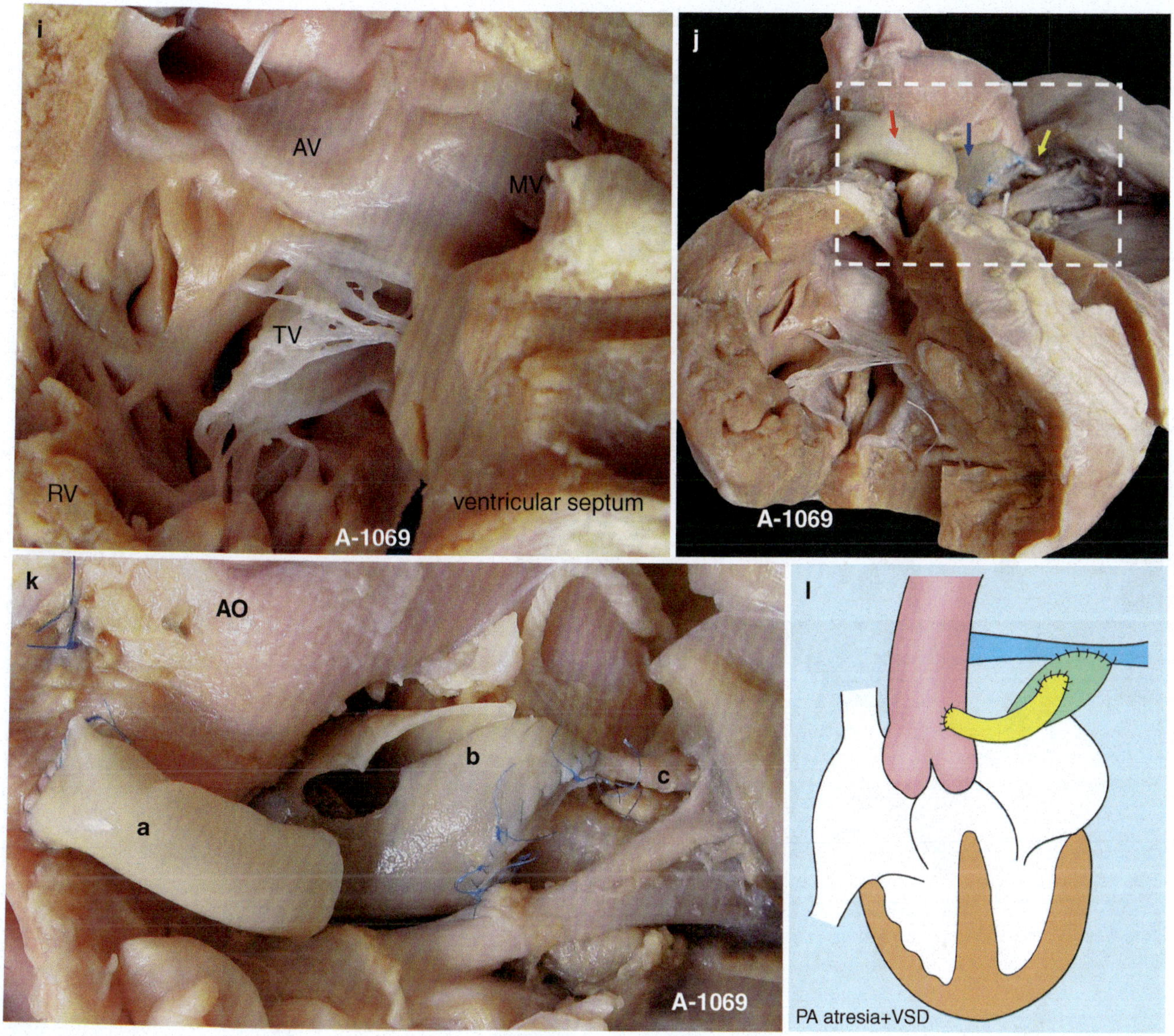

Fig.27.1 (continued)

27.4.2 Pulmonary Atresia Accompanied with PD

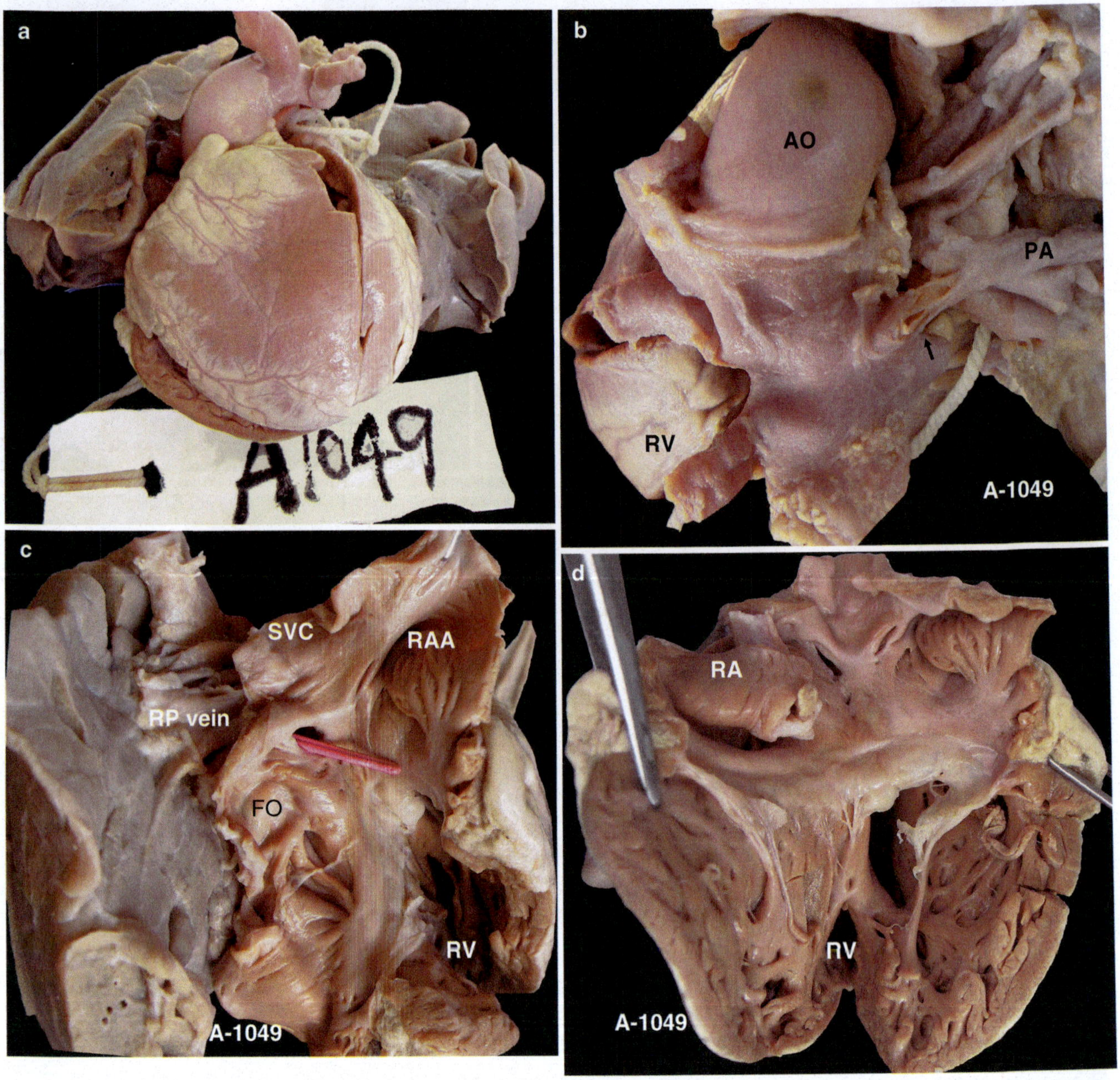

Fig. 27.2 (**a**) Specimen 1049. Specimen of pulmonary atresia. The thick ascending aorta can be seen in the front, but without the pulmonary artery; the anterior wall of the RV has been divided for pathological anatomy from the interventricular groove. (**b**) Specimen 1049. View from the top. The main pulmonary is blocked shaped like a strip (*arrow*). Only the thick ascending aorta and a trace of the slender main pulmonary can be seen. *PA* pulmonary artery, *AO* ascending aorta, *RV* right ventricle. (**c**) Specimen 1049. The right atrium unfolded. Figure reveals the SVC, the RAA, and the FO; one should make sure it is the morphological RA. Deformity of right pulmonary vein drains into the RA. The probe is in the doorway of the deformity. *RV* right ventricle. (**d**) Specimen 1049. The inflow tract of the right ventricle. Figure shows the moderate band and the TV. The muscular ridge and the lacuna in the muscular part prove this is the morphological RA. Hence, this is a case of normotopia of the atrium, with a normal connection of the atrium and the ventricle, and no deformity in the RVIT. *RV* right ventricle. (**e**) Both RVOT and AV are opened. The AV originates totally from the RV, the VSD (d) is under the AV, and the separation from the TV under the AV is conical muscle. *AO* ascending aorta. (**f**) Specimen 1049. Subarterial VSD. The leaflets of the AV have fibrous thickness, attributed to the reactive hyperplasia; the moderate band is thick, and the muscular ridge of the wall of the RV is obvious. *TV* tricuspid valve, *M* mitral valve. (**g**) Specimen 1049. PDA, The doorway of the arterial duct (*arrow*) can be seen in the far end of the ascending aorta; the blood supply to the lung is here. *AO* ascending aorta. (**h**) Specimen 1049. The doorway of the pulmonary vein and the left atrium. The *arrow* indicates the connection of the pulmonary vein and the LA, which is connected with the MV. The VSD (d) can be seen through the AV in the opposite side. The structure of the MV is normal. *LV* left ventricle. (**i**) Specimen 1049. The left atrium and the inflow tract of the left ventricle. The left atrium and ventricle divided, showing the LV, which has smooth walls, through the MV. *d* VSD; *LA* left atrium; *LV* left ventricle. (**j**) Specimen 1049. The outboard wall of the left ventricle opened to show the mitral valve and the outflow tract of the left ventricle. The three leaflets of the AV are completely in the RV. *LV* left ventricle, *D* ventricular septal defect. (**k**) Specimen 1049. The fibrous connection of the leaflets of aortic and mitral valves. Although the AV totally bestrides on the RV, we do not define it as a TGA but rather as a *dextropositioned aorta* because of the character of fibrous connection of the leaflets of the AV and the anterior leaflet of the MV. (**l**) Specimen 1049. Sketch map of structure. The subarterial VSD, the doorway of the aorta, is mostly in the TV, but there is still the fibrous connection of the leaflets of the AV and the MV and pulmonary atresia. *RV* right ventricle, *LV* left ventricle, *PDA* patent ductus arterious, *PA* pulmonary artery, *AO* ascending aorta

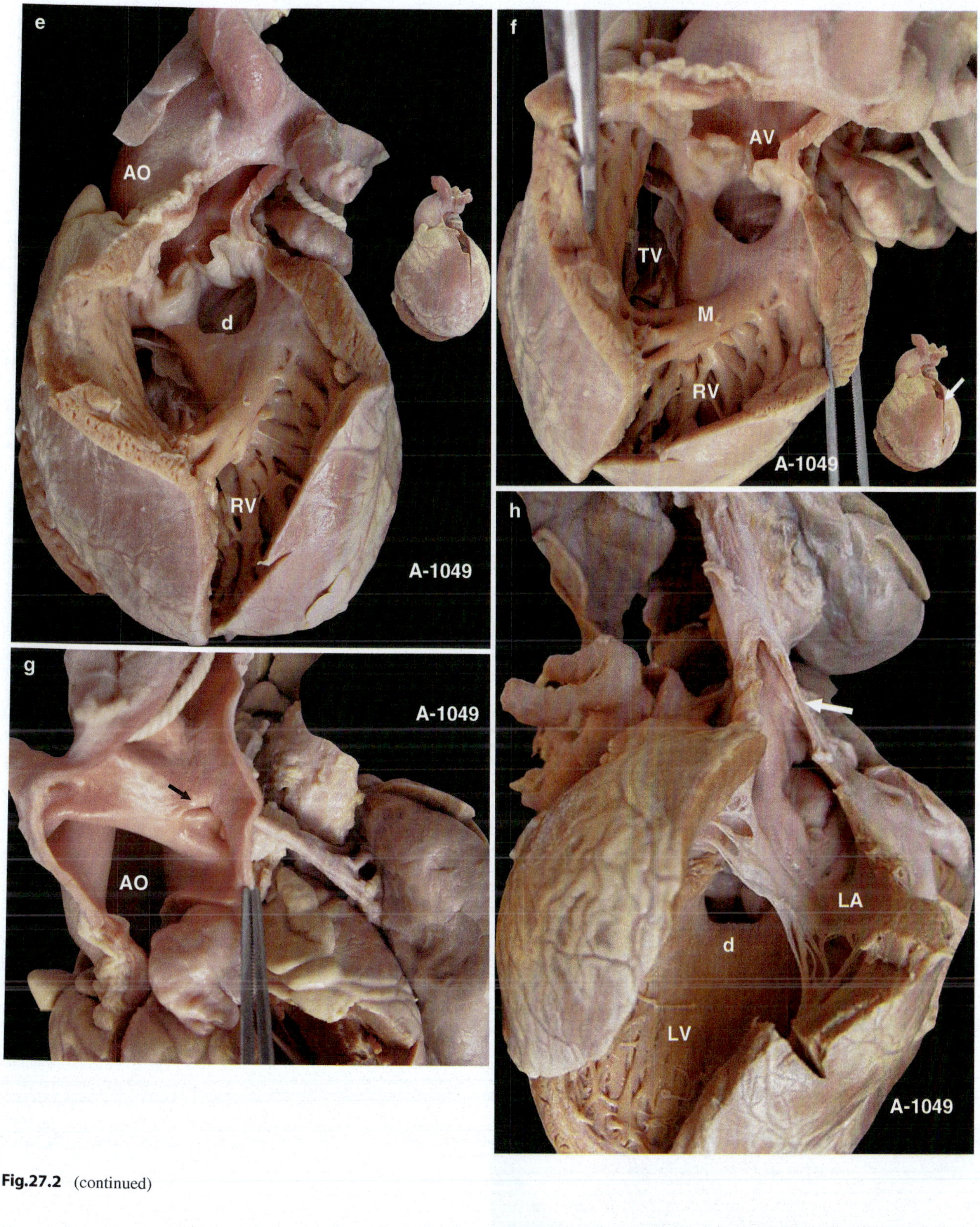

Fig.27.2 (continued)

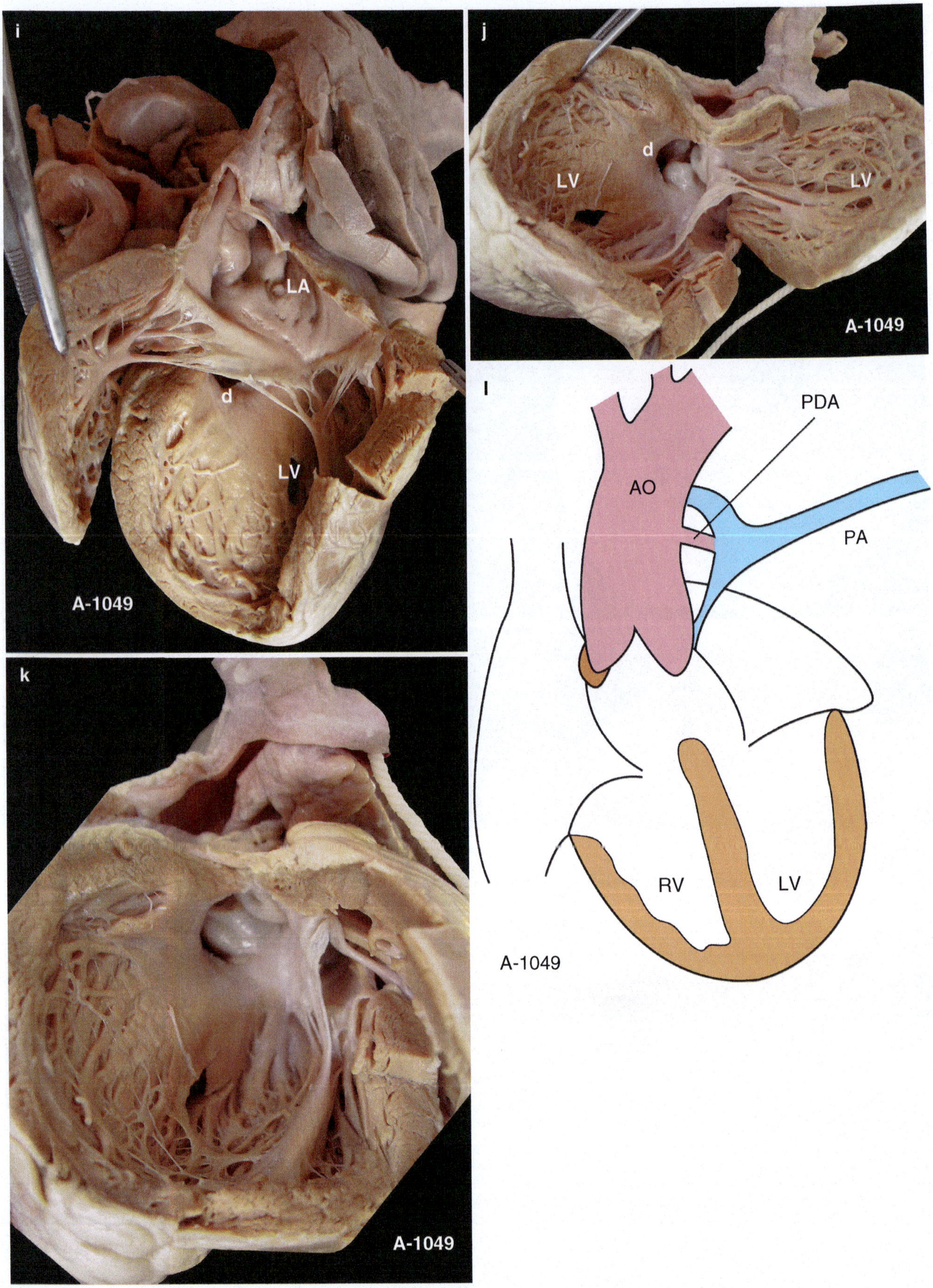

Fig.27.2 (continued)

27.4.3 Pulmonary Atresia, VSD

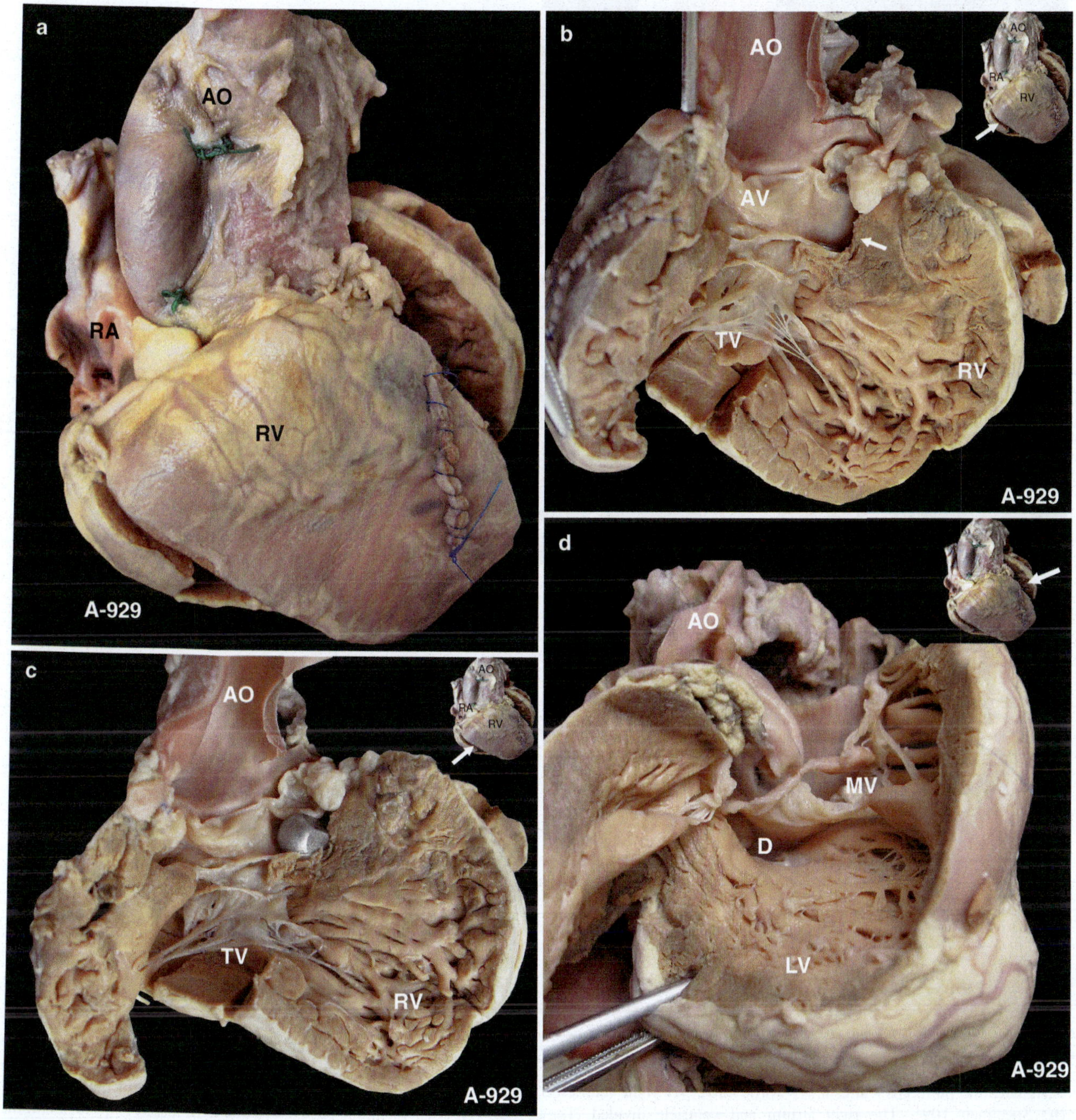

Fig. 27.3 (**a**) Specimen 929. Appearance of pulmonary atresia. The thick aorta is connected with the RV, but without the main pulmonary artery; there is an incision in the anterior wall of the RV. *AO* ascending aorta, *RA* right atrium. (**b**) Specimen 929. The AV originates from the RV, and there is still the fibrous connection of the leaflets of the AV and the TV. The *arrow* points to the inferior border of the VSD; the main pulmonary artery is absent. *AO* ascending aorta. (**c**) Specimen 929. VSD with opened RV. The thick probe is in the VSD, but without the main pulmonary artery. The *arrow* show the specimen with intact right ventricle. *AO* ascending aorta, *TV* tricuspid valve. (**d**) Specimen 929. The LV is connected with only the VSD; there is no LVOT or PV, proof of pulmonary atresia. The *arrow* show the specimen with intact right ventricle. *AO* ascending aorta, *MV* mitral valve, *D* ventricular septal defect. (**e**) Specimen 929. The left and right ventricle. The *red probe* runs from the LV to the aorta through the VSD; the main pulmonary artery is absent. *AO* ascending aorta, *RV* right ventricle

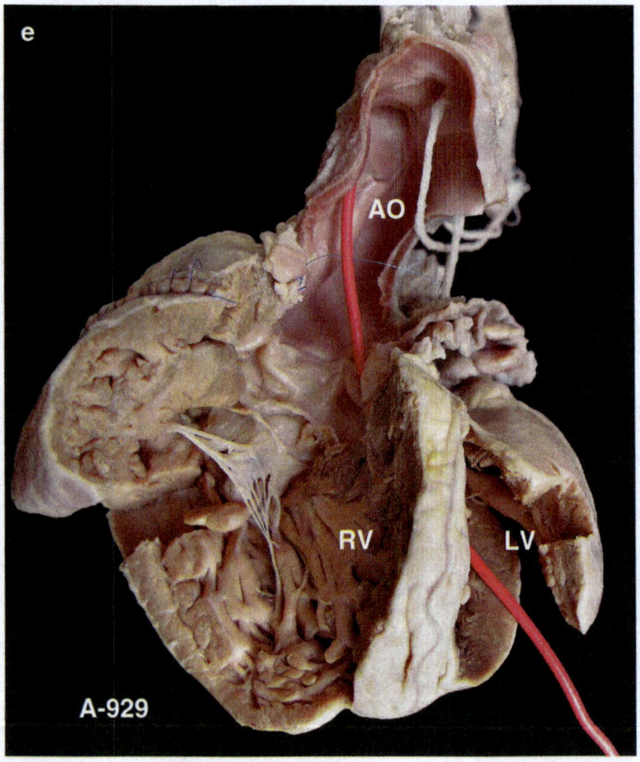

Fig.27.3 (continued)

Fig. 27.4 (**a**) Specimen 1091. The appearance of pulmonary atresia, accompanied by VSD, LSVC, left ventricular hypoplasia. The *arrow* designates the pulmonary atresia. *AV* aortic valve, *RV* right ventricle. (**b**) Specimen 1091. The right atrium and ventricle divided. The enlarged coronary sinus is indicated by the *red arrow*. There is LSVC present, and the TV is normal. *LSVC* left superior vena cava. *RA* right atrium, *RV* right ventricle. (**c**) Specimen 1091. The AV originates from the RV, the VSD is under the AV, and there remains a fibrous connection of the leaflets of the AV and the TV and the MV (*arrow*); there is no PV in the RVOT. *AO* ascending aorta. (**d**) Specimen 1091. Closer view of VSD and subarterial VSD (D). *AO* ascending aorta. (**e**) Specimen 1091. Opened anterior wall of the right ventricle. The AV and MV and a thick wall of the RV and a moderate band are exposed. The VSD is under the AV (i.e., a subarterial VSD); the PV cannot be seen in the view of the left or right ventricles. (**f**) Specimen 1091. The LV clearly is smaller than the RV, the MV is tiny, only part of the leaflets of the AV and the VSD (D) are present, the atrioventricular orifice also is quite small, the doorway of the aorta is on the right side of the RV, and the PV is absent. *AO* ascending aorta. (**g**) Specimen 1091. Typical pulmonary atresia. PA indicates the pulmonary atresia. There is no connection with the left or right ventricle. PDA indicates the big arterial duct, connected with the crotch of the PA. *AO* ascending aorta, *RV* right ventricle. (**h**) Specimen 1091. Sketch map, pulmonary atresia. VSD, left ventricular hypoplasia, right-aorta connection, and SLVC are seen. *PDA* patent ductus arteri-ous, *PA* pulmonary artery, *AO* ascending aorta, *RV* right ventricle, *RA* right atrium, *LV* left ventricle

27.4.4 Pulmonary Atresia, VSD, PDA

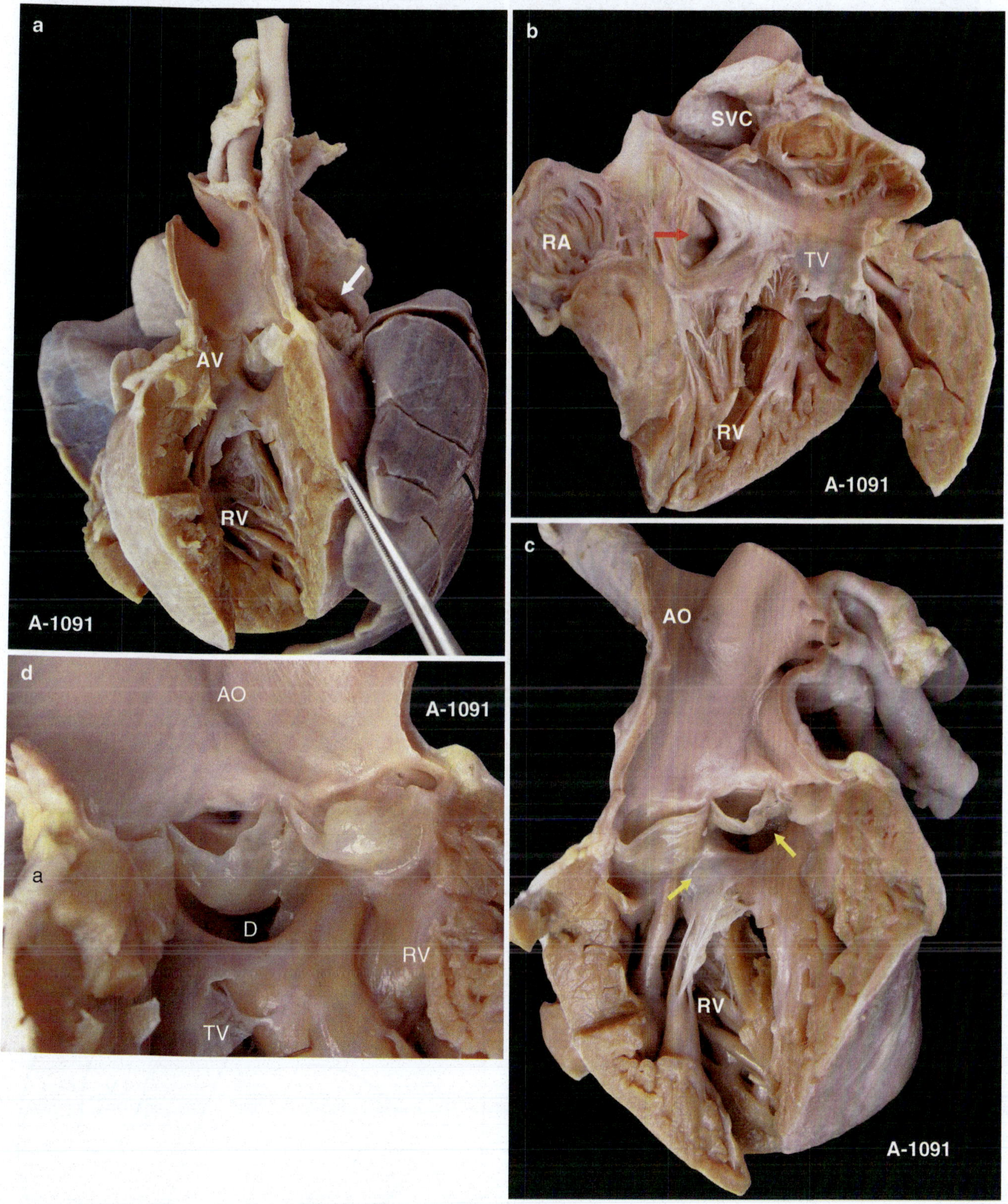

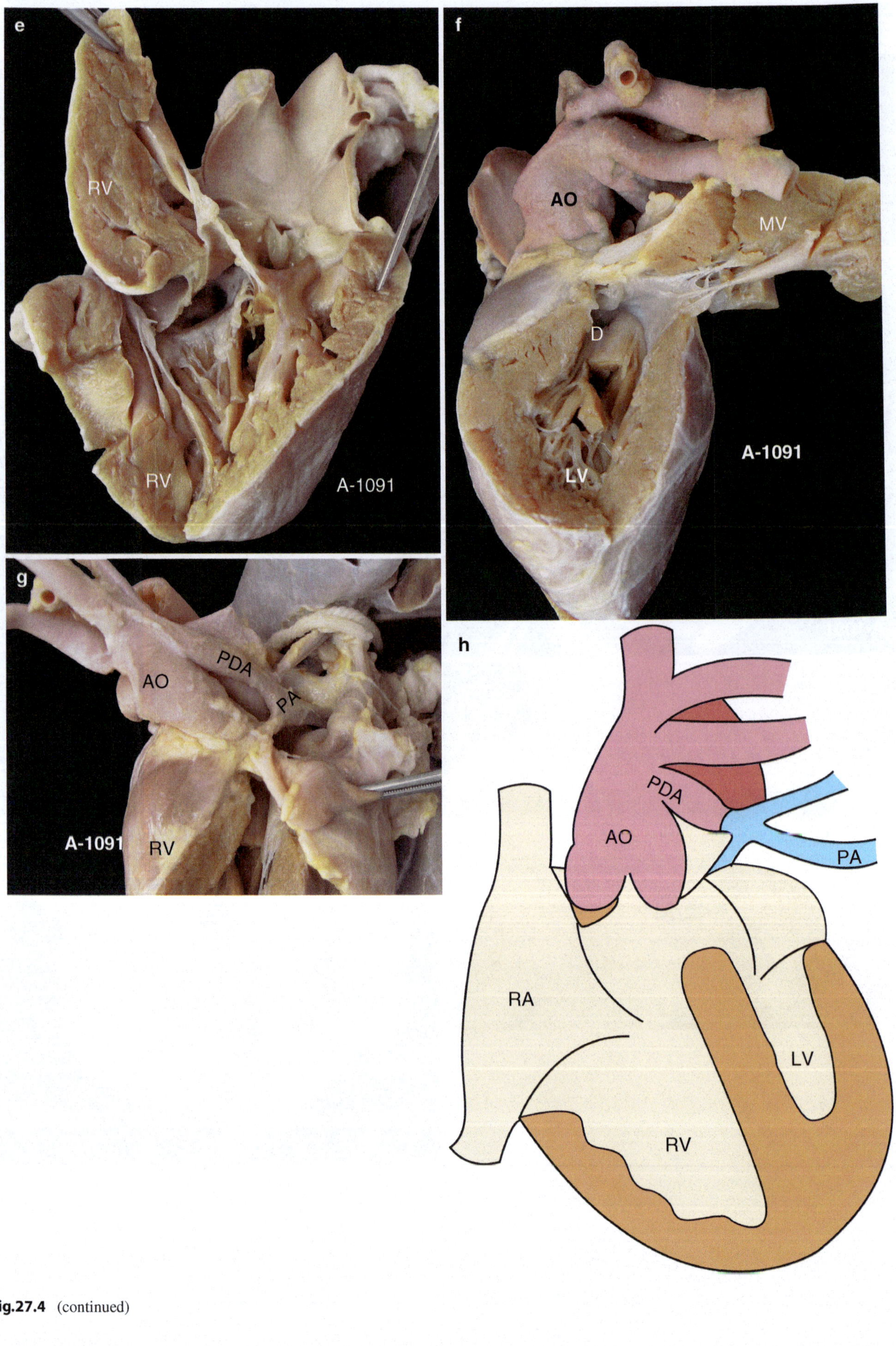

Fig.27.4 (continued)

27.4.5 Pulmonary Atresia Accompanied by VSD

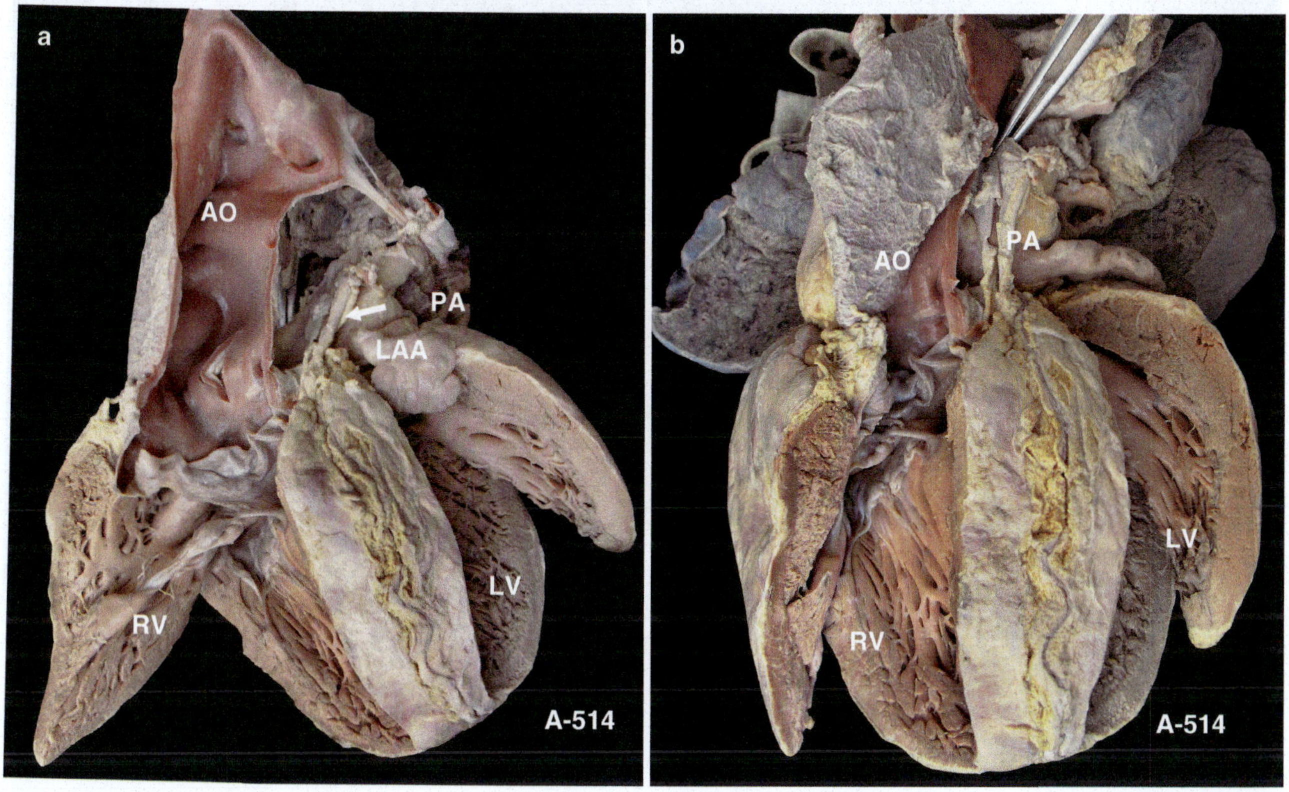

Fig. 27.5 (**a**) Specimen 514. Pulmonary atresia accompanied by VSD. The *arrow* points to the main pulmonary artery. The lumen is too thin to flow through the blood, the PV is absent, and the main pulmonary artery is on the left side of the aorta. *AO* ascending aorta; *LAA* left auricular appendix; *LV* left ventricle; *RV* right ventricle. (**b**) Specimen 514. In front of the heart. The aorta originates from the RV. The main pulmonary artery is thin and on the left side of the aorta, and the RVIT and TV are normal. *PA* pulmonary artery, *AO* ascending aorta, *LV* left ventricle. (**c**) Specimen 514. The large ascending aorta in view of the RV; the pulmonary artery cannot be seen. *AO* ascending aorta, *LV* left ventricle. (**d**) Specimen 514. The large ascending aorta originates from the RVOT. The VSD is under the AV, there is still a fibrous connection of the leaflets of the AV and the TV, and the papillary muscle of cone is under the VSD, which proved to be the structure of TV. *AO* ascending aorta, *D* ventricular septal defect, *RV* right ventricle. (**e**) Specimen 514. The *blue probe* is in the main pulmonary artery and connected with the RV, but without the PV. The RVOT is connected with the aorta, and the main pulmonary artery is on the left side of the aorta. Due to the small lumen of the main pulmonary artery, without leaflets or nearly blocked, this specimen was presented in the chapter of pulmonary atresia not DORV. *AO* ascending aorta, *LV* left ventricle, *D* ventricular septal defect. (**f**) Specimen 514. Left ventricular outflow tract. No half-moon valve originates from the LV. The *arrow* indicates the VSD; MV indicates the mitral valve. The muscular ridge is smashed in the LV, and the papillary muscle could be seen. The VSD (*arrow*) is under the AV, and the muscular ridge in the anterior wall of the LV is flat and manifests as the wall expands. *AO* ascending aorta. (**g**) Specimen 514. The LA and ventricular inflow tract are divided, showing the MV and two groups of papillary muscle. *LV* left ventricle. (**h**) Specimen 514. Opened dissociate wall of the left ventricle. The VSD is the only exit of the LV. *MV* mitral valve, *D* ventricular septal defect. (**i**) Specimen 514. Sketch map. *PA* pulmonary artery, *AO* ascending aorta, *RA* right atrium, *RV* right ventricle, *LA* left atrium, *LV* left ventricle

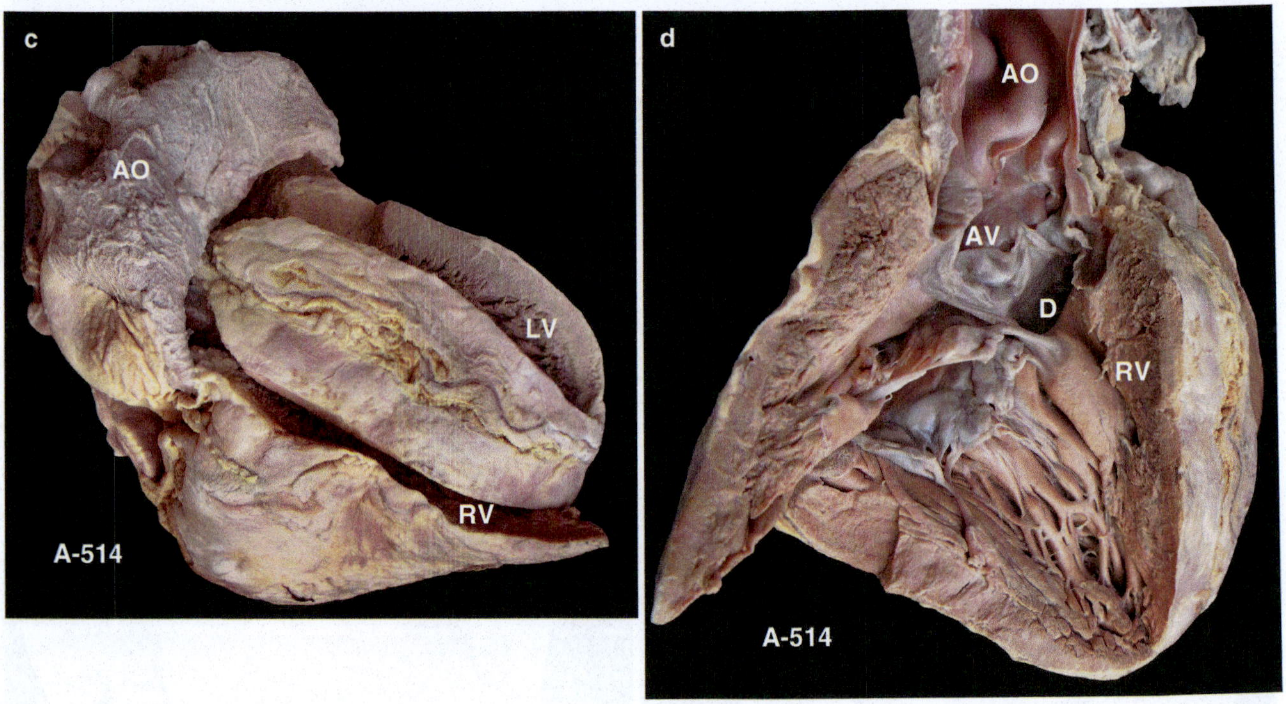

Fig.27.5 (continued)

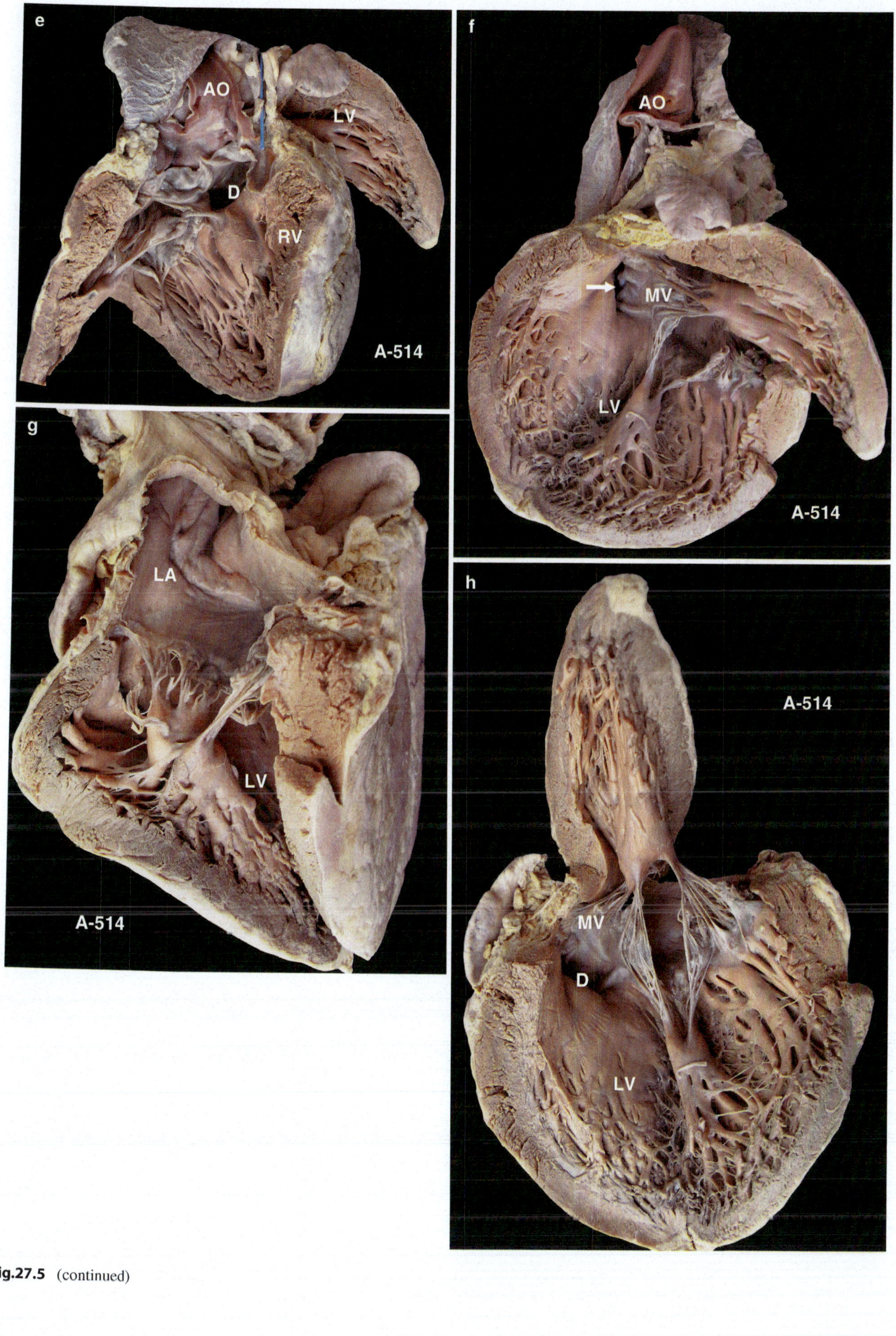

Fig.27.5 (continued)

i

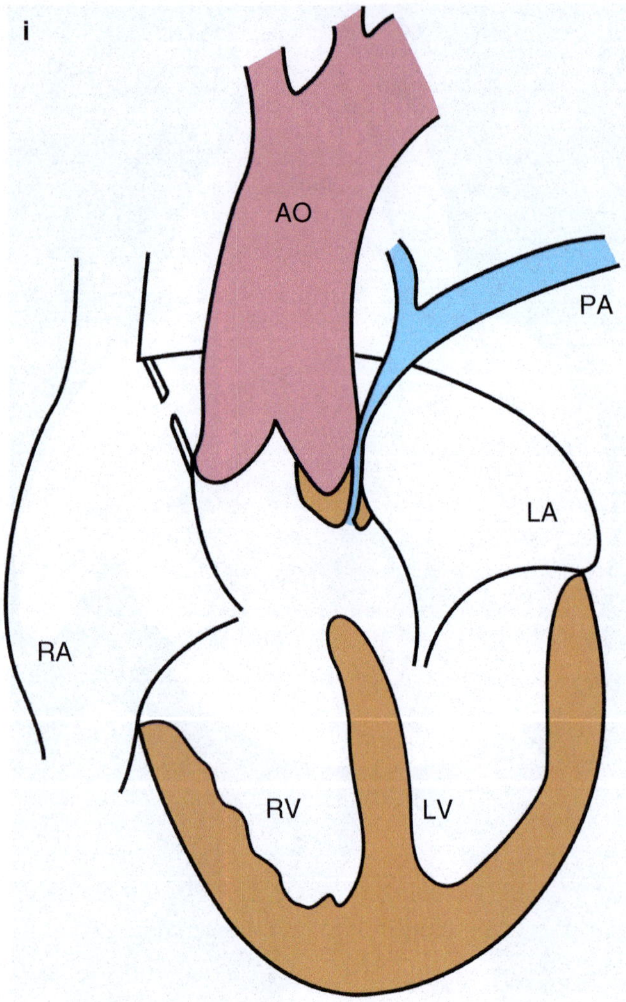

Fig.27.5 (continued)

27.4.6 Pulmonary Atresia Accompanied by Aortic Stenosis

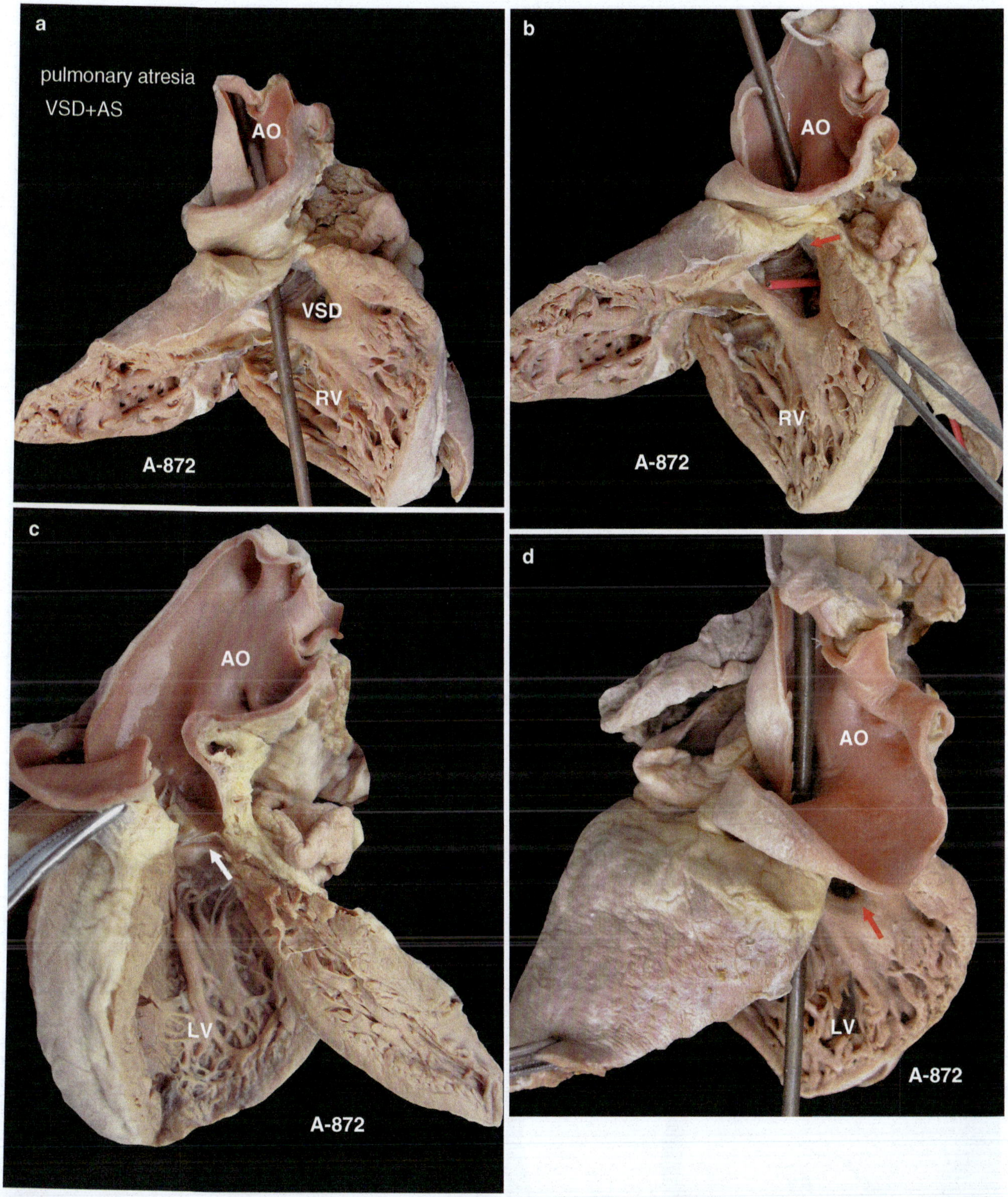

Fig. 27.6 (**a**) Specimen 872. Pulmonary atresia. There is only one exit in the view of the RVOT. The probe is inserted into the aorta through the RV, and the AV bestrides on the VSD. *AO* ascending aorta, *VSD* ventricular septal defect. (**b**) Specimen 872. The *red probe* is inserted from the LV into the RV through the VSD, which has no complete superior border. The leaflets of the AV can be seen gleaming. *AO* ascending aorta. (**c**) Specimen 872. The aortic stenosis can be seen from the left ventricle-aorta incision (*arrow*). *AO* ascending aorta, *LV* left ventricle. (**d**) Specimen 872. The LV and the ascending aorta are divided; the aortic ring has not been cut open. The probe runs into the aorta through the AV, and the expansion after stenosis of the aorta appears. The *arrow* points to the VSD. *AO* ascending aorta

27.4.7 Pulmonary Atresia, Paratactic Cardio-Auditory

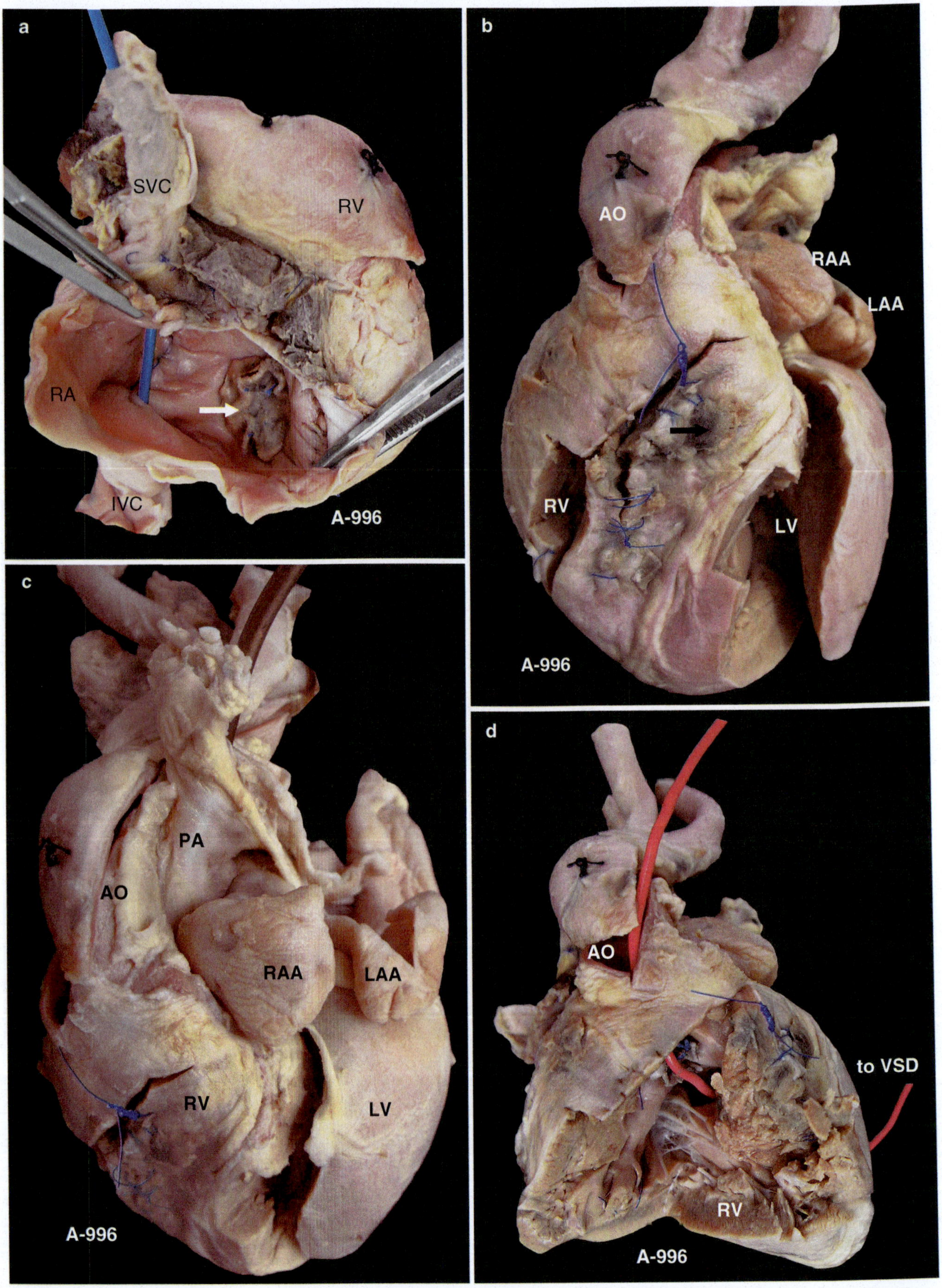

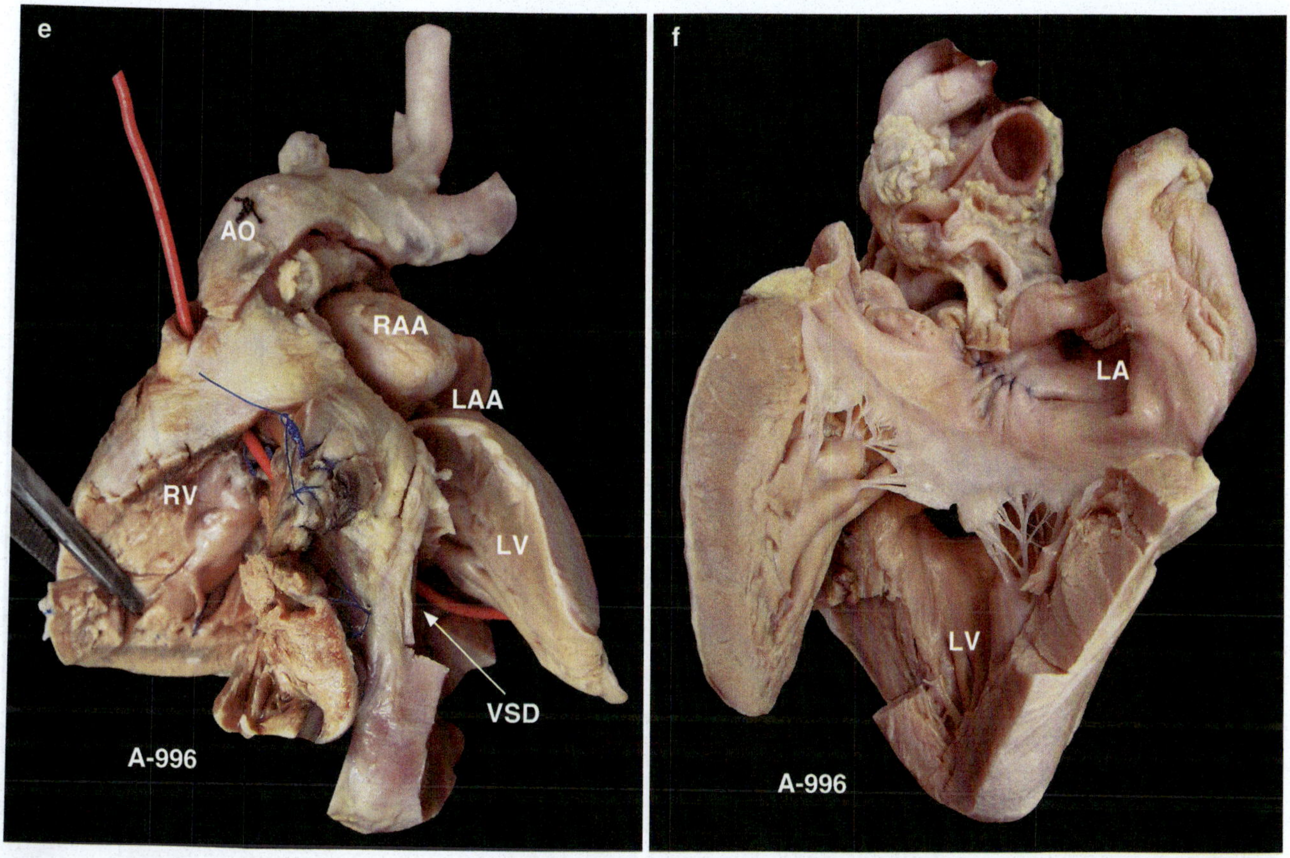

Fig.27.7 (continued)

Fig. 27.7 (**a**) Specimen 996. View of the right atrium. Figure shows a pulmonary artery and a large VSD. The aorta originates from the RV. The *arrow* points to the closed TV. The blood from the LV is excreted into the aorta through the VSD and the RV. *RA* right atrium, *SVC* superior vena cava, *IVC* inferior vena cava. (**b**) Front view of Specimen 996, the left and right ventricles have been cut open, and the paratactic cardio-auditory can be seen. *AO* ascending aorta, *RAA* right atrium auricle, *LAA* left atrium auricle, *RV* right ventricle, *LV* left ventricle. (**c**) Specimen 996. The atresic pulmonary artery. The probe could pass through the main pulmonary artery, but the root of pulmonary is total atresia, without being connected to anything. The aorta is in front of the main pulmonary artery, and the left atrial auditory (LAA) and right atrial auditory (RAA), which is paratactic cardio-auditory, can be seen. *PA* pulmonary artery, *AO* ascending aorta, *RV* right ventricle, *LV* left ventricle. (**d**) Specimen 996. The relationship of the left ventricle and aorta. The probe runs from the LV to the RV through the VSD and at last in the aorta, which shows the direction of blood. The TV demonstrates dysontogenesis. *AO* ascending aorta. (**e**) Specimen 996. The left ventricle and aorta. The *red probe* shows the direction of blood flow in the LV; there is no trace of the main pulmonary artery. *AO* ascending aorta, *RAA* right atrium auricle, *LAA* left atrium auricle, *RV* right ventricle, *VSD* ventricular septal defect. (**f**) Specimen 996. Inflow tract of left atrium and ventricle. The LV grows well, and there is a slightly fibrous thickness in the anterior leaflet of the MV; the FO has sutures. *LA* left atrium

27.4.8 Pulmonary Atresia, Intact Ventricular Septum

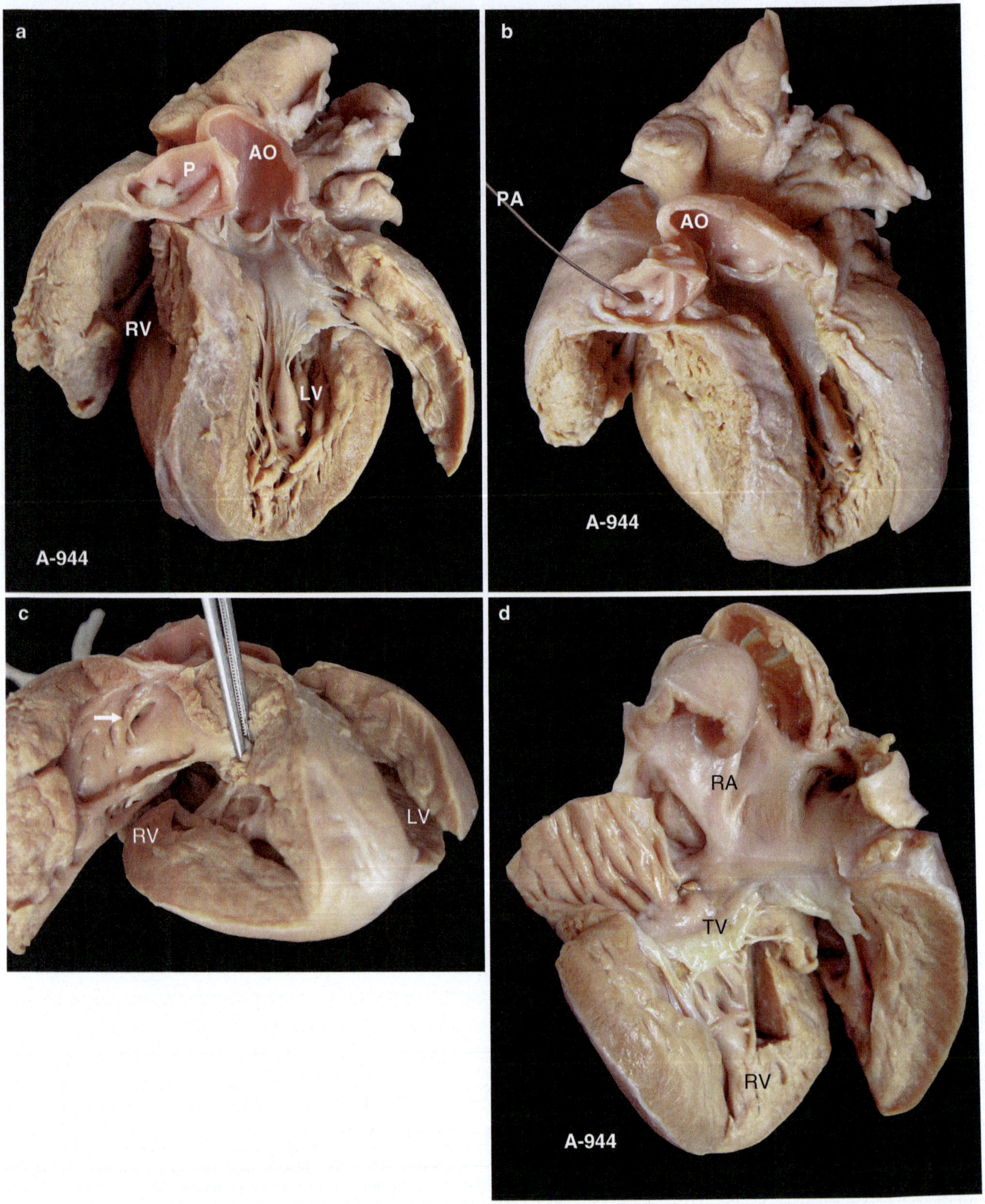

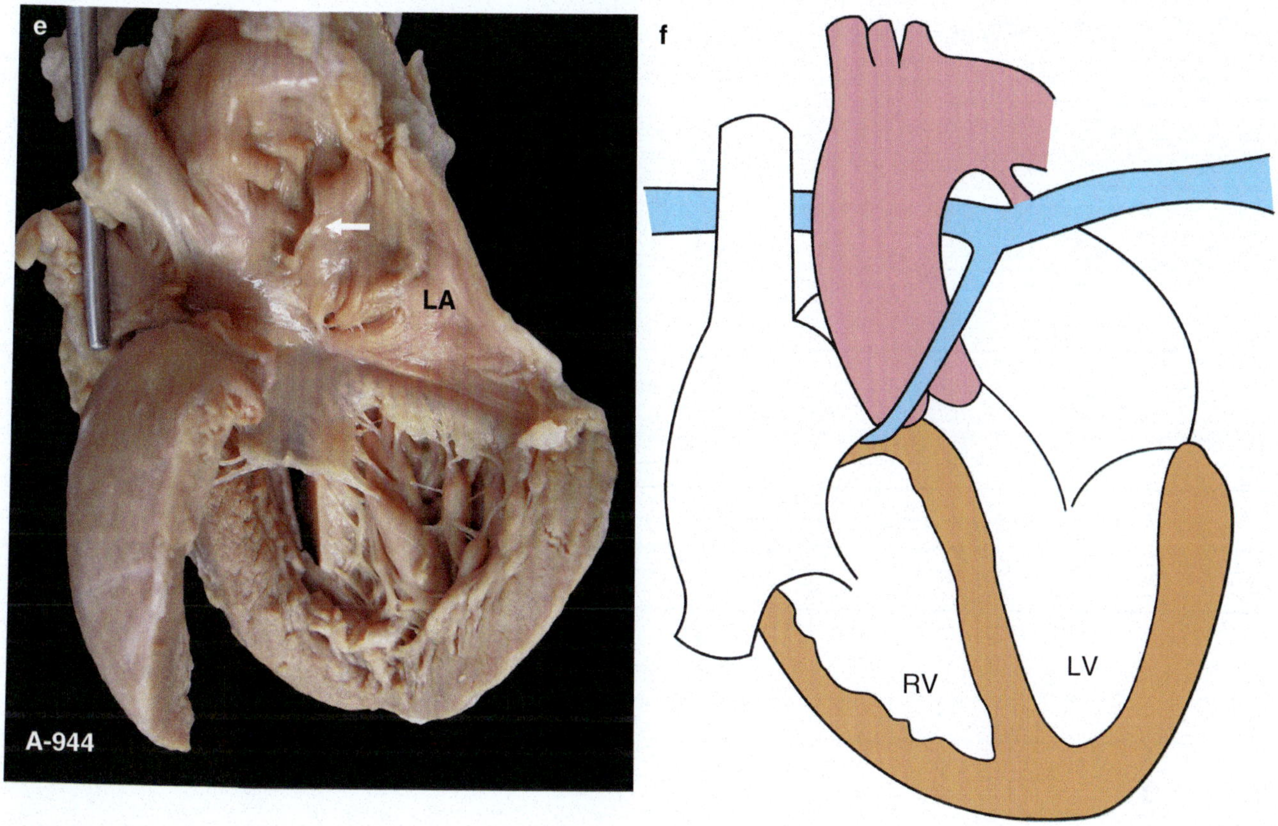

Fig. 27.8 (continued)

Fig. 27.8 (**a**) Specimen 944. In the front, with left and right ventricles divided along the VSD. The PDA cannot be seen. The LV is normal, the wall is a little thick, a piece of the anterior leaflet of the MV remains, the aorta originates from the LV, the AV is normal, and the root of the pulmonary artery is the caecum, instead of a membranous fibrous structure, without leaflets. *AO* ascending aorta, *RV* right ventricle. (**b**) Specimen 944. Outflow tract of left ventricle. The wall of the LV is a little thick. There is a fibrous connection of the MV and the AV, but the AV and the main pulmonary artery originate side by side. The PV is on the right, and its root is the caecum. A thin probe is inserted into the aorta, and the ventricular septum is intact. *PA* pulmonary artery, *AO* ascending aorta. (**c**) Specimen 944. The outflow tract of the right ventricle. The RV is not connected with the main artery. The *arrow* points to the lacuna where the PV is located. *LV* left ventricle. (**d**) Specimen 944. Divided right atrium and ventricle. The wall of the right atrium and ventricle is thick, the ventricular septum is intact, and the FO does not close completely. *RA* right atrium, *TV* tricuspid valve, *RV* right ventricle. (**e**) Specimen 944. View of the left atrium. The *arrow* points at the crack of the FO; the first and the second septum of the atrium have not merged together, and the MV is normal. *LA* left atrium. (**f**) Specimen 944. The aorta originates from the LV; the main pulmonary artery is thin; the PV is totally closed. Blood in the lung is from the PDA. The wall of the right atrium and ventricle are thick; the FO is open. Although it does not normally function, the wall of the RV does not show signs of disuse atrophy, due to the young age of the patient

Fig. [illegible]

Fig. 22.8 [illegible]

28 Double Outlet of Left Ventricle

28.1 General Considerations

DOLV is a rare deformity of the connection of the LV and the main artery. The two groups of main arteries both have origins from a morphological LV, generally accompanied by a VSD, usually the type under the AV.

The embryological theory is that the cause has something to do with abnormal absorption of the cone or another deformity such as pulmonary stenosis. The DOLV should be distinguished from the TGA in embryology, because both the AV and the PV originate from morphological LV. There is no fibrous connection of the two groups of main arteries and the MV, but there is fibrous connection of the AV, the PV, and the MV.

In this case, the DOLV is accompanied by a VSD under the AV and pulmonary stenosis, atrium and ventricle normotopia, a remaining muscular conical structure, a normal TV, and a small LV. Another case is ventricle inversus viscerum, DOLV. Surgical therapy is palliative operation or radical operation, according to the PAH or PS and the growth of the left and right ventricles (Fig. 28.1m).

X. Zhu (ed.), *Surgical Atlas of Cardiac Anatomy*,
DOI 10.1007/978-94-017-9409-1_28,

28.2 Specimen Demonstrations

28.2.1 DOLV, Ventricle Normotopia

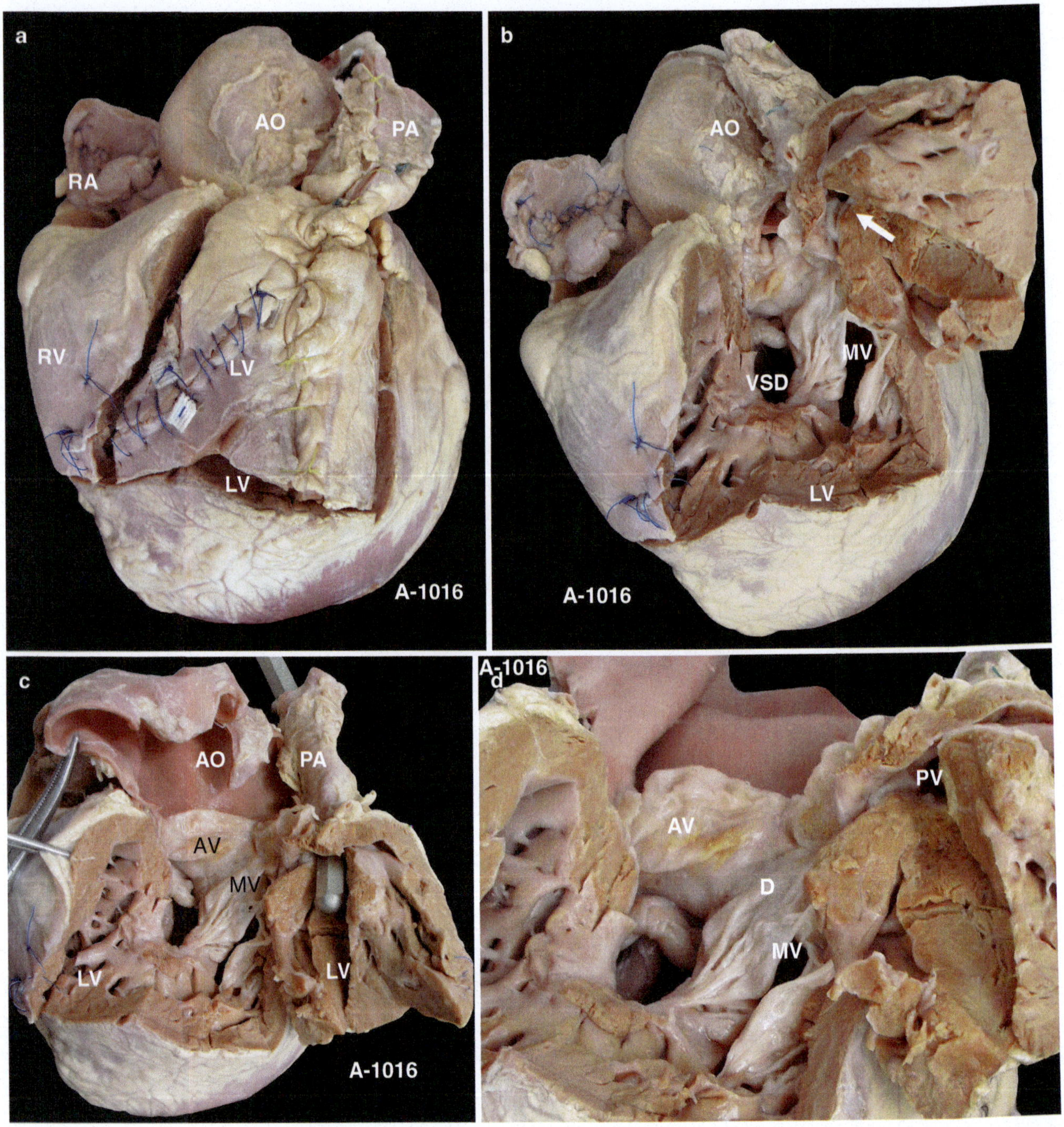

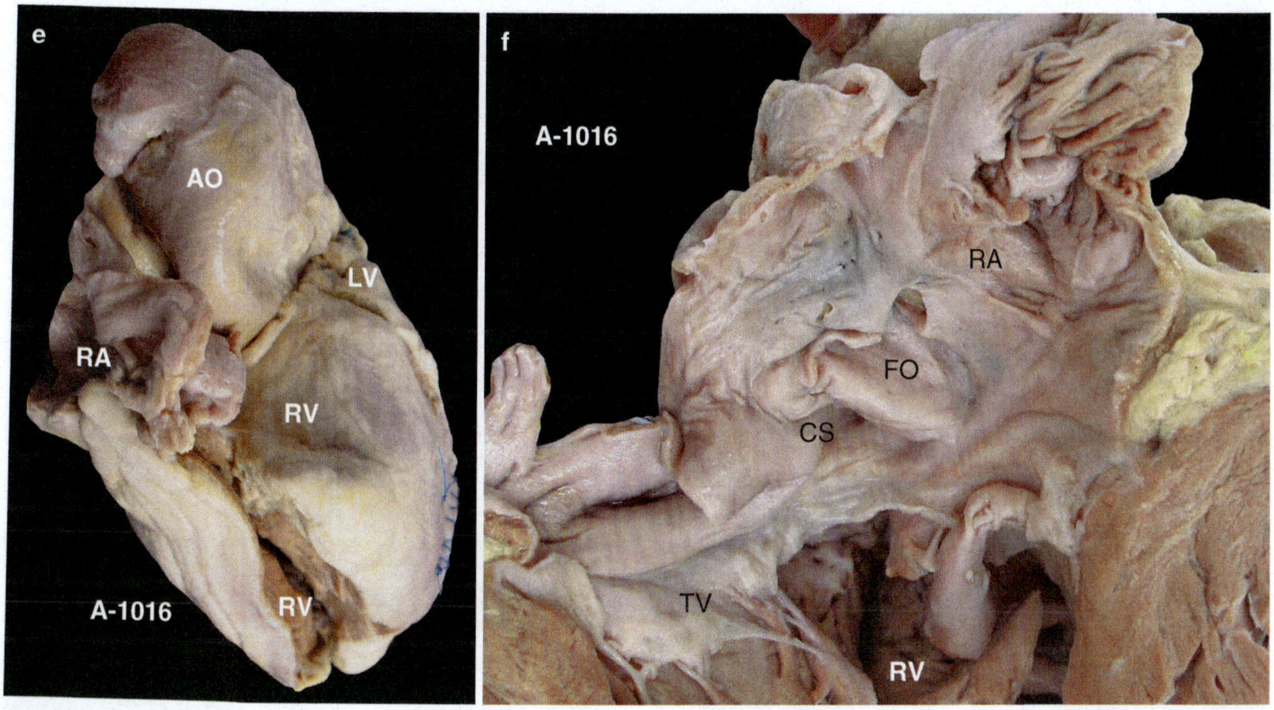

Fig. 28.1 (continued)

Fig. 28.1 (**a**) Specimen 1016. DOLV accompanied with muscular stenosis under pulmonary valve, rare in comparison. The anterior wall (dissociate wall) of the LV is straight ahead, the incision can be seen, and the main pulmonary artery is front and left of the aorta. *PA* pulmonary artery, *AO* ascending aorta, *RA* right atrium, *RV* right ventricle. (**b**) Specimen 1016. The dissociate wall of the LV, opened. Figure shows *first*, the diameter of the VSD is similar to the PV but much smaller than the AV; *second*, there is crack in the MV, and chordae tendineae of the anterior leaflet adhere to the edge of the VSD; *third*, there is a fibrous connection of the mitral, atrial, and pulmonary valves; *fourth*, the LV is small, and the enlarged RV is behind it; and *fifth*, the PV grows well, and there is mild pulmonary stenosis and a thick conical muscle under the pulmonary valve, which leads to obstruction under pulmonary artery (*arrow*). *MV* mitral valve. (**c**) Specimen 1016. The probe is inserted into the pulmonary artery, which reveals that both the pulmonary and aortic valves originate from the LV. There is fibrous connection of the mitral and aortic valves, and muscular pachynsis is present under PV. The LV is small, and chordae tendineae of the MV bestride on the VSD. *PA* pulmonary artery, *AO* ascending aorta. (**d**) Specimen 1016. Figure reveals the aortic, pulmonary, and mitral valves and the VSD (D) at close range. *AV* aortic valve, *PV* pulmonary valve, *MV* mitral valve. (**e**) Specimen 1016. From the right view. The right atrium and ventricle are divided at the same time, the aorta expands, the RV enlarges, and the main pulmonary artery is covered. *LV* left ventricle, *AO* ascending aorta, *RA* right atrium. (**f**) Specimen 1016. Figure reveals the coronary sinus and FO in the RA, which demonstrate a morphological RA. This clearly is atrium normotopia, in which the RA is on the right and the TV is normal. *CS* coronary sinus, *RV* right ventricle. (**g**) Specimen 1016. The right atrium and tricuspid valve viewed from the same angle as Fig. 28.1f. The *arrow* points to the anterior papillary muscle; there is no RVOT, rather a VSD. *RA* right atrium, *CS* coronary sinus, *TV* tricuspid valve, *RV* right ventricle. (**h**) Specimen 1016. A VSD observed in the right ventricle. There is no outflow tract or fibrous connection of the mitral and tricuspid valves, meaning there is no conjunction with the atrioventricular valve. Absence of an atrioventricular channel can be denied because the TV is normal. *RA* right atrium, *RV* right ventricle, *D* ventricular septal defect. (**i**) Specimen 1016. View of the left side of the heart. The main pulmonary artery grows well and is in the front and left of the aorta; the left atrium and auricle are enlarged. *PA* pulmonary artery, *AO* ascending aorta, *LA* left atrium, *LAA* left atrium auricle. (**j**) Specimen 1016. Figure shows that the doorway of the pulmonary vein is in the back (*arrow*), which proves the left side of the back wall is a morphological LA, another proof of atrium normotopia. *PA* pulmonary artery, *RA* right atrium. (**k**) Specimen 1016. Mitral valve in the left atrium. Thick leaflets, abnormal chordae tendineae, and papillary muscle can be seen. The chordae tendineae of the anterior leaflet adhere to the inferior border of the VSD; the "crack" could be considered as the crack of the anterior junction. *LA* left atrium, *MV* mitral valve. (**l**) Specimen 1016. Sketch map of the DOLV. The aorta and pulmonary artery both originate from the LV. The LV dysplasia is present: the diameter of the pulmonary artery is thinner than the aorta, and there is muscular conical stenosis under the PV, forming a VSD (*dotted line*). *PA* pulmonary artery, *AO* ascending aorta, *RV* right ventricle

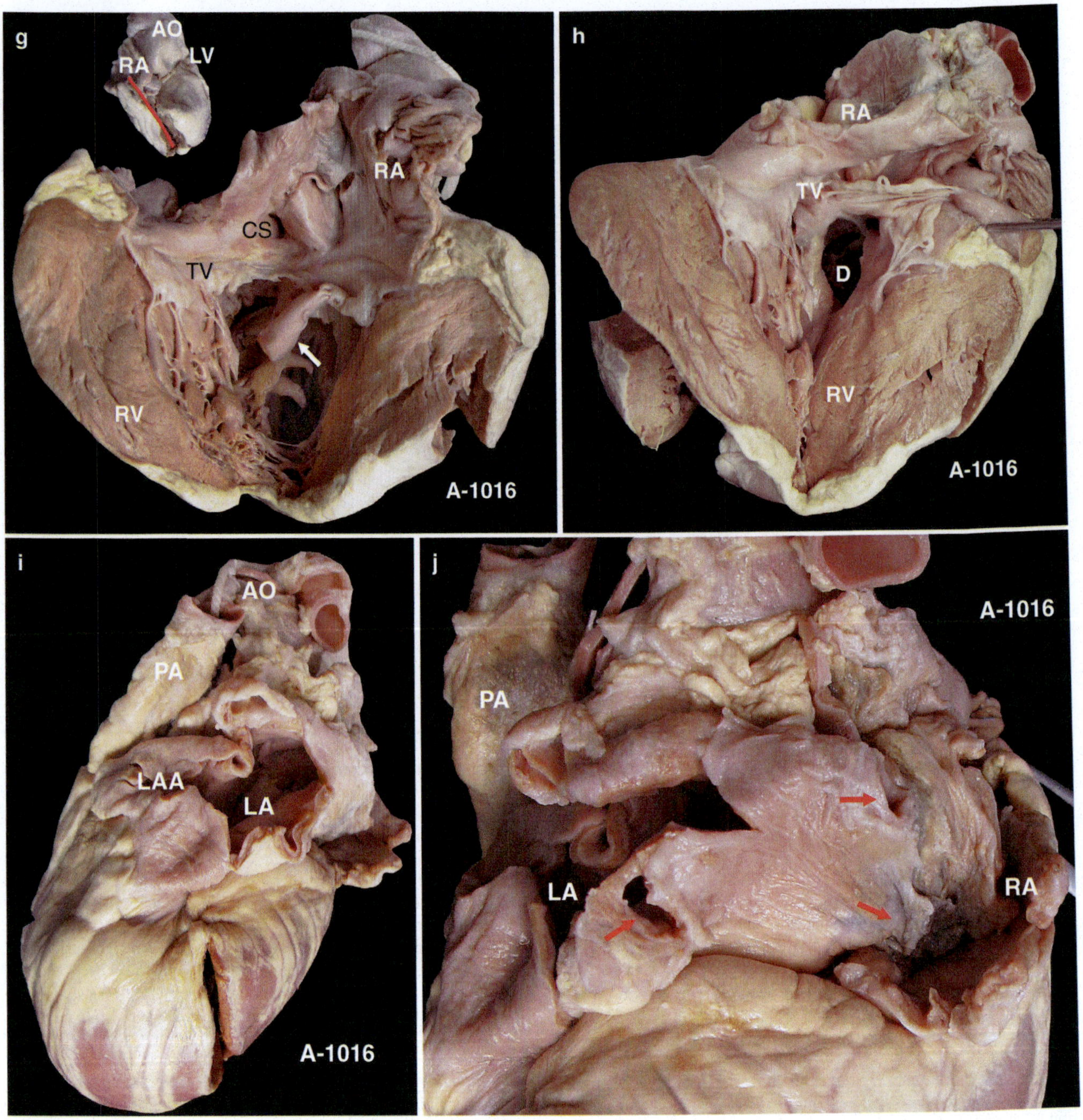

Fig. 28.1 (continued)

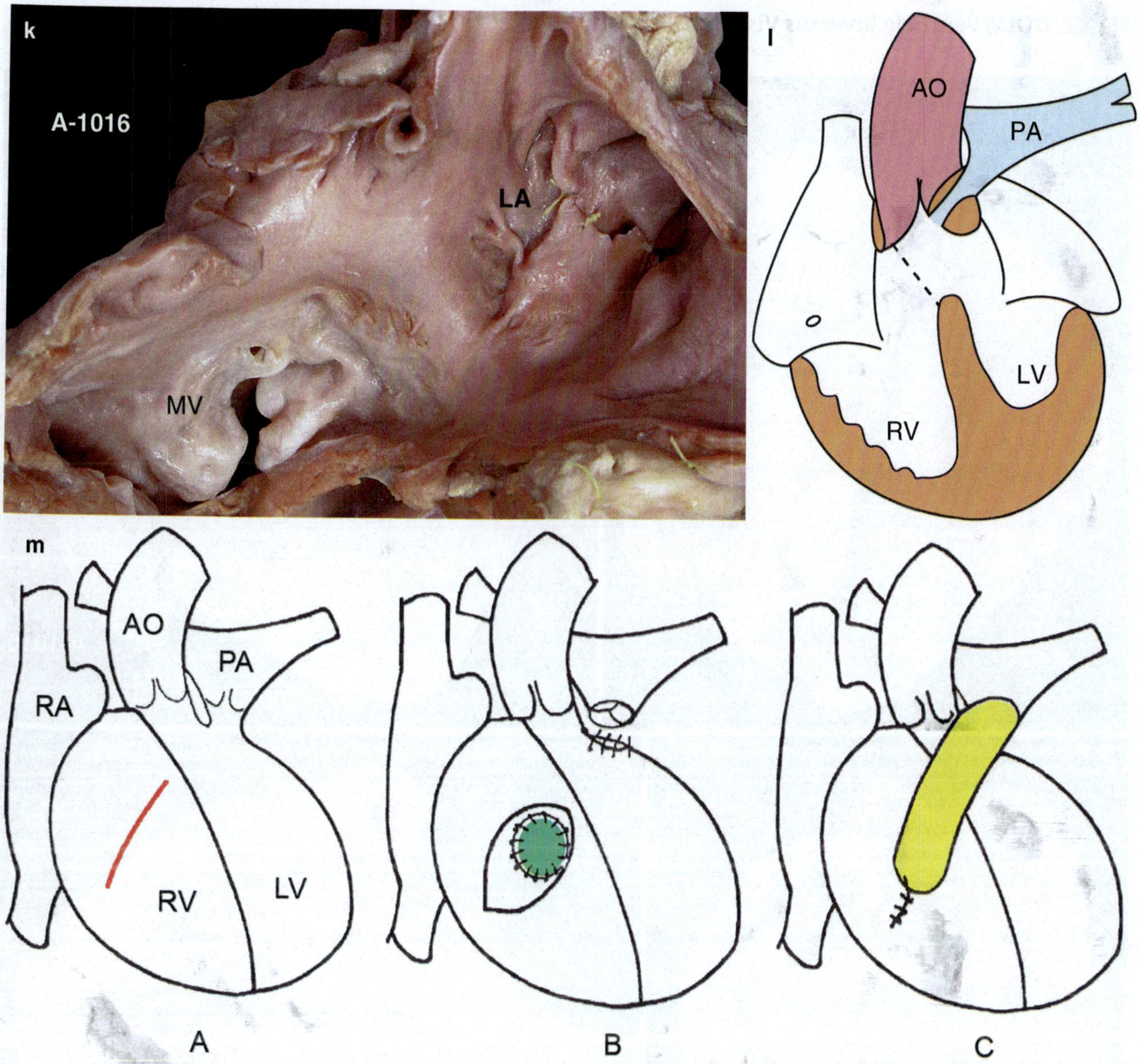

Fig. 28.1 (continued)

28.2.2 DOLV, Ventricle Inversus Viscerum, Outer Tunnel

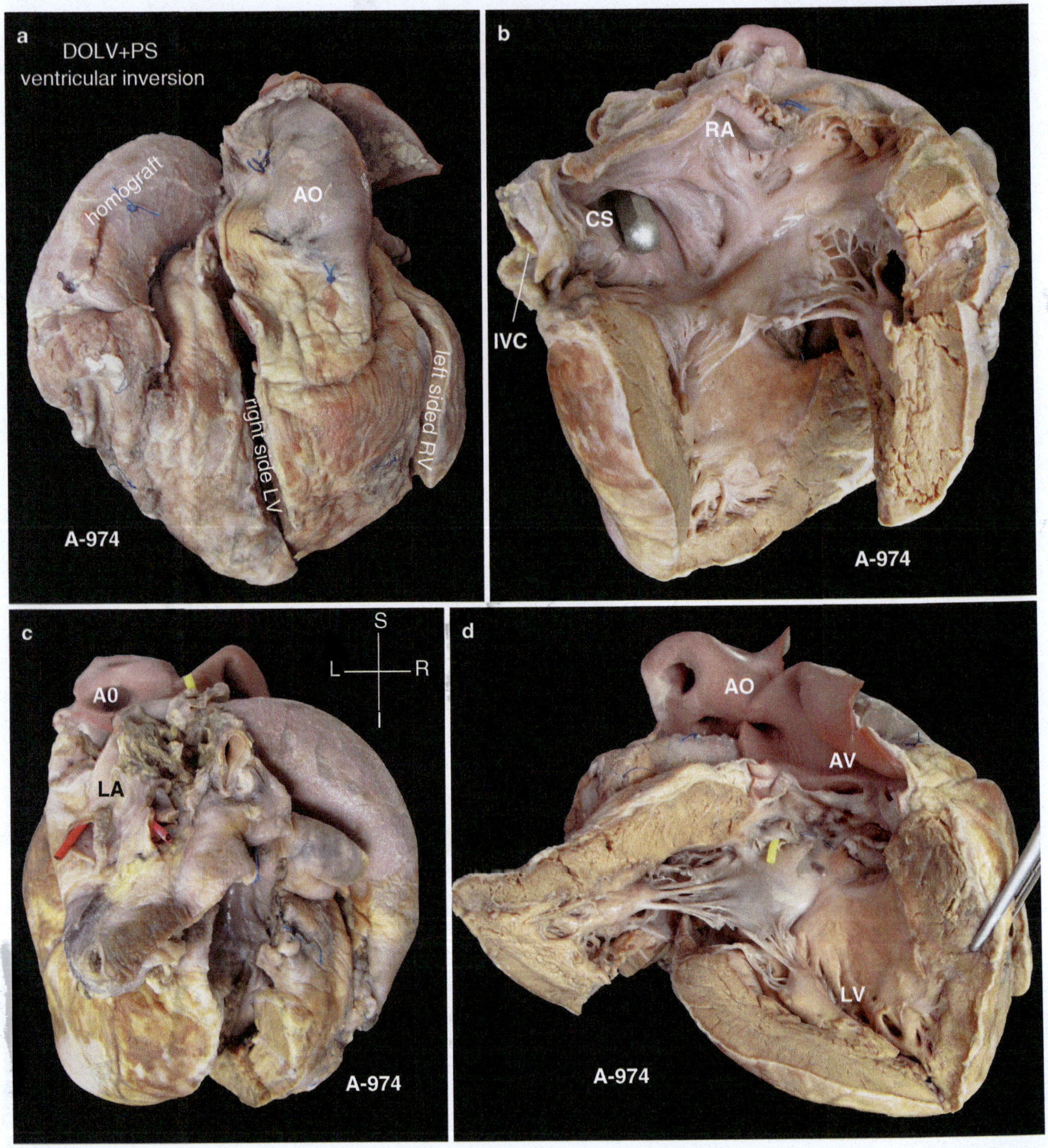

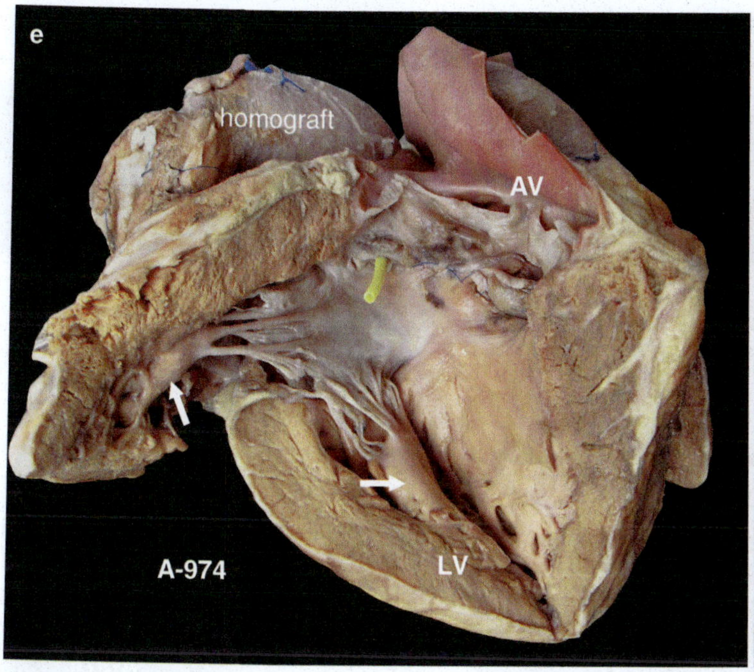

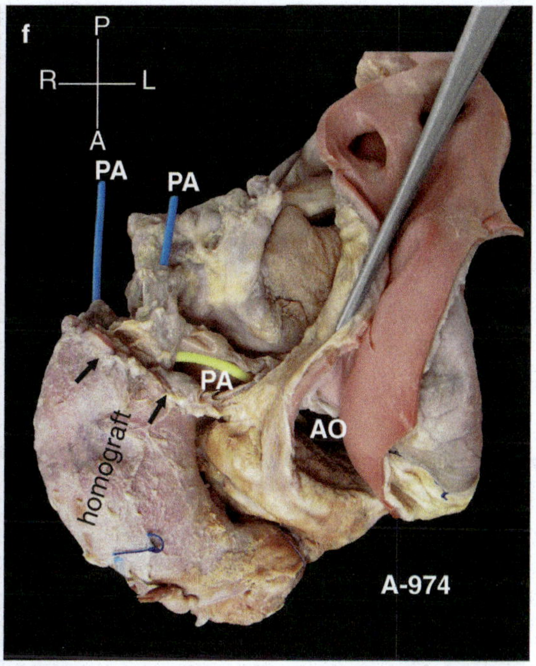

Fig. 28.2 (continued)

Fig. 28.2 (**a**) Specimen 974. Atrium normotopia, ventricle inversus viscerum, DOLV. The ascending aorta is in the front, the main pulmonary artery is in the back, and the RV is a morphological LV. The homograft should be made as the outer tunnel and connected with the narrow main pulmonary artery. *AO* ascending aorta. (**b**) Specimen 974. The doorway of the coronary sinus and the IVC are in the RA, which proved to be a morphological RA. The conjunction of the atrioventricular valve is the MV, and the anterior and back papillary muscle can be seen. *CS* coronary sinus. (**c**) Specimen 974. Back view of the heart. Two *red probes* are in the pulmonary vein, which proved to be the LA, so we are sure that the LA is on the left, rendering this atrium normotopia. *AO* ascending aorta. (**d**) Specimen 974. The structure of the left ventricle. The *yellow probe* is in the doorway of the PV. Notice: *first*, there is a totally fibrous connection of the leaflets of the pulmonary and atrioventricular valves, without muscle; *second*, there are no chordae tendineae of the atrioventricular valve on the ventricular septum; *third*, there are two thick groups of papillary muscle underneath. As a result, the ventricle on the right is the LV. *AO* ascending aorta, *AV* aortic valve. (**e**) Specimen 974. Papillary muscle of the mitral valve (*two arrows*). *AV* aortic valve, *LV* left ventricle. (**f**) Specimen 974. Overview shows that the aorta is in the front left, whereas the pulmonary artery is in the right diagonal back (*yellow probe*). The two *blue probes* are the branches of the pulmonary artery; the *arrow* points at the anastomosis of the pulmonary artery and homograft. *A* anterior, *P* posterior; *L* left; *R* right. *PA* pulmonary artery, *AO* ascending aorta. (**g**) Specimen 974. The thick aorta and small main pulmonary artery (*yellow probe*) both originate from the morphological LV. The VSD is under the AV; the metal probe is the VSD patch. *AO* ascending aorta. (**h**) Specimen 974. The morphological RV is on the left because there is muscular separation of the atrioventricular valve and main arterial valves. The TV has three leaflets, and the chordae tendineae adhere to the ventricular septum. Because there is no fibrous connection of the AV and TV, we are sure that the aorta originates from the morphological LV. (**i**) Specimen 974. Appearance in the front. The morphological RV on the left is smaller than the ventricle on the right, and the morphological LV is on the right, which emits both the aorta and the pulmonary artery. *AO* ascending aorta. (**j**) Specimen 974. Sketch map, SLL. The AV and PV both originate from the morphological LV, diagnosed as DOLV, pulmonary artery dysplasia, outer tunnel of LV-PA (*green*). The VSD patch is indicated by the *dotted line*. The proof of difference with congenitally corrected TGA is that it has no fibrous connection of the atrial and tricuspid valves in the morphological RV. The specimen uses a VSD patch and an outer tunnel at the same time; bilateral arterial switch operation could be considered. *PA* pulmonary artery, *AO* ascending aorta, *RV* right ventricle, *LA* left atrium

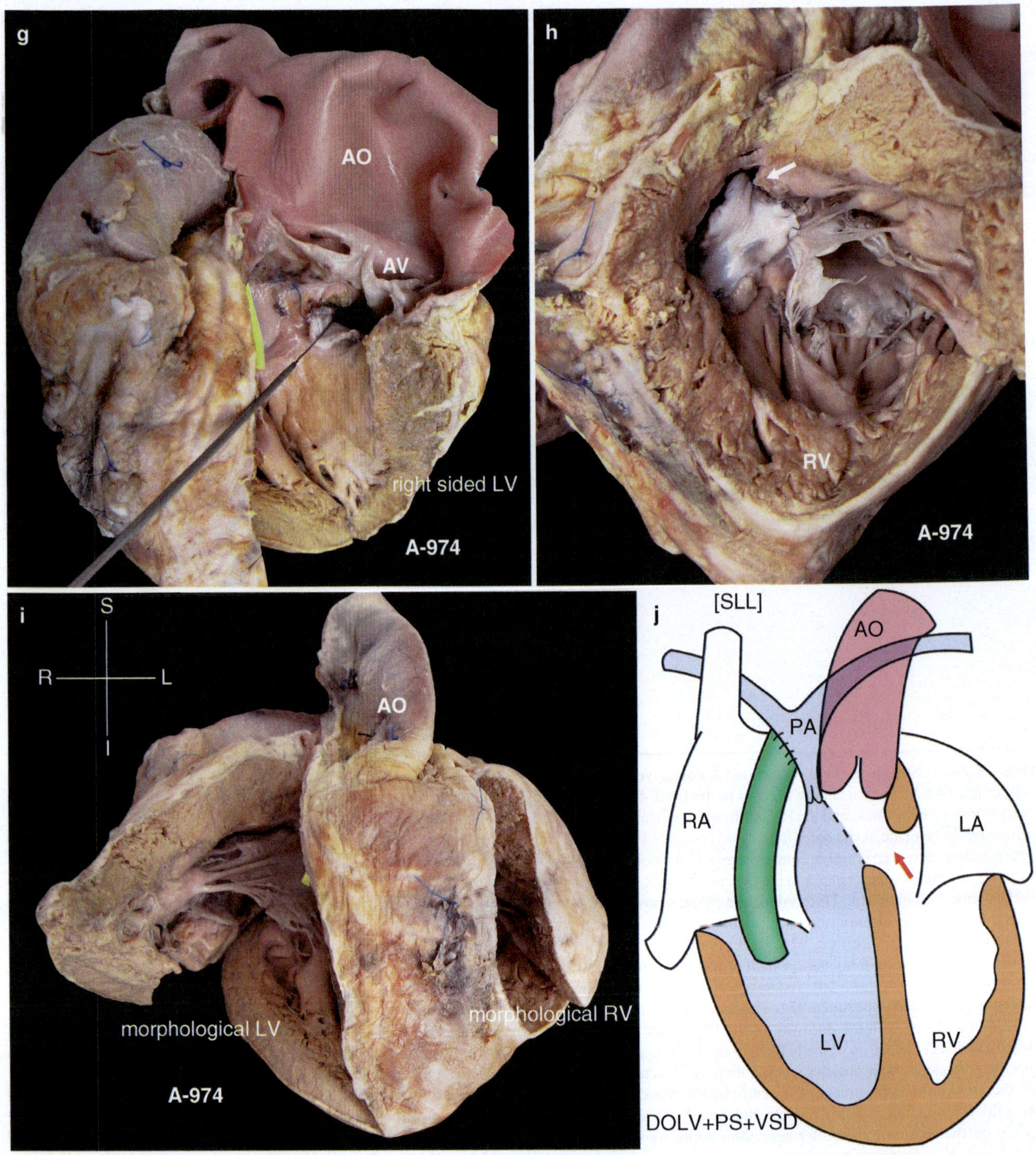

Fig. 28.2 (continued)

The manufacturer's authorised representative in the EU is Springer Nature Customer Service Centre GmbH, Europaplatz 3, 69115 Heidelberg, Germany. If you have any concerns regarding our products, please contact ProductSafety@springernature.com

Printed and bound by CPI Group (UK) Ltd, Croydon, CR0 4YY
30/09/2026
02225476-0006